W9-DHR-751

EMISSION SPECTRA

CONTINUOUS SPECTRUM (Incandescent solids or liquids and incandescent gases under high pressure give continuous spectra)   **INCANDESCENT LAMP**

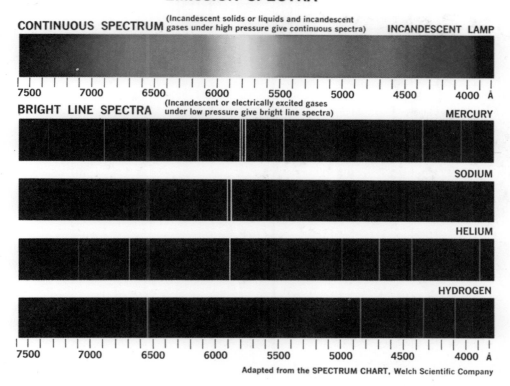

BRIGHT LINE SPECTRA (Incandescent or electrically excited gases under low pressure give bright line spectra)   **MERCURY**

SODIUM

HELIUM

HYDROGEN

Adapted from the SPECTRUM CHART, Welch Scientific Company

Useful Conversions and Combinations

$$1\text{ eV} = 1.602 \times 10^{-19}\text{ J}$$

$$1\text{ cal} = 4.184\text{ J}$$

$$1\text{ u} = 931.5\text{ MeV}/c^2$$

$$1\text{ MeV}/c^2 = 1.073 \times 10^{-3}\text{ u} = 1.783 \times 10^{-30}\text{ kg}$$

$$1\text{ Å} = 10^{-10}\text{ m} = 0.1\text{ nm}$$

$$1\text{ fm} = 10^{-15}\text{ m}$$

$$1\text{ in} = 2.540\text{ cm}$$

$$1\text{ mi} = 1609\text{ m}$$

$$hc = 1.240 \times 10^3\text{ eV}\cdot\text{nm} = 1.986 \times 10^{-25}\text{ J}\cdot\text{m}$$

$$\hbar c = 1.973 \times 10^2\text{ eV}\cdot\text{nm} = 3.162 \times 10^{-26}\text{ J}\cdot\text{m}$$

$$k_\text{B}T = 0.02525\text{ eV at }T = 300\text{ K}$$

$$ke^2 = e^2/4\pi\epsilon_0 = 1.440\text{ eV}\cdot\text{nm}$$

$$1\text{ barn} = 10^{-28}\text{ m}^2$$

$$1\text{ curie} = 3.7 \times 10^{10}\text{ decays/s}$$

$$1\text{ MeV}/c = 5.344 \times 10^{-22}\text{ kg}\cdot\text{m/s}$$

Modern Physics

Third Edition

RAYMOND A. SERWAY
Emeritus
James Madison University

CLEMENT J. MOSES
Emeritus
Utica College of Syracuse University

CURT A. MOYER
University of North Carolina-Wilmington

THOMSON
BROOKS/COLE

Australia • Canada • Mexico • Singapore • Spain
United Kingdom • United States

THOMSON

BROOKS/COLE

Physics Editor: Chris Hall
Development Editor: Jay Campbell
Editor-in-Chief: Michelle Julet
Publisher: David Harris
Editorial Assistant: Seth Dobrin
Technology Project Manager: Sam Subity
Marketing Manager: Kelley McAllister
Marketing Assistant: Leyla Jowza
Advertising Project Manager: Stacey Purviance
Project Manager, Editorial Production: Teri Hyde
Print/Media Buyer: Barbara Britton
Permissions Editor: Sarah Harkrader
Production Service: Progressive Publishing Alternatives
Text Designer: Patrick Devine

Art Director: Rob Hugel
Photo Researcher: Dena Digilio-Betz
Copy Editor: Progressive Publishing Alternatives
Illustrator: Rolin Graphics/Progressive Information Technologies
Cover Designer: Patrick Devine
Cover Image: Patrice Loiez, CERN/Science Photo Library, Artificially colored bubble chamber photo from CERN, the European particle physics laboratory outside Geneva (1984).
Cover Printer: Coral Graphic Services
Compositor: Progressive Information Technologies
Printer: Quebecor World, Taunton

COPYRIGHT © 2005, 1997, 1989 by Raymond A. Serway.

ALL RIGHTS RESERVED. No part of this work covered by the copyright hereon may be reproduced or used in any form or by any means—graphic, electronic, or mechanical, including but not limited to photocopying, recording, taping, Web distribution, information networks, or information storage and retrieval systems—without the written permission of the publisher.

Printed in the United States of America

1 2 3 4 5 6 7 08 07 06 05 04

For more information about our products, contact us at:
Thomson Learning Academic Resource Center
1-800-423-0563

For permission to use material from this text
or product, submit a request online at
http://www.thomsonrights.com.
Any additional questions about permissions can be
submitted by email to **thomsonrights@thomson.com.**

Library of Congress Control Number: 2004101232

Student's Edition: ISBN 0-534-49339-4

International Student Edition: ISBN 0-534-40624-6

Brooks/Cole—Thomson Learning
10 Davis Drive
Belmont, CA 94002
USA

Asia
Thomson Learning
5 Shenton Way #01-01
UIC Building
Singapore 068808

Australia/New Zealand
Thomson Learning
102 Dodds Street
Southbank, Victoria 3006
Australia

Canada
Nelson
1120 Birchmount Road
Toronto, Ontario M1K 5G4
Canada

Europe/Middle East/Africa
Thomson Learning
High Holborn House
50/51 Bedford Row
London WC1R 4LR
United Kingdom

Latin America
Thomson Learning
Seneca, 53
Colonia Polanco
11560 Mexico D.F.
Mexico

Spain/Portugal
Paraninfo
Calle/Magallanes, 25
28015 Madrid, Spain

About the Authors

Raymond A. Serway received his doctorate at Illinois Institute of Technology and is Professor Emeritus at James Madison University. Dr. Serway began his teaching career at Clarkson University, where he conducted research and taught from 1967 to 1980. His second academic appointment was at James Madison University as Professor of Physics and Head of the Physics Department from 1980 to 1986. He remained at James Madison University until his retirement in 1997. He was the recipient of the Madison Scholar Award at James Madison University in 1990, the Distinguished Teaching Award at Clarkson University in 1977, and the Alumni Achievement Award from Utica College in 1985. As Guest Scientist at the IBM Research Laboratory in Zurich, Switzerland, he worked with K. Alex Müller, 1987 Nobel Prize recipient. Dr. Serway also held research appointments at Rome Air Development center from 1961 to 1963, at IIT Research Institute from 1963 to 1967, and as a visiting scientist at Argonne National Laboratory, where he collaborated with his mentor and friend, Sam Marshall. In addition to earlier editions of this textbook, Dr. Serway is the co-author of *Physics for Scientists and Engineers,* 6th edition, *Principles of Physics,* 3rd edition, *College Physics,* 6th edition, and the high-school textbook *Physics,* published by Holt, Rinehart, and Winston. In addition, Dr. Serway has published more than 40 research papers in the field of condensed matter physics and has given more than 60 presentations at professional meetings. Dr. Serway and his wife Elizabeth enjoy traveling, golfing, fishing, and spending quality time with their four children and seven grandchildren.

Clement J. Moses is Emeritus Professor of Physics at Utica College. He was born and brought up in Utica, New York, and holds an A.B. from Hamilton College, an M.S. from Cornell University, and a Ph.D. from State University of New York at Binghamton. He has over 30 years of science writing and teaching experience at the college level, and is a co-author of *College Physics,* 6th edition, with Serway and Faughn. His research work, both in industrial and university settings, has dealt with defects in solids, solar cells, and the dynamics of atoms at surfaces. In addition to science writing, Dr. Moses enjoys reading novels, gardening, cooking, singing, and going to operas.

Curt A. Moyer has been Professor and Chair of the Department of Physics and Physical Oceanography at the University of North Carolina-Wilmington since 1999. Before his appointment to UNC-Wilmington, he taught in the Physics Department at Clarkson University from 1974 to 1999. Dr. Moyer earned a B.S. from Lehigh University and a Ph.D. from the State University of New York at Stony Brook. He has published more than 45 research articles in the fields of condensed matter physics and surface science. In addition to being an experienced teacher, Dr. Moyer is an advocate for the uses of computers in education and developed the Web-based *QMTools* software that accompanies this text. He and his wife, V. Sue, enjoy traveling and the special times they spend with their four children and three grandchildren.

Preface

This book is intended as a modern physics text for science majors and engineering students who have already completed an introductory calculus-based physics course. The contents of this text may be subdivided into two broad categories: an introduction to the theories of relativity, quantum and statistical physics (Chapters 1 through 10) and applications of elementary quantum theory to molecular, solid-state, nuclear, and particle physics (Chapters 11 through 16).

OBJECTIVES

Our basic objectives in this book are threefold:

1. To provide simple, clear, and mathematically uncomplicated explanations of physical concepts and theories of modern physics.
2. To clarify and show support for these theories through a broad range of current applications and examples. In this regard, we have attempted to answer questions such as: What holds molecules together? How do electrons tunnel through barriers? How do electrons move through solids? How can currents persist indefinitely in superconductors?
3. To enliven and humanize the text with brief sketches of the historical development of 20th century physics, including anecdotes and quotations from the key figures as well as interesting photographs of noted scientists and original apparatus.

COVERAGE

Topics. The material covered in this book is concerned with fundamental topics in modern physics with extensive applications in science and engineering. Chapters 1 and 2 present an introduction to the special theory of relativity. Chapter 2 also contains an introduction to general relativity. Chapters 3 through 5 present an historical and conceptual introduction to early developments in quantum theory, including a discussion of key experiments that show the quantum aspects of nature. Chapters 6 through 9 are an introduction to the real "nuts and bolts" of quantum mechanics, covering the Schrödinger equation, tunneling phenomena, the hydrogen atom, and multielectron atoms, while Chapter 10 contains an introduction to statistical physics. The remainder of the book consists mainly of applications of the theory set forth in earlier chapters to more specialized areas of modern physics. In particular, Chapter 11 discusses the physics of molecules, while Chapter 12 is an introduction to the physics of solids and electronic devices. Chapters 13 and 14 cover nuclear physics, methods of obtaining energy from nuclear reactions, and medical and other applications of nuclear processes. Chapter 15 treats elementary particle physics, and Chapter 16 (available online at **http://info. brookscole.com/mp3e**) covers cosmology.

CHANGES TO THE THIRD EDITION

The third edition contains two major changes from the second edition: *First,* this edition has been extensively rewritten in order to clarify difficult concepts, aid understanding, and bring the text up to date with rapidly developing technical applications of quantum physics. Artwork and the order of presentation of certain topics have been revised to help in this process. (Many new photos of physicists have been added to the text, and a new collection of color photographs of modern physics phenomena is also available on the Book Companion Web Site.) Typically, each chapter contains new worked examples and five new end-of-chapter questions and problems. Finally, the *Suggestions for Further Reading* have been revised as needed.

Second, this edition refers the reader to a new, online (platform independent) simulation package, *QMTools,* developed by one of the authors, Curt Moyer. We think these simulations clarify, enliven, and complement the analytical solutions presented in the text. Icons in the text highlight the problems designed for use with this software, which provides modeling tools to help students visualize abstract concepts. All instructions about the general use of the software as well as specific instructions for each problem are contained on the Book Companion Web Site, thereby minimizing interruptions to the logical flow of the text. The Book Companion Web Site at **http://info.brookscole. mp3e** also contains appendices and much supplemental information on current physics research and applications, allowing interested readers to dig deeper into many topics.

Specific changes by chapter in this third edition are as follows:

- Chapter 1 in the previous editions, "Relativity," has been extensively revised and divided into two chapters. The new **Chapter 1,** entitled "Relativity I," contains the history of relativity, new derivations of the Lorentz coordinate and velocity transformations, and a new section on spacetime and causality.
- **Chapter 2,** entitled "Relativity II," covers relativistic dynamics and energy and includes new material on general relativity, gravitational radiation, and the applications GPS (Global Positioning System) and LIGO (the Laser Interferometer Gravitational-wave Observatory).
- **Chapter 3** has been streamlined with a more concise treatment of the Rayleigh-Jeans and Planck blackbody laws. Material necessary for a complete derivation of these results has been placed on our Book Companion Web Site.
- **Chapter 5** contains a new section on the invention and principles of operation of transmission and scanning electron microscopes.

- **Chapter 6,** "Quantum Mechanics in One Dimension," features a new application on the principles of operation and utility of CCDs (Charge-Coupled Devices).
- **Chapter 8,** "Quantum Mechanics in Three Dimensions," includes a new discussion on the production and spectroscopic study of anti-hydrogen, a study which has important consequences for several fundamental physical questions.
- **Chapter 10** presents new material on the connection of wavefunction symmetry to the Bose-Einstein condensation and the Pauli exclusion principle, as well as describing potential applications of Bose-Einstein condensates.
- **Chapter 11** contains new material explaining Raman scattering, fluorescence, and phosphorescence, as well as giving applications of these processes to pollution detection and biomedical research. This chapter has also been streamlined with the discussion of overlap integrals being moved to the Book Companion Web Site.
- **Chapter 12** has been carefully revised for clarification and features new material on semiconductor devices, in particular MOSFETs and chips. In addition, the most important facts about superconductivity have been summarized, updated, and included in Chapter 12. For those desiring more material on superconductivity, the entire superconductivity chapter from previous editions is available at the Book Companion Web Site along with essays on the history of the laser and solar cells.
- **Chapter 13** contains new material on MRI (Magnetic Resonance Imaging) and an interesting history of the determination of the age of the Earth.
- **Chapter 14** presents updated sections on fission reactor safety and waste disposal, fusion reactor results, and applications of nuclear physics to tracing, neutron activation analysis, radiation therapy, and other areas.
- **Chapter 15** has been extensively rewritten in an attempt to convey the thrust toward unification in particle physics. By way of achieving this goal, new discussions of positrons, neutrino mass and oscillation, conservation laws, and grand unified theories, including supersymmetry and string theory, have been introduced.
- **Chapter 16** is a new chapter devoted exclusively to the exciting topic of the origin and evolution of the universe. Topics covered include the discovery of the expanding universe, primordial radiation, inflation, the future evolution of the universe, dark matter, dark energy, and the accelerating expansion of the universe. This cosmology chapter is available on our Book Companion Web Site.

FEATURES OF THIS TEXT

QMTools   Five chapters contain several new problems requiring the use of our simulation software, *QMTools*. *QMTools* is a sophisticated interactive learning tool with considerable flexibility and scope. Using *QMTools,* students can compose matter-wave packets and study their time evolution, find stationary state energies and wavefunctions, and determine the probability for particle transmission and reflection from nearly any potential well or barrier. Access to *QMTools* is available online at **http://info.brookscole.com/mp3e**.

Style.   We have attempted to write this book in a style that is clear and succinct yet somewhat informal, in the hope that readers will find the text appealing and enjoyable to read. All new terms have been carefully defined, and we have tried to avoid jargon.

Worked Examples.   A large number of worked examples of varying difficulty are presented as an aid in understanding both concepts and the chain of reasoning needed to solve realistic problems. In many cases, these examples will serve as models for solving some end-of-chapter problems. The examples are set off with colored bars for ease of location, and most examples are given titles to describe their content.

Exercises Following Examples.   As an added feature, many of the worked examples are followed immediately by exercises with answers. These exercises are intended to make the textbook more interactive with the student, and to test immediately the student's understanding of key concepts and problem-solving techniques. The exercises represent extensions of the worked examples and are numbered in case the instructor wishes to assign them for homework.

Problems and Questions.   An extensive set of questions and problems is included at the end of each chapter. Most of the problems are listed by section topic. Answers to all odd-numbered problems are given at the end of the book. Problems span a range of difficulty and more challenging problems have colored numbers. Most of the questions serve to test the student's understanding of the concepts presented in a given chapter, and many can be used to motivate classroom discussions.

Units.   The international system of units (SI) is used throughout the text. Occasionally, where common usage dictates, other units are used (such as the angstrom, Å, and cm^{-1}, commonly used by spectroscopists), but all such units are carefully defined in terms of SI units.

Chapter Format.   Each chapter begins with a **preview,** which includes a brief discussion of chapter objectives and content. **Marginal notes** set in color are used to locate important concepts and equations in the text. Important statements are *italicized* or **highlighted,** and important equations are set in a colored box for added emphasis and ease of review. Each chapter concludes with a **summary,** which reviews the important concepts and equations discussed in that chapter.

In addition, many chapters contain **special topic sections** which are clearly marked **optional.** These sections expose the student to slightly more advanced material either in the form of current interesting discoveries or as fuller developments of concepts or calculations discussed in that chapter. Many of these special topic sections will be of particular interest to certain student groups such as chemistry majors, electrical engineers, and physics majors.

Guest Essays.   Another feature of this text is the inclusion of interesting material in the form of essays by guest authors. These essays cover a wide range of topics and are intended to convey an insider's view of exciting current developments in modern physics. Furthermore, the essay topics present extensions and/or applications of the material discussed in specific chapters. Some of the essay topics covered are recent developments in general relativity, the scanning tunneling microscope, superconducting devices, the history of the laser, laser cooling of atoms, solar cells, and how the top quark was detected. The guest essays are either included in the text or referenced as being on our Web site at appropriate points in the text.

Mathematical Level.   Students using this text should have completed a comprehensive one-year calculus course, as calculus is used throughout the text. However, we have made an attempt to keep physical ideas foremost so as not to obscure our presentations with overly elegant mathematics. Most steps are shown when basic equations are developed, but exceptionally long and detailed proofs which interrupt the flow of physical arguments have been placed in appendices.

Appendices and Endpapers.   The appendices in this text serve several purposes. Lengthy derivations of important results needed in physical discussions have been placed on our Web site to avoid interrupting the main flow of arguments. Other appendices needed for quick reference are located at the end of the book. These contain physical constants, a table of atomic masses, and a list of Nobel prize winners. The endpapers inside the front cover of the book contain important physical constants and standard abbreviations of units used in the book, and conversion factors for quick reference, while a periodic table is included in the rear cover endpapers.

Ancillaries.   The ancillaries available with this text include a Student Solutions Manual, which has solutions to all odd-numbered problems in the book, an Instructor's Solutions Manual, consisting of solutions to all problems in the text, and a Multimedia Manager, a CD-ROM lecture tool that contains digital versions of all art and selected photographs in the text.

TEACHING OPTIONS

As noted earlier, the text may be subdivided into two basic parts: Chapters 1 through 10, which contain an introduction to relativity, quantum physics, and statistical physics, and Chapters 11 through 16, which treat applications to molecules, the solid state, nuclear physics, elementary particles, and cosmology. It is suggested that the first part of the book be covered sequentially. However, the relativity chapters may actually be covered at any time because $E^2 = p^2c^2 + m^2c^4$ is the only formula from these chapters which is essential for subsequent chapters. Chapters 11 through 16 are independent of one another and can be covered in any order with one exception: Chapter 14, "Nuclear Physics Applications," should follow Chapter 13, "Nuclear Structure."

A traditional sophomore or junior level modern physics course for science, mathematics, and engineering students should cover most of Chapters 1 through 10 and several of the remaining chapters, depending on the student major. For example, an audience consisting mainly of electrical engineering students might cover most of Chapters 1 through 10 with particular emphasis on tunneling and tunneling devices in Chapter 7, the Fermi-Dirac distribution in Chapter 10, semiconductors in Chapter 12, and radiation detectors in Chapter 14. Chemistry and chemical engineering majors could cover most of Chapters 1 through 10 with special emphasis on atoms in Chapter 9, classical and quantum statistics in Chapter 10, and molecular bonding and spectroscopy in Chapter 11. Mathematics and physics majors should pay special attention to the unique development of operator methods and the concept of sharp and fuzzy observables introduced in Chapter 6. The deep connection of sharp observables with classically conserved quantities and the powerful role of sharp observables in shaping the form of system wavefunctions is developed more fully in Chapter 8.

Our experience has shown that there is more material contained in this book than can be covered in a standard one semester three-credit-hour course. For this reason, one has to "pick-and-choose" from topics in the second part of the book as noted earlier. However, the text can also be used in a two-semester sequence with some supplemental material, such as one of many monographs on relativity, and/or selected readings in the areas of solid state, nuclear, and elementary particle physics. Some selected readings are suggested at the end of each chapter.

ACKNOWLEDGMENTS

We wish to thank the users and reviewers of the first and second editions who generously shared with us their comments and criticisms. In preparing this third edition we owe a special debt of gratitude to the following reviewers:

Melissa Franklin, Harvard University

Edward F. Gibson, California State University, Sacramento

Grant Hart, Brigham Young University

James Hetrick, University of the Pacific

Andres H. La Rosa, Portland State University

Pui-tak (Peter) Leung, Portland State University

Peter Moeck, Portland State University

Timothy S. Sullivan, Kenyon College

William R. Wharton, Wheaton College

We thank the professional staff at Brooks-Cole Publishing for their fine work during the development and production of this text, especially Jay Campbell, Chris Hall, Teri Hyde, Seth Dobrin, Sam Subity, Kelley McAllister, Stacey Purviance, Susan Dust Pashos, and Dena Digilio-Betz. We thank Suzon O. Kister for her helpful reference work, and all the authors of our guest essays: Steven Chu, Melissa Franklin, Roger A. Freedman, Clark A. Hamilton, Paul K. Hansma, David Kestenbaum, Sam Marshall, John Meakin, and Clifford M. Will.

Finally, we thank all of our families for their patience and continual support.

Raymond A. Serway
Leesburg, VA 20176

Clement J. Moses
Durham, NC 27713

Curt A. Moyer
Wilmington, NC 28403

December 2003

Contents Overview

Contents

QM Tools
Text References to the Software

Chapter 6
Section 6.2, after Example 6.4
Exercise 3, following Example 6.8

Problems 22, 27, 36

Chapter 7
Exercise 1, following Example 7.1
Section 7.2, after Example 7.6
Subsection on Ammonia Inversion in Section 7.2

Problems 8, 9, 10, 19, 20

Chapter 8
Problems 27, 28, 32, 33

Chapter 9
Problems 19, 20

Chapter 11
Subsection on The Hydrogen Molecular Ion in Section 11.4

Problems 16, 17, 22, 23

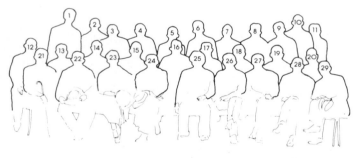

1. A. Piccard
2. E. Henriot
3. P. Ehrenfest
4. E. Herzen
5. Th. de Donder
6. E. Schroedinger
7. E. Verschaffelt
8. W. Pauli
9. W. Heisenberg
10. R.H. Fowler

11. L. Brillouin
12. P. Debye
13. M. Knudsen
14. W.L. Bragg
15. H.A. Kramers
16. P.A.M. Dirac
17. A.H. Compton
18. L.V. de Broglie
19. M. Born
20. N. Bohr

21. I. Langmuir
22. M. Planck
23. M. Curie
24. H.A. Lorentz
25. A. Einstein
26. P. Langevin
27. C.E. Guye
28. C.T.R. Wilson
29. O.W. Richardson

The "architects" of modern physics. This unique photograph shows many eminent scientists who participated in the Fifth International Congress of Physics held in 1927 by the Solvay Institute in Brussels. At this and similar conferences, held regularly from 1911 on, scientists were able to discuss and share the many dramatic developments in atomic and nuclear physics. This elite company of scientists includes fifteen Nobel prize winners in physics and three in chemistry. (*Photograph courtesy of AIP Niels Bohr Library*)

1

Relativity I

At the end of the 19th century, scientists believed that they had learned most of what there was to know about physics. Newton's laws of motion and his universal theory of gravitation, Maxwell's theoretical work in unifying electricity and magnetism, and the laws of thermodynamics and kinetic theory employed mathematical methods to successfully explain a wide variety of phenomena.

However, at the turn of the 20th century, a major revolution shook the world of physics. In 1900 Planck provided the basic ideas that led to the quantum theory, and in 1905 Einstein formulated his special theory of relativity. The excitement of the times is captured in Einstein's own words: "It was a marvelous time to be alive." Both ideas were to have a profound effect on our understanding of nature. Within a few decades, these theories inspired new developments and theories in the fields of atomic, nuclear, and condensed-matter physics.

Although modern physics has led to a multitude of important technological achievements, the story is still incomplete. Discoveries will continue to be made during our lifetime, many of which will deepen or refine our understanding of nature and the world around us. It is still a "marvelous time to be alive."

1.1 SPECIAL RELATIVITY

Light waves and other forms of electromagnetic radiation travel through free space at the speed $c = 3.00 \times 10^8$ m/s. As we shall see in this chapter, the speed of light sets an upper limit for the speeds of particles, waves, and the transmission of information.

Most of our everyday experiences deal with objects that move at speeds much less than that of light. Newtonian mechanics and early ideas on space and time were formulated to describe the motion of such objects, and this formalism is very successful in describing a wide range of phenomena. Although Newtonian mechanics works very well at low speeds, it fails when applied to particles whose speeds approach that of light. Experimentally, one can test the predictions of Newtonian theory at high speeds by accelerating an electron through a large electric potential difference. For example, it is possible to accelerate an electron to a speed of $0.99c$ by using a potential difference of several million volts. According to Newtonian mechanics, if the potential difference (as well as the corresponding energy) is increased by a factor of 4, then the speed of the electron should be doubled to $1.98c$. However, experiments show that the speed of the electron—as well as the speeds of all other particles in the universe—always remains less than the speed of light, regardless of the size of the accelerating voltage. In part because it places no upper limit on the speed that a particle can attain, Newtonian mechanics is contrary to modern experimental results and is therefore clearly a limited theory.

In 1905, at the age of 26, Albert Einstein published his *special theory of relativity*. Regarding the theory, Einstein wrote,

> The relativity theory arose from necessity, from serious and deep contradictions in the old theory from which there seemed no escape. The strength of the new theory lies in the consistency and simplicity with which it solves all these difficulties, using only a few very convincing assumptions. . . .[1]

Although Einstein made many important contributions to science, the theory of relativity alone represents one of the greatest intellectual achievements of the 20th century. With this theory, one can correctly predict experimental observations over the range of speeds from rest to speeds approaching the speed of light. Newtonian mechanics, which was accepted for over 200 years, is in fact a limiting case of Einstein's special theory of relativity. This chapter and the next give an introduction to the special theory of relativity, which deals with the analysis of physical events from coordinate systems moving with constant speed in straight lines with respect to one another. Chapter 2 also includes a short introduction to general relativity, which describes physical events from coordinate systems undergoing general or accelerated motion with respect to each other.

In this chapter we show that the special theory of relativity follows from two basic postulates:

1. The laws of physics are the same in all reference systems that move uniformly with respect to one another. That is, basic laws such as

---

[1]A. Einstein and L. Infeld, *The Evolution of Physics,* New York, Simon and Schuster, 1961.

$\Sigma \boldsymbol{F} = d\mathbf{p}/dt$ have the same mathematical form for all observers moving at constant velocity with respect to one another.

2. The speed of light in vacuum is always measured to be 3×10^8 m/s, and the measured value is independent of the motion of the observer or of the motion of the source of light. That is, the speed of light is the same for all observers moving at constant velocities.

Although it is well known that relativity plays an essential role in theoretical physics, it also has practical applications, for example, in the design of particle accelerators, global positioning system (GPS) units, and high-voltage TV displays. Note that these devices simply will not work if designed according to Newtonian mechanics! We shall have occasion to use the outcomes of relativity in many subsequent topics in this text.

1.2   THE PRINCIPLE OF RELATIVITY

To describe a physical event, it is necessary to establish a frame of reference, such as one that is fixed in the laboratory. Recall from your studies in mechanics that Newton's laws are valid in inertial frames of reference. *An inertial frame is one in which an object subjected to no forces moves in a straight line at constant speed*—thus the name "inertial frame" because an object observed from such a frame obeys Newton's first law, the law of inertia.[2] Furthermore, any frame or system moving with constant velocity with respect to an inertial system must also be an inertial system. Thus there is no single, preferred inertial frame for applying Newton's laws.

Inertial frame of reference

According to the **principle of Newtonian relativity,** the laws of mechanics must be the same in all inertial frames of reference. For example, if you perform an experiment while at rest in a laboratory, and an observer in a passing truck moving with constant velocity performs the same experiment, Newton's laws may be applied to both sets of observations. Specifically, in the laboratory or in the truck a ball thrown up rises and returns to the thrower's hand. Moreover, both events are measured to take the same time in the truck or in the laboratory, and Newton's second law may be used in both frames to compute this time. Although these experiments look different to different observers (see Fig. 1.1, in which the Earth observer sees a different path for the ball) and the observers measure different values of position and velocity for the ball at the same times, both observers agree on the validity of Newton's laws and principles such as conservation of energy and conservation of momentum. This implies that no experiment involving mechanics can detect any essential difference between the two inertial frames. The only thing that can be detected is the relative motion of one frame with respect to the other. That is, the notion of *absolute* motion through space is meaningless, as is the notion of a single, preferred reference frame. **Indeed, one of the firm philosophical principles of modern science is that all observers are equivalent and that the laws of nature must take the same mathematical form for all observers.** Laws of physics that exhibit the same mathematical form for observers with different motions at different locations are said to be *covariant*. Later in this section we will give specific examples of covariant physical laws.

---

[2]An example of a *noninertial frame* is a frame that accelerates in a straight line or rotates with respect to an inertial frame.

(a)   (b)

Figure 1.1   The observer in the truck sees the ball move in a vertical path when thrown upward. (b) The Earth observer views the path of the ball as a parabola.

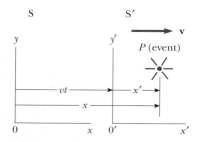

Figure 1.2   An event occurs at a point *P*. The event is observed by two observers in inertial frames S and S′, in which S′ moves with a velocity **v** relative to S.

In order to show the underlying equivalence of measurements made in different reference frames and hence the equivalence of different frames for doing physics, we need a mathematical formula that systematically relates measurements made in one reference frame to those in another. Such a relation is called a *transformation,* and the one satisfying Newtonian relativity is the so-called *Galilean transformation,* which owes its origin to Galileo. It can be derived as follows.

Consider two inertial systems or frames S and S′, as in Figure 1.2. The frame S′ moves with a constant velocity **v** along the *xx′* axes, where **v** is measured relative to the frame S. Clocks in S and S′ are synchronized, and the origins of S and S′ coincide at $t = t' = 0$. We assume that a point event, a physical phenomenon such as a lightbulb flash, occurs at the point *P*. An observer in the system S would describe the event with space–time coordinates (x, y, z, t), whereas an observer in S′ would use (x', y', z', t') to describe the same event. As we can see from Figure 1.2, these coordinates are related by the equations

$$
\begin{aligned}
x' &= x - vt \\
y' &= y \\
z' &= z \\
t' &= t
\end{aligned}
\tag{1.1}
$$

Galilean transformation of coordinates

These equations constitute what is known as a **Galilean transformation of coordinates.** Note that the fourth coordinate, time, is *assumed* to be the same in both inertial frames. That is, *in classical mechanics, all clocks run at the same rate regardless of their velocity,* so that the time at which an event occurs for an observer in S is the same as the time for the same event in S′. Consequently, the time interval between two successive events should be the same for both observers. Although this assumption may seem obvious, it turns out to be *incorrect* when treating situations in which v is comparable to the speed of light. In fact, this point represents one of the most profound differences between Newtonian concepts and the ideas contained in Einstein's theory of relativity.

Exercise 1   Show that although observers in S and S′ measure different coordinates for the ends of a stick at rest in S, they agree on the length of the stick. Assume the stick has end coordinates $x = a$ and $x = a + l$ in S and use the Galilean transformation.

An immediate and important consequence of the invariance of the distance between two points under the Galilean transformation is the invariance of force. For example if $F = \dfrac{kqQ}{(x_2 - x_1)^2}$ gives the electric force between two charges q,Q located at x_1 and x_2 on the x-axis in frame S, F', the force measured in S′, is given by $F' = \dfrac{kqQ}{(x_2' - x_1')^2} = F$ since $x_2' - x_1' = x_2 - x_1$. In fact *any* force would be invariant under the Galilean transformation as long as it involved only the relative positions of interacting particles.

Now suppose two events are separated by a distance dx and a time interval dt as measured by an observer in S. It follows from Equation 1.1 that the corresponding displacement dx' measured by an observer in S′ is given by $dx' = dx - v\,dt$, where dx is the displacement measured by an observer in S. Because $dt = dt'$, we find that

$$\frac{dx'}{dt'} = \frac{dx}{dt} - v$$

or

$$u_x' = u_x - v \tag{1.2}$$

Galilean addition law for velocities

where u_x and u_x' are the instantaneous velocities of the object relative to S and S′, respectively. This result, which is called the **Galilean addition law for velocities** (or Galilean velocity transformation), is used in everyday observations and is consistent with our intuitive notions of time and space.

To obtain the relation between the accelerations measured by observers in S and S′, we take a derivative of Equation 1.2 with respect to time and use the results that $dt = dt'$ and v is constant:

$$\frac{du_x'}{dt'} = a_x' = a_x \tag{1.3}$$

Thus observers in different inertial frames measure the same acceleration for an accelerating object. The mathematical terminology is to say that lengths (Δx), time intervals, and accelerations are *invariant* under a Galilean transformation. Example 1.1 points up the distinction between invariant and covariant and shows that **transformation equations, in addition to converting measurements made in one inertial frame to those in another, may be used to show the covariance of physical laws.**

EXAMPLE 1.1 $F_x = ma_x$ Is Covariant Under a Galilean Transformation

Assume that Newton's law $F_x = ma_x$ has been shown to hold by an observer in an inertial frame S. Show that Newton's law also holds for an observer in S′ or is covariant under the Galilean transformation, that is, has the form $F'_x = m'a'_x$. Note that inertial mass is an invariant quantity in Newtonian dynamics.

Solution   Starting with the established law $F_x = ma_x$, we use the Galilean transformation $a'_x = a_x$ and the fact that

$m' = m$ to obtain $F_x = m'a'_x$. If we now assume that F_x depends only on the relative positions of m and the particles interacting with m, that is, $F_x = f(x_2 - x_1, x_3 - x_1, \ldots)$, then $F_x = F'_x$, because the Δx's are invariant quantities. Thus we find $F'_x = m'a'_x$ and establish the covariance of Newton's second law in this simple case.

Exercise 2   *Conservation of Linear Momentum Is Covariant Under the Galilean Transformation.* Assume that two masses m'_1 and m'_2 are moving in the positive x direction with velocities v'_1 and v'_2 as measured by an observer in S′ before a collision. After the collision, the two masses stick together and move with a velocity v' in S′. Show that if an observer in S′ finds momentum to be conserved, so does an observer in S.

The Speed of Light

It is natural to ask whether the concept of Newtonian relativity and the Galilean addition law for velocities in mechanics also apply to electricity, magnetism, and optics. Recall that Maxwell in the 1860s showed that the speed of light in free space was given by $c = (\mu_0 \varepsilon_0)^{-1/2} = 3.00 \times 10^8$ m/s. Physicists of the late 1800s were certain that light waves (like familiar sound and water waves) required a definite medium in which to move, called the *ether*,[3] and that the speed of light was c only with respect to the ether or a frame fixed in the ether called the ether frame. In any other frame moving at speed v relative to the ether frame, the Galilean addition law was expected to hold. Thus, the speed of light in this other frame was expected to be $c - v$ for light traveling in the same direction as the frame, $c + v$ for light traveling opposite to the frame, and in between these two values for light moving in an arbitrary direction with respect to the moving frame.

Because the existence of the ether and a preferred ether frame would show that light was similar to other classical waves (in requiring a medium), considerable importance was attached to establishing the existence of the special ether frame. Because the speed of light is enormous, experiments involving light traveling in media moving at then attainable laboratory speeds had not been capable of detecting small changes of the size of $c \pm v$ prior to the late 1800s. Scientists of the period, realizing that the Earth moved rapidly around

[3]It was proposed by Maxwell that light and other electromagnetic waves were waves in a luminiferous ether, which was present everywhere, even in empty space. In addition to an overblown name, the ether had contradictory properties since it had to have great rigidity to support the high speed of light waves yet had to be tenuous enough to allow planets and other massive objects to pass freely through it, without resistance, as observed.

the Sun at 30 km/s, shrewdly decided to use the Earth itself as the moving frame in an attempt to improve their chances of detecting these small changes in light velocity.

From our point of view of observers fixed on Earth, we may say that we are stationary and that the special ether frame moves past us with speed v. Determining the speed of light under these circumstances is just like determining the speed of an aircraft in a moving air current or wind, and consequently we speak of an "ether wind" blowing through our apparatus fixed to the Earth. If **v** is the velocity of the ether relative to the Earth, then the speed of light should have its maximum value, $c + v$, when propagating downwind, as shown in Figure 1.3a. Likewise, the speed of light should have its minimum value, $c - v$, when propagating upwind, as in Figure 1.3b, and an intermediate value, $(c^2 - v^2)^{1/2}$, in the direction perpendicular to the ether wind, as in Figure 1.3c. *If the Sun is assumed to be at rest in the ether*, then the velocity of the ether wind would be equal to the orbital velocity of the Earth around the Sun, which has a magnitude of about 3×10^4 m/s compared to $c = 3 \times 10^8$ m/s. Thus, the change in the speed of light would be about 1 part in 10^4 for measurements in the upwind or downwind directions, and changes of this size should be detectable. However, as we show in the next section, all attempts to detect such changes and establish the existence of the ether proved futile!

1.3   THE MICHELSON–MORLEY EXPERIMENT

The famous experiment designed to detect small changes in the speed of light with motion of an observer through the ether was performed in 1887 by American physicist Albert A. Michelson (1852–1931) and the American chemist Edward W. Morley (1838–1923).[4] We should state at the outset that the outcome of the experiment was *negative*, thus contradicting the ether hypothesis. The highly accurate experimental tool perfected by these pioneers to measure small changes in light speed was the Michelson interferometer, shown in Figure 1.4. One of the arms of the interferometer was aligned along the direction of the motion of the Earth through the ether. The Earth moving through the ether would be equivalent to the ether flowing past the Earth in the opposite direction with speed v, as shown in Figure 1.4. This ether wind blowing in the opposite direction should cause the speed of light measured in the Earth's frame of reference to be $c - v$ as it approaches the mirror M_2 in Figure 1.4 and $c + v$ after reflection. The speed v is the speed of the Earth through space, and hence the speed of the ether wind, and c is the speed of light in the ether frame. The two beams of light reflected from M_1 and M_2 would recombine, and an interference pattern consisting of alternating dark and bright bands, or fringes, would be formed.

During the experiment, the interference pattern was observed while the interferometer was rotated through an angle of 90°. This rotation would change the speed of the ether wind along the direction of the arms of the interferometer. The effect of this rotation should have been to cause the fringe pattern to shift slightly but measurably. Measurements failed to show any change in the

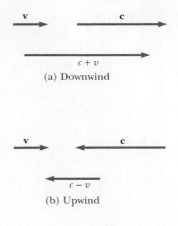
(a) Downwind

(b) Upwind

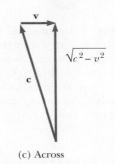
(c) Across

Figure 1.3 If the velocity of the ether wind relative to the Earth is **v,** and **c** is the velocity of light relative to the ether, the speed of light relative to the Earth is (a) $c + v$ in the downwind direction, (b) $c - v$ in the upwind direction, and (c) $(c^2 - v^2)^{1/2}$ in the direction perpendicular to the wind.

[4]A. A. Michelson and E. W. Morley, *Am. J. Sci.* 134:333, 1887.

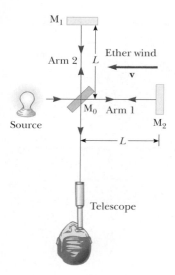

Figure 1.4 Diagram of the Michelson interferometer. According to the ether wind concept, the speed of light should be $c - v$ as the beam approaches mirror M_2 and $c + v$ after reflection.

interference pattern! The Michelson–Morley experiment was repeated by other researchers under various conditions and at different times of the year when the ether wind was expected to have changed direction and magnitude, but the results were always the same: *No fringe shift of the magnitude required was ever observed.*[5]

The negative results of the Michelson–Morley experiment not only meant that the speed of light does not depend on the direction of light propagation but also contradicted the ether hypothesis. The negative results also meant that it was impossible to measure the absolute velocity of the Earth with respect to the ether frame. As we shall see in the next section, Einstein's postulates compactly explain these and a host of other perplexing questions, relegating the idea of the ether to the ash heap of history. Light is now understood to be a phenomenon that *requires no medium for its propagation.* As a result, the idea of an ether in which these waves could travel became unnecessary.

Details of the Michelson–Morley Experiment

To understand the outcome of the Michelson–Morley experiment, let us assume that the interferometer shown in Figure 1.4 has two arms of equal length L. First consider the beam traveling parallel to the direction of the ether wind, which is taken to be horizontal in Figure 1.4. According to Newtonian mechanics, as the beam moves to the right, its speed is reduced by the wind and its speed with respect to the Earth is $c - v$. On its return journey, as the light beam moves to the left downwind, its speed with respect to the Earth is $c + v$. Thus, the time of travel to the right is $L/(c - v)$, and the time of travel to the left is $L/(c + v)$. The total time of travel for the round-trip along the horizontal path is

$$t_1 = \frac{L}{c + v} + \frac{L}{c - v} = \frac{2Lc}{c^2 - v^2} = \frac{2L}{c}\left(1 - \frac{v^2}{c^2}\right)^{-1}$$

Now consider the light beam traveling perpendicular to the wind, as shown in Figure 1.4. Because the speed of the beam relative to the Earth is $(c^2 - v^2)^{1/2}$ in this case (see Fig. 1.3c), the time of travel for each half of this trip is $L/(c^2 - v^2)^{1/2}$, and the total time of travel for the round-trip is

$$t_2 = \frac{2L}{(c^2 - v^2)^{1/2}} = \frac{2L}{c}\left(1 - \frac{v^2}{c^2}\right)^{-1/2}$$

Thus, the time difference between the light beam traveling horizontally and the beam traveling vertically is

$$\Delta t = t_1 - t_2 = \frac{2L}{c}\left[\left(1 - \frac{v^2}{c^2}\right)^{-1} - \left(1 - \frac{v^2}{c^2}\right)^{-1/2}\right]$$

[5]From an Earth observer's point of view, changes in the Earth's speed and direction in the course of a year are viewed as ether wind shifts. In fact, even if the speed of the Earth with respect to the ether were zero at some point in the Earth's orbit, six months later the speed of the Earth would be 60 km/s with respect to the ether, and one should find a clear fringe shift. None has ever been observed, however.

Because $v^2/c^2 \ll 1$, this expression can be simplified by using the following binomial expansion after dropping all terms higher than second order:

$$(1 - x)^n \approx 1 - nx \quad \text{(for } x \ll 1\text{)}$$

In our case, $x = v^2/c^2$, and we find

$$\Delta t = t_1 - t_2 \approx \frac{Lv^2}{c^3} \tag{1.4}$$

The two light beams start out in phase and return to form an interference pattern. Let us assume that the interferometer is adjusted for parallel fringes and that a telescope is focused on one of these fringes. The time difference between the two light beams gives rise to a phase difference between the beams, producing the interference fringe pattern when they combine at the position of the telescope. A difference in the pattern (Fig. 1.6) should be detected by rotating the interferometer through 90° in a horizontal plane, such that the two beams exchange roles. This results in a net time difference of twice that given by Equation 1.4. The path difference corresponding to this time difference is

$$\Delta d = c(2\Delta t) = \frac{2Lv^2}{c^2}$$

The corresponding fringe shift is equal to this path difference divided by the wavelength of light, λ, because a change in path of 1 wavelength corresponds to a shift of 1 fringe.

$$\text{Shift} = \frac{2Lv^2}{\lambda c^2} \tag{1.5}$$

In the experiments by Michelson and Morley, each light beam was reflected by mirrors many times to give an increased effective path length L of about 11 m. Using this value, and taking v to be equal to 3×10^4 m/s, the speed of the Earth about the Sun, gives a path difference of

$$\Delta d = \frac{2(11 \text{ m})(3 \times 10^4 \text{ m/s})^2}{(3 \times 10^8 \text{ m/s})^2} = 2.2 \times 10^{-7} \text{ m}$$

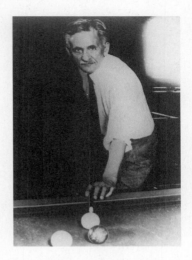

Figure 1.5 Albert A. Michelson (1852–1931). A German-American physicist, Michelson invented the interferometer and spent much of his life making accurate measurements of the speed of light. He was the first American to be awarded the Nobel prize (1907), which he received for his work in optics. His most famous experiment, conducted with Edward Morley in 1887, implied that it was impossible to measure the absolute velocity of the Earth with respect to the ether. Subsequent work by Einstein in his special theory of relativity eliminated the ether concept by assuming that the speed of light has the same value in all inertial reference frames. (*Nimitz Library, U.S.N.A./Courtesy of AIP Emilio Segre Visual Archive*)

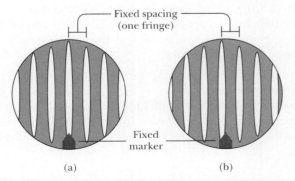

(a)   (b)

Figure 1.6 Interference fringe schematic showing (a) fringes before rotation and (b) expected fringe shift after a rotation of the interferometer by 90°.

This extra distance of travel should produce a noticeable shift in the fringe pattern. Specifically, using light of wavelength 500 nm, we find a fringe shift for rotation through 90° of

$$\text{Shift} = \frac{\Delta d}{\lambda} = \frac{2.2 \times 10^{-7} \text{ m}}{5.0 \times 10^{-7} \text{ m}} \approx 0.40$$

The precision instrument designed by Michelson and Morley had the capability of detecting a shift in the fringe pattern as small as 0.01 fringe. However, *they detected no shift in the fringe pattern.* Since then, the experiment has been repeated many times by various scientists under various conditions, and no fringe shift has ever been detected. Thus, it was concluded that one cannot detect the motion of the Earth with respect to the ether.

Many efforts were made to explain the null results of the Michelson–Morley experiment and to save the ether concept and the Galilean addition law for the velocity of light. Because all these proposals have been shown to be wrong, we consider them no further here and turn instead to an auspicious proposal made by George F. Fitzgerald and Hendrik A. Lorentz. In the 1890s, Fitzgerald and Lorentz tried to explain the null results by making the following ad hoc assumption. They proposed that the length of an object moving at speed v would contract along the direction of travel by a factor of $\sqrt{1 - v^2/c^2}$. The net result of this contraction would be a change in length of one of the arms of the interferometer such that no path difference would occur as the interferometer was rotated.

Never in the history of physics were such valiant efforts devoted to trying to explain the absence of an expected result as those directed at the Michelson–Morley experiment. The difficulties raised by this null result were tremendous, not only implying that light waves were a new kind of wave propagating without a medium but that the Galilean transformations were flawed for inertial frames moving at high relative speeds. The stage was set for Albert Einstein, who solved these problems in 1905 with his special theory of relativity.

1.4 POSTULATES OF SPECIAL RELATIVITY

In the previous section we noted the impossibility of measuring the speed of the ether with respect to the Earth and the failure of the Galilean velocity transformation in the case of light. In 1905, Albert Einstein (Fig. 1.7) proposed a theory that boldly removed these difficulties and at the same time completely altered our notion of space and time.[6] Einstein based his special theory of relativity on two postulates.

1. **The Principle of Relativity:** All the laws of physics have the same form in all inertial reference frames.

Postulates of special relativity

2. **The Constancy of the Speed of Light:** The speed of light in vacuum has the same value, $c = 3.00 \times 10^8$ m/s, in all inertial frames, regardless of the velocity of the observer or the velocity of the source emitting the light.

---

[6]A. Einstein, "On the Electrodynamics of Moving Bodies," *Ann. Physik* 17:891, 1905. For an English translation of this article and other publications by Einstein, see the book by H. Lorentz, A. Einstein, H. Minkowski, and H. Weyl, *The Principle of Relativity,* Dover, 1958.

A lbert Einstein, one of the greatest physicists of all time, was born in Ulm, Germany. As a child, Einstein was very unhappy with the discipline of German schools and completed his early education in Switzerland at age 16. Because he was unable to obtain an academic position following graduation from the Swiss Federal Polytechnic School in 1901, he accepted a job at the Swiss Patent Office in Berne. During his spare time, he continued his studies in theoretical physics. In 1905, at the age of 26, he published four scientific papers that

BIOGRAPHY

ALBERT EINSTEIN
(1879–1955)

revolutionized physics. One of these papers, which won him the Nobel prize in 1921, dealt with the photoelectric effect. Another was concerned with Brownian motion, the irregular motion of small particles suspended in a liquid. The remaining two papers were concerned with what is now considered his most important contribution of all, the

special theory of relativity. In 1915, Einstein published his work on the general theory of relativity, which relates gravity to the structure of space and time. One of the remarkable predictions of the theory is that strong gravitational forces in the vicinity of very massive objects cause light beams to deviate from straight-line paths. This and other predictions of the general theory of relativity have been experimentally verified (see the essay on our companion Web site by Clifford Will).

Einstein made many other important contributions to the development of modern physics, including the concept of the light quantum and the idea of stimulated emission of radiation, which led to the invention of the laser 40 years later. However, throughout his life, he rejected the probabilistic interpretation of quantum mechanics when describing events on the atomic scale in favor of a deterministic view. He is quoted as saying, "God does not play dice with the universe." This comment is reputed to have been answered by Niels Bohr, one of the founders of quantum mechanics, with "Don't tell God what to do!"

In 1933, Einstein left Germany (by then under Nazis control) and spent his remaining years at the Institute for Advanced Study in Princeton, New Jersey. He devoted most of his later years to an unsuccessful search for a unified theory of gravity and electromagnetism.

Figure 1.7   Albert Einstein playing his beloved fiddle, 1941. (*Hansel Micth, Life Magazine © Time, Inc.*)

The first postulate asserts that *all* the laws of physics, those dealing with electricity and magnetism, optics, thermodynamics, mechanics, and so on, will have the same mathematical form or be covariant in all coordinate frames moving with constant velocity relative to one another. This postulate is a sweeping generalization of Newton's principle of relativity, which refers only to the laws of mechanics. From an experimental point of view, Einstein's principle of relativity means that no experiment of any type can establish an absolute rest frame, and that all inertial reference frames are experimentally indistinguishable.

Note that postulate 2, the principle of the constancy of the speed of light, is consistent with postulate 1: If the speed of light was not the same in all inertial frames but was c in only one, it would be possible to distinguish between inertial frames, and one could identify a preferred, absolute frame in contradiction to postulate 1. Postulate 2 also does away with the problem of measuring the speed of the ether by essentially denying the existence of the ether and boldly asserting that light always moves with speed c with respect to any inertial observer. Postulate 2 was a brilliant theoretical insight on Einstein's part in 1905 and has since been directly confirmed experimentally in many ways. Perhaps the most direct demonstration involved measuring the speed of very high frequency electromagnetic waves (gamma rays) emitted by unstable particles (neutral pions) traveling at 99.975% of the speed of light with respect to the laboratory. The measured gamma ray speed relative to the laboratory agreed in this case to five significant figures with the speed of light in empty space.

The Michelson–Morley experiment was performed before Einstein published his work on relativity, and it is not clear that Einstein was aware of the details of the experiment. Nonetheless, the null result of the experiment can be readily understood within the framework of Einstein's theory. According to his principle of relativity, the premises of the Michelson–Morley experiment were incorrect. In the process of trying to explain the expected results, we stated that when light traveled against the ether wind its speed was $c - v$, in accordance with the Galilean addition law for velocities. However, if the state of motion of the observer or of the source has no influence on the value found for the speed of light, one will always measure the value to be c. Likewise, the light makes the return trip after reflection from the mirror at a speed of c, and not with the speed $c + v$. Thus, the motion of the Earth should not influence the fringe pattern observed in the Michelson–Morley experiment, and a null result should be expected.

Perhaps at this point you have rightly concluded that the Galilean velocity and coordinate transformations are incorrect; that is, the Galilean transformations do not keep all the laws of physics in the same form for different inertial frames. The correct coordinate and time transformations that preserve the covariant form of all physical laws in two coordinate systems moving uniformly with respect to each other are called *Lorentz transformations*. These are derived in Section 1.6. Although the Galilean transformation preserves the form of Newton's laws in two frames moving uniformly with respect to each other, Newton's laws of mechanics are limited laws that are valid only for low speeds. In general, Newton's laws must be replaced by Einstein's relativistic laws of mechanics, which hold for all speeds and are invariant, as are all physical laws, under the Lorentz transformations.

1.5   CONSEQUENCES OF SPECIAL RELATIVITY

Almost everyone who has dabbled even superficially with science is aware of some of the startling predictions that arise because of Einstein's approach to relative motion. As we examine some of the consequences of relativity in this section, we shall find that they conflict with our basic notions of space and time. We restrict our discussion to the concepts of length, time, and simultaneity, which are quite different in relativistic mechanics and Newtonian mechanics. For example, we will find that *the distance between two points and the time interval between two events depend on the frame of reference in which they are measured.* That is, *there is no such thing as absolute length or absolute time in relativity.* Furthermore, *events at different locations that occur simultaneously in one frame are not simultaneous in another frame moving uniformly past the first.*

Before we discuss the consequences of special relativity, we must first understand how an observer in an inertial reference frame describes an event. We define an event as an occurrence described by three space coordinates and one time coordinate. In general, different observers in different inertial frames would describe the same event with different spacetime coordinates.

The reference frame used to describe an event consists of a coordinate grid and a set of clocks situated at the grid intersections, as shown in Figure 1.8 in two dimensions. It is necessary that the clocks be synchronized. This can be accomplished in many ways with the help of light signals. For example, suppose an observer at the origin with a master clock sends out a pulse of light at $t = 0$. The light pulse takes a time r/c to reach a second clock, situated a distance r from the origin. Hence, the second clock will be synchronized with the clock at the origin if the second clock reads a time r/c at the instant the pulse reaches it. This procedure of synchronization assumes that the speed of light has the same value in all directions and in all inertial frames. Furthermore, the procedure concerns an event recorded by an observer in a specific inertial reference frame. Clocks in other inertial frames can be synchronized in a similar manner. An observer in some other inertial frame would assign different spacetime coordinates to events, using another coordinate grid with another array of clocks.

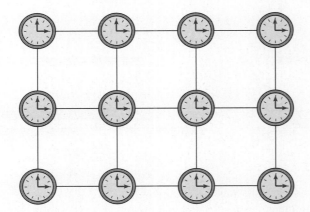

Figure 1.8   In relativity, we use a reference frame consisting of a coordinate grid and a set of synchronized clocks.

Simultaneity and the Relativity of Time

A basic premise of Newtonian mechanics is that a universal time scale exists that is the same for all observers. In fact, Newton wrote that "Absolute, true, and mathematical time, of itself, and from its own nature, flows equably without relation to anything external." Thus, Newton and his followers simply took simultaneity for granted. In his special theory of relativity, Einstein abandoned this assumption. According to Einstein, *a time interval measurement depends on the reference frame in which the measurement is made.*

Einstein devised the following thought experiment to illustrate this point. A boxcar moves with uniform velocity, and two lightning bolts strike the ends of the boxcar, as in Figure 1.9a, leaving marks on the boxcar and ground. The marks left on the boxcar are labeled A' and B'; those on the ground are labeled A and B. An observer at O' moving with the boxcar is midway between A' and B', and a ground observer at O is midway between A and B. The events recorded by the observers are the light signals from the lightning bolts.

The two light signals reach the observer at O at the same time, as indicated in Figure 1.9b. This observer realizes that the light signals have traveled at the same speed over equal distances. Thus, observer O concludes that the events at A and B occurred simultaneously. Now consider the same events as viewed by the observer on the boxcar at O'. By the time the light has reached observer O, observer O' has moved as indicated in Figure 1.9b. Thus, the light signal from B' has already swept past O', but the light from A' has not yet reached O'. According to Einstein, *observer O' must find that light travels at the same speed as that measured by observer O.* Therefore, observer O' concludes that the lightning struck the front of the boxcar before it struck the back. This thought experiment clearly demonstrates that the two events, which appear to O to be simultaneous, do not appear to O' to be simultaneous. In other words,

> Two events that are simultaneous in one frame are in general not simultaneous in a second frame moving with respect to the first. That is, simultaneity is not an absolute concept, but one that depends on the state of motion of the observer.

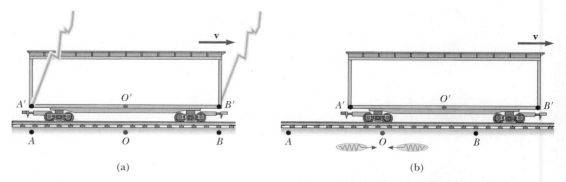

(a)   (b)

Figure 1.9 Two lightning bolts strike the ends of a moving boxcar. (a) The events appear to be simultaneous to the stationary observer at O, who is midway between A and B. (b) The events do not appear to be simultaneous to the observer at O', who claims that the front of the train is struck *before* the rear.

At this point, you might wonder which observer is right concerning the two events. The answer is that *both are correct,* because the principle of relativity states that *there is no preferred inertial frame of reference.* Although the two ob- servers reach different conclusions, both are correct in their own reference frame because the concept of simultaneity is not absolute. This, in fact, is the central point of relativity—any uniformly moving frame of reference can be used to describe events and do physics. However, observers in different inertial frames will always measure different time intervals with their clocks and differ- ent distances with their meter sticks. Nevertheless, they will both agree on the forms of the laws of physics in their respective frames, because these laws must be the same for all observers in uniform motion. It is the alteration of time and space that allows the laws of physics (including Maxwell's equations) to be the same for all observers in uniform motion.

Time Dilation

The fact that observers in different inertial frames always measure different time intervals between a pair of events can be illustrated in another way by consider- ing a vehicle moving to the right with a speed v, as in Figure 1.10a. A mirror is fixed to the ceiling of the vehicle, and observer O', at rest in this system, holds a laser a distance d below the mirror. At some instant the laser emits a pulse of light directed toward the mirror (event 1), and at some later time, after reflecting from the mirror, the pulse arrives back at the laser (event 2). Observer O' carries a clock, C', which she uses to measure the time interval $\Delta t'$ between these two events. Because the light pulse has the speed c, the time it takes to travel from O' to the mirror and back can be found from the definition of speed:

$$\Delta t' = \frac{\text{distance traveled}}{\text{speed of light}} = \frac{2d}{c} \qquad (1.6)$$

This time interval $\Delta t'$—measured by O', who, remember, is at rest in the mov- ing vehicle—requires only a *single* clock, C', in this reference frame.

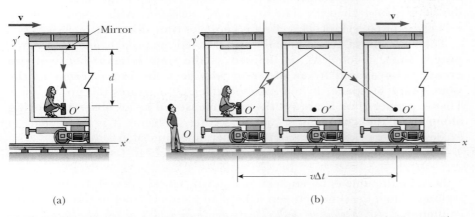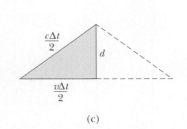

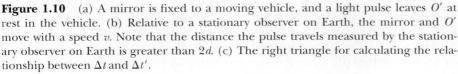

Figure 1.10   (a) A mirror is fixed to a moving vehicle, and a light pulse leaves O' at rest in the vehicle. (b) Relative to a stationary observer on Earth, the mirror and O' move with a speed v. Note that the distance the pulse travels measured by the station- ary observer on Earth is greater than $2d$. (c) The right triangle for calculating the rela- tionship between Δt and $\Delta t'$.

Now consider the same set of events as viewed by observer O in a second frame (Fig. 1.10b). According to this observer, the mirror and laser are moving to the right with a speed v, and as a result, the sequence of events appears different to this observer. By the time the light from the laser reaches the mirror, the mirror has moved to the right a distance $v\Delta t/2$, where Δt is the time interval required for the light pulse to travel from O' to the mirror and back as measured by O. In other words, O concludes that, because of the motion of the vehicle, if the light is to hit the mirror, it must leave the laser at an angle with respect to the vertical direction. Comparing Figures 1.10a and 1.10b, we see that the light must travel farther in (b) than in (a). (Note that neither observer "knows" that he or she is moving. Each is at rest in his or her own inertial frame.)

According to the second postulate of special relativity, both observers must measure c for the speed of light. Because the light travels farther according to O, it follows that the time interval Δt measured by O is longer than the time interval $\Delta t'$ measured by O'. To obtain a relationship between Δt and $\Delta t'$, it is convenient to use the right triangle shown in Figure 1.10c. The Pythagorean theorem gives

$$\left(\frac{c\Delta t}{2}\right)^2 = \left(\frac{v\Delta t}{2}\right)^2 + d^2$$

Solving for Δt gives

$$\Delta t = \frac{2d}{\sqrt{c^2 - v^2}} = \frac{2d}{c\sqrt{1 - v^2/c^2}} \tag{1.7}$$

Because $\Delta t' = 2d/c$, we can express Equation 1.7 as

Time dilation

$$\Delta t = \frac{\Delta t'}{\sqrt{1 - (v^2/c^2)}} = \gamma \Delta t' \tag{1.8}$$

where $\gamma = (1 - v^2/c^2)^{-1/2}$. Because γ is always greater than unity, this result says that the time interval Δt measured by the observer moving with respect to the clock is *longer* than the time interval $\Delta t'$ measured by the observer at rest

A moving clock runs slower

with respect to the clock. This effect is known as **time dilation.**

The time interval $\Delta t'$ in Equation 1.8 is called the proper time. In general, **proper time,** denoted Δt_p, **is defined as the time interval between two events as measured by an observer who sees the events occur at the same point in space.** In our case, observer O' measures the proper time. That is, **proper time is always the time measured by an observer moving along with the clock.** As an aid in solving problems it is convenient to express Equation 1.8 in terms of the proper time interval, Δt_p, as

$$\Delta t = \gamma \Delta t_p \tag{1.9}$$

Because the time between ticks of a moving clock, $\gamma(2d/c)$, is observed to be longer than the time between ticks of an identical clock at rest, $2d/c$, one commonly says, *"A moving clock runs slower than a clock at rest by a factor of γ."* This is true for ordinary mechanical clocks as well as for the light clock just described. In fact, we can generalize these results by stating that *all physical processes,* including chemical reactions and biological processes, slow down when observed from a reference frame in which they are moving. For example, the heartbeat of an astronaut moving through space would keep time with a clock inside the spaceship, but both the astronaut's clock and her heartbeat appear slow to an observer, with another clock, in any other reference frame. The astronaut would not have any sensation of life slowing down in her frame.

Time dilation is a very real phenomenon that has been verified by various experiments. For example, muons are unstable elementary particles that have a charge equal to that of an electron and a mass 207 times that of the electron. Muons are naturally produced by the collision of cosmic radiation with atoms at a height of several thousand meters above the surface of the Earth. Muons have a lifetime of only 2.2 μs when measured in a reference frame at rest with respect to them. If we take 2.2 μs (proper time) as the average lifetime of a muon and assume that its speed is close to the speed of light, we would find that these particles could travel a distance of about 650 m before they decayed. Hence, they could not reach the Earth from the upper atmosphere where they are produced. However, experiments show that a large number of muons *do* reach the Earth. The phenomenon of time dilation explains this effect (see Fig. 1.11a). Relative to an observer on Earth, the muons have a lifetime equal to $\gamma\tau$, where $\tau = 2.2$ μs is the lifetime in a frame of reference traveling with the muons. For example, for $v = 0.99c$, $\gamma \approx 7.1$ and $\gamma\tau \approx 16$ μs. Hence, the average distance traveled as measured by an observer on Earth is $\gamma v\tau \approx 4700$ m, as indicated in Figure 1.11b.

In 1976, experiments with muons were conducted at the laboratory of the European Council for Nuclear Research (CERN) in Geneva. Muons were injected into a large storage ring, reaching speeds of about $0.9994c$. Electrons produced by the decaying muons were detected by counters around the ring, enabling scientists to measure the decay rate, and hence the lifetime, of the muons. The lifetime of the moving muons was measured to be about 30 times as long as that of the stationary muon (see Fig. 1.12), in agreement with the prediction of relativity to within two parts in a thousand.

It is quite interesting that time dilation can be observed directly by comparing high-precision atomic clocks, one carried aboard a jet, the other

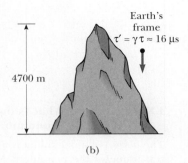

(a)

(b)

Figure 1.11 (a) Muons traveling with a speed of $0.99c$ travel only about 650 m as measured in the muons' reference frame, where their lifetime is about 2.2 μs. (b) The muons travel about 4700 m as measured by an observer on Earth. Because of time dilation, the muons' lifetime is longer as measured by the Earth observer.

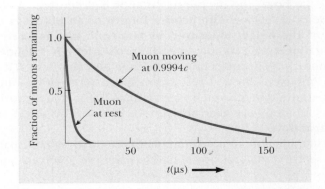

Figure 1.12 Decay curves for muons traveling at a speed of $0.9994c$ and for muons at rest.

remaining in a laboratory on Earth. The actual experiment involved the use of very stable cesium beam atomic clocks.[7] Time intervals measured with four such clocks in jet flight were compared with time intervals measured by reference atomic clocks located at the U.S. Naval Observatory. To compare these results with the theory, many factors had to be considered, including periods of acceleration and deceleration relative to the Earth, variations in direction of travel, and the weaker gravitational field experienced by the flying clocks compared with the Earth-based clocks. The results were in good agreement with the predictions of the special theory of relativity and can be completely explained in terms of the relative motion between the Earth and the jet aircraft.

EXAMPLE 1.2   What Is the Period of the Pendulum?

The period of a pendulum is measured to be 3.0 s in the rest frame of the pendulum. What is the period of the pendulum when measured by an observer moving at a speed of $0.95c$ with respect to the pendulum?

Solution   In this case, the proper time is equal to 3.0 s. From the point of view of the observer, the pendulum is moving at $0.95c$ past her. Hence the pendulum is an example of a moving clock. Because a moving clock runs slower than a stationary clock by γ, Equation 1.8 gives

$$T = \gamma T' = \frac{1}{\sqrt{1 - (0.95c)^2/c^2}} 3.0 \text{ s}$$

$$T = (3.2)(3.0 \text{ s}) = 9.6 \text{ s}$$

That is, a moving pendulum slows down or takes longer to complete one period.

Exercise 3   If the speed of the observer is increased by 5.0%, what is the period of the pendulum when measured by this observer?

Answer   43 s. Note that the 5.0% increase in speed causes more than a 300% increase in the dilated time.

Length Contraction

We have seen that measured time intervals are not absolute, that is, the time interval between two events depends on the frame of reference in which it is measured. Likewise, the measured distance between two points depends on the frame of reference. **The proper length of an object is defined as the length of the object measured by someone who is at rest with respect to the object.** You should note that proper length is defined similarly to proper time, in that proper time is the time between ticks of a clock measured by an observer who is at rest with respect to the clock. The length of an object measured by someone in a reference frame that is moving relative to the object is always less than the proper length. This effect is known as **length contraction.**

To understand length contraction quantitatively, consider a spaceship traveling with a speed v from one star to another and two observers, one on Earth

[7]J. C. Hafele and R. E. Keating, "Around the World Atomic Clocks: Relativistic Time Gains Observed," *Science,* July 14, 1972, p. 168.

and the other in the spaceship. The observer at rest on Earth (and also assumed to be at rest with respect to the two stars) measures the distance between the stars to be L_p, where L_p is the proper length. According to this observer, the time it takes the spaceship to complete the voyage is $\Delta t = L_p/v$. What does an observer in the moving spaceship measure for the distance between the stars? Because of time dilation, the space traveler measures a smaller time of travel: $\Delta t' = \Delta t/\gamma$. The space traveler claims to be at rest and sees the destination star as moving toward the spaceship with speed v. Because the space traveler reaches the star in the shorter time $\Delta t'$, he or she concludes that the distance, L, between the stars is shorter than L_p. This distance measured by the space traveler is given by

$$L = v\Delta t' = v\frac{\Delta t}{\gamma}$$

Because $L_p = v\Delta t$, we see that $L = L_p/\gamma$ or

$$L = L_p\left(1 - \frac{v^2}{c^2}\right)^{1/2}$$ (1.10)   **Length contraction**

where $(1 - v^2/c^2)^{1/2}$ is a factor less than 1. This result may be interpreted as follows:

> If an object has a proper length L_p when it is measured by an observer at rest with respect to the object, when it moves with speed v in a direction parallel to its length, its length L is measured to be shorter according to $L = L_p\left(1 - \frac{v^2}{c^2}\right)^{1/2}$.

Note that the length contraction takes place only along the direction of motion. For example, suppose a stick moves past a stationary Earth observer with a speed v, as in Figure 1.13b. The length of the stick as measured by an observer in the frame attached to it is the proper length L_p, as illustrated in Figure 1.13a. The length of the stick, L, as measured by the Earth observer is shorter than L_p by the factor $(1 - v^2/c^2)^{1/2}$. Note that length contraction is a symmetric effect: If the stick were at rest on Earth, an observer in a frame moving past the earth at speed v would also measure its length to be shorter by the same factor $(1 - v^2/c^2)^{1/2}$.

As we mentioned earlier, one of the basic tenets of relativity is that all inertial frames are equivalent for analyzing an experiment. Let us return to the example of the decaying muons moving at speeds close to the speed of light to see an example of this. An observer in the muon's reference frame would measure the proper lifetime, whereas an Earth-based observer measures the proper height of the mountain in Figure 1.11. In the muon's reference frame, there is no time dilation, but the distance of travel is observed to be shorter when measured from this frame. Likewise, in the Earth observer's reference frame, there is time dilation, but the distance of travel is measured to be the proper height of the mountain. Thus, when calculations on the muon are performed in both frames, one sees the effect of "offsetting penalties," and the outcome of the experiment is the same!

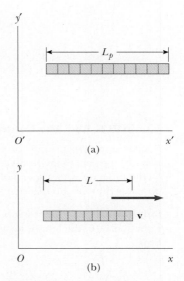

Figure 1.13 A stick moves to the right with a speed v. (a) The stick as viewed in a frame attached to it. (b) The stick as seen by an observer who sees it move past her at v. Any inertial observer finds that the length of a meter stick moving past her with speed v is less than the length of a stationary stick by a factor of $(1 - v^2/c^2)^{1/2}$.

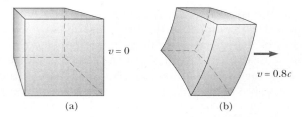

Figure 1.14   Computer-simulated photographs of a box (a) at rest relative to the camera and (b) moving at a speed $v = 0.8c$ relative to the camera.

Note that proper length and proper time are measured in *different* reference frames.

    If an object in the shape of a box passing by could be photographed, its image would show length contraction, but its shape would also be distorted. This is illustrated in the computer-simulated drawings shown in Figure 1.14 for a box moving past an observer with a speed $v = 0.8c$. When the shutter of the camera is opened, it records the shape of the object at a given instant of time. Because light from different parts of the object must arrive at the shutter at the same time (when the photograph is taken), light from more distant parts of the object must start its journey earlier than light from closer parts. Hence, the photograph records different parts of the object at different times. This results in a highly distorted image, which shows horizontal length contraction, vertical curvature, and image rotation.

EXAMPLE 1.3   The Contraction of a Spaceship

A spaceship is measured to be 100 m long while it is at rest with respect to an observer. If this spaceship now flies by the observer with a speed of $0.99c$, what length will the observer find for the spaceship?

Solution   The proper length of the ship is 100 m. From Equation 1.10, the length measured as the spaceship flies by is

$$L = L_p \sqrt{1 - \frac{v^2}{c^2}} = (100 \text{ m}) \sqrt{1 - \frac{(0.99c)^2}{c^2}} = 14 \text{ m}$$

Exercise 4   If the ship moves past the observer at $0.01000c$, what length will the observer measure?

Answer   99.99 m.

EXAMPLE 1.4   How High Is the Spaceship?

An observer on Earth sees a spaceship at an altitude of 435 m moving downward toward the Earth at $0.970c$. What is the altitude of the spaceship as measured by an observer in the spaceship?

Solution   The proper length here is the Earth–ship separation as seen by the Earth-based observer, or 435 m. The moving observer in the ship finds this separation (the altitude) to be

$$L = L_p \sqrt{1 - \frac{v^2}{c^2}} = (435 \text{ m}) \sqrt{1 - \frac{(0.970c)^2}{c^2}}$$
$$= 106 \text{ m}$$

EXAMPLE 1.5   The Triangular Spaceship

A spaceship in the form of a triangle flies by an observer at $0.950c$. When the ship is measured by an observer at rest with respect to the ship (Fig. 1.15a), the distances x and y are found to be 50.0 m and 25.0 m, respectively. What is the shape of the ship as seen by an observer who sees the ship in motion along the direction shown in Figure 1.15b?

Solution   The observer sees the horizontal length of the ship to be contracted to a length of

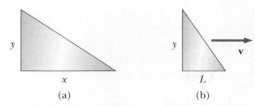

Figure 1.15 (Example 1.5) (a) When the spaceship is at rest, its shape is as shown. (b) The spaceship appears to look like this when it moves to the right with a speed v. Note that only its x dimension is contracted in this case.

$$L = L_p \sqrt{1 - \frac{v^2}{c^2}}$$

$$= (50.0 \text{ m}) \sqrt{1 - \frac{(0.950c)^2}{c^2}} = 15.6 \text{ m}$$

The 25-m vertical height is unchanged because it is perpendicular to the direction of relative motion between the observer and the spaceship. Figure 1.15b represents the shape of the spaceship as seen by the observer who sees the ship in motion.

THE TWINS PARADOX

OPTIONAL

If we placed a living organism in a box . . . one could arrange that the organism, after an arbitrary lengthy flight, could be returned to its original spot in a scarcely altered condition, while corresponding organisms which had remained in their original positions had long since given way to new generations. **(Einstein's original statement of the twins paradox in 1911)**

An intriguing consequence of time dilation is the so-called clock or twins paradox. Consider an experiment involving a set of identical 20-year-old twins named Speedo and Goslo. The twins carry with them identical clocks that have been synchronized. Speedo, the more adventuresome of the two, sets out on an epic journey to planet X, 10 lightyears from Earth. (Note that 1 lightyear (ly) is the distance light travels through free space in 1 year.) Furthermore, his spaceship is capable of a speed of $0.500c$ relative to the inertial frame of his twin brother. After reaching planet X, Speedo becomes homesick and impetuously sets out on a return trip to Earth at the same high speed of the outbound journey. On his return, Speedo is shocked to discover that many things have changed during his absence. To Speedo, the most significant change is that his twin brother Goslo *has aged more than he* and is now 60 years of age. Speedo, on the other hand, has aged by only 34.6 years.

At this point, it is fair to raise the following question—Which twin is the traveler and which twin would really be the younger of the two? If motion is relative, the twins are in a symmetric situation and either's point of view is equally valid. From Speedo's perspective, it is he who is at rest while Goslo is on a high-speed space journey. To Speedo, it is Goslo and the Earth that have raced away on a 17.3-year journey and then headed back for another 17.3 years. This leads to the paradox: Which twin will have developed the signs of excess aging?

To resolve this apparent paradox, recall that special relativity deals with inertial frames of reference moving with respect to one another at *uniform speed*. However, the trip situation is not symmetric. Speedo, the space traveler, must experience acceleration during his journey. As a result, his state of motion is not always uniform, and consequently Speedo is not in an inertial frame. He cannot regard himself to always be at rest and Goslo to be in uniform motion. Hence Speedo cannot apply simple time dilation to Goslo's motion, because to do so would be an incorrect application of special relativity. Therefore there is no paradox and Speedo will really be the younger twin at the end of the trip.

The conclusion that Speedo is not in a *single* inertial frame is inescapable. We may diminish the length of time needed to accelerate and decelerate Speedo's spaceship to an insignificant interval by using very large and expensive rockets and claim that he spends all but a negligible amount of time coasting to planet X at $0.500c$ in an inertial frame. However, to return to Earth, Speedo must slow down, reverse his motion, and return in a different inertial frame, one which is moving uniformly toward the Earth. At the very best, Speedo is in *two* different inertial frames. The important point is that even when we idealize Speedo's trip, it consists of motion in two different inertial frames and a very real lurch as he hops from the outbound ship to the returning Earth shuttle. Only Goslo remains in a single inertial frame, and so only he can correctly apply the simple time dilation formula of special relativity to Speedo's trip. Thus, Goslo finds that instead of aging 40 years $(20 \text{ ly}/0.500c)$, Speedo actually ages only $(\sqrt{1 - v^2/c^2})(40 \text{ yr})$, or 34.6 yr. Clearly, Speedo spends 17.3 years going to planet X and 17.3 years returning in agreement with our earlier statement.

The result that Speedo ages 34.6 yr while Goslo ages 40 yr can be confirmed in a very direct experimental way from Speedo's frame if we use the special theory of relativity but take into account the fact that *Speedo's idealized trip takes place in two different inertial frames.* In yet another flight of fancy, suppose that Goslo celebrates his birthday each year in a flashy way, sending a powerful laser pulse to inform his twin that Goslo is another year older and wiser. Because Speedo is in an inertial frame on the outbound trip in which the Earth appears to be receding at $0.500c$, the flashes occur at a rate of one every

$$\frac{1}{\sqrt{1 - (v^2/c^2)}} \text{ yr} = \frac{1}{\sqrt{1 - [(0.500c)^2/c^2]}} \text{ yr} = 1.15 \text{ yr}$$

This occurs because moving clocks run slower. Also, because the Earth is receding, each successive flash must travel an additional distance of $(0.500c)(1.15 \text{ yr})$ between flashes. Consequently, Speedo observes flashes to arrive with a total time between flashes of $1.15 \text{ yr} + (0.500c)(1.15 \text{ yr})/c = 1.73 \text{ yr}$. The total number of flashes seen by Speedo on his outbound voyage is therefore $(1 \text{ flash}/1.73 \text{ yr})(17.3 \text{ yr}) = 10$ flashes. This means that Speedo views the Earth clocks to run more slowly than his own on the outbound trip because he observes 17.3 years to have passed for him while only 10 years have passed on Earth.

On the return voyage, because the Earth is racing toward Speedo with speed $0.500c$, successive flashes have less distance to travel, and the total time Speedo sees between the arrival of flashes is drastically shortened: $1.15 \text{ yr} - (0.500)(1.15 \text{ yr}) = 0.577 \text{ yr/flash}$. Thus, during the return trip, Speedo sees $(1 \text{ flash}/0.577 \text{ yr})(17.3 \text{ yr}) = 30$ flashes in total. In sum, during his 34.6 years of travel, Speedo receives $(10 + 30)$ flashes, indicating that his twin has aged 40 years. Notice that there has been no failure of special relativity for Speedo as long as we take his *two* inertial frames into account and assume negligible acceleration and deceleration times. On both the outbound and inbound trips Speedo correctly judges the Earth clocks to run slower than his own, but on the return trip his rapid movement toward the light flashes more than compensates for the slower rate of flashing.

Figure 1.16 "I love hearing that lonesome wail of the train whistle as the frequency of the wave changes due to the Doppler effect."

The Relativistic Doppler Shift

Another important consequence of time dilation is the shift in frequency found for light emitted by atoms in motion as opposed to light emitted by atoms at rest. A similar phenomenon, the mournful drop in pitch of the sound of a passing train's whistle, known as the Doppler effect, is quite familiar to most cowboys (Fig. 1.16). The Doppler shift for sound is usually studied in introductory physics courses and is especially interesting because motion of the source with respect to the medium of propagation can be clearly distinguished from motion of the observer. This means that in the case of sound we can distinguish the "absolute motion" of frames moving with respect to the air, which is the medium of propagation for sound.

Light waves must be analyzed differently from sound, because light waves require no medium of propagation and no method exists of distinguishing the motion of the light source from the motion of the observer. Thus, we expect to find a different formula for the Doppler shift of light waves, one that is only sensitive to the *relative* motion of source and observer and that holds for relative speeds of source and observer approaching c.

Consider a source of light waves at rest in frame S, emitting waves of frequency f and wavelength λ as measured in S. We wish to find the frequency f' and wavelength λ' of the light as measured by an observer fixed in frame S', which is moving with speed v toward S, as shown in Figure 1.17a and b. In general, we expect f' to be greater than f if S' approaches S because more wave crests are crossed per unit time, and we expect f' to be less than f if S' recedes from S. In particular, consider the situation from the point of view of an observer fixed in S', as shown in Figure 1.18. This figure shows two successive wavefronts (color) emitted when the approaching source is at positions 1 and 2, respectively. If the time between the emission of these wavefronts as measured in S' is T', during this time front 1 will move a distance cT' from position 1. During this same time, the light source

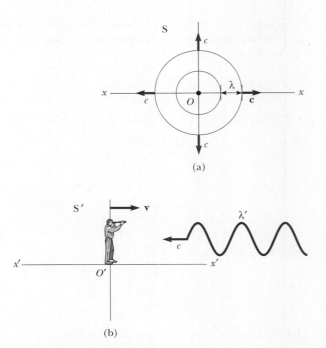

(a)

(b)

Figure 1.17   (a) A light source fixed in S emits wave crests separated in space by λ and moving outward at speed c as seen from S. (b) What wavelength λ' is measured by an observer at rest in S'? S' is a frame approaching S at speed v such that the x- and x'-axes coincide.

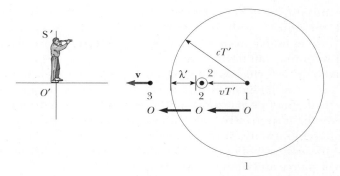

Figure 1.18   The view from S'. 1, 2, and 3 (in black) show three successive positions of O separated in time by T', the period of the light as measured from S'.

will advance a distance vT' to the left of position 1, and the distance between successive wavefronts will be measured in S' to be

$$\lambda' = cT' - vT' \tag{1.11}$$

Because we wish to obtain a formula for f' (the frequency measured in S') in terms of f (the frequency measured in S), we use the expression for λ' from Equation 1.11 in $f' = c/\lambda'$ to obtain

$$f' = \frac{c}{(c - v)\,T'} \tag{1.12}$$

To eliminate T' in favor of T, note that T is the proper time; that is, T is the time between two events (the emission of successive wavefronts) that occur at the same place in S, and consequently,

$$T' = \frac{T}{\sqrt{1 - (v^2/c^2)}}$$

Substituting for T' in Equation 1.12 and using $f = 1/T$ gives

$$f' = \frac{\sqrt{1 - (v^2/c^2)}}{1 - (v/c)}\,f \tag{1.13}$$

or

$$f' = \frac{\sqrt{1 + (v/c)}}{\sqrt{1 - (v/c)}}\,f \tag{1.14}$$

For clarity, this expression is often written

$$f_{\text{obs}} = \frac{\sqrt{1 + (v/c)}}{\sqrt{1 - (v/c)}}\,f_{\text{source}} \tag{1.15}$$

where f_{obs} is the frequency measured by an observer approaching a light source, and f_{source} is the frequency as measured in the source's rest frame.

Equation 1.15 is the relativistic Doppler shift formula, which, unlike the Doppler formula for sound, depends only on the relative speed v of the source and observer and holds for relative speeds as large as c. Equation 1.15 agrees with physical intuition in predicting f_{obs} to be greater than f_{source} for an approaching emitter and receiver. The expression for the case of a receding source is obtained by replacing v with $-v$ in Equation 1.15.

Although Christian Johann Doppler's name is most frequently associated with the effect in sound, he originally developed his ideas in an effort to understand the shift in frequency or wavelength of the light emitted by moving atoms and astronomical objects. The most spectacular and dramatic use of the Doppler effect has occurred in just this area in explaining the famous red shift of absorption lines (wavelengths) observed for most galaxies. (A galaxy is a cluster of millions of stars.) The term *redshift* refers to the shift of known absorption lines toward longer wavelengths, that is, toward the red end of the visible spectrum. For example, lines normally found in the extreme violet region for a galaxy at rest with respect to the Earth are shifted about 100 nm toward the red end of the spectrum for distant galaxies—indicating that these distant galaxies are rapidly *receding* from us. The American astronomer Edwin Hubble used this technique to confirm that most galaxies are moving away from us and that the Universe is *expanding*. (For more about the expanding Universe see Chapter 16, Cosmology, on our Web site.)

EXAMPLE 1.6   Determining the Speed of Recession of the Galaxy Hydra

The light emitted by a galaxy contains a continuous distribution of wavelengths because the galaxy is composed of millions of stars and other thermal emitters. However, some narrow gaps occur in the continuous spectrum where the radiation has been strongly absorbed by cooler gases in the galaxy. In particular, a cloud of ionized calcium atoms produces very strong absorption at 394 nm for a galaxy at rest with respect to the Earth. For the galaxy Hydra, which is 200 million ly away, this absorption is shifted to 475 nm. How fast is Hydra moving away from the Earth?

Solution   For an approaching source and observer, $f_{obs} > f_{source}$ and $\lambda_{obs} < \lambda_{source}$ because $f_{obs}\lambda_{obs} = c = f_{source}\lambda_{source}$. In the case of Hydra, $\lambda_{obs} > \lambda_{source}$, so Hydra must be receding and we must use

$$f_{obs} = \frac{\sqrt{1 - (v/c)}}{\sqrt{1 + (v/c)}} f_{source}$$

Substituting $f_{obs} = c/\lambda_{obs}$ and $f_{source} = c/\lambda_{source}$ into this equation gives

$$\lambda_{obs} = \frac{\sqrt{1 + (v/c)}}{\sqrt{1 - (v/c)}} \lambda_{source}$$

Finally, solving for v/c, we find

$$\frac{v}{c} = \frac{\lambda_{obs}^2 - \lambda_{source}^2}{\lambda_{obs}^2 + \lambda_{source}^2}$$

or

$$\frac{v}{c} = \frac{(475 \text{ nm})^2 - (394 \text{ nm})^2}{(475 \text{ nm})^2 + (394 \text{ nm})^2} = 0.185$$

Therefore, Hydra is receding from us at $v = 0.185c = 5.54 \times 10^7$ m/s.

1.6   THE LORENTZ TRANSFORMATION

We have seen that the Galilean transformation is not valid when v approaches the speed of light. In this section, we shall derive the correct coordinate and velocity transformation equations that apply for all speeds in the range of $0 \le v < c$. This transformation, known as the Lorentz transformation, was laboriously derived by Hendrik A. Lorentz (1853–1928, Dutch physicist) in 1890 as the transformation that made Maxwell's equations covariant. However, its real significance in a physical theory transcending electromagnetism was first recognized by Einstein.

The Lorentz coordinate transformation is a set of formulas that relates the space and time coordinates of two inertial observers moving with a relative speed v. We have already seen two consequences of the Lorentz transformation in the time dilation and length contraction formulas. The Lorentz velocity transformation is the set of formulas that relate the velocity components u_x, u_y, u_z of an object moving in frame S to the velocity components u'_x, u'_y, u'_z of the same object measured in frame S′, which is moving with a speed v relative to S. The Lorentz transformation formulas provide a formal, concise, and almost mechanical method of solution of relativity problems.

We start our derivation of the Lorentz transformation by noting that a reasonable guess (based on physical intuition) about the form of the coordinate equations can greatly reduce the algebraic complexity of the derivation. For simplicity, consider the standard frames, S and S′, with S′ moving at a speed v along the $+x$ direction (see Fig. 1.2). The origins of the two frames coincide at $t' = t = 0$. A reasonable guess about the dependence of x' on x and t is

$$x' = G(x - vt) \tag{1.16}$$

where G is a dimensionless factor that does not depend on x or t but is some function of v/c such that G is 1 in the limit as v/c approaches 0. The form of Equation 1.16 is suggested by the form of the Galilean transformation, $x' = x - vt$, which we know is correct in the limit as v/c approaches zero. The fact that Equation 1.16 is linear in x and t is also important because we require a single event in S (specified by x_1, t_1) to correspond to a single event in S′ (specified by x'_1, t'_1). Assuming that Equation 1.16 is correct, we can write the *inverse Lorentz coordinate transformation* for x in terms of x' and t' as

$$x = G(x' + vt') \tag{1.17}$$

This follows from Einstein's first postulate of relativity, which requires the laws of physics to have the same form in both S and S′ and where the sign of v has been changed to take into account the difference in direction of motion of the two frames. In fact, we should point out that this important technique for obtaining the inverse of a Lorentz transformation may be followed as a general rule:

To obtain the inverse Lorentz transformation of any quantity, simply interchange primed and unprimed variables and reverse the sign of the frame velocity.

Returning to our derivation of the Lorentz transformations, our argument will be to take the differentials of x' and t' and form an expression that relates the measured velocity of an object in S′, $u'_x = dx'/dt'$, to the measured velocity of that object in S, $u_x = dx/dt$. We then determine G by requiring that u'_x must equal c in the case that u_x, the velocity of an object in frame S, is equal to c, in accord with Einstein's second postulate of relativity. Once G has been determined, this simple algebraic argument conveniently provides both the Lorentz coordinate and velocity transformations. Following this plan, we first find

$$t' = G\left\{t + (1/G^2 - 1)\,\frac{x}{v}\right\}$$ (1.18)

by substituting Equation 1.16 into 1.17 and solving for t'. Taking differentials of Equations 1.16 and 1.18 yields

$$dx' = G(dx - v\,dt)$$ (1.19)

$$dt' = G\left\{dt + (1/G^2 - 1)\,\frac{dx}{v}\right\}$$ (1.20)

Forming $u'_x = dx'/dt'$ leads, after some simplification, to

$$u'_x = \frac{dx'}{dt'} = \frac{u_x - v}{1 + (1/G^2 - 1)(u_x/v)}$$ (1.21)

where $u_x = dx/dt$.

Postulate 2 requires that the velocity of light be c for any observer, so in the case $u_x = c$, we must also have $u'_x = c$. Using this condition in Equation 1.21 gives

$$c = \frac{c - v}{1 + (1/G^2 - 1)(c/v)}$$ (1.22)

Equation 1.22 may be solved to give

$$G \equiv \gamma = \frac{1}{\sqrt{1 - (v^2/c^2)}}$$

The direct coordinate transformation is thus $x' = \gamma(x - vt)$, and the inverse transformation is $x = \gamma(x' + vt')$. To get the time transformation (t' as a function of t and x), substitute $G = \gamma$ into Equation 1.18 to obtain

$$t' = \gamma\left(t - \frac{vx}{c^2}\right)$$

In summary, the complete coordinate transformations between an event found to occur at (x, y, z, t) in S and (x', y', z', t') in S' are

$$x' = \gamma(x - vt)$$ (1.23)

$$y' = y$$ (1.24)

$$z' = z$$ (1.25)

$$t' = \gamma\left(t - \frac{vx}{c^2}\right)$$ (1.26)

where

$$\gamma = \frac{1}{\sqrt{1 - (v^2/c^2)}}$$

If we wish to transform coordinates of an event in the S' frame to coordinates in the S frame, we simply replace v by $-v$ and interchange the primed

Lorentz transformation for S → S'

and unprimed coordinates in Equations 1.23 through 1.26. The resulting inverse transformation is given by

$$x = \gamma(x' - vt')$$
$$y = y'$$
$$z = z'$$
$$t = \gamma\left(t' + \frac{vx'}{c^2}\right)$$

(1.27)

Inverse Lorentz transformation for S' → S

where

$$\gamma = \frac{1}{\sqrt{1 - v^2/c^2}}$$

In the Lorentz transformation, note that t depends on both t' and x'. Likewise, t' depends on both t and x. This is unlike the case of the Galilean transformation, in which $t = t'$. When $v \ll c$, the Lorentz transformation should reduce to the Galilean transformation. To check this, note that as $v \to 0$, $v/c < 1$ and $v^2/c^2 \ll 1$, so that Equations 1.23–1.26 reduce in this limit to the Galilean coordinate transformation equations, given by

$$x' = x - vt \qquad y' = y \qquad z' = z \qquad t' = t$$

EXAMPLE 1.7   Time Dilation Is Contained in the Lorentz Transformation

Show that the phenomenon of time dilation is contained in the Lorentz coordinate transformation. A light located at (x_0, y_0, z_0) is turned abruptly on at t_1 and off at t_2 in frame S. (a) For what time interval is the light measured to be on in frame S'? (See Figure 1.2 for a picture of the two standard frames.) (b) What is the distance between where the light is turned on and off as measured by S'?

Solution   (a) The two events, the light turning on and the light turning off, are measured to occur in the two frames as follows:

	Event 1 (light on)	Event 2 (light off)
Frame S	x_0, t_1	x_0, t_2
Frame S'	$x'_1 = \gamma(x_0 - vt_1)$	$x'_2 = \gamma(x_0 - vt_2)$
	$t'_1 = \gamma\left(t_1 - \dfrac{vx_0}{c^2}\right)$	$t'_2 = \gamma\left(t_2 - \dfrac{vx_0}{c^2}\right)$

Note that the y and z coordinates are not affected because the motion of S' is along x. As measured by S', the light is on for a time interval

$$t'_2 - t'_1 = \gamma\left(t_2 - \frac{vx_0}{c^2}\right) - \gamma\left(t_1 - \frac{vx_0}{c^2}\right)$$
$$= \gamma(t_2 - t_1)$$

Because $\gamma > 1$ and $(t_2 - t_1)$ is the proper time, it follows that $(t'_2 - t'_1) > (t_2 - t_1)$, and we have recovered our previous result for time dilation, Equation 1.8.

(b) Although event 1 and event 2 occur at the same place in S, they are measured to occur at a separation of $x'_2 - x'_1$ in S' where

$$x'_2 - x'_1 = (\gamma x_0 - \gamma vt_2) - (\gamma x_0 - \gamma vt_1)$$
$$= \gamma v(t_1 - t_2)$$

This result is reasonable because it reduces to

$$v(t_1 - t_2) \quad \text{for } v/c \ll 1$$

Can you explain why $x'_2 - x'_1$ is negative?

Exercise 5   Use the Lorentz transformation to derive the expression for length contraction. Note that the length of a moving object is determined by measuring the positions of both ends simultaneously.

Lorentz Velocity Transformation

The explicit form of the Lorentz velocity transformation follows immediately upon substitution of $G \equiv \gamma = 1/\sqrt{1 - (v^2/c^2)}$ into Equation 1.21:

$$u'_x = \frac{u_x - v}{1 - (u_x v/c^2)} \qquad (1.28)$$

Lorentz velocity transformation for S → S′

where $u'_x = dx'/dt'$ is the instantaneous velocity in the x direction measured in S′ and $u_x = dx/dt$ is the velocity component u_x of the object as measured in S. Similarly, if the object has velocity components along y and z, the components in S′ are

$$u'_y = \frac{dy'}{dt'} = \frac{dy}{\gamma(dt - v\,dx/c^2)} = \frac{u_y}{\gamma[1 - (u_x v/c^2)]}$$

$$\text{and} \quad u'_z = \frac{u_z}{\gamma[1 - (u_x v/c^2)]} \qquad (1.29)$$

When u_x and v are both much smaller than c (the nonrelativistic case), we see that the denominator of Equation 1.28 approaches unity, and so $u'_x \approx u_x - v$. This corresponds to the Galilean velocity transformation. In the other extreme, when $u_x = c$, Equation 1.28 becomes

$$u'_x = \frac{c - v}{1 - (cv/c^2)} = \frac{c[1 - (v/c)]}{1 - (v/c)} = c$$

From this result, we see that an object moving with a speed c relative to an observer in S also has a speed c relative to an observer in S′—*independent* of the relative motion of S and S′. Note that this conclusion is consistent with Einstein's second postulate, namely, that the speed of light must be c with respect to all inertial frames of reference. Furthermore, the speed of an object can never exceed c. That is, the speed of light is the "ultimate" speed. We return to this point later in Chapter 2 when we consider the energy of a particle.

To obtain u_x in terms of u'_x, replace v by $-v$ in Equation 1.28 and interchange u_x and u'_x following the rule stated earlier for obtaining the inverse transformation. This gives

$$u_x = \frac{u'_x + v}{1 + (u'_x v/c^2)} \qquad (1.30)$$

Inverse Lorentz velocity transformation for S′ → S

EXAMPLE 1.8   Relative Velocity of Spaceships

Two spaceships A and B are moving in *opposite* directions, as in Figure 1.19. An observer on Earth measures the speed of A to be $0.750c$ and the speed of B to be $0.850c$. Find the velocity of B with respect to A.

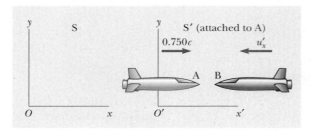

Figure 1.19 (Example 1.8) Two spaceships A and B move in *opposite* directions. The velocity of B relative to A is *less* than c and is obtained by using the relativistic velocity transformation.

Solution This problem can be solved by taking the S′ frame to be attached to spacecraft A, so that $v = 0.750c$ relative to an observer on Earth (the S frame). Spacecraft B can be considered as an object moving to the left with a velocity $u_x = -0.850c$ relative to the Earth observer. Hence, the velocity of B with respect to A can be obtained using Equation 1.28:

$$u'_x = \frac{u_x - v}{1 - \dfrac{u_x v}{c^2}} = \frac{-0.850c - 0.750c}{1 - \dfrac{(-0.850c)(0.750c)}{c^2}}$$

$$= -0.9771c$$

The negative sign for u'_x indicates that spaceship B is moving in the negative x direction as observed by A. Note that the result is less than c. That is, a body with speed less than c in one frame of reference must have a speed less than c in *any other* frame. If the incorrect Galilean velocity transformation were used in this example, we would find that $u'_x = u_x - v = -0.850c - 0.750c = -1.600c$, which is greater than the universal limiting speed c.

EXAMPLE 1.9   The Speeding Motorcycle

Imagine a motorcycle rider moving with a speed of $0.800c$ past a stationary observer, as shown in Figure 1.20. If the rider tosses a ball in the forward direction with a speed of $0.700c$ with respect to himself, what is the speed of the ball as seen by the stationary observer?

Solution In this situation, the velocity of the motorcycle with respect to the stationary observer is $v = 0.800c$. The velocity of the ball in the frame of reference of the motorcyclist is $u'_x = 0.700c$. Therefore, the velocity, u_x, of

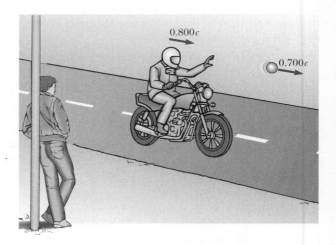

Figure 1.20 (Example 1.9) A motorcyclist moves past a stationary observer with a speed of $0.800c$ and throws a ball in the direction of motion with a speed of $0.700c$ relative to himself.

the ball relative to the stationary observer is

$$u_x = \frac{u'_x + v}{1 + (u'_x v / c^2)}$$

$$= \frac{0.700c + 0.800c}{1 + [(0.700c)(0.800c)/c^2]} = 0.9615c$$

Exercise 6 Suppose that the motorcyclist moving with a speed $0.800c$ turns on a beam of light that moves away from him with a speed of c in the same direction as the moving motorcycle. What would the stationary observer measure for the speed of the beam of light?

Answer c.

EXAMPLE 1.10   Relativistic Leaders of the Pack!

Imagine two motorcycle gang leaders racing at relativistic speeds along perpendicular paths from the local pool hall, as shown in Figure 1.21. How fast does pack leader Beta recede over Alpha's right shoulder as seen by Alpha?

Solution Figure 1.21 shows the situation as seen by a stationary police officer located in frame S, who observes the following:

Pack Leader Alpha	$u_x = 0.75c$	$u_y = 0$
Pack Leader Beta	$u_x = 0$	$u_y = -0.90c$

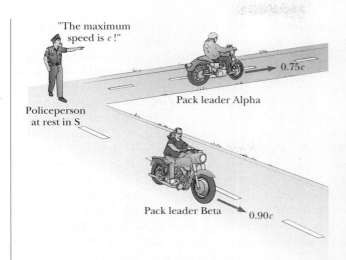

"The maximum speed is c!"

0.75c

Pack leader Alpha

Policeperson at rest in S

Pack leader Beta   0.90c

Figure 1.21 (Example 1.10) Two motorcycle pack leaders, Alpha and Beta, blaze past a stationary police officer. They are leading their respective gangs from the pool hall along perpendicular roads.

To get Beta's speed of recession as seen by Alpha, we take S′ to move along with Alpha, as shown in Figure 1.22, and we calculate u'_x and u'_y for Beta using Equations 1.28 and 1.29:

$$u'_x = \frac{u_x - v}{1 - (u_x v/c^2)} = \frac{0 - 0.75c}{1 - [(0)(0.75c)/c^2]} = -0.75c$$

$$u'_y = \frac{u_y}{\gamma[1 - (u_x v/c^2)]}$$

$$= \frac{\sqrt{1 - [(0.75c)^2/c^2]}(-0.90c)}{1 - [(0)(0.75c)/c^2]} = -0.60c$$

The speed of recession of Beta away from Alpha as observed by Alpha is then found to be less than c as required by relativity.

$$u' = \sqrt{(u'_x)^2 + (u'_y)^2} = \sqrt{(-0.75c)^2 + (-0.60c)^2} = 0.96c$$

Exercise 7 Calculate the classical speed of recession of Beta from Alpha using the incorrect Galilean transformation.

Answer $1.2c$

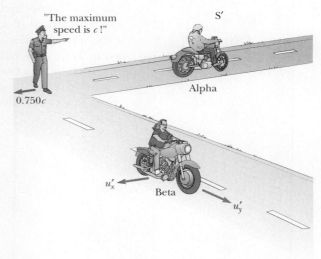

"The maximum speed is c!"   S′

0.750c

Alpha

u'_x Beta u'_y

Figure 1.22 (Example 1.10) Pack leader Alpha's view of things.

1.7   SPACETIME AND CAUSALITY

The views of space and time which I wish to lay before you have sprung from the soil of experimental physics, and therein lies their strength. They are radical. Henceforth space by itself, and time by itself, are doomed to fade away into mere shadows, and only a union of the two will preserve an independent reality. (Hermann Minkowski, 1908, in an address to the Assembly of German Natural Scientists and Physicians)

We have seen in relativity that space and time coordinates cannot be treated separately. This is apparent from both the combination of space and time coordinates required in the Lorentz coordinate transformation and in the variation of length and time intervals with inertial frame as shown in the time dilation and length contraction formulas. A convenient way to express the entanglement of space and time is with the concept of four-dimensional *spacetime* and *spacetime diagrams* introduced by the German mathematician Minkowski.[8] While classical mechanics uses vectors with three components, relativistic mechanics can be elegantly expressed in terms of four vectors, corresponding to the directions *x, y, z,* and *t.* However for simplicity we will confine our discussion to motion in one dimension along the *x*-axis.

A Minkowski or spacetime diagram showing the complete history or *world line* of a one-dimensional motion in frame S is shown in Figure 1.23. Note that the quantity *ct* is plotted on the *y*-axis and the coordinate *x* is plotted on the *x*-axis. The scale of distance is chosen to be the same for both axes. That is, both vertical and horizontal axis ticks occur every meter, so that a light signal starting out at $x = 0$, $t = 0$ follows a 45° line. Point E shows a point event described in frame S by the coordinates (x, t). Of course, other inertial frames (S′) may be used to describe the event or plot the world line and it is quite interesting that these other frames have nonorthogonal *ct′* and *x′* axes, as shown in Figure 1.23. (See Problem 40 for proof of this statement.) To find the space and time coordinates of a given event E in a specific frame, we draw lines parallel to the frame axes and measure the intercepts with the specific frame axes, as shown in the figure. Note too, that the velocity u_x of a particle is inversely proportional to the slope of its world line since

$$u_x = c\, \frac{\Delta x}{\Delta ct} = \frac{c}{\text{slope}} \tag{1.31}$$

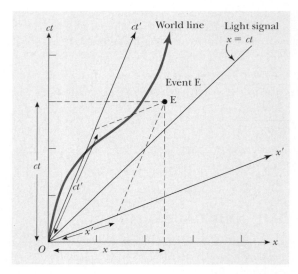

Figure 1.23   A spacetime diagram showing the position of a particle in one dimension at consecutive times. The path showing the complete history of the particle is called the world line of the particle. An event E has coordinates (x, t) in frame S and coordinates $(x′, t′)$ in S′.

---

[8]Minkowski was one of Einstein's teachers, who, commenting on Einstein's work on relativity, reputedly said something like, "I never would have expected that student to come up with anything so clever."

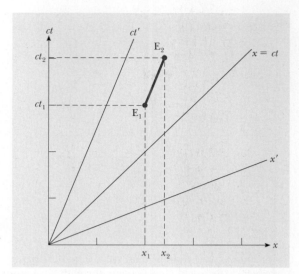

Figure 1.24   Two events, E_1 and E_2, with coordinates (x_1, t_1) and (x_2, t_2) in frame S.

We stated earlier in this section that neither lengths nor time intervals by themselves have any absolute meaning. Can we find a quantity that is absolute or invariant and represents the correct union of length and time? Figure 1.24 shows a spacetime graph with two events, E_1 and E_2 having coordinates (x_1, t_1) and (x_2, t_2) in frame S and coordinates (x_1', t_1') and (x_2', t_2') in frame S'. Let us define the quantity $(\Delta s)^2$ by

$$(\Delta s)^2 = (c\Delta t)^2 - (\Delta x)^2 = (c(t_2 - t_1))^2 - (x_2 - x_1)^2 \qquad (1.32)$$

where Δs has the dimension of length and is called *the spacetime interval* between two events; it is analogous to distance in classical mechanics. If we now evaluate the quantity

$$(\Delta s')^2 = (c\Delta t')^2 - (\Delta x')^2 = (c(t_2' - t_1'))^2 - (x_2' - x_1')^2$$

for the two events E_1 and E_2 whose coordinates in S and S' are connected by the Lorentz transforms $x_1' = \gamma(x_1 - vt_1)$, $t_1' = \gamma(t_1 - vx_1/c^2)$, and so on, we find after some algebra

$$(\Delta s')^2 = (c\Delta t)^2 - (\Delta x)^2 = (\Delta s)^2 \qquad (1.33)$$

This important result says that **the quantity Δs, the spacetime interval between two events, is an invariant and has the same value for all inertial observers.** We have found the quantity that correctly combines space and time in an invariant way.

Minkowski diagrams can be used to classify the entire universe of spacetime and clarify whether or not one event could be the cause of another. Figure 1.25 shows a spacetime diagram for one dimension with axes for two different inertial frames S and S', which share a common origin O at $x = x' = 0$ and $t = t' = 0$. The lines $x = \pm ct$ are world lines of light pulses passing through the origin and traveling in the positive or negative x direction. The regions labeled past and future correspond to negative and positive values of time as

The invariant spacetime interval Δs

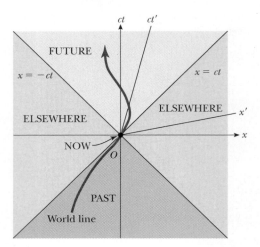

Figure 1.25   Classification of one-dimensional spacetime into past, future, and else-where regions. A particle with world line passing through O cannot reach regions marked elsewhere.

judged from the present moment (now), which occurs at the origin. Regions labeled "elsewhere" cannot be reached by an object whose world line passes through O since to get to them would require a spacetime slope <1 or speed greater than c.

The quantity $(\Delta s)^2 = (c\Delta t)^2 - (\Delta x)^2$ can be used to classify the interval between two events and determine whether one event *could* be caused by the other. To see this, consider the three pairs of events shown in Figure 1.26, where for simplicity the events V, A, and C have been taken to coincide with the origin. For the two events V, W, $(\Delta s)^2 > 0$ since $c\Delta t > |\Delta x|$. Event V *could* be the cause of event W because some signal or influence

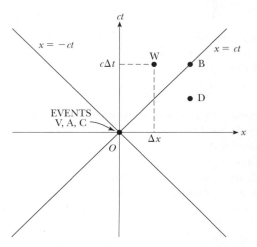

Figure 1.26   Three pairs of events in spacetime: V,W; A,B; C,D. V could cause W. A could cause B. C could not cause D.

could cover the distance Δx from V to W with a speed less than c and connect the two events. The interval between V and W is called "timelike" for reasons we won't go into here, but it is important to note that since $(\Delta s)^2$ is an invariant, if V causes W in frame S, it also causes W in any other inertial frame. Thus, events linked causally in one frame are linked causally in all other inertial frames.

For the two events A, B, $(\Delta s)^2 = 0$ because $c\Delta t = |\Delta x|$. In this case the world line of a light pulse connects point events A and B, and the spacetime interval Δs is said to be "lightlike."

In the final case of events C, D, $(\Delta s)^2 < 0$ because $c\Delta t < |\Delta x|$. This means that even a signal propagating at the speed of light can't cover the distance $|\Delta x|$ between the events C and D and so C cannot possibly be the cause of D in any inertial frame whatsoever.

SUMMARY

The two basic postulates of the **special theory of relativity** are as follows:

- The laws of physics must be the same for all observers moving at constant velocity with respect to one another.
- The speed of light must be the same for all inertial observers, independent of their relative motion.

To satisfy these postulates, the Galilean transformations must be replaced by the **Lorentz transformations** given by

$$x' = \gamma(x - vt) \tag{1.23}$$

$$y' = y \tag{1.24}$$

$$z' = z \tag{1.25}$$

$$t' = \gamma\left(t - \frac{v}{c^2}x\right) \tag{1.26}$$

where

$$\gamma = \frac{1}{\sqrt{1 - (v^2/c^2)}}$$

These equations relate an event with coordinates x, y, z, t measured in S to the same event with coordinates x', y', z', t' measured in S', where it is assumed that the primed system moves with a speed v along the xx'-axes.

The relativistic form of the **velocity transformation** is

$$u'_x = \frac{u_x - v}{1 - (u_x v/c^2)} \tag{1.28}$$

where u_x is the speed of an object as measured in the S frame and u'_x is its speed measured in the S' frame.

If the object has velocity components u_y and u_z along y and z respectively, the components in S' are

$$u'_y = \frac{u_y}{\gamma[1 - (u_x v/c^2)]} \quad \text{and} \quad u'_z = \frac{u_z}{\gamma[1 - (u_x v/c^2)]} \tag{1.29}$$

Some of the consequences of the special theory of relativity are as follows:

- Clocks in motion relative to an observer appear to be slowed down by a factor γ. This is known as **time dilation.**
- Lengths of objects in motion appear to be contracted in the direction of motion by a factor of $1/\gamma$. This is known as **length contraction.**
- Events that are simultaneous for one observer are not simultaneous for another observer in motion relative to the first. **This is known as the relativity of simultaneity.**

These three statements can be summarized by saying that duration, length, and simultaneity are not absolute concepts in relativity.

The relativistic Doppler shift for electromagnetic waves emitted by a moving source is given by

$$f_{obs} = \frac{\sqrt{1 + (v/c)}}{\sqrt{1 - (v/c)}} f_{source} \tag{1.15}$$

where f_{obs} is the frequency measured by an observer approaching a light source with relative speed v, and f_{source} is the frequency as measured in the source's rest frame. The expression for the case of a receding source is obtained by replacing v with $-v$ in Equation 1.15.

The quantity Δs, the spacetime interval between two events, is an invariant and has the same value for all inertial observers where Δs is defined by $(\Delta s)^2 = (c\Delta t)^2 - (\Delta x)^2$.

SUGGESTIONS FOR FURTHER READING

1. E. F. Taylor and J. A. Wheeler, *Spacetime Physics,* San Francisco, W. H. Freeman, 1963.
2. R. Resnick, *Introduction to Special Relativity,* New York, Wiley, 1968.
3. A. P. French, *Special Relativity,* New York, Norton, 1968.
4. H. Bondi, *Relativity and Common Sense,* Science Study Series, Garden City, N.Y., Doubleday, 1964.
5. J. Bronowski, "The Clock Paradox," *Sci. Amer.,* February, 1963, p. 134.
6. A. Einstein, *Out of My Later Years,* New York, World Publishing, 1971.
7. A. Einstein, *Ideas and Opinions,* New York, Crown, 1954.
8. G. Gamow, *Mr. Tompkins in Wonderland,* New York, Cambridge University Press, 1939.
9. L. Infeld, *Albert Einstein,* New York, Scribner's, 1950.
10. J. Schwinger, *Einstein's Legacy,* Scientific American Library, New York, W. H. Freeman, 1985.
11. R. S. Shankland, "The Michelson–Morley Experiment," *Sci. Amer.,* November 1964, p. 107.
12. R. Skinner, *Relativity for Scientists and Engineers,* New York, Dover Publications, 1982.
13. N. Mermin, *Space and Time in Special Relativity,* Prospect Heights, Illinois, Waveland Press, 1989.

QUESTIONS

1. What two measurements will two observers in relative motion *always* agree on?
2. A spaceship in the shape of a sphere moves past an observer on Earth with a speed of $0.5c$. What shape will the observer see as the spaceship moves past?
3. An astronaut moves away from Earth at a speed close to the speed of light. If an observer on Earth could make measurements of the astronaut's size and pulse rate, what changes (if any) would he or she measure? Would the astronaut measure any changes?
4. Two identically constructed clocks are synchronized. One is put in an eastward orbit around Earth while the other remains on Earth. Which clock runs slower? When the moving clock returns to Earth, will the two clocks still be synchronized?
5. Two lasers situated on a moving spacecraft are triggered simultaneously. An observer on the spacecraft claims to see the pulses of light simultaneously. What condition is necessary in order that another observer agrees that the two pulses are emitted simultaneously?

6. When we say that a moving clock runs slower than a stationary one, does this imply that there is something physically unusual about the moving clock?
7. When we speak of time dilation, do we mean that time passes more slowly in moving systems or that it simply appears to do so?
8. List some ways our day-to-day lives would change if the speed of light were only 50 m/s.
9. Give a physical argument to show that it is impossible to accelerate an object of mass m to the speed of light, even with a continuous force acting on it.

10. It is said that Einstein, in his teenage years, asked the question, "What would I see in a mirror if I carried it in my hands and ran at the speed of light?" How would you answer this question?
11. Suppose astronauts were paid according to the time spent traveling in space. After a long voyage at a speed near that of light, a crew of astronauts returns and opens their pay envelopes. What will their reaction be?
12. What happens to the density of an object as its speed increases, as measured by an Earth observer?

PROBLEMS

1.2 The Principle of Newtonian Relativity and the Galilean Transformation

1. In a lab frame of reference, an observer finds Newton's second law is valid in the form $\sum F = ma$. Show that

actual
physical
forces

Newton's second law is not valid in a reference frame moving past the laboratory frame of Problem 1 with a constant acceleration a_1. Assume that mass is an invariant quantity and is constant in time.
2. A 2000-kg car moving with a speed of 20 m/s collides with and sticks to a 1500-kg car at rest at a stop sign. Show that because momentum is conserved in the rest frame, momentum is also conserved in a reference frame moving with a speed of 10 m/s in the direction of the moving car.
3. A billiard ball of mass 0.3 kg moves with a speed of 5 m/s and collides elastically with a ball of mass 0.2 kg moving in the opposite direction with a speed of 3 m/s. Show that because momentum is conserved in the rest frame, it is also conserved in a frame of reference moving with a speed of 2 m/s in the direction of the second ball.

1.3 The Michelson–Morley Experiment

4. An airplane flying upwind, downwind, and crosswind shows the main principle of the Michelson–Morley experiment. A plane capable of flying at speed c in still air is flying in a wind of speed v. Suppose the plane flies upwind a distance L and then returns downwind to its starting point. (a) Find the time needed to make the round-trip and compare it with the time to fly crosswind a distance L and return. Before calculating these times, sketch the two situations. (b) Compute the time difference for the two trips if $L = 100$ mi, $c = 500$ mi/h, and $v = 100$ mi/h.

1.5 Consequences of Special Relativity

5. With what speed will a clock have to be moving in order to run at a rate that is one-half the rate of a clock at rest?

6. How fast must a meter stick be moving if its length is observed to shrink to 0.5 m?
7. A clock on a moving spacecraft runs 1 s slower per day relative to an identical clock on Earth. What is the relative speed of the spacecraft? (*Hint:* For $v/c \ll 1$, note that $\gamma \approx 1 + v^2/2c^2$.)
8. A meter stick moving in a direction parallel to its length appears to be only 75 cm long to an observer. What is the speed of the meter stick relative to the observer?
9. A spacecraft moves at a speed of $0.900c$. If its length is L as measured by an observer on the spacecraft, what is the length measured by a ground observer?
10. The average lifetime of a pi meson in its own frame of reference is 2.6×10^{-8} s. If the meson moves with a speed of $0.95c$, what is (a) its mean lifetime as measured by an observer on Earth and (b) the average distance it travels before decaying, as measured by an observer on Earth?
11. An atomic clock is placed in a jet airplane. The clock measures a time interval of 3600 s when the jet moves with a speed of 400 m/s. How much longer or shorter a time interval does an identical clock held by an observer on the ground measure? (*Hint:* For $v/c \ll 1$, $\gamma \approx 1 + v^2/2c^2$.)
12. An astronaut at rest on Earth has a heartbeat rate of 70 beats/min. What will this rate be when she is traveling in a spaceship at $0.90c$ as measured (a) by an observer also in the ship and (b) by an observer at rest on the Earth?
13. The muon is an unstable particle that spontaneously decays into an electron and two neutrinos. If the number of muons at $t = 0$ is N_0, the number at time t is given by $N = N_0 e^{-t/\tau}$, where τ is the mean lifetime, equal to 2.2 μs. Suppose the muons move at a speed of $0.95c$ and there are 5.0×10^4 muons at $t = 0$. (a) What is the observed lifetime of the muons? (b) How many muons remain after traveling a distance of 3.0 km?
14. A rod of length L_0 moves with a speed v along the horizontal direction. The rod makes an angle of θ_0 with

respect to the x'-axis. (a) Show that the length of the rod as measured by a stationary observer is given by $L = L_0[1 - (v^2/c^2)\cos^2 \theta_0]^{1/2}$. (b) Show that the angle that the rod makes with the x-axis is given by the expression $\tan \theta = \gamma \tan \theta_0$. These results show that the rod is both contracted and rotated. (Take the lower end of the rod to be at the origin of the primed coordinate system.)

15. *The classical Doppler shift for light.* A light source recedes from an observer with a speed v that is small compared with c. (a) Show that in this case, Equation 1.15 reduces to

$$\frac{\Delta f}{f} \approx -\frac{v}{c}$$

(b) Also show that in this case

$$\frac{\Delta \lambda}{\lambda} \approx \frac{v}{c}$$

(*Hint:* Differentiate $\lambda f = c$ to show that $\Delta\lambda/\lambda = -\Delta f/f$)
(c) Spectroscopic measurements of an absorption line normally found at $\lambda = 397$ nm reveal a redshift of 20 nm for light coming from a galaxy in Ursa Major. What is the recessional speed of this galaxy?

16. Calculate, for the judge, how fast you were going in miles per hour when you ran the red light because it appeared Doppler-shifted green to you. Take red light to have a wavelength of 650 nm and green to have a wavelength of 550 nm.

17. (a) How fast and in what direction must galaxy A be moving if an absorption line found at 550 nm (green) for a stationary galaxy is shifted to 450 nm (blue) for A? (b) How fast and in what direction is galaxy B moving if it shows the same line shifted to 700 nm (red)?

18. Police radar detects the speed of a car (Fig. P1.18) as follows: Microwaves of a precisely known frequency are broadcast toward the car. The moving car reflects the microwaves with a Doppler shift. The reflected waves are received and combined with an attenuated version of the transmitted wave. Beats occur between the two microwave signals. The beat frequency is measured. (a) For an electromagnetic wave reflected back to its source from a mirror approaching at speed v, show that the reflected wave has frequency

$$f = f_{\text{source}} \frac{c + v}{c - v}$$

where f_{source} is the source frequency. (b) When v is much less than c, the beat frequency is much smaller than the transmitted frequency. In this case use the approximation $f + f_{\text{source}} \approx 2 f_{\text{source}}$ and show that the beat frequency can be written as $f_{\text{beat}} = 2v/\lambda$. (c) What beat frequency is measured for a car speed of 30.0 m/s

if the microwaves have frequency 10.0 GHz? (d) If the beat frequency measurement is accurate to ± 5 Hz, how accurate is the velocity measurement?

Figure P1.18   (*Trent Steffler/David R. Frazier Photo Library*)

1.6 The Lorentz Transformation

19. Two spaceships approach each other, each moving with the *same* speed as measured by an observer on the Earth. If their *relative* speed is $0.70c$, what is the speed of each spaceship?

20. An electron moves to the right with a speed of $0.90c$ relative to the laboratory frame. A proton moves to the right with a speed of $0.70c$ relative to the electron. Find the speed of the proton relative to the laboratory frame.

21. An observer on Earth observes two spacecraft moving in the *same* direction toward the Earth. Spacecraft A appears to have a speed of $0.50c$, and spacecraft B appears to have a speed of $0.80c$. What is the speed of spacecraft A measured by an observer in spacecraft B?

22. *Speed of light in a moving medium.* The motion of a medium such as water influences the speed of light. This effect was first observed by Fizeau in 1851. Consider a light beam passing through a horizontal column of water moving with a speed v. (a) Show that if the beam travels in the same direction as the flow of water, the speed of light measured in the laboratory frame is given by

$$u = \frac{c}{n}\left(\frac{1 + nv/c}{1 + v/nc}\right)$$

where n is the index of refraction of the water. (*Hint:* Use the inverse Lorentz velocity transformation and note that the speed of light with respect to the moving frame is given by c/n.) (b) Show that for $v \ll c$, the preceding expression is in good agreement with Fizeau's experimental result:

$$u \approx \frac{c}{n} + v - \frac{v}{n^2}$$

This proves that the Lorentz velocity transformation and not the Galilean velocity transformation is correct for light.

23. An observer in frame S sees lightning simultaneously strike two points 100 m apart. The first strike occurs

ADDITIONAL PROBLEMS

25. In 1962, when Scott Carpenter orbited Earth 22 times, the press stated that for each orbit he aged 2 millionths of a second less than if he had remained on Earth. (a) Assuming that he was 160 km above Earth in an eastbound circular orbit, determine the time difference between someone on Earth and the orbiting astronaut for the 22 orbits. (b) Did the press report accurate information? Explain.

26. The proper length of one spaceship is three times that of another. The two spaceships are traveling in the same direction and, while both are passing overhead, an Earth observer measures the two spaceships to have the same length. If the slower spaceship is moving with a speed of $0.35c$, determine the speed of the faster spaceship.

27. The pion has an average lifetime of 26.0 ns when at rest. For it to travel 10.0 m, how fast must it move?

28. If astronauts could travel at $v = 0.95c$, we on Earth would say it takes $(4.2/0.95) = 4.4$ years to reach Alpha Centauri, 4.2 lightyears away. The astronauts disagree. (a) How much time passes on the astronauts' clocks? (b) What distance to Alpha Centauri do the astronauts measure?

29. A spaceship moves away from Earth at a speed v and fires a shuttle craft in the forward direction at a speed v relative to the ship. The pilot of the shuttle craft launches a probe at speed v relative to the shuttle craft. Determine (a) the speed of the shuttle craft relative to Earth, and (b) the speed of the probe relative to Earth.

30. An observer in a rocket moves toward a mirror at speed v relative to the reference frame labeled by S in Figure P1.30. The mirror is stationary with respect to S. A light pulse emitted by the rocket travels toward the mirror and is reflected back to the rocket. The front of the rocket is a distance d from the mirror (as measured by observers in S) at the moment the light pulse leaves the rocket. What is the total travel time of the pulse as measured by observers in (a) the S frame and (b) the front of the rocket?

at $x_1 = y_1 = z_1 = t_1 = 0$ and the second at $x_2 = 100$ m, $y_2 = z_2 = t_2 = 0$. (a) What are the coordinates of these two events in a frame S' moving in the standard configuration at $0.70c$ relative to S? (b) How far apart are the events in S'? (c) Are the events simultaneous in S'? If not, what is the difference in time between the events, and which event occurs first?

24. As seen from Earth, two spaceships A and B are approaching along perpendicular directions. If A is observed by an Earth observer to have velocity $u_y = -0.90c$ and B to have a velocity $u_x = +0.90c$, find the speed of ship A as measured by the pilot of B.

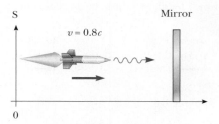

Figure P1.30

31. A physics professor on Earth gives an exam to her students who are on a spaceship traveling at speed v relative to Earth. The moment the ship passes the professor, she signals the start of the exam. If she wishes her students to have time T_0 (spaceship time) to complete the exam, show that she should wait a time (Earth time) of

$$T = T_0 \sqrt{\frac{1 - v/c}{1 + v/c}}$$

before sending a light signal telling them to stop. (*Hint:* Remember that it takes some time for the second light signal to travel from the professor to the students.)

32. A yet-to-be-built spacecraft starts from Earth moving at constant speed to the yet-to-be-discovered planet Retah, which is 20 lighthours away from Earth. It takes 25 h (according to an Earth observer) for a spacecraft to reach this planet. Assuming that the clocks are synchronized at the beginning of the journey, compare the time elapsed in the spacecraft's frame for this one-way journey with the time elapsed as measured by an Earth-based clock.

33. Suppose our Sun is about to explode. In an effort to escape, we depart in a spaceship at $v = 0.80c$ and head toward the star Tau Ceti, 12 lightyears away. When we reach the midpoint of our journey from the Earth, we see our Sun explode and, unfortunately, at the same instant we see Tau Ceti explode as well. (a) In the spaceship's frame of reference, should we conclude that the

two explosions occurred simultaneously? If not, which occurred first? (b) In a frame of reference in which the Sun and Tau Ceti are at rest, did they explode simultaneously? If not, which exploded first?

34. Two powerless rockets are on a collision course. The rockets are moving with speeds of 0.800c and 0.600c and are initially 2.52×10^{12} m apart as measured by Liz, an Earth observer, as shown in Figure P1.34. Both rockets are 50.0 m in length as measured by Liz. (a) What are their respective proper lengths? (b) What is the length of each rocket as measured by an observer in the other rocket? (c) According to Liz, how long before the rockets collide? (d) According to rocket 1, how long before they collide? (e) According to rocket 2, how long before they collide? (f) If both rocket crews are capable of total evacuation within 90 min (their own time), will there be any casualties?

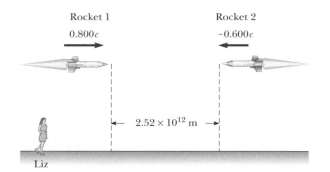

Figure P1.34

35. The identical twins Speedo and Goslo join a migration from Earth to Planet X. It is 20.0 ly away in a reference frame in which both planets are at rest. The twins, of the same age, depart at the same time on different spaceships. Speedo's ship travels steadily at 0.950c, and Goslo's at 0.750c. Calculate the age difference between the twins after Goslo's spaceship reaches Planet X. Which twin is the older?

36. Suzanne observes two light pulses to be emitted from the same location, but separated in time by 3.00 μs. Mark sees the emission of the same two pulses separated in time by 9.00 μs. (a) How fast is Mark moving relative to Suzanne? (b) According to Mark, what is the separation in space of the two pulses?

37. An observer in reference frame S sees two events as simultaneous. Event A occurs at the point (50.0 m, 0, 0) at the instant 9:00:00 Universal time, 15 January 2001. Event B occurs at the point (150 m, 0, 0) at the same moment. A second observer, moving past with a velocity of $0.800c\,\hat{\mathbf{i}}$, also observes the two events. In her reference frame S′, which event occurred first and what time elapsed between the events?

38. A spacecraft is launched from the surface of the Earth with a velocity of 0.600c at an angle of 50.0° above the horizontal, positive x-axis. Another spacecraft is moving past with a velocity of 0.700c in the negative x direction. Determine the magnitude and direction of the velocity of the first spacecraft as measured by the pilot of the second spacecraft.

39. An Earth satellite used in the Global Positioning System moves in a circular orbit with period 11 h 58 min. (a) Determine the radius of its orbit. (b) Determine its speed. (c) The satellite contains an oscillator producing the principal nonmilitary GPS signal. Its frequency is 1 575.42 MHz in the reference frame of the satellite. When it is received on the Earth's surface, what is the fractional change in this frequency due to time dilation, as described by special relativity? (d) The gravitational blueshift of the frequency according to general relativity is a separate effect. The magnitude of that fractional change is given by

$$\frac{\Delta f}{f} = \frac{\Delta U_g}{mc^2}$$

where $\Delta U_g/m$ is the change in gravitational potential energy per unit mass between the two points at which the signal is observed. Calculate this fractional change in frequency. (e) What is the overall fractional change in frequency? Superposed on both of these relativistic effects is a Doppler shift that is generally much larger. It can be a redshift or a blueshift, depending on the motion of a particular satellite relative to a GPS receiver (Fig. P1.39).

Figure P1.39   (*Photo courtesy of Garmin Ltd.*)

40. Show that the S′ axes, x′ and ct′, are nonorthogonal in a spacetime diagram. Assume that the S and S′ inertial frames move as shown in Figure 1.2 and that $t = t' = 0$ when $x = x' = 0$. (*Hint:* First use the fact that the ct′-axis is the world line of the origin of S′ to show that the ct′-axis is inclined with respect to the ct-axis. Next note that the world line of a light pulse moving in the +x direction starting out at x = 0 and ct = 0 is described by the equation $x = +ct$ in S and $x' = ct'$ in S′).

2

Relativity II

Chapter Outline

In this chapter we extend the theory of special relativity to classical mechanics, that is, we give relativistically correct expressions for momentum, Newton's second law, and the famous equivalence of mass and energy. The final section, on general relativity, deals with the physics of accelerating reference frames and Einstein's theory of gravitation.

2.1   RELATIVISTIC MOMENTUM AND THE RELATIVISTIC FORM OF NEWTON'S LAWS

The conservation of linear momentum states that when two bodies collide, the total momentum remains constant, assuming the bodies are isolated (that is, they interact only with each other). Suppose the collision is described in a reference frame S in which momentum is conserved. If the velocities of the colliding bodies are calculated in a second inertial frame S′ using the Lorentz transformation, and the classical definition of momentum $\mathbf{p} = m\mathbf{u}$ applied, one finds that momentum *is not* conserved in the second reference frame. However, because the laws of physics are the same in all inertial frames, momentum must be conserved in all frames if it is conserved in any one. This application of the principle of relativity demands that we modify the classical definition of momentum.

To see how the classical form $\mathbf{p} = m\mathbf{u}$ fails and to determine the correct relativistic definition of $\mathbf{p}$, consider the case of an inelastic collision

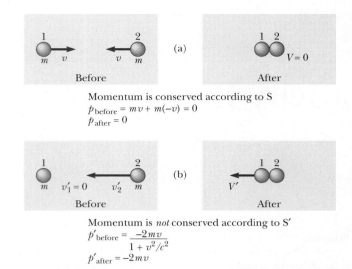

Momentum is conserved according to S
$p_{\text{before}} = mv + m(-v) = 0$
$p_{\text{after}} = 0$

Momentum is *not* conserved according to S'
$p'_{\text{before}} = \dfrac{-2mv}{1 + v^2/c^2}$
$p'_{\text{after}} = -2mv$

Figure 2.1   (a) An inelastic collision between two equal clay lumps as seen by an observer in frame S. (b) The same collision viewed from a frame S' that is moving to the right with speed v with respect to S.

between two particles of equal mass. Figure 2.1a shows such a collision for two identical particles approaching each other at speed v as observed in an inertial reference frame S. Using the classical form for momentum, $\mathbf{p} = m\mathbf{u}$ (we use the symbol $\mathbf{u}$ for particle velocity rather than $\mathbf{v}$, which is reserved for the relative velocity of two reference frames), the observer in S finds momentum is conserved as shown in Figure 2.1a. Suppose we now view things from an inertial frame S' moving to the right with speed v relative to S. In S' the new speeds are v'_1, v'_2 and V' (see Fig. 2.1b). If we use the Lorentz velocity transformation

$$u'_x = \frac{u_x - v}{1 - (u_x v / c^2)}$$

to find v'_1, v'_2 and V', and the classical form for momentum, $\mathbf{p} = m\mathbf{u}$, will momentum be conserved according to the observer in S'? To answer this question we first calculate the velocities of the particles in S' in terms of the given velocities in S.

$$v'_1 = \frac{v_1 - v}{1 - (v_1 v / c^2)} = \frac{v - v}{1 - (v^2/c^2)} = 0$$

$$v'_2 = \frac{v_2 - v}{1 - (v_2 v / c^2)} = \frac{-v - v}{1 - [(-v)(v)/c^2]} = \frac{-2v}{1 + (v^2/c^2)}$$

$$V' = \frac{V - v}{1 - (V v / c^2)} = \frac{0 - v}{1 - [(0)v/c^2]} = -v$$

Checking for momentum conservation in S', we have

$$p'_{\text{before}} = mv'_1 + mv'_2 = m(0) + m\left[\frac{-2v}{1 + (v^2/c^2)}\right] = \frac{-2mv}{1 + (v^2/c^2)}$$

$$p'_{\text{after}} = 2mV' = -2mv$$

Thus, *in the frame S'*, the momentum before the collision is not equal to the momentum after the collision, and *momentum is not conserved.*

It can be shown (see Example 2.6) that momentum is conserved in both S and S', (and indeed in all inertial frames), if we redefine momentum as

$$\mathbf{p} \equiv \frac{m\mathbf{u}}{\sqrt{1 - (u^2/c^2)}} \qquad (2.1)$$

Definition of relativistic momentum

where **u** is the velocity of the particle and *m is the proper mass,* that is, the mass measured by an observer at rest with respect to the mass.[1] Note that when u is much less than c, the denominator of Equation 2.1 approaches unity and **p** approaches $m\mathbf{u}$. Therefore, the relativistic equation for **p** reduces to the classical expression when u is small compared with c. Because it is a simpler expression, Equation 2.1 is often written $\mathbf{p} = \gamma m\mathbf{u}$, where $\gamma = 1/\sqrt{1 - (u^2/c^2)}$. Note that this γ has the same functional form as the γ in the Lorentz transformation, but here γ contains u, the particle speed, while in the Lorentz transformation, γ contains v, the relative speed of the two frames.

The **relativistic form of Newton's second law** is given by the expression

$$\mathbf{F} = \frac{d\mathbf{p}}{dt} = \frac{d}{dt}(\gamma m\mathbf{u}) \qquad (2.2)$$

where **p** is given by Equation 2.1. This expression is reasonable because it preserves classical mechanics in the limit of low velocities and requires **the momentum of an isolated system (F = 0) to be conserved relativistically as well as classically.** It is left as a problem (Problem 3) to show that the relativistic acceleration a of a particle *decreases* under the action of a constant force applied in the direction of **u**, as

$$a = \frac{F}{m}(1 - u^2/c^2)^{3/2}$$

From this formula we see that as the velocity approaches c, the acceleration caused by any finite force approaches zero. Hence, it is impossible to accelerate a particle from rest to a speed equal to or greater than c.

---

[1]In this book we shall always take m to be constant with respect to speed, and we call m the speed invariant mass, or proper mass. Some physicists refer to the mass in Equation 2.1 as the rest mass, m_0, and call the term $m_0/\sqrt{1 - (u^2/c^2)}$ the relativistic mass. Using this description, the relativistic mass is imagined to increase with increasing speed. We exclusively use the invariant mass m because we think it is a clearer concept and that the introduction of relativistic mass leads to no deeper physical understanding.

EXAMPLE 2.1 Momentum of an Electron

An electron, which has a mass of 9.11×10^{-31} kg, moves with a speed of $0.750c$. Find its relativistic momentum and compare this with the momentum calculated from the classical expression.

Solution Using Equation 2.1 with $u = 0.750c$, we have

$$
\begin{aligned}
p &= \frac{mu}{\sqrt{1 - (u^2/c^2)}} \\
&= \frac{(9.11 \times 10^{-31}\ \text{kg})(0.750 \times 3.00 \times 10^8\ \text{m/s})}{\sqrt{1 - [(0.750c)^2/c^2]}} \\
&= 3.10 \times 10^{-22}\ \text{kg·m/s}
\end{aligned}
$$

The incorrect classical expression would give

$$
\text{momentum} = mu = 2.05 \times 10^{-22}\ \text{kg·m/s}
$$

Hence, for this case the correct relativistic result is 50% greater than the classical result!

EXAMPLE 2.2 An Application of the Relativistic Form of $F = d\mathbf{p}/dt$: The Measurement of the Momentum of a High-Speed Charged Particle

Suppose a particle of mass m and charge q is injected with a relativistic velocity $\mathbf{u}$ into a region containing a magnetic field $\mathbf{B}$. The magnetic force $\mathbf{F}$ on the particle is given by $\mathbf{F} = q\mathbf{u} \times \mathbf{B}$. If $\mathbf{u}$ is perpendicular to $\mathbf{B}$, the force is radially inward, and the particle moves in a circle of radius R with $|\mathbf{u}|$ constant. From Equation 2.2 we have

$$
\mathbf{F} = \frac{d\mathbf{p}}{dt} = \frac{d}{dt}(\gamma m \mathbf{u})
$$

Solution Because the magnetic force is always perpendicular to the velocity, it does no work on the particle, and hence the speed, u, and γ are both constant with time. Thus, the magnitude of the force on the particle is

$$
F = \gamma m \left| \frac{d\mathbf{u}}{dt} \right| \tag{2.3}
$$

Substituting $F = quB$ and $|d\mathbf{u}/dt| = u^2/R$ (the usual definition of centripetal acceleration) into Equation 2.3, we can solve for $p = \gamma mu$. We find

$$
p = \gamma mu = qBR \tag{2.4}
$$

Equation 2.4 shows that the momentum of a relativistic particle of known charge q may be determined by measuring its radius of curvature R in a known magnetic field, $\mathbf{B}$. This technique is routinely used to determine the momentum of subatomic particles from photographs of their tracks in space.

2.2 RELATIVISTIC ENERGY

We have seen that the definition of momentum and the laws of motion required generalization to make them compatible with the principle of relativity. This implies that the relativistic form of the kinetic energy must also be modified.

To derive the relativistic form of the work–energy theorem, let us start with the definition of work done by a force F and make use of the definition of relativistic force, Equation 2.2. That is,

$$
W = \int_{x_1}^{x_2} F\,dx = \int_{x_1}^{x_2} \frac{dp}{dt}\,dx \tag{2.5}
$$

where we have assumed that the force and motion are along the x-axis. To perform this integration and find the work done on a particle or the relativistic kinetic energy as a function of the particle velocity u, we first evaluate dp/dt:

$$
\frac{dp}{dt} = \frac{d}{dt} \frac{mu}{\sqrt{1 - (u^2/c^2)}} = \frac{m\left(\dfrac{du}{dt}\right)}{[1 - (u^2/c^2)]^{3/2}} \tag{2.6}
$$

Substituting this expression for dp/dt and $dx = u\,dt$ into Equation 2.5 gives

$$W = \int_{x_1}^{x_2} \frac{m\left(\dfrac{du}{dt}\right)u\,dt}{[1 - (u^2/c^2)]^{3/2}} = m\int_0^u \frac{u\,du}{[1 - (u^2/c^2)]^{3/2}}$$

where we have assumed that the particle is accelerated from rest to some final velocity u. Evaluating the integral, we find that

$$W = \frac{mc^2}{\sqrt{1 - (u^2/c^2)}} - mc^2 \qquad (2.7)$$

Recall that the work–energy theorem states that the work done by all forces acting on a particle equals the change in kinetic energy of the particle. Because the initial kinetic energy is zero, we conclude that the work W in Eq. 2.7 is equal to the relativistic kinetic energy K, that is,

$$K = \frac{mc^2}{\sqrt{1 - (u^2/c^2)}} - mc^2 \qquad (2.8)$$

Relativistic kinetic energy

At low speeds, where $u/c \ll 1$, Equation 2.8 should reduce to the classical expression $K = \frac{1}{2}mu^2$. We can check this by using the binomial expansion $(1 - x^2)^{-1/2} \approx 1 + \frac{1}{2}x^2 + \cdots$, for $x \ll 1$, where the higher-order powers of x are ignored in the expansion. In our case, $x = u/c$, so that

$$\frac{1}{\sqrt{1 - (u^2/c^2)}} = \left(1 - \frac{u^2}{c^2}\right)^{-1/2} \approx 1 + \frac{1}{2}\frac{u^2}{c^2} + \cdots$$

Substituting this into Equation 2.8 gives

$$K \approx mc^2\left(1 + \frac{1}{2}\frac{u^2}{c^2} + \cdots\right) - mc^2 = \frac{1}{2}mu^2$$

which agrees with the classical result. A graph comparing the relativistic and nonrelativistic expressions for u as a function of K is given in Figure 2.2. Note that in the relativistic case, the particle speed never exceeds c, regard-

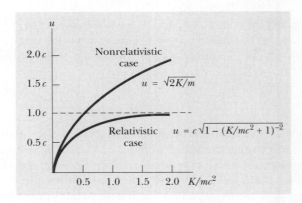

Figure 2.2   A graph comparing the relativistic and nonrelativistic expressions for speed as a function of kinetic energy. In the relativistic case, u is always *less* than c.

less of the kinetic energy, as is routinely confirmed in very high energy particle accelerator experiments. The two curves are in good agreement when $u \ll c$.

It is instructive to write the relativistic kinetic energy in the form

$$K = \gamma mc^2 - mc^2 \qquad (2.9)$$

where

$$\gamma = \frac{1}{\sqrt{1 - u^2/c^2}}$$

The constant term mc^2, which is independent of the speed, is called the **rest energy** of the particle. The term γmc^2, which depends on the particle speed, is therefore the sum of the kinetic and rest energies. We define γmc^2 to be the **total energy** E, that is,

Definition of total energy

$$E = \gamma mc^2 = K + mc^2 \qquad (2.10)$$

Mass–energy equivalence

The expression $E = \gamma mc^2$ is Einstein's famous mass–energy equivalence equation, which shows that mass is a measure of the total energy in all forms. Although we have been considering single particles for simplicity, Equation 2.10 applies to macroscopic objects as well. In this case it has the remarkable implication that any kind of energy added to a "brick" of matter—electric, magnetic, elastic, thermal, gravitational, chemical—actually increases the mass! Several end-of-chapter questions and problems explore this idea more fully. Another implication of Equation 2.10 is that a small mass corresponds to an enormous amount of energy because c^2 is a very large number. This concept has revolutionized the field of nuclear physics and is treated in detail in the next section.

In many situations, the momentum or energy of a particle is measured rather than its speed. It is therefore useful to have an expression relating the total energy E to the relativistic momentum p. This is accomplished using $E = \gamma mc^2$ and $p = \gamma mu$. By squaring these equations and subtracting, we can eliminate u (Problem 7). The result, after some algebra, is

Energy–momentum relation

$$E^2 = p^2c^2 + (mc^2)^2 \qquad (2.11)$$

When the particle is at rest, $p = 0$, and so we see that $E = mc^2$. That is, the total energy equals the rest energy. For the case of particles that have zero mass, such as photons (massless, chargeless particles of light), we set $m = 0$ in Equation 2.11, and find

$$E = pc \qquad (2.12)$$

This equation is an *exact* expression relating energy and momentum for photons, which always travel at the speed of light.

Finally, note that because the mass m of a particle is independent of its motion, m must have the same value in all reference frames. On the other hand, the total energy and momentum of a particle depend on the reference frame in which they are measured, because they both depend on velocity. Because m is a constant, then according to Equation 2.11 the quantity $E^2 - p^2c^2$ must have the same value in all reference frames. That is, $E^2 - p^2c^2$ is *invariant* under a Lorentz transformation.

When dealing with electrons or other subatomic particles, it is convenient to express their energy in **electron volts (eV)**, since the particles are usually given this energy by acceleration through a potential difference. The conversion factor is

$$1 \text{ eV} = 1.60 \times 10^{-19} \text{ J}$$

For example, the mass of an electron is 9.11×10^{-31} kg. Hence, the rest energy of the electron is

$$m_e c^2 = (9.11 \times 10^{-31} \text{ kg})(3.00 \times 10^8 \text{ m/s})^2 = 8.20 \times 10^{-14} \text{ J}$$

Converting this to electron volts, we have

$$m_e c^2 = (8.20 \times 10^{-14} \text{ J})(1 \text{ eV}/1.60 \times 10^{-19} \text{ J}) = 0.511 \text{ MeV}$$

where 1 MeV = 10^6 eV. Finally, note that because $m_e c^2 = 0.511$ MeV, the mass of the electron may be written $m_e = 0.511$ MeV/c^2, accounting for the practice of measuring particle masses in units of MeV/c^2.

EXAMPLE 2.3   The Energy of a Speedy Electron

An electron has a speed $u = 0.850c$. Find its total energy and kinetic energy in electron volts.

Solution Using the fact that the rest energy of the electron is 0.511 MeV together with $E = \gamma mc^2$ gives

$$E = \frac{m_e c^2}{\sqrt{1 - (u^2/c^2)}} = \frac{0.511 \text{ MeV}}{\sqrt{1 - [(0.85c)^2/c^2]}}$$
$$= 1.90(0.511 \text{ MeV}) = 0.970 \text{ MeV}$$

The kinetic energy is obtained by subtracting the rest energy from the total energy:

$$K = E - m_e c^2 = 0.970 \text{ MeV} - 0.511 \text{ MeV} = 0.459 \text{ MeV}$$

EXAMPLE 2.4   The Energy of a Speedy Proton

The total energy of a proton is three times its rest energy.

(a) Find the proton's rest energy in electron volts.

Solution

$$\text{rest energy} = m_p c^2$$
$$= (1.67 \times 10^{-27} \text{ kg})(3.00 \times 10^8 \text{ m/s})^2$$
$$= (1.50 \times 10^{-10} \text{ J})(1 \text{ eV}/1.60 \times 10^{-19} \text{ J})$$
$$= 938 \text{ MeV}$$

(b) With what speed is the proton moving?

Solution Because the total energy E is three times the rest energy, $E = \gamma mc^2$ gives

$$E = 3m_p c^2 = \frac{m_p c^2}{\sqrt{1 - (u^2/c^2)}}$$
$$3 = \frac{1}{\sqrt{1 - (u^2/c^2)}}$$

Solving for u gives

$$\left(1 - \frac{u^2}{c^2}\right) = \frac{1}{9} \quad \text{or} \quad \frac{u^2}{c^2} = \frac{8}{9}$$
$$u = \frac{\sqrt{8}}{3} c = 2.83 \times 10^8 \text{ m/s}$$

(c) Determine the kinetic energy of the proton in electron volts.

Solution

$$K = E - m_p c^2 = 3m_p c^2 - m_p c^2 = 2m_p c^2$$

Because $m_p c^2 = 938$ MeV, $K = 1876$ MeV.

(d) What is the proton's momentum?

Solution We can use Equation 2.11 to calculate the momentum with $E = 3m_p c^2$:

$$E^2 = p^2 c^2 + (m_p c^2)^2 = (3m_p c^2)^2$$
$$p^2 c^2 = 9(m_p c^2)^2 - (m_p c^2)^2 = 8(m_p c^2)^2$$
$$p = \sqrt{8} \frac{m_p c^2}{c} = \sqrt{8} \frac{(938 \text{ MeV})}{c} = 2650 \frac{\text{MeV}}{c}$$

Note that the unit of momentum is left as MeV/c for convenience.

2.3   MASS AS A MEASURE OF ENERGY

The equation $E = \gamma mc^2$ as applied to a particle suggests that even when a particle is at rest ($\gamma = 1$) it still possesses enormous energy through its mass. The clearest experimental proof of the equivalence of mass and energy occurs in nuclear and elementary particle interactions in which both the conversion of mass into energy and the conversion of energy into mass take place. Because of this convertibility from the currency of mass into the currency of energy, we can no longer accept the separate classical laws of the conservation of mass and the conservation of energy; we must instead speak of a single unified law, **the conservation of mass–energy.** Simply put, this law requires that **the sum of the mass–energy of a system of particles before interaction must equal the sum of the mass–energy of the system after interaction where the mass–energy of the ith particle is defined as the total relativistic energy**

Conservation of mass–energy

$$E_i = \frac{m_i c^2}{\sqrt{1 - (u_i^2/c^2)}}$$

To understand the conservation of mass–energy and to see how the relativistic laws possess more symmetry and wider scope than the classical laws of momentum and energy conservation, we consider the simple inelastic collision treated earlier.

As one can see in Figure 2.1a, *classically* momentum is conserved but kinetic energy is not because the total kinetic energy before collision equals mu^2 and the total kinetic energy after is zero (we have replaced the v shown in Figure 2.1 with u). Now consider the same two colliding clay lumps using the relativistic mass–energy conservation law. If the mass of each lump is m, and the mass of the composite object is M, we must have

$$E_{\text{before}} = E_{\text{after}}$$

$$\frac{mc^2}{\sqrt{1 - (u^2/c^2)}} + \frac{mc^2}{\sqrt{1 - (u^2/c^2)}} = Mc^2$$

or

$$M = \frac{2m}{\sqrt{1 - (u^2/c^2)}} \tag{2.13}$$

Because $\sqrt{1 - (u^2/c^2)} < 1$, the composite mass M is greater than the sum of the two individual masses! What's more, it is easy to show that the mass increase of the composite lump, $\Delta M = M - 2m$, is equal to the sum of the incident kinetic energies of the colliding lumps ($2K$) divided by c^2:

$$\Delta M = \frac{2K}{c^2} = \frac{2}{c^2}\left(\frac{mc^2}{\sqrt{1 - (u^2/c^2)}} - mc^2\right) \tag{2.14}$$

Thus, we have an example of the conversion of kinetic energy to mass, and the satisfying result that in relativistic mechanics, kinetic energy is not lost in an inelastic collision but shows up as an increase in the mass of the final composite object. In fact, the deeper symmetry of relativity theory shows that *both relativistic mass–energy and momentum are always conserved in a collision,* whereas classical methods show that momentum is conserved but kinetic energy is not unless the collision is perfectly elastic. Indeed, as we show in Example 2.6, relativistic momentum and energy are inextricably linked because momentum conservation only holds in all inertial frames if mass–energy conservation also holds.

EXAMPLE 2.5

(a) Calculate the mass increase for a completely inelastic head-on collision of two 5.0-kg balls each moving toward the other at 1000 mi/h (the speed of a fast jet plane). (b) Explain why measurements on macroscopic objects reinforce the relativistically *incorrect* beliefs that mass is conserved ($M = 2m$) and that kinetic energy is lost in an inelastic collision.

Solution (a) $u = 1000$ mi/h $= 450$ m/s, so

$$\frac{u}{c} = \frac{4.5 \times 10^2 \text{ m/s}}{3.0 \times 10^8 \text{ m/s}} = 1.5 \times 10^{-6}$$

Because $u^2/c^2 \ll 1$, substituting

$$\frac{1}{\sqrt{1 - (u^2/c^2)}} \approx 1 + \frac{1}{2}\frac{u^2}{c^2}$$

in Equation 2.14 gives

$$\Delta M = 2m \left(\frac{1}{\sqrt{1 - (u^2/c^2)}} - 1 \right)$$

$$\approx 2m \left(1 + \frac{1}{2}\frac{u^2}{c^2} - 1 \right) \approx \frac{mu^2}{c^2}$$

$$= (5.0 \text{ kg})(1.5 \times 10^{-6})^2 = 1.1 \times 10^{-11} \text{ kg}$$

(b) Because the mass increase of 1.1×10^{-11} kg is an unmeasurably minute fraction of $2m$ (10 kg), it is quite natural to believe that the mass remains constant when macroscopic objects suffer an inelastic collision. On the other hand, the change in kinetic energy from mu^2 to 0 is so large (10^6 J) that it is readily measured to be lost in an inelastic collision of macroscopic objects.

Exercise 1 Prove that $\Delta M = 2\Delta K/c^2$ for a completely inelastic collision, as stated.

EXAMPLE 2.6

Show that use of the relativistic definition of momentum

$$p = \frac{mu}{\sqrt{1 - (u^2/c^2)}}$$

leads to momentum conservation in both S and S′ for the inelastic collision shown in Figure 2.1.

Solution In frame S:

$$p_{\text{before}} = \gamma mv + \gamma m(-v) = 0$$

$$p_{\text{after}} = \gamma MV = (\gamma M)(0) = 0$$

Hence, momentum is conserved in S. Note that we have used M as the mass of the two combined masses after the collision and allowed for the possibility in relativity that M is not necessarily equal to $2m$.

In frame S′:

$$p'_{\text{before}} = \gamma mv'_1 + \gamma mv'_2 = \frac{(m)(0)}{\sqrt{1 - (0)^2/c^2}}$$

$$+ \frac{m}{\{\sqrt{1 - [-2v/1 + (v^2/c^2)]^2}\}(1/c^2)} \times \left(\frac{-2v}{1 + v^2/c^2} \right)$$

After some algebra, we find

$$\frac{m}{\{\sqrt{1 - [2v/1 + (v^2/c^2)]^2}\}(1/c^2)} = \frac{m(1 + v^2/c^2)}{(1 - v^2/c^2)}$$

and we obtain

$$p'_{\text{before}} = \frac{m(1 + v^2/c^2)}{(1 - v^2/c^2)} \left(\frac{-2v}{1 + v^2/c^2} \right) = \frac{-2mv}{(1 - v^2/c^2)}$$

$$p'_{\text{after}} = \gamma MV' = \frac{M(-v)}{\sqrt{1 - [(-v)^2/c^2]}} = \frac{-Mv}{\sqrt{1 - v^2/c^2}}$$

To show that momentum is conserved in S′, we use the fact that M is not simply equal to $2m$ in relativity. As shown, the combined mass, M, formed from the collision of two particles, each of mass m moving toward each other with speed v, is greater than $2m$. This occurs because of the equivalence of mass and energy, that is, the kinetic energy of the incident particles shows up in relativity theory as a tiny increase in mass, which can actually be measured as thermal energy. Thus, from Equation 2.13, which results from imposing the conservation of mass–energy, we have

$$M = \frac{2m}{\sqrt{1 - (v^2/c^2)}}$$

Substituting this result for M into p'_{after}, we obtain

$$p'_{\text{after}} = \frac{2m}{\sqrt{1 - (v^2/c^2)}} \frac{-v}{\sqrt{1 - (v^2/c^2)}}$$

$$= \frac{-2mv}{1 - (v^2/c^2)} = p'_{\text{before}}$$

Hence, momentum is conserved in both S and S′, provided that we use the correct relativistic definition of momentum, $p = \gamma mu$, and assume the conservation of mass–energy.

The absence of observable mass changes in inelastic collisions of macroscopic objects impels us to look for other areas to test this law, where particle velocities are higher, masses are more precisely known, and forces are stronger than electrical or mechanical forces. This leads us to consider nuclear reactions, because nuclear masses can be measured very precisely with a mass spectrometer, nuclear forces are much stronger than electrical forces, and decay products are often produced with extremely high velocities.

Perhaps the most direct confirmation of the conservation of mass–energy occurs in the decay of a heavy radioactive nucleus at rest into several lighter particles emitted with large kinetic energies. For such a nucleus **Fission** of mass M undergoing *fission* into particles with masses M_1, M_2, and M_3 and having speeds u_1, u_2, and u_3, conservation of total relativistic energy requires

$$Mc^2 = \frac{M_1c^2}{\sqrt{1-(u_1^2/c^2)}} + \frac{M_2c^2}{\sqrt{1-(u_2^2/c^2)}} + \frac{M_3c^2}{\sqrt{1-(u_3^2/c^2)}} \qquad (2.15)$$

Because the square roots are all less than 1, $M > M_1 + M_2 + M_3$ and the loss of mass, $M - (M_1 + M_2 + M_3)$, appears as energy of motion of the products. This **disintegration energy** released per fission is often denoted by the symbol Q and can be written for our case as

$$Q = [M - (M_1 + M_2 + M_3)]c^2 = \Delta mc^2 \qquad (2.16)$$

EXAMPLE 2.7   A Fission Reaction

An excited ${}_{92}^{236}\text{U}$ nucleus decays at rest into ${}_{37}^{90}\text{Rb}$, ${}_{55}^{143}\text{Cs}$, and several neutrons, ${}_0^1\text{n}$. (a) By conserving charge and the total number of protons and neutrons, write a balanced reaction equation and determine the number of neutrons produced. (b) Calculate by how much the combined "offspring" mass is less than the "parent" mass. (c) Calculate the energy released per fission. (d) Calculate the energy released in kilowatt hours when 1 kg of uranium undergoes fission in a power plant that is 40% efficient.

Solution   (a) In general, an element is represented by the symbol ${}_Z^A\text{X}$, where X is the symbol for the element, A is the number of neutrons plus protons in the nucleus (mass number), and Z is the number of protons in the nucleus (atomic number). Conserving charge and number of nucleons gives

$${}_{92}^{236}\text{U} \longrightarrow {}_{37}^{90}\text{Rb} + {}_{55}^{143}\text{Cs} + 3{}_0^1\text{n}$$

So three neutrons are produced per fission.
(b) The masses of the decay particles are given in Appendix B in terms of atomic mass units, u, where $1\ \text{u} = 1.660 \times 10^{-27}\ \text{kg} = 931.5\ \text{MeV}/c^2$.

$$\Delta m = M_\text{U} - (M_\text{Rb} + M_\text{Cs} + 3m_\text{n}) = 236.045563\ \text{u}$$
$$-(89.914811\ \text{u} + 142.927220\ \text{u}$$
$$+ (3)(1.008665)\ \text{u})$$
$$= 0.177537\ \text{u} = 2.9471 \times 10^{-28}\ \text{kg}$$

Therefore, the reaction products have a combined mass that is about 3.0×10^{-28} kg less than the initial uranium mass.
(c) The energy given off per fission event is just Δmc^2. This is most easily calculated if Δm is first converted to mass units of MeV/c^2. Because $1\ \text{u} = 931.5\ \text{MeV}/c^2$,

$$\Delta m = (0.177537\ \text{u})(931.5\ \text{MeV}/c^2)$$
$$= 165.4\ \text{MeV}/c^2$$

$$Q = \Delta mc^2 = 165.4\ \frac{\text{MeV}}{c^2}\ c^2 = 165.4\ \text{MeV}$$
$$= -165.4\ \text{MeV}$$

(d) To find the energy released by the fission of 1 kg of uranium we need to calculate the number of nuclei, N, contained in 1 kg of ${}^{236}\text{U}$.

$$N = \frac{(6.02 \times 10^{23} \text{ nuclei/mol})}{(236 \text{ g/mol})}(1000 \text{ g})$$

$$= 2.55 \times 10^{24} \text{ nuclei}$$

The total energy produced, E, is

$$E = (\text{efficiency})NQ$$

$$= (0.40)(2.55 \times 10^{24} \text{ nuclei})(165 \text{ MeV/nucleus})$$

$$= 1.68 \times 10^{26} \text{ MeV}$$

$$= (1.68 \times 10^{26} \text{ MeV})(4.45 \times 10^{-20} \text{ kWh/MeV})$$

$$= 7.48 \times 10^{6} \text{ kWh}$$

Exercise 2   How long will this amount of energy keep a 100-W lightbulb burning?

Answer ≈ 8500 years.

We have considered the simplest case showing the conversion of mass to energy and the release of this nuclear energy: the decay of a heavy unstable element into several lighter elements. However, the most common case is the one in which the mass of a composite particle is *less than* the sum of the particle masses composing it. By examining Appendix B, you can see that the mass of any nucleus is less than the sum of its component neutrons and protons by an amount Δm. This occurs because the nuclei are stable, *bound* systems of neutrons and protons (bound by strong attractive nuclear forces), and in order to disassociate them into separate nucleons an amount of energy Δmc^2 must be supplied to the nucleus. This energy or work required to pull a bound system apart, leaving its component parts free of attractive forces and at rest, is called the binding energy, *BE*. Thus, we describe the mass and energy of a bound system by the equation

$$Mc^2 + BE = \sum_{i=1}^{n} m_i c^2 \tag{2.17}$$

where M is the bound system mass, the m_i's are the free component particle masses, and n is the number of component particles. Two general comments are in order about Equation 2.17. First, it applies quite generally to any type of system bound by attractive forces, whether gravitational, electrical (chemical), or nuclear. For example, the mass of a water molecule is less than the combined mass of two free hydrogen atoms and a free oxygen atom, although the mass difference cannot be directly measured in this case. (The mass difference can be measured in the nuclear case because the forces and the binding energy are so much greater.) Second, Equation 2.17 shows the possibility of liberating huge quantities of energy, *BE*, if one reads the equation from right to left; that is, one collides nuclear particles with a small but sufficient amount of kinetic energy to overcome proton repulsion and fuse the particles into new elements with less mass. Such a process is called *fusion*, one example of which is a reaction in which two deuterium nuclei combine to form a helium nucleus, releasing 23.9 MeV per fusion. (See Chapter 14 for more on fusion processes.) We can write this reaction schematically as follows:

Fusion

$$^2_1\text{H} + ^2_1\text{H} \longrightarrow ^4_2\text{He} + 23.9 \text{ MeV}$$

EXAMPLE 2.8

(a) How much lighter is a molecule of water than two hydrogen atoms and an oxygen atom? The binding energy of water is about 3 eV. (b) Find the fractional loss of mass per gram of water formed. (c) Find the total energy released (mainly as heat and light) when 1 gram of water is formed.

Solution   (a) Equation 2.17 may be solved for the mass difference as follows:

$$\Delta m = (m_{\text{H}} + m_{\text{H}} + m_{\text{O}}) - M_{\text{H}_2\text{O}} = \frac{BE}{c^2} = \frac{3\text{ eV}}{c^2}$$

$$= \frac{(3.0\text{ eV})(1.6 \times 10^{-19}\text{ J/eV})}{(3.0 \times 10^8\text{ m/s})^2} = 5.3 \times 10^{-36}\text{ kg}$$

(b) To find the fractional loss of mass per molecule we divide Δm by the mass of a water molecule, $M_{\text{H}_2\text{O}} = 18\text{u} = 3.0 \times 10^{-26}$ kg:

$$\frac{\Delta m}{M_{\text{H}_2\text{O}}} = \frac{5.3 \times 10^{-36}\text{ kg}}{3.0 \times 10^{-26}\text{ kg}} = 1.8 \times 10^{-10}$$

Because the fractional loss of mass per molecule is the same as the fractional loss per gram of water formed, 1.8×10^{-10} g of mass would be lost for each gram of water formed. This is much too small a mass to be measured directly, and this calculation shows that nonconservation of mass does not generally show up as a measurable effect in chemical reactions.

(c) The energy released when 1 gram of H_2O is formed is simply the change in mass when 1 gram of water is formed times c^2:

$$E = \Delta mc^2 = (1.8 \times 10^{-13}\text{ kg})(3.0 \times 10^8\text{ m/s})^2 \approx 16\text{ kJ}$$

This energy change, as opposed to the decrease in mass, is easily measured, providing another case similar to Example 2.5 in which mass changes are minute but energy changes, amplified by a factor of c^2, are easily measured.

2.4   CONSERVATION OF RELATIVISTIC MOMENTUM AND ENERGY

So far we have considered only cases of the conservation of mass−energy. By far, however, the most common and strongest confirmation of relativity theory comes from the daily application of relativistic momentum and energy conservation to elementary particle interactions. Often the measurement of momentum (from the path curvature in a magnetic field—see Example 2.2) and kinetic energy (from the distance a particle travels in a known substance before coming to rest) are enough when combined with conservation of momentum and mass−energy to determine fundamental particle properties of mass, charge, and mean lifetime.

EXAMPLE 2.9   Measuring the Mass of the π^+ Meson

The π^+ meson (also called the *pion*) is a subatomic particle responsible for the strong nuclear force between protons and neutrons. It is observed to decay at rest into a μ^+ meson (muon) and a neutrino,[2] denoted v. Because the neutrino has no charge and little mass (talk about elusive!), it leaves no track in a bubble chamber. (A bubble chamber is a large chamber filled with liquid hydrogen that shows the tracks of charged particles as a series of tiny bubbles.) However, the track of the charged muon is visible as it loses kinetic energy and comes to rest (Fig. 2.3). If the mass of the muon is known to be 106 MeV/c^2, and the kinetic energy, K, of the muon is measured to be 4.6 MeV from its track length, find the mass of the π^+.

Solution   The decay equation is $\pi^+ \rightarrow \mu^+ + v$. Conserving energy gives

$$E_\pi = E_u + E_v$$

---

[2]Neutrino, from the Italian, means "little tiny neutral one." Following this practice, neutron should probably be neutrone (pronounced noo-trōn-eh) or "great big neutral one."

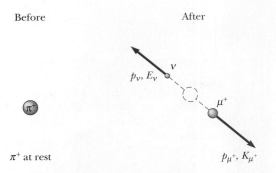

Figure 2.3 (Example 2.9) Decay of the pion at rest into a neutrino and a muon.

Because the pion is at rest when it decays, and the neutrino has negligible mass,

$$m_\pi c^2 = \sqrt{(m_\mu c^2)^2 + (p_v^2 c^2)} + p_v c \qquad (2.18)$$

Conserving momentum in the decay yields $p_\mu = p_v$. Substituting the muon momentum for the neutrino momentum in Equation 2.18 gives the following expression for the rest energy of the pion in terms of the muon's mass and momentum:

$$m_\pi c^2 = \sqrt{(m_\mu c^2)^2 + (p_\mu^2 c^2)} + p_\mu c \qquad (2.19)$$

Finally, to obtain p_μ from the measured value of the muon's kinetic energy, K_μ, we start with Equation 2.11, $E_\mu^2 = p_\mu^2 c^2 + (m_\mu c^2)^2$, and solve it for $p_\mu^2 c^2$:

$$p_\mu^2 c^2 = E_\mu^2 - (m_\mu c^2)^2 = (K_\mu + m_\mu c^2)^2 - (m_\mu c^2)^2$$
$$= K_\mu^2 + 2K_\mu m_\mu c^2$$

Substituting this expression for $p_\mu^2 c^2$ into Equation 2.19 yields the desired expression for the pion mass in terms of the muon's mass and kinetic energy:

$$m_\pi c^2 = \sqrt{(m_\mu^2 c^4 + K_\mu^2 + 2K_\mu m_\mu c^2} \\ + \sqrt{K_\mu^2 + 2K_\mu m_\mu c^2} \qquad (2.20)$$

Finally, substituting $m_\mu c^2 = 106$ MeV and $K_\mu = 4.6$ MeV into Equation 2.20 gives

$$m_\pi c^2 = 111 \text{ MeV} + 31 \text{ MeV} \approx 1.4 \times 10^2 \text{ MeV}$$

Thus, the mass of the pion is

$$m_\pi = 140 \text{ MeV}/c^2$$

This result shows why this particle is called a meson; it has an intermediate mass (from the Greek word *mesos* meaning "middle") between the light electron (0.511 MeV/c^2) and the heavy proton (938 MeV/c^2).

2.5 GENERAL RELATIVITY

Up to this point, we have sidestepped a curious puzzle. Mass has two seemingly different properties: a *gravitational attraction* for other masses and an *inertial* property that represents a resistance to acceleration. To designate these two attributes, we use the subscripts *g* and *i* and write

Gravitational property: $\qquad F_g = G\dfrac{m_g m_g'}{r^2}$

Inertial property: $\qquad \sum F = m_i a$

The value for the gravitational constant G was chosen to make the magnitudes of m_g and m_i numerically equal. Regardless of how G is chosen, however, the strict proportionality of m_g and m_i has been established experimentally to an extremely high degree: a few parts in 10^{12}. Thus, it appears that gravitational mass and inertial mass may indeed be exactly proportional.

But why? They seem to involve two entirely different concepts: a force of mutual gravitational attraction between two masses, and the resistance of a single mass to being accelerated. This question, which puzzled Newton and many other physicists over the years, was answered by Einstein in 1916 when he published his theory of gravitation, known as the *general theory of relativity*. Because it is a mathematically complex theory, we offer merely a hint of its elegance and insight.

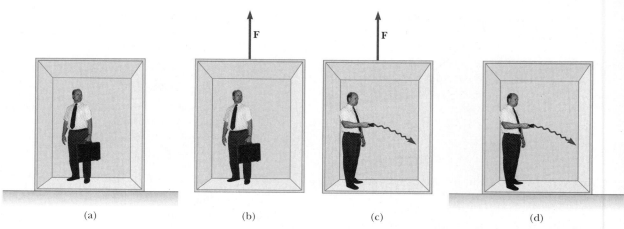

(a)   (b)   (c)   (d)

Figure 2.4 (a) The observer is at rest in a uniform gravitational field **g**, directed downward. (b) The observer is in a region where gravity is negligible, but the frame is accelerated by an external force **F** that produces an acceleration **g** directed upward. According to Einstein, the frames of reference in parts (a) and (b) are equivalent in every way. No local experiment can distinguish any difference between the two frames. (c) In the accelerating frame, a ray of light would appear to bend downward due to the acceleration of the elevator. (d) If parts (a) and (b) are truly equivalent, as Einstein proposed, then part (c) suggests that a ray of light would bend downward in a gravitational field.

In Einstein's view, the dual behavior of mass was evidence of a very intimate and basic connection between the two behaviors. He pointed out that no mechanical experiment (such as dropping an object) could distinguish between the two situations illustrated in Figures 2.4a and 2.4b. In Figure 2.4a, a person is standing in an elevator on the surface of a planet and feels pressed into the floor, due to the gravitational force. In Figure 2.4b, the person is in an elevator in empty space accelerating upward with $a = g$. The person feels pressed into the floor with the same force as in Figure 2.4a. In each case, an object released by the observer undergoes a downward acceleration of magnitude g relative to the floor. In Figure 2.4a, the person is in an inertial frame in a gravitational field. In Figure 2.4b, the person is in a noninertial frame accelerating in gravity-free space. Einstein's claim is that these two situations are *completely* equivalent. Because the two reference frames in relative acceleration can no longer be distinguished from one another, this extends the idea of complete physical equivalence to reference frames *accelerating translationally* with respect to each other. This solved another philosophical issue raised by Einstein, namely the artificiality of confining the principle of relativity to nonaccelerating frames.

Einstein carried his idea further and proposed that *no* experiment, mechanical or otherwise, could distinguish between the two cases. This extension to include all phenomena (not just mechanical ones) has interesting consequences. For example, suppose that a light pulse is sent horizontally across an elevator that is accelerating upward in empty space, as in Figure 2.4c. From the point of view of an observer in an inertial frame outside of the elevator, the light travels in a straight line while the floor of the elevator accelerates upward. According to the observer on the elevator, however, the trajectory of the light pulse bends downward as the floor of the elevator (and the observer) accelerates upward. Therefore, based on the equality of parts (a) and (b) of the figure for all phenomena, Einstein proposed that **a beam of light should also be defelected downward or fall in a gravitational field,** as in Figure 2.4d. Experiments have verified the effect, although the bending is small. A laser aimed at the horizon falls less than 1 cm after traveling 6000 km.

The two postulates of Einstein's **general theory of relativity** are

Postulates of general relativity

- The laws of nature have the same form for observers in any frame of reference, whether accelerated or not.
- In the vicinity of any point, a gravitational field is equivalent to an accelerated frame of reference in the absence of gravitational effects. (This is the **principle of equivalence.**)

An interesting effect predicted by the general theory is that time is altered by gravity. A clock in the presence of gravity runs slower than one located where gravity is negligible. Consequently, the frequencies of radiation emitted by atoms in the presence of a strong gravitational field are *redshifted* to lower frequencies when compared with the same emissions in the presence of a weak field. This gravitational redshift has been detected in spectral lines emitted by atoms in massive stars. It has also been verified on the Earth by comparing the frequencies of gamma rays (a high-energy form of electromagnetic radiation) emitted from nuclei separated vertically by about 20 m (see Section 3.7).

The second postulate suggests that a gravitational field may be "transformed away" at any point if we choose an appropriate accelerated frame of reference—a freely falling one. Einstein developed an ingenious method of describing the acceleration necessary to make the gravitational field "disappear." He specified a concept, the *curvature of spacetime,* that describes the gravitational effect at every point. In fact, the curvature of spacetime completely replaces Newton's gravitational theory. According to Einstein, there is no such thing as a gravitational force. Rather, the presence of a mass causes a curvature of spacetime in the vicinity of the mass, and this curvature dictates the spacetime path that all freely moving objects must follow. In 1979, John Wheeler (b. 1911, American theoretical physicist) summarized Einstein's general theory of relativity in a single sentence: "Space tells matter how to move and matter tells space how to curve."

As an example of the effects of curved spacetime, imagine two travelers moving on parallel paths a few meters apart on the surface of the Earth and maintaining an exact northward heading along two longitude lines. As they observe each other near the equator, they will claim that their paths are exactly parallel. As they approach the North Pole, however, they notice that they are moving closer together, and they will actually meet at the North Pole. Thus, they will claim that they moved along parallel paths, but moved toward each other, *as if there were an attractive force between them.* They will make this conclusion based on their everyday experience of moving on flat surfaces. From our perspective, however, we realize that they are walking on

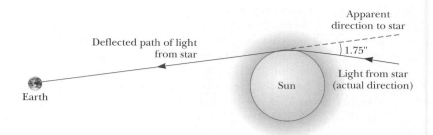

Apparent
direction to star

Deflected path of light
from star

1.75"

Sun

Light from star
(actual direction)

Earth

Figure 2.5 Deflection of starlight passing near the Sun. Because of this effect, the Sun or some other remote object can act as a *gravitational lens*. In his general theory of relativity, Einstein calculated that starlight just grazing the Sun's surface should be deflected by an angle of 1.75 s of arc.

a curved surface, and it is the geometry of the curved surface that causes them to converge, rather than an attractive force. In a similar way, general relativity replaces the notion of forces with the movement of objects through curved spacetime.

An important prediction of the general theory of relativity is that a light ray passing near the Sun should be deflected in the curved spacetime created by the Sun's mass. This prediction was confirmed when astronomers detected the bending of starlight near the Sun during a total solar eclipse that occurred shortly after World War I (Fig. 2.5). When this discovery was announced, Einstein became an international celebrity. (See the web essay by Clifford Will for other important tests and ramifications of general relativity at http://info.brookscole.com/mp3e.)

If the concentration of mass becomes very great, as is believed to occur when a large star exhausts its nuclear fuel and collapses to a very small volume, a **black hole** may form. Here, the curvature of spacetime is so extreme that, within a certain distance from the center of the black hole, all matter and light become trapped, as discussed in Section 3.7.

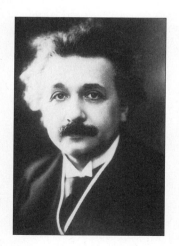

Figure 2.6 Albert Einstein. Gravity imaging was another triumph for Einstein since he pointed out that it might occur in 1936. (*Courtesy of AIP/Niels Bohr Library*).

Gravitational Radiation, or A Good Wave Is Hard to Find

Gravitational radiation is a subject almost as old as general relativity. By 1916, Einstein had succeeded in showing that the field equations of general relativity admitted wavelike solutions analogous to those of electromagnetic theory. For example, a dumbbell rotating about an axis passing at right angles through its handle will emit gravitational waves that travel at the speed of light. Gravitational waves also carry energy away from the dumbbell, just as electromagnetic waves carry energy away from a light source. Also, like electromagnetic (em) waves, gravity waves are believed to have a dual particle and wave nature. The gravitational particle, the graviton, is believed to have a mass of zero, to travel at the speed c, and to obey the relativistic equation $E = pc$.

In 1968, Joseph Weber initiated a program of gravitational-wave detection using as detectors massive aluminum bars, suspended in vacuum and isolated from outside forces. Gravity waves are notoriously more difficult to detect than

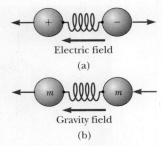

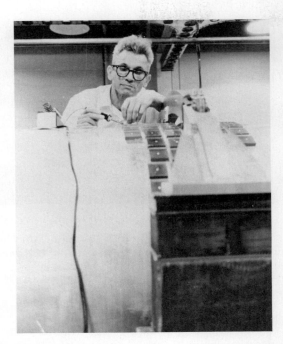

Figure 2.7   Joseph Weber working on a bar detector at the University of Maryland in the early 1970's. The fundamental frequency of the bar was 1660 Hz. Piezoelectric crystals around the center of the bar convert tiny mechanical vibrations to electrical signals. (*Courtesy of AIP Emilio Segre Visual Archives*)

Figure 2.8   Simple models of em and gravity wave detectors. The detectors are shown as two "charges" with a spring sandwiched in between, the idea being that the waves exert forces on the charges and set the spring vibrating in proportion to the wave intensity. The detector will be particularly sensitive when the wave frequency matches the natural frequency of the spring–mass system. (a) Equal and opposite electric charges move in opposite directions when subjected to an em wave and easily excite the spring. (b) A metal bar gravity wave detector can be modeled by a spring connecting two equal masses; however, a wave encountering both masses in phase will not cause the spring to vibrate.

em waves not only because gravitational forces are much weaker than electric forces but also because gravitational "charge" or mass only comes in one variety, positive. Figure 2.8a shows why a dipolar em wave detector is much more sensitive than a gravitational bar detector shown in Figure 2.8b. Nevertheless, as shown in Figure 2.9, if the distance between detecting masses is of the same order of magnitude as the wavelength of the gravity wave, passing gravitational waves exert a weak net oscillating force that alternately compresses and extends the bar lengthwise.

Tiny vibrations of the bar are detected by crystals attached to the bar that convert the vibrations to electrical signals. Currently, a dozen laboratories around the world are engaged in building and improving the basic "Weber bar" detector, striving to reduce noise from thermal, electrical, and environmental sources in order to detect the very weak oscillations produced by a gravitational wave. For a bar of 1 meter in length, the challenge is to detect a variation in length smaller than 10^{-20} m, or 10^{-5} of the radius of a proton. This sensitivity is predicated on a massive nearby *catastrophic* source of gravitational waves, such as the gravitational collapse of a star to form a black hole at the center of our galaxy. Thus, gravity waves are not only hard to detect but also hard to generate with great intensity. It is interesting that collapsing star models predict a collapse to take about a millisecond, with production of gravity waves of frequency around 1 kHz and wavelengths of several hundred km.

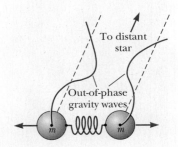

Figure 2.9   If the gravity wave detector is of the same size as the wavelength of the radiation detected, the waves arrive out-of-phase at the two masses and the system starts to vibrate.

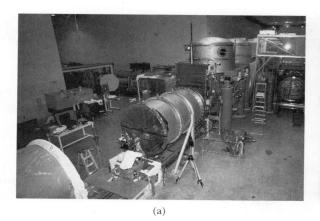

(a)

(b)

Figure 2.10   (a) Prototype LIGO apparatus with 40 m arms. (b) Photo of Louisiana LIGO with 4 km arms. (*a: Tony Tyson/Lucent Technologies/Bell Labs Innovations; b: Courtesy of California Institute of Technology*)

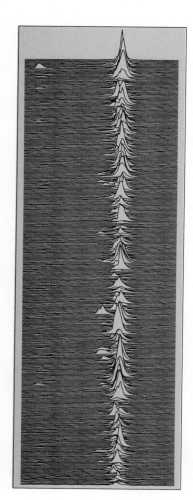

Figure 2.11   400 consecutive radio pulses from pulsar PSR 0950+08. Each line of the 400 represents a consecutive time interval of 0.253 s.

At this time, several "**l**aser-**i**nterferometric" **g**ravitational-wave **o**bservatories (LIGO) are in operation or under construction in the United States and Europe. These reflect laser beams along perpendicular arms to monitor tiny variations in length between mirrors spaced several kilometers apart in a giant Michelson–Morley apparatus. (See Figures 2.10a and b.) The variations in arm length should occur when a gravitational wave passes through the apparatus. Two LIGO sites with 4-km arms are currently in operation in the United States in Livingston, Louisiana and Hanford, Washington. The two sites, separated by about 2000 miles, search for signals that appear simultaneously at both sites. Such coincidences are more likely to be gravity waves from a distant star rather than local noise signals.

Although gravitational radiation has not been detected directly, we know that it exists through the observations of a remarkable system known as the binary pulsar. Discovered in 1974 by radio astronomers Russell Hulse and Joseph Taylor, it consists of a pulsar (which is a rapidly spinning neutron star) and a companion star in orbit around each other. Although the companion has not been seen directly, it is also believed to be a neutron star. The pulsar acts as an extremely stable clock, its pulse period of approximately 59 milliseconds drifting by only 0.25 ns/year. Figure 2.11 shows the remarkable regularity of 400 consecutive radio pulses from a pulsar. By measuring the arrival times of radio pulses at Earth, observers were able to determine the motion of the pulsar about its companion with amazing accuracy. For example, the accurate value for the orbital period is 27906.980 895 s, and the orbital eccentricity is 0.617|131. Like a rotating dumbbell, an orbiting binary system should emit gravitational radiation and, in the process, lose some of its orbital energy. This energy loss will cause the pulsar and its companion to spiral in toward each other and the orbital period to shorten. According to general relativity, the predicted decrease in the orbital period is 75.8 μs/year. The observed decrease in orbital period is in agreement with the prediction to better than 0.5%. This confirms the existence of gravitational radiation and the general relativistic equations that describe it.

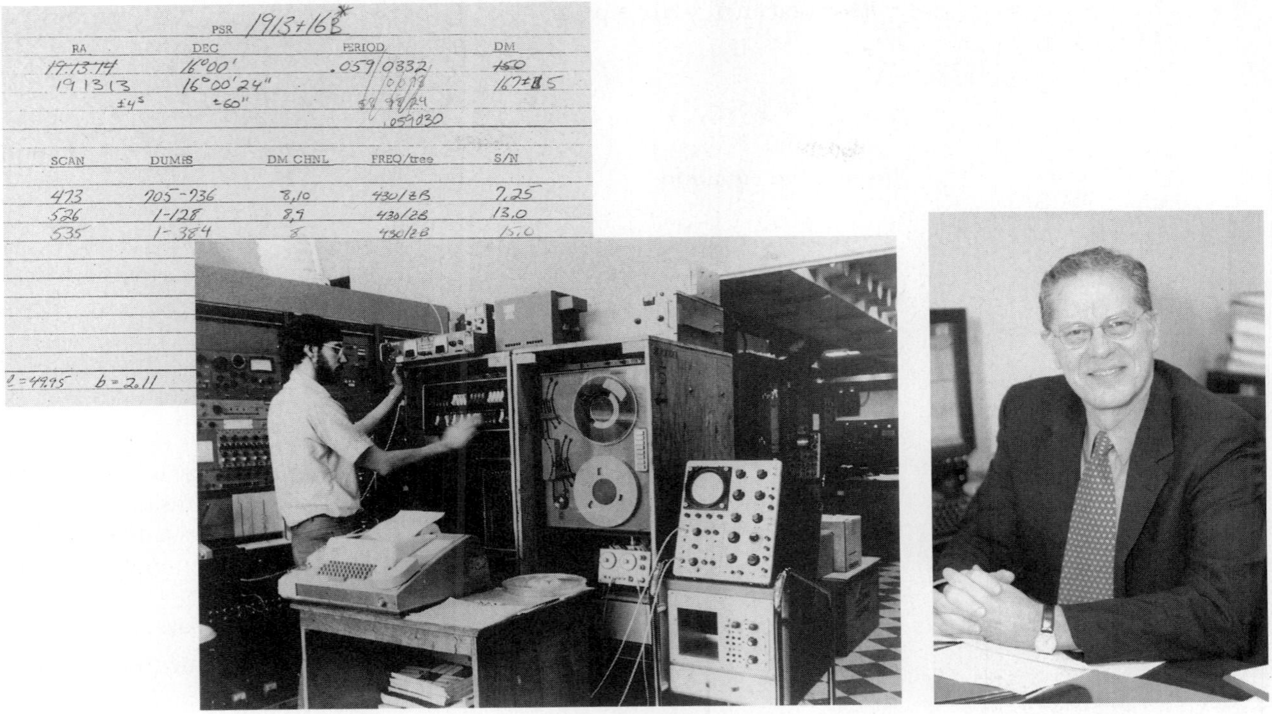

Figure 2.12   (a) Russell Hulse shown in 1974 operating his computer and teletype at Arecibo observatory in Puerto Rico. The form records the "fantastic" detection of PSR 1913+16, with its ever-changing periods scratched out by Hulse in frustration. (b) Joseph Taylor found evidence for gravity waves in the motions of PSR 1913+16, a pair of neutron stars. (*a: Photo Courtesy of Russell Hulse. © The Nobel Foundation, 1993; b: Courtesy of Office of Communications, Princeton University*)

Hulse and Taylor (Figure 2.12) received the Nobel prize in 1993 for this discovery.

SUMMARY

The relativistic expression for the **linear momentum** of a particle moving with a velocity **u** is

$$\mathbf{p} \equiv \frac{m\mathbf{u}}{\sqrt{1 - (u^2/c^2)}} = \gamma m\mathbf{u} \tag{2.1}$$

where γ is given by

$$\gamma = \frac{1}{\sqrt{1 - (u^2/c^2)}}$$

The relativistic expression for the **kinetic energy** of a particle is

$$K = \gamma mc^2 - mc^2 \tag{2.9}$$

where mc^2 is called the **rest energy** of the particle.

The total energy E of a particle is related to the mass through the expression

$$E = \gamma mc^2 = \frac{mc^2}{\sqrt{1 - (u^2/c^2)}} \qquad (2.10)$$

The total energy of a particle of mass m is related to the momentum through the equation

$$E^2 = p^2c^2 + (mc^2)^2 \qquad (2.11)$$

Finally, the law of the conservation of mass—energy states that *the sum of the mass—energy of a system of particles before interaction must equal the sum of the mass—energy of the system after interaction where the mass—energy of the ith particle is defined as*

$$E_i = \frac{m_ic^2}{\sqrt{1 - (u_i^2/c^2)}}$$

Application of the principle of conservation of mass—energy to the specific cases of (1) the fission of a heavy nucleus at rest and (2) the fusion of several particles into a composite nucleus with less total mass allows us to define (1) the energy released per fission, Q, and (2) the binding energy of a composite system, *BE*.

The two postulates of Einstein's **general theory of relativity** are

- The laws of nature have the same form for observers in any frame of reference, whether accelerated or not.
- In the vicinity of any point, a gravitational field is equivalent to an accelerated frame of reference in the absence of gravitational effects. (This is the **principle of equivalence**.)

The field equations of general relativity predict gravitational waves, and a worldwide search is currently in progress to detect these elusive waves.

(© *S. Harris*)

SUGGESTIONS FOR FURTHER READING

1. E. F. Taylor and J. A. Wheeler, *Spacetime Physics*, San Francisco, W. H. Freeman, 1963.

2. R. Resnick, *Introduction to Special Relativity*, New York, Wiley, 1968.

3. A. P. French, *Special Relativity*, New York, Norton, 1968.

4. H. Bondi, *Relativity and Common Sense*, Science Study Series, Garden City, NY, Doubleday, 1964.

5. A. Einstein, *Out of My Later Years*, New York, World Publishing, 1971.

6. A. Einstein, *Ideas and Opinions*, New York, Crown, 1954.

7. G. Gamow, *Mr. Tompkins in Wonderland*, New York, Cambridge University Press, 1939.

8. L. Infeld, *Albert Einstein*, New York, Scribner's, 1950.

9. J. Schwinger, *Einstein's Legacy*, Scientific American Library, New York, W. H. Freeman, 1985.

10. R. S. Shankland, "The Michelson-Morley Experiment," *Sci. Amer.*, November 1964, p. 107.

11. R. Skinner, *Relativity for Scientists and Engineers*, New York, Dover Publications, 1982.

12. N. Mermin, *Space and Time in Special Relativity*, Prospect Heights, IL, Waveland Press, 1989.

13. M. Bartusiak, *Einstein's Unfinished Symphony*, New York, Berkley Books, 2000. (A nonmathematical history and explanation of the search for gravity waves.)

QUESTIONS

1. A particle is moving at a speed of less than $c/2$. If the speed of the particle is doubled, what happens to its momentum?

2. Give a physical argument showing that it is impossible to accelerate an object of mass m to the speed of light, even with a continuous force acting on it.

3. The upper limit of the speed of an electron is the speed of light, c. Does that mean that the momentum of the electron has an upper limit?

4. Because mass is a measure of energy, can we conclude that the mass of a compressed spring is greater than the mass of the same spring when it is not compressed?

5. Photons of light have zero mass. How is it possible that they have momentum?

6. "Newtonian mechanics correctly describes objects moving at ordinary speeds, and relativistic mechanics correctly describes objects moving very fast." "Relativistic mechanics must make a smooth transition as it reduces to Newtonian mechanics in a case where the speed of an object becomes small compared to the speed of light." Argue for or against each of these two statements.

7. Two objects are identical except that one is hotter than the other. Compare how they respond to identical forces.

8. With regard to reference frames, how does general relativity differ from special relativity?

9. Two identical clocks are in the same house, one upstairs in a bedroom, and the other downstairs in the kitchen. Which clock runs more slowly? Explain.

10. *A thought experiment.* Imagine ants living on a merry-go-round, which is their two-dimensional world. From measurements on small circles they are thoroughly familiar with the number π. When they measure the circumference of their world, and divide it by the diameter, they expect to calculate the number $\pi = 3.14159. \ldots$ We see the merry-go-round turning at relativistic speed. From our point of view, the ants' measuring rods on the circumference are experiencing Lorentz contraction in the tangential direction; hence the ants will need some extra rods to fill that entire distance. The rods measuring the diameter, however, do not contract, because their motion is perpendicular to their lengths. As a result, the computed ratio does not agree with the number π. If you were an ant, you would say that the rest of the universe is spinning in circles, and your disk is stationary. What possible explanation can you then give for the discrepancy, in view of the general theory of relativity?

PROBLEMS

2.1 Relativistic Momentum and the Relativistic Form of Newton's Laws

1. Calculate the momentum of a proton moving with a speed of (a) $0.010c$, (b) $0.50c$, (c) $0.90c$. (d) Convert the answers of (a)–(c) to MeV/c.

2. An electron has a momentum that is 90% larger than its classical momentum. (a) Find the speed of the electron. (b) How would your result change if the particle were a proton?

3. Consider the relativistic form of Newton's second law. Show that when **F** is parallel to **v**,

$$F = m \left(1 - \frac{v^2}{c^2} \right)^{-3/2} \frac{dv}{dt}$$

where m is the mass of an object and v is its speed.

4. A charged particle moves along a straight line in a uniform electric field E with a speed v. If the motion and the electric field are both in the x direction, (a) show

that the magnitude of the acceleration of the charge q is given by

$$a = \frac{dv}{dt} = \frac{qE}{m}\left(1 - \frac{v^2}{c^2}\right)^{3/2}$$

(b) Discuss the significance of the dependence of the acceleration on the speed. (c) If the particle starts from rest at $x = 0$ at $t = 0$, find the speed of the particle and its position after a time t has elapsed. Comment on the limiting values of v and x as $t \rightarrow \infty$.

5. Recall that the magnetic force on a charge q moving with velocity $\mathbf{v}$ in a magnetic field $\mathbf{B}$ is equal to $q\mathbf{v} \times \mathbf{B}$. If a charged particle moves in a circular orbit with a fixed speed v in the presence of a constant magnetic field, use the relativistic form of Newton's second law to show that the frequency of its orbital motion is

$$f = \frac{qB}{2\pi m}\left(1 - \frac{v^2}{c^2}\right)^{1/2}$$

6. Show that the momentum of a particle having charge e moving in a circle of radius R in a magnetic field B is given by $p = 300BR$, where p is in MeV/c, B is in teslas, and R is in meters.

2.2   Relativistic Energy

7. Show that the energy–momentum relationship given by $E^2 = p^2c^2 + (mc^2)^2$ follows from the expressions $E = \gamma mc^2$ and $p = \gamma mu$.

8. A proton moves at a speed of $0.95c$. Calculate its (a) rest energy, (b) total energy, and (c) kinetic energy.

9. An electron has a kinetic energy 5 times greater than its rest energy. Find (a) its total energy and (b) its speed.

10. Find the speed of a particle whose total energy is 50% greater than its rest energy.

11. A proton in a high-energy accelerator is given a kinetic energy of 50 GeV. Determine the (a) momentum and (b) speed of the proton.

12. An electron has a speed of $0.75c$. Find the speed of a proton that has (a) the same kinetic energy as the electron and (b) the same momentum as the electron.

13. Protons in an accelerator at the Fermi National Laboratory near Chicago are accelerated to an energy of 400 times their rest energy. (a) What is the speed of these protons? (b) What is their kinetic energy in MeV?

14. *How long will the Sun shine, Nellie?* The Sun radiates about 4.0×10^{26} J of energy into space each second. (a) How much mass is released as radiation each second? (b) If the mass of the Sun is 2.0×10^{30} kg, how long can the Sun survive if the energy release continues at the present rate?

15. Electrons in projection television sets are accelerated through a total potential difference of 50,000 V. (a) Calculate the speed of the electrons using the relativistic form of kinetic energy assuming the electrons start from rest. (b) Calculate the speed of the electrons using the classical form of kinetic energy. (c) Is the difference in speed significant in the design of this set in your opinion?

16. As noted in Section 2.2, the quantity $E - p^2c^2$ is an *invariant* in relativity theory. This means that the quantity $E^2 - p^2c^2$ has the same value in all inertial frames even though E and p have different values in different frames. Show this explicitly by considering the following case. A particle of mass m is moving in the $+x$ direction with speed u and has momentum p and energy E in the frame S. (a) If S′ is moving at speed v in the standard way, find the momentum p' and energy E' observed in S′. (*Hint:* Use the Lorentz velocity transformation to find p' and E'. Does $E = E'$ and $p = p'$? (b) Show that $E^2 - p^2c^2$ is equal to $E'^2 - p'^2c^2$.

2.3   Mass as a Measure of Energy

17. A radium isotope decays to a radon isotope, ^{222}Rn, by emitting an α particle (a helium nucleus) according to the decay scheme ^{226}Ra $\rightarrow$ ^{222}Rn + ^{4}He. The masses of the atoms are 226.0254 (Ra), 222.0175 (Rn), and 4.0026 (He). How much energy is released as the result of this decay?

18. Consider the decay $^{55}_{24}$Cr $\rightarrow$ $^{55}_{25}$Mn + e$^-$, where e$^-$ is an electron. The ^{55}Cr nucleus has a mass of 54.9279 u, and the ^{55}Mn nucleus has a mass of 54.9244 u. (a) Calculate the mass difference in MeV. (b) What is the maximum kinetic energy of the emitted electron?

19. Calculate the binding energy in MeV per nucleon in the isotope $^{12}_{6}$C. Note that the mass of this isotope is exactly 12 u, and the masses of the proton and neutron are 1.007276 u and 1.008665 u, respectively.

20. The free neutron is known to decay into a proton, an electron, and an antineutrino $\bar{v}$ (of negligible rest mass), according to

$$n \longrightarrow p + e^- + \bar{v}$$

This is called *beta decay* and will be discussed further in Chapter 13. The decay products are measured to have a total kinetic energy of 0.781 MeV $\pm$ 0.005 MeV. Show that this observation is consistent with the excess energy predicted by the Einstein mass–energy relationship.

2.4   Conservation of Relativistic Momentum and Energy

21. An electron having kinetic energy $K = 1.000$ MeV makes a head-on collision with a positron at rest. (A positron is an antimatter particle that has the same mass as the electron but opposite charge.) In the collision the two particles annihilate each other and are replaced by two γ rays of equal energy, each traveling at equal angles θ with the electron's direction of motion. (Gamma rays are massless particles of elec-

tromagnetic radiation having energy $E = pc$.) Find the energy E, momentum p, and angle of emission θ of the γ rays.

22. The K^0 meson is an uncharged member of the particle "zoo" that decays into two charged pions according to $K^0 \rightarrow \pi^+ + \pi^-$. The pions have opposite charges, as indicated, and the same mass, $m_\pi = 140$ MeV/c^2. Suppose that a K^0 at rest decays into two pions in a bubble chamber in which a magnetic field of 2.0 T is present (see Fig. P2.22). If the radius of curvature of the pions is 34.4 cm, find (a) the momenta and speeds of the pions and (b) the mass of the K^0 meson.

23. An unstable particle having a mass of 3.34×10^{-27} kg is initially at rest. The particle decays into two fragments that fly off with velocities of $0.987c$ and $-0.868c$. Find the rest masses of the fragments.

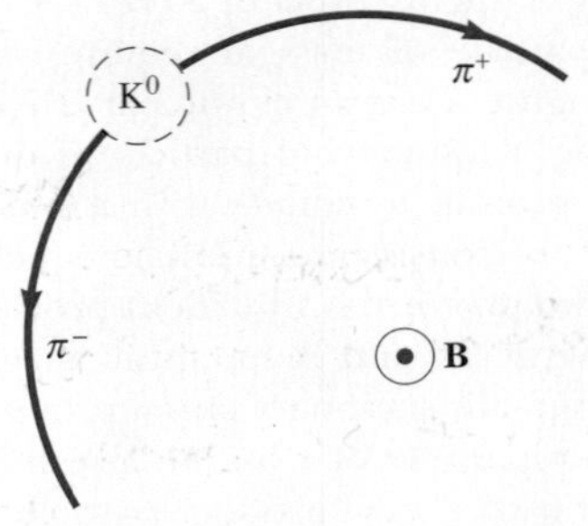

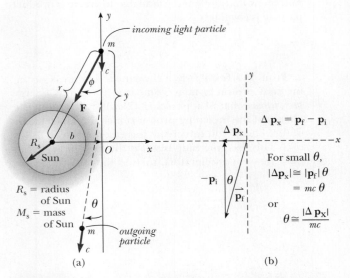

Figure P2.22 A sketch of the tracks made by the π^+ and π^- in the decay of the K^0 meson at rest. The pion motion is perpendicular to **B**. (**B** is directed out of the page.)

ADDITIONAL PROBLEMS

24. As measured by observers in a reference frame S, a particle having charge q moves with velocity **v** in a magnetic field **B** and an electric field **E**. The resulting force on the particle is then measured to be $\mathbf{F} = q(\mathbf{E} + \mathbf{v} \times \mathbf{B})$. Another observer moves along with the charged particle and measures its charge to be q also but measures the electric field to be **E'**. If both observers are to measure the same force, **F**, show that $\mathbf{E'} = \mathbf{E} + \mathbf{v} \times \mathbf{B}$.

25. *Classical deflection of light by the Sun* Estimate the deflection of starlight grazing the surface of the Sun. Assume that light consists of particles of mass m traveling with velocity c and that the deflection is small. (a) Use $\Delta p_x = \int_{-\infty}^{+\infty} F_x \, dt$ to show that the angle of deflection θ is given by $\theta \cong \dfrac{2GM_s}{bc^2}$ where Δp_x is the total change in momentum of a light particle grazing the Sun. See Figures P2.25a and b. (b) For $b = R_s$, show that $\theta = 4.2 \times 10^{-6}$ rad.

26. An object having mass of 900 kg and traveling at a speed of $0.850c$ collides with a stationary object having mass 1400 kg. The two objects stick together. Find (a) the speed and (b) the mass of the composite object.

27. Imagine that the entire Sun collapses to a sphere of radius R_g such that the work required to remove a small mass m from the surface would be equal to its rest energy mc^2. This radius is called the *gravitational radius* for the Sun. Find R_g. (It is believed that the ultimate fate of very massive stars is to collapse beyond their gravitational radii into black holes.)

28. A rechargeable AA battery with a mass of 25.0 g can supply a power of 1.20 W for 50.0 min. (a) What is the difference in mass between a charged and an un-

Figure P2.25 The classical deflection of starlight grazing the sun.

charged battery? (b) What fraction of the total mass is this mass difference?

29. An object disintegrates into two fragments. One of the fragments has mass 1.00 MeV/c^2 and momentum 1.75 MeV/c in the positive x direction. The other fragment has mass 1.50 MeV/c^2 and momentum 2.005 MeV/c in the positive y direction. Find (a) the mass and (b) the speed of the original object.

30. The creation and study of new elementary particles is an important part of contemporary physics. Especially

interesting is the discovery of a very massive particle. To create a particle of mass M requires an energy Mc^2. With enough energy, an exotic particle can be created by allowing a fast-moving particle of ordinary matter, such as a proton, to collide with a similar target particle. Let us consider a perfectly inelastic collision between two protons: An incident proton with mass m, kinetic energy K, and momentum magnitude p joins with an originally stationary target proton to form a single product particle of mass M. You might think that the creation of a new product particle, 9 times more massive than in a previous experiment, would require just 9 times more energy for the incident proton. Unfortunately, not all of the kinetic energy of the incoming proton is available to create the product particle, since conservation of momentum requires that after the collision the system as a whole still must have some kinetic energy. Only a fraction of the energy of the incident particle is thus available to create a new particle. You will determine how the energy available for particle creation depends on the energy of the moving proton. Show that the energy available to create a product particle is given by

$$Mc^2 = 2mc^2 \sqrt{1 + \frac{K}{2mc^2}}$$

From this result, when the kinetic energy K of the incident proton is large compared to its rest energy mc^2, we see that M approaches $(2mK)^{1/2}/c$. Thus if the energy of the incoming proton is increased by a factor of 9, the mass you can create increases only by a factor of 3. This disappointing result is the main reason that most modern accelerators, such as those at CERN (in Europe), at Fermilab (near Chicago), at SLAC (at Stanford), and at DESY (in Germany), use *colliding beams*. Here the total momentum of a pair of interacting particles can be zero. The center of mass can be at rest after the collision, so in principle all of the initial kinetic energy can be used for particle creation, according to

$$Mc^2 = 2mc^2 + K = 2mc^2 \left(1 + \frac{K}{2mc^2}\right)$$

where K is the total kinetic energy of two identical colliding particles. Here, if $K \gg mc^2$, we have M directly proportional to K, as we would desire. These machines are difficult to build and to operate, but they open new vistas in physics.

31. A particle of mass m moving along the x-axis with a velocity component $+u$ collides head-on and sticks to a particle of mass $m/3$ moving along the x-axis with the velocity component $-u$. What is the mass M of the resulting particle?

32. Compact high-power lasers can produce a 2.00-J light pulse of duration 100 fs focused to a spot 1 μm in diameter. (See Mourou and Umstader, "Extreme Light," *Scientific American*, May 2002, p. 81.) The electric field in the light accelerates electrons in the target material to near the speed of light. (a) What is the average power of the laser during the pulse? (b) How many electrons can be accelerated to $0.9999c$ if 0.0100% of the pulse energy is converted into energy of electron motion?

33. Energy reaches the upper atmosphere of the Earth from the Sun at the rate of 1.79×10^{17} W. If all of this energy were absorbed by the Earth and not re-emitted, how much would the mass of the Earth increase in 1.00 yr?

3

The Quantum Theory of Light

Chapter Outline

At the beginning of the 20th century, following the lead of Newton and Maxwell, physicists might have rewritten the biblical story of creation as follows:

In the beginning He created the heavens and the earth—

$$F = G \frac{mm'}{r^2} = ma$$

and He said, "Let there be light"—

$$\oint \mathbf{E} \cdot d\mathbf{A} = \frac{Q}{\varepsilon_0} \qquad \oint \mathbf{E} \cdot d\mathbf{s} = -\frac{d\Phi_B}{dt}$$

$$\oint \mathbf{B} \cdot d\mathbf{A} = 0 \qquad \oint \mathbf{B} \cdot d\mathbf{s} = \mu_0 I + \varepsilon_0 \mu_0 \frac{d\Phi_E}{dt}$$

Actually, in addition to the twin pillars of mechanics and electromagnetism erected by the giants Newton and Maxwell, there was a third sturdy support for physics in 1900—thermodynamics and statistical mechanics. Classical thermodynamics was the work of many men (Carnot, Mayer, Helmholtz, Clausius, Lord Kelvin). It is especially notable because it starts with two simple propositions and gives solid and conclusive results independent of detailed physical mechanisms. Statistical mechanics, founded by Maxwell, Clausius,

Boltzmann,[1] and Gibbs, uses the methods of probability theory to calculate averages and fluctuations from the average for systems containing many particles or modes of vibration. It is interesting that quantum physics started not with a breakdown of Maxwell's or Newton's laws applied to the atom, but with a problem of classical statistical mechanics—that of calculating the intensity of radiation at a given wavelength from a heated cavity. The desperate solution to this radiation problem was found by a thoroughly classical thermodynamicist, Max Planck, in 1900. Indeed, it is significant that both Planck and Einstein returned again and again to the simple and general foundation of thermodynamics and statistical mechanics as the only certain bases for the new quantum theory. Although we shall not follow the original thermodynamic arguments completely, we shall see in this chapter how Planck arrived at the correct spectral distribution for cavity radiation by allowing only certain energies for the radiation-emitting oscillators in the cavity walls. We shall also see how Einstein extended this quantization of energy to light itself, thereby brilliantly explaining the photoelectric effect. We conclude our brief history of the quantum theory of light with a discussion of the scattering of light by electrons (Compton effect), which showed conclusively that the light quantum carried momentum as well as energy. Finally, we describe the pull of gravity on light in Section 3.7.

3.1 HERTZ'S EXPERIMENTS—LIGHT AS AN ELECTROMAGNETIC WAVE

It is ironic that the same experimentalist who so carefully confirmed that the "newfangled" waves of Maxwell actually existed and possessed the same properties as light also undermined the electromagnetic wave theory as the complete explanation of light. To understand this irony, let us briefly review the theory of electromagnetism developed by the great Scottish physicist James Clerk Maxwell between 1865 and 1873.

Maxwell was primarily interested in the effects of electric current oscillations in wires. According to his theory, an alternating current would set up fluctuating electric and magnetic fields in the region surrounding the original disturbance. Moreover, these waves were predicted to have a frequency equal to the frequency of the current oscillations. *In addition, and most importantly, Maxwell's theory predicted that the radiated waves would behave in every way like light:* electromagnetic waves would be reflected by metal mirrors, would be refracted by dielectrics like glass, would exhibit polarization and interference, and would travel outward from the wire through a vacuum with a speed of 3.0×10^8 m/s. Naturally this led to the unifying and simplifying postulate that light was also a type of Maxwell wave or electromagnetic disturbance, created by extremely high frequency electric oscillators in matter. At the end of the 19th century the precise nature of these charged submicroscopic oscillators was unknown (Planck called them resonators), but physicists assumed that somehow they were able to emit light waves whose frequency was equal to the oscillator's frequency of motion.

Even at this time, however, it was apparent that this model of light emission was incapable of direct experimental verification, because the highest

---

[1]On whose tombstone is written $S = k_B \log W$, a basic formula of statistical mechanics attributed to Boltzmann.

electrical frequencies then attainable were about 10^9 Hz and visible light was known to possess a frequency a million times higher. But Heinrich Hertz (Fig. 3.1) did the next best thing. In a series of brilliant and exhaustive experiments, he showed that Maxwell's theory was correct and that an oscillating electric current does indeed radiate electromagnetic waves that possess every characteristic of light except the same wavelength as light. Using a simple spark gap oscillator consisting of two short stubs terminated in small metal spheres separated by an air gap of about half an inch, he applied pulses of high voltage, which caused a spark to jump the gap and produce a high-frequency electric oscillation of about 5×10^8 Hz. This oscillation, or ringing, occurs while the air gap remains conducting, and charge surges back and forth between the spheres until electrical equilibrium is established. Using a simple loop antenna with a small spark gap as the receiver, Hertz very quickly succeeded in detecting the radiation from his spark gap oscillator, even at distances of several hundred meters. Moreover, he found the detected radiation to have a wavelength of about 60 cm, corresponding to the oscillator frequency of 5×10^8 Hz. (Recall that $c = \lambda f$, where λ is the wavelength and f is the frequency.)

Figure 3.1 Heinrich Hertz (1857–1894), an extraordinarily gifted German experimentalist. (©*Bettmann/Corbis*)

In an exhaustive tour de force, Hertz next proceeded to show that these electromagnetic waves could be reflected, refracted, focused, polarized, and made to interfere—in short, he convinced physicists of the period that Hertzian waves and light waves were one and the same. The classical model for light emission was an idea whose time had come. It spread like wildfire. The idea that light was an electromagnetic wave radiated by oscillating submicroscopic electric charges (now known to be atomic electrons) was applied in rapid succession to the transmission of light through solids, to reflection from metal surfaces, and to the newly discovered Zeeman effect. In 1896, Pieter Zeeman, a Dutch physicist, discovered that a strong magnetic field changes the frequency of the light emitted by a glowing gas. In an impressive victory for Maxwell, it was found that Maxwell's equations correctly predicted (in most cases) the change of vibration of the electric oscillators and hence, the change in frequency of the light emitted. (See Problem 1.) Maxwell, with Hertz behind the throne, reigned supreme, for he had united the formerly independent kingdoms of electricity, magnetism, and light! (See Fig. 3.2.)

A terse remark made by Hertz ends our discussion of his confirmation of the electromagnetic wave nature of light. In describing his spark gap transmitter, he emphasizes that "it is essential that the pole surfaces of the spark gap

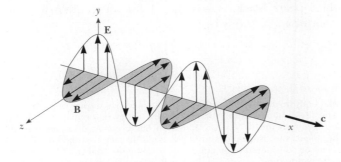

Figure 3.2   A light or radio wave far from the source according to Maxwell and Hertz.

should be frequently repolished" to ensure reliable operation of the spark.[2] Apparently this result was initially quite mysterious to Hertz. In an effort to resolve the mystery, he later investigated this side effect and concluded that it was the ultraviolet light from the initial spark acting on a clean metal surface that caused current to flow more freely between the poles of the spark gap. In the process of verifying the electromagnetic wave theory of light, Hertz had discovered the photoelectric effect, a phenomenon that would undermine the priority of the wave theory of light and establish the particle theory of light on an equal footing.

3.2 BLACKBODY RADIATION

The tremendous success of Maxwell's theory of light emission immediately led to attempts to apply it to a long-standing puzzle about radiation—the so-called "blackbody" problem. The problem is to predict the radiation intensity at a given wavelength emitted by a hot glowing solid at a specific temperature. Instead of launching immediately into Planck's solution of this problem, let us develop a feeling for its importance to classical physics by a quick review of its history.

Thomas Wedgwood, Charles Darwin's relative and a renowned maker of china, seems to have been the first to note the universal character of all heated objects. In 1792, he observed that all the objects in his ovens, regardless of their chemical nature, size, or shape, became red at the same temperature. This crude observation was sharpened considerably by the advancing state of spectroscopy, so that by the mid-1800s it was known that glowing solids emit continuous spectra rather than the bands or lines emitted by heated gases. (See Fig. 3.3.) In 1859, Gustav Kirchhoff proved a theorem as important as his circuit loop theorem when he showed by arguments based on thermodynamics that for any body in thermal equilibrium with radiation[3] the emitted power is proportional to the power absorbed. More specifically,

$$e_f = J(f, T)A_f \tag{3.1}$$

where e_f is the power emitted per unit area per unit frequency by a particular heated object, A_f is the absorption power (fraction of the incident power absorbed per unit area per unit frequency by the heated object), and $J(f, T)$ is a universal function (the same for all bodies) that depends only on f, the light frequency, and T, the absolute temperature of the body. A *blackbody* is defined as an object that absorbs all the electromagnetic radiation falling on it and consequently appears black. It has $A_f = 1$ for all frequencies and so Kirchhoff's theorem for a blackbody becomes

$$e_f = J(f, T) \tag{3.2}$$

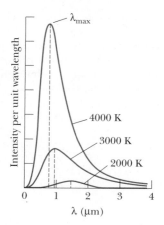

Figure 3.3 Emission from a glowing solid. Note that the amount of radiation emitted (the area under the curve) increases rapidly with increasing temperature.

Blackbody

[2]H. Hertz, *Ann. Physik* (Leipzig), 33:983, 1887.

[3]An example of a body in equilibrium with radiation would be an oven with closed walls at a fixed temperature and the radiation within the oven cavity. To say that radiation is in thermal equilibrium with the oven walls means that the radiation has exchanged energy with the walls many times and is homogeneous, isotropic, and unpolarized. In fact, thermal equilibrium of radiation within a cavity can be considered to be quite similar to the thermal equilibrium of a fluid within a container held at constant temperature—both will cause a thermometer in the center of the cavity to achieve a final stationary temperature equal to that of the container.

Equation 3.2 shows that the power emitted per unit area per unit frequency by a blackbody depends only on temperature and light frequency and not on the physical and chemical makeup of the blackbody, in agreement with Wedgwood's early observation.

Because absorption and emission are connected by Kirchhoff's theorem, we see that a blackbody or perfect absorber is also an ideal radiator. In practice, a small opening in any heated cavity, such as a port in an oven, behaves like a blackbody because such an opening traps all incident radiation (Fig. 3.4). If the direction of the radiation is reversed in Figure 3.4, the light emitted by a small opening is in thermal equilibrium with the walls, because it has been absorbed and re-emitted many times.

The next important development in the quest to understand the universal character of the radiation emitted by glowing solids came from the Austrian physicist Josef Stefan (1835–1893) in 1879. He found experimentally that the total power per unit area emitted at all frequencies by a hot solid, e_{total}, was proportional to the fourth power of its absolute temperature. Therefore, Stefan's law may be written as

$$e_{\text{total}} = \int_0^\infty e_f \, df = \sigma T^4 \tag{3.3}$$

where e_{total} is the power per unit area emitted at the surface of the blackbody at all frequencies, e_f is the power per unit area per unit frequency emitted by the blackbody, T is the absolute temperature of the body, and σ is the Stefan–Boltzmann constant, given by $\sigma = 5.67 \times 10^{-8} \, \text{W} \cdot \text{m}^{-2} \cdot \text{K}^{-4}$. A body that is not an ideal radiator will obey the same general law but with a coefficient, a, less than 1:

$$e_{\text{total}} = a\sigma T^4 \tag{3.4}$$

Only 5 years later another impressive confirmation of Maxwell's electromagnetic theory of light occurred when Boltzmann derived Stefan's law from a combination of thermodynamics and Maxwell's equations.

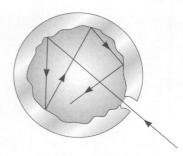

Figure 3.4   The opening to the cavity inside a body is a good approximation of a blackbody. Light entering the small opening strikes the far wall, where some of it is absorbed but some is reflected at a random angle. The light continues to be reflected, and at each reflection a portion of the light is absorbed by the cavity walls. After many reflections essentially all of the incident energy is absorbed.

Stefan's law

EXAMPLE 3.1   Stefan's Law Applied to the Sun

Estimate the surface temperature of the Sun from the following information. The Sun's radius is given by $R_s = 7.0 \times 10^8$ m. The average Earth–Sun distance is $R = 1.5 \times 10^{11}$ m. The power per unit area (at all frequencies) from the Sun is measured at the Earth to be 1400 W/m². Assume that the Sun is a blackbody.

Solution   For a black body, we take $a = 1$, so Equation 3.4 gives

$$e_{\text{total}}(R_s) = \sigma T^4 \tag{3.5}$$

where the notation $e_{\text{total}}(R_s)$ stands for the total power per unit area at the surface of the Sun. Because the problem gives the total power per unit area at the Earth, $e_{\text{total}}(R)$, we need the connection between $e_{\text{total}}(R)$ and

$e_{\text{total}}(R_s)$. This comes from the conservation of energy:

$$e_{\text{total}}(R_s) \cdot 4\pi R_s^2 = e_{\text{total}}(R) \cdot 4\pi R^2$$

or

$$e_{\text{total}}(R_s) = e_{\text{total}}(R) \cdot \frac{R^2}{R_s^2}$$

Using Equation 3.5, we have

$$T = \left[\frac{e_{\text{total}}(R) \cdot R^2}{\sigma R_s^2} \right]^{1/4}$$

or

$$T = \left[\frac{(1400 \, \text{W/m}^2)(1.5 \times 10^{11} \, \text{m})^2}{(5.6 \times 10^{-8} \, \text{W/m}^2 \cdot \text{K}^4)(7.0 \times 10^8 \, \text{m})^2} \right]^{1/4}$$

$$= 5800 \, \text{K}$$

As can be seen in Figure 3.3, *the wavelength marking the maximum power emission of a blackbody*, λ_{max}, shifts toward shorter wavelengths as the blackbody gets hotter. This agrees with Wedgwood's general observation that objects in his kiln progressed from dull red to orange to white in color as the temperature was raised. This simple effect of $\lambda_{max} \propto T^{-1}$ was not definitely established, however, until about 20 years after Kirchhoff's seminal paper had started the search to find the form of the universal function $J(f, T)$. In 1893, Wilhelm Wien proposed a general form for the blackbody distribution law $J(f, T)$ that gave the correct experimental behavior of λ_{max} with temperature. This law is called *Wien's displacement law* and may be written

$$\lambda_{max} T = 2.898 \times 10^{-3} \text{ m} \cdot \text{K} \tag{3.6}$$

where λ_{max} is the wavelength in meters corresponding to the blackbody's maximum intensity and T is the absolute temperature of the surface of the object emitting the radiation. Assuming that the peak sensitivity of the human eye (which occurs at about 500 nm—blue-green light) coincides with λ_{max} for the Sun (a blackbody), we can check the consistency of Wien's displacement law with Stefan's law by recalculating the Sun's surface temperature:

$$T = \frac{2.898 \times 10^{-3} \text{ m} \cdot \text{K}}{500 \times 10^{-9} \text{ m}} = 5800 \text{ K}$$

Thus we have good agreement between measurements made at all wavelengths (Example 3.1) and at the maximum-intensity wavelength.

Exercise 1   How convenient that the Sun's emission peak is at the same wavelength as our eyes' sensitivity peak! Can you account for this?

Spectral energy density of a blackbody

So far, the power radiated per unit area per unit frequency by the blackbody, $J(f, T)$ has been discussed. However, it is more convenient to consider the spectral energy density, or *energy per unit volume per unit frequency of the radiation within the blackbody cavity, $u(f, T)$*. For light in equilibrium with the walls, the power emitted per square centimeter of opening is simply proportional to the energy density of the light in the cavity. Because the cavity radiation is isotropic and unpolarized, one can average over direction to show that the constant of proportionality between $J(f, T)$ and $u(f, T)$ is $c/4$, where c is the speed of light. Therefore,

$$J(f, T) = u(f, T)c/4 \tag{3.7}$$

An important guess as to the form of the universal function $u(f, T)$ was made in 1893 by Wien and had the form

$$u(f, T) = Af^3 e^{-\beta f/T} \tag{3.8}$$

where A and β are constants. This result was known as Wien's exponential law; it resembles and was loosely based on Maxwell's velocity distribution for gas molecules. Within a year the great German spectroscopist Friedrich Paschen

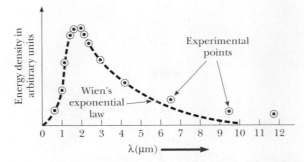

Figure 3.5 Discrepancy between Wien's law and experimental data for a blackbody at 1500 K.

had confirmed Wien's guess by working in the then difficult infrared range of 1 to 4 μm and at temperatures of 400 to 1600 K.[4]

As can be seen in Figure 3.5, Paschen had made most of his measurements in the maximum energy region of a body heated to 1500 K and had found good agreement with Wien's exponential law. In 1900, however, Lummer and Pringsheim extended the measurements to 18 μm, and Rubens and Kurlbaum went even farther—to 60 μm. Both teams concluded that Wien's law failed in this region (see Fig. 3.5). The experimental setup used by Rubens and Kurlbaum is shown in Figure 3.6. It is interesting to note that these historic

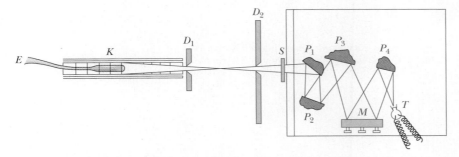

Figure 3.6 Apparatus for measuring blackbody radiation at a single wavelength in the far infrared region. The experimental technique that disproved Wien's law and was so crucial to the discovery of the quantum theory was the method of residual rays (*Restrahlen*). In this technique, one isolates a narrow band of far infrared radiation by causing white light to undergo multiple reflections from alkalide halide crystals (P_1–P_4). Because each alkali halide has a maximum reflection at a characteristic wavelength, quite pure bands of far infrared radiation may be obtained with repeated reflections. These pure bands can then be directed onto a thermopile (T) to measure intensity. E is a thermocouple used to measure the temperature of the blackbody oven, K.

[4]We should point out the great difficulty in making blackbody radiation measurements and the singular advances made by German spectroscopists in the crucial areas of blackbody sources, sensitive detectors, and techniques for operating far into the infrared region. In fact, it is dubious whether Planck would have found the correct blackbody law as quickly without his close association with the experimentalists at the Physikalisch Technische Reichsanstalt of Berlin (a sort of German National Bureau of Standards)—Otto Lummer, Ernst Pringsheim, Heinrich Rubens, and Ferdinand Kurlbaum.

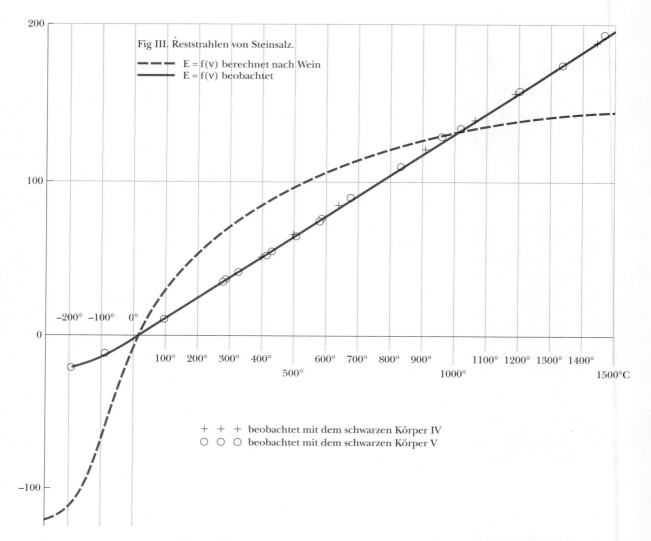

Fig III. Reststrahlen von Steinsalz.

– – – – E = f(ν) berechnet nach Wein
———— E = f(ν) beobachtet

+ + + beobachtet mit dem schwarzen Körper IV
O O O beobachtet mit dem schwarzen Körper V

Figure 3.7   Comparison of theoretical and experimental blackbody emission curves at 51.2 μm and over the temperature range of −188° to 1500°C. The title of this modified figure is "Residual Rays from Rocksalt." *Berechnet nach* means "calculated according to," and *beobachtet* means "observed." The vertical axis is emission intensity in arbitrary units. (*From H. Rubens and S. Kurlbaum,* Ann. Physik, *4:649, 1901.*)

experiments involved the measurement of blackbody radiation intensity at a fixed wavelength and variable temperature. Typical results measured at λ = 51.2 μm and over the temperature range of −200° to +1500°C are shown in Figure 3.7, from the paper by Rubens and Kurlbaum.

Enter Planck

On a Sunday evening early in October of 1900, Max Planck discovered the famous blackbody formula, which truly ushered in the quantum theory. Planck's proximity to the Reichsanstalt experimentalists was extremely important for his discovery—earlier in the day he had heard from Rubens that his latest measurements showed that $u(f, T)$, the spectral energy density, was proportional to T for long wavelengths or low frequency. Planck knew that Wien's law agreed well with the data at high frequency and indeed had been working hard for several years to derive Wien's exponential law from the principles of statistical mechanics and Maxwell's laws. Interpolating between the two limiting forms (Wien's exponential law and an energy density proportional to temperature), he immediately found a general formula, which he sent to Rubens, on a postcard, the same evening. His formula was[5]

$$u(f, T) = \frac{8\pi h f^3}{c^3} \left(\frac{1}{e^{hf/k_B T} - 1} \right) \tag{3.9}$$

where h is Planck's constant $= 6.626 \times 10^{-34}\,\text{J}\cdot\text{s}$, and k_B is Boltzmann's constant $= 1.380 \times 10^{-23}\,\text{J/K}$. We can see that Equation 3.9 has the correct limiting behavior at high and low frequencies with the help of a few approximations. At high frequencies, where $hf/k_B T \gg 1$,

$$\frac{1}{e^{hf/k_B T} - 1} \approx e^{-hf/k_B T}$$

so that

$$u(f, T) = \frac{8\pi h f^3}{c^3} \left(\frac{1}{e^{hf/k_B T} - 1} \right) \approx \frac{8\pi h f^3}{c^3} e^{-hf/k_B T}$$

and we recover Wien's exponential law, Equation 3.8. At low frequencies, where $hf/k_B T \ll 1$,

$$\frac{1}{e^{hf/k_B T} - 1} = \frac{1}{1 + \dfrac{hf}{k_B T} + \cdots - 1} \approx \frac{k_B T}{hf}$$

and

$$u(f, T) = \frac{8\pi h f^3}{c^3} \left(\frac{1}{e^{hf/k_B T} - 1} \right) \approx \frac{8\pi f^2}{c^3} k_B T$$

This result shows that the spectral energy density is proportional to T in the low-frequency or so-called classical region, as Rubens had found.

We should emphasize that Planck's work entailed much more than clever mathematical manipulation. For more than six years Planck (Fig. 3.8) labored to find a rigorous derivation of the blackbody distribution curve. He was driven, in his own words, by the fact that the emission problem "represents something absolute, and since I had always regarded the search for the absolute as the loftiest goal of all scientific activity, I eagerly set to work." This work was to occupy most of his life as he strove to give his formula an ever deeper physical interpretation and to reconcile discrete quantum energies with classical theory.

Figure 3.8 Max Planck (1858–1947). The work leading to the "lucky" blackbody radiation formula was described by Planck in his Nobel prize acceptance speech (1920): "But even if the radiation formula proved to be perfectly correct, it would after all have been only an interpolation formula found by lucky guess-work and thus, would have left us rather unsatisfied. I therefore strived from the day of its discovery, to give it a real physical interpretation and this led me to consider the relations between entropy and probability according to Boltzmann's ideas. After some weeks of the most intense work of my life, light began to appear to me and unexpected views revealed themselves in the distance." (*AIP Niels Bohr Library, W. F. Meggers Collection*)

---

[5]Planck originally published his formula as $u(\lambda, T) = \dfrac{C_1}{\lambda^5} \left(\dfrac{1}{e^{C_2/\lambda T} - 1} \right)$, where $C_1 = 8\pi ch$ and $C_2 = hc/k_B$. He then found best-fit values to the experimental data for C_1 and C_2 and evaluated $h = 6.55 \times 10^{-34}\,\text{J}\cdot\text{s}$ and $k_B = N_A/R = 1.345 \times 10^{-23}\,\text{J/K}$. As R, the universal gas constant, was fairly well known at the time, this technique also resulted in another method for finding N_A, Avogadro's number.

The Quantum of Energy

Planck's original theoretical justification of Equation 3.9 is rather abstract because it involves arguments based on entropy, statistical mechanics, and several theorems proved earlier by Planck concerning matter and radiation in equilibrium.[6] We shall give arguments that are easier to visualize physically yet attempt to convey the spirit and revolutionary impact of Planck's original work.

Planck was convinced that blackbody radiation was produced by vibrating submicroscopic electric charges, which he called resonators. He assumed that the walls of a glowing cavity were composed of literally billions of these resonators (whose exact nature was unknown at the time), all vibrating at different frequencies. Hence, according to Maxwell, each oscillator should emit radiation with a frequency corresponding to its vibration frequency. **Also according to classical Maxwellian theory, an oscillator of frequency f could have any value of energy and could change its amplitude continuously as it radiated any fraction of its energy.** This is where Planck made his revolutionary proposal. To secure agreement with experiment, **Planck had to assume that the total energy of a resonator with mechanical frequency f could only be an integral multiple of hf or**

$$E_{\text{resonator}} = nhf \qquad n = 1, 2, 3, \ldots \tag{3.10}$$

where h is a fundamental constant of quantum physics, $h = 6.626 \times 10^{-34}\,\text{J}\cdot\text{s}$, known as Planck's constant. In addition, he concluded that emission of radiation of frequency f occurred when a resonator dropped to the next lowest energy state. *Thus the resonator can change its energy only by the difference ΔE according to*

$$\Delta E = hf \tag{3.11}$$

That is, it cannot lose just any amount of its total energy, but only a finite amount, hf, the so-called quantum of energy. Figure 3.9 shows the quantized energy levels and allowed transitions proposed by Planck.

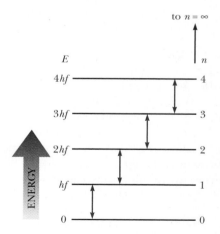

Figure 3.9   Allowed energy levels according to Planck's original hypothesis for an oscillator with frequency f. Allowed transitions are indicated by the double-headed arrows.

[6]M. Planck, *Ann. Physik*, 4:553, 1901.

EXAMPLE 3.2   A Quantum Oscillator versus a Classical Oscillator

Consider the implications of Planck's conjecture that *all* oscillating systems of natural frequency f have discrete allowed energies $E = nhf$ and that the smallest change in energy of the system is given by $\Delta E = hf$.

(a) First compare an atomic oscillator sending out 540-nm light (green) to one sending out 700-nm light (red) by calculating the minimum energy change of each. For the green quantum,

$$\Delta E_{\text{green}} = hf = \frac{hc}{\lambda}$$

$$= \frac{(6.63 \times 10^{-34}\,\text{J}\cdot\text{s})(3.00 \times 10^{8}\,\text{m/s})}{540 \times 10^{-9}\,\text{m}}$$

$$= 3.68 \times 10^{-9}\,\text{J}$$

Actually, the joule is much too large a unit of energy for describing atomic processes; a more appropriate unit of energy is the electron volt (eV). The electron volt takes the charge on the electron as its unit of charge. By definition, an electron accelerated through a potential difference of 1 volt has an energy of 1 eV. An electron volt may be converted to joules by noting that

$$E = V \cdot q = 1\,\text{eV} = (1.602 \times 10^{-19}\,\text{C})(1\,\text{J/C})$$

$$= 1.602 \times 10^{-19}\,\text{J}$$

It is also useful to have expressions for h and hc in terms of electron volts. These are

$$h = 4.136 \times 10^{-15}\,\text{eV}\cdot\text{s}$$

$$hc = 1.240 \times 10^{-6}\,\text{eV}\cdot\text{m} = 1240\,\text{eV}\cdot\text{nm}$$

Returning to our example, we see that the minimum energy change of an atomic oscillator sending out green light is

$$\Delta E_{\text{green}} = \frac{3.68 \times 10^{-19}\,\text{J}}{1.602 \times 10^{-19}\,\text{J/eV}} = 2.30\,\text{eV}$$

For the red quantum the minimum energy change is

$$\Delta E_{\text{red}} = \frac{hc}{\lambda} = \frac{(6.63 \times 10^{-34}\,\text{J}\cdot\text{s})(3.00 \times 10^{8}\,\text{m/s})}{700 \times 10^{-9}\,\text{m}}$$

$$= 2.84 \times 10^{-19}\,\text{J} = 1.77\,\text{eV}$$

Note that the minimum allowed amount or "quantum" of energy is not uniform under all conditions as is the quantum of charge—the quantum of energy is proportional to the natural frequency of the oscillator. Note, too, that the high frequency of atomic oscillators produces a measurable quantum of energy of several electron volts.

(b) Now consider a pendulum undergoing small oscillations with length $\ell = 1$ m. According to classical theory, if air friction is present, the amplitude of swing and

consequently the energy decrease *continuously* with time, as shown in Figure 3.10a. Actually, *all* systems vibrating with frequency f are quantized (according to Equation 3.10) and lose energy in discrete packets or quanta, hf. This would lead to a decrease of the pendulum's energy in a stepwise manner, as shown in Figure 3.10b. We shall show that there is no contradiction between quantum theory and the observed behavior of laboratory pendulums and springs.

An energy change of one quantum corresponds to

$$\Delta E = hf$$

where the pendulum frequency f is

$$f = \frac{1}{2\pi}\sqrt{\frac{g}{\ell}} = 0.50\,\text{Hz}$$

Thus,

$$\Delta E = (6.63 \times 10^{-34}\,\text{J}\cdot\text{s})(0.50\,\text{s}^{-1})$$

$$= 3.3 \times 10^{-34}\,\text{J}$$

$$= 2.1 \times 10^{-15}\,\text{eV}$$

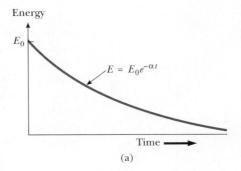

(a)

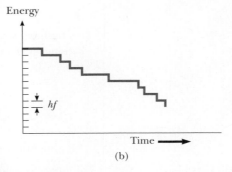

(b)

Figure 3.10   (Example 3.2) (a) Observed classical behavior of a pendulum. (b) Predicted quantum behavior of a pendulum.

Because the total energy of a pendulum of mass m and length ℓ displaced through an angle θ is

$$E = mg\ell(1 - \cos\theta)$$

we have for a typical pendulum with $m = 100$ g, $\ell = 1.0$ m, and $\theta = 10°$,

$$E = (0.10 \text{ kg})(9.8 \text{ m/s}^2)(1.0 \text{ m})(1 - \cos 10°) = 0.015 \text{ J}$$

Therefore, the fractional change in energy, $\Delta E/E$, is unobservably small:

$$\frac{\Delta E}{E} = \frac{3.3 \times 10^{-34}\text{J}}{1.5 \times 10^{-2}\text{J}} = 2.2 \times 10^{-32}$$

Note that the energy quantization of large vibrating systems is unobservable because of their low frequencies compared to the high frequencies of atomic oscillators. Hence there is no contradiction between Planck's quantum postulate and the behavior of macroscopic oscillators.

Exercise 2   Calculate the quantum number, n, for this pendulum with $E = 1.5 \times 10^{-2}$ J.

Answer 4.6×10^{31}

Exercise 3   An object of mass m on a spring of stiffness k oscillates with an amplitude A about its equilibrium position. Suppose that $m = 300$ g, $k = 10$ N/m, and $A = 10$ cm. (a) Find the total energy. (b) Find the mechanical frequency of vibration of the mass. (c) Calculate the change in amplitude when the system loses one quantum of energy.

Answer (a) $E_{\text{total}} = 0.050$ J; (b) $f = 0.92$ Hz; (c) $\Delta E_{\text{quantum}} = 6.1 \times 10^{-34}$ J, so

$$\Delta A \approx -\frac{\Delta E}{\sqrt{2Ek}} = -6.1 \times 10^{-34} \text{ m}$$

Until now we have been concentrating on the remarkable quantum properties of single oscillators of frequency f. Planck explained the continuous spectrum of the blackbody by assuming that the heated walls contained resonators vibrating at many different frequencies, each emitting light at the same frequency as its vibration frequency. By considering the conditions leading to equilibrium between the wall resonators and the radiation in the blackbody cavity, he was able to show that the spectral energy density $u(f, T)$ could be expressed as the product of the number of oscillators having frequency between f and $f + df$, denoted by $N(f)\,df$, and the average energy emitted per oscillator, $\bar{E}$. Thus we have the important result

$$u(f, T)\,df = \bar{E}N(f)\,df \tag{3.12}$$

Furthermore, Planck showed that the number of oscillators with frequency between f and $f + df$ was proportional to f^2 or

$$N(f)\,df = \frac{8\pi f^2}{c^3}\,df \tag{3.13}$$

(See Appendix 1 on our book Web site at http://info.brookscole.com/mp3e for details.)

Substituting Equation 3.13 into Equation 3.12 gives

$$u(f, T)\,df = \bar{E}\,\frac{8\pi f^2}{c^3}\,df \tag{3.14}$$

This result shows that the spectral energy density is proportional to the product of the frequency squared and the average oscillator energy. Also, since $u(f, T)$ approaches zero at high frequencies (see Fig. 3.5), $\overline{E}$ must tend to zero at high frequencies faster than $1/f^2$. The fact that the mean oscillator energy must become extremely small when the frequency becomes high guided Planck in the development of his theory. In the next section we shall see that the failure of $\overline{E}$ to become small at high frequencies in the classical Rayleigh–Jeans theory led to the "ultraviolet catastrophe"—the prediction of an infinite spectral energy density at high frequencies in the ultraviolet region.

3.3   THE RAYLEIGH–JEANS LAW AND PLANCK'S LAW

O P T I O N A L

Rayleigh–Jeans Law

Both Planck's law and the Rayleigh–Jeans law (the classical theory of blackbody radiation formulated by Lord Rayleigh, John William Strutt, 1842–1919, English physicist, and James Jeans, 1887–1946, English astronomer and physicist) may be derived using the idea that the blackbody radiation energy per unit volume with frequency between f and $f + df$ can be expressed as the product of the number of oscillators per unit volume in this frequency range and the average energy per oscillator:

$$u(f, T)\,df = \overline{E}\,N(f)\,df \qquad (3.12)$$

It is instructive to perform both the Rayleigh–Jeans and Planck calculations to see the effect on $u(f, T)$ of calculating $\overline{E}$ from a *continuous* distribution of classical oscillator energies (Rayleigh–Jeans) as opposed to a *discrete* set of quantum oscillator energies (Planck). We discuss Lord Rayleigh's derivation first because it is a more direct classical calculation.

While Planck concentrated on the thermal equilibrium of cavity radiation with oscillating electric charges in the cavity walls, Rayleigh concentrated directly on the electromagnetic waves in the cavity. Rayleigh and Jeans reasoned that the standing electromagnetic waves in the cavity could be considered to have a temperature T, because they constantly exchanged energy with the walls and caused a thermometer within the cavity to reach the same temperature as the walls. Further, they considered a standing polarized electromagnetic wave to be equivalent to a one-dimensional oscillator (Fig. 3.11). Using the same general idea as Planck, they expressed the energy density as a product of the number of standing waves (oscillators) and the average energy per oscillator. They found the average oscillator energy $\overline{E}$ to be independent of frequency and equal to $k_B T$ from the Maxwell-Boltzmann distribution law (see Chapter 10). According to this distribution law, the probability P of finding an individual system (such as a molecule or an atomic oscillator) with energy E above some minimum energy, E_0, in a large group of systems at temperature T is

$$P(E) = P_0 e^{-(E-E_0)/k_B T} \qquad (3.15)$$

where P_0 is the probability that a system has the minimum energy. In the case of a *discrete* set of allowed energies, the average energy, $\overline{E}$, is given by

$$\overline{E} = \frac{\sum E \cdot P(E)}{\sum P(E)} \qquad (3.16)$$

where division by the sum in the denominator serves to normalize the total probability to 1. *In the classical case considered by Rayleigh,* an oscillator could have any

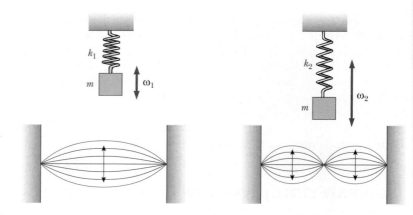

Figure 3.11 A one-dimensional harmonic oscillator is equivalent to a plane-polarized electromagnetic standing wave.

energy E in a continuous range from 0 to ∞. Thus the sums in Equation 3.16 must be replaced with integrals, and the expression for $\bar{E}$ becomes

$$\bar{E} = \frac{\displaystyle\int_0^\infty E e^{-E/k_B T} \, dE}{\displaystyle\int_0^\infty e^{-E/k_B T} \, dE} = k_B T$$

The calculation of $N(f)$ is a bit more complicated but is of importance here as well as in the free electron model of metals. Appendix 1 on our Web site gives the derivation of the density of modes, $N(f) \, df$. One finds

$$N(f) \, df = \frac{8\pi f^2}{c^3} \, df \tag{3.45}$$

or in terms of wavelength,

$$N(\lambda) \, d\lambda = \frac{8\pi}{\lambda^4} \, d\lambda \tag{3.46}$$

Density of standing waves in a cavity

The spectral energy density is simply the density of modes multiplied by $k_B T$, or

$$u(f, T) \, df = \frac{8\pi f^2}{c^3} \, k_B T \, df \tag{3.17}$$

In terms of wavelength,

Rayleigh–Jeans blackbody law

$$u(\lambda, T) \, d\lambda = \frac{8\pi}{\lambda^4} \, k_B T \, d\lambda \tag{3.18}$$

However, as one can see from Figure 3.12, this classical expression, known as the Rayleigh–Jeans law, does not agree with the experimental results in the short wavelength region. Equation 3.18 diverges as $\lambda \to 0$, predicting unlimited energy emission in the ultraviolet region, which was dubbed the "ultraviolet catastrophe." One is forced to conclude that classical theory fails miserably to explain blackbody radiation.

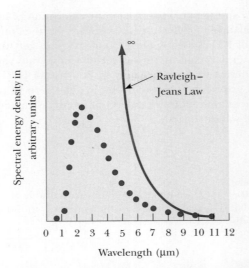

Figure 3.12   The failure of the classical Rayleigh–Jeans law (Equation 3.18) to fit the observed spectrum of a blackbody heated to 1000 K.

Planck's Law

As mentioned earlier, Planck concentrated on the energy states of resonators in the cavity walls and used the condition that the resonators and cavity radiation were in equilibrium to determine the spectral quality of the radiation. By thermodynamic reasoning (and apparently unaware of Rayleigh's derivation), he arrived at the same expression for $N(f)$ as Rayleigh. However, Planck arrived at a different form for $\overline{E}$ by allowing only discrete values of energy for his resonators. He found, using the Maxwell-Boltzmann distribution law,

$$\overline{E} = \frac{hf}{e^{hf/k_B T} - 1} \qquad (3.19)$$

(See the book Web site at http://info.brookscole.com/mp3e for Planck's derivation of $\overline{E}$.)

Multiplying $\overline{E}$ by $N(f)$ gives the Planck distribution formula:

$$u(f, T)\, df = \frac{8\pi f^2}{c^3}\left(\frac{hf}{e^{hf/k_B T} - 1}\right) df \qquad (3.9)$$

Planck blackbody law

or in terms of wavelength, λ,

$$u(\lambda, T)\, d\lambda = \frac{8\pi hc\, d\lambda}{\lambda^5 (e^{hc/\lambda k_B T} - 1)} \qquad (3.20)$$

Equation 3.9 shows that the ultraviolet catastrophe is avoided because the $\overline{E}$ term dominates the f^2 term at high frequencies. One can qualitatively understand why $\overline{E}$ tends to zero at high frequencies by noting that the first allowed oscillator level (hf) is so large for large f compared to the average thermal energy available ($k_B T$) that Boltzmann's law predicts almost zero probability that the first excited state is occupied.

In summary, Planck arrived at his blackbody formula by making two startling assumptions: (1) the energy of a charged oscillator of frequency f is limited to discrete values nhf and (2) during emission or absorption of light, the change in energy of an oscillator is hf. But Planck was every bit the "unwilling revolutionary." From most of Planck's early correspondence one gets the impression that his concept of energy quantization was really a desperate calculational device, and moreover a device that applied only in the case of blackbody radiation. It remained for the great Albert Einstein, the popular icon of physics in the 20th century, to elevate quantization to the level of a universal phenomenon by showing that light itself was quantized.

EXAMPLE 3.3   Derivation of Stefan's Law from the Planck Distribution

In this example, we show that the Planck spectral distribution formula leads to the experimentally observed Stefan law for the total radiation emitted by a blackbody at all wavelengths,

$$e_{total} = 5.67 \times 10^{-8}\, T^4\ \text{W}\cdot\text{m}^{-2}\cdot\text{K}^{-4}$$

Solution   Since Stefan's law is an expression for the total power per unit area radiated at all wavelengths, we must integrate the expression for $u(\lambda, T)\, d\lambda$ given by Equation 3.20 over λ and use Equation 3.7 for the connection between the energy density inside the blackbody cavity and the power emitted per unit area of blackbody surface. We find

$$e_{total} = \frac{c}{4}\int_0^\infty u(\lambda, T)\, d\lambda = \int_0^\infty \frac{2\pi hc^2}{\lambda^5(e^{hc/\lambda k_B T} - 1)}\, d\lambda$$

If we make the change of variable $x = hc/\lambda k_B T$, the integral assumes a form commonly found in tables:

$$e_{total} = \frac{2\pi k_B^4 T^4}{c^2 h^3}\int_0^\infty \frac{x^3}{(e^x - 1)}\, dx$$

Using

$$\int_0^\infty \frac{x^3}{(e^x - 1)}\, dx = \frac{\pi^4}{15}$$

we find

$$e_{total} = \frac{2\pi^5 k_B^4}{15 c^2 h^3}\, T^4 = \sigma T^4$$

Finally, substituting for k_B, c, and h, we have

$$\sigma = \frac{(2)(3.141)^5(1.381 \times 10^{-23}\ \text{J/K})^4}{(15)(2.998 \times 10^8\ \text{m/s})^2(6.626 \times 10^{-34}\ \text{J}\cdot\text{s})^3}$$
$$= 5.67 \times 10^{-8}\ \text{W}\cdot\text{m}^{-2}\cdot\text{K}^{-4}$$

Exercise 4   Show that

$$\int_0^\infty \frac{2\pi hc^2}{\lambda^5(e^{hc/\lambda k_B T} - 1)}\, d\lambda = \frac{2\pi k_B^4 T^4}{h^3 c^2}\int_{x=0}^\infty \frac{x^3}{(e^x - 1)}\, dx$$

3.4   LIGHT QUANTIZATION AND THE PHOTOELECTRIC EFFECT

We now turn to the year 1905, in which the next major development in quantum theory took place. The year 1905 was an incredible one for the "willing revolutionary" Albert Einstein (Fig. 3.13). In this year Einstein produced three immortal papers on three different topics, each revolutionary and each worthy of a Nobel prize. All three papers contained balanced, symmetric, and unifying new results achieved by spare and clean logic and simple mathematics. The first work, entitled "A Heuristic[7] Point of View

[7]A heuristic argument is one that is plausible and enlightening but not rigorously justified.

About the Generation and Transformation of Light," formulated the theory of light quanta and explained the photoelectric effect.[8] The second paper was entitled "On the Motion of Particles Suspended in Liquids as Required by the Molecular-Kinetic Theory of Heat." It explained Brownian motion and provided strong proof of the reality of atoms.[9] The third paper, which is perhaps his most famous, contained the invention of the theory of special relativity[10] and was entitled "On the Electrodynamics of Moving Bodies." It is interesting to note that when Einstein was awarded the Nobel prize in 1922, the Swedish Academy judged his greatest contribution to physics to have been the theory of the photoelectric effect. No mention was made at all of his theory of relativity!

Let us turn now to the paper concerning the light quantum, in which Einstein crossed swords with Maxwell and challenged the unqualified successes of the classical wave theory of light. Einstein recognized an inconsistency between Planck's quantization of oscillators in the walls of the blackbody and Planck's insistence that the cavity radiation consisted of classical electromagnetic waves. By showing that the change in entropy of blackbody radiation was like the change in entropy of an ideal gas consisting of independent particles, Einstein reached the conclusion that light itself is composed of "grains," irreducible finite amounts, or quanta of energy.[11] Furthermore, he asserted that light interacting with matter also consists of quanta, and he worked out the implications for photoelectric and photochemical processes. His explanation of the photoelectric effect offers such convincing proof that light consists of energy packets that we shall describe it in more detail. First, however, we need to consider the main experimental features of the photoelectric effect and the failure of classical theory to explain this effect.

As noted earlier, Hertz first established that clean metal surfaces emit charges when exposed to ultraviolet light. In 1888 Hallwachs discovered that the emitted charges were negative, and in 1899 J. J. Thomson showed that the emitted charges were electrons, now called photoelectrons. He did this by measuring the charge-to-mass ratio of the particles produced by ultraviolet light and even succeeded in measuring e separately by a cloud chamber technique (see Chapter 4).

The last crucial discovery before Einstein's explanation was made in 1902 by Philip Lenard, who was studying the photoelectric effect with intense carbon arc light sources. He found that electrons are emitted from the metal with a range of velocities and that the maximum kinetic energy of photoelectrons, K_{max}, *does not* depend on the *intensity* of the exciting light. Although he

Figure 3.13 Albert Einstein shown in a playful mood riding his bicycle. The photograph was taken in 1933 in Santa Barbara. (*California Institute of Technology Archives*)

---

[8]A. Einstein, *Ann. Physik*, 17:132, 1905 (March).

[9]A. Einstein, *Ann. Physik*, 17:549, 1905 (May).

[10]A. Einstein, *Ann. Physik*, 17:891, 1905 (June).

[11]Einstein, as Planck before him, fell back on the unquestionable solidity of thermodynamics and statistical mechanics to derive his revolutionary results. At the time it was well known that the probability, W, for n independent gas atoms to be in a partial volume V of a larger volume V_0 is $(V/V_0)^n$. Einstein showed that light of frequency f and total energy E enclosed in a cavity obeys an identical law, where in this case W is the probability that all the radiation is in the partial volume and $n = E/hf$.

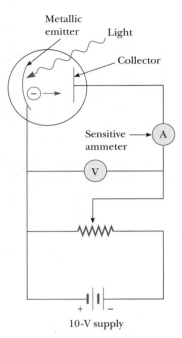

Metallic
emitter   Light

Collector

Sensitive —→ (A)
ammeter

(V)

www

+ |⊢| −

10-V supply

Figure 3.14 Photoelectric effect apparatus.

was unable to establish the precise relationship, Lenard also indicated that K_{max} increases with light *frequency*. A typical apparatus used to measure the maximum kinetic energy of photoelectrons is shown in Figure 3.14. K_{max} is easily measured by applying a retarding voltage and gradually increasing it until the most energetic electrons are stopped and the photocurrent becomes zero. At this point,

$$K_{max} = \tfrac{1}{2}m_e v_{max}^2 = eV_s \qquad (3.21)$$

where m_e is the mass of the electron, v_{max} is the maximum electron speed, e is the electronic charge, and V_s is the stopping voltage. A plot of the type found by Lenard is shown in Figure 3.15a; it illustrates that K_{max} or V_s is independent of light intensity I. The increase in current (or number of electrons per second) with increasing light intensity shown in Figure 3.15a was expected and could be explained classically. However, the result that K_{max} *does not depend on the intensity* was completely unexpected.

Two other experimental results were completely unexpected classically as well. One was *the linear dependence of K_{max} on light frequency*, shown in Figure 3.15b. Note that Figure 3.15b also shows the existence of a threshold frequency, f_0, below which no photoelectrons are emitted. (Actually, a threshold energy called the *work function*, ϕ, is associated with the binding energy of an electron in a metal and is expected classically. That there is an energy barrier holding electrons in a solid is evident from the fact that electrons are not spontaneously emitted from a metal in a vacuum, but require high temperatures or incident light to provide an energy of ϕ and cause emission.) The other interesting result impossible to explain classically is that *there is no time lag between the start of illumination and the start of the photocurrent*. Measurements have shown that if there is a time lag, it is less than 10^{-9} s. In summary, as shown in detail in the following example, classical electromagnetic theory has major difficulties explaining the independence of K_{max} and light intensity, the linear dependence of K_{max} on light frequency, and the instantaneous response of the photocurrent.

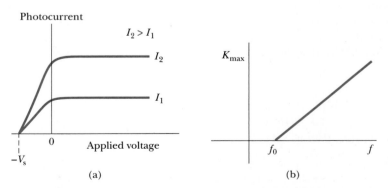

Figure 3.15 (a) A plot of photocurrent versus applied voltage. The graph shows that K_{max} is independent of light intensity I for light of fixed frequency. (b) A graph showing the dependence of K_{max} on light frequency.

EXAMPLE 3.4   Maxwell Takes a Licking

For a typical case of photoemission from sodium, show that *classical theory* predicts that (a) $K_{\max}$ depends on the incident light intensity, I, (b) $K_{\max}$ does not depend on the frequency of the incident light, and (c) there is a long time lag between the start of illumination and the beginning of the photocurrent. The work function for sodium is $\phi = 2.28$ eV and an absorbed power per unit area of 1.00×10^{-7} mW/cm^2 produces a measurable photocurrent in sodium.

Solution   (a) *According to classical theory, the energy in a light wave is spread out uniformly and continuously over the wavefront.* Assuming that all absorption of light occurs in the top atomic layer of the metal, that each atom absorbs an equal amount of energy proportional to its cross-sectional area, A, and that each atom somehow funnels this energy into one of its electrons, we find that each electron absorbs an energy K in time t given by

$$K = CIAt$$

where C is a fraction accounting for less than 100% light absorption. Because the most energetic electrons are held in the metal by a surface energy barrier or work function of ϕ, these electrons will be emitted with $K_{\max}$ once they have absorbed enough energy to overcome the barrier ϕ. We can express this as

$$K_{\max} = CIAt - \phi \tag{3.22}$$

Thus, classical theory predicts that for a fixed absorption period, t, at low light intensities when $CIAt < \phi$, no electrons ought to be emitted. At higher intensities, when $CIAt > \phi$, electrons should be emitted with higher kinetic energies the higher the light intensity. Therefore, classical predictions contradict experiment at both very low and very high light intensities.

(b) According to classical theory, *the intensity of a light wave is proportional to the square of the amplitude of the electric field, E_0^2, and it is this electric field amplitude that increases with increasing intensity and imparts an increasing acceleration and kinetic energy to an electron.* Replacing I with a quantity proportional to E_0^2 in Equation 3.22 shows that $K_{\max}$ should not depend at all on the frequency of the classical light wave, again contradicting the experimental results.

(c) To estimate the time lag between the start of illumination and the emission of electrons, we assume that an electron must accumulate just enough light energy to overcome the work function. Setting $K_{\max} = 0$ in Equation 3.22 gives

$$0 = CIAt - \phi$$

or

$$t = \frac{\phi}{CIA} = \frac{\phi}{IA}$$

assuming that I is the actual absorbed intensity. Because ϕ and I are given, we need A, the cross-sectional area of an atom, to calculate the time. As an estimate of A we simply use $A = \pi r^2$, where r is a typical atomic radius. Taking $r = 1.0 \times 10^{-8}$ cm, we find $A = \pi \times 10^{-16}$ cm^2. Finally, substituting this value into the expression for t, we obtain

$$t = \frac{2.28 \text{ eV} \times 1.60 \times 10^{-16} \text{ mJ/eV}}{(10^{-7} \text{ mJ/s} \cdot \text{cm}^2)(\pi \times 10^{-16} \text{ cm}^2)}$$

$$= 1.2 \times 10^7 \text{ s} \approx 130 \text{ days}$$

Thus we see that the classical calculation of the time lag for photoemission does not agree with the experimental result, disagreeing by a factor of 10^{16}!

Exercise 5   Why do the I–V curves in Figure 3.15a rise gradually between $-V_s$ and 0, that is, why do they not rise abruptly upward at $-V_s$? What statistical information about the conduction electrons inside the metal is contained in the slope of the I–V curve?

Einstein's explanation of the puzzling photoelectric effect was as brilliant for what it focused on as for what it omitted. For example, he stressed that Maxwell's classical theory had been immensely successful in describing the progress of light through space *over long time intervals* but that a different theory might be needed to describe *momentary interactions* of light and matter, as in light emission by oscillators or the transformation of light energy to kinetic energy of the electron in the photoelectric effect. He also focused only on the energy aspect of the light and avoided models or mechanisms concerning the conversion of the quantum of light energy to kinetic energy

Table 3.1   Work Functions of Selected Metals

Metal	Work Function, ϕ, (in eV)
Na	2.28
Al	4.08
Cu	4.70
Zn	4.31
Ag	4.73
Pt	6.35
Pb	4.14
Fe	4.50

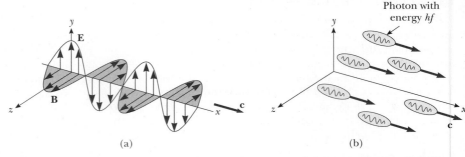

Figure 3.16   (a) A classical view of a traveling light wave. (b) Einstein's photon picture of "a traveling light wave."

of the electron. In short, he introduced only those ideas necessary to explain the photoelectric effect. **He maintained that the energy of light is not distributed evenly over the classical wavefront, but is concentrated in discrete regions (or in "bundles"), called quanta, each containing energy, hf.** A suggestive image, not to be taken too literally, is shown in Figure 3.16b. Einstein's picture was that a light quantum was so localized that it gave all its energy, *hf*, directly to a single electron in the metal. Therefore, according to Einstein, the maximum kinetic energy for emitted electrons is

Einstein's theory of the photoelectric effect

$$K_{max} = hf - \phi \tag{3.23}$$

where ϕ is the work function of the metal, which corresponds to the minimum energy with which an electron is bound in the metal. Table 3.1 lists values of work functions measured for different metals.

Equation 3.23 beautifully explained the puzzling independence of K_{max} and intensity found by Lenard. For a fixed light frequency *f*, an increase in light intensity means more photons and more photoelectrons per second, although K_{max} remains unchanged according to Equation 3.23. In addition, Equation 3.23 explained the phenomenon of threshold frequency. Light of threshold frequency f_0, which has just enough energy to knock an electron out of the metal surface, causes the electron to be released with zero kinetic energy. Setting $K_{max} = 0$ in Equation 3.23 gives

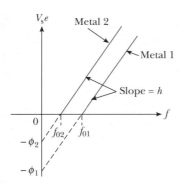

Figure 3.17   Universal characteristics of all metals undergoing the photoelectric effect.

$$f_0 = \frac{\phi}{h} \tag{3.24}$$

Thus the variation in threshold frequency for different metals is produced by the variation in work function. Note that light with $f < f_0$ has insufficient energy to free an electron. Consequently, the photocurrent is zero for $f < f_0$.

With any theory, one looks not only for explanations of previously observed results but also for new predictions. This was indeed the case here, as Equation 3.23 predicted the result (new in 1905) that K_{max} should vary linearly with *f* for any material and that the slope of the K_{max} versus *f* plot should yield the universal constant h (see Fig. 3.17). In 1916, the American physicist Robert Millikan (1868–1953) reported photoelectric measurement data, from which he substantiated the linear relation between K_{max} and f and determined h with a precision of about 0.5%.[12]

EXAMPLE 3.5   The Photoelectric Effect in Zinc

Philip Lenard determined that photoelectrons released from zinc by ultraviolet light were stopped by a voltage of 4.3 V. Find K_{max} and v_{max} for these electrons.

Solution

$$K_{max} = eV_s = (1.6 \times 10^{-19}\ C)(4.3\ V) = 6.9 \times 10^{-19}\ J$$

To find v_{max}, we set the work done by the electric field equal to the change in the electron's kinetic energy, to obtain

$$\tfrac{1}{2}m_e v_{max}^2 = eV_s$$

or

$$v_{max} = \sqrt{\frac{2eV_s}{m_e}} = \sqrt{\frac{2(6.9 \times 10^{-19}\ J)}{9.11 \times 10^{-31}\ kg}}$$
$$= 1.2 \times 10^6\ m/s$$

Therefore, a 4.3-eV electron is rather energetic and moves with a speed of about a million meters per second. Note, however, that this is still only about 0.4% of the speed of light, so relativistic effects are negligible in this case.

EXAMPLE 3.6   The Photoelectric Effect for Iron

Suppose that light of total intensity 1.0 $\mu W/cm^2$ falls on a clean iron sample 1.0 cm^2 in area. Assume that the iron sample reflects 96% of the light and that only 3.0% of the absorbed energy lies in the violet region of the spectrum above the threshold frequency.

(a) What intensity is actually available for the photoelectric effect?

Because only 4.0% of the incident energy is absorbed, and only 3.0% of this energy is able to produce photoelectrons, the intensity available is

$$I = (0.030)(0.040)I_0 = (0.030)(0.040)(1.0\ \mu W/cm^2)$$
$$= 1.2\ nW/cm^2$$

(b) Assuming that all the photons in the violet region have an effective wavelength of 250 nm, how many electrons will be emitted per second?

For an efficiency of 100%, one photon of energy, hf, will produce one electron, so

Number of electrons/s

$$= \frac{1.2 \times 10^{-9}\ W}{hf} = \frac{\lambda(1.2 \times 10^{-9})}{hc}$$
$$= \frac{(250 \times 10^{-9}\ m)(1.2 \times 10^{-9}\ J/s)}{(6.6 \times 10^{-34}\ J \cdot s)(3.0 \times 10^8\ m/s)}$$
$$= 1.5 \times 10^9$$

(c) Calculate the current in the phototube in amperes.

$$i = (1.6 \times 10^{-19}\ C)(1.5 \times 10^9\ electrons/s)$$
$$= 2.4 \times 10^{-10}\ A$$

A sensitive electrometer is needed to detect this small current.

(d) If the cutoff frequency is $f_0 = 1.1 \times 10^{15}$ Hz, find the work function, ϕ, for iron.

From Equation 3.24, we have

$$\phi = hf_0 = (4.14 \times 10^{-15}\ eV \cdot s)(1.1 \times 10^{15}\ s^{-1})$$
$$= 4.5\ eV$$

(e) Find the stopping voltage for iron if photoelectrons are produced by light with $\lambda = 250$ nm.

From the photoelectric equation,

$$eV_s = hf - \phi = \frac{hc}{\lambda} - \phi$$
$$= \frac{(4.14 \times 10^{-15}\ eV \cdot s)(3.0 \times 10^8\ m/s)}{250 \times 10^{-9}\ m} - 4.5\ eV$$
$$= 0.46\ eV$$

Thus the stopping voltage is 0.46 V.

---

[12]R. A. Millikan, *Phys. Rev.*, 7:355, 1916. Some of the experimental difficulties in the photoelectric effect were the lack of strong monochromatic uv sources, small photocurrents, and large effects of rough and impure metal surfaces on f_0 and K_{max}. Millikan cleverly circumvented these difficulties by using alkali metal cathodes, which are sensitive in the visible to about 600 nm (thus making it possible to use the strong visible lines of the mercury arc), and machining fresh alkali surfaces while the metal sample was held under high vacuum. Also when the phototube emitter and collector are composed of different metals, the work function ϕ determined from plots of V_s vs. f is actually that of the collector. See J. Rudnick and D. S. Tannhauser, AJP 44, 796, 1976.

3.5   THE COMPTON EFFECT AND X-RAYS

Although Einstein introduced the concept that light consists of pointlike quanta of *energy* in 1905, he did not directly treat the *momentum* carried by light until 1906. In that year, in a paper treating a molecular gas in thermal equilibrium with electromagnetic radiation (statistical mechanics again!), Einstein concluded that a light quantum of energy E travels in a single direction (unlike a spherical wave) and carries a momentum directed along its line of motion of E/c, or hf/c. In his own words, "If a bundle of radiation causes a molecule to emit or absorb an energy packet hf, then momentum of quantity hf/c is transferred to the molecule, directed along the line of motion of the bundle for absorption and opposite the bundle for emission."

After developing the first theoretical justification for photon momentum, and treating the photoelectric effect much earlier, it is curious that Einstein carried the treatment of photon momentum no further. The theoretical treatment of photon–particle collisions had to await the insight of Peter Debye (1884–1966, Dutch physical chemist), and Arthur Holly Compton (1892–1962, American physicist). In 1923, both men independently realized that the scattering of x-ray photons from electrons could be explained by treating photons as pointlike particles with energy hf and momentum hf/c and by conserving relativistic energy and momentum of the photon–electron pair in a collision.[13,14] This remarkable development completed the particle picture of light by showing that photons, in addition to carrying energy, hf, carry momentum, hf/c, and scatter like particles. Before treating this in detail, a brief introduction to the important topic of x-rays will be given.

X-Rays

X-rays were discovered in 1895 by the German physicist Wilhelm Roentgen. He found that a beam of high-speed electrons striking a metal target produced a new and extremely penetrating type of radiation (Fig. 3.18). Within months of Roentgen's discovery the first medical x-ray pictures were taken, and within several years it became evident that x-rays were electromagnetic vibrations similar to light but with extremely short wavelengths and great penetrating power (see Fig. 3.19). Rough estimates obtained from the diffraction of x rays by a narrow slit showed x-ray wavelengths to be about 10^{-10} m, which is of the same order of magnitude as the atomic spacing in crystals. Because the best artificially ruled gratings of the time had spacings of 10^{-7} m, Max von Laue in Germany and William Henry Bragg and William Lawrence Bragg (a father and son team) in England suggested using single crystals such as calcite as natural three-dimensional gratings, the periodic atomic arrangement in the crystals constituting the grating rulings.

A particularly simple method of analyzing the scattering of x-rays from parallel crystal planes was proposed by W. L. Bragg in 1912. Consider two successive planes of atoms as shown in Figure 3.20. Note that adjacent atoms *in a single plane*, A, will scatter constructively if the angle of incidence, θ_i,

[13]P. Debye, *Phys. Zeitschr.*, 24:161, 1923. In this paper, Debye acknowledges Einstein's pioneering work on the quantum nature of light.

[14]A. H. Compton, *Phys. Rev.*, 21:484, 1923.

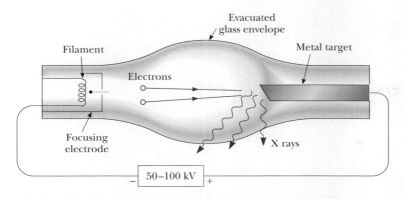

Figure 3.18   X-rays are produced by bombarding a metal target (copper, tungsten, and molybdenum are common) with energetic electrons having energies of 50 to 100 keV.

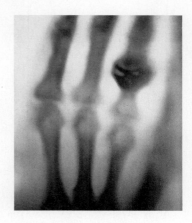

Figure 3.19   One of the first images made by Roentgen using x-rays (December 22, 1895).

equals the angle of reflection, θ_r. Atoms in *successive planes* (A and B) will scatter constructively at an angle θ if the path length difference for rays (1) and (2) is a whole number of wavelengths, $n\lambda$. From the diagram, constructive interference will occur when

$$AB + BC = n\lambda \qquad n = 1, 2, 3, \ldots$$

and because $AB = BC = d \sin \theta$, it follows that

$$n\lambda = 2d \sin \theta \qquad n = 1, 2, 3, \ldots \qquad (3.25a)$$

Bragg equation

where n is the order of the intensity maximum, λ is the x-ray wavelength, d is the spacing between planes, and θ is the angle of the intensity maximum measured from plane A. Note that there are several maxima at different angles for a fixed d and λ corresponding to $n = 1, 2, 3, \ldots$. Equation 3.25a is known as the Bragg equation; it was used with great success by the Braggs to determine atomic positions in crystals. A diagram of a Bragg x-ray spectrometer is shown in Figure 3.21a. The crystal is slowly rotated until a strong reflection is

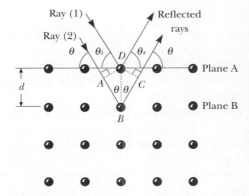

Figure 3.20   Bragg scattering of x-rays from successive planes of atoms. Constructive interference occurs for *ABC* equal to an integral number of wavelengths.

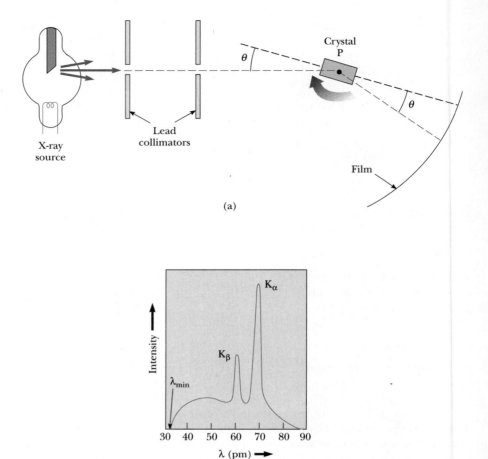

Figure 3.21 (a) A Bragg crystal x-ray spectrometer. The crystal is rotated about an axis through *P*. (b) The x-ray spectrum of a metal target consists of a broad, continuous spectrum plus a number of sharp lines, which are due to the characteristic x-rays. Those shown were obtained when 35-keV electrons bombarded a molybdenum target. Note that $1 \text{ pm} = 10^{-12} \text{ m} = 10^{-3} \text{ nm}$.

observed, which means that Equation 3.25a holds. If λ is known, *d* can be calculated and, from the series of *d* values found, the crystal structure may be determined. (See Problem 38.) If measurements are made with a crystal with known *d*, the x-ray intensity vs. wavelength may be determined and the x-ray emission spectrum examined.

The actual x-ray emission spectrum produced by a metal target bombarded by electrons is interesting in itself and is shown in Figure 3.21b. Although the broad, continuous spectrum is well explained by classical electromagnetic theory, a feature of Figure 3.21b, λ_{min}, shows proof of the photon theory. The broad continuous x-ray spectrum shown in Figure 3.21b results from glancing or indirect scattering of electrons from metal atoms. In such collisions only part of the electron's energy is converted to electromagnetic radiation. This radiation is called *bremsstrahlung* (German for braking radiation), which refers to the radiation given off by any charged particle when it is decelerated. The minimum continuous x-ray wavelength, λ_{min}, is found to be independent of target composition and depends only on the tube voltage, V. It may be explained by attributing it to the case of a head-on electron–atom collision in which all of the incident electron's kinetic energy is converted to electromagnetic energy in the form of a single x-ray photon. For this case we have

$$eV = hf = \frac{hc}{\lambda_{min}}$$

or

$$\lambda_{min} = \frac{hc}{eV} \qquad (3.26)$$

where V is the x-ray tube voltage.

Superimposed on the continuous spectrum are sharp x-ray lines labeled K_α and K_β, which are like sharp lines emitted in the visible light spectrum. The sharp lines depend on target composition and provide evidence for discrete atomic energy levels separated by thousands of electron volts, as explained in Chapter 9.

The Compton Effect

Let us now turn to the year 1922 and the experimental confirmation by Arthur Holly Compton that x-ray photons behave like particles with momentum hf/c. For some time prior to 1922, Compton and his coworkers had been accumulating evidence that showed that classical wave theory failed to explain the scattering of x-rays from free electrons. In particular, classical theory predicted that incident radiation of frequency f_0 should accelerate an electron in the direction of propagation of the incident radiation, and that it should cause forced oscillations of the electron and reradiation at frequency f', where $f' \leq f_0$ (see Fig. 3.22a).[15] Also, according to classical theory, the frequency or wavelength of the scattered radiation should depend on the length of time the electron was exposed to the incident radiation as well as on the intensity of the incident radiation.

Imagine the surprise when Compton showed experimentally that the wavelength shift of x-rays scattered at a given angle is absolutely independent of the intensity of radiation and the length of exposure, and depends only on the scattering angle. Figure 3.22b shows the quantum model of the transfer of momentum and energy between an individual x-ray photon and an electron. Note that the quantum model easily explains the lower scattered frequency f', because the incident photon gives some of its original energy hf to the recoiling electron.

A schematic diagram of the apparatus used by Compton is shown in Figure 3.23a. In the original experiment, Compton measured the dependence of scattered x-ray intensity on wavelength at three different scattering angles

---

[15]This decrease in frequency of the reradiated wave is caused by a double Doppler shift, first because the electron is receding from the incident radiation, and second because the electron is a moving radiator as viewed from the fixed lab frame. See D. Bohm, *Quantum Theory*, Upper Saddle River, NJ, Prentice-Hall, 1961, p. 35.

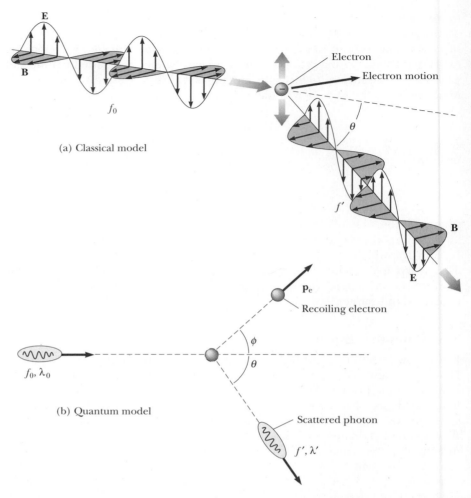

Figure 3.22   X-ray scattering from an electron: (a) the classical model, (b) the quantum model.

of 45°, 90°, and 135°. The wavelength was measured with a rotating crystal spectrometer, and the intensity was determined by an ionization chamber that generated a current proportional to the x-ray intensity. Monochromatic x-rays of wavelength $\lambda_0 = 0.71$ Å constituted the incident beam. A carbon target with a low atomic number, $Z = 12$, was used because atoms with small Z have a higher percentage of loosely bound electrons. The experimental intensity versus wavelength plots observed by Compton for scattering angles of 0°, 45°, 90°, and 135° are shown in Figure 3.23b. They show two peaks, one at λ_0 and a shifted peak at a longer wavelength λ'. The shifted peak at λ' is caused by the scattering of x-rays from nearly free electrons. Assuming that x-rays behave like particles, λ' was predicted by Compton to depend on scattering angle as

Compton effect

$$\lambda' - \lambda_0 = \frac{h}{m_e c}(1 - \cos\theta) \tag{3.27}$$

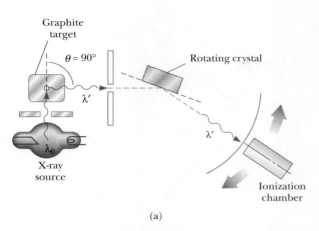

Graphite target

$\theta = 90°$

Rotating crystal

λ'

λ'

X-ray source

λ_0

Ionization chamber

(a)

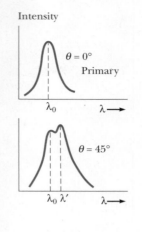

Intensity

$\theta = 0°$
Primary

λ_0 $\lambda \longrightarrow$

$\theta = 45°$

λ_0 λ' $\lambda \longrightarrow$

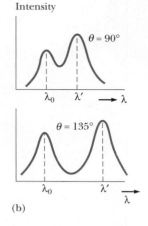

Intensity

$\theta = 90°$

λ_0 λ' $\longrightarrow \lambda$

$\theta = 135°$

λ_0 λ'
$\longrightarrow \lambda$

(b)

Figure 3.23   (a) Schematic diagram of Compton's apparatus. The wavelength was measured with a rotating crystal spectrometer using graphite (carbon) as the target. The intensity was determined by a movable ionization chamber that generated a current proportional to the x-ray intensity. (b) Scattered x-ray intensity versus wavelength of Compton scattering at $\theta = 0°$, $45°$, $90°$, and $135°$.

where m_e = electron mass; the combination of constants $h/m_e c$ is called the Compton wavelength of the electron and has a currently accepted value of

$$\frac{h}{m_e c} = 0.0243 \text{ Å} = 0.00243 \text{ nm}$$

Compton's careful measurements completely confirmed the dependence of λ' on scattering angle θ and determined the Compton wavelength of the electron to be 0.0242 Å, in excellent agreement with the currently accepted value. It is fair to say that these results were the first to really convince most American physicists of the basic validity of the quantum theory!

The unshifted peak at λ_0 in Figure 3.23 is caused by x-rays scattered from electrons tightly bound to carbon atoms. This unshifted peak is actually predicted by Equation 3.27 if the electron mass is replaced by the mass of a carbon atom, which is about 23,000 times the mass of an electron.

Let us now turn to the derivation of Equation 3.27 assuming that the photon exhibits particle-like behavior and collides elastically like a billiard ball with a free electron initially at rest. Figure 3.24 shows the photon–electron collision for which energy and momentum are conserved. Because the electron typically recoils at high speed, we treat the collision relativistically. The expression for conservation of energy gives

$$E + m_e c^2 = E' + E_e \qquad (3.28)$$

Energy conservation

where E is the energy of the incident photon, E' is the energy of the scattered photon, $m_e c^2$ is the rest energy of the electron, and E_e is the total relativistic energy of the electron after the collision. Likewise, from momentum conservation we have

$$p = p' \cos \theta + p_e \cos \phi \qquad (3.29)$$

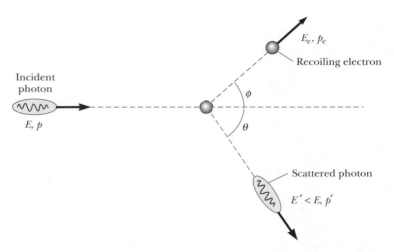

Figure 3.24 Diagram representing Compton scattering of a photon by an electron. The scattered photon has less energy (or longer wavelength) than the incident photon.

$$p' \sin \theta = p_e \sin \phi \qquad (3.30)$$

where p is the momentum of the incident photon, p' is the momentum of the scattered photon, and p_e is the recoil momentum of the electron. Equations 3.29 and 3.30 may be solved simultaneously to eliminate ϕ, the electron scattering angle, to give the following expression for p_e^2:

$$p_e^2 = (p')^2 + p^2 - 2pp' \cos \theta \qquad (3.31)$$

At this point it is necessary, paradoxically, to use the wave nature of light to explain the particle-like behavior of photons. We have already seen that the energy of a photon and the frequency of the associated light wave are related by $E = hf$. If we assume that a photon obeys the relativistic expression $E^2 = p^2c^2 + m^2c^4$ and that a photon has a mass of zero, we have

$$p_{\text{photon}} = \frac{E}{c} = \frac{hf}{c} = \frac{h}{\lambda} \qquad (3.32)$$

Here again we have a paradoxical situation; a particle property, the photon momentum, is given in terms of a wave property, λ, of an associated light wave. If the relations $E = hf$ and $p = hf/c$ are substituted into Equations 3.28 and 3.31, these become respectively

$$E_e = hf - hf' + m_ec^2 \qquad (3.33)$$

and

$$p_e^2 = \left(\frac{hf'}{c}\right)^2 + \left(\frac{hf}{c}\right)^2 - \frac{2h^2ff'}{c^2} \cos \theta \qquad (3.34)$$

Because the Compton measurements do not concern the total energy and momentum of the electron, we eliminate E_e and p_e by substituting Equations 3.33 and 3.34 into the expression for the electron's relativistic energy,

$$E_e^2 = p_e^2 c^2 + m_e^2 c^4$$

After some algebra (see Problem 33), one obtains Compton's result for the increase in a photon's wavelength when it is scattered through an angle θ:

$$\lambda' - \lambda_0 = \frac{h}{m_e c}(1 - \cos\theta) \qquad (3.27)$$

EXAMPLE 3.7   The Compton Shift for Carbon

X-rays of wavelength $\lambda = 0.200$ nm are aimed at a block of carbon. The scattered x-rays are observed at an angle of 45.0° to the incident beam. Calculate the increased wavelength of the scattered x-rays at this angle.

Solution   The shift in wavelength of the scattered x-rays is given by Equation 3.27. Taking $\theta = 45.0°$, we find

$$\Delta\lambda = \frac{h}{m_e c}(1 - \cos\theta)$$

$$= \frac{6.63 \times 10^{-34}\,\text{J}\cdot\text{s}}{(9.11 \times 10^{-31}\,\text{kg})(3.00 \times 10^8\,\text{m/s})}(1 - \cos 45.0°)$$

$$= 7.11 \times 10^{-13}\,\text{m} = 0.00071\,\text{nm}$$

Hence, the wavelength of the scattered x-ray at this angle is

$$\lambda = \Delta\lambda + \lambda_0 = 0.200711\,\text{nm}$$

Exercise 6   Find the fraction of energy lost by the photon in this collision.

Answer   Fraction $= \Delta E/E = 0.00355$.

EXAMPLE 3.8   X-ray Photons versus Visible Photons

(a) Why are x-ray photons used in the Compton experiment, rather than visible-light photons? To answer this question, we shall first calculate the Compton shift for scattering at 90° from graphite for the following cases: (1) very high energy γ-rays from cobalt, $\lambda = 0.0106$ Å; (2) x-rays from molybdenum, $\lambda = 0.712$ Å; and (3) green light from a mercury lamp, $\lambda = 5461$ Å.

Solution   In all cases, the Compton shift formula gives $\Delta\lambda = \lambda' - \lambda_0 = (0.0243\,\text{Å})(1 - \cos 90°) = 0.0243\,\text{Å} = 0.00243$ nm. That is, regardless of the incident wavelength, the same small shift is observed. However, the fractional change in wavelength, $\Delta\lambda/\lambda_0$, is quite different in each case:

γ-rays from cobalt:

$$\frac{\Delta\lambda}{\lambda_0} = \frac{0.0243\,\text{Å}}{0.0106\,\text{Å}} = 2.29$$

X-rays from molybdenum:

$$\frac{\Delta\lambda}{\lambda_0} = \frac{0.0243\,\text{Å}}{0.712\,\text{Å}} = 0.0341$$

Visible light from mercury:

$$\frac{\Delta\lambda}{\lambda_0} = \frac{0.0243\,\text{Å}}{5461\,\text{Å}} = 4.45 \times 10^{-6}$$

Because both incident and scattered wavelengths are simultaneously present in the beam, they can be easily resolved only if $\Delta\lambda/\lambda_0$ is a few percent or if $\lambda_0 \lesssim 1$ Å.

(b) The so-called free electrons in carbon are actually electrons with a binding energy of about 4 eV. Why may this binding energy be ignored for x-rays with $\lambda_0 = 0.712$ Å?

Solution   The energy of a photon with this wavelength is

$$E = hf = \frac{hc}{\lambda} = \frac{12\,400\,\text{eV}\cdot\text{Å}}{0.712\,\text{Å}} = 17\,400\,\text{eV}$$

Therefore, the electron binding energy of 4 eV is negligible in comparison with the incident x-ray energy.

3.6 PARTICLE–WAVE COMPLEMENTARITY

As we have seen, the Compton effect offers ironclad evidence that when light interacts with matter it behaves as if it were composed of particles with energy hf and momentum h/λ. Yet the very success of Compton's theory raises many questions. If the photon is a particle, what can be the meaning of the "frequency" and "wavelength" of the particle, which determine its energy and momentum? Is light in some sense simultaneously a wave and a particle? Although photons have zero mass, is there a simple expression for an effective gravitational photon mass that determines a photon's gravitational attraction? What is the spatial extent of a photon, and how does an electron absorb or scatter a photon?

Although answers to some of these questions are possible, it is well to be aware that some demand a view of atomic processes that is too pictorial and literal. Many of these questions issue from the viewpoint of classical mechanics, in which all matter and energy are seen in the context of colliding billiard balls or water waves breaking on a shore. Quantum theory gives light a more flexible nature by implying that different experimental conditions elicit either the wave properties or particle properties of light. In fact, *both views are necessary and complementary*. Neither model can be used exclusively to describe electromagnetic radiation adequately. A complete understanding is obtained only if the two models are combined in a complementary manner.

The physicist Max Born, an important contributor to the foundations of quantum theory, had this to say about the particle–wave dilemma:

> The ultimate origin of the difficulty lies in the fact (or philosophical principle) that we are compelled to use the words of common language when we wish to describe a phenomenon, not by logical or mathematical analysis, but by a picture appealing to the imagination. Common language has grown by everyday experience and can never surpass these limits. Classical physics has restricted itself to the use of concepts of this kind; by analyzing visible motions it has developed two ways of representing them by elementary processes: moving particles and waves. There is no other way of giving a pictorial description of motions—we have to apply it even in the region of atomic processes, where classical physics breaks down.
>
> Every process can be interpreted either in terms of corpuscles or in terms of waves, but on the other hand it is beyond our power to produce proof that it is actually corpuscles or waves with which we are dealing, for we cannot simultaneously determine all the other properties which are distinctive of a corpuscle or of a wave, as the case may be. We can therefore say that the wave and corpuscular descriptions are only to be regarded as complementary ways of viewing one and the same objective process, a process which only in definite limiting cases admits of complete pictorial interpretation.[16]

Thus we are left with an uneasy compromise between wave and particle concepts and must accept, at this point, that both are necessary to explain the observed behavior of light. Further considerations of the dual nature of light and indeed of all matter will be taken up again in Chapters 4 and 5.

[16]M. Born, *Atomic Physics*, fourth edition, New York, Hafner Publishing Co., 1946, p. 92.

3.7   DOES GRAVITY AFFECT LIGHT?

It is interesting to speculate on how far the particle model of light may be carried. Encouraged by the successful particle explanation of the photoelectric and Compton effects, one may ask whether the photon possesses an effective gravitational-mass, and whether photons will be attracted gravitationally by large masses, such as those of the Sun or Earth, and experience an observable change in energy.

To investigate these questions, recall that the photon has zero mass, but its effective inertial mass, m_i, may reasonably be taken to be the mass equivalent of the photon energy, E, or

$$m_i = \frac{E}{c^2} = \frac{hf}{c^2} \tag{3.35}$$

The same result is obtained if we divide the photon momentum by the photon speed c:

$$m_i = \frac{p}{c} = \frac{hf}{c^2}$$

Recall that the *effective inertial mass* determines how the photon responds to an applied force such as that exerted on it during a collision with an electron. The *gravitational mass* of an object determines the force of gravitational attraction of that object to another, such as the Earth. Although it is a remarkable unexplained fact in Newtonian mechanics that the inertial mass of all material bodies is equal to the gravitational mass to within one part in 10^{12}, Einstein's Equivalence Principle of general relativity requires this result as mentioned in Chapter 2.

Let us assume that the photon, like other objects, also has a gravitational mass equal to its inertial mass. In this case a photon falling from a height H should increase in energy by mgH and therefore increase in frequency, although its speed cannot increase and remains at c. In fact, experiments have been carried out that show this increase in frequency and confirm that the photon indeed has an effective gravitational mass of hf/c^2. Figure 3.25 shows a schematic representation of the experiment. An expression for f' in terms of f may be derived by applying conservation of energy to the photon at points A and B.

$$KE_B + PE_B = KE_A + PE_A$$

Because the photon's kinetic energy is $E = pc = hf$ and its potential energy is mgH, where $m = hf/c^2$, we have

$$hf' + 0 = hf + \left(\frac{hf}{c^2}\right)gH$$

or

$$f' = f\left(1 + \frac{gH}{c^2}\right) \tag{3.36}$$

The fractional change in frequency, $\Delta f/f$, is given by

$$\frac{\Delta f}{f} = \frac{f' - f}{f} = \frac{gH}{c^2} \tag{3.37}$$

For $H = 50$ m, we find

$$\frac{\Delta f}{f} = \frac{(9.8 \text{ m/s}^2)(50 \text{ m})}{(3.0 \times 10^8 \text{ m/s})^2} = 5.4 \times 10^{-15}$$

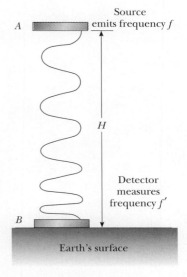

Figure 3.25 Schematic diagram of the falling-photon experiment.

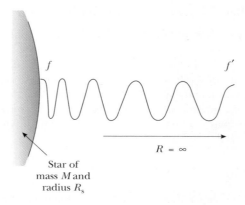

Figure 3.26   Gravitational redshift from a high-density star.

This incredibly small increase in frequency has actually been measured (with difficulty)![17] The shift amounts to only about $1/250$ of the line width of the monochromatic γ-ray photons used in the falling-photon experiment.

The increase in frequency for a photon falling inward suggests a decrease in frequency for a photon that escapes outward to infinity against the gravitational pull of a star (see Fig. 3.26). This effect, known as "gravitational redshift," would cause an emitted photon to be shifted in frequency toward the red end of the spectrum. An expression for the redshift may be derived once again by conserving photon energy:

$$[KE + PE]_{R=\infty} = [KE + PE]_{R=R_s}$$

Using hf for the photon's kinetic energy and $-GMm/R$ for its potential energy, with m equal to hf/c^2 and R_s equal to the star's radius, yields

$$hf' - 0 = hf - \frac{GM}{R_s}\left(\frac{hf}{c^2}\right) \tag{3.38}$$

or

$$f' = f\left(1 - \frac{GM}{R_s c^2}\right) \tag{3.39}$$

EXAMPLE 3.9   The Gravitational Redshift for a White Dwarf

White dwarf stars are extremely massive, compact stars that have a mass on the order of the Sun's mass concentrated in a volume similar to that of the Earth. Calculate the gravitational redshift for 300-nm light emitted from such a star.

Solution   We can write Equation 3.39 in the alternate form

$$\frac{f' - f}{f} = \frac{\Delta f}{f} = \frac{GM}{R_s c^2}$$

Using the values

$$M = \text{mass of Sun} = 1.99 \times 10^{30}\text{ kg}$$

---

[17]R. V. Pound and G. A. Rebka, Jr., *Phys. Rev. Lett.*, 4:337, 1960.

$$R_s = \text{radius of Earth} = 6.37 \times 10^6 \text{ m}$$

$$G = 6.67 \times 10^{-11} \text{ N} \cdot \text{m}^2/\text{kg}^2$$

we find

$$\frac{\Delta f}{f} = \frac{(6.67 \times 10^{-11} \text{ N} \cdot \text{m}^2/\text{kg}^2)(1.99 \times 10^{30} \text{ kg})}{(6.37 \times 10^6 \text{ m})(3.00 \times 10^8 \text{ m/s})^2}$$

$$= 2.31 \times 10^{-4}$$

Because $\Delta f/f \approx df/f$, and $df = -(c/\lambda^2) \ d\lambda$ (from $f = c/\lambda$), we find $|df/f| = d\lambda/\lambda$. Therefore, the shift in wavelength is

$$\Delta\lambda = (300 \text{ nm})(2.31 \times 10^{-4}) = 0.0695 \text{ nm} \approx 0.7 \text{ Å}$$

Note that this is a redshift, so the observed wavelength would be 300.07 nm.

One more observation about Equation 3.39 is irresistible. Is it possible for a very massive star in the course of its life cycle to become so dense that the term $GM/R_s c^2$ becomes greater than 1? In that case Equation 3.38 suggests that the photon cannot escape from the star, because escape requires more energy than the photon initially possesses. Such a star is called a *black hole* because it emits no light and acts like a celestial vacuum cleaner for all nearby matter and radiation. Even though the black hole itself is not luminous, it may be possible to observe it indirectly in two ways. One way is through the gravitational attraction the black hole would exert on a normal luminous star if the two constituted a binary star system. In this case the normal star would orbit the center of mass of the black hole/normal star pair, and the orbital motion might be detectable. A second indirect technique for "viewing" a black hole would be to search for x-rays produced by inrushing matter attracted to the black hole. Although the black hole itself would not emit x-rays, an x-ray-emitting region of roughly stellar diameter should be observable, as shown in Figure 3.27. X-rays are produced by the

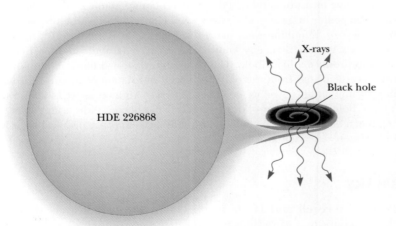

Figure 3.27   The Cygnus X-1 black hole. The stellar wind from HDE 226868 pours matter onto a huge disk around its black hole companion. The infalling gases are heated to enormous temperatures as they spiral toward the black hole. The gases are so hot that they emit vast quantities of x-rays.

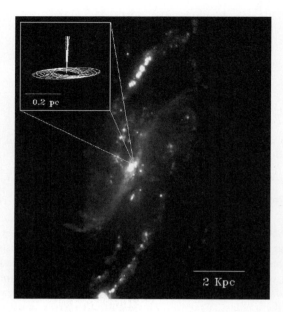

Figure 3.28   A background radio telescope image of NGC 4258 and an inset image generated by a computer model of the accretion disk surrounding the black hole in the galaxy's center. A parsec (pc) is a distance of about 3.3 light years. (*Gerald Cecil (UNC) and Holland Ford (JHU), Composite image courtesy of Lincoln Greenhill and James Herrnstein*)

heating of the infalling matter as it circulates, is compressed, and eventually falls into the black hole. Such an intense nonluminous point source of x-rays has been detected in the constellation of the Swan. This source, designated Cygnus X-1, is believed by most astronomers to be a black hole; it possesses a luminosity, or power output, of 10^{30} W in the 2- to 10-keV x-ray range.

Recently, even more convincing evidence of a black hole has been obtained from radio telescope measurements of a dust torus rotating rapidly around a huge central mass at the center of galaxy NGC 4258. (See Figure 3.28.) These observations pinpoint a mass of 39 million solar masses within a radius of 4.0×10^{15} m, a density 10,000 times greater than any known cluster of stars and almost certainly high enough to produce a black hole. The central gravitational mass of 39 million solar masses was calculated from the observed speed of rotation of the dust torus, which is about 1 million m/s. And we needn't even go so far away as NGC 4258. Evidence of a black hole at the center of our own galaxy has been rapidly accumulating, indicating that a black hole of about 3 million solar masses, concealed by dust, is located in the constellation Sagittarius.[18]

SUMMARY

The work of Maxwell and Hertz in the late 1800s conclusively showed that light, heat radiation, and radio waves were all electromagnetic waves differing only in frequency and wavelength. Thus it astonished scientists to find that the

[18]See the interesting book *The Black Hole at the Center of our Galaxy*, by Fulvio Melia, Princeton University Press, 2003.

spectral distribution of radiation from a heated cavity could not be explained in terms of classical wave theory. Planck was forced to introduce the concept of the quantum of energy in order to derive the correct blackbody formula. According to Planck, the atomic oscillators responsible for blackbody radiation can have only *discrete*, or quantized, *energies* given by

$$E = nhf \qquad (3.10)$$

where n is an integer, h is Planck's constant, and f is the oscillator's natural frequency. Using general thermodynamic arguments, Planck was able to show that $u(f, T)$, the blackbody radiation energy per unit volume with frequency between f and $f + df$, could be expressed as the product of the number of oscillators per unit volume in this frequency range, $N(f)\, df$, and the average energy emitted per oscillator, $\bar{E}$. That is,

$$u(f, T)\, df = \bar{E}\, N(f)\, df \qquad (3.12)$$

If $\bar{E}$ is calculated by allowing a *continuous* distribution of oscillator energies, the incorrect Rayleigh–Jeans law is obtained. If $\bar{E}$ is calculated from a *discrete* set of oscillator energies (following Planck), the correct blackbody radiation formula is obtained:

$$u(f, T)\, df = \frac{8\pi f^2}{c^3} \left(\frac{hf}{e^{hf/k_B T} - 1} \right) df \qquad (3.9)$$

Planck quantized the energy of atomic oscillators, but Einstein extended the concept of quantization to light itself. In Einstein's view, light of frequency f consists of a stream of particles, called *photons*, each with energy $E = hf$. The photoelectric effect, a process in which electrons are ejected from a metallic surface when light of sufficiently high frequency is incident on the surface, can be simply explained with the photon theory. According to this theory, the maximum kinetic energy of the ejected photoelectron, K_{max}, is given by

$$K_{max} = hf - \phi \qquad (3.23)$$

where ϕ is the work function of the metal.

Although the idea that light consists of a stream of photons with energy hf was put forward in 1905, the idea that these photons also carry momentum was not experimentally confirmed until 1923. In that year it was found that x-rays scattered from free electrons suffer a simple shift in wavelength with scattering angle, known as the Compton shift. When an x-ray of frequency f is viewed as a *particle* with *energy* hf and *momentum* hf/c, x-ray–electron scattering can be simply analyzed to yield the Compton shift formula:

$$\Delta\lambda = \frac{h}{m_e c} (1 - \cos\theta) \qquad (3.27)$$

Here, m_e is the mass of the electron and θ is the x-ray scattering angle.

The striking success of the photon theory in explaining interactions between light and electrons contrasts sharply with the success of classical wave theory in explaining the polarization, reflection, and interference of light. This leaves us with the dilemma of whether light is a wave or a parti-

cle. The currently accepted view suggests that light has both wave and particle characteristics and that these characteristics together constitute a complementary view of light.

QUESTIONS

1. What assumptions were made by Planck in dealing with the problem of blackbody radiation? Discuss the consequences of these assumptions.
2. If the photoelectric effect is observed for one metal, can you conclude that the effect will also be observed for another metal under the same conditions? Explain.
3. Suppose the photoelectric effect occurs in a gaseous target rather than a solid. Will photoelectrons be produced at *all* frequencies of the incident photon? Explain.
4. How does the Compton effect differ from the photoelectric effect?
5. What assumptions were made by Compton in dealing with the scattering of a photon from an electron?
6. An x-ray photon is scattered by an electron. What happens to the frequency of the scattered photon relative to that of the incident photon?
7. Why does the existence of a cutoff frequency in the photoelectric effect favor a particle theory for light rather than a wave theory?
8. All objects radiate energy. Why, then, are we not able to see all objects in a dark room?

9. Which has more energy, a photon of ultraviolet radiation or a photon of yellow light?
10. What effect, if any, would you expect the temperature of a material to have on the ease with which electrons can be ejected from it in the photoelectric effect?
11. Some stars are observed to be reddish, and some are blue. Which stars have the higher surface temperature? Explain.
12. When wood is stacked on a special elevated grate (which is commercially available) in a fireplace, a pocket of burning wood forms beneath the grate whose temperature is higher than that of the burning wood at the top of the stack. Explain how this device provides more heat to the room than a conventional fire does and thus increases the efficiency of the fireplace.
13. What physical process described in this chapter might reasonably be called "the inverse photoelectric effect"? Can you account for this process classically or must it be accounted for by viewing light as a collection of many little particles each with energy hf? Explain.

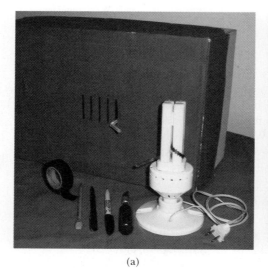

(a)

(b)

Figure Q3.15

14. In the photoelectric effect, if the intensity of incident light is very low, then the number of photons per second striking the metal surface will be small and the probability per second of electron emission *per surface atom* will also be small. Account for the observed instantaneous emission of photoelectrons under these conditions.

15. *Blacker than black, brighter than white.* (a) Take a large, closed, empty cardboard box. Cut a slot a few millimeters wide in one side. Use black pens, markers, and soot to make some stripes next to the slot, as shown in Figure Q3.15a. Inspect them with care and choose which is blackest—the figure does not show enough contrast to reveal which it is. Explain why it is blackest. (b) Locate an intricately shaped compact fluorescent light fixture, as in Figure Q3.15b. Look at it through dark glasses and describe where it appears brightest. Explain why it is brightest there. *Suggestion:*

Gustav Kirchhoff, professor at Heidelberg and master of the obvious, gave the same answer to part (a) as you likely will. His answer to part (b) would begin like this: When electromagnetic radiation falls on its surface, an object reflects some fraction r of the energy and absorbs the rest. Whether the fraction reflected is 0.8 or 0.001, the fraction absorbed is $a = 1 - r$. Suppose the object and its surroundings are at the same temperature. The energy the object absorbs joins its fund of internal energy, but the second law of thermodynamics implies that the absorbed energy cannot raise the object's temperature. It does not produce a temperature increase because the object's energy budget has one more term: energy radiated . . . . You still have to make the observations and answer questions (a) and (b), but you can incorporate some of Kirchhoff's ideas into your answer if you wish. *(Alexandra Héder)*

PROBLEMS

3.1 Light as an Electromagnetic Wave

1. *Classical Zeeman effect or the triumph of Maxwell's equations!* As pointed out in Section 3.1, Maxwell's equations may be used to predict the change in emission frequency when gas atoms are placed in a magnetic field. Consider the situation shown in Figure P3.1. Note that the application of a magnetic field perpendicular to the orbital plane of the electron induces an electric field, which changes the direction of the velocity vector. (a) Using

$$\oint \mathbf{E} \cdot d\mathbf{s} = -\frac{d\Phi_B}{dt}$$

show that the magnitude of the electric field is given by

$$E = \frac{r}{2}\frac{dB}{dt}$$

(b) Using $F\,dt = m\,dv$, calculate the change in speed, Δv, of the electron. Show that if r remains constant,

$$\Delta v = \frac{erB}{2m_e}$$

(c) Find the change in angular frequency, $\Delta\omega$, of the electron and calculate the numerical value of $\Delta\omega$ for B equal to 1 T. Note that this is also the change in frequency of the light emitted according to Maxwell's equations. Find the fractional change in frequency, $\Delta\omega/\omega$, for an ordinary emission line of 500 nm. (d) Actually, the original emission line at ω_0 is split into three components at $\omega_0-\Delta\omega$, ω_0, and $\omega_0 + \Delta\omega$. The line at

$\omega_0 + \Delta\omega$ is produced by atoms with electrons rotating as shown in Figure P3.1, whereas the line at $\omega_0 - \Delta\omega$ is produced by atoms with electrons rotating in the opposite sense. The line at ω_0 is produced by atoms with electronic planes of rotation oriented parallel to **B**. Explain.

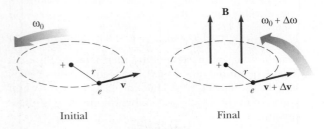

Figure P3.1

3.2 Blackbody Radiation

2. The temperature of your skin is approximately 35°C. What is the wavelength at which the peak occurs in the radiation emitted from your skin?

3. A 2.0-kg mass is attached to a massless spring of force constant $k = 25$ N/m. The spring is stretched 0.40 m from its equilibrium position and released. (a) Find the total energy and frequency of oscillation according to classical calculations. (b) Assume that the energy is quantized and find the quantum number, n, for the sys-

tem. (c) How much energy would be carried away in a 1-quantum change?

4. (a) Use Stefan's law to calculate the total power radiated per unit area by a tungsten filament at a temperature of 3000 K. (Assume that the filament is an ideal radiator.) (b) If the tungsten filament of a lightbulb is rated at 75 W, what is the surface area of the filament? (Assume that the main energy loss is due to radiation.)

5. Consider the problem of the distribution of black-body radiation described in Figure 3.3. Note that as *T increases*, the wavelength λ_{max} at which $u(\lambda, T)$ reaches a maximum shifts toward shorter wavelengths. (a) Show that there is a general relationship between temperature and λ_{max} stating that $T\lambda_{max}$ = constant (Wien's displacement law). (b) Obtain a numerical value for this constant. (*Hint:* Start with Planck's radiation law and note that the slope of $u(\lambda, T)$ is zero when $\lambda = \lambda_{max}$.)

6. *Planck's fundamental constant, h.* Planck ultimately realized the great and fundamental importance of *h*, which, much more than a curve-fitting parameter, is actually the measure of all quantum phenomena. In fact, Planck suggested using the universal constants *h*, *c* (the velocity of light), and *G* (Newton's gravitational constant) to construct "natural" or universal units of length, time, and mass. He reasoned that the current units of length, time, and mass were based on the accidental size, motion, and mass of our particular planet, but that truly universal units should be based on the quantum theory, the speed of light in a vacuum, and the law of gravitation—all of which hold anywhere in the universe and at all times. Show that the expressions $\left(\dfrac{hG}{c^3}\right)^{1/2}$, $\left(\dfrac{hG}{c^5}\right)^{1/2}$, and $\left(\dfrac{hc}{G}\right)^{1/2}$ have dimensions of length, time, and mass and find their numerical values. These quantities are called, respectively, the Planck length, the Planck time, and the Planck mass. Would you care to speculate on the physical meaning of these quantities?

3.3   Derivation of the Rayleigh–Jeans Law and Planck's Law (Optional)

7. *Density of modes.* The essentials of calculating the number of modes of vibration of waves confined to a cavity may be understood by considering a one-dimensional example. (a) Calculate the number of modes (standing waves of different wavelength) with wavelengths between 2.0 cm and 2.1 cm that can exist on a string with fixed ends that is 2 m long. (*Hint:* use $n(\lambda/2) = L$, where $n = 1, 2, 3, 4, 5 \ldots$. Note that a specific value of *n* defines a specific mode or standing wave with different wavelength.) (b) Calculate, in analogy to our three-dimensional calculation, the number of modes per unit

wavelength per unit length, $\dfrac{\Delta n}{L\,\Delta\lambda}$. (c) Show that in general the number of modes per unit wavelength per unit length for a string of length *L* is given by

$$\frac{1}{L}\left|\frac{dn}{d\lambda}\right| = \frac{2}{\lambda^2}$$

Does this expression yield the same numerical answer as found in (a)? (d) Under what conditions is it justified to replace $\left|\left(\dfrac{\Delta n}{L\,\Delta\lambda}\right)\right|$ with $\left|\left(\dfrac{dn}{L\,d\lambda}\right)\right|$? Is the expression $n = 2L/\lambda$ a continuous function?

3.4   Light Quantization and the Photoelectric Effect

8. Calculate the energy of a photon whose frequency is (a) 5×10^{14} Hz, (b) 10 GHz, (c) 30 MHz. Express your answers in electron volts.

9. Determine the corresponding wavelengths for the photons described in Problem 8.

10. An FM radio transmitter has a power output of 100 kW and operates at a frequency of 94 MHz. How many photons per second does the transmitter emit?

11. The average power generated by the Sun has the value 3.74×10^{26} W. Assuming the average wavelength of the Sun's radiation to be 500 nm, find the number of photons emitted by the Sun in 1 s.

12. A sodium-vapor lamp has a power output of 10 W. Using 589.3 nm as the average wavelength of the source, calculate the number of photons emitted per second.

13. The photocurrent of a photocell is cut off by a retarding potential of 2.92 V for radiation of wavelength 250 nm. Find the work function for the material.

14. The work function for potassium is 2.24 eV. If potassium metal is illuminated with light of wavelength 350 nm, find (a) the maximum kinetic energy of the photoelectrons and (b) the cutoff wavelength.

15. Molybdenum has a work function of 4.2 eV. (a) Find the cutoff wavelength and threshold frequency for the photoelectric effect. (b) Calculate the stopping potential if the incident light has a wavelength of 200 nm.

16. When cesium metal is illuminated with light of wavelength 300 nm, the photoelectrons emitted have a maximum kinetic energy of 2.23 eV. Find (a) the work function of cesium and (b) the stopping potential if the incident light has a wavelength of 400 nm.

17. Consider the metals lithium, beryllium, and mercury, which have work functions of 2.3 eV, 3.9 eV, and 4.5 eV, respectively. If light of wavelength 300 nm is incident on each of these metals, determine (a) which metals exhibit the photoelectric effect and (b) the

maximum kinetic energy for the photoelectron in each case.

18. Light of wavelength 500 nm is incident on a metallic surface. If the stopping potential for the photoelectric effect is 0.45 V, find (a) the maximum energy of the emitted electrons, (b) the work function, and (c) the cutoff wavelength.

19. The active material in a photocell has a work function of 2.00 eV. Under reverse-bias conditions (where the polarity of the battery in Figure 3.14 is reversed), the cutoff wavelength is found to be 350 nm. What is the value of the bias voltage?

20. A light source of wavelength λ illuminates a metal and ejects photoelectrons with a maximum kinetic energy of 1.00 eV. A second light source with half the wavelength of the first ejects photoelectrons with a maximum kinetic energy of 4.00 eV. What is the work function of the metal?

21. Figure P3.21 shows the stopping potential versus incident photon frequency for the photoelectric effect for sodium. Use these data points to find (a) the work function, (b) the ratio h/e, and (c) the cutoff wavelength. (d) Find the percent difference between your answer to (b) and the accepted value. (Data taken from R. A. Millikan, *Phys. Rev.*, 7:362, 1916.)

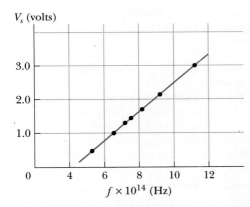

Figure P3.21   Some of Millikan's original data for sodium.

22. Photons of wavelength 450 nm are incident on a metal. The most energetic electrons ejected from the metal are bent into a circular arc of radius 20 cm by a magnetic field whose strength is equal to 2.0×10^{-5} T. What is the work function of the metal?

2.5   The Compton Effect and X-Rays

23. Calculate the energy and momentum of a photon of wavelength 500 nm.

24. X-rays of wavelength 0.200 nm are scattered from a block of carbon. If the scattered radiation is detected at 90° to the incident beam, find (a) the Compton shift, $\Delta\lambda$, and (b) the kinetic energy imparted to the recoiling electron.

25. X-rays with an energy of 300 keV undergo Compton scattering from a target. If the scattered rays are detected at 30° relative to the incident rays, find (a) the Compton shift at this angle, (b) the energy of the scattered x-ray, and (c) the energy of the recoiling electron.

26. X-rays with a wavelength of 0.040 nm undergo Compton scattering. (a) Find the wavelength of photons scattered at angles of 30°, 60°, 90°, 120°, 150°, 180°, and 210°. (b) Find the energy of the scattered electrons corresponding to these scattered x-rays. (c) Which one of the given scattering angles provides the electron with the greatest energy?

27. Show that a photon cannot transfer all of its energy to a free electron. (*Hint:* Note that energy and momentum must be conserved.)

28. In the Compton scattering event illustrated in Figure 3.24, the scattered photon has an energy of 120 keV and the recoiling electron has an energy of 40 keV. Find (a) the wavelength of the incident photon, (b) the angle θ at which the photon is scattered, and (c) the recoil angle ϕ of the electron.

29. Gamma rays (high-energy photons) of energy 1.02 MeV are scattered from electrons that are initially at rest. If the scattering is *symmetric*, that is, if $\theta = \phi$ in Figure 3.24, find (a) the scattering angle θ and (b) the energy of the scattered photons.

30. If the maximum energy given to an electron during Compton scattering is 30 keV, what is the wavelength of the incident photon? (*Hint:* What is the scattering angle for maximum energy transfer?)

31. A photon of initial energy 0.1 MeV undergoes Compton scattering at an angle of 60°. Find (a) the energy of the scattered photon, (b) the recoil kinetic energy of the electron, and (c) the recoil angle of the electron.

32. An excited iron (Fe) nucleus (mass 57 u) decays to its ground state with the emission of a photon. The energy available from this transition is 14.4 keV. (a) By how much is the photon energy reduced from the full 14.4 keV as a result of having to share energy with the recoiling atom? (b) What is the wavelength of the emitted photon?

33. Show that the Compton formula

$$\lambda' - \lambda_0 = \frac{h}{m_e c}(1 - \cos\theta)$$

results when expressions for the electron energy (Equation 3.33) and momentum (Equation 3.34) are substituted into the relativistic energy expression,

$$E_e^2 = p_e^2 c^2 + m_e^2 c^4$$

34. Find the energy of an x-ray photon that can impart a maximum energy of 50 keV to an electron by Compton collision.

35. Compton used photons of wavelength 0.0711 nm. (a) What is the energy of these photons? (b) What is the wavelength of the photons scattered at an angle of 180° (backscattering case)? (c) What is the energy of the backscattered photons? (d) What is the recoil energy of the electrons in this case?

36. A photon undergoing Compton scattering has an energy after scattering of 80 keV, and the electron recoils with an energy of 25 keV. (a) Find the wavelength of the incident photon. (b) Find the angle at which the photon is scattered. (c) Find the angle at which the electron recoils.

37. X-radiation from a molybdenum target (0.626 Å) is incident on a crystal with adjacent atomic planes spaced 4.00×10^{-10} m apart. Find the three smallest angles at which intensity maxima occur in the diffracted beam.

38. As a single crystal is rotated in an x-ray spectrometer (Fig. 3.22a), many parallel planes of atoms besides AA and BB produce strong diffracted beams. Two such planes are shown in Figure P3.38. (a) Determine geometrically the interplanar spacings d_1 and d_2 in terms of d_0. (b) Find the angles (with respect to the surface plane AA) of the $n = 1$, 2, and 3 intensity maxima from planes with spacing d_1. Let $\lambda = 0.626$ Å and $d_0 = 4.00$ Å. Note that a given crystal structure (for example, cubic) has interplanar spacings with characteristic ratios, which produce characteristic diffraction patterns. In this way, measurement of the angular position of diffracted x-rays may be used to infer the crystal structure.

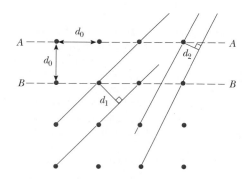

Figure P3.38 Atomic planes in a cubic lattice.

39. *The determination of Avogadro's number with x-rays.* X-rays from a molybdenum target (0.626 Å) are incident on an NaCl crystal, which has the atomic arrangement shown in Figure P3.39. If NaCl has a density of 2.17 g/cm³ and the $n = 1$ diffraction maximum from planes separated by d is found at $\theta = 6.41°$, compute

Avogadro's number. (*Hint:* First determine d. Using Figure P3.39, determine the number of NaCl molecules per primitive cell and set the mass per unit volume of the primitive cell equal to the density.)

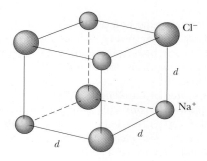

Figure P3.39 The primitive cell of NaCl.

3.7 Does Gravity Affect Light? (Optional)

40. In deriving expressions for the change in frequency of a photon falling or rising in a gravitational field, we have assumed a small change in frequency and a constant photon mass of hf/c^2. Suppose that a star is so dense that Δf is not small. (a) Show that f', the photon frequency at ∞, is related to f, the photon frequency at the star's surface, by

$$f' = fe^{-GM_s/R_s c^2}$$

(b) Show that this expression reduces to Equation 3.39 for small M_s/R_s. (*Hint:* The decrease in photon energy, $h\,df$, as the photon moves dr away from the star is equal to the work done against gravity, $F_G\,dr$.)

41. If the Sun were to contract and become a black hole, (a) what would its approximate radius be and (b) by what factor would its density increase?

Additional Problems

42. Two light sources are used in a photoelectric experiment to determine the work function for a particular metal surface. When green light from a mercury lamp ($\lambda = 546.1$ nm) is used, a retarding potential of 1.70 V reduces the photocurrent to zero. (a) Based on this measurement, what is the work function for this metal? (b) What stopping potential would be observed when using the yellow light from a helium discharge tube ($\lambda = 587.5$ nm)?

43. In a Compton collision with an electron, a photon of violet light ($\lambda = 4000$ Å) is backscattered through an angle of 180°. (a) How much energy (eV) is transferred to the electron in this collision? (b) Compare your result with the energy this electron would acquire in a photoelectric process with the same photon. (c) Could

violet light eject electrons from a metal by Compton collision? Explain.

44. Ultraviolet light is incident normally on the surface of a certain substance. The work function of the electrons in this substance is 3.44 eV. The incident light has an intensity of 0.055 W/m^2. The electrons are photoelectrically emitted with a maximum speed of $4.2 \times 10^5 \text{ m/s}$. What is the maximum number of electrons emitted from a square centimeter of the surface per second? Pretend that none of the photons are reflected or heat the surface.

45. The following data are found for photoemission from calcium:

$\lambda \ (nm)$	253.6	313.2	365.0	404.7
$V_s \ (V)$	1.95	0.98	0.50	0.14

Plot V_s versus f, and from the graph obtain Planck's constant, the threshold frequency, and the work function for calcium.

46. A 0.500-nm x-ray photon is deflected through 134° in a Compton scattering event. At what angle (with respect to the incident beam) is the recoiling electron found?

47. An electron initially at rest recoils from a head-on collision with a photon. Show that the kinetic energy acquired by the electron is $2hf\alpha/(1 + 2\alpha)$, where α is the ratio of the photon's initial energy to the rest energy of the electron.

48. In a Compton scattering experiment, an x-ray photon scatters through an angle of 17.4° from a free electron that is initially at rest. The electron recoils with a speed of 2180 km/s. Calculate (a) the wavelength of the incident photon and (b) the angle through which the electron scatters.

4

The Particle Nature
of Matter

Chapter Outline

In Chapter 3 we reviewed the evidence for the wave nature of electromagnetic radiation and dealt with major experimental puzzles of the first quarter of the 20th century, which required a particle-like behavior of radiation for their solution. In particular, we discussed Planck's revolutionary idea of energy quantization of oscillators in the walls of a perfect radiator, Einstein's extension of energy quantization to light in the photoelectric effect, and Compton's further confirmation of the existence of the photon as a particle carrying momentum in x-ray scattering experiments.

In this chapter we shall examine the evidence for the particle nature of matter. We only mention briefly the early atomists and concentrate instead on the developments from 1800 onward that dealt with the composition of atoms. In particular, we review the ingenious and fascinating experiments that led to the discoveries of the electron, the proton, the nucleus, and the important Rutherford–Bohr planetary model of the atom.

4.1 THE ATOMIC NATURE OF MATTER

To say that the world is made up of atoms is, today, commonplace. Because the atomic picture of reality is often accepted without question, students can miss out on the rich and fascinating story of how atoms were shown to be real. The discovery and proof of the graininess of the world seem especially fascinating for two reasons. First, because of the size of individual atoms, measurements of atomic properties are usually indirect and necessarily involve clever manipulations of large-scale measurements to infer properties of microscopic particles. Second, the historical evolution of ideas about atomicity shows clearly the real way in which science progresses. This progression is often nonlinear and involves an interdependence of physics, chemistry, and mathematics, and the convergence of many different lines of investigation.

There is also an exalted romance in honoring the great atomists who were able to pick out organizing principles from the confusing barrage of marketplace ideas of their time: Democritus and Leucippus, who speculated that the unchanging substratum of the world was atoms in motion; the debonair French chemist Lavoisier and his wife (see Fig. 4.1), who established the conservation of matter in many careful chemical experiments; Dalton, who perceived the atomicity of nature in the law of multiple proportions of compounds; Avogadro, who in a most obscure and little-appreciated paper, postulated that all pure gases at the same temperature and pressure have the same number of molecules per unit volume; and Maxwell,[1] who showed with his molecular-kinetic theory of gases how macroscopic quantities, such as pressure and temperature, could be derived from averages over distributions

Figure 4.1   Antoine Lavoisier (French chemist, 1743–1794) and Madame Lavoisier who together established the principle of conservation of mass in chemical reactions. In this painting they appear to have matters other than chemistry on the mind. (*© Bettmann/CORBIS*)

---

[1]Maxwell was a genius twice over. Either his theory of electricity and magnetism or his kinetic theory of gases would qualify him for that rank.

of molecular properties. The list could run on and on. We abbreviate it by naming Jean Perrin and the ubiquitous Albert Einstein,[2] who carried on very important theoretical and experimental work concerning Brownian motion, the zigzag movement of small suspended particles caused by molecular impacts. Their work produced additional confirmation of the atomic-molecular hypothesis and resulted in improved values of Avogadro's number as late as the early 1900s.

4.2   THE COMPOSITION OF ATOMS

We now turn our attention to answering the rather dangerous question, "If matter is primarily composed of atoms, what are atoms composed of?" Again, we can point to some primary discoveries that showed that atoms are composed of light, negatively charged particles orbiting a heavy, positively charged nucleus. These were

- *The discovery of the law of electrolysis in 1833 by Michael Faraday.* Through careful experimental work on electrolysis, Faraday showed that the mass of an element liberated at an electrode is directly proportional to the charge transferred and to the atomic weight of the liberated material but is inversely proportional to the valence of the freed material.
- *The identification of cathode rays as electrons and the measurement of the charge-to-mass ratio (e/m_e) of these particles by Joseph John (J. J.) Thomson in 1897.* Thomson measured the properties of negative particles emitted from different metals and found that the value of e/m_e was always the same. He thus came to the conclusion that the electron is a constituent of all matter!
- *The precise measurement of the electronic charge (e) by Robert Millikan in 1909.* By combining his result for (e) with Thomson's e/m_e value, Millikan showed unequivocally that particles about 1000 times less massive than the hydrogen atom exist.
- *The establishment of the nuclear model of the atom by Ernest Rutherford and coworkers Hans Geiger and Ernest Marsden in 1913.* By scattering fast-moving α particles (charged nuclei of helium atoms emitted spontaneously in radioactive decay processes) from metal foil targets, Rutherford established that atoms consist of a compact positively charged nucleus (diameter $\approx 10^{-14}$ m) surrounded by a swarm of orbiting electrons (electron cloud diameter $\approx 10^{-10}$ m).

Let us describe these developments in more detail. We start with a brief example of Faraday's experiments, in particular the electrolysis of molten common salt (NaCl). Faraday found that if 96,500 C of charge (1 faraday) is passed through such a molten solution, 23.0 g of Na will deposit on the cathode and 35.5 g of chlorine gas will bubble off the anode (Fig. 4.2). In this case, exactly 1 gram atomic weight or mole of each element is released because both are monovalent. For divalent and trivalent elements, exactly $\frac{1}{2}$ and $\frac{1}{3}$ of a mole, respectively, would be released. As expected, doubling the

---

[2]Much of Einstein's earliest work was concerned with the molecular analysis of solutions and determinations of molecular radii and Avogadro's number. See A. Pais, "*Subtle is the Lord . . .*" *The Science and the Life of Albert Einstein*, New York, Oxford University Press, 1982, Chapter 5.

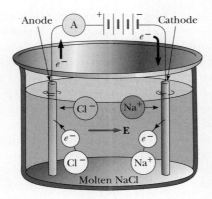

Figure 4.2   Electrolysis of molten NaCl.

quantity of charge passed doubles the mass of the neutral element liberated. Faraday's results may be given in equation form as

$$m = \frac{(q)\,(\text{molar mass})}{(96,500\ \text{C})\,(\text{valence})}$$

(4.1)   **Faraday's law of electrolysis**

where m is the mass of the liberated substance in grams, q is the total charge passed in coulombs, the molar mass is in grams, and the valence is dimensionless.

EXAMPLE 4.1   The Electrolysis of BaCl$_2$

How many grams of barium and chlorine (cough!) will you get if you pass a current of 10.0 A through molten BaCl$_2$ for 1 h? Barium has a molar weight of 137 g and a valence of 2. Chlorine has a molar weight of 35.5 g and a valence of 1.

Solution   Using Equation 4.1 and $q = It$, where I is the current and t is the time, we have

$$m_{\text{Ba}} = \frac{(q)\,(\text{molar mass})}{(96,500\ \text{C})\,(\text{valence})}$$

$$= \frac{(10.0\ \text{C/s})\,(3600\ \text{s})\,(137\ \text{g})}{(96,500\ \text{C})\,(2)} = 25.6\ \text{g}$$

$$m_{\text{Cl}} = \frac{(10.0\ \text{C/s})\,(3600\ \text{s})\,(35.5\ \text{g})}{(96,500\ \text{C})\,(1)} = 13.2\ \text{g}$$

Faraday's law of electrolysis is explained in terms of an atomic picture shown in Figure 4.2. Charge passes through the molten solution in the form of ions, which possess an excess or deficiency of one or more electrons. Under the influence of the electric field produced by the battery, these ions move to the anode or cathode, where they respectively lose or gain electrons and are liberated as neutral atoms.

Although it was far from clear in 1833, Faraday's law of electrolysis confirmed three important parts of the atomic picture. First, it offered proof that matter consists of molecules and that molecules consist of atoms. Second, it showed that charge is quantized, because only integral numbers of charges are transferred at the electrodes. Third, it showed that the subatomic parts of atoms are positive and negative charges, although the mass and the size of the charge of these subatomic particles remained unknown.

Figure 4.3 J. J. Thomson. (*AIP Emilio Segrè Visual Archives/W. F. Meggers Collection*)

The next major step explaining the composition of atoms was taken by Joseph John (J. J.) Thomson (see Figure 4.3). His discovery in 1897[3] that the "rays" seen in low-pressure gas discharges were actually caused by negative particles (electrons) ended a debate dating back nearly 30 years: Were cathode rays material particles or waves? Contrary to our rather blasé present acceptance of the electron, many of Thomson's distinguished contemporaries responded with utter disbelief to the idea that electrons were a constituent of all matter. Much of the opposition to Thomson's discovery stemmed from the fact that it required the abandonment of the recently established concept of the atom as an indivisible entity. Thomson's discovery of the electron disturbed this newly established order in atomic theory and provoked startling new developments—Rutherford's nuclear model and the first satisfactory theory of the emission of light by atomic systems, the Bohr model of the atom.

Figure 4.4 shows the original vacuum tube used by Thomson in his e/m_e experiments. Figure 4.5 shows the various parts of the Thomson apparatus for easy reference. Electrons are accelerated from the cathode to the anode, collimated by slits in the anodes, and then allowed to drift into a region of crossed (perpendicular) electric and magnetic fields. The simultaneously applied **E** and **B** fields are first adjusted to produce an undeflected beam. If the **B** field is then turned off, the **E** field alone produces a measurable beam deflection on the phosphorescent screen. From the size of the deflection and the measured values of **E** and **B**, the charge-to-mass ratio, e/m_e, may be determined. The truly ingenious feature of this experiment is the manner in which Thomson measured v_x, the horizontal velocity component of the beam. He did this by balancing the magnetic and electric forces. In effect, he created a *velocity selector*, which could select out of the beam those particles having a velocity within a narrow range of values. This device was extensively used in the first quarter

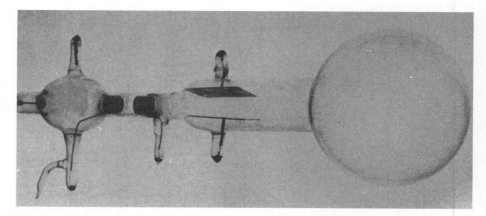

Figure 4.4 The original e/m_e tube used by J. J. Thomson. (*After Figure 1.3, p. 7, R. L. Sproull and W. A. Phillips*, Modern Physics, *3rd ed., New York, John Wiley & Sons, 1980*).

[3]J. J. Thomson, *Phil. Mag.* 44:269, 1897.

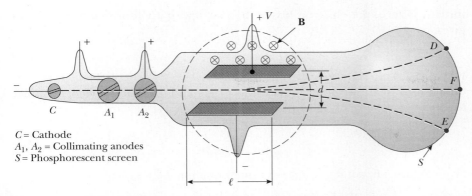

Figure 4.5   A diagram of Thomson's e/m_e tube (patterned after J. J. Thomson, *Philosophical Magazine* (5)44:293, 1897). Electrons subjected to an electric field alone land at *D*, while those subjected to a magnetic field alone land at *E*. When both electric and magnetic fields are present and properly adjusted, the electrons experience no net deflection and land at *F*.

of the 20th century in charge-to-mass measurements (q/m) on many particles and in early mass spectrometers.

To gain a clearer picture of the Thomson experiment, let us analyze the electron's motion in his apparatus. Figure 4.6 shows the trajectory of a beam of negative particles entering the **E** and **B** field regions with horizontal velocity v_x. Consider first only an **E** field between the plates. For this case, v_x remains constant throughout the motion because there is no force acting in the *x* direction. The *y* component of velocity, v_y, is constant everywhere except between the plates, where the electron experiences a constant upward acceleration due to the electric force and follows a parabolic path. To solve for the deflection angle, θ, we must solve for v_x and v_y. Because v_y

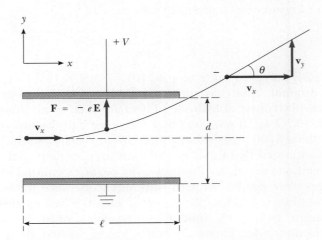

Figure 4.6   Deflection of negative particles by an electric field.

initially is zero, the electron leaves the plates with a y component of velocity given by

$$v_y = a_y t \qquad (4.2)$$

Because $a_y = F/m_e = Ee/m_e = Ve/m_e d$, and $t = \ell/v_x$, where d and ℓ are the dimensions of the region between the plates and V is the applied potential, we obtain

$$v_y = \frac{V\ell e}{m_e v_x d} \qquad (4.3)$$

From Figure 4.6, $\tan \theta = v_y/v_x$, so using Equation 4.3 we obtain

$$\tan \theta = \frac{V\ell}{v_x^2 d} \left(\frac{e}{m_e} \right) \qquad (4.4)$$

Assuming small deflections, $\tan \theta \approx \theta$, so we have

$$\theta \approx \frac{V\ell}{v_x^2 d} \left(\frac{e}{m_e} \right) \qquad (4.5)$$

Note that θ, the beam deflection, V, the voltage applied to the horizontal deflecting plates, and d and ℓ, the spacing and length, respectively, of the horizontal deflecting plates can all be measured. Hence, one only needs to measure v_x to determine e/m_e. Thomson determined v_x by applying a **B** field and adjusting its magnitude to just balance the deflection of the still present **E** field. Equating the magnitudes of the electric and magnetic forces gives

$$qE = qv_x B$$

or

$$v_x = \frac{E}{B} = \frac{V}{Bd} \qquad (4.6)$$

Substituting Equation 4.6 into Equation 4.5 immediately yields a formula for e/m_e entirely in terms of measurable quantities:

$$\frac{e}{m_e} = \frac{V\theta}{B^2 \ell d} \qquad (4.7)$$

The currently accepted value of e/m_e is 1.758803×10^{11} C/kg. Although Thomson's original value was only about 1.0×10^{11} C/kg, prior experiments on the electrolysis of hydrogen ions had given q/m values for hydrogen of about 10^8 C/kg. It was clear that Thomson had discovered a particle with a mass about 1000 times smaller than the smallest atom! In his observations, Thomson noted that the e/m_e ratio was independent of the discharge gas and the cathode metal. Furthermore, the particles emitted when electrical discharges were passed through different gases were found to be the same as those observed in the photoelectric effect. Based on these observations, Thomson concluded that these particles must be a universal constituent of all matter. Humanity had achieved its first glimpse into the subatomic world!

EXAMPLE 4.2   Deflection of an Electron Beam by E and B Fields

Using the accepted e/m_e value, calculate the magnetic field required to produce a deflection of 0.20 rad in Thomson's experiment, assuming the values $V = 200$ V, $\ell = 5.0$ cm, and $d = 1.5$ cm (the approximate values used by Thomson). Compare this value of B to the Earth's magnetic field.

Solution   Because $e/m_e = V\theta/B^2\ell d$, solving for B gives

$$B = \sqrt{\frac{V\theta}{\ell d(e/m_e)}}$$

so

$$B = \left[\frac{(200\text{ V})(0.20\text{ rad})}{(0.050\text{ m})(0.015\text{ m})(1.76 \times 10^{11}\text{ C/kg})} \right]^{1/2}$$

$$= [3.03 \times 10^{-7}\text{ V}\cdot\text{kg/m}^2\text{ C}]^{1/2}$$

$$= [3.03 \times 10^{-7}\text{ N}^2/(\text{m/s})^2\text{ C}^2]^{1/2}$$

$$= 5.5 \times 10^{-4}\text{ N}/(\text{m/s})\cdot\text{C} = 5.5 \times 10^{-4}\text{ T}$$

As the Earth's magnetic field has a magnitude of about 0.5×10^{-4} T, we require a field 11 times as strong as the Earth's field.

Exercise 1   Find the horizontal speed v_x for this case.

Answer 2.4×10^7 m/s $= 0.080c$, where c is the speed of light.

Millikan's Value of the Elementary Charge

In 1897, Thomson had been unable to determine e or m_e separately. However, about two years later this great British experimentalist had bracketed the accepted value of e (1.602×10^{-19} C) with values of 2.3×10^{-19} C for charges emitted from zinc illuminated by ultraviolet light and 1.1×10^{-19} C for charges produced by ionizing x rays and radium emissions. He was also able to conclude that "e is the same in magnitude as the charge carried by the hydrogen atom in the electrolysis of solutions." The technique used by Thomson and his students to measure e is especially interesting because it represents the first use of the cloud chamber technique in physics and also formed the starting point for the famous Millikan oil-drop experiment. Charles Wilson, one of Thomson's students, had discovered that ions act as condensation centers for water droplets when damp air is cooled by a rapid expansion. Thomson used this idea to form charged clouds by using the apparatus shown in Figure 4.7a. Here Q is the measured total charge of the cloud, W is the measured weight of the cloud, and v is the rate of fall or terminal speed. Thomson assumed that the cloud was composed of spherical droplets having a constant mass (no evaporation) and that the magnitude of the drag force D on a single falling droplet was given by Stokes's law,

$$D = 6\pi a\eta v \tag{4.8}$$

where a is the droplet radius, η is the viscosity of air, and v is the terminal speed of the droplet. The following procedure was used to find a and w, the weight of a single drop. Because v is constant, the droplet is in equilibrium under the combined action of its weight, w, and the drag force, D, as shown in Figure 4.7b. Hence, we require that $w = D$, or

$$w = \tfrac{4}{3}\pi a^3 \rho g = D = 6\pi a\eta v$$

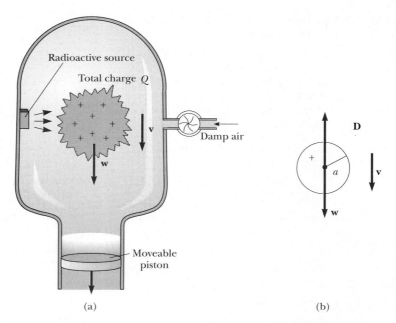

Radioactive source

Total charge Q

Damp air

w

v

Moveable piston

(a)

D

v

w

a

(b)

Figure 4.7 (a) A diagram of Thomson's apparatus for determining e. (b) A single droplet in the cloud.

so

$$a = \sqrt{\frac{9\eta v}{2\rho g}} \qquad (4.9)$$

where ρ is the mass density of the droplet and g is the free-fall acceleration. From the droplet radius and the known density we can find w, the weight. Once w is obtained, the number of drops n (or number of ions) is given by W/w and the electronic charge e is equal to Q/n, assuming that each droplet carries only one electronic charge. Although ingenious, this method is inaccurate because the theory applies only to a single particle and the particles are all assumed to be identical in order to compare the theory to experiments performed on a cloud.

The tremendous advance of Millikan was made possible by his clever idea of making the experiment "fit" the theory. By observing single droplets he eliminated the problems of assuming all particles to be identical and of making uncertain measurements on a cloud. Millikan's basic idea was to measure the rate of fall of a single drop acted on by gravity and drag forces, apply Stokes's law to determine the drop radius and mass, then to measure its upward velocity in an opposing electric field, and hence determine the total charge on an individual drop.[4] A schematic of the Millikan apparatus is shown in Figure 4.8. Oil droplets charged by an atomizer are allowed to pass through a small hole in the upper plate of a parallel-plate capacitor. If these droplets are illuminated from the side, they appear as brilliant stars against a dark background, and the rate of fall of

---

[4]Actually, the idea of allowing charges to "fall" under a combined gravitational and electric field was first applied to charged clouds of water vapor by H. A. Wilson in 1903. Millikan switched from water to oil to avoid the problems of a changing droplet mass and radius caused by water evaporation.

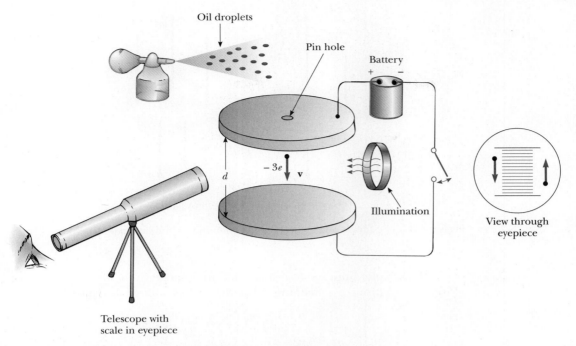

Figure 4.8   A schematic view of the Millikan oil-drop apparatus.

individual drops may be determined.[5] If an electrostatic field of several thousand volts per meter is applied to the capacitor plates, the drop may move slowly upward, typically at rates of *hundredths* of a centimeter per second. Because the rate of fall is comparable, a single droplet with constant mass and radius may be followed for hours, alternately rising and falling, by simply turning the electric field on and off. The atomicity of charge is shown directly by the observation that after a long series of measurements of constant upward velocities one observes a discontinuous change or jump to a different *upward* velocity (higher or lower). This discontinuous change is caused by the attraction of an ion to the charged droplet and a consequent change in droplet charge. Such changes become more frequent when a source of ionizing radiation is placed between the plates.

The quantitative analysis of the Millikan experiment starts with Newton's second law applied to the oil drop, $\Sigma F_y = ma_y$. Because the drag force D is large, a constant velocity of fall is quickly achieved, and all measurements are made for the case $a_y = 0$, or $\Sigma F_y = 0$. If we assume that the magnitude of the drag force is proportional to the speed ($D = Cv$), and refer to Figure 4.9, we find

Millikan's determination of the electronic charge

$$Cv - mg = 0 \qquad \text{(field off)}$$

$$q_1 E - mg - Cv_1' = 0 \qquad \text{(field on)}$$

Eliminating C from these expressions gives

$$q_1 = \frac{mg}{E}\left(\frac{v + v_1'}{v}\right) \qquad (4.10)$$

---

[5]Perhaps the reason for the failure of "Millikan's Shining Stars" as a poetic and romantic image has something to do with the generations of physics students who have experienced hallucinations, near blindness, migraine attacks, etc. while repeating his experiment!

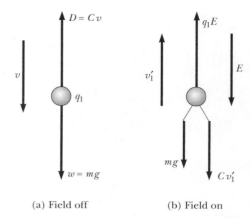

(a) Field off   (b) Field on

Figure 4.9   The forces on a charged oil droplet in the Millikan experiment.

When the droplet undergoes a discontinuous change in its *upward speed* from v_1' to v_2' (m, g, E, and v remaining constant), its new charge q_2 is given by

$$q_2 = \frac{mg}{E}\left(\frac{v + v_2'}{v}\right) \tag{4.11}$$

Dividing Equation 4.10 by Equation 4.11 gives

$$\frac{q_1}{q_2} = \frac{v + v_1'}{v + v_2'} \tag{4.12}$$

Robert Millikan (1868–1953). Although Millikan *(left)* studied Greek as an undergraduate at Oberlin College, he fell in love with physics during his graduate training after teaching school for a few years. He was the first student to receive a Ph.D. in physics from Columbia University in 1895. Following postdoctoral work in Germany under Planck and Nernst, Millikan obtained an academic appointment at the University of Chicago in 1910, where he worked with Michelson. He received the Nobel prize in 1923 for his famous experimental determination of the electronic charge. He is also remembered for his careful experimental work to verify the theory of the photoelectric effect deduced by Einstein. Following World War I, he transferred to the California Institute of Technology, where he worked in atmospheric physics and remained until his retirement. (*Courtesy AIP Emilio Segrè Visual Archives*)

Equation 4.12 constitutes a remarkably direct and powerful proof of the quantization of charge, because if successive speed ratios are ratios of whole numbers, successive charges on the drop must be multiples of the same elementary charge! Millikan's experimental measurements of speed ratios beautifully confirmed this quantization of charge to within about 1% accuracy.[6]

Up to this point our arguments have been quite general and have assumed only that the drag force on the droplet is proportional to its velocity. To determine the actual value of the electronic charge, e, the mass of the drop must be determined, as can be seen from Equation 4.10. As noted earlier, the droplet radius a may be determined from the application of Stokes's law. This value of a, in turn, can be used to find m from the oil density, ρ. In this procedure, a is

$$a = \sqrt{\frac{9\eta v}{2\rho g}} \tag{4.9}$$

and the mass of the droplet can be expressed as

$$m = \rho \cdot \text{volume} = \rho \tfrac{4}{3}\pi a^3 \tag{4.13}$$

An example of this technique for determining e using Stokes's law is given in Example 4.3.

EXAMPLE 4.3   Experimental Determination of e

In a Millikan experiment the distance of rise or fall of a droplet is 0.600 cm and the average time of fall (field off) is 21.0 s. The observed successive rise times are 46.0, 15.5, 28.1, 12.9, 45.3, and 20.0 s.

(a) Prove that charge is quantized.

Solution   Charge is quantized if q_1/q_2, q_2/q_3, q_3/q_4, and so on are ratios of small whole numbers. Because

$$\frac{q_1}{q_2} = \frac{v + v_1'}{v + v_2'}, \quad \frac{q_2}{q_3} = \frac{v + v_2'}{v + v_3'}, \quad \text{etc.}$$

we must find the speeds. Thus,

$$v = \frac{\Delta y}{\Delta t} = \frac{0.600 \text{ cm}}{21.0 \text{ s}} = 0.0286 \text{ cm/s}$$

$$v_1' = \frac{\Delta y}{\Delta t} = \frac{0.600 \text{ cm}}{46.0 \text{ s}} = 0.0130 \text{ cm/s}$$

$$v_2' = 0.600/15.5 = 0.0387 \text{ cm/s}$$

$$v_3' = 0.600/28.1 = 0.0214 \text{ cm/s}$$

$$v_4' = 0.600/12.9 = 0.0465 \text{ cm/s}$$

$$v_5' = 0.600/45.3 = 0.0132 \text{ cm/s}$$

$$v_6' = 0.600/20.0 = 0.0300 \text{ cm/s}$$

so

$$\frac{q_1}{q_2} = \frac{v + v_1'}{v + v_2'} = \frac{0.0286 + 0.0130}{0.0286 + 0.0387} = 0.618 \approx \frac{3}{5}$$

$$\frac{q_2}{q_3} = \frac{v + v_2'}{v + v_3'} = \frac{0.0286 + 0.0387}{0.0286 + 0.0214} = 1.35 \approx \frac{4}{3}$$

$$\frac{q_3}{q_4} = \frac{v + v_3'}{v + v_4'} = \frac{0.0286 + 0.0214}{0.0286 + 0.0465} = 0.666 \approx \frac{2}{3}$$

$$\frac{q_4}{q_5} = \frac{v + v_4'}{v + v_5'} = \frac{0.0286 + 0.0465}{0.0286 + 0.0132} = 1.80 \approx \frac{9}{5}$$

$$\frac{q_5}{q_6} = \frac{v + v_5'}{v + v_6'} = \frac{0.0286 + 0.0132}{0.0286 + 0.0300}$$

$$= 0.713 \approx 8/11 \quad \text{or} \quad 7/10$$

(b) If the oil density is 858 kg/m^3 and the viscosity of air is 1.83×10^{-5} kg/m·s, find the radius, volume, and mass of the drop used in this experiment.

Solution   The radius of the drop is

$$a = \sqrt{\frac{9\eta v}{2\rho g}}$$

---

[6]R. A. Millikan, *Phys. Rev.* 1911, p. 349.

$$= \left[\frac{9(1.83 \times 10^{-5}\,\text{kg/m·s})(0.0286 \times 10^{-2}\,\text{m/s})}{2(858\,\text{kg/m}^3)(9.81\,\text{m/s}^2)} \right]^{1/2}$$

$$= [2.80 \times 10^{-12}\,\text{m}^2]^{1/2}$$

$$= 1.67 \times 10^{-6}\,\text{m} \quad \text{or } 1.67\ \mu\text{m}$$

The volume is

$$V = \tfrac{4}{3}\pi a^3 = 1.95 \times 10^{-17}\,\text{m}^3$$

The mass is

$$m = \rho V = (858\,\text{kg/m}^3)(1.95 \times 10^{-17}\,\text{m}^3)$$

$$= 1.67 \times 10^{-14}\,\text{kg}$$

(c) Calculate the successive charges on the drop, and from these results determine the electronic charge. Assume a plate separation of 1.60 cm and a potential difference of 4550 V for the parallel-plate capacitor.

Solution   To calculate the charge on each drop using Equation 4.10, we must first calculate the magnitude of the electric field, E. Thus,

$$E = \frac{V}{d} = \frac{4550\,\text{V}}{0.0160\,\text{m}} = 2.84 \times 10^5\,\text{V/m}$$

Now we can find the charges on the drop:

$$q_1 = \left(\frac{mg}{E} \right)\left(\frac{v + v_1'}{v} \right)$$

$$= \frac{(1.67 \times 10^{-14}\,\text{kg})(9.81\,\text{m/s}^2)}{(2.84 \times 10^5\,\text{V/m})}\left(\frac{0.0286 + 0.0130}{0.0286} \right)$$

$$= 8.39 \times 10^{-19}\,\text{C}$$

Likewise, $q_2 = 13.6 \times 10^{-19}$ C, $q_3 = 10.1 \times 10^{-19}$ C, $q_4 = 15.2 \times 10^{-19}$ C, $q_5 = 8.43 \times 10^{-19}$ C, and $q_6 = 11.8 \times 10^{-19}$ C.

To find the average value of e, we shall use the fact that at the time of Millikan's work e was known to be between 1.5×10^{-19} C and 2.0×10^{-19} C. Dividing q_1 through q_6 by these values gives the range of integral charges on each drop. Thus, the range of q_1 is $8.39/1.5 = 5.6$ electronic charges to $8.39/2.0 = 4.2$ electronic charges. Similarly, q_2 has 9.1 to 5.8 charges, q_3 has 6.7 to 5.1 charges, q_4 has 10.1 to 7.6 charges, q_5 has 5.6 to 4.2 charges, and q_6 has 7.9 to 5.9 charges. Because there must be an integral number of charges on each drop, we pick an integer in the middle of the allowed range. Therefore, in terms of the electronic charge e, we conclude that $q_1 = 5e$, $q_2 = 8e$, $q_3 = 6e$, $q_4 = 9e$, $q_5 = 5e$, and $q_6 = 7e$. Using the preceding values, we find

$$e_1 = q/5 = 1.68 \times 10^{-19}\,\text{C}$$

$$e_2 = q_1/8 = 1.70 \times 10^{-19}\,\text{C}$$

$$e_3 = q_2/6 = 1.68 \times 10^{-19}\,\text{C}$$

$$e_4 = q_3/9 = 1.69 \times 10^{-19}\,\text{C}$$

$$e_5 = q_4/5 = 1.69 \times 10^{-19}\,\text{C}$$

$$e_6 = q_5/7 = 1.69 \times 10^{-19}\,\text{C}$$

Taking the average of these values, we find the value of the electronic charge to be $e = 1.688 \times 10^{-19}$ C for this data set.

Stokes's law, as Millikan was aware, is only approximately correct for tiny spheres moving through a gas. The expression $D = 6\pi a\eta v$ holds quite accurately for a 0.1-cm radius sphere moving through a liquid or for any case where the moving-object radius, a, is large compared with the mean free path, L, of the surrounding molecules. (The mean free path is essentially the average distance between molecules.) In the Millikan experiment, however, a is of the same order of magnitude as the mean free path of air at STP. Consequently, Stokes's law overestimates the drag force, because the droplet actually moves for appreciable times through a frictionless "vacuum." Millikan corrected Stokes's law by using a drag force whose magnitude is

$$D = \frac{6\pi a\eta v}{1 + \alpha(L/a)} \tag{4.14}$$

and found that $\alpha = 0.81$ gave the most consistent values of e for drops of different radii. Further corrections to Stokes's law were made by Perrin and Roux, and corrections to Stokes's law and the correct value of e remained a controversial issue for more than 20 years. The currently accepted value of the magnitude of the electronic charge is

$$e = 1.60217733 \times 10^{-19} \text{ C}$$

Rutherford's Model of the Atom

The early years of the 20th century were generally a period of incredible ferment and change in physics, including the advent of relativity, quantum theory, and atomic and subatomic physics. Hardly had the reality of pristine, indivisible atoms been established ("They are the only material things which still remain in the precise condition in which they first began to exist" wrote Maxwell in 1872), when Thomson announced their divisibility in 1899: "Electrification essentially involves the splitting of the atom, a part of the mass of the atom getting free and becoming detached from the original atom." Further daring assaults on the indivisibility of the chemical atom came from the experimental work on radioactivity by Marie Curie (1867–1934, a Polish physicist–chemist), and by Ernest Rutherford (1871–1937, New Zealand physicist), and Frederick Soddy, a British physicist, who explained radioactive transformations of elements in terms of the emission of subatomic particles. The porosity of atoms was also known before 1910 from Lenard's experiments, which showed that electrons are easily transmitted through thin metal and mica foils. All these discoveries, plus the suspicion that the intricate atomic spectral lines (light emitted at a discrete set of frequencies characteristic of each element) must be produced by charge rattling around inside the atom, led to various proposals concerning the internal structure of atoms. The most famous of these early atomic models was the Thomson "plum-pudding" model (1898). This proposal viewed the atom as a homogeneous sphere of uniformly distributed mass and positive charge in which were embedded, like raisins in a plum pudding, negatively charged electrons, which just balanced the positive charge to produce electrically neutral atoms. Although such models possessed electrical stability against collapse or explosion of the atom, they failed to explain the rich line spectra of even the simplest atom, hydrogen.

The key to understanding the mysterious line spectra and the correct model of the atom were both furnished by Ernest Rutherford and his students Hans Geiger (1882–1945, German physicist) and Ernest Marsden (1899–1970, British physicist) through a series of experiments conducted from 1909 to 1914. Noticing that a beam of collimated α particles broadened on passing through a metal foil yet easily penetrated the thin film of metal, they embarked on experiments to probe the distribution of mass within the atom by observing in detail the scattering of α particles from foils. These experiments ultimately led Rutherford to the discovery that most of the atomic mass and all of the positive charge lie in a minute central nucleus of the atom. The accidental chain of events and the clever capitalization on the accidental discoveries leading up to Rutherford's monumental nuclear theory of the atom are nowhere better described than in Rutherford's own essay summarizing the development of the theory of atomic structure:

Rutherford's nuclear model

> . . . I would like to use this example to show how you often stumble upon facts by accident. In the early days I had observed the scattering of α-particles, and Dr. Geiger in my laboratory had examined it in detail. He found, in thin pieces of heavy metal, that the scattering was usually small, of the order of one degree. One

day Geiger came to me and said, "Don't you think that young Marsden, whom I am training in radioactive methods, ought to begin a small research?" Now I had thought that, too, so I said, "Why not let him see if any α-particles can be scattered through a large angle?" I may tell you in confidence that I did not believe that they would be, since we knew that the α-particle was a very fast, massive particle, with a great deal of energy, and you could show that if the scattering was due to the accumulated effect of a number of small scatterings the chance of an α-particle's being scattered backwards was very small. Then I remember two or three days later Geiger coming to me in great excitement and saying, "We have been able to get some of the α-particles coming backwards. . . ." It was quite the most incredible event that has ever happened to me in my life. It was almost as incredible as if you fired a 15-inch shell at a piece of tissue paper and it came back and hit you. On consideration, I realized that this scattering backwards must be the result of a single collision, and when I made calculations I saw that it was impossible to get anything of that order of magnitude unless you took a system in which the greater part of the mass of the atom was concentrated in a minute nucleus. It was then that I had the idea of an atom with a minute massive center carrying a charge. I worked out mathematically what laws the scattering should obey, and I found that the number of particles scattered through a given angle should be proportional to the thickness of the scattering foil, the square of the nuclear charge, and inversely proportional to the fourth power of the velocity. These deductions were later verified by Geiger and Marsden in a series of beautiful experiments.[7]

The essential experimental features of Rutherford's apparatus are shown in Figure 4.10. A finely collimated beam of α particles emitted with speeds of about 2×10^7 m/s struck a thin gold foil several thousand atomic layers thick. Most of the α's passed straight through the foil along the line DD' (again showing the porosity of the atom), but some were scattered at an angle ϕ. The number of scattered α's at each angle per unit detector area and per unit time

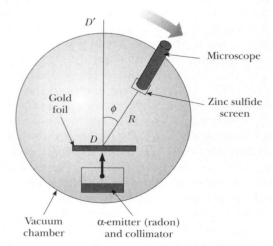

Figure 4.10   A schematic view of Rutherford's α scattering apparatus.

[7]An essay on "The Development of the Theory of Atomic Structure," 1936, Lord Rutherford, published in *Background to Modern Science*, New York, Macmillan Company, 1940.

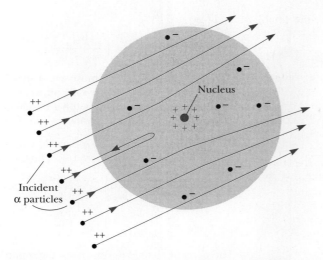

Figure 4.11   Scattering of α particles by a dense, positively charged nucleus.

was measured by counting the scintillations produced by scattered α's on the ZnS screen. These scintillations were counted with the aid of the microscope. The distance from the point where α particles strike the foil to the zinc sulfide screen is denoted R in Figure 4.10.

Rutherford's basic insight was that because the mass and kinetic energy of the α's are large, even a nearly head-on collision with a particle with the mass of a hydrogen atom would deflect the α particle only slightly and knock the hydrogen atom straight ahead. Multiple scattering of the α particles in the foil accounted for the small broadening (about 1°) originally observed by Rutherford, but it could not account for the occasional large-scale deflections. On the other hand,[8] if all of the positive charge in an atom is assumed to be concentrated at a single central point and not spread out throughout the atom, the electric repulsion experienced by an incident α particle in a head-on collision becomes much greater. Because the charge and mass of the gold atom are concentrated at the nucleus, large deflections of the α particle could be experienced in a *single* collision with the massive nucleus. This situation is shown in Figure 4.11.

EXAMPLE 4.4   Collision of an α Particle with a Proton

(a) An α particle of mass m_α and speed v_α strikes a stationary proton with mass m_p. If the collision is elastic and head-on, show that the speed of the proton after the collision, v_p, and the speed of the α particle after the collision, v'_α, are given by

$$v_p = \left(\frac{2m_\alpha}{m_\alpha + m_p}\right) v_\alpha$$

and

$$v'_\alpha = \left(\frac{m_\alpha - m_p}{m_\alpha + m_p}\right) v_\alpha$$

(b) Calculate the percent change in velocity for an α particle colliding with a proton.

---

[8]A dangerous expression that never fails to bring to mind President Truman's infamous request: "If you know of any one-handed economists, bring them to me."

Solution   (a) Because the collision is elastic, the total kinetic energy is conserved; therefore,

$$\tfrac{1}{2} m_\alpha v_\alpha^2 = \tfrac{1}{2} m_\alpha v_\alpha'^2 + \tfrac{1}{2} m_p v_p^2 \qquad (1)$$

Conservation of momentum for this one-dimensional collision yields

$$m_\alpha v_\alpha = m_p v_p + m_\alpha v_\alpha' \qquad (2)$$

Solving Equation 1 for $(m_\alpha v_\alpha')^2$ yields

$$(m_\alpha v_\alpha')^2 = m_\alpha(m_\alpha v_\alpha^2 - m_p v_p^2) \qquad (3)$$

Solving Equation 2 for $(m_\alpha v_\alpha')^2$ and equating this to Equation 3 gives

$$(m_\alpha v_\alpha)^2 + (m_p v_p)^2 - 2 m_\alpha m_p v_\alpha v_p = (m_\alpha v_\alpha)^2 - m_\alpha m_p v_p^2$$

or

$$(m_p v_p)(m_p v_p - 2 m_\alpha v_\alpha + m_\alpha v_p) = 0$$

The solutions to this equation are

$$v_p = 0$$

and

$$v_p = \left(\frac{2 m_\alpha}{m_\alpha + m_p} \right) v_\alpha \qquad (4)$$

Because the proton must move when struck by the heavy α particle, Equation 4 is the only physically reasonable solution for v_p. The solution for v_α' follows immediately from the substitution of Equation 4 into Equation 2.

Solution   (b) Because an α particle consists of two protons and two neutrons, $m_\alpha = 4 m_p$. Thus,

$$v_p = \left(\frac{2 m_\alpha}{m_\alpha + m_p} \right) v_\alpha = \left(\frac{8 m_p}{5 m_p} \right) v_\alpha = 1.60 v_\alpha$$

$$v_\alpha' = \left(\frac{m_\alpha - m_p}{m_\alpha + m_p} \right) v_\alpha = \left(\frac{3 m_p}{5 m_p} \right) v_\alpha = 0.60 v_\alpha$$

The percent change in velocity of the α particle is

$$\% \text{ change in } v_\alpha = \left(\frac{v_\alpha' - v_\alpha}{v_\alpha} \right) \times 100\% = -40\%$$

Exercise 2   An α particle with initial velocity v_α undergoes an elastic, head-on collision with an electron initially at rest. Using the fact that an electron's mass is about $1/2000$ of the proton mass, calculate the final velocities of the electron and α particle and the percent change in velocity of the α particle.

Answers $v_e = 1.998 v_\alpha$, $v_\alpha' = 0.9998 v_\alpha$, and the percent change in $v_\alpha = -0.02\%$.

In his analysis, Rutherford assumed that large-angle scattering is produced by a single nuclear collision and that the repulsive force between an α particle and a nucleus separated by a distance r is given by Coulomb's law,

$$F = k \frac{(2e)(Ze)}{r^2} \qquad (4.15)$$

where $+2e$ is the charge on the α, $+Ze$ is the nuclear charge, and k is the Coulomb constant. With this assumption, Rutherford was able to show that the number of α particles entering the detector per unit time, Δn, at an angle ϕ is given by

$$\Delta n = \frac{k^2 Z^2 e^4 N n A}{4 R^2 (\tfrac{1}{2} m_\alpha v_\alpha^2)^2 \sin^4(\phi/2)} \qquad (4.16)$$

Here R and ϕ are defined in Figure 4.10, N is the number of nuclei per unit area of the foil (and is thus proportional to the foil thickness), n is the total number of α particles incident on the target per unit time, and A is the area of the detector. The dependence of scattering on foil thickness, α particle speed, and scattering angle was confirmed experimentally by Geiger and Marsden.[9]

---

[9]H. Geiger and E. Marsden, "Deflection of α-Particles through Large Angles," *Phil. Mag.* (6)25:605, 1913.

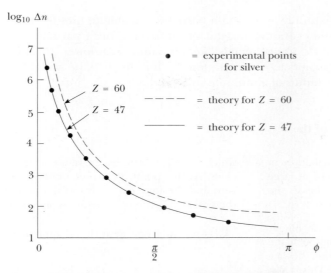

Figure 4.12   Comparison of theory and experiment for α particle scattering from a silver foil. (*From E. Rutherford, J. Chadwick, and J. Ellis,* Radiations from Radioactive Substances, *Cambridge, Cambridge University Press, 1951.*)

Because the values of atomic number (Z) were uncertain at the time, the scattering dependence on Z could not be directly checked. Turning the argument around, however, and assuming the correctness of Equation 4.16, one could find the value of Z that gave the best fit of this equation to the experimental data points. An illustration of this sensitive technique for determining Z for a silver foil is shown in Figure 4.12. Note that changing Z produces only a vertical shift in the graph and not a change in shape.

Much of the remarkable experimental work of the ingenious Lord Rutherford can be credited to an ability to use his current discoveries to probe even deeper into nature's mysteries. For example, he turned his studies of the transmission of radioactive particles through matter into a sensitive and delicate technique for probing the atom. Another example was his clever technique for measuring the size of the nucleus. Realizing that Equation 4.15 would hold only if the α particle did not have enough energy to deform or penetrate the scattering nucleus, he systematically looked for the threshold α energy at which departures from his scattering equation occurred, the idea being that at this threshold energy the α should be just penetrating the nuclear radius at its distance of closest approach. Following Rutherford, we set the kinetic energy of the α at infinity equal to the potential energy of the system (α + target nucleus) at the distance of closest approach, $d_{\min}$, or

$$\tfrac{1}{2} m_\alpha v_\alpha^2 = k\, \frac{(Ze)(2e)}{d_{\min}} \tag{4.17}$$

Equation 4.17 may then be solved for $d_{\min}$ to determine the distance of closest approach. In the case when the kinetic energy of the α is so high that Equation 4.16 begins to fail, this distance of closest approach is approximately equal to the nuclear radius.

Rutherford was confronted with the experimental dilemma that no failures of Equations 4.15 or 4.16 were found for heavy metal foils with the most energetic naturally occurring α particles available to him (≈ 8 MeV). Showing characteristic economy, instead of embarking on a particle accelerator program, he made use of metals like aluminum with lower Z's and hence lower Coulomb barriers to α penetration.[10] Thus, in 1919, he was able to determine the nuclear radius of aluminum to be about 5×10^{-15} m.

EXAMPLE 4.5   Estimate of the Radius of the Aluminum Nucleus

In 1919, Rutherford was able to show a breakdown in Equation 4.16 for 7.7-MeV α particles scattered at large angles from aluminum nuclei ($Z = 13$). Estimate the radius of the aluminum nucleus from these facts.

Solution   Rutherford's scattering formula is no longer valid when α particles begin to penetrate or touch the nucleus. When the α particle is very far from the aluminum nucleus, its kinetic energy is 7.7 MeV. This is also the total energy of the system (α particle plus aluminum nucleus), because the aluminum nucleus is at rest and the potential energy is zero for an infinite separation of particles. When the α particle is at the point of closest approach to the aluminum nucleus in a head-on collision, its kinetic energy is zero and it is at a distance d_{min},

which we may take to be the radius of the aluminum nucleus. At this point, the kinetic energy of the system is zero and the total energy is just the potential energy of the system. Applying conservation of energy gives

$$K_\alpha = \text{potential energy at closest approach} = \frac{k(Ze)(2e)}{d_{min}}$$

or

$$d_{min} = k\,\frac{2Ze^2}{K_\alpha}$$
$$= \frac{2(13)(1.60 \times 10^{-19}\text{ C})^2\,(8.99 \times 10^9\text{ N·m}^2/\text{C}^2)}{(7.7 \times 10^6\text{ eV})(1.60 \times 10^{-19}\text{ J/eV})}$$
$$= 4.9 \times 10^{-15}\text{ m}$$

The overall success of the Rutherford nuclear model was striking. Rutherford and his students had shown that all the mass and positive charge Ze were concentrated in a minute nucleus of the atom of diameter 10^{-14} m and that Z electrons must circle the nucleus in some way. As with all great discoveries, however, the idea of the nuclear atom raised a swarm of questions at the next deeper level: (1) If there are only Z protons in the nucleus, what composes the other half of the nuclear mass? (2) What provides the cohesive force to keep many protons confined in the incredibly small distance of 10^{-14} m? (3) How do the electrons move around the nucleus to form a stable atom, and how does their motion account for the observed spectral lines?

Rutherford had no precise answer to the first question. He speculated that the difference between the mass of Z protons and the total nuclear mass could be accounted for by additional groupings of neutral particles, each consisting of a bound electron–proton pair. This conjecture seemed especially satisfying because it built the atom out of the most fundamental particles then known to exist.

In answer to the second question, Rutherford cautiously held that electrical forces provided the cement to hold the nucleus together. He wrote, "The nucleus, though of minute dimensions, is in itself a very complex system

---

[10]Rutherford was famous for the remark to his graduate students, "There is no money for apparatus—we shall have to use our heads" (A. Keller, *Infancy of Atomic Physics: Hercules in His Cradle*, Oxford, Clarendon Press, 1983, p. 215).

Figure 4.13   Bohr (on the right) and Rutherford (on the left) were literal as well as intellectual supports for each other. This photograph of Bohr and Rutherford sitting back to back was taken at a rowing regatta in June, 1923 at Cambridge University. (*AIP Niels Bohr Library, and* Physics Today *October 1985, an issue devoted to Bohr.*)

consisting of positively and negatively charged bodies bound closely together by intense electrical forces." In fact, it was not until 1921 that it was clearly recognized that the Coulomb force did not hold the nucleus together and that a completely new and very strong type of force binds protons together. Interestingly it was James Chadwick, the discoverer of the neutron, who first recognized that a new force of much more than electric intensity was at work in the nucleus.[11] Perhaps Rutherford's magnificent achievement of explaining α scattering with the Coulomb law blinded him to the possibility that this was not the ultimate law at work within the nucleus.

The answer to the third question was not to be given by Rutherford. That was to be the masterwork of Niels Bohr (Fig. 4.13). Even so, with characteristic insight, Rutherford mentioned a planetary model of the atom or, more precisely, that negative charges revolved around the dense positive core as the planets revolved around the Sun.[12]

4.3   THE BOHR ATOM

Bohr's original quantum theory of spectra was one of the most revolutionary, I suppose, that was ever given to science, and I do not know of any theory that has been more successful . . . . I consider the work of Bohr one of the greatest triumphs of the human mind. (Lord Rutherford)

Then it is one of the greatest discoveries. (Albert Einstein, on hearing of Bohr's theoretical calculation of the Rydberg constants for hydrogen and singly ionized helium)

---

[11]J. Chadwick and E. S. Biele, *Phil. Mag.* 42:923, 1921.

[12]Thomson and Hantaro Nagaoka, a Japanese physicist, had worked even earlier with planetary atomic models in 1904.

Spectral Series

Before looking in detail at the first successful theory of atomic dynamics, we review the experimental work on line spectra that served as the impetus for and the clear confirmation of the first quantum theory of the atom. As already pointed out in Chapter 3, glowing solids and liquids (and even gases at the high densities found in stars) emit a continuous distribution of wavelengths. This distribution exhibits a common shape for the intensity-versus-wavelength curve, and the peak in this curve shifts toward shorter wavelengths with increasing temperature. This universal "blackbody" curve is shown in Figure 4.14.

In sharp contrast to this continuous spectrum is the discrete line spectrum emitted by a low-pressure gas subject to an electric discharge. When the light from such a low-pressure gas discharge is examined with a spectroscope, it is found to consist of a few bright lines of pure color on a dark background. This contrasts sharply with the continuous rainbow of colors seen when a glowing solid is viewed through a spectroscope. Furthermore, as can be seen from Figure 4.15, the wavelengths contained in a given line spectrum are character-istic of the particular element emitting the light. (Also see the inside front cover.) The simplest line spectrum is observed for atomic hydrogen, and we shall describe this spectrum in detail. Other atoms, such as mercury, helium, and neon, give completely different line spectra. Because no two elements emit the same line spectrum, this phenomenon represents a practical and sensitive technique for identifying the elements present in unknown samples. In fact, by 1860 spectroscopy had advanced so far in the hands of Gustav Robert Kirchhoff (Fig. 4.16) and Robert Wilhelm von Bunsen (Fig. 4.17) at the University of Heidelberg that they were able to discover two new elements, rubidium and cesium, by observing new sequences of spectral lines in mineral samples. Improvements in instruments and techniques resulted in an enormous growth in spectral analysis in Europe from 1860 to 1900. Even the European public imagination was captured by spectroscopy when spectro-scopic techniques showed that "celestial" meteorites consisted only of known Earth elements after all.

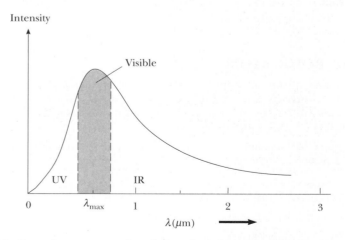

Figure 4.14   Intensity versus wavelength for a body heated to 6000 K.

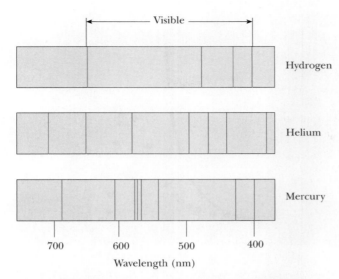

Figure 4.15   Emission line spectra of a few representative elements.

Figure 4.16 Gustav Robert Kirchhoff (1824–1887). Yes, this is the same fellow who brought us the circuit loop theorem and established the connection between the absorption and emission of an object (see Section 3.2). (*AIP Emilio Segrè Visual Archives, W. F. Meggers Collection*)

Kirchhoff's immense contribution to spectroscopy is also shown by another advance he made in 1859—the foundation of **absorption spectroscopy** and the explanation of Fraunhofer's dark D-lines in the solar spectrum.[13] In 1814, Joseph Fraunhofer had passed the continuous spectrum from the Sun through a narrow slit and then through a prism. He observed the surprising result of nearly 1000 fine dark lines, or gaps, in the continuous rainbow spectrum of the Sun, and he assigned the letters A, B, C, D . . . to the most

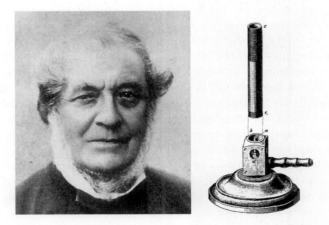

Figure 4.17   Robert Wilhelm von Bunsen (1811–1899). Bunsen is pictured with his most famous invention, the gas laboratory burner named for him. The greatest achievement of this fine chemist, however, was the development, with Kirchhoff, of the powerful analytical method of spectral analysis. (*AIP Emilio Segrè Visual Archives, E. Scott Barr Collection*)

[13]G. Kirchhoff, *Monatsber.*, Berlin, 1859, p. 662.

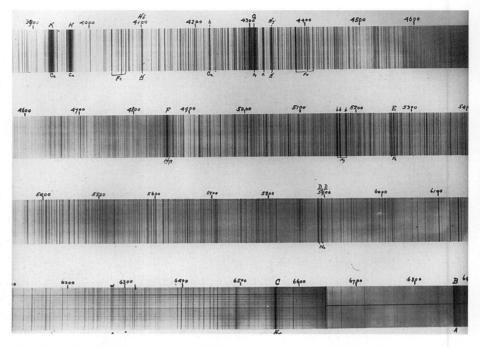

Figure 4.18   The spectrum of the Sun as viewed with a high-resolution spectrograph, showing many dark Fraunhofer lines. Wavelengths are indicated above the spectrum in angstrom units (10^{-8} cm). (*Palomar Observatory photograph, Cal Tech*)

prominent dark lines. These lines and many more are shown in Figure 4.18. Kirchhoff correctly deduced that the mysterious dark lines are produced by a cloud of vaporized atoms in the Sun's outer, cooler layers, which absorb at discrete frequencies the intense continuous radiation from the center of the Sun. Further, he showed that the Fraunhofer D-lines were produced by vaporized sodium and that they had the same wavelengths as the strong yellow lines in the emission spectrum of sodium. Kirchhoff also correctly deduced that all of Fraunhofer's dark lines should be attributable to absorption by different elements present in the Sun. In a single stroke he opened the way to determining the elemental composition of stars trillions of miles from the Earth. His elegant yet simple method for demonstrating the presence of sodium vapor in the solar atmosphere is shown in schematic form in Figure 4.19.

Today, absorption spectroscopy is certainly as important as emission spectroscopy for qualitative and quantitative analyses of elements and molecular groups. In general, one obtains an absorption spectrum by passing light from a continuous source [whether in the ultraviolet (uv), visible (vis), or infrared (IR) regions] through a gas of the element being analyzed. The absorption spectrum consists of a series of dark lines superimposed on the otherwise continuous spectrum emitted by the source. Each line in the absorption spectrum of an element coincides with a line in the emission spectrum of that same element; however, not all of the emission lines are present in an absorption spectrum. The differences between emission and absorption spectra are complicated in general and depend on the temperature of the absorbing vapor.

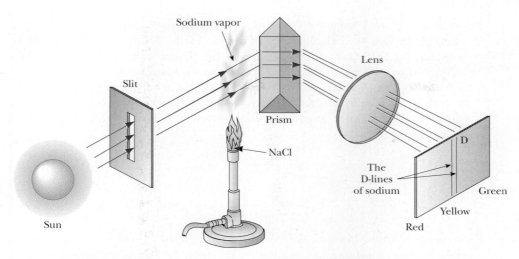

Figure 4.19 Kirchhoff's experiment explaining the Fraunhofer D-lines. The D-lines darken noticeably when sodium vapor is introduced between the slit and the prism.

An interesting use of the coincidence of absorption and emission lines is made in the atomic absorption spectrometer. This device is routinely used to measure parts per million (ppm) of metals in unknowns. For example, if sodium is to be measured, a sodium lamp emitting a line spectrum is chosen as the light source. The unknown is heated in a hot flame (usually oxyacetylene) to vaporize the sample, to break the chemical bonds of sodium to other elements, and to produce a gas of elemental sodium. The spectrometer is then tuned to a wavelength for which both absorption and emission lines exist (say, one of the D-lines at 588.99 or 589.59 nm), and the amount of darkening or decrease in intensity is measured with a sensitive photomultiplier. The decrease in intensity is a measure of the sodium concentration. With proper calibration, concentrations of 0.1 ppm can be measured with this extremely selective technique. Atomic absorption spectroscopy has been a useful technique in analyzing heavy-metal contamination of the food chain. For example, the first determinations of high levels of mercury in tuna fish were made with atomic absorption.

From 1860 to 1885 spectroscopic measurements accumulated voluminously, burying frenzied theoreticians under a mountain of data. Accurate measurements of four visible emission lines of hydrogen had recently been made by Anders Ångström, a Swedish physicist, when in 1885 a Swiss schoolteacher, Johann Jakob Balmer, published a paper with the unpretentious title "Notice Concerning the Spectral Lines of Hydrogen." By trial and error Balmer had found a formula that correctly predicted the wavelengths of Ångström's four visible lines: H_α (red), H_β (green), H_γ (blue), and H_δ (violet). Figure 4.20 shows these and other lines in the emission spectrum of hydrogen. Balmer gave his formula in the form

$$\lambda(\text{cm}) = C_2\left(\frac{n^2}{n^2 - 2^2}\right) \qquad n = 3, 4, 5, \ldots \qquad (4.18)$$

where λ is the wavelength emitted in cm and $C_2 = 3645.6 \times 10^{-8}$ cm, a constant called the **convergence limit** because it gave the wavelength of

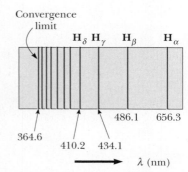

Figure 4.20 The Balmer series of spectral lines for hydrogen (emission spectrum).

the line with the largest n value ($n = \infty$). Also, note that $n = 3, 4, 5, \ldots$, where H_α has $n = 3$, H_β has $n = 4$, and so forth. Although only four lines were known to Balmer when he started his paper, by the time he had finished, ten more lines in the violet and ultraviolet had been measured. Much to his delight and satisfaction, these lines agreed with his empirical formula to within 0.1%! Encouraged by his success and because he was a bit of a numerologist, Balmer suggested that other hydrogen series might exist of the form

$$\lambda = C_3 \left(\frac{n^2}{n^2 - 3^2} \right) \qquad n = 4, 5, 6, \ldots \qquad (4.19)$$

$$\lambda = C_4 \left(\frac{n^2}{n^2 - 4^2} \right) \qquad n = 5, 6, 7, \ldots \qquad (4.20)$$

As we now know, his speculations were correct, and these series do indeed exist. In today's notation, all of these series are given by a single formula:

$$\frac{1}{\lambda} = R \left(\frac{1}{n_f^2} - \frac{1}{n_i^2} \right) \qquad (4.21)$$

where n_f and n_i are *integers*. The Rydberg constant, R, is the same for all series and has the value

$$R = 1.0973732 \times 10^7 \text{ m}^{-1} \qquad (4.22)$$

Note that for a given series, n_f has a constant value. Furthermore, for a given series $n_i = n_f + 1, n_f + 2, \ldots$. Table 4.1 lists the name of each series (named after their discoverers) and the integers that define the series.

Bohr's Quantum Model of the Atom

In April of 1913, a young Danish physicist, Niels Bohr (who had recently been working with both Thomson and Rutherford), published a three-part paper that shook the world of physics to its foundations.[14] Not only did this young rebel give the first successful theory of atomic line spectra but in the process he overthrew some of the most cherished principles of the reigning king of electromagnetism, James Clerk Maxwell.

Table 4.1   Some Spectral Series for the Hydrogen Atom

Lyman Series (uv)	$n_f = 1$	$n_i = 2, 3, 4, \ldots$
Balmer Series (vis–uv)	$n_f = 2$	$n_i = 3, 4, 5, \ldots$
Paschen Series (IR)	$n_f = 3$	$n_i = 4, 5, 6, \ldots$
Brackett Series IR)	$n_f = 4$	$n_i = 5, 6, 7, \ldots$
Pfund Series (IR)	$n_f = 5$	$n_i = 6, 7, 8, \ldots$

[14] N. Bohr, "On the Constitution of Atoms and Molecules," *Phil. Mag.* 26:1, 1913. Also, N. Bohr, *Nature* 92:231, 1913.

From our point of view, Bohr's model may seem only a reasonable next step, but it appeared astounding, confounding, and incredibly bold to his contemporaries. As mentioned earlier, both Thomson and Rutherford realized that the electrons must revolve about the nucleus in order to avoid falling into it. They, along with Bohr, realized that according to Maxwell's theory, accelerated charges revolving with orbital frequency f should radiate light waves of frequency f. Unfortunately, pushed to its logical conclusion, this classical model leads to disaster. As the electron radiates energy, its orbit radius steadily decreases and its frequency of revolution increases. This leads to an ever-increasing frequency of emitted radiation and an ultimate catastrophic collapse of the atom as the electron plunges into the nucleus (Fig. 4.21).

These deductions of electrons falling into the nucleus and a continuous emission spectrum from elements were boldly circumvented by Bohr. He simply postulated that classical radiation theory, which had been confirmed by Hertz's detection of radio waves using large circuits, did not hold for atomic-sized systems. Moreover, he drew on the work of Planck and Einstein as sources of the correct theory of atomic systems. He overcame the problem of a classical electron that continually lost energy by applying Planck's ideas of quantized energy levels to orbiting atomic electrons. Thus he postulated that electrons in atoms are generally confined to certain stable, nonradiating energy levels and orbits known as **stationary states.**[15] He applied Einstein's concept of the photon to arrive at an expression for the frequency of the light emitted when the electron jumps from one stationary state to another. Thus, if ΔE is the separation of two possible electronic stationary states, then $\Delta E = hf$, where h is Planck's constant and f is the frequency of the emitted light regardless of the frequency of the electron's orbital motion. In this way, by combining certain principles of classical mechanics with new quantum principles of light emission, Bohr arrived at a theory of the atom that agreed remarkably with experiment.

Stationary states

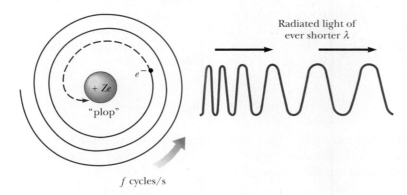

Figure 4.21   The classical model of the nuclear atom.

[15]*Stationary state* was a term used by Bohr to mean a state of an atom that was stable, nonradiating, and had an energy constant with time. It does not mean "fixed in position" or "without motion," since electrons in stationary orbits move with high speed.

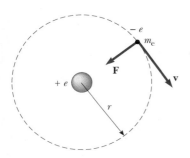

Figure 4.22 Diagram representing Bohr's model of the hydrogen atom.

Now that we have looked at the general principles of Bohr's model of hydrogen and at the detailed experimental spectra already discovered by 1913, let us examine Bohr's quantum theory in detail. The basic ideas of the Bohr theory as it applies to an atom of hydrogen are as follows:

- The electron moves in circular orbits about the proton under the influence of the Coulomb force of attraction, as in Figure 4.22. So far nothing new!
- Only certain orbits are stable. These stable orbits are ones in which the electron does not radiate. Hence the energy is fixed or stationary in time, and ordinary classical mechanics may be used to describe the electron's motion in these stable orbits.
- Radiation is emitted by the atom when the electron "jumps" from a more energetic initial stationary state to a less energetic lower state. This "jump" cannot be visualized or treated classically. In particular, **the frequency f of the photon emitted in the jump is independent of the frequency of the electron's orbital motion.** Instead, the frequency of the light emitted is related to the change in the atom's energy and is given by the Planck–Einstein formula

$$E_i - E_f = hf \qquad (4.23)$$

where E_i is the energy of the initial state, E_f is the energy of the final state, and $E_i > E_f$.

- The size of the allowed electron orbits is determined by an additional quantum condition imposed on the electron's orbital angular momentum. Namely, the allowed orbits are those for which the electron's orbital angular momentum about the nucleus is an integral multiple of $\hbar = h/2\pi$,

$$m_e vr = n\hbar \qquad n = 1, 2, 3, \ldots \qquad (4.24)$$

Using these four assumptions, we can now calculate the allowed energy levels and emission wavelengths of the hydrogen atom. Recall that the electrical potential energy of the system shown in Figure 4.22 is given by $U = qV = -ke^2/r$, where k (the Coulomb constant) has the value $1/4\pi\varepsilon_0$. Thus, the total energy of the atom, which contains both kinetic and potential energy terms, is

$$E = K + U = \tfrac{1}{2}m_e v^2 - k\frac{e^2}{r} \qquad (4.25)$$

Applying Newton's second law to this system, we see that the Coulomb attractive force on the electron, ke^2/r^2, must equal the mass times the centripetal acceleration of the electron, or

$$\frac{ke^2}{r^2} = \frac{m_e v^2}{r}$$

From this expression, we immediately find the kinetic energy to be

$$K = \frac{m_e v^2}{2} = \frac{ke^2}{2r} \qquad (4.26)$$

Substituting this value of K into Equation 4.25 gives the total energy of the atom as

$$E = -\frac{ke^2}{2r} \qquad (4.27)$$

Note that the total energy is negative, indicating a *bound* electron–proton system. This means that energy in the amount of $ke^2/2r$ must be added to the atom to remove the electron to infinity and leave it motionless. An expression for r, the radius of the electron orbit, may be obtained by eliminating v between Equations 4.24 and 4.26:

$$r_n = \frac{n^2\hbar^2}{m_e ke^2} \qquad n = 1, 2, 3, \ldots \qquad (4.28)$$

Radii of Bohr orbits in hydrogen

Equation 4.28 shows that only certain orbits are allowed and that these preferred orbits follow from the nonclassical step of requiring the electron's angular momentum to be an integral multiple of $\hbar$. The smallest radius occurs for $n = 1$, is called the **Bohr radius,** and is denoted a_0. The value for the Bohr radius is

$$a_0 = \frac{\hbar^2}{m_e ke^2} = 0.529 \text{ Å} = 0.0529 \text{ nm} \qquad (4.29)$$

The fact that Bohr's theory gave a value for a_0 in good agreement with the experimental size of hydrogen without any empirical calibration of orbit size was considered a striking triumph for this theory. The first three Bohr orbits are shown to scale in Figure 4.23.

The quantization of the orbit radii immediately leads to energy quantization. This can be seen by substituting $r_n = n^2 a_0$ into Equation 4.27, giving for the allowed energy levels

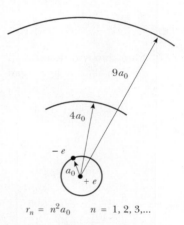

$$r_n = n^2 a_0 \qquad n = 1, 2, 3, \ldots$$

Figure 4.23 The first three Bohr orbits for hydrogen.

$$E_n = -\frac{ke^2}{2a_0}\left(\frac{1}{n^2}\right) \qquad n = 1, 2, 3, \ldots \qquad (4.30)$$

Energy levels of hydrogen

Inserting numerical values into Equation 4.30 gives

$$E_n = -\frac{13.6}{n^2} \text{ eV} \qquad n = 1, 2, 3, \ldots \qquad (4.31)$$

The integers n corresponding to the discrete, or quantized, values of the atom's energy have the special name **quantum numbers.** Quantum numbers are central to quantum theory and in general refer to the set of integers that label the discrete values of important atomic quantities, such as energy and angular momentum. The lowest stationary, or nonradiating, state is called the **ground state,** has $n = 1$, and has an energy $E_1 = -13.6$ eV. The next state, or **first excited state,** has $n = 2$ and an energy $E_2 = E_1/2^2 = -3.4$ eV. An energy-level diagram showing the energies of these discrete energy states and the corresponding quantum numbers is shown in Figure 4.24. The uppermost level, corresponding to $n = \infty$ (or $r = \infty$) and $E = 0$, represents the state for which the electron is removed from the atom and is motionless. The minimum energy required to ionize

Quantum numbers

the atom (that is, to completely remove an electron in the ground state from the proton's influence) is called the **ionization energy.** As can be seen from Figure 4.24, the ionization energy for hydrogen based on Bohr's calculation is 13.6 eV. This constituted another major achievement for the Bohr theory, because the ionization energy for hydrogen had already been measured to be precisely 13.6 eV.

Equation 4.30 together with Bohr's third postulate can be used to calculate the frequency of the photon emitted when the electron jumps from an outer orbit to an inner orbit:

$$f = \frac{E_i - E_f}{h} = \frac{ke^2}{2a_0 h}\left(\frac{1}{n_f^2} - \frac{1}{n_i^2}\right) \tag{4.32}$$

Because the quantity actually measured is wavelength, it is convenient to convert frequency to wavelength using $c = f\lambda$ to get

Emission wavelengths of hydrogen

$$\frac{1}{\lambda} = \frac{f}{c} = \frac{ke^2}{2a_0 hc}\left(\frac{1}{n_f^2} - \frac{1}{n_i^2}\right) \tag{4.33}$$

The remarkable fact is that the *theoretical* expression, Equation 4.33, is identical to Balmer's *empirical* relation

$$\frac{1}{\lambda} = R\left(\frac{1}{n_f^2} - \frac{1}{n_i^2}\right) \tag{4.34}$$

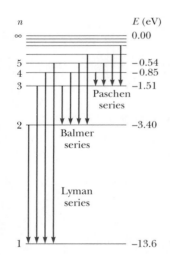

provided that the combination of constants $ke^2/2a_0 hc$ is equal to the experimentally determined Rydberg constant, $R = 1.0973732 \times 10^7 \text{ m}^{-1}$. When Bohr demonstrated the agreement of these two quantities to a precision of about 1% late in 1913, it was recognized as the crowning achievement of his quantum theory of hydrogen. Furthermore, Bohr showed that all of the observed spectral series for hydrogen mentioned previously in this section have a natural interpretation in his theory. These spectral series are shown as transitions between energy levels in Figure 4.24.

Bohr immediately extended his model for hydrogen to other elements in which all but one electron had been removed. Ionized elements such as He^+, Li^{2+}, and Be^{3+} were suspected to exist in hot stellar atmospheres, where frequent atomic collisions occurred with enough energy to completely remove one or more atomic electrons. Bohr showed that several mysterious lines observed in the Sun and stars could not be due to hydrogen, but were correctly predicted by his theory if attributed to singly ionized helium. In general, to describe a single electron orbiting a fixed nucleus of charge $+Ze$, Bohr's theory gives

$$r_n = (n^2)\frac{a_0}{Z} \tag{4.35}$$

and

$$E_n = -\frac{ke^2}{2a_0}\left(\frac{Z^2}{n^2}\right) \qquad n = 1, 2, 3, \ldots \tag{4.36}$$

Figure 4.24 An energy-level diagram for hydrogen. In such diagrams the allowed energies are plotted on the vertical axis. Nothing is plotted on the horizontal axis, but the horizontal extent of the diagram is made large enough to show allowed transitions. Note that the quantum numbers are given on the left.

EXAMPLE 4.6   Spectral Lines from the Star ζ-Puppis

The mysterious lines observed by the American astronomer Edward Charles Pickering in 1896 in the spectrum of the star ζ-Puppis fit the empirical formula

$$\frac{1}{\lambda} = R\left(\frac{1}{(n_f/2)^2} - \frac{1}{(n_i/2)^2}\right)$$

where R is, again, the Rydberg constant. Show that these lines can be explained by the Bohr theory as originating from He^+.

Solution He^+ has $Z = 2$. Thus, the allowed energy levels are given by Equation 4.36 as

$$E_n = \frac{ke^2}{2a_0}\left(\frac{4}{n^2}\right)$$

Using $hf = E_i - E_f$ we find

$$f = \frac{E_i - E_f}{h} = \frac{ke^2}{2a_0 h}\left(\frac{4}{n_f^2} - \frac{4}{n_i^2}\right)$$

$$= \frac{ke^2}{2a_0 h}\left(\frac{1}{(n_f/2)^2} - \frac{1}{(n_i/2)^2}\right)$$

or

$$\frac{1}{\lambda} = \frac{f}{c} = \frac{ke^2}{2a_0 hc}\left(\frac{1}{(n_f/2)^2} - \frac{1}{(n_i/2)^2}\right)$$

This is the desired solution, because $R \equiv ke^2/2a_0 hc$.

EXAMPLE 4.7   An Electronic Transition in Hydrogen

The electron in a hydrogen atom at rest makes a transition from the $n = 2$ energy state to the $n = 1$ ground state.

(a) Find the wavelength, frequency, and energy (eV) of the emitted photon.

Solution   We can use Equation 4.34 directly to obtain λ, with $n_i = 2$ and $n_f = 1$:

$$\frac{1}{\lambda} = R\left(\frac{1}{n_f^2} - \frac{1}{n_i^2}\right) = R\left(\frac{1}{1^2} - \frac{1}{2^2}\right) = \frac{3R}{4}$$

$$\lambda = \frac{4}{3R} = \frac{4}{3(1.097 \times 10^7\ \text{m}^{-1})}$$

$$= 1.215 \times 10^{-7}\ \text{m} = 121.5\ \text{nm}$$

This wavelength lies in the ultraviolet region.

Because $c = f\lambda$, the frequency of the photon is

$$f = \frac{c}{\lambda} = \frac{3.00 \times 10^8\ \text{m/s}}{1.215 \times 10^{-7}\ \text{m}} = 2.47 \times 10^{15}\ \text{Hz}$$

The energy of the photon is given by $E = hf$, so

$$E = hf$$
$$= (4.136 \times 10^{-15}\ \text{eV} \cdot \text{s})(2.47 \times 10^{15}\ \text{Hz}) = 10.2\ \text{eV}$$

(b) The calculation of part (a) assumes that *all* of the $n = 2$ to $n = 1$ transition energy is carried off by the photon; however, this is technically incorrect because some of this energy must go into the recoil motion of the atom. Using conservation of momentum as it applies to the system (atom + photon), and assuming that the recoil energy of the atom is small compared with the $n = 2$ to $n = 1$ energy-level separation, find the momentum and energy of the recoiling hydrogen atom.

Solution   Because momentum is conserved, and the total momentum before emission is zero, the total momentum after emission must also be zero. The photon and atom therefore move off in opposite directions, with

$$mv = \frac{E_{photon}}{c}$$

where m and v are the mass and recoil speed of the hydrogen atom, E_{photon} is the actual energy of the photon (less than 10.2 eV), and c is the speed of light. Because the energy difference between the $n = 2$ and $n = 1$ levels, E, is the source of both the photon energy and the recoil kinetic energy of the atom, we can write

$$E = E_{photon} + \tfrac{1}{2}mv^2$$

Because the atom is massive, we can assume that its recoil speed v and kinetic energy are so small that $E \approx E_{photon}$. Substituting $E_{photon} = 10.2$ eV into the expression for mv yields

$$mv = 10.2\ \text{eV}/c$$

The (approximate) recoil kinetic energy of the hydrogen atom can now be calculated:

$$K = \frac{1}{2}mv^2 = \frac{1}{2}\frac{(mv)^2}{m} = (0.5)\frac{(10.2\ \text{eV})^2}{mc^2}$$

$$= \frac{(0.5)(10.2\ \text{eV})^2}{938.8 \times 10^6\ \text{eV}} = 5.56 \times 10^{-8}\ \text{eV}$$

Thus the fraction of the energy difference between the $n = 2$ and $n = 1$ levels that goes into atomic recoil energy is *very* small, approximately 5 parts per billion:

$$\frac{K}{E} = \frac{5.56 \times 10^{-8}\ \text{eV}}{10.2\ \text{eV}} = 5.4 \times 10^{-9}$$

Evidently the process of simply equating the photon's energy to the atomic energy-level separation yields accurate answers because little energy is needed to conserve momentum.

Exercise 3   Check the approximation that $E \approx E_{photon}$ made in Example 4.7 by directly calculating the recoil kinetic energy of the hydrogen atom, $1/2 mv^2$. (*Hint:* Solve $mv = E_{photon}/c$ and $E = E_{photon} + 1/2 mv^2$ simultaneously to show $v \approx E/mc$, calculate the numerical value of $1/2 mv^2$, and compare this answer to the result given in Example 4.7.)

Exercise 4   What is the wavelength of the photon emitted by hydrogen when the electron makes a transition from the $n = 3$ state to the $n = 1$ state?

Answer $\dfrac{9}{8R} = 102.6$ nm.

EXAMPLE 4.8   The Balmer Series for Hydrogen

The Balmer series for the hydrogen atom corresponds to electronic transitions that terminate in the state of quantum number $n = 2$, as shown in Figure 4.24.

(a) Find the longest-wavelength photon emitted and determine its energy.

Solution   The longest-wavelength (least-energetic) photon in the Balmer series results from the transition from $n = 3$ to $n = 2$. Using Equation 4.34 gives

$$\frac{1}{\lambda} = R\left(\frac{1}{n_f^2} - \frac{1}{n_i^2}\right)$$

$$\frac{1}{\lambda_{max}} = R\left(\frac{1}{2^2} - \frac{1}{3^2}\right) = \frac{5}{36} R$$

$$\lambda_{max} = \frac{36}{5R} = \frac{36}{5(1.097 \times 10^7 \text{ m}^{-1})} = 656.3 \text{ nm}$$

This wavelength is in the red region of the visible spectrum.

The energy of this photon is

$$E_{photon} = hf = \frac{hc}{\lambda_{max}}$$

$$= \frac{(6.626 \times 10^{-34} \text{ J} \cdot \text{s})(3.00 \times 10^8 \text{ m/s})}{656.3 \times 10^{-9} \text{ m}}$$

$$= 3.03 \times 10^{-19} \text{ J} = 1.89 \text{ eV}$$

We could also obtain the energy of the photon by using the expression $hf = E_3 - E_2$, where E_2 and E_3 are the energy levels of the hydrogen atom, which can be calculated from Equation 4.31. Note that this is the lowest-energy photon in this series because it involves the smallest energy change.

(b) Find the shortest-wavelength photon emitted in the Balmer series.

Solution   The shortest-wavelength (most-energetic) photon in the Balmer series is emitted when the electron makes a transition from $n = \infty$ to $n = 2$. Therefore,

$$\frac{1}{\lambda_{min}} = R\left(\frac{1}{2^2} - \frac{1}{\infty}\right) = \frac{R}{4}$$

$$\lambda_{min} = \frac{4}{R} = \frac{4}{1.097 \times 10^7 \text{ m}^{-1}} = 364.6 \text{ nm}$$

This wavelength is in the ultraviolet region and corresponds to the series limit.

Exercise 5   Find the energy of the shortest-wavelength photon emitted in the Balmer series for hydrogen.

Answer   3.40 eV.

Although the theoretical derivation of the line spectrum was a remarkable feat in itself, the scope and impact of Bohr's monumental achievement is truly seen only when it is realized what else he treated in his three-part paper of 1913:

- He explained why fewer lines are seen in the absorption spectrum of hydrogen than in the emission spectrum.
- He explained the emission of x rays from atoms.
- He explained the nuclear origin of β particles.

- He explained the chemical properties of atoms in terms of the electron shell model.
- He explained how atoms associate to form molecules.

Two of these topics, the comparison of absorption and emission in hydrogen and the shell structure of atoms, are of such general importance that they deserve more explanation.

We have already pointed out that a gas will absorb at wavelengths that correspond exactly to some emission lines, but not every line present in emission is seen as a dark absorption line. Bohr explained absorption as the reverse of emission; that is, an electron in a given energy state can only absorb a photon of the exact frequency required to produce a "jump" from a lower energy state to a higher energy state. Ordinarily, hydrogen atoms are in the ground state ($n = 1$) and so only the high-energy Lyman series corresponding to transitions from the ground state to higher energy states is seen in absorption. The longer-wavelength Balmer series corresponding to transitions originating in the first excited state ($n = 2$) is not seen because the average thermal energy of each atom is insufficient to raise the electron to the first excited state. That is, the number of electrons in the first excited state is insufficient at ordinary temperatures to produce measurable absorption.

EXAMPLE 4.9   Hydrogen in Its First Excited State

Calculate the temperature at which many hydrogen atoms will be in the first excited state ($n = 2$). What series should be prominent in absorption at this temperature? (Calculate both from $N_2/N_1 = \exp(-\Delta E/k_B T)$ and from $\frac{3}{2}k_B T$ = average thermal energy.)

Solution   At room temperature almost all hydrogen atoms are in the ground state with an energy of -13.6 eV. The first excited state ($n = 2$) has an energy equal to $E_2 = -3.4$ eV. Therefore, each hydrogen atom must gain an energy of 10.2 eV to reach the first excited state. If the atoms are to obtain this energy from heat, we must have

$$\frac{3}{2}k_B T = \frac{\text{average thermal energy}}{\text{atom}} \approx 10.2 \text{ eV}$$

or

$$T = \frac{10.2 \text{ eV}}{(3/2)k_B} = \frac{10.2 \text{ eV}}{(1.5)(8.62 \times 10^{-5} \text{ eV/K})}$$
$$= 79{,}000 \text{ K}$$

Let us check this result by using the Boltzmann distribution. In Section 3.3 we saw that the probability of finding an atom with energy E at temperature T is

$$P(E) = P_0 e^{-(E-E_0)/k_B T}$$

where P_0 is the probability of finding the atom in the ground state of energy, E_0. From this expression, it follows that the ratio of the number of atoms in two different energy levels in thermal equilibrium at temperature T is

$$\frac{N_2}{N_1} = \frac{P(E_2)}{P(E_1)} = e^{-(E_2-E_1)/k_B T}$$

where N_2 is the number in the upper level, N_1 is the number in the lower level, and ΔE is the energy separation of the two levels. Let us use this equation to determine the temperature at which approximately 10% of the hydrogen atoms are in the $n = 2$ state.

$$\frac{N_2}{N_1} = 0.10 = e^{-(10.2 \text{ eV})/k_B T}$$

or

$$\ln(0.10) = -\frac{10.2 \text{ eV}}{k_B T}$$

Solving for T gives

$$T = -\frac{10.2 \text{ eV}}{k_B \ln(0.10)} = -\frac{10.2 \text{ eV}}{(8.62 \times 10^{-5} \text{ eV/K})\ln(0.10)}$$
$$= 51{,}000 \text{ K}$$

Thus the two estimates agree in order of magnitude and show that the Balmer series will only be seen in absorption if the absorbing gas is quite hot, as in a stellar atmosphere.

The final comment on Bohr's work concerns his development of electronic shell theory to treat multielectron atoms. In part II of his paper he attempted to find stable electronic arrangements subject to the conditions that the total angular momentum of all the electrons is quantized and, simultaneously, that the total energy is a minimum. This is a difficult problem, and one that becomes more difficult as more electrons are introduced into a system. Nevertheless, Bohr had considerable success in explaining the chemical activity of multielectron atoms. For example, he was able to show that neutral hydrogen could add another electron to become H^-, and that neutral helium was particularly stable with a closed innermost shell of two electrons and a high ionization potential. He also proposed that lithium ($Z = 3$) had an electronic arrangement consisting of two electrons in one orbit near the nucleus and the third in a large, loosely bound outer orbit. This explains the tendency of lithium atoms to lose an electron and "take a positive charge in chemical combinations with other elements." Although we cannot afford the luxury of looking in detail at all of Bohr's predictions about multielectron atoms, his basic ideas of shell structure are as follows:

- Electrons of elements with higher atomic number form stable concentric rings, with definite numbers of electrons allowed for each ring or shell.
- The number of electrons in the outermost ring determines the valency.[16]

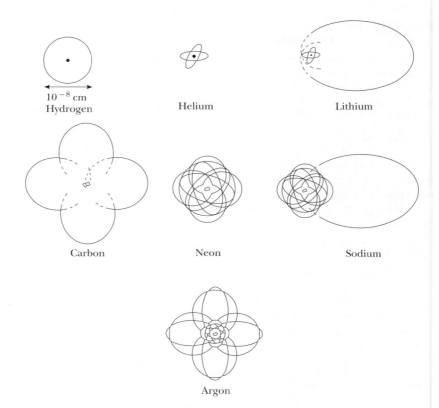

Figure 4.25 Bohr's sketches of electronic orbits.

[16]G. N. Lewis, an American chemist, contributed much to our understanding of shell structure in 1916, building on Bohr's remarkable foundation.

With almost magical insight, Bohr ended part II of his classic paper with the explanation of the similar chemical properties of the iron group (Fe, Co, Ni) and the rare earths, which have atomic numbers that progressively increase by 1 and would not normally be expected to be chemically alike. The answer, according to Bohr, is that the configuration of electrons in the outermost ring of these elements is identical and that it is energetically more favorable to add electrons to inner rings. At the risk of encouraging some to take the idea of electronic orbits too seriously, Figure 4.25 shows some sketches of electronic orbits as drawn by Bohr in the early 1900s.

4.4   BOHR'S CORRESPONDENCE PRINCIPLE, OR WHY IS ANGULAR MOMENTUM QUANTIZED?

Where others might have left a wild and lawless gap between the revolutionary new laws that apply to atomic systems and those that hold for classical systems, Bohr provided a gentle and refined continuum in the form of the **correspondence principle.** This principle states that **predictions of quantum theory must correspond to the predictions of classical physics in the region of sizes where classical theory is known to hold.** These classical sizes for length, mass, and time are on the order of centimeters, grams, and seconds and typically involve very large quantum numbers, as can be seen by calculating n for a hydrogen atom with a radius of 1 cm. If the quantum number becomes large because of increased size or mass, we may state the correspondence principle symbolically as

$$\lim_{n \to \infty} [\text{quantum physics}] = [\text{classical physics}]$$

where n is a typical quantum number of the system such as the quantum number for hydrogen. In the hands of Bohr, the correspondence principle became a masterful tool to test new quantum results as well as a source of fundamental postulates about atomic systems. In fact, Bohr used reasoning of this type to arrive at the concept of the quantization of the electron's orbital angular momentum. Both Bohr's idea of discrete, nonradiating energy states and the emission postulate for atoms were foreshadowed by Planck's quantization of the energy of blackbody oscillators and by Einstein's treatment of the photoelectric effect. However, the concept of angular momentum quantization seems to have sprung full blown from Bohr's *Gedankenkuche* (thought kitchen), as so aptly expressed by Einstein. Indeed, in some of his later writings Bohr emphasized the point of view that the quantization of angular momentum was a postulate, underivable from any deeper law, and that its validity depended simply on the agreement of his model with experimental spectra.

What is most interesting is that in his 1913 paper Bohr ingeniously showed that the quantization of angular momentum is a consequence of the smooth and gradual emergence of classical results from quantum theory in the limit of large quantum number. In particular, Bohr argued that according to his correspondence principle, the quantum condition for emission ($\Delta E = hf$) and Maxwell's classical radiation theory (electronic charges with orbital frequency f radiate light waves of frequency f) *must simultaneously hold for the case of extremely large electronic orbits.* This case is

Correspondence principle

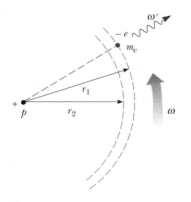

Figure 4.26 The classical limit of the Bohr atom. Note that r_1 and r_2 are the radii of the two adjacent quantum orbits, in which the electron has orbital angular frequencies of ω_1 and ω_2. We assume $r_1 \approx r_2 \approx r$ and $\omega_1 \approx \omega_2 \approx \omega$; ω' is the angular frequency of a photon emitted in a transition from r_1 to r_2.

shown in Figure 4.26. In this figure, r_1 and r_2 are the radii of two large adjacent orbits that are separated in energy by an amount dE and ω is the orbital angular frequency of the electron, where ω is approximately constant in a transition between large orbits. (The algebra is simpler if we use angular frequency instead of frequency. Recall that the connection is $\omega = 2\pi f$.) Because we want to determine the allowed values of the angular momentum from the known change in the atom's energy when light is emitted, we need the relation between the total energy of the atom, $E = -ke^2/2r$ (Equation 4.27) and the magnitude of the total angular momentum of the atom, $L = m_e v r = m_e \omega r^2$. Using the fact that the electron is kept in orbit by the Coulomb force, it is not difficult to show that $1/r = m_e ke^2/L^2$ (see Problem 30), so Equation 4.27 becomes

$$E = -\frac{1}{2} \frac{m_e k^2 e^4}{L^2} \tag{4.37}$$

Taking a derivative of Equation 4.37 gives the desired connection between the change in energy and the change in angular momentum for the Bohr atom.

$$\frac{dE}{dL} = \frac{m_e k^2 e^4}{L^3} \tag{4.38}$$

Finally, we obtain dE/dL in terms of ω, the electron orbital angular frequency, by using $L^3 = m_e k^2 e^4/\omega$ (see Problem 30). Thus,

$$\frac{dE}{dL} = \frac{m_e k^2 e^4}{(m_e k^2 e^4/\omega)} = \omega \tag{4.39}$$

Now consider the emission of a photon of energy $dE = \hbar\omega'$ when the electron makes a transition from r_1 to r_2. Equation 4.39 becomes

$$dE = \omega\, dL$$

or

$$\hbar\omega' = \omega\, dL \tag{4.40}$$

where ω' is the photon angular frequency and ω is the electron orbital angular frequency. Ordinarily, ω' and ω are not simply related. However, because we are dealing with large orbits in this situation, the correspondence principle tells us that the quantum theory must predict the same frequency for the emitted light as Maxwell's law of radiation. Because Maxwell's classical theory requires the electron to radiate light of the same frequency as its orbital motion frequency, $\omega' = \omega$, and Equation 4.40 becomes

$$\hbar\omega = \omega\, dL$$

or

$$dL = \hbar \tag{4.41}$$

Equation 4.41 shows that the change in electronic angular momentum for a transition between adjacent, large electronic orbits is *always* $\hbar$. This means that the magnitude of total angular momentum of the electron in a specific orbit may be taken to have a value equal to an integral multiple of $\hbar$, or

$$L = m_e v r = n\hbar \tag{4.42}$$

for n = large integers.

Bohr realized that although Equation 4.42 was derived for the case of large electron orbits, it was a universal quantum principle applicable to all systems and of wider applicability than Maxwell's law of radiation. Such a bold and far-seeing vision was characteristic of the man so aptly described by Einstein in the following quote: "That this insecure and contradictory foundation [physics from 1910 to 1920] was sufficient to enable a man of Bohr's unique instinct and tact to discover the major laws of the spectral lines and of the electron shells of the atoms together with their significance for chemistry appeared to me like a miracle—and appears to me as a miracle even today. This is the highest form of musicality in the sphere of thought."

4.5   DIRECT CONFIRMATION OF ATOMIC ENERGY LEVELS: THE FRANCK–HERTZ EXPERIMENT

In the preceding sections we have shown the involved trail of reasoning indirectly proving the existence of quantized energy levels in atoms from observations of the optical line spectra emitted by different elements. Now we turn to a simpler and more direct experimental proof of the existence of discrete energy levels in atoms involving their excitation by collision with low-energy electrons. The first experiment of this type was performed by German physicists James Franck and Gustav Hertz (a nephew of Heinrich Hertz) in 1914 on mercury (Hg) atoms. It provided clear experimental proof of the existence of quantized energy levels in atoms and showed that the levels deduced from electron bombardment agreed with those deduced from optical line spectra. Furthermore, it confirmed the universality of energy quantization in atoms, because the quite different physical processes of photon emission and electron bombardment yielded the same energy levels.

Figure 4.27 shows a schematic of a typical college laboratory device similar to the Franck–Hertz apparatus. Electrons emitted by the filament are accelerated over a relatively long region (≈ 1 cm) by the positive potential on the

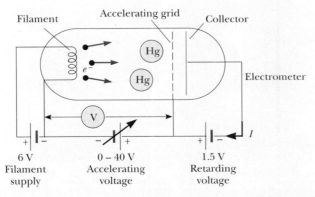

Figure 4.27 Franck–Hertz apparatus. A drop of pure mercury is sealed into an evacuated tube. The tube is heated to 185°C during measurements to provide a high-enough density of mercury to ensure many electron–atom collisions.

grid, V. The electrons can reach the collector and be registered on the electrometer (a sensitive ammeter) if they have sufficient energy to overcome the retarding potential of about 1.5 V set up over the short distance (≈ 1 mm) between grid and collector. At low electron energies or accelerating voltages, perfectly elastic collisions occur between the electrons and Hg atoms in which the *sum* of the kinetic energies of both electron and atom are conserved. Because the Hg atom is much more massive than the electron, the electron transfers very little kinetic energy to the atom in a collision (see Problem 38). Even after multiple collisions the electron reaches the grid with a kinetic energy of approximately e times V and will be collected if the accelerating voltage V is greater than 1.5 V. When V is modestly increased, more electrons reach the collector and the current, I, rises.

As the accelerating voltage is increased further, a threshold voltage is reached at which inelastic collisions occur at the grid, where the electrons reach an energy of e times V. In these inelastic collisions, electrons can transfer almost all of their kinetic energy to the atom, raising it to its first excited state (see Problem 39 and Question 9). Electrons that have collided inelastically are unable to overcome the retarding potential and consequently I decreases for this threshold voltage. Figure 4.28 shows a typical plot of current versus accelerating voltage, with the first weak current dip (A) occurring at a threshold voltage of slightly more than 7 V. When the voltage is increased once again, the inelastic collision region moves closer to the filament and the electrons that were stopped by an inelastic collision are reaccelerated, reaching the collector and causing another rise in current (B). Another dip (C) occurs when V is increased enough for an electron to have two successive inelastic collisions: An electron excites an atom halfway between filament and grid, loses all its energy, and is then reaccelerated to excite another atom at the grid, finally ending up with insufficient energy to be collected. This process takes place periodically with increasing grid voltage, giving rise to equally spaced maxima and minima in the I–V curve, as shown in Figure 4.28.

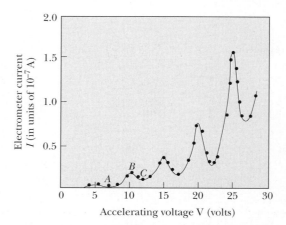

Figure 4.28 Current as a function of voltage in the Franck–Hertz experiment. To obtain these data, the filament voltage was set at 6.0 V and the tube heated to 185°C. (*Data taken by Bob Rodick, Utica College, class of 1992*)

If the adjacent maxima and minima separations of Figure 4.28 are carefully averaged, one finds an average of 4.9 ± 0.1 V, or a ground to first excited state separation of 4.9 ± 0.1 eV. Note, however, that the first minimum does not occur at 4.9 V but at about 7.1 V. The extra energy $(7.1 - 4.9 = 2.2 \text{ eV})$ is required because the filament and collector are made of different metals with different work functions. (Recall that the work function is the energy needed to pull an electron out of a metal—see Chapter 3.) Although the filament, like all good emitters, has a low work function, the collector has a high work function, and this work function energy must be supplied to extract an electron from the collector so a current can flow in the circuit.

As we have seen Franck and Hertz used simple ammeter and voltmeter measurements to show that atoms can only accept discrete amounts of energy from an electron beam. In addition, they showed that the energy levels obtained from electron bombardment agreed with the spectroscopic results. Reasoning that an Hg atom actually excited to an energy level 4.9 eV above its ground state could return to its ground state by emitting a single photon (as Bohr had just postulated), they calculated the wavelength of such a photon to be

$$\Delta E = hf = \frac{hc}{\lambda}$$

or

$$\lambda = \frac{hc}{\Delta E} = \frac{1240 \text{ eV} \cdot \text{nm}}{4.9 \text{ eV}} = 253 \text{ nm} \qquad (4.43)$$

Because glass is not transparent to such ultraviolet radiation, they constructed a quartz apparatus and carefully measured the radiation emitted, finding radiation of wavelength 254 nm to be emitted as soon as the accelerating voltage exceeded 4.9 V. For this direct experimental confirmation of Bohr's basic ideas of discrete energy levels in atoms and the process of photon emission, Franck and Hertz were awarded the Nobel prize in 1925.

SUMMARY

The determination of the composition of atoms relies heavily on four classic experiments:

- **Faraday's law of electrolysis,** which may be stated as

$$m = \frac{(q)(\text{molar mass})}{(96,500 \text{ C})(\text{valence})} \qquad (4.1)$$

  where m is the mass liberated at an electrode and q is the total charge passed through the solution. Faraday's law shows that atoms are composed of positive and negative charges and that atomic charges always consist of multiples of some unit charge.

- **J. J. Thomson's determination of e/m_e** and that the electron is a part of all atoms. Thomson measured e/m_e of electrons from a variety of elements by measuring the deflection of an electron beam by an electric field. He then applied a magnetic field to just cancel the electric deflection in order to determine the electron velocity. The charge-to-mass ratio of the electron in the Thomson experiment is

$$\frac{e}{m_e} = \frac{V\theta}{B^2 \ell d} \tag{4.7}$$

where V/d is the magnitude of the applied electric field, ℓ is the length of the vertical deflecting plates, θ is the deflection produced by the electric field, and B is the applied magnetic field.

- **Millikan's determination of the fundamental charge, _e_.** By balancing the electric and gravitational force on individual oil drops, Millikan was able to determine the fundamental electric charge and to show that charges always occur in multiples of e. The quantum of charge may be determined from the relation

$$ne = \left(\frac{mg}{E}\right)\left(\frac{v + v_1'}{v}\right) \tag{4.11}$$

where n is an integer, m is the mass of the drop, E is the magnitude of the electric field, v is the terminal speed of the drop with field off (falling), and v_1' is the terminal speed of the drop with field on (rising).

- **Rutherford's scattering of α particles from gold atoms,** which established the nuclear model of the atom. By measuring the rate of scattering of α particles into an angle ϕ, Rutherford was able to establish that most of the mass and all of the positive charge of an atom, $+Ze$, are concentrated in a minute volume of the atom with a diameter of about 10^{-14} m.

The explanation of the motion of electrons within the atom and of the rich and elaborate series of spectral lines emitted by the atom was given by Bohr. Bohr's theory was based partly on classical mechanics and partly on some startling new quantum ideas. **Bohr's postulates** were

- Electrons move about the nucleus in circular orbits determined by Coulomb's and Newton's laws.
- Only certain orbits are stable. The electron does not radiate electromagnetic energy in these special orbits, and because the energy is constant with time these are called **stationary states.**
- A spectral line of frequency f is emitted when an electron jumps from an initial orbit of energy E_i to a final orbit of energy E_f, where

$$hf = E_i - E_f \tag{4.23}$$

- The sizes of the stable electron orbits are determined by requiring the electron's angular momentum to be an integral multiple of $\hbar$:

$$m_e vr = n\hbar \qquad n = 1, 2, 3, \ldots \tag{4.24}$$

These postulates lead to quantized orbits and quantized energies for a single electron orbiting a nucleus with charge $+Ze$, given by

$$r_n = \frac{n^2 a_0}{Z} \tag{4.35}$$

and

$$E_n = -\frac{ke^2}{2a_0}\frac{Z^2}{n^2} = -\frac{13.6\ Z^2}{n^2}\ \text{eV} \tag{4.36}$$

where n is an integer and $a_0 = (\hbar^2/m_e k e^2) = 0.529$ Å $= 0.0529$ nm is the Bohr radius.

As a bridge between the familiar domain of classical physics and the more uncertain domain of atomic systems and quantum theory, Bohr provided the **correspondence principle.** This principle states that predictions of quantum theory must correspond to the predictions of classical physics in the region of sizes where classical theory is known to hold.

Direct experimental evidence of the quantized energy of atoms is provided by the Franck–Hertz experiment. This experiment shows that mercury atoms can only accept discrete amounts of energy from a bombarding electron beam.

SUGGESTIONS FOR FURTHER READING

1. G. Holton, *Introduction to Concepts and Theories in Physical Science*, Reading, MA, Addison-Wesley, 1952. This is an accurate, humane, and eminently readable overview of the history and progression of scientific concepts from the Greeks to quantum theory.
2. J. Perrin, *Atoms*, translated by D. L. Hammick, New York, D. Van Nostrand Co., 1923. This is a superb account at first hand of the evidence for atoms at the beginning of the 20th century.
3. G. Thomson, *J. J. Thomson and the Cavendish Laboratory in His Day*, New York, Doubleday and Co., 1965. This book

offers a fascinating account by Thomson's son of the *e/m* experiment and other works.
4. R. A. Millikan, *Electrons (+ and −), Protons, Neutrons, Mesotrons, and Cosmic Rays*, Chicago, University of Chicago Press, 1947. This book gives detailed accounts of the determination of *e*.
5. *Physics Today*, October 1985, Special Issue: Niels Bohr Centennial. This magazine contains three articles dealing with Bohr's scientific, social, and cultural contributions to the physics community. Many interesting photos of the key players are included.

QUESTIONS

1. The Bohr theory of the hydrogen atom is based on several assumptions. Discuss these assumptions and their significance. Do any of these assumptions contradict classical physics?
2. Suppose that the electron in the hydrogen atom obeyed classical mechanics rather than quantum mechanics. Why should such a *hypothetical* atom emit a continuous spectrum rather than the observed line spectrum?
3. Can the electron in the ground state of the hydrogen atom absorb a photon of energy (a) *less* than 13.6 eV and (b) *greater* than 13.6 eV?
4. Explain the concept of an atomic stationary state. Why is this idea of central importance in explaining the stability of the Bohr atom?
5. Does Bohr's correspondence principle apply only to quantum theory? Can you give an example of the application of this principle to relativity theory?
6. On the basis of Bohr's ideas, explain why all emission lines are not seen in absorption.
7. The results of classical measurements and calculations are sometimes called *classical numbers*. Contrast and explain the differences between quantum numbers and classical numbers.
8. What factor causes the finite width of the peaks in the *I–V* curve of the Franck–Hertz experiment?

9. An electron with a kinetic energy of 4.9 eV (mass $= 5.49 \times 10^{-4}$ u) collides inelastically with a stationary mercury atom (mass $= 201$ u). Explain qualitatively why almost 100% of the electron's energy can go into raising the atom to its first excited state.
10. Why don't other current dips corresponding to excitation of the mercury atom's second excited state, third excited state, and so forth show up in the Franck–Hertz experiment? (*Hint:* At the high density of mercury vapor used in the experiment, the probability of a 4.9-eV electron experiencing an inelastic collision is approximately 1.)
11. Four possible transitions for a hydrogen atom are listed here.

$$\text{(A) } n_i = 2; \; n_f = 5$$

$$\text{(B) } n_i = 5; \; n_f = 3$$

$$\text{(C) } n_i = 7; \; n_f = 4$$

$$\text{(D) } n_i = 4; \; n_f = 7$$

(a) Which transition emits the photons having the shortest wavelength? (b) For which transition does the atom gain the most energy? (c) For which transition(s) does the atom lose energy?

PROBLEMS

4.2 The Composition of Atoms

1. Using the Faraday (96,500 C) and Avogadro's number, determine the electronic charge. Explain your reasoning.

2. *Weighing a copper atom in an electrolysis experiment.* A standard experiment involves passing a current of several amperes through a copper sulfate solution ($CuSO_4$) for a period of time and determining the mass of copper plated onto the cathode. If it is found that a current of 1.00 A flowing for 3600 s deposits 1.185 g of copper, find (a) the number of copper atoms deposited, (b) the weight of a copper atom, and (c) the molar mass of copper.

3. A mystery particle enters the region between the plates of a Thomson apparatus as shown in Figure 4.6. The deflection angle θ is measured to be 0.20 radians (downwards) for this particle when $V = 2000$ V, $\ell = 10.0$ cm, and $d = 2.00$ cm. If a perpendicular magnetic field of magnitude 4.57×10^{-2} T is applied simultaneously with the electric field, the particle passes through the plates without deflection. (a) Find q/m for this particle. (b) Identify the particle. (c) Find the horizontal speed with which the particle entered the plates. (d) Must we use relativistic mechanics for this particle?

4. Figure P4.4 shows a cathode ray tube for determining e/m_e without applying a magnetic field. In this case v_x may be found by measuring the rise in temperature when a known amount of charge is stopped in a target. If V, ℓ, d, D, and y are measured, e/m_e may be found. Show that

$$\frac{e}{m_e} = \frac{y v_x^2 d}{V\ell[(\ell/2) + D]}$$

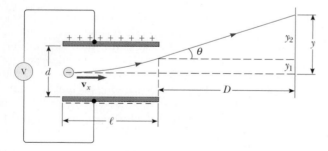

Figure P4.4   Deflection of a charged particle by an electric field.

5. *A Thomson-type experiment with relativistic electrons.* One of the earliest experiments to show that $p = \gamma mv$ (rather than $p = mv$) was that of Neumann. [G. Neumann, *Ann. Physik* 45:529 (1914)]. The apparatus shown in Figure P4.5 is identical to Thomson's except that the source of high-speed electrons is a radioactive radium source and the magnetic field **B** is arranged to act on the electron over its entire trajectory from source to detector. The combined electric and magnetic fields act as a *velocity selector*, only passing electrons with speed v, where $v = V/Bd$ (Equation 4.6), while in the region where there is only a magnetic field the electron moves in a circle of radius r, with r given by $p = Bre$. This latter region ($\mathbf{E} = 0$, $\mathbf{B} = $ constant) acts as a *momentum selector* because electrons with larger momenta have paths with larger radii. (a) Show that the radius of the circle described by the electron is given by $r = (l^2 + y^2)/2y$. (b) Typical values for the Neumann experiment were $d = 2.51 \times 10^{-4}$ m, $B = 0.0177$ T, and $l = 0.0247$ m. For $V = 1060$ V, y, the most critical value, was measured to be 0.0024 ± 0.0005 m. Show that these values disagree with the y value calculated from $p = mv$ but agree with the y value calculated from $p = \gamma mv$ within experimental error. (*Hint:* Find v from Equation 4.6, use $mv = Bre$ or $\gamma mv = Bre$ to find r, and use r to find y.)

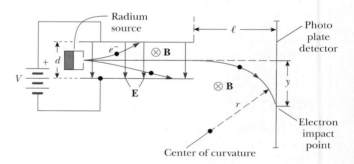

Figure P4.5   The Neumann apparatus.

6. In a Millikan oil-drop experiment, the condenser plates are spaced 2.00 cm apart, the potential across the plates is 4000 V, the rise or fall distance is 4.00 mm, the density of the oil droplets is 0.800 g/cm³, and the viscosity of the air is 1.81×10^{-5} kg·m⁻¹s⁻¹. The average time of fall in the absence of an electric field is 15.9 s. The following different rise times in seconds are observed when the field is turned on: 36.0, 17.3, 24.0, 11.4, 7.54. (a) Find the radius and mass of the drop used in this experiment. (b) Calculate the charge on each drop, and show that charge is quantized by considering both the size of each charge and the amount of charge gained (lost) when the rise time changes. (c) Determine the electronic charge from these data. You may assume that e lies between 1.5 and 2.0×10^{-19} C.

7. Actual data from one of Millikan's early experiments are as follows:

$a = 0.000276$ cm
$\rho = 0.9561$ g/cm^3
Average time of fall = 11.894 s
Rise or fall distance = 10.21 nm
Plate separation = 16.00 mm
Average potential difference between plates = 5085 V
Sequential rise times in seconds: 80.708, 22.336, 22.390, 22.368, 140.566, 79.600, 34.748, 34.762, 29.286, 29.236

Find the average value of e by requiring that the difference in charge for drops with different rise times be equal to an *integral* number of elementary charges.

8. A parallel beam of α particles with fixed kinetic energy is normally incident on a piece of gold foil. (a) If 100 α particles per minute are detected at 20°, how many will be counted at 40°, 60°, 80°, and 100°? (b) If the kinetic energy of the incident α particles is doubled, how many scattered α particles will be observed at 20°? (c) If the original α particles were incident on a copper foil of the same thickness, how many scattered α particles would be detected at 20°? Note that $\rho_{Cu} = 8.9$ g/cm^3 and $\rho_{Au} = 19.3$ g/cm^3.

9. It is observed that α particles with kinetic energies of 13.9 MeV and higher, incident on Cu foils, do not obey Rutherford's $(\sin \phi/2)^{-4}$ law. Estimate the nuclear size of copper from this observation, assuming that the Cu nucleus remains fixed in a head-on collision with an α particle.

10. A typical Rutherford scattering apparatus consists of an evacuated tube containing a polonium-210 α source (5.2-MeV α's), collimators, a gold foil target, and a special alpha-detecting film. The detecting film simultaneously measures all the alphas scattered over a range from $\phi = 2.5°$ to 12.5°. (See Fig. P4.10.) The total number of counts measured over a week's time falling in a specific ring (denoted by its average scattering angle) and the corresponding ring area are given in Table 4.2. (a) Find the counts per area at each angle and correct these values for the angle-independent background. The background correction may be found from a seven-day count taken with the beam blocked with a metal shutter in which 72 counts were measured evenly distributed over the total detector area of 8.50 cm^2. (b) Show that the corrected counts per unit area are proportional to $\sin^{-4}(\phi/2)$ or, in terms of the Rutherford formula, Equation 4.16,

$$\frac{\Delta n}{A} = \frac{C}{\sin^4(\phi/2)}$$

Notes: If a plot of $(\Delta n/A)$ versus ϕ will not fit on a single sheet of graph paper, try plotting $\log(\Delta n/A)$ versus $\log[1/(\sin \phi/2)^4]$. This plot should yield a straight line with a slope of 1 and an intercept that gives C. Explain why this technique works.

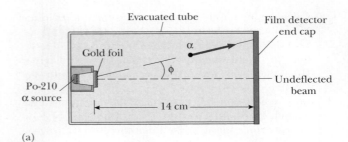

(a)

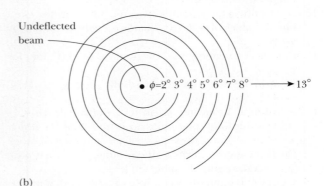

(b)

Figure P4.10   (a) Side view of Rutherford's scattering apparatus: ϕ is the scattering angle. (b) End view of the Rutherford apparatus showing the film detector end cap with grid marking the angle ϕ. The α particles damage the film emulsion and after development show up as dots within the rings.

Table 4.2   Data to Be Used in Problem 10

Angle (degrees)	Counts/Ring	Ring Area (cm^2)	Counts/Area
2.5	605	0.257	
3.5	631	0.360	
4.5	520	0.463	
5.5	405	0.566	
6.5	301	0.669	
7.5	201	0.772	
8.5	122	0.875	
9.5	78	0.987	
10.5	65	1.08	
11.5	66	1.18	
12.5	44	1.29	

4.3   The Bohr Atom

11. Calculate the wavelengths of the first three lines in the Balmer series for hydrogen.

12. Calculate the wavelengths of the first three lines in the Lyman series for hydrogen.

13. (a) What value of n is associated with the Lyman series line in hydrogen whose wavelength is 102.6 nm? (b) Could this wavelength be associated with the Paschen or Brackett series?

14. (a) Use Equation 4.35 to calculate the radii of the first, second, and third Bohr orbits of hydrogen. (b) Find the electron's speed in the same three orbits. (c) Is a relativistic correction necessary? Explain.

15. (a) Construct an energy-level diagram for the He^+ ion, for which $Z = 2$. (b) What is the ionization energy for He^+?

16. Construct an energy level diagram for the Li^{2+} ion, for which $Z = 3$.

17. What is the radius of the first Bohr orbit in (a) He^+, (b) Li^{2+}, and (c) Be^{3+}?

18. A hydrogen atom initially in its ground state ($n = 1$) absorbs a photon and ends up in the state for which $n = 3$. (a) What is the energy of the absorbed photon? (b) If the atom returns to the ground state, what photon energies could the atom emit?

19. A photon is emitted from a hydrogen atom that undergoes an electronic transition from the state $n = 3$ to the state $n = 2$. Calculate (a) the energy, (b) the wavelength, and (c) the frequency of the emitted photon.

20. What is the energy of the photon that could cause (a) an electronic transition from the $n = 4$ state to the $n = 5$ state of hydrogen and (b) an electronic transition from the $n = 5$ state to the $n = 6$ state?

21. (a) Calculate the longest and shortest wavelengths for the Paschen series. (b) Determine the photon energies corresponding to these wavelengths.

22. Find the potential energy and kinetic energy of an electron in the ground state of the hydrogen atom.

23. A hydrogen atom is in its ground state ($n = 1$). Using the Bohr theory of the atom, calculate (a) the radius of the orbit, (b) the linear momentum of the electron, (c) the angular momentum of the electron, (d) the kinetic energy, (e) the potential energy, and (f) the total energy.

24. A hydrogen atom initially at rest in the $n = 3$ state decays to the ground state with the emission of a photon. (a) Calculate the wavelength of the emitted photon. (b) Estimate the recoil momentum of the atom and the kinetic energy of the recoiling atom. Where does this energy come from?

25. Calculate the frequency of the photon emitted by a hydrogen atom making a transition from the $n = 4$ to the $n = 3$ state. Compare your result with the frequency of revolution for the electron in these two Bohr orbits.

26. Calculate the longest and shortest wavelengths in the Lyman series for hydrogen, indicating the underlying electronic transition that gives rise to each. Are any of the Lyman spectral lines in the visible spectrum? Explain.

27. Show that Balmer's formula, $\lambda = C_2 \left(\dfrac{n^2}{n^2 - 2^2} \right)$, reduces to the Rydberg formula, $\dfrac{1}{\lambda} = R \left(\dfrac{1}{2^2} - \dfrac{1}{n^2} \right)$, provided that $(2^2/C_2) = R$. Check that $(2^2/C_2)$ has the same numerical value as R.

28. *The Auger process.* An electron in chromium makes a transition from the $n = 2$ state to the $n = 1$ state without emitting a photon. Instead, the excess energy is transferred to an outer electron (in the $n = 4$ state), which is ejected by the atom. (This is called an *Auger process*, and the ejected electron is referred to as an *Auger electron*.) Use the Bohr theory to find the kinetic energy of the Auger electron.

29. An electron initially in the $n = 3$ state of a one-electron atom of mass M at rest undergoes a transition to the $n = 1$ ground state. (a) Show that the recoil speed of the atom from emission of a photon is given approximately by

$$v = \frac{8hR}{9M}$$

(b) Calculate the percent of the $3 \rightarrow 1$ transition energy that is carried off by the recoiling atom if the atom is deuterium.

30. Apply classical mechanics to an electron in a stationary state of hydrogen to show that $L^2 = m_e k e^2 r$ and $L^3 = m_e k^2 e^4 / \omega$. Here k is the Coulomb constant, L is the magnitude of the orbital angular momentum of the electron, and m_e, e, r, and ω are the mass, charge, orbit radius, and orbital angular frequency of the electron, respectively.

31. (a) Find the frequency of the electron's orbital motion, f_e, around a fixed nucleus of charge $+Ze$ by using Equation 4.24 and $f_e = (v/2\pi r)$ to obtain

$$f_e = \frac{m_e k^2 Z^2 e^4}{2\pi \hbar^3} \left(\frac{1}{n^3} \right)$$

(b) Show that the frequency of the photon emitted when an electron jumps from an outer to an inner orbit can be written

$$f_{photon} = \frac{kZ^2 e^2}{2a_0 h} \left(\frac{1}{n_f^2} - \frac{1}{n_i^2} \right)$$

$$= \frac{m_e k^2 e^4 Z^2}{2\pi \hbar^3} \left(\frac{n_i + n_f}{2 n_i^2 n_f^2} \right) (n_i - n_f)$$

For an electronic transition between adjacent orbits, $n_i - n_f = 1$ and

$$f_{photon} = \frac{m_e k^2 Z^2 e^4}{2\pi \hbar^3} \left(\frac{n_i + n_f}{2 n_i^2 n_f^2} \right)$$

Now examine the factor

$$\left(\frac{n_i + n_f}{2n_i^2 n_f^2}\right)$$

and use $n_i > n_f$ to argue that

$$\frac{1}{n_i^3} < \frac{n_i + n_f}{2n_i^2 n_f^2} < \frac{1}{n_f^3}$$

(c) What do you conclude about the frequency of emitted radiation compared with the frequencies of orbital revolution in the initial and final states? What happens as $n_i \rightarrow \infty$?

32. Wavelengths of spectral lines depend to some extent on the nuclear mass. This occurs because the nucleus is not an infinitely heavy stationary mass and both the electron and nucleus actually revolve around their common center of mass. It can be shown that a system of this type is entirely equivalent to a single particle of reduced mass μ that revolves around the position of the heavier particle at a distance equal to the electron–nucleus separation. See Figure P4.32. Here, $\mu = m_e M/(m_e + M)$, where m_e is the electron mass and M is the nuclear mass. To take the moving nucleus into account in the Bohr theory we replace m_e with μ. Thus Equation 4.30 becomes

$$E_n = \frac{-\mu k e^2}{2m_e a_0}\left(\frac{1}{n^2}\right)$$

and Equation 4.33 becomes

$$\frac{1}{\lambda} = \frac{\mu k e^2}{2m_e a_0 hc}\left(\frac{1}{n_f^2} - \frac{1}{n_i^2}\right) = \left(\frac{\mu}{m_e}\right)R\left(\frac{1}{n_f^2} - \frac{1}{n_i^2}\right)$$

Determine the corrected values of wavelength for the first Balmer line ($n = 3$ to $n = 2$ transition) taking nuclear motion into account for (a) hydrogen, ^{1}H, (b) deuterium, ^{2}H, and (c) tritium, ^{3}H. (Deuterium, was actually discovered in 1932 by Harold Urey, who measured the small wavelength difference between ^{1}H and ^{2}H.)

(a)   (b)

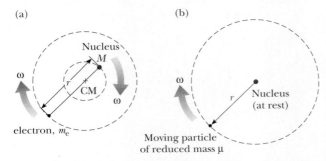

Figure P4.32   (a) Both the electron and the nucleus actually revolve around the center of mass. (b) To calculate the effect of nuclear motion, the nucleus can be considered to be at rest and m_e is replaced by the reduced mass μ.

33. A *muon* is a particle with a charge equal to that of an electron and a mass equal to 207 times the mass of an electron. Muonic lead is formed when ^{208}Pb captures a muon to replace an electron. Assume that the muon moves in such a small orbit that it "sees" a nuclear charge of $Z = 82$. According to the Bohr theory, what are the radius and energy of the ground state of muonic lead? Use the concept of reduced mass introduced in Problem 32.

34. A muon (Problem 33) is captured by a deuteron (an ^{2}H nucleus) to form a muonic atom. (a) Find the energy of the ground state and the first excited state. (b) What is the wavelength of the photon emitted when the atom makes a transition from the first excited state to the ground state? Use the concept of reduced mass introduced in Problem 32.

35. *Positronium* is a hydrogen-like atom consisting of a positron (a positively charged electron) and an electron revolving around each other. Using the Bohr model, find the allowed radii (relative to the center of mass of the two particles) and the allowed energies of the system. Use the concept of reduced mass introduced in Problem 32.

4.4 The Correspondence Principle

36. (a) Calculate the frequency of revolution and the orbit radius of the electron in the Bohr model of hydrogen for $n = 100$, 1000, and 10,000. (b) Calculate the photon frequency for transitions from the n to $n - 1$ states for the same values of n as in part (a) and compare with the revolution frequencies found in part (a). (c) Explain how your results verify the correspondence principle.

37. Use Bohr's model of the hydrogen atom to show that when the atom makes a transition from the state n to the state $n - 1$, the frequency of the emitted light is given by

$$f = \frac{2\pi^2 m_e k^2 e^4}{h^3}\left[\frac{2n - 1}{(n - 1)^2 n^2}\right]$$

Show that as $n \rightarrow \infty$, the preceding expression varies as $1/n^3$ and reduces to the classical frequency one would expect the atom to emit. (*Hint:* To calculate the classical frequency, note that the frequency of revolution is $v/2\pi r$, where r is given by Equation 4.28.) This is an example of the correspondence principle, which requires that the classical and quantum models agree for large values of n.

4.5 The Franck–Hertz Experiment

38. An electron with kinetic energy less than 100 eV collides head-on in an elastic collision with a massive mercury atom at rest. (a) If the electron reverses direction in the collision (like a ball hitting a wall), show that the electron loses only a tiny fraction of its initial kinetic energy, given by

$$\frac{\Delta K}{K} = \frac{4M}{m_e(1 + M/m_e)^2}$$

where m_e is the electron mass and M is the mercury atom mass. (b) Using the accepted values for m_e and M, show that

$$\frac{\Delta K}{K} \approx \frac{4m_e}{M}$$

and calculate the numerical value of $\Delta K/K$.

ADDITIONAL PROBLEMS

39. An electron collides inelastically and head-on with a mercury atom at rest. (a) If the separation of the first excited state and the ground state of the atom is exactly 4.9 eV, what is the minimum initial electron kinetic energy needed to raise the atom to its first excited state and also conserve momentum? Assume that the collision is completely inelastic. (b) What is the initial speed of the electron in this case? (c) What is the speed of the electron and atom after the collision? (d) What is the kinetic energy (in electron volts) of the electron after collision? Is the approximation that the electron loses all of its kinetic energy in an inelastic collision justified?

40. If the Franck–Hertz experiment could be performed with high-density monatomic hydrogen, at what voltage separations would the current dips appear? Take the separation between ground state and first excited state to be exactly 10.2 eV and be sure to consider momentum as well as energy in arriving at your answer.

41. Liquid oxygen has a bluish color, meaning that it preferentially absorbs light toward the red end of the visible spectrum. Although the oxygen molecule (O_2) does not strongly absorb visible radiation, it does absorb strongly at 1269 nm, which is in the infrared region of the spectrum. Research has shown that it is possible for two colliding O_2 molecules to absorb a single photon, sharing its energy equally. The transition that both molecules undergo is the same transition that results when they absorb 1269-nm radiation. What is the wavelength of the single photon that causes this double transition? What is the color of this radiation?

42. Two hydrogen atoms collide head-on and end up with zero kinetic energy. Each then emits a photon with a wavelength of 121.6 nm ($n = 2$ to $n = 1$ transition). At what speed were the atoms moving before the collision?

43 Steven Chu, Claude Cohen-Tannoudji, and William Phillips received the 1997 Nobel prize in physics for "the development of methods to cool and trap atoms with laser light." One part of their work was with a beam of atoms (mass $\sim 10^{-25}$ kg) that move at a speed on the order of 1 km/s, similar to the speed of molecules in air at room temperature. An intense laser light beam tuned to a visible atomic transition (assume 500 nm) is directed straight into the atomic beam. That is, the atomic beam and light beam are traveling in opposite directions. An atom in the ground state immediately absorbs a photon. Total system momentum is conserved in the absorption process. After a lifetime on the order of 10^{-8} s, the excited atom radiates by spontaneous emission. It has an equal probability of emitting a photon in any direction. Thus, the average "recoil" of the atom is zero over many absorption and emission cycles. (a) Estimate the average deceleration of the atomic beam. (b) What is the order of magnitude of the distance over which the atoms in the beam will be brought to a halt?

44. In a hot star, a multiply ionized atom with a single remaining electron produces a series of spectral lines as described by the Bohr model. The series corresponds to electronic transitions that terminate in the same final state. The longest and shortest wavelengths of the series are 63.3 nm and 22.8 nm, respectively. (a) What is the ion? (b) Find the wavelengths of the next three spectral lines nearest to the line of longest wavelength.

5

Matter Waves

In the previous chapter we discussed some important discoveries and theoretical concepts concerning the *particle nature of matter*. We now point out some of the shortcomings of these theories and introduce the fascinating and bizarre *wave properties of particles*. Especially notable are Count Louis de Broglie's remarkable ideas about how to represent electrons (and other particles) as waves and the experimental confirmation of de Broglie's hypothesis by the electron diffraction experiments of Davisson and Germer. We shall also see how the notion of representing a particle as a localized wave or wave group leads naturally to limitations on simultaneously measuring position and momentum of the particle. Finally, we discuss the passage of electrons through a double slit as a way of "understanding" the wave–particle duality of matter.

Figure 5.1 Louis de Broglie was a member of an aristocratic French family that produced marshals, ambassadors, foreign ministers, and at least one duke, his older brother Maurice de Broglie. Louis de Broglie came rather late to theoretical physics, as he first studied history. Only after serving as a radio operator in World War I did he follow the lead of his older brother and begin his studies of physics. Maurice de Broglie was an outstanding experimental physicist in his own right and conducted experiments in the palatial family mansion in Paris. (*AIP Meggers Gallery of Nobel Laureates*)

De Broglie wavelength

5.1   THE PILOT WAVES OF DE BROGLIE

By the early 1920s scientists recognized that the Bohr theory contained many inadequacies:

- It failed to predict the observed intensities of spectral lines.
- It had only limited success in predicting emission and absorption wavelengths for multielectron atoms.
- It failed to provide an equation of motion governing the time development of atomic systems starting from some initial state.
- It overemphasized the particle nature of matter and could not explain the newly discovered wave–particle duality of light.
- It did not supply a general scheme for "quantizing" other systems, especially those without periodic motion.

The first bold step toward a new mechanics of atomic systems was taken by Louis Victor de Broglie in 1923 (Fig. 5.1). In his doctoral dissertation he postulated that *because photons have wave and particle characteristics, perhaps all forms of matter have wave as well as particle properties*. This was a radical idea with no experimental confirmation at that time. According to de Broglie, electrons had a dual particle–wave nature. Accompanying every electron was a wave (not an electromagnetic wave!), which guided, or "piloted," the electron through space. He explained the source of this assertion in his 1929 Nobel prize acceptance speech:

> On the one hand the quantum theory of light cannot be considered satisfactory since it defines the energy of a light corpuscle by the equation $E = hf$ containing the frequency f. Now a purely corpuscular theory contains nothing that enables us to define a frequency; for this reason alone, therefore, we are compelled, in the case of light, to introduce the idea of a corpuscle and that of periodicity simultaneously. On the other hand, determination of the stable motion of electrons in the atom introduces integers, and up to this point the only phenomena involving integers in physics were those of interference and of normal modes of vibration. This fact suggested to me the idea that electrons too could not be considered simply as corpuscles, but that periodicity must be assigned to them also.

Let us look at de Broglie's ideas in more detail. He concluded that the wavelength and frequency of a *matter wave* associated with any moving object were given by

$$\lambda = \frac{h}{p} \tag{5.1}$$

and

$$f = \frac{E}{h} \tag{5.2}$$

where h is Planck's constant, p is the relativistic momentum, and E is the total relativistic energy of the object. Recall from Chapter 2 that p and E can be written as

$$p = \gamma mv \tag{5.3}$$

and

$$E^2 = p^2c^2 + m^2c^4 = \gamma^2 m^2 c^4 \tag{5.4}$$

where $\gamma = (1 - v^2/c^2)^{-1/2}$ and v is the object's speed. Equations 5.1 and 5.2 immediately suggest that it should be easy to calculate the speed of a de Broglie wave from the product λf. However, as we will show later, this is not the speed of the particle. Since the correct calculation is a bit complicated, we postpone it to Section 5.3. Before taking up the question of the speed of matter waves, we prefer first to give some introductory examples of the use of $\lambda = h/p$ and a brief description of how de Broglie waves provide a physical picture of the Bohr theory of atoms.

De Broglie's Explanation of Quantization in the Bohr Model

Bohr's model of the atom had many shortcomings and problems. For example, as the electrons revolve around the nucleus, how can one understand the fact that only certain electronic energies are allowed? Why do all atoms of a given element have precisely the same physical properties regardless of the infinite variety of starting velocities and positions of the electrons in each atom?

De Broglie's great insight was to recognize that although these are deep problems for particle theories, wave theories of matter handle these problems neatly by means of interference. For example, a plucked guitar string, although initially subjected to a wide range of wavelengths, supports only standing wave patterns that have nodes at each end. Thus only a discrete set of wavelengths is allowed for standing waves, while other wavelengths not included in this discrete set rapidly vanish by destructive interference. This same reasoning can be applied to electron matter waves bent into a circle around the nucleus. Although initially a continuous distribution of wavelengths may be present, corresponding to a distribution of initial electron velocities, most wavelengths and velocities rapidly die off. The residual standing wave patterns thus account for the identical nature of all atoms of a given element and show that atoms are more like vibrating drum heads with discrete modes of vibration than like miniature solar systems. This point of view is emphasized in Figure 5.2, which shows the standing wave pattern of the electron in the hydrogen atom corresponding to the $n = 3$ state of the Bohr theory.

Another aspect of the Bohr theory that is also easier to visualize physically by using de Broglie's hypothesis is the quantization of angular momentum. One simply assumes that **the allowed Bohr orbits arise because the electron matter waves interfere constructively when an integral number of wavelengths exactly fits into the circumference of a circular orbit.** Thus

$$n\lambda = 2\pi r \tag{5.5}$$

where r is the radius of the orbit. From Equation 5.1, we see that $\lambda = h/m_e v$. Substituting this into Equation 5.5, and solving for $m_e vr$, the angular momentum of the electron, gives

$$m_e vr = n\hbar \tag{5.6}$$

Note that this is precisely the Bohr condition for the quantization of angular momentum.

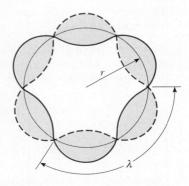

Figure 5.2 Standing waves fit to a circular Bohr orbit. In this particular diagram, three wavelengths are fit to the orbit, corresponding to the $n = 3$ energy state of the Bohr theory.

EXAMPLE 5.1 Why Don't We See the Wave Properties of a Baseball?

An object will appear "wavelike" if it exhibits interference or diffraction, both of which require scattering objects or apertures of about the same size as the wavelength. A baseball of mass 140 g traveling at a speed of 60 mi/h (27 m/s) has a de Broglie wavelength given by

$$\lambda = \frac{h}{p} = \frac{6.63 \times 10^{-34}\,\text{J}\cdot\text{s}}{(0.14\,\text{kg})(27\,\text{m/s})} = 1.7 \times 10^{-34}\,\text{m}$$

Even a nucleus (whose size is $\approx 10^{-15}$ m) is much too large to diffract this incredibly small wavelength! This explains why all macroscopic objects appear particle-like.

EXAMPLE 5.2 What Size "Particles" Do Exhibit Diffraction?

A particle of charge q and mass m is accelerated from rest through a small potential difference V. (a) Find its de Broglie wavelength, assuming that the particle is non-relativistic.

Solution When a charge is accelerated from rest through a potential difference V, its gain in kinetic energy, $\frac{1}{2}mv^2$, must equal the loss in potential energy qV. That is,

$$\tfrac{1}{2}mv^2 = qV$$

Because $p = mv$, we can express this in the form

$$\frac{p^2}{2m} = qV \quad \text{or} \quad p = \sqrt{2mqV}$$

Substituting this expression for p into the de Broglie relation $\lambda = h/p$ gives

$$\lambda = \frac{h}{p} = \frac{h}{\sqrt{2mqV}}$$

(b) Calculate λ if the particle is an electron and $V = 50$ V.

Solution The de Broglie wavelength of an electron accelerated through 50 V is

$$\lambda = \frac{h}{\sqrt{2m_e qV}}$$

$$= \frac{6.63 \times 10^{-34}\,\text{J}\cdot\text{s}}{\sqrt{2(9.11 \times 10^{-31}\,\text{kg})(1.6 \times 10^{-19}\,\text{C})(50\,\text{V})}}$$

$$= 1.7 \times 10^{-10}\,\text{m} = 1.7\,\text{Å}$$

This wavelength is of the order of atomic dimensions and the spacing between atoms in a solid. Such low-energy electrons are routinely used in electron diffraction experiments to determine atomic positions on a surface.

Exercise 1 (a) Show that the de Broglie wavelength for an electron accelerated from rest through a *large* potential difference, V, is

$$\lambda = \frac{12.27}{V^{1/2}}\left(\frac{Ve}{2m_e c^2} + 1\right)^{-1/2} \tag{5.7}$$

where λ is in angstroms (Å) and V is in volts. (b) Calculate the percent error introduced when $\lambda = 12.27/V^{1/2}$ is used instead of the correct relativistic expression for 10 MeV electrons.

Answer (b) 230%.

5.2 THE DAVISSON–GERMER EXPERIMENT

Direct experimental proof that electrons possess a wavelength $\lambda = h/p$ was furnished by the diffraction experiments of American physicists Clinton J. Davisson (1881–1958) and Lester H. Germer (1896–1971) at the Bell Laboratories in New York City in 1927 (Fig. 5.3).[1] In fact, de Broglie had already suggested in 1924 that a stream of electrons traversing a small aperture should exhibit diffraction phenomena. In 1925, Einstein was led to the necessity of postulating matter waves from an analysis of fluctuations of a molecular gas. In addition, he noted that a molecular beam should show small but measurable diffraction effects. In the same year, Walter Elsasser pointed out that the slow

---

[1] C. J. Davisson and L. H. Germer, *Phys. Rev.* 30:705, 1927.

THE DAVISSON–GERMER EXPERIMENT

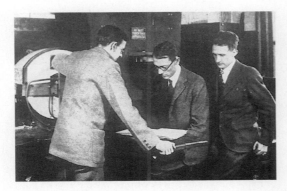

Figure 5.3   Clinton J. Davisson (left) and Lester H. Germer (center) at Bell Laboratories in New York City. (*Bell Laboratories, courtesy AIP Emilio Segrè Visual Archives*)

electron scattering experiments of C. J. Davisson and C. H. Kunsman at the Bell Labs could be explained by electron diffraction.

Clear-cut proof of the wave nature of electrons was obtained in 1927 by the work of Davisson and Germer in the United States and George P. Thomson (British physicist, 1892–1975, the son of J. J. Thomson) in England. Both cases are intriguing not only for their physics but also for their human interest. The first case was an accidental discovery, and the second involved the discovery of the particle properties of the electron by the father and the wave properties by the son.

The crucial experiment of Davisson and Germer was an offshoot of an attempt to understand the arrangement of atoms on the surface of a nickel sample by elastically scattering a beam of low-speed electrons from a polycrystalline nickel target. A schematic drawing of their apparatus is shown in Figure 5.4. Their device allowed for the variation of three experimental parameters—electron energy; nickel target orientation, α; and scattering angle, ϕ. Before a fortunate accident occurred, the results seemed quite pedestrian. For constant electron energies of about 100 eV, the scattered intensity rapidly decreased as ϕ increased. But then someone dropped a flask of liquid air on the glass vacuum system, rupturing the vacuum and oxidizing the nickel target, which had been at high temperature. To remove the oxide, the sample was reduced by heating it cautiously[2] in a flowing stream of hydrogen. When the apparatus was reassembled, quite different results were found: Strong variations in the intensity of scattered electrons with angle were observed, as shown in Figure 5.5. The prolonged heating had evidently annealed the nickel target, causing large single-crystal regions to develop in the polycrystalline sample. These crystalline regions furnished the extended regular lattice needed to observe electron diffraction. Once Davisson and Germer realized that it was the elastic scattering from *single crystals* that produced such unusual results (1925), they initiated a thorough investigation of elastic scattering from large single crystals

[2]At present this can be done without the slightest fear of "stinks or bangs," because 5% hydrogen–95% argon safety mixtures are commercially available.

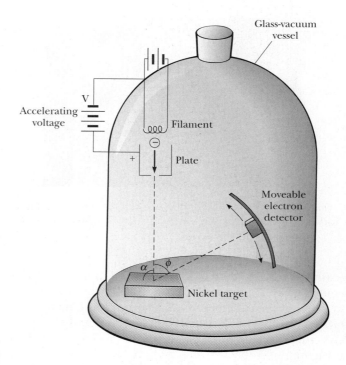

Figure 5.4   A schematic diagram of the Davisson–Germer apparatus.

with predetermined crystallographic orientation. Even these experiments were not conducted at first as a test of de Broglie's wave theory, however. Following discussions with Richardson, Born, and Franck, the experiments and their analysis finally culminated in 1927 in the proof that electrons experience diffraction with an electron wavelength that is given by $\lambda = h/p$.

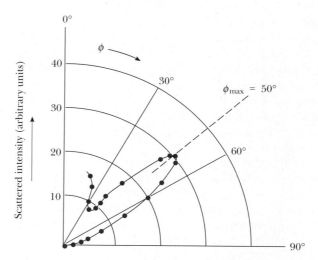

Figure 5.5   A polar plot of scattered intensity versus scattering angle for 54-eV electrons, based on the original work of Davisson and Germer. The scattered intensity is proportional to the distance of the point from the origin in this plot.

The idea that electrons behave like waves when interacting with the atoms of a crystal is so striking that Davisson and Germer's proof deserves closer scrutiny. In effect, they calculated the wavelength of electrons from a simple diffraction formula and compared this result with de Broglie's formula $\lambda = h/p$. Although they tested this result over a wide range of target orientations and electron energies, we consider in detail only the simple case shown in Figures 5.4 and 5.5 with $\alpha = 90.0°$, $V = 54.0$ V, and $\phi = 50.0°$, corresponding to the $n = 1$ diffraction maximum. In order to calculate the de Broglie wavelength for this case, we first obtain the velocity of a nonrelativistic electron accelerated through a potential difference V from the energy relation

$$\tfrac{1}{2} m_e v^2 = eV$$

Substituting $v = \sqrt{2Ve/m_e}$ into the de Broglie relation gives

$$\lambda = \frac{h}{m_e v} = \frac{h}{\sqrt{2Vem_e}} \tag{5.8}$$

Thus the wavelength of 54.0-V electrons is

$$\lambda = \frac{6.63 \times 10^{-34}\,\text{J·s}}{\sqrt{2(54.0\ \text{V})(1.60 \times 10^{-19}\ \text{C})(9.11 \times 10^{-31}\ \text{kg})}}$$
$$= 1.67 \times 10^{-10}\ \text{m} = 1.67\ \text{Å}$$

The experimental wavelength may be obtained by considering the nickel atoms to be a reflection diffraction grating, as shown in Figure 5.6. Only the surface layer of atoms is considered because low-energy electrons, unlike x-rays, do not penetrate deeply into the crystal. Constructive interference occurs when the path length difference between two adjacent rays is an integral number of wavelengths or

$$d \sin \phi = n\lambda \tag{5.9}$$

As d was known to be 2.15 Å from x-ray diffraction measurements, Davisson and Germer calculated λ to be

$$\lambda = (2.15\ \text{Å})(\sin 50.0°) = 1.65\ \text{Å}$$

in excellent agreement with the de Broglie formula.

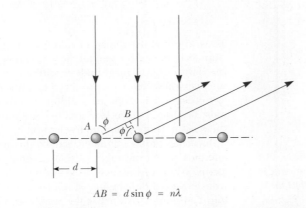

$$AB = d \sin \phi = n\lambda$$

Figure 5.6   Constructive interference of electron matter waves scattered from a single layer of atoms at an angle ϕ.

Figure 5.7   Diffraction of 50-kV electrons from a film of Cu_3Au. The alloy film was 400 Å thick. (*Courtesy of the late Dr. L. H. Germer*)

Figure 5.8 Diffraction pattern of neutrons produced by a single crystal of NaCl. (*Courtesy of Dr. E. O. Wollan*)

It is interesting to note that while the diffraction lines from low-energy reflected electrons are quite broad (see Fig. 5.5), the lines from high-energy electrons transmitted through metal foils are quite sharp (see Fig. 5.7). This effect occurs because hundreds of atomic planes are penetrated by high-energy electrons, and consequently Equation 5.9, which treats diffraction from a surface layer, no longer holds. Instead, the Bragg law, $2d \sin \theta = n\lambda$, applies to high-energy electron diffraction. The maxima are extremely sharp in this case because if $2d \sin \theta$ is not exactly equal to $n\lambda$, there will be no diffracted wave. This occurs because there are scattering contributions from so many atomic planes that eventually the path length difference between the wave from the first plane and some deeply buried plane will be *an odd multiple of $\lambda/2$*, resulting in complete cancellation of these waves (see Problem 13).

If de Broglie's postulate is true for all matter, then any object of mass m has wavelike properties and a wavelength $\lambda = h/p$. In the years following Davisson and Germer's discovery, experimentalists tested the universal character of de Broglie's postulate by searching for diffraction of other "particle" beams. In subsequent experiments, diffraction was observed for helium atoms (Estermann and Stern in Germany) and hydrogen atoms (Johnson in the United States). Following the discovery of the neutron in 1932, it was shown that neutron beams of the appropriate energy also exhibit diffraction when incident on a crystalline target (Fig. 5.8).

EXAMPLE 5.3   Thermal Neutrons

What kinetic energy (in electron volts) should neutrons have if they are to be diffracted from crystals?

Solution Appreciable diffraction will occur if the de Broglie wavelength of the neutron is of the same order of magnitude as the interatomic distance. Taking $\lambda = 1.00$ Å, we find

$$p = \frac{h}{\lambda} = \frac{6.63 \times 10^{-34}\, \text{J} \cdot \text{s}}{1.00 \times 10^{-10}\, \text{m}} = 6.63 \times 10^{-24}\, \text{kg} \cdot \text{m/s}$$

The kinetic energy is given by

$$K = \frac{p^2}{2m_n} = \frac{(6.63 \times 10^{-24}\, \text{J} \cdot \text{s})^2}{2(1.66 \times 10^{-27}\, \text{kg})}$$

$$= 1.32 \times 10^{-20}\, \text{J} = 0.0825\, \text{eV}$$

Note that these neutrons are nonrelativistic because K is much less than the neutron rest energy of 940 MeV, and so our use of the classical expression $K = p^2/2m_n$ is justified. Because the average thermal energy of a par-

ticle in thermal equilibrium is $\frac{1}{2}k_B T$ for each independent direction of motion, neutrons at room temperature (300 K) possess a kinetic energy of

$$K = \frac{3}{2}k_B T = (1.50)(8.62 \times 10^{-5}\, \text{eV/K})(300\, \text{K})$$

$$= 0.0388\, \text{eV}$$

Thus "thermal neutrons," or neutrons in thermal equilibrium with matter at room temperature, possess energies of the right order of magnitude to diffract appreciably from single crystals. Neutrons produced in a nuclear reactor are far too energetic to produce diffraction from crystals and must be slowed down in a graphite column as they leave the reactor. In the graphite moderator, repeated collisions with carbon atoms ultimately reduce the average neutron energies to the average thermal energy of the carbon atoms. When this occurs, these so-called thermalized neutrons possess a distribution of velocities and a corresponding distribution of de Broglie wavelengths with average wavelengths comparable to crystal spacings.

Exercise 2   *Monochromatic Neutrons.* A beam of neutrons with a single wavelength may be produced by means of a mechanical velocity selector of the type shown in Figure 5.9. (a) Calculate the speed of neutrons with a wavelength of 1.00 Å. (b) What rotational speed (in rpm) should the shaft have in order to pass neutrons with wavelength of 1.00 Å?

Answers   (a) 3.99×10^3 m/s. (b) 13,300 rev/min.

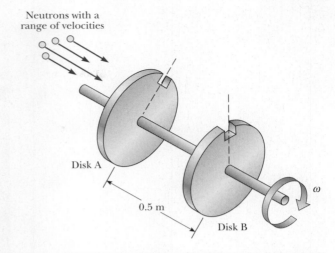

Neutrons with a range of velocities

Disk A

0.5 m

Disk B

ω

Figure 5.9   A neutron velocity selector. The slot in disk B lags the slot in disk A by 10°.

The Electron Microscope

The idea that electrons have a controllable wavelength that can be made much shorter than visible light wavelengths and, accordingly, possess a much better ability to resolve fine details was only one of the factors that led to the development of the electron microscope. In fact, ideas of such a device were tossed about in the cafés and bars of Paris and Berlin as early as 1928. What really made the difference was the coming together of several lines of development—electron tubes and circuits, vacuum technology, and electron beam control—all pioneered in the development of the cathode ray tube (CRT). These factors led to the construction of the first transmission electron microscope (TEM) with magnetic lenses by electrical engineers Max Knoll and Ernst Ruska in Berlin in 1931. The testament to the fortitude and brilliance of Knoll and Ruska in overcoming the "cussedness of objects" and building and getting such a complicated experimental device to work for the first time is shown in Figure 5.10. It is remarkable that although the overall performance of the TEM has been improved thousands of times since its invention, it is basically the same in principle as that first designed by Knoll and Ruska: a device that focuses electron beams with magnetic lenses and creates a flat-looking two-dimensional shadow pattern on its screen, the result of varying degrees of electron transmission through the object. Figure 5.11a is a diagram showing this basic design and Figure 5.11b shows, for comparison, an optical projection microscope. The best optical microscopes using ultraviolet light have a magnification of about 2000 and can resolve two objects separated by 100 nm, but a TEM using electrons accelerated through 100 kV has a magnification of as much as 1,000,000 and a maximum resolution of 0.2 nm. In practice, magnifications of 10,000 to 100,000 are easier to use. Figure 5.12 shows typical TEM micrographs of microbes, Figure 5.12b showing a microbe and its DNA strands magnified 40,000 times. Although it would seem that increasing electron energy should lead to shorter electron wavelength and increased resolution, imperfections or aberrations in the magnetic lenses actually set the limit of resolution at about 0.2 nm. Increasing electron energy above 100 keV does not

Ernst Ruska played a major role in the invention of the TEM. He was awarded the Nobel prize in physics for this work in 1986. (*AIP Emilio Segre Visual Archives, W. F. Meggers Gallery of Nobel Laureates*)

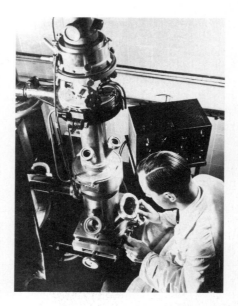

Figure 5.10   Ruska's 1934 electron microscope. (*Siemens Archives Munich*)

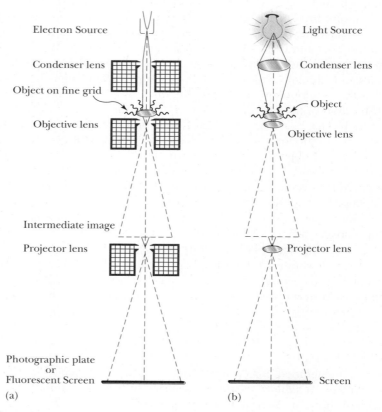

Figure 5.11   (a) Schematic drawing of a transmission electron microscope with magnetic lenses. (b) Schematic of a light-projection microscope.

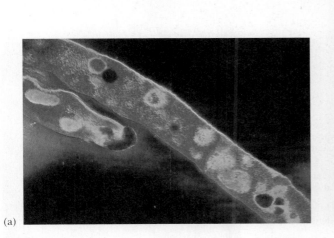

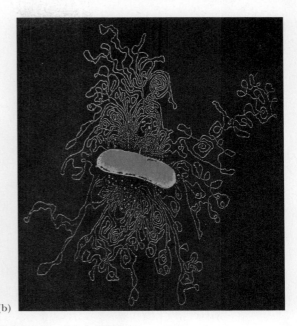

(a)   (b)

Figure 5.12 (a) A false-color TEM micrograph of tuberculosis bacteria. (b) A TEM micrograph of a microbe leaking DNA (×40,000). (*CNRI/Photo Researchers, Inc., Dr. Gopal Murti/Photo Researchers, Inc.*)

improve resolution—it only permits electrons to sample regions deeper inside an object. Figures 5.13a and 5.13b show, respectively, a diagram of a modern TEM and a photo of the same instrument.

A second type of electron microscope with less resolution and magnification than the TEM, but capable of producing striking three-dimensional images, is the scanning electron microscope (SEM). Figure 5.14 shows dramatic three-dimensional SEM micrographs made possible by the large range of focus (depth of field) of the SEM, which is several hundred times better than that of a light microscope. The SEM was the brainchild of the same Max Knoll who helped invent the TEM. Knoll had recently moved to the television department at Telefunken when he conceived of the idea in 1935. The SEM produces a sort of giant television image by collecting electrons scattered from an object, rather than light. The first operating scanning microscope was built by M. von Ardenne in 1937, and it was extensively developed and perfected by Vladimir Zworykin and collaborators at RCA Camden in the early 1940s.

Figure 5.15 shows how a typical SEM works. Such a device might be operated with 20-keV electrons and have a resolution of about 10 nm and a magnification ranging from 10 to 100,000. As shown in Figure 5.15, an electron beam is sharply focused on a specimen by magnetic lenses and then scanned (rastered) across a tiny region on the surface of the specimen. The high-energy primary beam scatters lower-energy secondary electrons out of the object depending on specimen composition and surface topography. These secondary electrons are detected by a plastic scintillator coupled to a photomultiplier, amplified, and used to modulate the brightness of a simultaneously

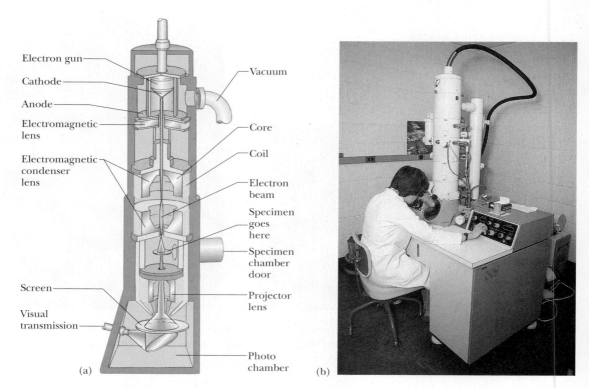

Figure 5.13 (a) Diagram of a transmission electron microscope. (b) A photo of the same TEM. (*W. Ormerod/Visuals Unlimited*)

rastered display CRT. The ratio of the display raster size to the microscope electron beam raster size determines the magnification. Modern SEM's can also collect x-rays and high-energy electrons from the specimen to detect chemical elements at certain locations on the specimen's surface, thus answering the bonus question, "Is the bitty bump on the bilayer boron or bismuth?"

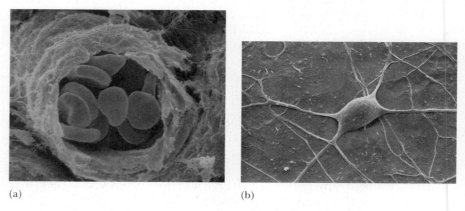

Figure 5.14 (a) A SEM micrograph showing blood cells in a tiny artery. (b) A SEM micrograph of a single neuron (×4000). (*P. Motta & S. Correr/Photo Researchers, Inc., David McCarthy/Photo Researchers, Inc.*)

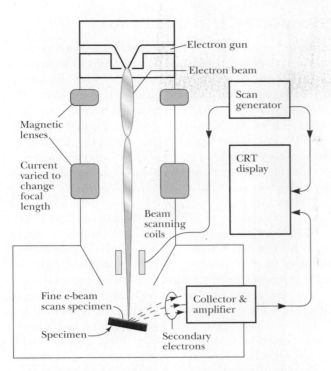

Figure 5.15   The working parts of a scanning electron microscope.

The newer, higher-resolution scanning tunneling microscope (STM) and atomic force microscope (AFM), which can image individual atoms and molecules, are discussed in Chapter 7. These instruments are exciting not only for their superb pictures of surface topography and individual atoms (see Figure 5.16 for an AFM picture) but also for their potential as microscopic machines capable of detecting and moving a few atoms at a time in proposed microchip terabit memories and mass spectrometers.

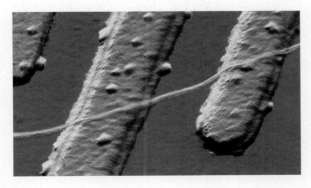

Figure 5.16   World's smallest electrical wire. An AFM image of a carbon nanotube wire on platinum electrodes. The wire is 1.5 nm wide, a mere 10 atoms. The magnification is 120,000. (*Delft University of Technology/Photo Researchers, Inc.*)

5.3 WAVE GROUPS AND DISPERSION

The matter wave representing a moving particle must reflect the fact that the particle has a large probability of being found in a small region of space only at a specific time. This means that a traveling sinusoidal matter wave of infinite extent and constant amplitude cannot properly represent a localized moving particle. What is needed is a pulse, or "wave group," of limited spatial extent. Such a pulse can be formed by adding sinusoidal waves with different wavelengths. The resulting wave group can then be shown to move with a speed v_g (the group speed) identical to the classical particle speed. This argument is shown schematically in Figure 5.17 and will be treated in detail after the introduction of some general ideas about wave groups.

Actually, all observed waves are limited to definite regions of space and are called *pulses, wave groups,* or *wave packets* in the case of matter waves. The plane wave with an exact wavelength and infinite extension is an abstraction. Water waves from a stone dropped into a pond, light waves emerging from a briefly opened shutter, a wave generated on a taut rope by a single flip of one end, and a sound wave emitted by a discharging capacitor must all be modeled by wave groups. A wave group consists of a superposition of waves with *different wavelengths*, with the amplitude and phase of each component wave adjusted so that the waves interfere constructively over a small region of space. Outside of this region the combination of waves produces a net amplitude that approaches zero rapidly as a result of destructive interference. Perhaps the most familiar physical example in which wave groups arise is the phenomenon of beats. Beats occur when two sound waves of slightly different wavelength (and hence different frequency) are combined. The resultant sound wave has a frequency equal to the average of the two combining waves and an amplitude that fluctuates, or "beats," at a rate

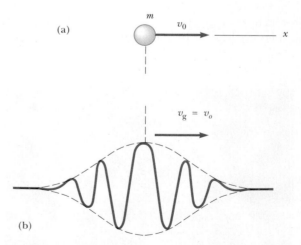

Figure 5.17 Representing a particle with matter waves: (a) particle of mass m and speed v_0; (b) superposition of many matter waves with a spread of wavelengths centered on $\lambda_0 = h/mv_0$ correctly represents a particle.

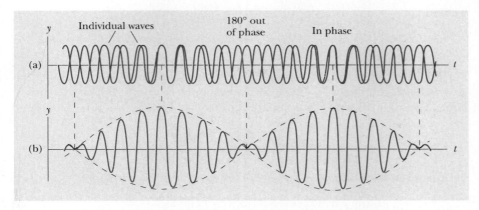

Figure 5.18   Beats are formed by the combination of two waves of slightly different frequency traveling in the same direction. (a) The individual waves. (b) The combined wave has an amplitude (broken line) that oscillates in time.

given by the difference of the two original frequencies. This case is illustrated in Figure 5.18.

Let us examine this situation mathematically. Consider a one-dimensional wave propagating in the positive x direction with a *phase speed* v_p. Note that v_p is the speed of a point of constant phase on the wave, such as a wave crest or trough. This traveling wave with wavelength λ, frequency f, and amplitude A may be described by

$$y = A \cos\left(\frac{2\pi x}{\lambda} - 2\pi f t\right) \tag{5.10}$$

where λ and f are related by

$$v_p = \lambda f \tag{5.11}$$

A more compact form for Equation 5.10 results if we take $\omega = 2\pi f$ (where ω is *the angular frequency*) and $k = 2\pi/\lambda$ (where k *is the wavenumber*). With these substitutions the infinite wave becomes

$$y = A \cos(kx - \omega t) \tag{5.12}$$

with

$$v_p = \frac{\omega}{k} \tag{5.13}$$ **Phase velocity**

Let us now form the superposition of two waves of equal amplitude both traveling in the positive x direction but with slightly different wavelengths, frequencies, and phase velocities. The resultant amplitude y is given by

$$y = y_1 + y_2 = A \cos(k_1 x - \omega_1 t) + A \cos(k_2 x - \omega_2 t)$$

Using the trigonometric identity

$$\cos a + \cos b = 2 \cos \tfrac{1}{2}(a - b) \cdot \cos \tfrac{1}{2}(a + b)$$

we find

$$y = 2A \cos \tfrac{1}{2}\{(k_2 - k_1)x - (\omega_2 - \omega_1)t\} \cdot \cos \tfrac{1}{2}\{(k_1 + k_2)x - (\omega_1 + \omega_2)t\} \tag{5.14}$$

For the case of two waves with slightly different values of k and ω, we see that $\Delta k = k_2 - k_1$ and $\Delta \omega = \omega_2 - \omega_1$ are small, but $(k_1 + k_2)$ and $(\omega_1 + \omega_2)$ are large. Thus, Equation 5.14 may be interpreted as a broad sinusoidal envelope

$$2A \cos\left(\frac{\Delta k}{2}x - \frac{\Delta \omega}{2}t\right)$$

limiting or modulating a high-frequency wave within the envelope

$$\cos[\tfrac{1}{2}(k_1 + k_2)x - \tfrac{1}{2}(\omega_1 + \omega_2)t]$$

This superposition of two waves is shown in Figure 5.19.

Although our model is primitive and does not represent a pulse limited to a small region of space, it shows several interesting features common to more complicated models. For example, the envelope and the wave within the envelope move at different speeds. The speed of either the high-frequency wave or the envelope is given by dividing the coefficient of the t term by the coefficient of the x term as was done in Equations 5.12 and 5.13. For the wave *within the envelope*,

$$v_{\mathrm{p}} = \frac{(\omega_1 + \omega_2)/2}{(k_1 + k_2)/2} \approx \frac{\omega_1}{k_1} = v_1$$

Thus, the high-frequency wave moves at the phase velocity v_1 of one of the waves or at v_2 because $v_1 \approx v_2$. The envelope or group described by $2A \cos[(\Delta k/2)x - (\Delta \omega/2)t]$ moves with a different velocity however, the group velocity given by

$$v_{\mathrm{g}} = \frac{(\omega_2 - \omega_1)/2}{(k_2 - k_1)/2} = \frac{\Delta \omega}{\Delta k} \tag{5.15}$$

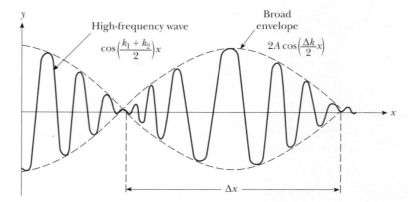

Figure 5.19 Superposition of two waves of slightly different wavelengths resulting in primitive wave groups; t has been set equal to zero in Equation 5.14.

Another general characteristic of wave groups for waves of any type is both a limited duration in time, Δt, and a limited extent in space, Δx. It is found that the smaller the spatial width of the pulse, Δx, the larger the range of wavelengths or wavenumbers, Δk, needed to form the pulse. This may be stated mathematically as

$$\Delta x \, \Delta k \approx 1 \qquad (5.16)$$

Likewise, if the time duration, Δt, of the pulse is small, we require a wide spread of frequencies, $\Delta \omega$, to form the group. That is,

$$\Delta t \, \Delta \omega \approx 1 \qquad (5.17)$$

In pulse electronics, this condition is known as the "response time–bandwidth formula."[3] In this situation Equation 5.17 shows that in order to amplify a voltage pulse of time width Δt without distortion, a pulse amplifier must equally amplify all frequencies in a frequency band of width $\Delta \omega$.

Equations 5.16 and 5.17 are important because they constitute "uncertainty relations," or "reciprocity relations," for pulses of any kind—electromagnetic, sound, or even matter waves. In particular, Equation 5.16 shows that Δx, the uncertainty in spatial extent of a pulse, is inversely proportional to Δk, the range of wavenumbers making up the pulse: **both Δx and Δk cannot become arbitrarily small, but as one decreases the other must increase.**

It is interesting that our simple two-wave model also shows the general principles given by Equations 5.16 and 5.17. If we call (rather artificially) the spatial extent of our group the distance between adjacent minima (labeled Δx in Figure 5.12), we find from the envelope term $2A\cos(\frac{1}{2}\Delta kx)$ the condition $\frac{1}{2}\Delta k \, \Delta x = \pi$ or

$$\Delta k \, \Delta x = 2\pi \qquad (5.18)$$

Here, $\Delta k = k_2 - k_1$ is the range of wavenumbers present. Likewise, if x is held constant and t is allowed to vary in the envelope portion of Equation 5.14, the result is $\frac{1}{2}(\omega_2 - \omega_1)\,\Delta t = \pi$, or

$$\Delta \omega \, \Delta t = 2\pi \qquad (5.19)$$

Therefore, Equations 5.18 and 5.19 agree with the general principles, respectively, of $\Delta k \, \Delta x \approx 1$ and $\Delta \omega \, \Delta t \approx 1$.

The addition of only two waves with discrete frequencies is instructive but produces an infinite wave instead of a true pulse. In the general case, many waves having a continuous distribution of wavelengths must be added to form a packet that is finite over a limited range and really zero everywhere else. In this case Equation 5.15 for the **group velocity,** v_{g} becomes

$$v_{\mathrm{g}} = \left. \frac{d\omega}{dk} \right|_{k_0} \qquad (5.20) \qquad \textbf{Group velocity}$$

---

[3]It should be emphasized that Equations 5.16 and 5.17 are true in general and that the quantities Δx, Δk, Δt, and $\Delta \omega$ represent the spread in values present in an *arbitrary* pulse formed from the superposition of two or *more* waves.

where the derivative is to be evaluated at k_0, the central wavenumber of the many waves present. The connection between the group velocity and the phase velocity of the composite waves is easily obtained. Because $\omega = kv_\mathrm{p}$, we find

$$v_\mathrm{g} = \left.\frac{d\omega}{dk}\right|_{k_0} = \left.v_\mathrm{p}\right|_{k_0} + \left.k\frac{dv_\mathrm{p}}{dk}\right|_{k_0} \tag{5.21}$$

where v_p is the phase velocity and is, in general, a function of k or λ. Materials in which the phase velocity varies with wavelength are said to exhibit **dispersion.** An example of a dispersive medium is glass, in which the index of refraction varies with wavelength and different colors of light travel at different speeds. Media in which the phase velocity does not vary with wavelength (such as vacuum for electromagnetic waves) are termed *nondispersive.* The term *dispersion* arises from the fact that the individual harmonic waves that form a pulse travel at different phase velocities and cause an originally sharp pulse to change shape and become spread out, or dispersed. As an example, dispersion of a laser pulse after traveling 1 km along an optical fiber is shown in Figure 5.20. In a nondispersive medium where all waves have the same velocity, the group velocity is equal to the phase velocity. In a dispersive medium the group velocity can be less than or greater than the phase velocity, depending on the sign of dv_p/dk, as shown by Equation 5.21.

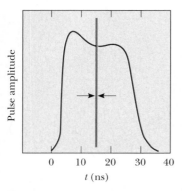

Figure 5.20 Dispersion in a 1-ns laser pulse. A pulse that starts with the width shown by the vertical lines has a time width of approximately 30 ns after traveling 1 km along an optical fiber.

EXAMPLE 5.4   Group Velocity in a Dispersive Medium

In a particular substance the phase velocity of waves doubles when the wavelength is halved. Show that wave groups in this system move at twice the *central* phase velocity.

Solution   From the given information, the dependence of phase velocity on wavelength must be

$$v_\mathrm{p} = \frac{A'}{\lambda} = Ak$$

for some constants A' and A. From Equation 5.21 we obtain

$$v_\mathrm{g} = \left.v_\mathrm{p}\right|_{k_0} + \left.k\frac{dv_\mathrm{p}}{dk}\right|_{k_0} = Ak_0 + Ak_0 = 2Ak_0$$

Thus,

$$v_\mathrm{g} = \left.2v_\mathrm{p}\right|_{k_0}$$

EXAMPLE 5.5   Group Velocity in Deep Water Waves

Newton showed that the phase velocity of deep water waves having wavelength λ is given by

$$v_\mathrm{p} = \sqrt{\frac{g\lambda}{2\pi}}$$

where g is the acceleration of gravity and where the minor contribution of surface tension has been ignored. Show that in this case the velocity of a group of these waves is one-half of the phase velocity of the central wavelength.

Solution   Because $k = 2\pi/\lambda$, we can write v_p as

$$v_\mathrm{p} = \left(\frac{g}{k}\right)^{1/2}$$

Therefore, we find

$$v_\mathrm{g} = \left.v_\mathrm{p}\right|_{k_0} + \left.k\frac{dv_\mathrm{p}}{dk}\right|_{k_0} = \left(\frac{g}{k_0}\right)^{1/2} - \frac{1}{2}\left(\frac{g}{k_0}\right)^{1/2}$$

$$= \frac{1}{2}\left(\frac{g}{k_0}\right)^{1/2} \equiv \left.\frac{1}{2}v_\mathrm{p}\right|_{k_0}$$

Matter Wave Packets

We are now in a position to apply our general theory of wave groups to electrons. We shall show both the dispersion of de Broglie waves and the satisfying result that the wave packet and the particle move at the same velocity. According to de Broglie, *individual* matter waves have a frequency f and a wavelength λ given by

$$f = \frac{E}{h} \quad \text{and} \quad \lambda = \frac{h}{p}$$

where E and p are the relativistic energy and momentum of the particle, respectively. The phase speed of these matter waves is given by

$$v_p = f\lambda = \frac{E}{p} \tag{5.22}$$

The phase speed can be expressed as a function of p or k alone by substituting $E = (p^2c^2 + m^2c^4)^{1/2}$ into Equation 5.22:

$$v_p = c\sqrt{1 + \left(\frac{mc}{p}\right)^2} \tag{5.23}$$

The dispersion relation for de Broglie waves can be obtained as a function of k by substituting $p = h/\lambda = \hbar k$ into Equation 5.23. This gives

$$v_p = c\sqrt{1 + \left(\frac{mc}{\hbar k}\right)^2} \tag{5.24}$$

Phase velocity of matter waves

Equation 5.24 shows that individual de Broglie waves representing a particle of mass m show dispersion even in *empty space* and always travel at a speed that is *greater than* or at least equal to c. Because these component waves travel at different speeds, the width of the wave packet, Δx, spreads or disperses as time progresses, as will be seen in detail in Chapter 6. To obtain the group speed, we use

$$v_g = \left[v_p + k\,\frac{dv_p}{dk} \right]_{k_0}$$

and Equation 5.24. After some algebra, we find

$$v_g = \frac{c}{\left[1 + \left(\dfrac{mc}{\hbar k_0}\right)^2 \right]^{1/2}} = \left.\frac{c^2}{v_p}\right|_{k_0} \tag{5.25}$$

Solving for the phase speed from Equation 5.22, we find

$$v_p = \frac{E}{p} = \frac{\gamma mc^2}{\gamma mv} = \frac{c^2}{v}$$

where v is the particle's speed. Finally, substituting $v_p = c^2/v$ into Equation 5.25 for v_g shows that the group velocity of the matter wave packet is the same as the particle speed. This agrees with our intuition that the matter wave envelope should move at the same speed as the particle.

O P T I O N A L

5.4   FOURIER INTEGRALS

In this section we show in detail how to construct wave groups, or pulses, that are truly localized in space or time and also show that very general reciprocity relations of the type $\Delta k \Delta x \approx 1$ and $\Delta \omega \Delta t \approx 1$ hold for these pulses.

To form a true pulse that is zero everywhere outside of a finite spatial range Δx requires adding together an infinite number of harmonic waves with continuously varying wavelengths and amplitudes. This addition can be done with a Fourier integral, which is defined as follows:

$$f(x) = \frac{1}{\sqrt{2\pi}} \int_{-\infty}^{+\infty} a(k) e^{ikx} dk \tag{5.26}$$

Here $f(x)$ is a spatially localized wave group, $a(k)$ gives the amount or amplitude of the wave with wavenumber $k = (2\pi/\lambda)$ to be added, and $e^{ikx} = \cos kx + i \sin kx$ is Euler's compact expression for a harmonic wave. The amplitude distribution function $a(k)$ can be obtained if $f(x)$ is known by using the symmetric formula

$$a(k) = \frac{1}{\sqrt{2\pi}} \int_{-\infty}^{+\infty} f(x) e^{-ikx} dk \tag{5.27}$$

Equations 5.26 and 5.27 apply to the case of a spatial pulse at fixed time, but it is important to note that they are mathematically identical to the case of a time pulse passing a fixed position. This case is common in electrical engineering and involves adding together a continuously varying set of frequencies:

$$V(t) = \frac{1}{\sqrt{2\pi}} \int_{-\infty}^{+\infty} g(\omega) e^{i\omega t} d\omega \tag{5.28}$$

$$g(\omega) = \frac{1}{\sqrt{2\pi}} \int_{-\infty}^{+\infty} V(t) e^{-i\omega t} dt \tag{5.29}$$

where $V(t)$ is the strength of a signal as a function of time, and $g(\omega)$ is the *spectral content* of the signal and gives the amount of the harmonic wave with frequency ω that is present.

Let us now consider several examples of how to use Equations 5.26 through 5.29 and how they lead to uncertainty relationships of the type $\Delta \omega \, \Delta t \approx 1$ and $\Delta k \, \Delta x \approx 1$.

EXAMPLE 5.6

This example compares the spectral contents of infinite and truncated sinusoidal waves. A truncated sinusoidal wave is a wave cut off or truncated by a shutter, as shown in Figure 5.21. (a) What is the spectral content of an infinite sinusoidal wave $e^{i\omega_0 t}$? (b) Find and sketch the spectral content of a truncated sinusoidal wave given by

$$V(t) = e^{i\omega_0 t} \qquad -T < t < +T$$

$$V(t) = 0 \qquad \text{otherwise}$$

(c) Show that for this truncated sinusoid $\Delta t \, \Delta \omega = \pi$, where Δt and $\Delta \omega$ are the half-widths of $v(t)$ and $g(\omega)$, respectively.

Solution   (a) The spectral content consists of a single strong contribution at the frequency ω_0.

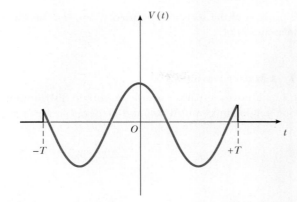

Figure 5.21   (Example 5.6) The real part of a truncated sinusoidal wave.

(b)
$$g(\omega) = \frac{1}{\sqrt{2\pi}} \int_{-\infty}^{+\infty} V(t)e^{-i\omega t}dt = \frac{1}{\sqrt{2\pi}} \int_{-T}^{+T} e^{i(\omega_0 - \omega)t}\, dt$$

$$= \frac{1}{\sqrt{2\pi}} \int_{-T}^{+T} [\cos(\omega_0 - \omega)t + i\sin(\omega_0 - \omega)t]\, dt$$

Because the sine term is an odd function and the cosine is even, the integral reduces to

$$g(\omega) = \frac{2}{\sqrt{2\pi}} \int_0^T \cos(\omega_0 - \omega)t\, dt = \sqrt{\frac{2}{\pi}} \frac{\sin(\omega_0 - \omega)T}{(\omega_0 - \omega)} = \sqrt{\frac{2}{\pi}}(T)\frac{\sin(\omega_0 - \omega)T}{(\omega_0 - \omega)T}$$

A sketch of $g(\omega)$ (Figure 5.22) shows a typical $\sin Z/Z$ profile centered on ω_0. Note that both positive and negative amounts of different frequencies must be added to produce the truncated sinusoid. Furthermore, the strongest frequency contribution comes from the frequency region near $\omega = \omega_0$, as expected.

(c) Δt clearly equals T and $\Delta \omega$ may be taken to be half the width of the main lobe of $g(\omega)$, $\Delta \omega = \pi/T$. Thus, we get

$$\Delta \omega \, \Delta t = \frac{\pi}{T} \times T = \pi$$

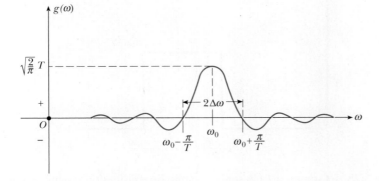

Figure 5.22   (Example 5.6) The Fourier transform of a truncated sinusoidal wave. The curve shows the amount of a given frequency that must be added to produce the truncated wave.

We see that the product of the spread in frequency, $\Delta\omega$, and the spread in time, Δt, is a constant independent of T.

EXAMPLE 5.7   A Matter Wave Packet

(a) Show that the matter wave packet whose amplitude distribution $a(k)$ is a rectangular pulse of height unity, width Δk, and centered at k_0 (Fig. 5.23) has the form

$$f(x) = \frac{\Delta k}{\sqrt{2\pi}} \frac{\sin(\Delta k \cdot x/2)}{(\Delta k \cdot x/2)} e^{ik_0 x}$$

Solution

$$f(x) = \frac{1}{\sqrt{2\pi}} \int_{-\infty}^{+\infty} a(k) e^{ikx}\, dk = \frac{1}{\sqrt{2\pi}} \int_{k_0-(\Delta k/2)}^{k_0+(\Delta k/2)} e^{ikx}\, dk = \frac{1}{\sqrt{2\pi}} \frac{e^{ik_0 x}}{x} 2 \sin(\Delta k \cdot x/2)$$

$$= \frac{\Delta k}{\sqrt{2\pi}} \frac{\sin(\Delta k \cdot x/2)}{(\Delta k \cdot x/2)} e^{ik_0 x}$$

(b) Observe that this wave packet is a complex function. Later in this chapter we shall see how the definition of probability density results in a real function, but for the time being consider only the real part of $f(x)$ and make a sketch of its behavior, showing its envelope and the cosine function within. Determine Δx, and show that an uncertainty relation of the form $\Delta x\, \Delta k \approx 1$ holds.

Solution   The real part of the wave packet is shown in Figure 5.24 where the full width of the main lobe is $\Delta x = 4\pi/\Delta k$. This immediately gives the uncertainty relation $\Delta x\, \Delta k = 4\pi$. Note that the constant on the right-hand side of the uncertainty relation depends on the shape chosen for $a(k)$ and the precise definition of Δx and Δk.

Exercise 3   Assume that a narrow triangular voltage pulse $V(t)$ arises in some type of radar system (see Fig. 5.25). (a) Find and sketch the spectral content $g(\omega)$. (b) Show that a relation of the type $\Delta\omega\, \Delta t \approx 1$ holds. (c) If the width of the pulse is

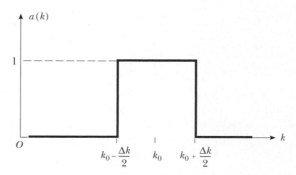

Figure 5.23   (Example 5.7) A simple amplitude distribution specifying a uniform contribution of all wavenumbers from $k_0 - \Delta k/2$ to $k_0 + \Delta k/2$. Although we have used only positive k's here, both positive and negative k values are allowed, in general corresponding to waves traveling to the right ($k > 0$) or left ($k < 0$).

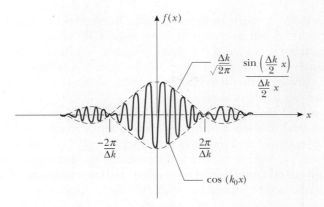

Figure 5.24   (Example 5.7) The real part of the wave packet formed by the uniform amplitude distribution shown in Figure 5.23.

$2\tau = 10^{-9}$ s, what range of frequencies must this system pass if the pulse is to be undistorted? Take $\Delta t = \tau$ and define $\Delta \omega$ similarly.

Answer   (a) $g(\omega) = (\sqrt{2/\pi})(1/\omega^2 \tau)(1 - \cos \omega \tau)$.   (b) $\Delta \omega \, \Delta t = 2\pi$.   (c) $2\Delta f = 4.00 \times 10^9$ Hz.

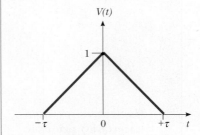

Figure 5.25   (Exercise 3).

Constructing Moving Wave Packets

Figure 5.24 represents a snapshot of the wave packet at $t = 0$. To construct a moving wave packet representing a moving particle, we replace kx in Equation 5.26 with $(kx - \omega t)$. Thus, the representation of the moving wave packet becomes

$$f(x, t) = \frac{1}{\sqrt{2\pi}} \int_{-\infty}^{+\infty} a(k) e^{i(kx - \omega t)} \, dk \qquad (5.30)$$

It is important to realize that here $\omega = \omega(k)$, that is, ω is a function of k and therefore depends on the type of wave and the medium traversed. In general, it is difficult to solve this integral analytically. For matter waves, the QMTools software available from our companion Web site (http://info.brookscole.com/mp3e) produces the same result by solving numerically a certain differential equation that governs the behavior of such waves. This approach will be explored further in the next chapter.

5.5   THE HEISENBERG UNCERTAINTY PRINCIPLE

In the period 1924–25, Werner Heisenberg, the son of a professor of Greek and Latin at the University of Munich, invented a complete theory of quantum mechanics called matrix mechanics. This theory overcame some of the problems with the Bohr theory of the atom, such as the postulate of "unobservable" electron orbits. Heisenberg's formulation was based primarily on measurable quantities such as the transition probabilities for electronic jumps between quantum states. Because transition probabilities depend on the initial and final states, Heisenberg's mechanics used variables labeled by two subscripts. Although at first Heisenberg presented his theory in the form of noncommuting algebra, Max Born quickly realized that this theory could be more elegantly described by matrices. Consequently, Born, Heisenberg, and Pascual Jordan soon worked out a comprehensive theory of matrix mechanics. Although the matrix formulation was quite elegant, it attracted little attention outside of a small group of gifted physicists because it was difficult to apply in specific cases, involved mathematics unfamiliar to most physicists, and was based on rather vague physical concepts.

Although we will investigate this remarkable form of quantum mechanics no further, we shall discuss another of Heisenberg's discoveries, the **uncertainty principle,** elucidated in a famous paper in 1927. In this paper Heisenberg introduced the notion that **it is impossible to determine simultaneously with unlimited precision the position and momentum of a particle.** In words we may state the uncertainty principle as follows:

Momentum–position uncertainty principle

> If a measurement of position is made with precision Δx and a simultaneous measurement of momentum in the x direction is made with precision Δp_x, then the product of the two uncertainties can never be smaller than $\hbar/2$. That is,
>
> $$\Delta p_x \Delta x \geq \frac{\hbar}{2} \qquad (5.31)$$

In his paper of 1927, Heisenberg was careful to point out that the inescapable uncertainties Δp_x and Δx do not arise from imperfections in practical measuring instruments. Rather, they arise from the need to use a large range of wavenumbers, Δk, to represent a matter wave packet localized in a small region, Δx. The uncertainty principle represents a sharp break with the ideas of classical physics, in which it is assumed that, with enough skill and ingenuity, it is possible to simultaneously measure a particle's position and momentum to any desired degree of precision. As shown in Example 5.8, however, there is no contradiction between the uncertainty principle and classical laws for macroscopic systems because of the small value of $\hbar$.

One can show that $\Delta p_x \Delta x \geq \hbar/2$ comes from the uncertainty relation governing any type of wave pulse formed by the superposition of waves with different wavelengths. In Section 5.3 we found that to construct a wave group localized in a small region Δx, we had to add up a large range of wavenumbers Δk, where $\Delta k \, \Delta x \approx 1$ (Eq. 5.16). The precise value of the number on the right-hand side of Equation 5.16 depends on the functional form $f(x)$ of the wave group as well as on the specific definition of Δx and Δk. A different choice of $f(x)$ or a different rule for defining Δx and Δk (or both) will give a slightly different number. With Δx and Δk defined as standard deviations, it can be shown that the smallest number, $\frac{1}{2}$, is obtained for a Gaussian wavefunction.[4] In this minimum uncertainty case we have

$$\Delta x \Delta k = \tfrac{1}{2}$$

---

[4]See Section 6.7 for a definition of the standard deviation and Problem 6.34 for a complete mathematical proof of this statement.

This photograph of Werner Heisenberg was taken around 1924. Heisenberg obtained his Ph.D. in 1923 at the University of Munich where he studied under Arnold Sommerfeld and became an enthusiastic mountain climber and skier. Later, he worked as an assistant to Max Born at Göttingen and Niels Bohr in Copenhagen. While physicists such as de Broglie and Schrödinger tried to develop visualizable models of the atom, Heisenberg, with the help of Born and Pascual Jordan, developed an abstract mathematical model called matrix mechanics to explain the wavelengths of spectral lines. The more successful wave mechanics of Schrödinger an-

B I O G R A P H Y

_____

WERNER HEISENBERG
(1901–1976)

nounced a few months later was shown to be equivalent to Heisenberg's approach. Heisenberg made many other significant contributions to physics, including his famous uncertainty principle, for which he received the Nobel prize in 1932, the prediction of two forms of molecular hydrogen, and theoretical models of the nucleus. During World War II he was director of the Max Planck Institute at Berlin where he was in charge of German research on atomic weapons. Following the war, he moved to West Germany and became director of the Max Planck Institute for Physics at Göttingen.

(Photo courtesy of the University of Hamburg)

For any other choice of $f(x)$,

$$\Delta x \Delta k \geq \tfrac{1}{2} \tag{5.32}$$

and using $\Delta p_x = \hbar \Delta k$, $\Delta x \, \Delta k \geq \tfrac{1}{2}$ immediately becomes

$$\Delta p_x \, \Delta x \geq \frac{\hbar}{2} \tag{5.33}$$

The basic meaning of $\Delta p \, \Delta x \geq \hbar/2$ is that as one uncertainty increases the other decreases. In the extreme case as one uncertainty approaches ∞, the other must approach zero. This extreme case is illustrated by a plane wave $e^{ik_0 x}$ that has a precise momentum $\hbar k_0$ and an infinite extent—that is, the wavefunction is not concentrated in any segment of the x axis.

Another important uncertainty relation involves the uncertainty in energy of a wave packet, ΔE, and the time, Δt, taken to measure that energy. Starting with $\Delta \omega \Delta t \geq \tfrac{1}{2}$ as the minimum form of the time–frequency uncertainty principle, and using the de Broglie relation for the connection between the matter wave energy and frequency, $E = \hbar \omega$, we immediately find the **energy–time uncertainty principle**

$$\Delta E \Delta t \geq \frac{\hbar}{2} \tag{5.34}$$

Energy–time uncertainty principle

Equation 5.34 states that the precision with which we can know the energy of some system is limited by the time available for measuring the energy. A common application of the energy–time uncertainty is in calculating the lifetimes of very short-lived subatomic particles whose lifetimes cannot be measured directly, but whose uncertainty in energy or mass can be measured. (See Problem 26.)

A Different View of the Uncertainty Principle

Although we have indicated that $\Delta p_x \Delta x \geq \hbar/2$ arises from the theory of forming pulses or wave groups, there is a more physical way to view the origin of the un-

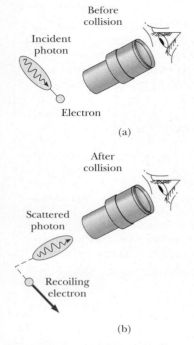

Before collision

Incident photon

Electron

(a)

After collision

Scattered photon

Recoiling electron

(b)

Figure 5.26 A thought experiment for viewing an electron with a powerful microscope. (a) The electron is shown before colliding with the photon. (b) The electron recoils (is disturbed) as a result of the collision with the photon.

certainty principle. We consider certain idealized experiments (called *thought experiments*) and show that it is impossible to carry out an experiment that allows the position and momentum of a particle to be simultaneously measured with an accuracy that violates the uncertainty principle. The most famous thought experiment along these lines was introduced by Heisenberg himself and involves the measurement of an electron's position by means of a microscope (Fig. 5.26), which forms an image of the electron on a screen or the retina of the eye.

Because light can scatter from and perturb the electron, let us minimize this effect by considering the scattering of only a single light quantum from an electron initially at rest (Fig. 5.27). To be collected by the lens, the photon must be scattered through an angle ranging from $-\theta$ to $+\theta$, which consequently imparts to the electron an x momentum value ranging from $+(h \sin \theta)/\lambda$ to $-(h \sin \theta)/\lambda$. Thus the uncertainty in the electron's momentum is $\Delta p_x = (2h \sin \theta)/\lambda$. After passing through the lens, the photon lands somewhere on the screen, but the image and consequently the position of the electron is "fuzzy" because the photon is diffracted on passing through the lens aperture. According to physical optics, the resolution of a microscope or the uncertainty in the image of the electron, Δx, is given by $\Delta x = \lambda/(2 \sin \theta)$. Here 2θ is the angle subtended by the objective lens, as shown in Figure 5.27.[5] Multiplying the expressions for Δp_x and Δx, we find for the electron

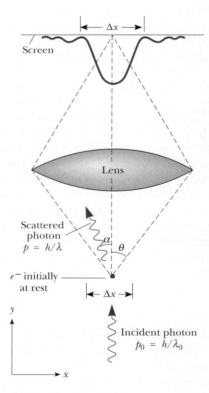

Screen

Δx

Lens

Scattered photon
$p = h/\lambda$

α θ

e^- initially at rest

Δx

y

Incident photon
$p_0 = h/\lambda_0$

x

Figure 5.27 The Heisenberg microscope.

[5]The resolving power of the microscope is treated clearly in F. A. Jenkins and H. E. White, *Fundamentals of Optics*, 4th ed., New York, McGraw-Hill Book Co., 1976, pp. 332–334.

$$\Delta p_x \Delta x \approx \left(\frac{2h}{\lambda} \sin \theta \right) \left(\frac{\lambda}{2 \sin \theta} \right) = h$$

in agreement with the uncertainty relation. Note also that this principle is inescapable and relentless! If Δx is reduced by increasing θ or the lens size, there is an equivalent increase in the uncertainty of the electron's momentum.

Examination of this simple experiment shows several key physical properties that lead to the uncertainty principle:

- The indivisible nature of light particles or quanta (nothing less than a single photon can be used!).
- The wave property of light as shown in diffraction.
- The impossibility of predicting or measuring the precise classical path of a single scattered photon and hence of knowing the precise momentum transferred to the electron.[6]

We conclude this section with some examples of the types of calculations that can be done with the uncertainty principle. In the spirit of Fermi or Heisenberg, these "back-of-the-envelope calculations" are surprising for their simplicity and essential description of quantum systems of which the details are unknown.

EXAMPLE 5.8   The Uncertainty Principle Changes Nothing for Macroscopic Objects

(a) Show that the spread of velocities caused by the uncertainty principle does not have measurable consequences for macroscopic objects (objects that are large compared with atoms) by considering a 100-g racquetball confined to a room 15 m on a side. Assume the ball is moving at 2.0 m/s along the x axis.

Solution

$$\Delta p_x \geq \frac{\hbar}{2 \, \Delta x} = \frac{1.05 \times 10^{-34} \, \text{J} \cdot \text{s}}{2 \times 15 \, \text{m}} = 3.5 \times 10^{-36} \, \text{kg} \cdot \text{m/s}$$

Thus the minimum spread in velocity is given by

$$\Delta v_x = \frac{\Delta p_x}{m} = \frac{3.05 \times 10^{-36} \, \text{kg} \cdot \text{m/s}}{0.100 \, \text{kg}} = 3.5 \times 10^{-35} \, \text{m/s}$$

This gives a relative uncertainty of

$$\frac{\Delta v_x}{v_x} = \frac{3.5 \times 10^{-35}}{2.0} = 1.8 \times 10^{-35}$$

which is certainly not measurable.

(b) If the ball were to suddenly move along the y axis perpendicular to its well-defined classical trajectory along x, how far would it move in 1 s? Assume that the ball moves in the y direction with the top speed in the spread Δv_y produced by the uncertainty principle.

Solution   It is important to realize that uncertainty relations hold in the y and z directions as well as in the x direction. This means that $\Delta p_x \, \Delta x \geq \hbar/2$, $\Delta p_y \, \Delta y \geq \hbar/2$, and $\Delta p_z \, \Delta z \geq \hbar/2$ and because all the position uncertainties are equal, all of the velocity spreads are equal. Consequently, we have $\Delta v_y = 3.5 \times 10^{-35} \, \text{m/s}$ and the ball moves 3.5×10^{-35} m in the y direction in 1 s. This distance is again an immeasurably small quantity, being 10^{-20} times the size of a nucleus!

Exercise 4   How long would it take the ball to move 50 cm in the y direction? (The age of the universe is thought to be 15 billion years, give or take a few billion).

---

[6]Attempts to measure the photon's position by scattering electrons from it in a Compton process only serve to make its path to the lens more uncertain.

EXAMPLE 5.9 Do Electrons Exist Within the Nucleus?

Estimate the kinetic energy of an electron confined within a nucleus of size 1.0×10^{-14} m by using the uncertainty principle.

Solution Taking Δx to be the half-width of the confinement length in the equation $\Delta p_x \geq \dfrac{\hbar}{2\,\Delta x}$, we have

$$\Delta p_x \geq \frac{6.58 \times 10^{-16}\ \text{eV·s}}{1.0 \times 10^{-14}\ \text{m}} \times \frac{3.00 \times 10^8\ \text{m/s}}{c}$$

or

$$\Delta p_x \geq 2.0 \times 10^7\ \frac{\text{eV}}{c}$$

This means that measurements of the component of momentum of electrons trapped inside a nucleus would range from less than -20 MeV/c to greater than $+20$ MeV/c and that some electrons would have momentum at least as large as 20 MeV/c. Because this appears to be a large momentum, to be safe we calculate the electron's energy relativistically.

$$\begin{aligned}E^2 &= p^2c^2 + (m_ec^2)^2 \\ &= (20\ \text{MeV}/c)^2c^2 + (0.511\ \text{MeV})^2 \\ &= 400(\text{MeV})^2\end{aligned}$$

or

$$E \geq 20\ \text{MeV}$$

Finally, the kinetic energy of an intranuclear electron is

$$K = E - m_ec^2 \geq 19.5\ \text{MeV}$$

Since electrons emitted in radioactive decay of the nucleus (beta decay) have energies much less than 19.5 MeV (about 1 MeV or less) and it is known that no other mechanism could carry off an intranuclear electron's energy during the decay process, we conclude that electrons observed in beta decay do not come from within the nucleus but are actually created at the instant of decay.

EXAMPLE 5.10 The Width of Spectral Lines

Although an excited atom can radiate at any time from $t = 0$ to $t = \infty$, the average time after excitation at which a group of atoms radiates is called the **lifetime,** τ, of a particular excited state. (a) If $\tau = 1.0 \times 10^{-8}$ s (a typical value), use the uncertainty principle to compute the line width Δf of light emitted by the decay of this excited state.

Solution We use $\Delta E\,\Delta t \approx \hbar/2$, where ΔE is the uncertainty in energy of the excited state, and $\Delta t = 1.0 \times 10^{-8}$ s is the average time available to measure the excited state. Thus,

$$\Delta E \approx \hbar/2\,\Delta t = \hbar/(2.0 \times 10^{-8}\ \text{s})$$

Since ΔE is also the uncertainty in energy of a photon emitted when the excited state decays, and $\Delta E = h\Delta f$ for a photon,

$$h\,\Delta f = \hbar/(2.0 \times 10^{-8}\ \text{s})$$

or

$$\Delta f = \frac{1}{4\pi \times 10^{-8}\ \text{s}} = 8.0 \times 10^6\ \text{Hz}$$

(b) If the wavelength of the spectral line involved in this process is 500 nm, find the fractional broadening $\Delta f/f$.

Solution First, we find the center frequency of this line as follows:

$$f_0 = \frac{c}{\lambda} = \frac{3.0 \times 10^8\ \text{m/s}}{500 \times 10^{-9}\ \text{m}} = 6.0 \times 10^{14}\ \text{Hz}$$

Hence,

$$\frac{\Delta f}{f_0} = \frac{8.0 \times 10^6\ \text{Hz}}{6.0 \times 10^{14}\ \text{Hz}} = 1.3 \times 10^{-8}$$

This narrow natural line width can be seen with a sensitive interferometer. Usually, however, temperature and pressure effects overshadow the natural line width and broaden the line through mechanisms associated with the Doppler effect and atomic collisions.

Exercise 5 Using the nonrelativistic Doppler formula, calculate the Doppler broadening of a 500-nm line emitted by a hydrogen atom at 1000 K. Do this by considering the atom to be moving either directly toward or away from an observer with an energy of $\frac{3}{2}k_BT$.

Answer 0.0083 nm, or 0.083 Å.

5.6 IF ELECTRONS ARE WAVES, WHAT'S WAVING?

Although we have discussed in some detail the notion of de Broglie matter waves, we have not discussed the precise nature of the field $\Psi(x, y, z, t)$ or **wavefunction** that represents the matter waves. We have delayed this discussion because Ψ

(Greek letter psi) is rather abstract. Ψ is definitely *not* a measurable disturbance requiring a medium for propagation like a water wave or a sound wave. Instead, the stuff that is waving requires no medium. Furthermore, Ψ is in general represented by a complex number and is used to calculate the probability of finding the particle at a given time in a small volume of space. If any of this seems confusing, you should not lose heart, as the nature of the wavefunction has been confusing people since its invention. It even confused its inventor, Erwin Schrödinger, who incorrectly interpreted $\Psi^*\Psi$ as the electric charge density.[7] The great philosopher of the quantum theory, Bohr, immediately objected to this interpretation. Subsequently, Max Born offered the currently accepted statistical view of $\Psi^*\Psi$ in late 1926. The confused state of affairs surrounding Ψ at that time was nicely described in a poem by Walter Huckel:

> Erwin with his psi can do
> Calculations quite a few.
> But one thing has not been seen
> Just what does psi really mean?
> (*English translation by Felix Bloch*)

The currently held view is that a particle is described by a function $\Psi(x, y, z, t)$ called the **wavefunction.** The quantity $\Psi^*\Psi = |\Psi|^2$ represents the probability per unit volume of finding the particle at a time t in a small volume of space centered on (x, y, z). We will treat methods of finding Ψ in much more detail in Chapter 6, but for now all we require is the idea that **the probability of finding a particle is directly proportional to $|\Psi|^2$.**

5.7 THE WAVE–PARTICLE DUALITY

The Description of Electron Diffraction in Terms of Ψ

In this chapter and previous chapters we have seen evidence for both the wave properties and the particle properties of electrons. Historically, the particle properties were first known and connected with a definite mass, a discrete charge, and detection or localization of the electron in a small region of space. Following these discoveries came the confirmation of the wave nature of electrons in scattering at low energy from metal crystals. In view of these results and because of the everyday experience of seeing the world in terms of *either* grains of sand *or* diffuse water waves, it is no wonder that we are tempted to simplify the issue and ask, "Well, is the electron a wave or a particle?" The answer is that **electrons are very delicate and rather plastic—they behave like either particles or waves, depending on the kind of experiment performed on them. In any case, it is impossible to measure both the wave and particle properties simultaneously.**[8] The view of Bohr was expressed in an idea known as **complementarity.** As different as they are, both wave and particle views are needed and they complement each other to fully describe the electron. The view of

Complementarity

---

[7]Ψ^* represents the complex conjugate of Ψ. Thus, if $\Psi = a + ib$, then $\Psi^* = a - ib$. In exponential form, if $\Psi = Ae^{i\theta}$, then $\Psi^* = Ae^{-i\theta}$. Note that $\Psi^*\Psi = |\Psi|^2$; a, b, A, and θ are all real quantities.

[8]Many feel that the elder Bragg's remark, originally made about light, is a more satisfying answer: Electrons behave like waves on Mondays, Wednesdays, and Fridays, like particles on Tuesdays, Thursdays, and Saturdays, and like nothing at all on Sundays.

Feynman[9] was that both electrons and photons behave in their own inimitable way. This is like nothing we have seen before, because we do not live at the very tiny scale of atoms, electrons, and photons.

Perhaps the best way to crystallize our ideas about the wave–particle duality is to consider a "simple" double-slit electron diffraction experiment. This experiment highlights much of the mystery of the wave–particle paradox, shows the impossibility of measuring *simultaneously* both wave and particle properties, and illustrates the use of the wavefunction, Ψ, in determining interference effects. A schematic of the experiment with monoenergetic (single-wavelength) electrons is shown in Figure 5.28. A parallel beam of electrons falls on a double slit, which has individual openings much smaller than D so that single-slit diffraction effects are negligible. At a distance from the slits much greater than D is an electron detector capable of detecting individual electrons. It is important to note that the detector always registers discrete particles localized in space and time. In a real experiment this can be achieved if the electron source is weak enough (see Fig. 5.29): **In all cases if the detector collects electrons at different positions for a long enough time, a typical wave interference pattern for the counts per minute or probability of arrival of electrons is found** (see Fig. 5.28). If one imagines a single electron to produce in-phase "wavelets" at the slits, standard wave theory can be used to find the angular separation, θ, of the

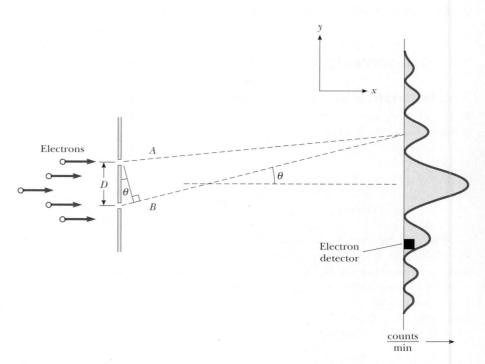

Figure 5.28   Electron diffraction. D is much greater than the individual slit widths and much less than the distance between the slits and the detector.

[9]R. Feynman, *The Character of Physical Law*, Cambridge, MA, MIT Press, 1982.

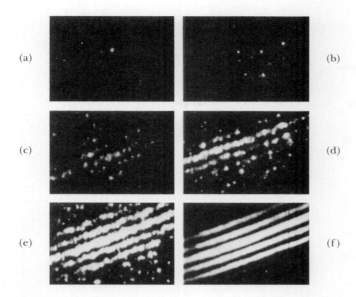

Figure 5.29   Gradual accumulation of interference fringes after the diffraction of an increasing number of electrons. (*From P. G. Medi, G. F. Missiroli, and G. Pozzi*, Am. J. of Phys. *44:306, 1976*).

central probability maximum from its neighboring minimum. The minimum occurs when the path length difference between A and B in Figure 5.28 is half a wavelength, or

$$D \sin \theta = \lambda/2$$

As the electron's wavelength is given by $\lambda = h/p_x$, we see that

$$\sin \theta \approx \theta = \frac{h}{2p_x D} \qquad (5.35)$$

for small θ. Thus we can see that the dual nature of the electron is clearly shown in this experiment: **although the electrons are detected as particles at a localized spot at some instant of time, the probability of arrival at that spot is determined by finding the intensity of two interfering matter waves.**

But there is more. What happens if one slit is covered during the experiment? In this case one obtains a symmetric curve peaked around the center of the open slit, much like the pattern formed by bullets shot through a hole in armor plate. Plots of the counts per minute or probability of arrival of electrons with the lower or upper slit *closed* are shown in Figure 5.30. These are expressed as the appropriate square of the absolute value of some wavefunction, $|\Psi_1|^2 = \Psi_1{}^*\Psi_1$ or $|\Psi_2|^2 = \Psi_2{}^*\Psi_2$, where Ψ_1 and Ψ_2 represent the cases of the electron passing through slit 1 and slit 2, respectively. If an experiment is now performed with slit 1 open and slit 2 blocked for time T and then slit 1 blocked and slit 2 open for time T, the accumulated pattern of counts per minute is completely different from the case with

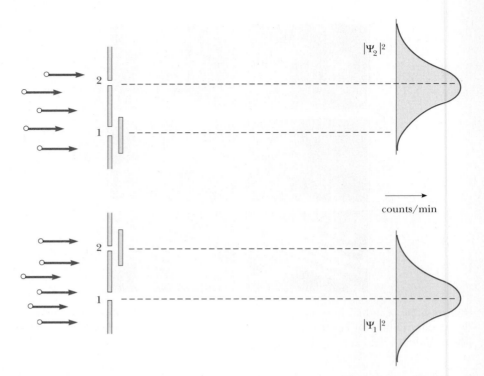

Figure 5.30   The probability of finding electrons at the screen with either the lower or upper slit closed.

both slits open. Note in Figure 5.31 that there is no longer a maximum probability of arrival of an electron at $\theta = 0$. In fact, **the interference pattern has been lost and the accumulated result is simply the sum of the individual results.** The results shown by the black curves in Figure 5.31 are easier to understand and more reasonable than the interference effects seen with both slits open (blue curve). When only one slit is open at a time, we know the electron has the same localizability and indivisibility at the slits as we measure at the detector, because the electron clearly goes through slit 1 or slit 2. Thus, the total must be analyzed as the sum of those electrons that come through slit 1, $|\Psi_1|^2$, and those that come through slit 2, $|\Psi_2|^2$. When both slits are open, it is tempting to assume that the electron goes through either slit 1 or slit 2 and that the counts per minute are again given by $|\Psi_1|^2 + |\Psi_2|^2$. We know, however, that the experimental results contradict this. Thus, our assumption that the electron is localized and goes through only one slit when both slits are open must be wrong (a painful conclusion!). Somehow the electron must be simultaneously present at both slits in order to exhibit interference.

To find the probability of detecting the electron at a particular point on the screen with both slits open, we may say that the electron is in a *superposition state* given by

$$\Psi = \Psi_1 + \Psi_2$$

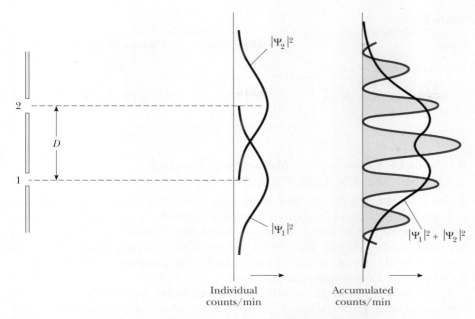

Figure 5.31 Accumulated results from the two-slit electron diffraction experiment with each slit closed half the time. For comparison, the results with both slits open are shown in color.

Thus, the probability of detecting the electron at the screen is equal to the quantity $|\Psi_1 + \Psi_2|^2$ and not $|\Psi_1|^2 + |\Psi_2|^2$. Because matter waves that start out in phase at the slits in general travel different distances to the screen (see Fig. 5.28), Ψ_1 and Ψ_2 will possess a relative phase difference ϕ at the screen. Using a phasor diagram (Fig. 5.32) to find $|\Psi_1 + \Psi_2|^2$ immediately yields

$$|\Psi|^2 = |\Psi_1 + \Psi_2|^2 = |\Psi_1|^2 + |\Psi_2|^2 + 2|\Psi_1||\Psi_2|\cos\phi$$

Note that the term $2|\Psi_1||\Psi_2|\cos\phi$ is an interference term that predicts the interference pattern actually observed in this case. For ease of comparison, a summary of the results found in both cases is given in Table 5.1.

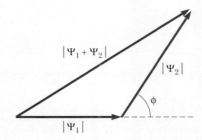

Figure 5.32 Phasor diagram to represent the addition of two complex wavefunctions, Ψ_1 and Ψ_2, differing in phase by ϕ.

Table 5.1

Case	Wavefunction	Counts/Minute at Screen
Electron is measured to pass through slit 1 or slit 2	Ψ_1 or Ψ_2	$|\Psi_1|^2 + |\Psi_2|^2$
No measurements made on electron at slits	$\Psi_1 + \Psi_2$	$|\Psi_1|^2 + |\Psi_2|^2 + 2|\Psi_1||\Psi_2|\cos\phi$

A Thought Experiment: Measuring Through Which Slit the Electron Passes

Another way to view the electron double-slit experiment is to say that the electron passes through the upper or lower slit only when one *measures* the electron to do so. Once one measures unambiguously which slit the electron passes through (yes, you guessed it . . . here comes the uncertainty principle again . . .), the act of measurement disturbs the electron's path enough to destroy the delicate interference pattern.

Let us look again at our two-slit experiment to see in detail how the interference pattern is destroyed.[10] To determine which slit the electron goes through, imagine that a group of particles is placed right behind the slits, as shown in Figure 5.33. If we use the recoil of a small particle to determine

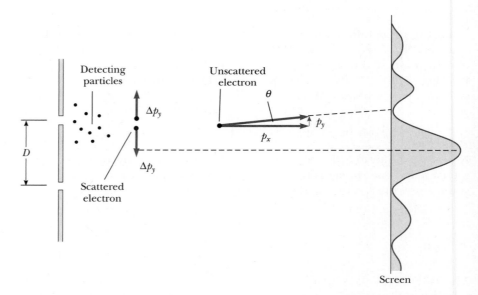

Figure 5.33   A thought experiment to determine through which slit the electron passes.

[10]Although we shall use the uncertainty principle in its standard form, it is worth noting that an alternative statement of the uncertainty principle involves this pivotal double-slit experiment: *It is impossible to design any device to determine through which slit the electron passes that will not at the same time disturb the electron and destroy the interference pattern.*

which slit the electron goes through, we must have the uncertainty in the detecting particle's position, $\Delta y \ll D$. Also, during the collision the detecting particle suffers a change in momentum, Δp_y, equal and opposite to the change in momentum experienced by the electron, as shown in Figure 5.33. An undeviated electron landing at the first minimum *and producing an interference pattern* has

$$\tan \theta \approx \theta = \frac{p_y}{p_x} = \frac{h}{2p_x D}$$

from Equation 5.35. Thus, we require that an electron scattered by a detecting particle have

$$\frac{\Delta p_y}{p_x} \ll \theta = \frac{h}{2p_x D}$$

or

$$\Delta p_y \ll \frac{h}{2D}$$

if the interference pattern is not to be distorted. Because the change in momentum of the scattered electron is equal to the change in momentum of the detecting particle, $\Delta p_y \ll h/2D$ also applies to the detecting particle. Thus, we have for the detecting particle

$$\Delta p_y \Delta y \ll \frac{h}{2D} \cdot D$$

or

$$\Delta p_y \Delta y \ll \frac{h}{2}$$

This is a clear violation of the uncertainty principle. Hence we see that **the small uncertainties needed, both to observe interference and to know which slit the electron goes through, are impossible, because they violate the uncertainty principle.** If Δy is small enough to determine which slit the electron goes through, Δp_y is so large that electrons heading for the first minimum are scattered into adjacent maxima and the interference pattern is destroyed.

Exercise 6   In a real experiment it is likely that some electrons would miss the detecting particles. Thus, we would really have two categories of electrons arriving at the detector: those measured to pass through a definite slit and those not observed, or just missed, at the slits. In this case what kind of pattern of counts per minute would be accumulated by the detector?

Answer   A mixture of an interference pattern $|\Psi_1 + \Psi_2|^2$ (those not measured) and $|\Psi_1|^2 + |\Psi_2|^2$ (those measured) would result.

5.8 A FINAL NOTE

Scientists once viewed the world as being made up of distinct and unchanging parts that interact according to strictly deterministic laws of cause and effect. In the classical limit this is fundamentally correct because classical processes involve so many quanta that deviations from the average are imperceptible. At the atomic level, however, we have seen that a given piece of matter (an electron, say) is not a distinct and unchanging part of the universe obeying completely deterministic laws. Such a particle exhibits wave properties when it interacts with a metal crystal and particle properties a short while later when it registers on a Geiger counter. Thus, rather than viewing the electron as a distinct and separate part of the universe with an intrinsic particle nature, we are led to the view that the electron and indeed all particles are amorphous entities possessing the potential to cycle endlessly between wave and particle behavior. We also find that it is much more difficult to separate the object measured from the measuring instrument at the atomic level, because the type of measuring instrument determines whether wave properties or particle properties are observed.

SUMMARY

Every lump of matter of mass m and momentum p has wavelike properties with wavelength given by the de Broglie relation

$$\lambda = \frac{h}{p} \tag{5.1}$$

By applying this wave theory of matter to electrons in atoms, de Broglie was able to explain the appearance of integers in certain Bohr orbits as a natural consequence of electron wave interference. In 1927, Davisson and Germer demonstrated directly the wave nature of electrons by showing that low-energy electrons were diffracted by single crystals of nickel. In addition, they confirmed Equation 5.1.

Although the wavelength of matter waves can be experimentally determined, it is important to understand that they are not just like other waves because their frequency and phase velocity cannot be directly measured. In particular, the phase velocity of an individual matter wave is greater than the velocity of light and varies with wavelength or wavenumber as

$$v_{\mathrm{p}} = f\lambda = \left(\frac{E}{h}\right)\left(\frac{h}{p}\right) = c\left[1 + \left(\frac{mc}{\hbar k}\right)^2\right]^{1/2} \tag{5.24}$$

To represent a particle properly, a superposition of matter waves with different wavelengths, amplitudes, and phases must be chosen to interfere constructively over a limited region of space. The resulting wave packet or group can then be shown to travel with the same speed as the classical particle. In addition, a wave packet localized in a region Δx contains a range of wavenumbers Δk, where $\Delta x\,\Delta k \geq \frac{1}{2}$. Because $p_x = \hbar k$, this implies that there is an uncertainty principle for position and momentum:

$$\Delta p_x\,\Delta x \geq \frac{\hbar}{2} \tag{5.31}$$

In a similar fashion one can show that an energy–time uncertainty relation exists, given by

$$\Delta E \Delta t \geq \frac{\hbar}{2} \qquad (5.34)$$

In quantum mechanics matter waves are represented by a wavefunction $\Psi(x, y, z, t)$. The probability of finding a particle represented by Ψ in a small volume centered at (x, y, z) at time t is proportional to $|\Psi|^2$. The wave–particle duality of electrons may be seen by considering the passage of electrons through two narrow slits and their arrival at a viewing screen. We find that although the electrons are detected as particles at a localized spot on the screen, the probability of arrival at that spot is determined by finding the intensity of two interfering matter waves.

Although we have seen the importance of matter waves or wavefunctions in this chapter, we have provided no method of finding Ψ for a given physical system. In the next chapter we introduce the Schrödinger wave equation. The solutions to this important differential equation will provide us with the wavefunctions for a given system.

SUGGESTIONS FOR FURTHER READING

1. D. Bohm, *Quantum Theory*, Englewood Cliffs, NJ, Prentice-Hall, 1951. Chapters 3 and 6 in this book give an excellent account of wave packets and the wave–particle duality of matter at a more advanced level.
2. R. Feynman, *The Character of Physical Law*, Cambridge, MA, The MIT Press, 1982, Chapter 6. This monograph is an incredibly lively and readable treatment of the double-slit experiment presented in Feynman's inimitable fashion.
3. B. Hoffman, *The Strange Story of the Quantum*, New York, Dover Publications, 1959. This short book presents a beautifully written nonmathematical discussion of the history of quantum mechanics.

QUESTIONS

1. Is light a wave or a particle? Support your answer by citing specific experimental evidence.
2. Is an electron a particle or a wave? Support your answer by citing some experimental results.
3. An electron and a proton are accelerated from rest through the same potential difference. Which particle has the longer wavelength?
4. If matter has a wave nature, why is this wavelike character not observable in our daily experiences?
5. In what ways does Bohr's model of the hydrogen atom violate the uncertainty principle?
6. Why is it impossible to measure the position and speed of a particle simultaneously with infinite precision?
7. Suppose that a beam of electrons is incident on three or more slits. How would this influence the interference pattern? Would the state of an electron depend on the number of slits? Explain.
8. In describing the passage of electrons through a slit and arriving at a screen, Feynman said that "electrons arrive in lumps, like particles, but the probability of arrival of these lumps is determined as the intensity of the waves would be. It is in this sense that the electron behaves sometimes like a particle and sometimes like a wave." Elaborate on this point in your own words. (For a further discussion of this point, see R. Feynman, *The Character of Physical Law*, Cambridge, MA, MIT Press, 1982, Chapter 6.)
9. Do you think that most major experimental discoveries are made by careful planning or by accident? Cite examples.
10. In the case of accidental discoveries, what traits must the experimenter possess to capitalize on the discovery?
11. Are particles even things? An extreme view of the plasticity of electrons and other particles is expressed in this fa-

mous quote of Heisenberg: "The invisible elementary particle of modern physics does not have the property of occupying space any more than it has properties like color or solidity. Fundamentally, it is not a material structure in space and time but only a symbol that allows the laws of nature to be expressed in especially simple form."

Are you satisfied with viewing science as a set of predictive rules or do you prefer to see science as a description of an objective world of things—in the case of particle physics, tiny, scaled-down things? What problems are associated with each point of view?

PROBLEMS

5.1   The Pilot Waves of de Broglie

1. Calculate the de Broglie wavelength for a proton moving with a speed of 10^6 m/s.
2. Calculate the de Broglie wavelength for an electron with kinetic energy (a) 50 eV and (b) 50 keV.
3. Calculate the de Broglie wavelength of a 74-kg person who is running at a speed of 5.0 m/s.
4. The "seeing" ability, or resolution, of radiation is determined by its wavelength. If the size of an atom is of the order of 0.1 nm, how fast must an electron travel to have a wavelength small enough to "see" an atom?
5. To "observe" small objects, one measures the diffraction of particles whose de Broglie wavelength is approximately equal to the object's size. Find the kinetic energy (in electron volts) required for electrons to resolve (a) a large organic molecule of size 10 nm, (b) atomic features of size 0.10 nm, and (c) a nucleus of size 10 fm. Repeat these calculations using alpha particles in place of electrons.
6. An electron and a photon each have kinetic energy equal to 50 keV. What are their de Broglie wavelengths?
7. Calculate the de Broglie wavelength of a proton that is accelerated through a potential difference of 10 MV.
8. Show that the de Broglie wavelength of an electron accelerated from rest through a small potential difference V is given by $\lambda = 1.226/\sqrt{V}$, where λ is in nanometers and V is in volts.
9. Find the de Broglie wavelength of a ball of mass 0.20 kg just before it strikes the Earth after being dropped from a building 50 m tall.
10. An electron has a de Broglie wavelength equal to the diameter of the hydrogen atom. What is the kinetic energy of the electron? How does this energy compare with the ground-state energy of the hydrogen atom?
11. For an electron to be confined to a nucleus, its de Broglie wavelength would have to be less than 10^{-14} m. (a) What would be the kinetic energy of an electron confined to this region? (b) On the basis of this result, would you expect to find an electron in a nucleus? Explain.
12. Through what potential difference would an electron have to be accelerated to give it a de Broglie wavelength of 1.00×10^{-10} m?

5.2   The Davisson–Germer Experiment

13. Figure P5.13 shows the top three planes of a crystal with planar spacing d. If $2d \sin \theta = 1.01\lambda$ for the two waves shown, and high-energy electrons of wavelength λ penetrate many planes deep into the crystal, which atomic plane produces a wave that cancels the surface reflection? This is an example of how extremely narrow maxima in high-energy electron diffraction are formed—that is, there are no diffracted beams unless $2d \sin \theta$ is equal to an integral number of wavelengths.

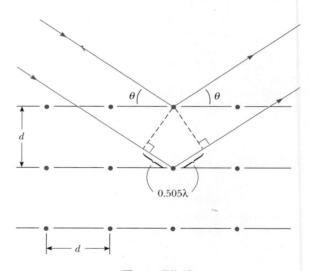

Figure P5.13

14. (a) Show that the formula for low-energy electron diffraction (LEED), when electrons are incident perpendicular to a crystal surface, may be written as

$$\sin \phi = \frac{nhc}{d(2m_e c^2 K)^{1/2}}$$

where n is the order of the maximum, d is the atomic spacing, m_e is the electron mass, K is the electron's kinetic energy, and ϕ is the angle between the incident and diffracted beams. (b) Calculate the atomic spacing in a crystal that has consecutive diffraction maxima at $\phi = 24.1°$ and $\phi = 54.9°$ for 100-eV electrons.

5.3 Wave Groups and Dispersion

15. Show that the group velocity for a nonrelativistic free electron is also given by $v_g = p/m_e = v_0$, where v_0 is the electron's velocity.

16. When a pebble is tossed into a pond, a circular wave pulse propagates outward from the disturbance. If you are alert (and it's not a sleepy afternoon in late August), you will see a fine structure in the pulse consisting of surface ripples moving inward through the circular disturbance. Explain this effect in terms of group and phase velocity if the phase velocity of ripples is given by $v_p = \sqrt{2\pi S/\lambda\rho}$, where S is the surface tension and ρ is the density of the liquid.

17. The dispersion relation for free relativistic electron waves is

$$\omega(k) = \sqrt{c^2 k^2 + (m_e c^2/\hbar)^2}$$

Obtain expressions for the phase velocity v_p and group velocity v_g of these waves and show that their product is a constant, independent of k. From your result, what can you conclude about v_g if $v_p > c$?

5.5 The Heisenberg Uncertainty Principle

18. A ball of mass 50 g moves with a speed of 30 m/s. If its speed is measured to an accuracy of 0.1%, what is the minimum uncertainty in its position?

19. A proton has a kinetic energy of 1.0 MeV. If its momentum is measured with an uncertainty of 5.0%, what is the minimum uncertainty in its position?

20. We wish to measure simultaneously the wavelength and position of a photon. Assume that the wavelength measurement gives $\lambda = 6000$ Å with an accuracy of one part in a million, that is, $\Delta\lambda/\lambda = 10^{-6}$. What is the minimum uncertainty in the position of the photon?

21. A woman on a ladder drops small pellets toward a spot on the floor. (a) Show that, according to the uncertainty principle, the miss distance must be at least

$$\Delta x = \left(\frac{\hbar}{2m}\right)^{1/2}\left(\frac{H}{2g}\right)^{1/4}$$

where H is the initial height of each pellet above the floor and m is the mass of each pellet. (b) If $H = 2.0$ m and $m = 0.50$ g, what is Δx?

22. A beam of electrons is incident on a slit of variable width. If it is possible to resolve a 1% difference in momentum, what slit width would be necessary to resolve the interference pattern of the electrons if their kinetic energy is (a) 0.010 MeV, (b) 1.0 MeV, and (c) 100 MeV?

23. Suppose Fuzzy, a quantum-mechanical duck, lives in a world in which $h = 2\pi$ J·s. Fuzzy has a mass of 2.0 kg and is initially known to be within a region 1.0 m wide.

(a) What is the minimum uncertainty in his speed?
(b) Assuming this uncertainty in speed to prevail for 5.0 s, determine the uncertainty in position after this time.

24. An electron of momentum p is at a distance r from a stationary proton. The system has a kinetic energy $K = p^2/2m_e$ and potential energy $U = -ke^2/r$. Its total energy is $E = K + U$. If the electron is bound to the proton to form a hydrogen atom, its average position is at the proton but the uncertainty in its position is approximately equal to the radius, r, of its orbit. The electron's average momentum will be zero, but the uncertainty in its momentum will be given by the uncertainty principle. Treat the atom as a one-dimensional system in the following: (a) Estimate the uncertainty in the electron's momentum in terms of r. (b) Estimate the electron's kinetic, potential, and total energies in terms of r. (c) The actual value of r is the one that minimizes the total energy, resulting in a stable atom. Find that value of r and the resulting total energy. Compare your answer with the predictions of the Bohr theory.

25. An excited nucleus with a lifetime of 0.100 ns emits a γ ray of energy 2.00 MeV. Can the energy width (uncertainty in energy, ΔE) of this 2.00-MeV γ emission line be directly measured if the best gamma detectors can measure energies to ±5 eV?

26. Typical measurements of the mass of a subatomic delta particle ($m \approx 1230$ MeV/c^2) are shown in Figure P5.26. Although the lifetime of the delta is much too short to measure directly, it can be calculated from the energy–time uncertainty principle. Estimate the lifetime from the full width at half-maximum of the mass measurement distribution shown.

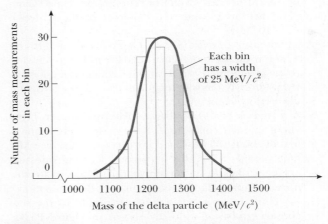

Figure P5.26 Histogram of mass measurements of the delta particle.

5.7 The Wave–Particle Duality

27. A monoenergetic beam of electrons is incident on a single slit of width 0.50 nm. A diffraction pattern is

formed on a screen 20 cm from the slit. If the distance between successive minima of the diffraction pattern is 2.1 cm, what is the energy of the incident electrons?

28. A neutron beam with a selected speed of 0.40 m/s is directed through a double slit with a 1.0-mm separation. An array of detectors is placed 10 m from the slit. (a) What is the de Broglie wavelength of the neutrons? (b) How far off axis is the first zero-intensity point on the detector array? (c) Can we say which slit any particular neutron passed through? Explain.

29. A two-slit electron diffraction experiment is done with slits of *unequal* widths. When only slit 1 is open, the number of electrons reaching the screen per second is 25 times the number of electrons reaching the screen per second when only slit 2 is open. When both slits are open, an interference pattern results in which the destructive interference is not complete. Find the ratio of the probability of an electron arriving at an interference maximum to the probability of an electron arriving at an adjacent interference minimum. (*Hint:* Use the superposition principle).

Additional Problems

30. Robert Hofstadter won the 1961 Nobel prize in physics for his pioneering work in scattering 20-GeV electrons from nuclei. (a) What is the γ factor for a 20-GeV electron, where $\gamma = (1 - v^2/c^2)^{-1/2}$? What is the momentum of the electron in kg·m/s? (b) What is the wavelength of a 20-GeV electron and how does it compare with the size of a nucleus?

31. An air rifle is used to shoot 1.0-g particles at 100 m/s through a hole of diameter 2.0 mm. How far from the rifle must an observer be to see the beam spread by 1.0 cm because of the uncertainty principle? Compare this answer with the diameter of the Universe (2×10^{26} m).

32. An atom in an excited state 1.8 eV above the ground state remains in that excited state 2.0 μs before moving to the ground state. Find (a) the frequency of the emitted photon, (b) its wavelength, and (c) its approximate uncertainty in energy.

33. A π^0 meson is an unstable particle produced in high-energy particle collisions. It has a mass–energy equivalent of about 135 MeV, and it exists for an average lifetime of only 8.7×10^{-17} s before decaying into two γ rays. Using the uncertainty principle, estimate the fractional uncertainty $\Delta m/m$ in its mass determination.

34. (a) Find and sketch the spectral content of the rectangular pulse of width 2τ shown in Figure P5.34. (b) Show that a reciprocity relation $\Delta\omega \, \Delta t \approx \pi$ holds

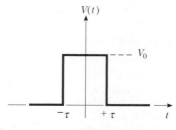

Figure P5.34

in this case. Take $\Delta t = \tau$ and define $\Delta\omega$ similarly. (c) What range of frequencies is required to compose a pulse of width $2\tau = 1$ μs? A pulse of width $2\tau = 1$ ns?

35. *A matter wave packet.* (a) Find and sketch the real part of the matter wave pulse shape $f(x)$ for a Gaussian amplitude distribution $a(k)$, where

$$a(k) = Ae^{-\alpha^2(k-k_0)^2}$$

Note that $a(k)$ is peaked at k_0 and has a width that decreases with increasing α. (*Hint:* In order to put $f(x) = (2\pi)^{-1/2} \int_{-\infty}^{+\infty} a(k)e^{ikx} \, dk$ into the standard form $\int_{-\infty}^{+\infty} e^{-az^2} \, dz$, complete the square in k.) (b) By comparing the result for the real part of $f(x)$ to the standard form of a Gaussian function with width Δx, $f(x) \propto Ae^{-(x/2\Delta x)^2}$, show that the width of the matter wave pulse is $\Delta x = \alpha$. (c) Find the width Δk of $a(k)$ by writing $a(k)$ in standard Gaussian form and show that $\Delta x \, \Delta k = \frac{1}{2}$, independent of α.

36. Consider a freely moving quantum particle with mass m and speed v. Its energy is $E = K + U = \frac{1}{2}mv^2 + 0$. Determine the phase speed of the quantum wave representing the particle and show that it is different from the speed at which the particle transports mass and energy.

37. In a vacuum tube, electrons are boiled out of a hot cathode at a slow, steady rate and accelerated from rest through a potential difference of 45.0 V. Then they travel altogether 28.0 cm as they go through an array of slits and fall on a screen to produce an interference pattern. Only one electron at a time will be in flight in the tube, provided the beam current is below what value? In this situation the interference pattern still appears, showing that each individual electron can interfere with itself.

6

Quantum Mechanics in One Dimension

Chapter Outline

We have seen that associated with any particle is a matter wave called the **wavefunction.** How this wavefunction affects our description of a particle and its behavior is the subject of **quantum mechanics,** or wave mechanics. This scheme, developed from 1925 to 1926 by Schrödinger, Heisenberg, and others, makes it possible to understand a host of phenomena involving elementary particles, atoms, molecules, and solids. In this and subsequent chapters, we shall describe the basic features of wave mechanics and its application to simple systems. The relevant concepts for particles confined to motion along a straight line (the *x*-axis) are developed in the present chapter.

6.1 THE BORN INTERPRETATION

The wavefunction Ψ contains within it all the information that can be known about the particle. That basic premise forms the cornerstone of our investigation: One of our objectives will be to discover how information may be extracted from the wavefunction; the other, to learn how to obtain this wavefunction for a given system.

The currently held view connects the wavefunction Ψ with *probabilities* in the manner first proposed by Max Born in 1925:

Max Born was a German theoretical physicist who made major contributions in many areas of physics, including relativity, atomic and solid-state physics, matrix mechanics, the quantum mechanical treatment of particle scattering ("Born approximation"), the foundations of quantum mechanics (Born interpretation of Ψ), optics, and the kinetic theory of liquids. Born received the doctorate in physics from the University of Göttingen in 1907, and he acquired an extensive knowledge of mathematics as the private assistant to the great German mathematician David Hilbert. This strong mathematical background proved a great asset when he was quickly able to reformulate Heisenberg's quantum theory in a more consistent way with matrices.

In 1921, Born was offered a post at the University of Göttingen, where he helped build one of the strongest physics centers of the 20th century. This group consisted, at one time

B I O G R A P H Y

MAX BORN
(1882–1970)

or another, of the mathematicians Hilbert, Courant, Klein, and Runge and the physicists Born, Jordan, Heisenberg, Franck, Pohl, Heitler, Herzberg, Nordheim, and Wigner,

among others. In 1926, shortly after Schrödinger's publication of wave mechanics, Born applied Schrödinger's methods to atomic scattering and developed the Born approximation method for carrying out calculations of the probability of scattering of a particle into a given solid angle. This work furnished the basis for Born's startling (in 1926) interpretation of $|\Psi|^2$ as the probability density. For this so-called statistical interpretation of $|\Psi|^2$ he was awarded the Nobel prize in 1954.

Fired by the Nazis, Born left Germany in 1933 for Cambridge and eventually the University of Edinburgh, where he again became the leader of a large group investigating the statistical mechanics of condensed matter. In his later years, Born campaigned against atomic weapons, wrote an autobiography, and translated German humorists into English.

(AIP Emilio Segrè Visual Archives)

Born interpretation of Ψ

The probability that a particle will be found in the infinitesimal interval dx about the point x, denoted by $P(x)\,dx$, is

$$P(x)\,dx = |\Psi(x,t)|^2\,dx \qquad (6.1)$$

Therefore, although it is not possible to specify with certainty the location of a particle, it is possible to assign probabilities for observing it at any given position. The quantity $|\Psi|^2$, the square of the absolute value of Ψ, represents the intensity of the matter wave and is computed as the product of Ψ with its complex conjugate, that is, $|\Psi|^2 = \Psi^*\Psi$. Notice that Ψ itself is not a measurable quantity; however, $|\Psi|^2$ is measurable and is just the probability per unit length, or **probability density** $P(x)$, for finding the particle at the point x at time t. For example, the intensity distribution in a light diffraction pattern is a measure of the probability that a photon will strike a given point within the pattern. Because of its relation to probabilities, we insist that $\Psi(x, t)$ be a *single-valued and continuous function of x* and t so that no ambiguities can arise concerning the predictions of the theory. The wavefunction Ψ also should be *smooth*, a condition that will be elaborated later as it is needed.

Because the particle must be somewhere along the *x*-axis, the probabilities summed over all values of *x* must add to 1:

$$\int_{-\infty}^{\infty} |\Psi(x, t)|^2 \, dx = 1 \qquad (6.2)$$

Any wavefunction satisfying Equation 6.2 is said to be **normalized.** Normalization is simply a statement that the particle can be found somewhere with certainty. The probability of finding the particle in any finite interval $a \leq x \leq b$ is

$$P = \int_{a}^{b} |\Psi(x, t)|^2 \, dx \qquad (6.3)$$

That is, the probability is just the area included under the curve of probability density between the points $x = a$ and $x = b$ (Fig. 6.1).

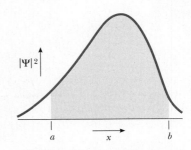

Figure 6.1   The probability for a particle to be in the interval $a \leq x \leq b$ is the area under the curve from *a* to *b* of the probability density function $|\Psi(x, t)|^2$.

EXAMPLE 6.1   Normalizing the Wavefunction

The initial wavefunction of a particle is given as $\Psi(x, 0) = C \exp(-|x|/x_0)$, where *C* and x_0 are constants. Sketch this function. Find *C* in terms of x_0 such that $\Psi(x, 0)$ is normalized.

Solution   The given wavefunction is symmetric, decaying exponentially from the origin in either direction, as shown in Figure 6.2. The decay length x_0 represents the distance over which the wave amplitude is diminished by the factor $1/e$ from its maximum value $\Psi(0, 0) = C$.

The normalization requirement is

$$1 = \int_{-\infty}^{\infty} |\Psi(x, 0)|^2 \, dx = C^2 \int_{-\infty}^{\infty} e^{-2|x|/x_0} \, dx$$

Because the integrand is unchanged when *x* changes sign (it is an even function), we may evaluate the integral over the whole axis as twice that over the half-axis $x > 0$, where $|x| = x$. Then,

$$1 = 2C^2 \int_{0}^{\infty} e^{-2x/x_0} \, dx = 2C^2 \left(\frac{x_0}{2}\right) = C^2 x_0$$

Thus, we must take $C = 1/\sqrt{x_0}$ for normalization.

EXAMPLE 6.2   Calculating Probabilities

Calculate the probability that the particle in the preceding example will be found in the interval $-x_0 \leq x \leq x_0$.

Solution   The probability is the area under the curve of $|\Psi(x, 0)|^2$ from $-x_0$ to $+x_0$ and is obtained by integrating the probability density over the specified interval:

$$P = \int_{-x_0}^{x_0} |\Psi(x, 0)|^2 \, dx = 2 \int_{0}^{x_0} |\Psi(x, 0)|^2 \, dx$$

where the second step follows because the integrand is an even function, as discussed in Example 6.1. Thus,

$$P = 2C^2 \int_{0}^{x_0} e^{-2x/x_0} \, dx = 2C^2(x_0/2)(1 - e^{-2})$$

Substituting $C = 1/\sqrt{x_0}$ into this expression gives for the probability $P = 1 - e^{-2} = 0.8647$, or about 86.5%, independent of x_0.

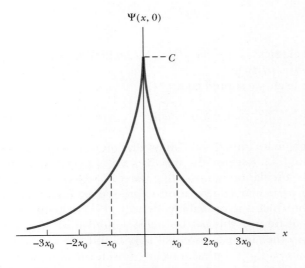

Ψ(x, 0)

- - - C

−3x₀   −2x₀   −x₀   x₀   2x₀   3x₀   x

Figure 6.2   (Example 6.1) The symmetric wavefunction $\Psi(x, 0) = C \exp(-|x|/x_0)$. At $x = \pm x_0$ the wave amplitude is down by the factor $1/e$ from its peak value $\Psi(0, 0) = C$. *C* is a normalizing constant whose proper value is $C = 1/\sqrt{x_0}$.

The fundamental problem of quantum mechanics is this: Given the wavefunction at some initial instant, say $t = 0$, find the wavefunction at any subsequent time t. The wavefunction $\Psi(x, 0)$ represents the initial information that must be specified; once this is known, however, the wave propagates according to prescribed laws of nature.

Because it describes how a given system evolves, quantum mechanics is a dynamical theory much like Newtonian mechanics. There are, of course, important differences. In Newton's mechanics, the state of a particle at $t = 0$ is specified by giving its initial position $x(0)$ and velocity $v(0)$—just two numbers; quantum mechanics demands an entire wavefunction $\Psi(x, 0)$—an infinite set of numbers corresponding to the wavefunction value at every point x. But both theories describe how this state changes with time when the forces acting on the particle are known. In Newton's mechanics $x(t)$ and $v(t)$ are calculated from Newton's second law; in quantum mechanics $\Psi(x, t)$ must be calculated from another law—*Schrödinger's equation.*

6.2 WAVEFUNCTION FOR A FREE PARTICLE

A free particle is one subject to no force. This special case can be studied using prior assumptions without recourse to the Schrödinger equation. The development underscores the role of the initial conditions in quantum physics.

The wavenumber k and frequency ω of free particle matter waves are given by the de Broglie relations

$$k = \frac{p}{\hbar} \qquad \text{and} \qquad \omega = \frac{E}{\hbar} \tag{6.4}$$

For nonrelativistic particles ω is related to k as

$$\omega(k) = \frac{\hbar k^2}{2m} \tag{6.5}$$

which follows from the classical connection $E = p^2/2m$ between the energy E and momentum p for a free particle.[1]

For the wavefunction itself, we should take

Plane wave representation for a free particle

$$\Psi_k(x, t) = Ae^{i(kx - \omega t)} = A\{\cos(kx - \omega t) + i\sin(kx - \omega t)\} \tag{6.6}$$

where $i = \sqrt{-1}$ is the imaginary unit. This is an oscillation with wavenumber k, frequency ω, and amplitude A. Because the variables x and t occur only in the combination $kx - \omega t$, the oscillation is a traveling wave, as befits a free particle in motion. Further, the particular combination expressed by Equation 6.6 is that of a plane wave,[2] for which the probability density $|\Psi|^2 (= A^2)$ is *uniform*. That is, the probability of finding this particle in any interval of the x-axis is the same as that for any other interval of equal length and does not change with time. The plane wave is the simplest traveling waveform with this prop-

---

[1]The functional form for $\omega(k)$ was discussed in Section 5.3 for relativistic particles, where $E = \sqrt{(cp)^2 + (mc^2)^2}$. In the nonrelativistic case ($v \ll c$), this reduces to $E = p^2/2m + mc^2$. The rest energy $E_0 = mc^2$ can be disregarded in this case if we agree to make E_0 our energy reference. By measuring all energies from this level, we are in effect setting E_0 equal to zero.

[2]For a plane wave, the wave fronts (points of constant phase) constitute planes perpendicular to the direction of wave propagation. In the present case the constant phase requirement $kx - \omega t = constant$ demands only that x be fixed, so the wave fronts occupy the y–z planes.

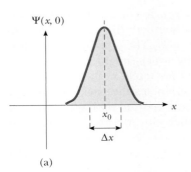

(a)

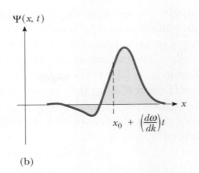

(b)

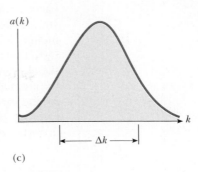

(c)

Figure 6.3   (a) A wave packet $\Psi(x, 0)$ formed from a superposition of plane waves. (b) The same wave packet some time t later (real part only). Because $v_p = \omega/k = \hbar k/2m$, the plane waves with smaller wavenumber move at slower speeds, and the packet becomes distorted. The body of the packet propagates with the group speed $d\omega/dk$ of the plane waves. (c) The amplitude distribution function $a(k)$ for this packet, indicating the amplitude of each plane wave in the superposition. A narrow wave packet requires a broad spectral content, and vice versa. That is, the widths Δx and Δk are inversely related as $\Delta x \Delta k \approx 1$.

erty—it expresses the reasonable notion that there are no special places for a free particle to be found. The particle's location is completely unknown at $t = 0$ and remains so for all time; however, its momentum and energy are known precisely as $p = \hbar k$ and $E = \hbar \omega$, respectively.

But not all free particles are described by Equation 6.6. For instance, we may establish (by measurement) that our particle initially is in some range Δx about x_0. In that case, $\Psi(x, 0)$ must be a *wave packet* concentrated in this interval, as shown in Figure 6.3a. The plane wave description is inappropriate now because the initial wave shape is not given correctly by $\Psi_k(x, 0) = e^{ikx}$. Instead, a sum of plane waves with different wavenumbers must be used to represent the packet. Because k is unrestricted, the sum actually is an integral here and we write

$$\Psi(x, 0) = \int_{-\infty}^{\infty} a(k) e^{ikx} \, dk \qquad (6.7)$$

Representing a particle with a wave group

The coefficients $a(k)$ specify the amplitude of the plane wave with wavenumber k in the mixture and are chosen to reproduce the initial wave shape. For a given $\Psi(x, 0)$, the required $a(k)$ can be found from the theory of Fourier integrals. We shall not be concerned with the details of this analysis here; the essential point is that it can be done for a packet of any shape (see optional Section 5.4). If each plane wave constituting the packet is assumed to propagate independently of the others according to Equation 6.6, the packet at any time t is

$$\Psi(x, t) = \int_{-\infty}^{\infty} a(k) e^{i\{kx - \omega(k)t\}} \, dk \qquad (6.8)$$

Notice that the initial data are used only to establish the amplitudes $a(k)$; subsequently, the packet develops according to the evolution of its plane wave constituents. Because each of these constituents moves with a different velocity $v_p = \omega/k$ (the phase velocity), the wave packet undergoes dispersion (see Section 5.3) and **the packet changes its shape as it propagates** (Fig. 6.3b). The speed of propagation of the wave packet as a whole is given by the group velocity $d\omega/dk$ of the plane waves forming the packet. Equation 6.8 no longer describes a particle with precise values of momentum and energy. To construct a wave packet (that is, localize the particle), a mixture of wavenumbers (hence, particle momenta) is necessary, as indicated by the different $a(k)$. The amplitudes $a(k)$ furnish the so-called *spectral content* of the packet, which might look like that sketched in Figure 6.3c. The narrower the desired packet $\Psi(x, 0)$, the broader is the function $a(k)$ representing that packet. If Δx denotes the packet width and Δk the extent of the corresponding $a(k)$, one finds that the product always is a number of order unity, that is, $\Delta x \, \Delta k \approx 1$. Together with $p = \hbar k$, this implies an uncertainty principle:

$$\Delta x \, \Delta p \sim \hbar \tag{6.9}$$

EXAMPLE 6.3 Constructing a Wave Packet

Find the wavefunction $\Psi(x, 0)$ that results from taking the function $a(k) = (C\alpha/\sqrt{\pi})\exp(-\alpha^2 k^2)$, where C and α are constants. Estimate the product $\Delta x \, \Delta k$ for this case.

Solution   The function $\Psi(x, 0)$ is given by the integral of Equation 6.7 or

$$\Psi(x, 0) = \int_{-\infty}^{\infty} a(k) e^{ikx} \, dk = \frac{C\alpha}{\sqrt{\pi}} \int_{-\infty}^{\infty} e^{(ikx - \alpha^2 k^2)} \, dk$$

To evaluate the integral, we first complete the square in the exponent as

$$ikx - \alpha^2 k^2 = -\left(\alpha k - \frac{ix}{2\alpha}\right)^2 - \frac{x^2}{4\alpha^2}$$

The second term on the right is constant for the integration over k; to integrate the first term we change variables with the substitution $z = \alpha k - ix/2\alpha$, obtaining

$$\Psi(x, 0) = \frac{C}{\sqrt{\pi}} e^{-x^2/4\alpha^2} \int_{-\infty}^{\infty} e^{-z^2} \, dz$$

The integral now is a standard one whose value is known to be $\sqrt{\pi}$. Then,

$$\Psi(x, 0) = Ce^{-x^2/4\alpha^2} = Ce^{-(x/2\alpha)^2}$$

This function $\Psi(x, 0)$, called a *Gaussian* function, has a single maximum at $x = 0$ and decays smoothly to zero on either side of this point (Fig. 6.4a). The width of this Gaussian packet becomes larger with increasing α. Accordingly, it is reasonable to identify α with Δx, the initial degree of localization. By the same token, $a(k)$ also is a Gaussian function, but with amplitude $C\alpha/\sqrt{\pi}$ and width $1/2\alpha$ (since $\alpha^2 k^2 = (k/2[1/2\alpha])^2$). Thus, $\Delta k = 1/2\alpha$ and $\Delta x \Delta k = 1/2$, *independent of α*. The multiplier C is a scale factor chosen to normalize Ψ.

Because our Gaussian packet is made up of many individual waves all moving with different speeds, the shape of the packet changes over time. In Problem 4 it is shown that the packet *disperses*, its width growing ever larger with the passage of time as

$$\Delta x(t) = \sqrt{[\Delta x(0)]^2 + \left[\frac{\hbar t}{2m\Delta x(0)}\right]^2}$$

Similarly, the peak amplitude diminishes steadily in order to keep the waveform normalized for all times (Fig. 6.4b). The wave as a whole does not propagate, because for every wavenumber k present in the wave group there is an equal admixture of the plane wave with the opposing wavenumber $-k$.

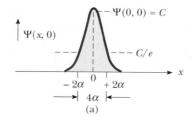
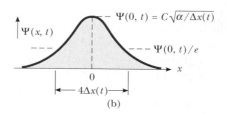

(a)   (b)

Figure 6.4 (Example 6.3) (a) The Gaussian wavefunction $\Psi(x, 0) = C\exp\{-(x/2\alpha)^2\}$, representing a particle initially localized around $x = 0$. C is the amplitude. At $x = \pm 2\alpha$, the amplitude is down from its maximum value by the factor $1/e$; accordingly, α is identified as the width of the Gaussian, $\alpha = \Delta x$. (b) The Gaussian wavefunction of Figure 6.4a at time t (apart from a phase factor). The width has increased to $\Delta x(t) = \sqrt{\alpha^2 + (\hbar t/2m\alpha)^2}$ and the amplitude is reduced by the factor $\sqrt{\alpha/\Delta x(t)}$.

EXAMPLE 6.4   Dispersion of Matter Waves

An atomic electron initially is localized to a region of space 0.10 nm wide (atomic size). How much time elapses before this localization is destroyed by dispersion? Repeat the calculation for a 1.0-g marble initially localized to 0.10 mm.

Solution   Taking for the initial state a Gaussian wave shape, we may use the results of the previous example. In particular, the extent of the matter wave after a time t has elapsed is

$$\Delta x(t) = \sqrt{[\Delta x(0)]^2 + \left[\frac{\hbar t}{2m\Delta x(0)}\right]^2}$$

where $\Delta x(0)$ is its initial width. The packet has effectively dispersed when $\Delta x(t)$ becomes appreciable compared to $\Delta x(0)$, say, $\Delta x(t) = 10\ \Delta x(0)$. This happens when $\hbar t/2m = \sqrt{99}\ [\Delta x(0)]^2$, or $t = \sqrt{99}\ (2m/\hbar)[\Delta x(0)]^2$.

The electron is initially localized to 0.10 nm $(= 10^{-10}$ m$)$, and its mass is $m_e = 9.11 \times 10^{-31}$ kg. Thus, the electron wave packet disperses after a time

$$t = \sqrt{99} \left\{\frac{(2)(9.11 \times 10^{-31}\ \text{kg})}{1.055 \times 10^{-34}\ \text{J}\cdot\text{s}}\right\} (1.00 \times 10^{-10}\ \text{m})^2$$
$$= 1.7 \times 10^{-15}\ \text{s}$$

The same calculation for a 1.0-g marble localized to 0.10 mm $= 10^{-4}$ m gives

$$t = \sqrt{99} \left\{\frac{(2)(10^{-3}\ \text{kg})}{1.055 \times 10^{-34}\ \text{J}\cdot\text{s}}\right\} (10^{-4}\ \text{m})^2$$
$$= 1.9 \times 10^{24}\ \text{s}$$

or about 6.0×10^{16} years! This is nearly 10 million times the currently accepted value for the age of the Universe. With its much larger mass, the marble does not show the quantum effects of dispersion on any measurable time scale and will, for all practical purposes, remain localized "forever." By contrast, the localization of an atomic electron is destroyed in a time that is very short, on a par with the time it takes the electron to complete one Bohr orbit.

In closing this section, we note that *in principle* Equations 6.7 and 6.8 solve the fundamental problem of quantum mechanics for free particles subject to any initial condition $\Psi(x, 0)$. Because of its mathematical simplicity, the Gaussian wave packet is commonly used to represent the initial system state, as in the previous examples. However, the Gaussian form is often only an approximation to reality. Yet even in this simplest of cases, the mathematical challenge of obtaining $\Psi(x, t)$ from $\Psi(x, 0)$ tends to obscure the important results. Numerical simulation affords a convenient alternative to analytical calculation that also aids in visualizing the important phenomena of wave packet propagation and dispersion. To "see" quantum waveforms in action and further explore their time evolution, go to our companion Web site http://info.brookscole.com/mp3e, select QMTools Simulations → Evolution of Free Particle Wave Packets (Tutorial), and follow the on-site instructions.

6.3   WAVEFUNCTIONS IN THE PRESENCE OF FORCES

For a particle acted on by a force F, $\Psi(x, t)$ must be found from **Schrödinger's equation:**

$$-\frac{\hbar^2}{2m}\frac{\partial^2\Psi}{\partial x^2} + U(x)\Psi = i\hbar\frac{\partial\Psi}{\partial t} \qquad (6.10)$$

The Schrödinger wave equation

Again, we assume knowledge of the initial wavefunction $\Psi(x, 0)$. In this expression, $U(x)$ is the potential energy function for the force F; that is,

$F = -dU/dx$. Schrödinger's equation is not derivable from any more basic principle, but is one of the laws of quantum physics. As with any law, its "truth" must be gauged ultimately by its ability to make predictions that agree with experiment.

Erwin Schrödinger was an Austrian theoretical physicist best known as the creator of wave mechanics. As a young man he was a good student who liked mathematics and physics, but also Latin and Greek for their logical grammar. He received a doctorate in physics from the University of Vienna. Although his work in physics was interrupted by World War I, Schrödinger had by 1920 produced important papers on statistical mechanics, color vision, and general relativity, which he at first found quite difficult to understand. Expressing his feelings about a scientific theory in the remarkably open and outspoken way he maintained throughout his life, Schrödinger found general relativity initially "depressing" and "unnecessarily complicated." Other Schrödinger remarks in this vein, with which some readers will enthusiastically agree, are as follows: The Bohr–Sommerfeld quantum theory was "unsatisfactory, even disagreeable." "I . . . feel intimidated, not to say repelled, by what seem to me the very difficult methods [of matrix mechanics] and by the lack of clarity."

Shortly after de Broglie introduced the concept of matter waves in 1924, Schrödinger began to develop a new relativistic atomic theory based on de Broglie's ideas, but his failure to include electron spin led to the failure of this theory for hydrogen. By January of 1926, however, by treating the electron as a nonrelativistic particle, Schrödinger had introduced his famous wave equation and successfully obtained the energy values and wavefunctions for hydrogen. As Schrödinger himself pointed out, an outstanding fea-

BIOGRAPHY

ERWIN SCHRÖDINGER
(1887–1961)

ture of his approach was that the discrete energy values emerged from his wave equation in a natural way (as in the case of standing waves on a string), and in a way superior to the artificial postulate approach of Bohr. Another outstanding feature of Schrödinger's wave mechanics was that it was easier to apply to physical problems than Heisenberg's matrix mechanics, because it involved a partial differential equation very similar to the classical wave equation. Intrigued by the remarkable differences in conception and mathematical method of wave and matrix mechanics, Schrödinger did much to hasten the universal acceptance of all of quantum theory by demonstrating the mathematical equivalence of the two theories in 1926.

Although Schrödinger's wave theory was generally based on clear physical ideas, one of its major problems in 1926 was the physical interpretation of the wavefunction Ψ. Schrödinger felt that the electron was ultimately a wave, Ψ was the vibration amplitude of this wave, and $\Psi^*\Psi$ was the electric charge density. As mentioned in Chapter 4, Born, Bohr, Heisenberg, and others pointed out the problems with this interpretation and presented the currently accepted view that $\Psi^*\Psi$ is a probability and that the electron is ultimately no more a wave than a particle. Schrödinger never accepted this view, but registered his "concern and disappointment" that this "transcendental, almost psychical interpretation" had become "universally accepted dogma."

In 1927, Schrödinger, at the invitation of Max Planck, accepted the chair of theoretical physics at the University of Berlin, where he formed a close friendship with Planck and experienced six stable and productive years. In 1933, disgusted with the Nazis like so many of his colleagues, he left Germany. After several moves reflecting the political instability of Europe, he eventually settled at the Dublin Institute for Advanced Studies. Here he spent 17 happy, creative years working on problems in general relativity, cosmology, and the application of quantum physics to biology. This last effort resulted in a fascinating short book, *What is Life?*, which induced many young physicists to investigate biological processes with chemical and physical methods. In 1956, he returned home to his beloved Tyrolean mountains. He died there in 1961.

(AIP Emilio Segrè Visual Archives)

The Schrödinger equation propagates the initial wave forward in time. To see how this works, suppose $\Psi(x, 0)$ has been given. Then the left-hand side (LHS) of Schrödinger's equation can be evaluated and Equation 6.10 gives $\partial\Psi/\partial t$ at $t = 0$, the initial rate of change of the wavefunction. From this we compute the wavefunction a short time, δt, later as $\Psi(x, \delta t) = \Psi(x, 0) + [\partial\Psi/\partial t]_0 \delta t$. This allows the LHS to be re-evaluated, now at $t = \delta t$. With each such repetition, Ψ is advanced another step δt into the future. Continuing the process generates Ψ at any later time t. Such repetitious calculations are ideally suited to computers, and the method just outlined may be used to solve the Schrödinger equation numerically.[3]

But how can we obtain an explicit mathematical expression for $\Psi(x, t)$? Returning to the free particle case, we see that the plane waves $\Psi_k(x, t)$ of Equation 6.6 serve a dual purpose: On the one hand, they represent particles whose momentum (hence, energy) is known precisely; on the other, they become the building blocks for constructing wavefunctions satisfying any initial condition. From this perspective, the question naturally arises: Do analogous functions exist when forces are present? The answer is yes! To obtain them we look for solutions to the Schrödinger equation having the separable form[4]

$$\Psi(x, t) = \psi(x)\phi(t) \tag{6.11}$$

where $\psi(x)$ is a function of x only and $\phi(t)$ is a function of t only. (Note that the plane waves have just this form, with $\psi(x) = e^{ikx}$ and $\phi(t) = e^{-i\omega t}$.) Substituting Equation 6.11 into Equation 6.10 and dividing through by $\psi(x)\phi(t)$ gives

$$-\frac{\hbar^2}{2m}\frac{\psi''(x)}{\psi(x)} + U(x) = i\hbar\frac{\phi'(t)}{\phi(t)}$$

where primes denote differentiation with respect to the arguments. Now the LHS of this equation is a function of x only,[5] and the RHS is a function of t only. Since we can assign any value of x independently of t, the two sides of the equation can be equal only if each is equal to the same constant, which we call E.[6] This yields two equations determining the unknown functions $\psi(x)$ and $\phi(t)$. The resulting equation for the time-

---

[3]This straightforward approach suffers from numerical instabilities and does not, for example, conserve probability. In practice, a more sophisticated discretization scheme is usually employed, such as that provided by the Crank–Nicholson method. See, for example, section 17.2 of *Numerical Recipes* by W. H. Press, B. P. Flannery, S. A. Teukolsky, W. T. Vetterling, Cambridge, U.K., Cambridge University Press, 1986.

[4]Obtaining solutions to partial differential equations in separable form is called *separation of variables*. On separating variables, a partial differential equation in, say, N variables is reduced to N ordinary differential equations, each involving only a single variable. The technique is a general one which may be applied to many (but not all!) of the partial differential equations encountered in science and engineering applications.

[5]Implicitly we have assumed that the potential energy $U(x)$ is a function of x only. For potentials that also depend on t (for example, those arising from a time-varying electric field), solutions to the Schrödinger equation in separable form generally do not exist.

[6]More explicitly, changing t cannot affect the LHS because this depends only on x. Since the two sides of the equation are equal, we conclude that changing t cannot affect the RHS either. It follows that the RHS must reduce to a constant. The same argument with x replacing t shows the LHS also must reduce to this same constant.

dependent function $\phi(t)$ is

$$i\hbar \frac{d\phi}{dt} = E\phi(t) \qquad (6.12)$$

This can be integrated immediately to give $\phi(t) = e^{-i\omega t}$ with $\omega = E/\hbar$. Thus, the time dependence is the same as that obtained for free particles! The equation for the space function $\psi(x)$ is

Wave equation for matter waves in separable form

$$-\frac{\hbar^2}{2m} \frac{d^2\psi}{dx^2} + U(x)\psi(x) = E\psi(x) \qquad (6.13)$$

Equation 6.13 is called the **time-independent Schrödinger equation.** Explicit solutions to this equation cannot be written for an arbitrary potential energy function $U(x)$. But whatever its form, $\psi(x)$ must be well behaved because of its connection with probabilities. In particular, $\psi(x)$ must be everywhere finite, single-valued, and continuous. Furthermore, $\psi(x)$ must be "smooth," that is, the slope of the wave $d\psi/dx$ also must be continuous wherever $U(x)$ has a finite value.[7]

For free particles we take $U(x) = 0$ in Equation 6.13 (to give $F = -dU/dx = 0$) and find that $\psi(x) = e^{ikx}$ is a solution with $E = \hbar^2 k^2/2m$. Thus, for free particles the separation constant E becomes the total particle energy; this identification continues to be valid when forces are present. The wavefunction $\psi(x)$ will change, however, with the introduction of forces, because particle momentum (hence, k) is no longer constant.

The separable solutions to Schrödinger's equation describe conditions of particular physical interest. One feature shared by all such wavefunctions is especially noteworthy: Because $|e^{-i\omega t}|^2 = e^{+i\omega t}e^{-i\omega t} = e^0 = 1$, we have

$$|\Psi(x, t)|^2 = |\psi(x)|^2 \qquad (6.14)$$

This equality expresses the time independence of all probabilities calculated from $\Psi(x, t)$. For this reason, solutions in separable form are called **stationary states.** Thus, **for stationary states all probabilities are static** and can be calculated from the time-independent wavefunction $\psi(x)$.

6.4 THE PARTICLE IN A BOX

Of the problems involving forces, the simplest is that of particle confinement. Consider a particle moving along the x-axis between the points $x = 0$ and $x = L$, where L is the length of the "box." Inside the box the particle is free; at the endpoints, however, it experiences strong forces that serve to

---

[7]On rearrangement, the Schrödinger equation specifies the second derivative of the wavefunction $d^2\psi/dx^2$ at any point as

$$\frac{d^2\psi}{dx^2} = \frac{2m}{\hbar^2}[U(x) - E]\psi(x)$$

It follows that if $U(x)$ is finite at x, the second derivative also is finite here and the slope $d\psi/dx$ will be continuous.

contain it. A simple example is a ball bouncing elastically between two impenetrable walls (Fig. 6.5). A more sophisticated one is a charged particle
moving along the axis of aligned metallic tubes held at different potentials,
as shown in Figure 6.6a. The central tube is grounded, so a test charge inside this tube has zero electric potential energy and experiences no electric
force. When both outer tubes are held at a high electric potential V, there
are no electric fields within them, but strong repulsive fields arise in the
gaps at 0 and L. The potential energy $U(x)$ for this situation is sketched in
Figure 6.6b. As V is increased without limit and the gaps are simultaneously
reduced to zero, we approach the idealization known as the *infinite square
well*, or "box" potential (Fig. 6.6c).

From a classical viewpoint, our particle simply bounces back and forth between the confining walls of the box. Its speed remains constant, as does its kinetic energy. Furthermore, classical physics places no restrictions on the values
of its momentum and energy. The quantum description is quite different and
leads to the interesting phenomenon of energy quantization.

We are interested in the time-independent wavefunction $\psi(x)$ of our particle. Because it is confined to the box, the particle can never be found outside,
which requires ψ to be zero in the exterior regions $x < 0$ and $x > L$. Inside the
box, $U(x) = 0$ and Equation 6.13 for $\psi(x)$ becomes, after rearrangement,

$$\frac{d^2\psi}{dx^2} = -k^2\psi(x) \qquad \text{with} \qquad k^2 = \frac{2mE}{\hbar^2}$$

Independent solutions to this equation are $\sin kx$ and $\cos kx$, indicating that
k is the wavenumber of oscillation. The most general solution is a linear

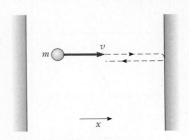

Figure 6.5 A particle of mass m and speed v bouncing elastically between two impenetrable walls.

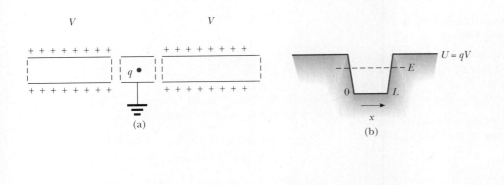

(a)

(b)

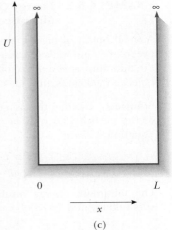

(c)

Figure 6.6 (a) Aligned metallic cylinders serve to confine a charged particle. The inner cylinder is grounded, while the outer ones are held at some high electric potential
V. A charge q moves freely within the cylinders, but encounters electric forces in the
gaps separating them. (b) The electric potential energy seen by this charge. A charge
whose total energy is less than qV is confined to the central cylinder by the strong repulsive forces in the gaps at $x = 0$ and $x = L$. (c) As V is increased and the gaps between cylinders are narrowed, the potential energy approaches that of the infinite
square well.

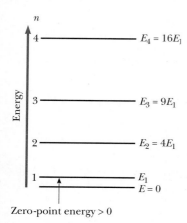

Figure 6.7 Energy-level diagram for a particle confined to a one-dimensional box of width L. The lowest allowed energy is E_1, with value $\pi^2\hbar^2/2mL^2$.

Allowed energies for a particle in a box

combination of these two,

$$\psi(x) = A \sin kx + B \cos kx \qquad \text{for } 0 < x < L \qquad (6.15)$$

This interior wave must match the exterior wave at the walls of the box for $\psi(x)$ to be continuous everywhere.[8] Thus, we require the interior wave to vanish at $x = 0$ and $x = L$:

$$\psi(0) = B = 0 \qquad \text{(continuity at } x = 0)$$
$$\psi(L) = A \sin kL = 0 \qquad \text{(continuity at } x = L) \qquad (6.16)$$

The last condition requires that $kL = n\pi$, where n is any positive integer.[9] Because $k = 2\pi/\lambda$, this is equivalent to fitting an integral number of half-wavelengths into the box (see Fig. 6.9a). Using $k = n\pi/L$, we find that the particle energies are *quantized*, being restricted to the values

$$E_n = \frac{\hbar^2 k^2}{2m} = \frac{n^2\pi^2\hbar^2}{2mL^2} \qquad n = 1, 2, \ldots \qquad (6.17)$$

The lowest allowed energy is given by $n = 1$ and is $E_1 = \pi^2\hbar^2/2mL^2$. This is the **ground state.** Because $E_n = n^2E_1$, the **excited states** for which $n = 2, 3, 4, \ldots$ have energies $4E_1, 9E_1, 16E_1, \ldots$ An energy-level diagram is given in Figure 6.7. Notice that $E = 0$ is not allowed; that is, *the particle can never be at rest.* The least energy the particle can have, E_1, is called the **zero-point energy.** This result clearly contradicts the classical prediction, for which $E = 0$ is an acceptable energy, as are all positive values of E. The following example illustrates how this contradiction is reconciled with our everyday experience.

EXAMPLE 6.5   Energy Quantization for a Macroscopic Object

A small object of mass 1.00 mg is confined to move between two rigid walls separated by 1.00 cm. (a) Calculate the minimum speed of the object. (b) If the speed of the object is 3.00 cm/s, find the corresponding value of n.

Solution   Treating this as a particle in a box, the energy of the particle can only be one of the values given by Equation 6.17, or

$$E_n = \frac{n^2\pi^2\hbar^2}{2mL^2} = \frac{n^2h^2}{8mL^2}$$

The minimum energy results from taking $n = 1$. For $m = 1.00$ mg and $L = 1.00$ cm, we calculate

$$E_1 = \frac{(6.626 \times 10^{-34}\,\text{J·s})^2}{8.00 \times 10^{-10}\,\text{kg·m}^2} = 5.49 \times 10^{-58}\,\text{J}$$

Because the energy is all kinetic, $E_1 = mv_1^2/2$ and the minimum speed v_1 of the particle is

$$v_1 = \sqrt{2(5.49 \times 10^{-58}\,\text{J})/(1.00 \times 10^{-6}\,\text{kg})}$$
$$= 3.31 \times 10^{-26}\,\text{m/s}$$

This speed is immeasurably small, so that for practical purposes the object can be considered to be at rest. Indeed, the time required for an object with this speed to move the 1.00 cm separating the walls is about

---

[8]Although $\psi(x)$ must be continuous everywhere, the slope of $d\psi/dx$ is *not* continuous at the walls of the box, where $U(x)$ becomes infinite (cf. footnote 7).

[9]For $n = 0$ ($E = 0$), Schrödinger's equation requires $d^2\psi/dx^2 = 0$, whose solution is given by $\psi(x) = Ax + B$ for some choice of constants A and B. For this wavefunction to vanish at $x = 0$ and $x = L$, both A and B must be zero, leaving $\psi(x) = 0$ everywhere. In such a case the particle is nowhere to be found; that is, no description is possible when $E = 0$. Also, the inclusion of negative integers $n < 0$ produces no new states, because changing the sign of n merely changes the sign of the wavefunction, leading to the same probabilities as for positive integers.

3×10^{23} s, or about 1 million times the present age of the Universe! It is reassuring to verify that quantum mechanics applied to macroscopic objects does not contradict our everyday experiences.

If, instead, the speed of the particle is $v = 3.00$ cm/s, then its energy is

$$E = \frac{mv^2}{2} = \frac{(1.00 \times 10^{-6} \text{ kg})(3.00 \times 10^{-2} \text{ m/s})^2}{2}$$

$$= 4.50 \times 10^{-10} \text{ J}$$

This, too, must be one of the special values E_n. To find which one, we solve for the quantum number n, obtaining

$$n = \frac{\sqrt{8mL^2E}}{h}$$

$$= \frac{\sqrt{(8.00 \times 10^{-10} \text{ kg} \cdot \text{m}^2)(4.50 \times 10^{-10} \text{ J})}}{6.626 \times 10^{-34} \text{ J} \cdot \text{s}}$$

$$= 9.05 \times 10^{23}$$

Notice that the quantum number representing a typical speed for this ordinary-size object is enormous. In fact, the value of n is so large that we would never be able to distinguish the quantized nature of the energy levels. That is, the difference in energy between two consecutive states with quantum numbers $n_1 = 9.05 \times 10^{23}$ and $n_2 = 9.05 \times 10^{23} + 1$ is only about 10^{-33} J, much too small to be detected experimentally. This is another example that illustrates the working of Bohr's correspondence principle, which asserts that quantum predictions must agree with classical results for large masses and lengths.

EXAMPLE 6.6   Model of an Atom

An atom can be viewed as a number of electrons moving around a positively charged nucleus, where the electrons are subject mainly to the Coulombic attraction of the nucleus (which actually is partially "screened" by the intervening electrons). The potential well that each electron "sees" is sketched in Figure 6.8. Use the model of a particle in a box to estimate the energy (in eV) required to raise an atomic electron from the state $n = 1$ to the state $n = 2$, assuming the atom has a radius of 0.100 nm.

Solution   Taking the length L of the box to be 0.200 nm (the diameter of the atom), $m_e = 511$ keV/c^2, and $\hbar c = 197.3$ eV · nm for the electron, we calculate

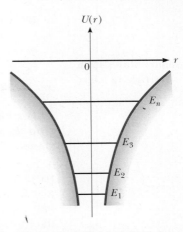

Figure 6.8   (Example 6.6) Model of the potential energy versus r for the one-electron atom.

$$E_1 = \frac{\pi^2 \hbar^2}{2m_e L^2}$$

$$= \frac{\pi^2 (197.3 \text{ eV} \cdot \text{nm}/c)^2}{2(511 \times 10^3 \text{ eV}/c^2)(0.200 \text{ nm})^2}$$

$$= 9.40 \text{ eV}$$

and

$$E_2 = (2)^2 E_1 = 4(9.40 \text{ eV}) = 37.6 \text{ eV}$$

Therefore, the energy that must be supplied to the electron is

$$\Delta E = E_2 - E_1 = 37.6 \text{ eV} - 9.40 \text{ eV} = 28.2 \text{ eV}$$

We could also calculate the wavelength of the photon that would cause this transition by identifying ΔE with the photon energy hc/λ, or

$$\lambda = hc/\Delta E = (1.24 \times 10^3 \text{ eV} \cdot \text{nm})/(28.2 \text{ eV}) = 44.0 \text{ nm}$$

This wavelength is in the far ultraviolet region, and it is interesting to note that the result is roughly correct. Although this oversimplified model gives a good estimate for transitions between lowest-lying levels of the atom, the estimate gets progressively worse for higher-energy transitions.

Exercise 1   Calculate the minimum speed of an atomic electron modeled as a particle in a box with walls that are 0.200 nm apart.

Answer 1.82×10^6 m/s.

Returning to the wavefunctions, we have from Equation 6.15 (with $k = n\pi/L$ and $B = 0$)

Stationary states for a particle in a box

$$\psi_n(x) = A\sin\left(\frac{n\pi x}{L}\right) \qquad \text{for } 0 < x < L \text{ and } n = 1, 2, \ldots \qquad (6.18)$$

For each value of the quantum number n there is a specific wavefunction $\psi_n(x)$ describing the state of the particle with energy E_n. Figure 6.9 shows plots of ψ_n versus x and of the probability density $|\psi_n|^2$ versus x for $n = 1, 2,$ and 3, corresponding to the three lowest allowed energies for the particle. For $n = 1$, the probability of finding the particle is largest at $x = L/2$—this is the *most probable position* for a particle in this state. For $n = 2$, $|\psi|^2$ is a maximum at $x = L/4$ and again at $x = 3L/4$: Both points are equally likely places for a particle in this state to be found.

There are also points within the box where it is impossible to find the particle. Again for $n = 2$, $|\psi|^2$ is zero at the midpoint, $x = L/2$; for $n = 3$, $|\psi|^2$ is zero at $x = L/3$ and at $x = 2L/3$, and so on. But this raises an interesting question: How does our particle get from one place to another when there is no probability for its ever being at points in between? It is as if there were *no path at all*, and not just that the probabilities $|\psi|^2$ express our ignorance about a world somehow hidden from view. Indeed, what is at stake here is the very essence of a particle as something that gets from one place to another by occupying all intervening positions. The objects of quantum mechanics are not particles, but more complicated things having both particle *and wave* attributes.

Actual probabilities can be computed only after ψ_n is normalized, that is, we must be sure that all probabilities sum to unity:

$$1 = \int_{-\infty}^{\infty} |\psi_n(x)|^2\, dx = A^2 \int_0^L \sin^2\left(\frac{n\pi x}{L}\right) dx$$

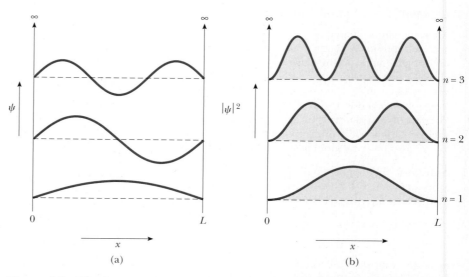

Figure 6.9   The first three allowed stationary states for a particle confined to a one-dimensional box. (a) The wavefunctions for $n = 1, 2,$ and 3. (b) The probability distributions for $n = 1, 2,$ and 3.

The integral is evaluated with the help of the trigonometric identity $2 \sin^2 \theta = 1 - \cos 2\theta$:

$$\int_0^L \sin^2\left(\frac{n\pi x}{L}\right) dx = \frac{1}{2}\int_0^L [1 - \cos(2n\pi x/L)]\, dx$$

Only the first term contributes to the integral, because the cosine integrates to $\sin(2n\pi x/L)$, which vanishes at the limits 0 and L. Thus, normalization requires $1 = A^2 L/2$, or

$$A = \sqrt{\frac{2}{L}} \tag{6.19}$$

EXAMPLE 6.7   Probabilities for a Particle in a Box

A particle is known to be in the ground state of an infinite square well with length L. Calculate the probability that this particle will be found in the middle half of the well, that is, between $x = L/4$ and $x = 3L/4$.

Solution   The probability density is given by $|\psi_n|^2$ with $n = 1$ for the ground state. Thus, the probability is

$$P = \int_{L/4}^{3L/4} |\psi_1|^2\, dx = \left(\frac{2}{L}\right)\int_{L/4}^{3L/4} \sin^2(\pi x/L)\, dx$$

$$= \left(\frac{1}{L}\right)\int_{L/4}^{3L/4} [1 - \cos(2\pi x/L)]\, dx$$

$$= \left(\frac{1}{L}\right)\left[\frac{L}{2} - \left(\frac{L}{2\pi}\right)\sin(2\pi x/L)\,\bigg|_{L/4}^{3L/4}\right]$$

$$= \frac{1}{2} - \left(\frac{1}{2\pi}\right)[-1 - 1] = 0.818$$

Notice that this is considerably larger than $\frac{1}{2}$, which would be expected for a classical particle that spends equal time in all parts of the well.

Exercise 2   Repeat the calculation of Example 6.7 for a particle in the nth state of the infinite square well, and show that the result approaches the classical value $\frac{1}{2}$ in the limit $n \rightarrow \infty$.

Charge-Coupled Devices (CCDs)

Potential wells are essential to the operation of many modern electronic devices, though rarely is the well shape so simple that it can be accurately modeled by the infinite square well discussed in this section. The charge-coupled device, or CCD, uses potential wells to trap electrons and create a faithful electronic reproduction of light intensity across the active surface.

For more than two decades now, CCDs have been helping astronomers see amazing detail in distant galaxies using much shorter exposure times than with traditional photographic emulsions (Fig. 6.10). These devices consist of a two-dimensional array of moveable electron boxes (or wells) created beneath a set of electrodes formed on the surface of a thin silicon chip (Fig. 6.11). The silicon serves the dual purpose of emitting an electron when struck by a photon and acting as a local trap for electrons. The potential energy seen by an electron in this environment is shown by the curve on the right in Figure 6.11, with the depth coordinate increasing downward. Though far removed from a

Figure 6.10 Researchers at Arizona State University, using NASA's Hubble Space Telescope, believe they are seeing the conclusion of the cosmic epoch where the young galaxies started to shine in significant numbers, about 13 billion years ago. The image shows some of the objects that the team discovered using Hubble's new Advanced Camera for Surveys (ACS), based on CCD technology. Astronomers believe that these numerous objects are faint young star-forming galaxies seen when the universe was seven times smaller than it is today (at redshifts of about 6) and less than a billion years old. (*H.-J. Yan, R. Windhorst and S. Cohen, Arizona State University and NASA*).

"box" potential, the well shape nevertheless serves to confine the emitted electrons in the depth dimension. [Each well or picture element (pixel) in the array also is isolated electrically from its neighbors, in effect confining the electrons in the remaining two dimensions perpendicular to the figure.] The number of electrons in a given well, and consequently the number of photons striking a particular point on the chip, may be read out electronically and the signal processed by computer to enhance the image. The name "charge-coupled device" was coined to describe the way the signals are read from the individual wells. A row of wells containing trapped electrons is moved vertically one step at a time by changing the voltage on the vertical electrodes in a progressive manner. When a row reaches the output register, the pixels are moved horizontally by systematically changing the voltage on the horizontal electrodes. In this way an entire row is read out in serial fashion by an amplifier at the end of the output register. Figure 6.12 illustrates the operating principle. CCD development has been impressive over the past two decades, and currently square arrays of over 4 million pixels (2048 pixels on a side) packed into a chip of several square centimeters are available. An entire CCD sensor is shown in Figure 6.13a; Figure 6.13b shows the cross section of a single pixel in a CCD image sensor, enlarged 5000 times.

CCD imagers possess several advantages over other light detectors. Because CCDs detect as many as 90% of the photons hitting their surface, they are far more sensitive than the best photographic emulsions, which can detect only 2–3% of those bone-weary photons that have traveled millions of lightyears from distant galaxies. In addition, CCDs can accurately measure the exact brightness of an object, since their voltage output is directly proportional to light input over a very wide brightness range. Another great feature of CCDs is their ability to measure accurately both faint and bright objects in the same frame. This is not true for photographic emulsions, where bright objects wash out faint details. Faint objects are recorded by cooling the CCD with liquid nitrogen to keep competing thermally generated electrons (noise) to a minimum. The simultaneous measurement of bright images is limited only by the filling of potential wells with electrons. State-of-the-art CCDs can hold as many as 100,000 electrons in a single well and are about 100 times better than photographic plates at simultaneously recording bright and faint objects. The ability to record where an incident photon strikes also is important for locating the exact position of a faint star. CCDs afford exceptional geometric accuracy because each pixel position is defined by the rigid physical structure of the

Figure 6.11   Structure of a single picture element (pixel) in a CCD array. The sketch on the right shows how the potential energy of an electron varies with depth in the device.

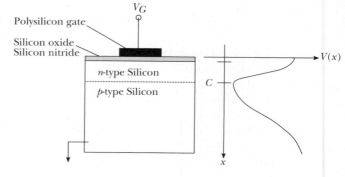

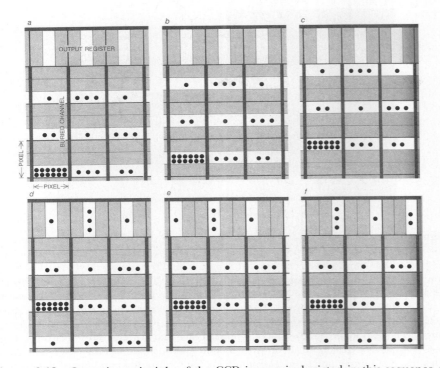

Figure 6.12   Operating principle of the CCD imager is depicted in this sequence of schematic diagrams, each of which corresponds to a small segment near the top edge of the device. The strips of gray and white bars function as a system of electronic conveyor belts. The white bars represent zones of low electric potential, called potential wells, in which the photoelectrons (*colored dots*) are collected; the gray bars are zones of higher electric potential that act as barriers to keep the electrons in the potential wells. The three vertical strips in each diagram are electron-conducting channels "buried" in the body of the device's imaging section; the horizontal strip across the top is the serial output register. Three pixels are shown in each channel. Each pixel is in turn subdivided into three parts: one part low (the potential well) and two parts high (the potential barriers). The heights of the three parts can be changed by means of three sets of electrodes called gates (not shown here), which run across the surface of the chip at right angles to the channels and work in concert to move the electrons along the channels. The electrons are kept from moving sideways out of the channels by permanent barriers called channel stops (*thick black lines*). In (*a*) the CCD is being exposed. Photons, or light quanta, enter the chip from the rear. Each photon can liberate one electron from the regular crystalline structure of the silicon. The electrons are promptly stored in the nearest potential well. After the exposure is finished the image is read out by moving the potential wells with their trapped charge packets in a systematic fashion. First the level of the next barrier toward the output register is lowered to the same level as the well. The electrons then divide between the two wells. Finally, the level of the original well is raised so that it becomes a barrier (*b*). The effect of this operation is to move the electrons one-third of a pixel upward. After two more shifts (*c*, *d*) the entire pattern of charge has been moved one full pixel upward and the electrons that were in the top row of pixels have been deposited in the output register. The same technique is now applied to move this row of pixels along the output register toward the left (*e*, *f*). An amplifier at the end of the output register measures each charge packet in turn, thereby reading out an entire row. The process is then repeated, with each row read out until the entire chip has been emptied of information. (*From "Charged-coupled Devices in Astronomy," by Jerome Kristian and Morley Blouke. © Oct. 1982 by Scientific American, Inc. All rights reserved.*)

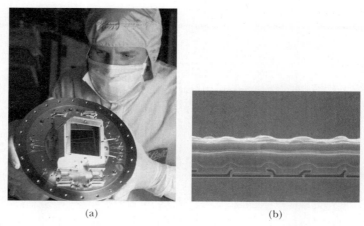

(a)   (b)

Figure 6.13   (a) A Ball Aerospace-built CCD used on the ACS (Advanced Camera for Surveys) instrument onboard the Hubble Space Telescope. The flat square in the center is the CCD chip. (*Photo courtesy Ball Aerospace and Technologies Corp.*) (b) Cross section of a pixel in the imaging section of a Scientific Imaging Technologies' 1024 × 1024 Charge-Coupled Device image sensor. The image is enlarged about 5,000 diameters in this SEM photo. The cross section is along a vertical cut in the chip and shows about $1\frac{1}{3}$ pixels. The characteristic step like structure is formed by the three overlapping layers of poly crystalline silicon (polysilicon) that serve as the electrodes. Each first-level electrode is a poly silicon ribbon eight microns wide (about a twentieth of the thickness of a human hair) and 24 millimeters long, running in and out of the plane of the page. On a single chip there are altogether about 84 feet of such ribbon in each of the three sets of electrodes. (*Photo courtesy Scientific Imaging Technologies.*)

chip. (Because of their high resolution and geometric accuracy, CCDs also are used to record the paths of energetic elementary particles by collecting the electrons generated along their tracks.) Finally, overall noise and signal

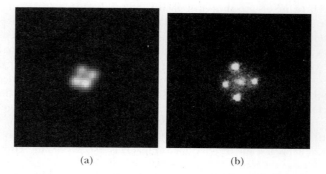

(a)   (b)

Figure 6.14   The "clover leaf," the quadruply lensed quasar H1413+117. The four images of comparable brightness are only 1 arcsec apart. The spectra of two of the images are identical, except for some absorption lines in one that presumably come from different gas clouds that are in the other's line of sight. The redshift is 2.55. The rare configuration and identical spectra show that we are seeing gravitational lensing rather than a cluster of quasars. (a) Ground-based image taken with a CCD camera attached to the 2.2m ESO/MPI telescope on La Silla (Chile). (*P. Magain, European Space Observatory. Reprinted by permission from Nature (334:325) copyright (1988) Macmillan Publishers Ltd.*) (b) A Hubble Space Telescope view, in which the lensing galaxy is revealed. (*NASA/ESA*)

degradation have decreased so markedly in CCDs that as many as 99.9999% of the electrons are transferred in each well shift. This is crucial since image readout involves thousands of such transfers.

Figure 6.14a shows a remarkable quadruply lensed quasar. The multiple images result when light from a single quasar is deflected by gravitational forces as it passes near an intervening galaxy on its journey to Earth. Figure 16.14b shows the lensing galaxy, beautifully resolved by the CCD imager on board the Hubble Space Telescope. These, and similar images offer conclusive proof of the superior ability of CCDs to make extremely accurate position measurements of faint objects in the presence of much brighter ones.

6.5   THE FINITE SQUARE WELL

O P T I O N A L

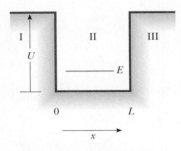

Figure 6.15 Potential-energy diagram for a well of finite height U and width L. The energy E of the particle is less than U.

The "box" potential is an oversimplification that is never realized in practice. Given sufficient energy, a particle can escape the confines of any well. The potential energy for a more realistic situation—the finite square well—is shown in Figure 6.15, and essentially is that depicted in Figure 6.6b before taking the limit $V \to \infty$. A classical particle with energy E greater than the well height U can penetrate the gaps at $x = 0$ and $x = L$ to enter the outer region. Here it moves freely, but with reduced speed corresponding to a diminished kinetic energy $E - U$.

A classical particle with energy E less than U is permanently bound to the region $0 < x < L$. Quantum mechanics asserts, however, that there is some probability that the particle can be found *outside* this region! That is, the wavefunction generally is nonzero outside the well, and so the probability of finding the particle here also is nonzero. For stationary states, the wavefunction $\psi(x)$ is found from the time-independent Schrödinger equation. Outside the well where $U(x) = U$, this is

$$\frac{d^2\psi}{dx^2} = \alpha^2\psi(x) \qquad x < 0 \text{ and } x > L$$

with $\alpha^2 = 2m(U - E)/\hbar^2$ a constant. Because $U > E$, α^2 necessarily is positive and the independent solutions to this equation are the *real* exponentials $e^{+\alpha x}$ and $e^{-\alpha x}$. The positive exponential must be rejected in region III where $x > L$ to keep $\psi(x)$ finite as $x \to \infty$; likewise, the negative exponential must be rejected in region I where $x < 0$ to keep $\psi(x)$ finite as $x \to -\infty$. Thus, the exterior wave takes the form

$$\psi(x) = Ae^{+\alpha x} \qquad \text{for } x < 0$$
$$\psi(x) = Be^{-\alpha x} \qquad \text{for } x > L \tag{6.20}$$

The coefficients A and B are determined by matching this wave smoothly onto the wavefunction in the well interior. Specifically, we require $\psi(x)$ and its first derivative $d\psi/dx$ to be continuous at $x = 0$ and again at $x = L$. This can be done only for certain values of E, corresponding to the allowed energies for the bound particle. For these energies, the matching conditions specify the entire wavefunction except for a multiplicative constant, which then is determined by normalization. Figure 6.16 shows the wavefunctions and probability densities that result for the three lowest allowed particle energies. Note that in each case the waveforms join smoothly at the boundaries of the potential well.

The fact that ψ is nonzero at the walls *increases* the de Broglie wavelength in the well (compared with that in the infinite well), and this in turn lowers the energy and momentum of the particle. This observation can be used to approximate the

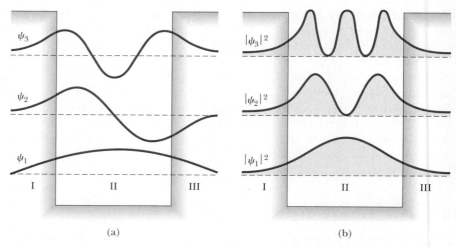

Figure 6.16   (a) Wavefunctions for the lowest three energy states for a particle in a potential well of finite height. (b) Probability densities for the lowest three energy states for a particle in a potential well of finite height.

allowed energies for the bound particle.[10] The wavefunction penetrates the exterior region on a scale of length set by the **penetration depth** δ, given by

Penetration depth

$$\delta = \frac{1}{\alpha} = \frac{\hbar}{\sqrt{2m(U-E)}} \tag{6.21}$$

Specifically, at a distance δ beyond the well edge, the wave amplitude has fallen to $1/e$ of its value at the edge and approaches zero exponentially in the exterior region. That is, the exterior wave is essentially zero beyond a distance δ on either side of the potential well. If it were truly zero beyond this distance, the allowed energies would be those for an infinite well of length $L + 2\delta$ (compare Equation 6.17), or

Approximate energies for a particle in a well of finite height

$$E_n \approx \frac{n^2 \pi^2 \hbar^2}{2m(L+2\delta)^2} \qquad n = 1, 2, \ldots \tag{6.22}$$

The allowed energies for a particle bound to the finite well are given approximately by Equation 6.22 so long as δ is small compared with L. But δ itself is energy dependent according to Equation 6.21. Thus, Equation 6.22 becomes an implicit relation for E that must be solved numerically for a given value of n. The approximation is best for the lowest-lying states and breaks down completely as E approaches U, where δ becomes infinite. From this we infer (correctly) that the number of bound states is limited by the height U of our potential well. Particles with energies E exceeding U are not bound to the well, that is, they may be found with comparable probability in the exterior regions. The case of unbound states will be taken up in the following chapter.

[10]This specific approximation method was reported by S. Garrett in the *Am. J. Phys.* 47:195–196, 1979.

EXAMPLE 6.8   A Bound Electron

Estimate the ground-state energy for an electron confined to a potential well of width 0.200 nm and height 100 eV.

Solution   We solve Equations 6.21 and 6.22 together, using an iterative procedure. Because we expect $E \ll U(= 100 \text{ eV})$, we estimate the decay length δ by first neglecting E to get

$$\delta \approx \frac{\hbar}{\sqrt{2mU}} = \frac{(197.3 \text{ eV} \cdot \text{nm}/c)}{\sqrt{2(511 \times 10^3 \text{ eV}/c^2)(100 \text{ eV})}}$$

$$= 0.0195 \text{ nm}$$

Thus, the effective width of the (infinite) well is $L + 2\delta = 0.239$ nm, for which we calculate the ground-state energy:

$$E \approx \frac{\pi^2 (197.3 \text{ eV} \cdot \text{nm}/c)^2}{2(511 \times 10^3 \text{ eV}/c^2)(0.239 \text{ nm})^2} = 6.58 \text{ eV}$$

From this E we calculate $U - E = 93.42$ eV and a new decay length

$$\delta \approx \frac{(197.3 \text{ eV} \cdot \text{nm}/c)}{\sqrt{2(511 \times 10^3 \text{ eV}/c^2)(93.42 \text{ eV})}} = 0.0202 \text{ nm}$$

This, in turn, increases the effective well width to 0.240 nm and lowers the ground-state energy to $E = 6.53$ eV. The iterative process is repeated until the desired accuracy is achieved. Another iteration gives the same result to the accuracy reported. This is in excellent agreement with the exact value, about 6.52 eV for this case.

Exercise 3   Bound-state waveforms and allowed energies for the finite square well also can be found using purely numerical methods. Go to our companion Web site (http://info.brookscole.com/mp3e) and select QMTools Simulations → Exercise 6.3. The applet shows the potential energy for an electron confined to a finite well of width 0.200 nm and height 100 eV. Follow the on-site instructions to add a stationary wave and determine the energy of the ground state. Repeat the procedure for the first excited state. Compare the symmetry and the number of nodes for these two wavefunctions. Find the highest-lying bound state for this finite well. Count nodes to determine which excited state this is, and thus deduce the total number of bound states this well supports.

EXAMPLE 6.9   Energy of a Finite Well: Exact Treatment

Impose matching conditions on the interior and exterior wavefunctions and show how these lead to energy quantization for the finite square well.

Solution   The exterior wavefunctions are the decaying exponential functions given by Equation 6.20 with decay constant $\alpha = [2m(U - E)/\hbar^2]^{1/2}$. The interior wave is an oscillation with wavenumber $k = (2mE/\hbar^2)^{1/2}$ having the same form as that for the infinite well, Equation 6.15; here we write it as

$$\psi(x) = C \sin kx + D \cos kx \qquad \text{for } 0 < x < L$$

To join this smoothly onto the exterior wave, we insist that the wavefunction and its slope be continuous at the well edges $x = 0$ and $x = L$. At $x = 0$ the conditions for smooth joining require

$$A = D \qquad \text{(continuity of } \psi\text{)}$$

$$\alpha A = kC \qquad \left(\text{continuity of } \frac{d\psi}{dx}\right)$$

Dividing the second equation by the first eliminates A, leaving

$$\frac{C}{D} = \frac{\alpha}{k}$$

In the same way, smooth joining at $x = L$ requires

$$C \sin kL + D \cos kL = Be^{-\alpha L} \qquad \text{(continuity of } \psi\text{)}$$

$$kC \cos kL - kD \sin kL = -\alpha Be^{-\alpha L} \qquad \left(\text{continuity of } \frac{d\psi}{dx}\right)$$

Again dividing the second equation by the first eliminates B. Then replacing C/D with α/k gives

$$\frac{(\alpha/k)\cos kL - \sin kL}{(\alpha/k)\sin kL + \cos kL} = -\frac{\alpha}{k}$$

For a specified well height U and width L, this last relation can only be satisfied for special values of E (E is contained in both k and α). For any other energies, the waveform will not match smoothly at the well edges, leaving a wavefunction that is physically inadmissable. (Note that the equation cannot be solved explicitly for E; rather, solutions must be obtained using numerical or graphical methods.)

Exercise 4   Use the result of Example 6.9 to verify that the ground-state energy for an electron confined to a square well of width 0.200 nm and height 100 eV is about 6.52 eV.

6.6   THE QUANTUM OSCILLATOR

As a final example of a potential well for which exact results can be obtained, let us examine the problem of a particle subject to a linear restoring force $F = -Kx$. Here x is the displacement of the particle from equilibrium ($x = 0$) and K is the force constant. The corresponding potential energy is given by $U(x) = \frac{1}{2}Kx^2$. The prototype physical system fitting this description is a mass on a spring, but the mathematical description actually applies to any object limited to small excursions about a point of stable equilibrium.

Consider the general potential function sketched in Figure 6.17. The positions a, b, and c all label equilibrium points where the force $F = -dU/dx$ is zero. Further, positions a and c are examples of *stable* equilibria, but b is *unstable*. The stability of equilibrium is decided by examining the forces in the immediate neighborhood of the equilibrium point. Just to the left of a, for example, $F = -dU/dx$ is positive, that is, the force is directed to the right; conversely, to the right of a the force is directed to the left. Therefore, a particle displaced slightly from equilibrium at a encounters a force driving it back to the equilibrium point (restoring force). Similar arguments show that the equilibrium at c also is stable. On the other hand, a particle displaced in either direction from point b experiences a force that drives it further away from equilibrium—an unstable condition. In general, stable and unstable equilibria are marked by potential curves that are concave or convex, respectively, at

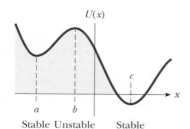

Figure 6.17 A general potential function $U(x)$. The points labeled a and c are positions of stable equilibrium, for which $dU/dx = 0$ and $d^2U/dx^2 > 0$. Point b is a position of unstable equilibrium, for which $dU/dx = 0$ and $d^2U/dx^2 < 0$.

the equilibrium point. To put it another way, **the curvature of $U(x)$ is positive ($d^2U/dx^2 > 0$) at a point of stable equilibrium, and negative ($d^2U/dx^2 < 0$) at a point of unstable equilibrium.**

Near a point of stable equilibrium such as a (or c), $U(x)$ can be fit quite well by a parabola:

$$U(x) = U(a) + \tfrac{1}{2}K(x - a)^2 \qquad (6.23)$$

Of course, the curvature of this parabola ($= K$) must match that of $U(x)$ at the equilibrium point $x = a$:

$$K = \frac{d^2U}{dx^2}\bigg|_a \qquad (6.24)$$

Further, $U(a)$, the potential energy at equilibrium, may be taken as zero if we agree to make this our energy reference, that is, if we subsequently measure all energies from this level. In the same spirit, the coordinate origin may be placed at $x = a$, in effect allowing us to set $a = 0$. With $U(a) = 0$ and $a = 0$, Equation 6.23 becomes the spring potential once again; in other words, **a particle limited to small excursions about any stable equilibrium point behaves as if it were attached to a spring with a force constant K prescribed by the curvature of the true potential at equilibrium.** In this way the oscillator becomes a first approximation to the vibrations occurring in many real systems.

Harmonic approximation to vibrations occurring in real systems

The motion of a classical oscillator with mass m is simple harmonic vibration at the angular frequency $\omega = \sqrt{K/m}$. If the particle is removed from equilibrium a distance A and released, it oscillates between the points $x = -A$ and $x = +A$ (A is the amplitude of vibration), with total energy $E = \tfrac{1}{2}KA^2$. By changing the initial point of release A, the classical particle can in principle be given any (nonnegative) energy whatsoever, including zero.

The quantum oscillator is described by the potential energy $U(x) = \tfrac{1}{2}Kx^2 = \tfrac{1}{2}m\omega^2x^2$ in the Schrödinger equation. After a little rearrangement we get

$$\frac{d^2\psi}{dx^2} = \frac{2m}{\hbar^2}\left(\frac{1}{2}m\omega^2x^2 - E\right)\psi(x) \qquad (6.25)$$

as the equation for the stationary states of the oscillator. The mathematical technique for solving this equation is beyond the level of this text. (The exponential and trigonometric forms for ψ employed previously will not work here because of the presence of x^2 in the potential.) It is instructive, however, to make some intelligent guesses and verify their accuracy by direct substitution. The ground-state wavefunction should possess the following attributes:

1. ψ should be *symmetric* about the midpoint of the potential well $x = 0$.
2. ψ should be *nodeless*, but approaching zero for $|x|$ large.

Both expectations are derived from our experience with the lowest energy states of the infinite and finite square wells, which you might want to review at this time. The symmetry condition (1) requires ψ to be some function of x^2; further, the function must have no zeros (other than at infinity) to meet the nodeless requirement (2). The simplest choice fulfilling both demands is the Gaussian form

$$\psi(x) = C_0 e^{-\alpha x^2} \qquad (6.26)$$

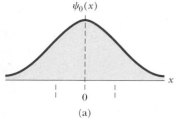

$\psi_0(x)$

(a)

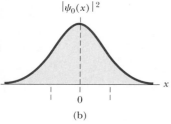

$|\psi_0(x)|^2$

(b)

Figure 6.18 (a) Wavefunction for the ground state of a particle in the oscillator potential well. (b) The probability density for the ground state of a particle in the oscillator potential well. The dashed vertical lines mark the limits of vibration for a classical particle with the same energy, $x = \pm A = \pm\sqrt{\hbar/m\omega}$.

for some as-yet-unknown constants C_0 and α. Taking the second derivative of $\psi(x)$ in Equation 6.26 gives (as you should verify)

$$\frac{d^2\psi}{dx^2} = \{4\alpha^2 x^2 - 2\alpha\}C_0 e^{-\alpha x^2} = \{4\alpha^2 x^2 - 2\alpha\}\psi(x)$$

which has the same structure as Equation 6.25. Comparing like terms between them, we see that we have a solution provided that both

$$4\alpha^2 = \frac{2m}{\hbar^2}\frac{1}{2}m\omega^2 \qquad \text{or} \qquad \alpha = \frac{m\omega}{2\hbar} \qquad (6.27)$$

and

$$\frac{2mE}{\hbar^2} = 2\alpha = \frac{m\omega}{\hbar} \qquad \text{or} \qquad E = \tfrac{1}{2}\hbar\omega \qquad (6.28)$$

In this way we discover that the oscillator ground state is described by the wavefunction $\psi_0(x) = C_0\exp(-m\omega x^2/2\hbar)$ and that the energy of this state is $E_0 = \tfrac{1}{2}\hbar\omega$. The constant C_0 is reserved for normalization (see Example 6.10). The ground-state wave ψ_0 and associated probability density $|\psi_0|^2$ are illustrated in Figure 6.18. The dashed vertical lines mark the limits of vibration for a classical oscillator with the same energy. Note the considerable penetration of the wave into the classically forbidden regions $x > A$ and $x < -A$. A detailed analysis shows that the particle can be found in these nonclassical regions about 16% of the time (see Example 6.12).

EXAMPLE 6.10   Normalizing the Oscillator Ground State Wavefunction

Normalize the oscillator ground-state wavefunction found in the preceding paragraph.

Solution With $\psi_0(x) = C_0 e^{-m\omega x^2/2\hbar}$, the integrated probability is

$$\int_{-\infty}^{\infty}|\psi_0(x)|^2\,dx = C_0^2\int_{-\infty}^{\infty}e^{-m\omega x^2/\hbar}\,dx$$

Evaluation of the integral requires advanced techniques. We shall be content here simply to quote the formula

$$\int_{-\infty}^{\infty}e^{-ax^2}\,dx = \sqrt{\frac{\pi}{a}} \qquad a > 0$$

In our case we identify a with $m\omega/\hbar$ and obtain

$$\int_{-\infty}^{\infty}|\psi_0(x)|^2\,dx = C_0^2\sqrt{\frac{\pi\hbar}{m\omega}}$$

Normalization requires this integrated probability to be 1, leading to

$$C_0 = \left(\frac{m\omega}{\pi\hbar}\right)^{1/4}$$

EXAMPLE 6.11   Limits of Vibration for a Classical Oscillator

Obtain the limits of vibration for a classical oscillator having the same total energy as the quantum oscillator in its ground state.

Solution The ground-state energy of the quantum oscillator is $E_0 = \tfrac{1}{2}\hbar\omega$. At its limits of vibration $x = \pm A$, the classical oscillator has transformed all this energy into elastic potential energy of the spring, given by $\tfrac{1}{2}KA^2 = \tfrac{1}{2}m\omega^2 A^2$. Therefore,

$$\tfrac{1}{2}\hbar\omega = \tfrac{1}{2}m\omega^2 A^2 \qquad \text{or} \qquad A = \sqrt{\frac{\hbar}{m\omega}}$$

The classical oscillator vibrates in the interval given by $-A \le x \le A$, having insufficient energy to exceed these limits.

EXAMPLE 6.12   The Quantum Oscillator in the Nonclassical Region

Calculate the probability that a quantum oscillator in its ground state will be found outside the range permitted for a classical oscillator with the same energy.

Solution Because the classical oscillator is confined to the interval $-A \leq x \leq A$, where A is its amplitude of vibration, the question is one of finding the quantum oscillator outside this interval. From the previous example we have $A = \sqrt{\hbar/m\omega}$ for a classical oscillator with energy $\frac{1}{2}\hbar\omega$. The quantum oscillator with this energy is described by the wavefunction $\psi_0(x) = C_0\exp(-m\omega x^2/2\hbar)$, with $C_0 = (m\omega/\pi\hbar)^{1/4}$ from Example 6.10. The probability in question is found by integrating the probability density $|\psi_0|^2$ in the region beyond the classical limits of vibration, or

$$P = \int_{-\infty}^{-A} |\psi_0|^2 \, dx + \int_{A}^{\infty} |\psi_0|^2 \, dx$$

From the symmetry of ψ_0, the two integrals contribute equally to P, so

$$P = 2\left(\frac{m\omega}{\pi\hbar}\right)^{1/2} \int_{A}^{\infty} e^{-m\omega x^2/\hbar} \, dx$$

Changing variables from x to $z = \sqrt{m\omega/\hbar}\ x$ and using $A = \sqrt{\hbar/m\omega}$ (corresponding to $z = 1$) leads to

$$P = \frac{2}{\sqrt{\pi}} \int_{1}^{\infty} e^{-z^2} \, dz$$

Expressions of this sort are encountered frequently in probability studies. With the lower limit of integration changed to a variable—say, y—the result for P defines the *complementary error function* erfc(y). Values of the error function may be found in tables. In this way we obtain $P = \text{erfc}(1) = 0.157$, or about 16%.

To obtain excited states of the oscillator, a procedure can be followed similar to that for the ground state. The first excited state should be antisymmetric about the midpoint of the oscillator well ($x = 0$) and display exactly one node. By virtue of the antisymmetry, this node must occur at the origin, so that a suitable trial solution would be $\psi(x) = x\exp(-\alpha x^2)$. Substituting this form into Equation 6.25 yields the same α as before, along with the first excited-state energy $E_1 = \frac{3}{2}\hbar\omega$.

Continuing in this manner, we could generate ever-higher-lying oscillator states with their respective energies, but the procedure rapidly becomes too laborious to be practical. What is needed is a systematic approach, such as that provided by the method of power series expansion.[11] Pursuing this method would take us too far afield, but the result for the allowed oscillator energies is quite simple and sufficiently important that it be included here:

$$E_n = (n + \tfrac{1}{2})\hbar\omega \qquad n = 0, 1, 2, \ldots \qquad (6.29)$$

The energy-level diagram following from Equation 6.29 is given in Figure 6.19. Note the uniform spacing of levels, widely recognized as the hallmark of the harmonic oscillator spectrum. The energy difference between adjacent levels is just $\Delta E = \hbar\omega$. In these results we find the quantum justification for Planck's revolutionary hypothesis concerning his cavity resonators (see Section 3.2). In deriving his blackbody radiation formula, Planck assumed that these resonators (oscillators), which made up the cavity walls, could possess only those energies that were multiples of $hf = \hbar\omega$. Although Planck could not have foreseen the zero-point energy $\hbar\omega/2$, it would make no difference: His resonators still would emit or absorb light energy in the bundles $\Delta E = hf$ necessary to reproduce the blackbody spectrum.

Energy levels for the harmonic oscillator

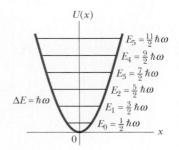

Figure 6.19 Energy-level diagram for the quantum oscillator. Note that the levels are equally spaced, with a separation equal to $\hbar\omega$. The ground state energy is E_0.

---

[11]The method of power series expansion as applied to the problem of the quantum oscillator is developed in any more advanced quantum mechanics text. See, for example, E. E. Anderson, *Modern Physics and Quantum Mechanics*, Philadelphia, W. B. Saunders Company, 1971.

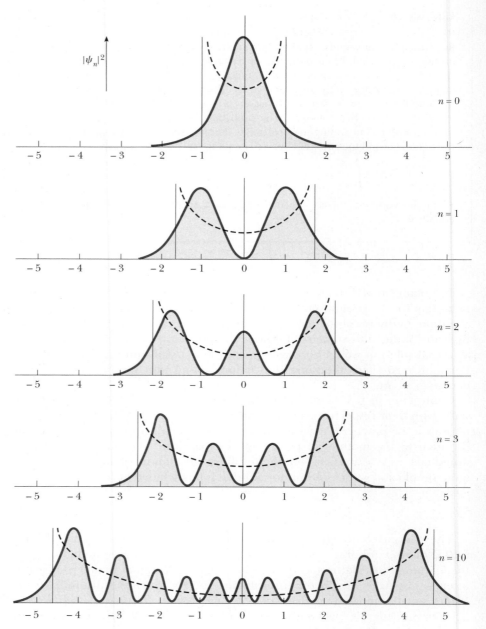

Figure 6.20 Probability densities for a few states of the quantum oscillator. The dashed curves represent the classical probabilities corresponding to the same energies.

The probability densities for some of the oscillator states are plotted in Figure 6.20. The dashed lines, representing the classical probability densities for the same energy, are provided for comparison (see Problem 28 for the calculation of classical probabilities). Note that as n increases, agreement between the classical and quantum probabilities improves, as expected from the correspondence principle.

EXAMPLE 6.13   Quantization of Vibrational Energy

The energy of a quantum oscillator is restricted to be one of the values $(n + \frac{1}{2})\hbar\omega$. How can this quantization apply to the motion of a mass on a spring, which seemingly can vibrate with any amplitude (energy) whatever?

Solution   The discrete values for the allowed energies of the oscillator would go unnoticed if the spacing between adjacent levels were too small to be detected. At the macroscopic level, a laboratory mass m of, say, 0.0100 kg on a spring having force constant $K = 0.100$ N/m (a typical value) would oscillate with angular frequency $\omega = \sqrt{K/m} = 3.16$ rad/s. The corresponding period of vibration is $T = 2\pi/\omega = 1.99$ s. In this case the quantum level spacing is only

$$\Delta E = \hbar\omega = (6.582 \times 10^{-16} \text{ eV·s})(3.16 \text{ rad/s})$$
$$= 2.08 \times 10^{-15} \text{ eV}$$

Such small energies are far below present limits of detection.

At the atomic level, however, much higher frequencies are commonplace. Consider the vibrational frequency of the hydrogen molecule. This behaves as an oscillator with $K = 510.5$ N/m and reduced mass $\mu = 8.37 \times 10^{-28}$ kg. The angular frequency of oscillation is therefore

$$\omega = \sqrt{\frac{K}{\mu}} = \sqrt{\frac{510.5 \text{ N/m}}{8.37 \times 10^{-28} \text{ kg}}}$$
$$= 7.81 \times 10^{14} \text{ rad/s}$$

At such frequencies, the quantum of energy $\hbar\omega$ is 0.513 eV, which can be measured easily!

6.7   EXPECTATION VALUES

It should be evident by now that two distinct types of measurable quantities are associated with a given wavefunction $\Psi(x, t)$. One type—like the energy E for the stationary states—is fixed by the quantum number labeling the wave. Therefore, every measurement of this quantity performed on the system described by Ψ yields the *same* value. Quantities such as E we call **sharp** to distinguish them from others—like the position x—for which the wavefunction Ψ furnishes only probabilities. We say x is an example of a dynamic quantity that is **fuzzy.** In the following paragraphs we discuss what more can be learned about these "fuzzy" quantities.

Sharp and fuzzy variables

A particle described by the wavefunction Ψ may occupy various places x with probability given by the wave intensity there, $|\Psi(x)|^2$. Predictions made this way from Ψ can be tested by making repeated measurements of the particle position. Table 6.1 shows results that might be obtained in a hypothetical experiment of this sort. The table consists of 18 entries, each one representing the actual position of the particle recorded in that particular measurement. We see that the

Table 6.1   Hypothetical Data Set for Position of a Particle as Recorded in Repeated Trials

Trial	Position (arbitrary units)	Trial	Position (arbitrary units)	Trial	Position (arbitrary units)
1	$x_1 = 2.5$	7	$x_7 = 8.0$	13	$x_{13} = 4.2$
2	$x_2 = 3.7$	8	$x_8 = 6.4$	14	$x_{14} = 8.8$
3	$x_3 = 1.4$	9	$x_9 = 4.1$	15	$x_{15} = 6.2$
4	$x_4 = 7.9$	10	$x_{10} = 5.4$	16	$x_{16} = 7.1$
5	$x_5 = 6.2$	11	$x_{11} = 7.0$	17	$x_{17} = 5.4$
6	$x_6 = 5.4$	12	$x_{12} = 3.3$	18	$x_{18} = 5.3$

entry 5.4 occurs most often (in 3 of the 18 trials); it represents the most probable position based on the data available. The probability associated with this position, again based on the available data, is $3/18 = 0.167$. These numbers will fluctuate as additional measurements are taken, but they should approach limiting values. The theoretical predictions refer to these limiting values. A good test of the theory would require much more data than we have shown in this illustration.

The information in Table 6.1 also can be used to find the average position of the particle:

$$\bar{x} = \frac{(2.5 + 3.7 + 1.4 + \cdots + 5.4 + 5.3)}{18} = 5.46$$

This same number can be found in a different way. First, order the table entries by value, starting with the smallest: 1.4, 2.5, 3.3, . . . , 5.4, 6.2, . . . , 8.0, 8.8. Now take each value, multiply by its frequency of occurrence, and sum the results:

$$1.4 \left(\frac{1}{18}\right) + 2.5 \left(\frac{1}{18}\right) + \cdots + 5.4 \left(\frac{3}{18}\right)$$
$$+ 6.2 \left(\frac{2}{18}\right) + \cdots + 8.8 \left(\frac{1}{18}\right) = 5.46$$

The two procedures are equivalent, but the latter involves a sum over *ordered values* rather than individual table entries. We may generalize this last expression to include other values for the position of the particle, provided we weight each one by its observed frequency of occurrence (in this case, zero). This allows us to write a general prescription to calculate the average particle position from any data set:

$$\bar{x} = \sum x P_x \tag{6.30}$$

The sum now includes all values of x, each weighted by its frequency or probability of occurrence P_x. Because the possible values of x are distributed continuously over the entire range of real numbers, the sum in Equation 6.30 really should be an integral and P_x should refer to the probability of finding the particle in the infinitesimal interval dx about the point x; that is, the probability $P_x \rightarrow P(x)\,dx$, where $P(x)$ is the probability density. In quantum mechanics, $P(x) = |\Psi|^2$ and the average value of x, written in quantum mechanics as $\langle x \rangle$, is called the **expectation value.** Then,

Average position of a particle

$$\langle x \rangle = \int_{-\infty}^{\infty} x |\Psi(x, t)|^2 \, dx \tag{6.31}$$

Notice that $\langle x \rangle$ may be a function of time. For a stationary state, however, $|\Psi|^2$ is static and, as a consequence, $\langle x \rangle$ is independent of t.

In similar fashion we find that the average or expectation value for any function of x, say $f(x)$, is

$$\langle f \rangle = \int_{-\infty}^{\infty} f(x) |\Psi|^2 \, dx \tag{6.32}$$

With $f(x) = U(x)$, Equation 6.32 becomes $\langle U \rangle$, the average potential energy of the particle. With $f(x) = x^2$, the quantum uncertainty in particle position may be found. To see how this is done, we return to Table 6.1 and notice that the entries scatter about the average value. The amount of scatter is measured by the standard deviation, σ, of the data, defined as

$$\sigma = \sqrt{\frac{\Sigma(x_i - \bar{x})^2}{N}} \tag{6.33}$$

where N is the number of data points—in this case, 18. Writing out the square under the radical gives

$$\frac{\Sigma(x_i)^2}{N} - 2(\bar{x})\frac{\Sigma(x_i)}{N} + (\bar{x})^2\Sigma\left(\frac{1}{N}\right) = \overline{(x^2)} - 2(\bar{x})(\bar{x}) + (\bar{x})^2$$

$$= \overline{(x^2)} - (\bar{x})^2$$

and so

$$\sigma = \sqrt{\overline{(x^2)} - (\bar{x})^2}$$

From Equation 6.33 we see that if the standard deviation were zero, all data entries would be identical and equal to the average. In that case the distribution is *sharp;* otherwise, the data exhibit some spread (as in Table 6.1) and the standard deviation is greater than zero. In quantum mechanics the standard deviation, written Δx, is often called the *uncertainty* in position. The preceding development implies that the quantum uncertainty in position can be calculated from expectation values as

$$\Delta x = \sqrt{\langle x^2 \rangle - \langle x \rangle^2} \tag{6.34}$$

The degree to which particle position is fuzzy is given by the magnitude of Δx; note that the position is sharp only if $\Delta x = 0$.

EXAMPLE 6.14   Standard Deviation from Averages

Compute $\overline{(x^2)}$ and the standard deviation for the data given in Table 6.1.

Solution Squaring the data entries of Table 6.1 and adding the results gives $\Sigma(x_i)^2 = 603.91$. Dividing this by the number of data points, $N = 18$, we find $\overline{(x^2)} = 603.91/18 = 33.55$. Then,

$$\sigma = \sqrt{33.55 - (5.46)^2} = 1.93$$

for this case.

EXAMPLE 6.15   Location of a Particle in a Box

Compute the average position $\langle x \rangle$ and the quantum uncertainty in this value, Δx, for the particle in a box, assuming it is in the ground state.

Solution The possible particle positions within the box are weighted according to the probability density given by $|\Psi|^2 = (2/L)\sin^2(n\pi x/L)$, with $n = 1$ for the ground state. The average position is calculated as

$$\langle x \rangle = \int_{-\infty}^{\infty} x|\Psi|^2\,dx = \left(\frac{2}{L}\right)\int_0^L x\sin^2\left(\frac{\pi x}{L}\right)dx$$

Making the change of variable $\theta = \pi x/L$ (so that $d\theta = \pi\,dx/L$) gives

$$\langle x \rangle = \frac{2L}{\pi^2}\int_0^\pi \theta\sin^2\theta\,d\theta$$

The integral is evaluated with the help of the trigonometric identity $2\sin^2\theta = 1 - \cos 2\theta$, giving

$$\langle x \rangle = \frac{L}{\pi^2}\left(\int_0^\pi \theta\,d\theta - \int_0^\pi \theta\cos 2\theta\,d\theta\right)$$

An integration by parts shows that the second integral vanishes, whereas the first integrates to $\pi^2/2$. Thus, the average particle position is the midpoint $\langle x \rangle = L/2$ as expected, because there is equal probability of finding the particle in the left half or the right half of the box.

$\langle x^2 \rangle$ is computed in much the same way, but with an extra factor of x in the integrand. After changing variables to $\theta = \pi x/L$, we get

$$\langle x^2 \rangle = \frac{L^2}{\pi^3}\left(\int_0^\pi \theta^2\,d\theta - \int_0^\pi \theta^2\cos 2\theta\,d\theta\right)$$

The first integral evaluates to $\pi^3/3$; the second may be integrated twice by parts to get

$$\int_0^\pi \theta^2 \cos 2\theta \, d\theta = -\int_0^\pi \theta \sin 2\theta \, d\theta$$

$$= \tfrac{1}{2}\theta \cos 2\theta \big|_0^\pi = \pi/2$$

Then,

$$\langle x^2 \rangle = \frac{L^2}{\pi^3}\left(\frac{\pi^3}{3} - \frac{\pi}{2}\right) = \frac{L^2}{3} - \frac{L^2}{2\pi^2}$$

Finally, the uncertainty in position for this particle is

$$\Delta x = \sqrt{\langle x^2 \rangle - \langle x \rangle^2} = L\sqrt{\frac{1}{3} - \frac{1}{2\pi^2} - \frac{1}{4}} = 0.181L$$

This is an appreciable figure, amounting to nearly one-fifth the size of the box. Consequently, the whereabouts of such a particle are largely unknown. With some confidence, we may assert only that the particle is likely to be in the range $L/2 \pm 0.181L$.

Finally, notice that none of these results depends on the time, because in a stationary state t enters only through the exponential factor $e^{-i\omega t}$, which cancels when Ψ is combined with Ψ^* in the calculation of averages. Therefore, it is generally true that, **in a stationary state, all averages, as well as probabilities, are time independent.**

We have learned how to predict the average position of a particle, $\langle x \rangle$; the uncertainty in this position, Δx; the average potential energy of the particle, $\langle U \rangle$; and so on. But what about the average momentum $\langle p \rangle$ of the particle or its average kinetic energy $\langle K \rangle$? These could be calculated if $p(x)$, the momentum as a function of x, were known. In classical mechanics, $p(x)$ may be obtained from the equation for the classical path taken by the particle, $x(t)$. Differentiating this function once gives the velocity $v(t)$. Then inverting $x(t)$ to get t as a function of x, and substituting this result into $v(t)$, gives $v(x)$ and the desired relation $p(x) = mv(x)$. In quantum mechanics, however, x and t are independent variables—*there is no path*, nor any function connecting p with x! If there were, then p could be found from x using $p(x)$ and both x and p would be known precisely, in violation of the uncertainty principle.

To obtain $\langle p \rangle$ we must try a different approach: We identify the time derivative of the average particle position with the average velocity of the particle. After multiplication by m, this gives the average momentum $\langle p \rangle$:

Average momentum of a particle

$$\langle p \rangle = m \, \frac{d\langle x \rangle}{dt} \tag{6.35}$$

Equation 6.35 cannot be derived from anything we have said previously. When applied to macroscopic objects where the quantum uncertainties in position and momentum are small, the averages $\langle x \rangle$ and $\langle p \rangle$ become indistinguishable from "the" position and "the" momentum of the object, and Equation 6.35 reduces to the classical definition of momentum.

An equivalent expression for $\langle p \rangle$ follows from Equation 6.35 by substituting $\langle x \rangle$ from Equation 6.31 and differentiating under the integral sign. Using Schrödinger's equation to eliminate time derivatives of Ψ and its conjugate Ψ^* gives (after much manipulation!)

$$\langle p \rangle = \int_{-\infty}^{\infty} \Psi^* \left(\frac{\hbar}{i}\right) \frac{\partial \Psi}{\partial x} \, dx \tag{6.36}$$

Exercise 5   Show that $\langle p \rangle = 0$ for *any* state of a particle in a box.

6.8 OBSERVABLES AND OPERATORS

An observable is any particle property that can be measured. The position and momentum of a particle are observables, as are its kinetic and potential energies.[12] In quantum mechanics, we associate an *operator* with each of these observables. Using this operator, one can calculate the average value of the corresponding observable. An operator here refers to an operation to be performed on whatever function follows the operator. The quantity operated on is called the *operand*. In this language a constant c becomes an operator, whose meaning is understood by supplying any function $f(x)$ to obtain $cf(x)$. Here the operator c means "multiplication by the constant c." A more complicated operator is d/dx, which, after supplying an operand $f(x)$, means "take the derivative of $f(x)$ with respect to x." Still another example is $(d/dx)^2 = (d/dx)(d/dx)$. Supplying the operand $f(x)$ gives $(d/dx)^2 f(x) = (d/dx)(df/dx) = d^2f/dx^2$. Hence, $(d/dx)^2$ means "take the second derivative with respect to x, that is, take the indicated derivative twice."

The operator concept is useful in quantum mechanics because all expectation values we have encountered so far can be written in the same general form, namely,

$$\langle Q \rangle = \int_{-\infty}^{\infty} \Psi^* [Q] \Psi \, dx \qquad (6.37)$$

Operators in quantum mechanics

In this expression, Q is the observable and $[Q]$ is the associated operator. *The order of terms in Equation 6.37 is important;* it indicates that the operand for $[Q]$ always is Ψ. Comparing the general form with that for $\langle p \rangle$ in Equation 6.36 shows that the momentum operator is $[p] = (\hbar/i)(\partial/\partial x)$. Similarly, writing $x|\Psi|^2 = \Psi^* x \Psi$ in Equation 6.31 implies that the operator for position is $[x] = x$. From $[x]$ and $[p]$ the operator for any other observable can be found. For instance, the operator for x^2 is just $[x^2] = [x]^2 = x^2$. For that matter, the operator for potential energy is simply $[U] = U([x]) = U(x)$, meaning that average potential energy is computed as

$$\langle U \rangle = \int_{-\infty}^{\infty} \Psi^* [U] \Psi \, dx = \int_{-\infty}^{\infty} \Psi^* U(x) \Psi \, dx$$

Still another example is the kinetic energy K. Classically, K is a function of p: $K = p^2/2m$. Then the kinetic energy operator is $[K] = ([p])^2/2m = (-\hbar/2m)\partial^2/\partial x^2$, and average kinetic energy is found from

$$\langle K \rangle = \int_{-\infty}^{\infty} \Psi^* [K] \Psi \, dx = \int_{-\infty}^{\infty} \Psi^* \left(-\frac{\hbar^2}{2m} \frac{\partial^2 \Psi}{\partial x^2} \right) dx$$

To find the average total energy for a particle, we sum the average kinetic and potential energies to get

$$\langle E \rangle = \langle K \rangle + \langle U \rangle = \int_{-\infty}^{\infty} \Psi^* \left\{ -\frac{\hbar^2}{2m} \frac{\partial^2}{\partial x^2} + U(x) \right\} \Psi \, dx \qquad (6.38)$$

---

[12]By contrast, the wavefunction Ψ, although clearly indispensable to the quantum description, is not directly measurable and so is *not* an observable.

Table 6.2   Common Observables and Associated Operators

Observable	Symbol	Associated Operator
Position	x	x
Momentum	p	$\dfrac{\hbar}{i}\dfrac{\partial}{\partial x}$
Potential energy	U	$U(x)$
Kinetic energy	K	$-\dfrac{\hbar^2}{2m}\dfrac{\partial^2}{\partial x^2}$
Hamiltonian	H	$-\dfrac{\hbar^2}{2m}\dfrac{\partial^2}{\partial x^2}+U(x)$
Total energy	E	$i\hbar\dfrac{\partial}{\partial t}$

The form of this result suggests that the term in the braces is the operator for total energy. This operator is called the *Hamiltonian,* symbolized by $[H]$:

$$[H] = -\frac{\hbar^2}{2m}\frac{\partial^2}{\partial x^2} + U(x) \tag{6.39}$$

The designation $[E]$ is reserved for another operator, which arises as follows: Inspection of Schrödinger's equation (Equation 6.10) shows that it can be written neatly as $[H]\Psi = i\hbar\partial\Psi/\partial t$. Using this in Equation 6.38 gives an equivalent expression for $\langle E\rangle$ and leads to the identification of the *energy operator:*

$$[E] = i\hbar\,\frac{\partial}{\partial t} \tag{6.40}$$

Notice that $[H]$ is an operation involving only the spatial coordinate x, whereas $[E]$ depends only on the time t. That is, $[H]$ and $[E]$ really are two different operators, but they produce identical results when applied to any solution of Schrödinger's equation. This is because the LHS of Schrödinger's equation is simply $[H]\Psi$, while the RHS is none other than $[E]\Psi$ (compare Equation 6.10)! Table 6.2 summarizes the observables we have discussed and their associated operators.

O P T I O N A L

QUANTUM UNCERTAINTY AND THE EIGENVALUE PROPERTY

In Section 6.7 we showed how Δx, the quantum uncertainty in position, could be found from the expectation values $\langle x^2\rangle$ and $\langle x\rangle$. But the argument given there applies to *any* observable, that is, the quantum uncertainty ΔQ for any observable Q is calculated as

Quantum uncertainty for any observable Q

$$\Delta Q = \sqrt{\langle Q^2\rangle - \langle Q\rangle^2} \tag{6.41}$$

Again, if $\Delta Q = 0$, Q is said to be a sharp observable and all measurements of Q yield the same value. More often, however, $\Delta Q > 0$ and repeated measurements reveal a distribution of values—as in Table 6.1 for the observable x. In such cases, we say the observable is *fuzzy,* suggesting that, prior to actual measurement, the particle cannot be said to possess a unique value of Q.

In classical physics all observables are sharp.[13] The extent to which sharp observables can be specified in quantum physics is limited by uncertainty principles, such as

$$\Delta x \, \Delta p \geq \frac{1}{2} \, \hbar \qquad (6.42)$$

The uncertainties here are to be calculated from Equation 6.41. Equation 6.42 says that no matter what the state of the particle, the spread in distributions obtained in measurements of x and of p will be inversely related: when one is small, the other will be large. Alternatively, if the position of the particle is quite "fuzzy," its momentum can be relatively "sharp," and vice versa. The degree to which both may be simultaneously sharp is limited by the size of $\hbar$. The incredibly small value of $\hbar$ in SI units is an indication that quantum ideas are unnecessary at the macroscopic level.

Despite restrictions imposed by uncertainty principles, some observables in quantum physics may still be sharp. The energy E of all stationary states is one example. In the free particle plane waves of Section 6.2 we have another: The plane wave with wavenumber k,

$$\Psi_k(x, t) = e^{i(kx - \omega t)}$$

describes a particle with momentum $p = \hbar k$. Evidently, momentum is a sharp observable for this wavefunction. We find that the action of the momentum operator in this instance is especially simple:

$$[p]\Psi_k(x, t) = \left(\frac{\hbar}{i} \frac{\partial}{\partial x}\right) e^{i(kx - \omega t)} = \hbar k \Psi_k(x, t)$$

that is, the operation $[p]$ returns the original function multiplied by a constant. This is an example of an **eigenvalue problem** for the operator $[p]$.[14] The wavefunction Ψ_k is the *eigenfunction*, and the constant, in this case $\hbar k$, is the *eigenvalue*. Notice that the eigenvalue is just the sharp value of particle momentum for this wave. This connection between sharp observables and eigenvalues is a general one: **For an observable Q to be sharp, the wavefunction must be an eigenfunction of the operator for Q. Further, the sharp value for Q in this state is the eigenvalue.** In this way the eigenvalue property can serve as a simple test for sharp observables, as the following examples illustrate.

Eigenfunctions and eigenvalues

EXAMPLE 6.16   Plane Waves and Sharp Observables

Use the eigenvalue test to show that the plane wave $\Psi_k(x, t) = e^{i(kx - \omega t)}$ is one for which total energy is a sharp observable. What value does the energy take in this case?

Solution   To decide the issue we examine the action of the energy operator $[E]$ on the candidate function $e^{i(kx - \omega t)}$. Since taking a derivative with respect to t of this function is equivalent to multiplying the function by $-i\omega$, we have

$$[E]e^{i(kx - \omega t)} = \left(i\hbar \frac{\partial}{\partial t}\right) e^{i(kx - \omega t)} = \hbar\omega e^{i(kx - \omega t)}$$

---

[13]We discount in this discussion any random errors of measurement. In principle at least, the imprecision resulting from such errors can be reduced to arbitrarily low levels.

[14]The eigenvalue problem for any operator $[Q]$ is $[Q]\psi = q\psi$; that is, the result of the operation $[Q]$ on some function ψ is simply to return a multiple q of the same function. This is possible only for certain special functions ψ, the *eigenfunctions*, and then only for certain special values of q, the *eigenvalues*. Generally, $[Q]$ is known; the eigenfunctions and eigenvalues are found by imposing the eigenvalue condition.

showing that $e^{i(kx-\omega t)}$ is an eigenfunction of the energy operator $[E]$ and the eigenvalue is $\hbar\omega$. Thus, energy is a sharp observable and has the value $\hbar\omega$ in this state.

It is instructive to compare this result with the outcome found by using the other energy operator, $[H]$. The Hamiltonian for a free particle is simply the kinetic energy operator $[K]$, because the potential energy is zero in this case. Then

$$[H]e^{i(kx-\omega t)} = \left(-\frac{\hbar^2}{2m}\frac{\partial^2}{\partial x^2}\right)e^{i(kx-\omega t)}$$

$$= \left(-\frac{\hbar^2}{2m}\right)(ik)^2 e^{i(kx-\omega t)}$$

Again, the operation returns the original function with a multiplier, so that $e^{i(kx-\omega t)}$ also is an eigenfunction of $[H]$. The eigenvalue in this case is $\hbar^2 k^2/2m$, which also must be the sharp value of particle energy. The equivalence with $\hbar\omega$ follows from the dispersion relation for free particles (see footnote 1).

Exercise 6   Show that total energy is a sharp observable for *any* stationary state.

EXAMPLE 6.17   Sharp Observables for a Particle in a Box

Are the stationary states of the infinite square well eigenfunctions of $[p]$? of $[p]^2$? If so, what are the eigenvalues? Discuss the implications of these results.

Solution   The candidate function in this case is any one of the square well wavefunctions $\Psi(x, t) = \sqrt{2/L}\sin(n\pi x/L)e^{-iE_n t/\hbar}$. Because the first derivative gives $(d/dx)\sin(n\pi x/L) = (n\pi/L)\cos(n\pi x/L)$, we see at once that the operator $[p]$ will *not* return the original function Ψ, and so these are not eigenfunctions of the momentum operator. They *are*, however, eigenfunctions of $[p]^2$. In particular, we have $(d^2/dx^2)\sin(n\pi x/L) = -(n\pi/L)^2\sin(n\pi x/L)$, so that

$$[p]^2\Psi(x, t) = -(\hbar/i)^2\left(\frac{n\pi}{L}\right)^2\Psi(x, t)$$

$$= \left(\frac{n\pi\hbar}{L}\right)^2\Psi(x, t)$$

The eigenvalue is the multiplier $(n\pi\hbar/L)^2$. Thus, the squared momentum (or magnitude of momentum) is sharp for such states, and repeated measurements of p^2 (or $|p|$) for the state labeled by n will give identical results equal to $(n\pi\hbar/L)^2$ (or $n\pi\hbar/L$). By contrast, the momentum itself is not sharp, meaning that different values for p will be obtained in successive measurements. In particular, it is the sign or direction of momentum that is fuzzy, consistent with the classical notion of a particle bouncing back and forth between the walls of the "box."

SUMMARY

In quantum mechanics, matter waves (or de Broglie waves) are represented by a wavefunction $\Psi(x, t)$. The probability that a particle constrained to move along the x-axis will be found in an interval dx at time t is given by $|\Psi|^2 dx$. These probabilities summed over all values of x must total 1 (certainty). That is,

$$\int_{-\infty}^{\infty} |\Psi|^2 \, dx = 1 \qquad (6.2)$$

This is called the **normalization condition.** Furthermore, the probability that the particle will be found in any interval $a \leq x \leq b$ is obtained by integrating the **probability density** $|\Psi|^2$ over this interval.

Aside from furnishing probabilities, the wavefunction can be used to find the average, or **expectation value,** of any dynamical quantity. The average position of a particle at any time t is

$$\langle x \rangle = \int_{-\infty}^{\infty} \Psi^* x \Psi \, dx \qquad (6.31)$$

In general, the average value of any observable Q at time t is

$$\langle Q \rangle = \int_{-\infty}^{\infty} \Psi^* [Q] \Psi \, dx \qquad (6.37)$$

where $[Q]$ is the associated operator. The operator for position is just $[x] = x$, and that for particle momentum is $[p] = (\hbar / i) \partial / \partial x$.

The wavefunction Ψ must satisfy the **Schrödinger equation,**

$$-\frac{\hbar^2}{2m} \frac{\partial^2 \Psi}{\partial x^2} + U(x) \Psi(x, t) = i\hbar \frac{\partial \Psi}{\partial t} \qquad (6.10)$$

Separable solutions to this equation, called **stationary states,** are $\Psi(x, t) = \psi(x) e^{-i\omega t}$, with $\psi(x)$ a time-independent wavefunction satisfying the **time-independent Schrödinger equation**

$$-\frac{\hbar^2}{2m} \frac{d^2 \psi}{dx^2} + U(x) \psi(x) = E\psi(x) \qquad (6.13)$$

The approach of quantum mechanics is to solve Equation 6.13 for ψ and E, given the potential energy $U(x)$ for the system. In doing so, we must require

- that $\psi(x)$ be continuous
- that $\psi(x)$ be finite for all x, including $x = \pm\infty$
- that $\psi(x)$ be single valued
- that $d\psi/dx$ be continuous wherever $U(x)$ is finite

Explicit solutions to Schrödinger's equation can be found for several potentials of special importance. For a free particle the stationary states are the **plane waves** $\psi(x) = e^{ikx}$ of wavenumber k and energy $E = \hbar^2 k^2 / 2m$. The particle momentum in such states is $p = \hbar k$, but the location of the particle is completely unknown. A free particle known to be in some range Δx is described not by a plane wave, but by a **wave packet,** or **group,** formed from a superposition of plane waves. The momentum of such a particle is not known precisely, but only to some accuracy Δp that is related to Δx by the **uncertainty principle,**

$$\Delta x \, \Delta p \geq \tfrac{1}{2}\hbar \qquad (6.42)$$

For a particle confined to a one-dimensional box of length L, the stationary-state waves are those for which an integral number of half-wavelengths can be fit inside, that is, $L = n\lambda/2$. In this case the energies are **quantized** as

$$E_n = \frac{n^2 \pi^2 \hbar^2}{2mL^2} \qquad n = 1, 2, 3, \ldots \qquad (6.17)$$

and the wavefunctions within the box are given by

$$\psi_n(x) = \sqrt{\frac{2}{L}} \sin\left(\frac{n\pi x}{L}\right) \qquad n = 1, 2, 3, \ldots \qquad (6.18)$$

For the **harmonic oscillator** the potential energy function is $U(x) = \frac{1}{2}m\omega^2 x^2$, and the total particle energy is quantized according to the relation

$$E_n = \left(n + \frac{1}{2}\right)\hbar\omega \qquad n = 0, 1, 2, \ldots \qquad (6.29)$$

The lowest energy is $E_0 = \frac{1}{2}\hbar\omega$; the separation between adjacent energy levels is uniform and equal to $\hbar\omega$. The wavefunction for the oscillator ground state is

$$\psi_0(x) = C_0 e^{-\alpha x^2} \qquad (6.26)$$

where $\alpha = m\omega/2\hbar$ and C_0 is a normalizing constant. The oscillator results apply to any system executing small-amplitude vibrations about a point of stable equilibrium. The effective spring constant in the general case is

$$K = m\omega^2 = \left.\frac{d^2U}{dx^2}\right|_a \qquad (6.24)$$

with the derivative of the potential evaluated at the equilibrium point a.

The **stationary state** waves for any potential share the following attributes:

- Their time dependence is $e^{-i\omega t}$.
- They yield probabilities that are time independent.
- All average values obtained from stationary states are time independent.
- The energy in any stationary state is a **sharp observable;** that is, repeated measurements of particle energy performed on identical systems always yield the same result, $E = \hbar\omega$.

For other observables, such as position, repeated measurements usually yield different results. We say these observables are **fuzzy.** Their inherent "fuzziness" is reflected by the spread in results about the average value, as measured by the standard deviation, or **uncertainty.** The uncertainty in any observable Q can be calculated from expectation values as

$$\Delta Q = \sqrt{\langle Q^2 \rangle - \langle Q \rangle^2} \qquad (6.41)$$

SUGGESTIONS FOR FURTHER READING

1. L. de Broglie, *New Perspectives in Physics*, New York, Basic Books Inc., 1962. A series of essays by Louis de Broglie on various aspects of theoretical physics and on the history and philosophy of science. A large part of this work is devoted to de Broglie's developing attitudes toward the interpretation of wave mechanics and wave–particle duality.

2. For an in-depth look at the problems of interpretation and measurement surrounding the formalism of quantum theory, see M. Jammer, *The Philosophy of Quantum Mechanics*, New York, John Wiley and Sons, Inc., 1974.

3. A concise, solid introduction to the basic principles of quantum physics with applications may be found in Chapter 41 of *Physics for Scientists and Engineers with Mod-* ern Physics, 6th ed., by R. Serway and J. Jewett, Jr., Belmont, CA, Brooks/Cole–Thomson Learning, 2004.

4. A novel but delightfully refreshing exposition of quantum theory is presented by R. Feynman, R. Leighton, and M. Sands in *The Feynman Lectures on Physics Vol. III, Modern Physics*, Reading, MA, Addison-Wesley Publishing Co., 1965. This work is more advanced and sophisticated than other introductory texts in the field, though still somewhat below the intermediate level.

5. For a very readable introduction to charge-coupled devices, see "Charge-coupled Devices in Astronomy", *Sci. Am.*, volume 247(4), pp 66–74, Oct. 1982, by Jerome Kristian and Morley Blouke.

QUESTIONS

1. The probability density at certain points for a particle in a box is zero, as seen in Figure 6.9. Does this imply that the particle cannot move across these points? Explain.
2. Discuss the relation between the zero-point energy and the uncertainty principle.
3. Consider a square well with one finite wall and one infinite wall. Compare the energy and momentum of a particle trapped in this well to the energy and momentum of an identical particle trapped in an infinite well with the same width.
4. Explain why a wave packet moves with the group velocity rather than with the phase velocity.
5. According to Section 6.2, a free particle can be represented by any number of waveforms, depending on the values chosen for the coefficients $a(k)$. What is the source of this ambiguity, and how is it resolved?
6. Because the Schrödinger equation can be formulated in terms of operators as $[H]\Psi = [E]\Psi$, is it incorrect to conclude from this the operator equivalence $[H] = [E]$?
7. For a particle in a box, the squared momentum p^2 is a sharp observable, but the momentum itself is fuzzy. Explain how this can be so, and how it relates to the classical motion of such a particle.
8. A philosopher once said that "it is necessary for the very existence of science that the same conditions always produce the same results." In view of what has been said in this chapter, present an argument showing that this statement is false. How might the statement be reworded to make it true?

PROBLEMS

6.1 The Born Interpretation

1. Of the functions graphed in Figure P6.1, which are candidates for the Schrödinger wavefunction of an actual physical system? For those that are not, state why they fail to qualify.
2. A particle is described by the wavefunction

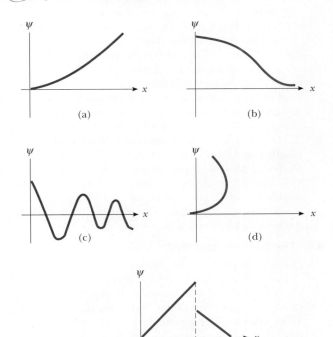

(a)   (b)   (c)   (d)   (e)

Figure P6.1

$$\psi(x) = \begin{cases} A\cos\left(\dfrac{2\pi x}{L}\right) & \text{for } -\dfrac{L}{4} \le x \le \dfrac{L}{4} \\ 0 & \text{otherwise} \end{cases}$$

(a) Determine the normalization constant A. (b) What is the probability that the particle will be found between $x = 0$ and $x = L/8$ if a measurement of its position is made?

6.2 Wavefunction for a Free Particle

3. A free electron has a wavefunction

$$\psi(x) = A\sin(5 \times 10^{10}\, x)$$

where x is measured in meters. Find (a) the electron's de Broglie wavelength, (b) the electron's momentum, and (c) the electron's energy in electron volts.
4. *Spreading of a Gaussian wave packet.* The Gaussian wave packet $\Psi(x, 0)$ of Example 6.3 is built out of plane waves according to the amplitude distribution function $a(k) = (C\alpha/\sqrt{\pi})\exp(-\alpha^2 k^2)$. Calculate $\Psi(x, t)$ for this packet and describe its evolution.

6.3 Wavefunctions in the Presence of Forces

5. In a region of space, a particle with zero energy has a wavefunction

$$\psi(x) = Axe^{-x^2/L^2}$$

(a) Find the potential energy U as a function of x.
(b) Make a sketch of $U(x)$ versus x.
6. The wavefunction of a particle is given by

$$\psi(x) = A\cos(kx) + B\sin(kx)$$

where A, B, and k are constants. Show that ψ is a solution of the Schrödinger equation (Eq. 6.13), assuming the particle is free ($U = 0$), and find the corresponding energy E of the particle.

6.4 The Particle in a Box

7. Show that allowing the state $n = 0$ for a particle in a one-dimensional box violates the uncertainty principle, $\Delta x \, \Delta p \geq \hbar/2$.

8. A bead of mass 5.00 g slides freely on a wire 20.0 cm long. Treating this system as a particle in a one-dimensional box, calculate the value of n corresponding to the state of the bead if it is moving at a speed of 0.100 nm per year (that is, apparently at rest).

9. The nuclear potential that binds protons and neutrons in the nucleus of an atom is often approximated by a square well. Imagine a proton confined in an infinite square well of length 10^{-5} nm, a typical nuclear diameter. Calculate the wavelength and energy associated with the photon that is emitted when the proton undergoes a transition from the first excited state ($n = 2$) to the ground state ($n = 1$). In what region of the electromagnetic spectrum does this wavelength belong?

10. An electron is contained in a one-dimensional box of width 0.100 nm. (a) Draw an energy-level diagram for the electron for levels up to $n = 4$. (b) Find the wavelengths of *all* photons that can be emitted by the electron in making transitions that would eventually get it from the $n = 4$ state to the $n = 1$ state.

11. Consider a particle moving in a one-dimensional box with walls at $x = -L/2$ and $x = L/2$. (a) Write the wavefunctions and probability densities for the states $n = 1$, $n = 2$, and $n = 3$. (b) Sketch the wavefunctions and probability densities. (*Hint:* Make an analogy to the case of a particle in a box with walls at $x = 0$ and $x = L$.)

12. A ruby laser emits light of wavelength 694.3 nm. If this light is due to transitions from the $n = 2$ state to the $n = 1$ state of an electron in a box, find the width of the box.

13. A proton is confined to moving in a one-dimensional box of width 0.200 nm. (a) Find the lowest possible energy of the proton. (b) What is the lowest possible energy of an electron confined to the same box? (c) How do you account for the large difference in your results for (a) and (b)?

14. A particle of mass m is placed in a one-dimensional box of length L. The box is so small that the particle's motion is *relativistic*, so that $E = p^2/2m$ is *not valid*. (a) Derive an expression for the energy levels of the particle using the relativistic energy–momentum relation and the quantization of momentum that derives from confinement. (b) If the particle is an electron in a box of length $L = 1.00 \times 10^{-12}$ m, find its lowest possible kinetic energy. By what percent is the nonrelativistic formula for the energy in error?

15. Consider a "crystal" consisting of two nuclei and two electrons, as shown in Figure P6.15. (a) Taking into account all the pairs of interactions, find the potential energy of the system as a function of d. (b) Assuming the electrons to be restricted to a one-dimensional box of length $3d$, find the minimum kinetic energy of the two electrons. (c) Find the value of d for which the total energy is a *minimum*. (d) Compare this value of d with the spacing of atoms in lithium, which has a density of 0.53 g/cm^3 and an atomic weight of 7. (This type of calculation can be used to estimate the densities of crystals and certain stars.)

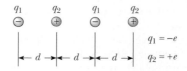

Figure P6.15

16. An electron is trapped in an infinitely deep potential well 0.300 nm in width. (a) If the electron is in its ground state, what is the probability of finding it within 0.100 nm of the left-hand wall? (b) Repeat (a) for an electron in the 99th excited state ($n = 100$). (c) Are your answers consistent with the correspondence principle?

17. An electron is trapped at a defect in a crystal. The defect may be modeled as a one-dimensional, rigid-walled box of width 1.00 nm. (a) Sketch the wavefunctions and probability densities for the $n = 1$ and $n = 2$ states. (b) For the $n = 1$ state, find the probability of finding the electron between $x_1 = 0.15$ nm and $x_2 = 0.35$ nm, where $x = 0$ is the left side of the box. (c) Repeat (b) for the $n = 2$ state. (d) Calculate the energies in electron volts of the $n = 1$ and $n = 2$ states.

18. Find the points of maximum and minimum probability density for the nth state of a particle in a one-dimensional box. Check your result for the $n = 2$ state.

19. A 1.00-g marble is constrained to roll inside a tube of length $L = 1.00$ cm. The tube is capped at both ends. Modeling this as a one-dimensional infinite square well, find the value of the quantum number n if the marble is initially given an energy of 1.00 mJ. Calculate the *excitation energy* required to promote the marble to the next available energy state.

6.5 The Finite Square Well

20. Consider a particle with energy E bound to a *finite* square well of height U and width $2L$ situated on $-L \leq x \leq +L$. Because the potential energy is symmetric about the midpoint of the well, the stationary state waves will be either symmetric or antisymmetric about this point. (a) Show that for $E < U$, the conditions for smooth joining of the interior and exterior waves lead to the following equation for the allowed energies of the symmetric waves:

$$k \tan kL = \alpha \quad \text{(symmetric case)}$$

where $\alpha = \sqrt{(2m/\hbar^2)(U - E)}$ and $k = \sqrt{2mE/\hbar^2}$ is the wavenumber of oscillation in the interior. (b) Show that the energy condition found in (a) can be rewritten as

$$k \sec kL = \frac{\sqrt{2mU}}{\hbar}$$

Apply the result in this form to an electron trapped at a defect site in a crystal, modeling the defect as a square well of height 5 eV and width 0.2 nm. Solve the equation numerically to find the ground-state energy for the electron, accurate to ±0.001 eV.

21. Sketch the wavefunction $\psi(x)$ and the probability density $|\psi(x)|^2$ for the $n = 4$ state of a particle in a *finite* potential well.

22. 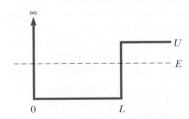 The potential energy of a proton confined to an atomic nucleus can be modeled as a square well of width 1.00×10^{-5} nm and height 26.0 MeV. Determine the energy of the proton in the ground state and first excited state for this case, using the Java applet available at our companion Website (http://info. brookscole.com/mp3e QMTools Simulations → Problem 6.22). Refer to Exercise 3 of Example 6.8 for details. Calculate the wavelength of the photon emitted when the proton undergoes a transition from the first excited state to the ground state, and compare your result with that found using the infinite-well model of Problem 9.

23. Consider a square well having an infinite wall at $x = 0$ and a wall of height U at $x = L$ (Fig. P6.23). For the case $E < U$, obtain solutions to the Schrödinger equation inside the well $(0 \le x \le L)$ and in the region beyond $(x > L)$ that satisfy the appropriate boundary conditions at $x = 0$ and $x = \infty$. Enforce the proper matching conditions at $x = L$ to find an equation for the allowed energies of this system. Are there conditions for which no solution is possible? Explain.

Figure P6.23

6.6   The Quantum Oscillator

24. The wavefunction

$$\psi(x) = Cxe^{-\alpha x^2}$$

also describes a state of the quantum oscillator, provided the constant α is chosen appropriately. (a) Using Schrödinger's equation, obtain an expression for α in terms of the oscillator mass m and the classical frequency of vibration ω. What is the energy of this state? (b) Normalize this wave. (*Hint:* See the integral of Problem 32.)

25. Show that the oscillator energies in Equation 6.29 correspond to the classical amplitudes

$$A_n = \sqrt{\frac{(2n + 1)\hbar}{m\omega}}$$

26. Obtain an expression for the probability density $P_c(x)$ of a *classical* oscillator with mass m, frequency ω, and amplitude A. (*Hint:* See Problem 28 for the calculation of classical probabilities.)

27. 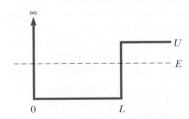 *Coherent states.* Use the Java applet available at our companion website (http://info. brookscole.com/mp3e QMTools Simulations → Problem 6.27) to explore the time development of a Gaussian waveform confined to the oscillator well. The default settings for the initial wave describe a Gaussian centered in the well with an adjustable width set by the value of the parameter a. Describe the time evolution of this wavefunction. Is it what you expected? Account for your observations. Now displace the initial waveform off of center by increasing the parameter d from zero to $d = 1$. Again describe the time evolution of the resulting wavefunction. What is remarkable about this case? Such wavefunctions, called *coherent states*, are important in the quantum theory of radiation.

6.7   Expectation Values

28. *Classical probabilities.* (a) Show that the classical probability density describing a particle in an infinite square well of dimension L is $P_c(x) = 1/L$. (*Hint:* The classical probability for finding a particle in dx—$P_c(x)\,dx$—is proportional to the *time* the particle spends in this interval.) (b) Using $P_c(x)$, determine the *classical* averages $\langle x \rangle$ and $\langle x^2 \rangle$ for a particle confined to the well, and compare with the quantum results found in Example 6.15. Discuss your findings in light of the correspondence principle.

29. An electron is described by the wavefunction

$$\psi(x) = \begin{cases} 0 & \text{for } x < 0 \\ Ce^{-x}(1 - e^{-x}) & \text{for } x > 0 \end{cases}$$

where x is in nanometers and C is a constant. (a) Find the value of C that normalizes ψ. (b) Where is the electron most likely to be found; that is, for what value of x is the probability for finding the electron largest? (c) Calculate $\langle x \rangle$ for this electron and compare your result with its most likely position. Comment on any differences you find.

30. For any eigenfunction ψ_n of the infinite square well, show that $\langle x \rangle = L/2$ and that

$$\langle x^2 \rangle = \frac{L^2}{3} - \frac{L^2}{2(n\pi)^2}$$

where L is the well dimension.

31. An electron has a wavefunction

$$\psi(x) = Ce^{-|x|/x_0}$$

where x_0 is a constant and $C = 1/\sqrt{x_0}$ for normalization (see Example 6.1). For this case, obtain expressions for $\langle x \rangle$ and Δx in terms of x_0. Also calculate the probability that the electron will be found within a standard deviation of its average position, that is, in the range $\langle x \rangle - \Delta x$ to $\langle x \rangle + \Delta x$, and show that this is independent of x_0.

32. Calculate $\langle x \rangle$, $\langle x^2 \rangle$, and Δx for a quantum oscillator in its ground state. *Hint:* Use the integral formula

$$\int_0^\infty x^2 e^{-ax^2}\, dx = \frac{1}{4a}\sqrt{\frac{\pi}{a}} \qquad a > 0$$

33. (a) What value do you expect for $\langle p \rangle$ for the quantum oscillator? Support your answer with a symmetry argument rather than a calculation. (b) Energy principles for the quantum oscillator can be used to relate $\langle p^2 \rangle$ to $\langle x^2 \rangle$. Use this relation, along with the value of $\langle x^2 \rangle$ from

Problem 32, to find $\langle p^2 \rangle$ for the oscillator ground state. (c) Evaluate Δp, using the results of (a) and (b).

34. From the results of Problems 32 and 33, evaluate $\Delta x\, \Delta p$ for the quantum oscillator in its ground state. Is the result consistent with the uncertainty principle? (Note that your computation verifies the minimum uncertainty product; furthermore, the harmonic oscillator ground state is the *only* quantum state for which this minimum uncertainty is realized.)

6.8   Observables and Operators

35. Which of the following functions are eigenfunctions of the momentum operator $[p]$? For those that are eigenfunctions, what are the eigenvalues?
(a) $A \sin(kx)$ (c) $A \cos(kx) + iA \sin(kx)$
(b) $A \sin(kx) - A \cos(kx)$ (d) $A e^{ik(x-a)}$

ADDITIONAL PROBLEMS

36. 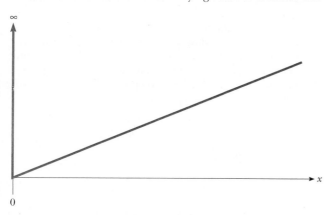 *The quantum bouncer.* The bouncer is the quantum analog to the classical problem of a ball bouncing vertically (and elastically) on a level surface and is modeled by the potential energy shown in Figure P6.36. The coordinate normal to the surface is denoted by x, and the surface itself is located at $x = 0$. Above the surface, the potential energy for the bouncer is linear, representing the attractive force of a uniform field—in this case the gravity field near the Earth. Below the surface, the potential energy rises abruptly to a very large value consistent with the bouncer's inability to penetrate this region. Obtaining stationary states for the bouncer from the Schrödinger equation using analytical techniques requires knowledge of special functions. Numerical solution furnishes a simpler alternative and allows for effortless study of the bound-state waveforms, once they are found. The Java applet for the quantum bouncer can be found at http://info.brookscole.com/mp3e QMTools Simulations → Problem 6.36. Use the applet as described there to find the three lowest-lying states of a tennis ball

(mass ≈ 50 g) bouncing on a hard floor. Count the number of nodes for each wavefunction to verify the general rule that the nth excited state exhibits exactly n nodes. For each state, determine the most probable distance above the floor for the bouncing ball and compare with the maximum height reached in the classical case. (Classically, the ball is most likely to be found at the top of its flight, where its speed drops to zero—see Problem 28.)

37. *Nonstationary states.* Consider a particle in an infinite square well described initially by a wave that is a superposition of the ground and first excited states of the well:

$$\Psi(x, 0) = C[\psi_1(x) + \psi_2(x)]$$

(a) Show that the value $C = 1/\sqrt{2}$ normalizes this wave, assuming ψ_1 and ψ_2 are themselves normalized. (b) Find $\psi(x, t)$ at any later time t. (c) Show that the superposition is *not* a stationary state, but that the average energy in this state is the arithmetic mean $(E_1 + E_2)/2$ of the ground- and first excited-state energies E_1 and E_2.

38. For the nonstationary state of Problem 37, show that the average particle position $\langle x \rangle$ oscillates with time as

$$\langle x \rangle = x_0 + A \cos(\Omega t)$$

where

$$x_0 = \frac{1}{2}\left(\int x |\psi_1|^2\, dx + \int x |\psi_2|^2\, dx \right)$$

$$A = \int x \psi_1^* \psi_2\, dx$$

and $\Omega = (E_2 - E_1)/\hbar$. Evaluate your results for the mean position x_0 and amplitude of oscillation A for an electron in a well 1 nm wide. Calculate the time for the electron to shuttle back and forth in the well once. Calculate the same time classically for an electron with energy equal to the average, $(E_1 + E_2)/2$.

Figure P6.36

7

Tunneling Phenomena

Chapter Outline

In this chapter the principles of wave mechanics are applied to particles striking a potential *barrier*. Unlike potential wells that attract and trap particles, barriers repel them. Because barriers have no bound states, the emphasis shifts to determining whether a particle incident on a barrier is reflected or transmitted.

In the course of this study we shall encounter a peculiar phenomenon called *tunneling*. A purely wave-mechanical effect, tunneling nevertheless is essential to the operation of many modern-day devices and shapes our world on a scale from atomic all the way up to galactic proportions. The chapter includes a discussion of the role played by tunneling in several phenomena of practical interest, such as field emission, radioactive decay, and the operation of the ammonia maser. Finally, the chapter is followed by an essay on the scanning tunneling microscope, or STM, a remarkable device that uses tunneling to make images of surfaces with resolution comparable to the size of a single atom.

7.1 THE SQUARE BARRIER

The square barrier is represented by a potential energy function $U(x)$ that is constant at U in the barrier region, say between $x = 0$ and $x = L$, and zero outside this region. One method for producing a square barrier potential using charged hollow cylinders is shown in Figure 7.1a. The outer cylinders are grounded while the central one is held at some positive potential V. For a particle with charge q, the barrier potential energy is $U = qV$. The

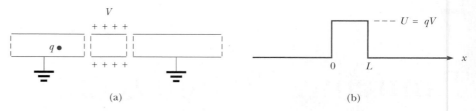

Figure 7.1 (a) Aligned metallic cylinders serve as a potential barrier to charged particles. The central cylinder is held at some positive electric potential V, and the outer cylinders are grounded. A charge q whose total energy is less than qV is unable to penetrate the central cylinder classically, but can do so quantum mechanically by a process called tunneling. (b) The potential energy seen by this charge in the limit where the gaps between the cylinders have shrunk to zero size. The result is the square barrier potential of height U.

charge experiences no electric force except in the gaps separating the cylinders. The force in the gaps is repulsive, tending to expel a positive charge q from the central cylinder. The electric potential energy for the idealized case in which the gaps have shrunk to zero size is the **square barrier,** sketched in Figure 7.1b.

A classical particle incident on the barrier, say from the left, experiences a retarding force on arriving at $x = 0$. Particles with energies E greater than U are able to overcome this force, but suffer a reduction in speed to a value commensurate with their diminished kinetic energy $(E - U)$ in the barrier region. Such particles continue moving to the right with reduced speed until they reach $x = L$, where they receive a "kick" accelerating them back to their original speed. Thus, particles having energy $E > U$ are able to cross the barrier with their speed restored to its initial value. By contrast, particles with energy $E < U$ are turned back (reflected) by the barrier, having insufficient energy to cross or even penetrate it. In this way the barrier divides the space into classically allowed and forbidden regions determined by the particle energy: If $E > U$, the whole space is accessible to the particle; for $E < U$ only the interval to the side of the barrier in which the particle originates is accessible—the barrier region itself is forbidden, and this precludes particle motion on the far side as well.

According to quantum mechanics, however, **there is no region inaccessible to our particle, regardless of its energy,** since the matter wave associated with the particle is nonzero everywhere. A typical wavefunction for this case, illustrated in Figure 7.2a, clearly shows the penetration of the wave into the barrier and beyond. This barrier penetration is in complete disagreement with classical physics. The process of penetrating the barrier is called **tunneling:** we say the particle has tunneled through the barrier.

The mathematical expression for Ψ on either side of the barrier is easily found. To the left of the barrier the particle is free, so the wavefunction here is composed of the free particle plane waves introduced in Chapter 6:

$$\Psi(x, t) = Ae^{i(kx - \omega t)} + Be^{i(-kx - \omega t)} \tag{7.1}$$

This wavefunction $\Psi(x, t)$ is actually the sum of two plane waves. Both have frequency ω and energy $E = \hbar\omega = \hbar^2 k^2/2m$, but the first moves from left to right (wavenumber k), the second from right to left (wavenumber $-k$). Thus,

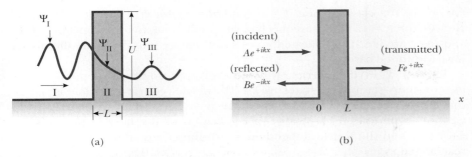

Figure 7.2   (a) A typical stationary-state wave for a particle in the presence of a square barrier. The energy E of the particle is less than the barrier height U. Since the wave amplitude is nonzero in the barrier, there is some probability of finding the particle there. (b) Decomposition of the stationary wave into incident, reflected, and transmitted waves.

that part of Ψ proportional to A is interpreted as a wave incident on the barrier from the left; that proportional to B as a wave reflected from the barrier and moving from right to left (Fig. 7.2b). The **reflection coefficient** R for the barrier is calculated as the ratio of the reflected probability density to the incident probability density:

$$R = \frac{(\Psi^*\Psi)_{\text{reflected}}}{(\Psi^*\Psi)_{\text{incident}}} = \frac{B^*B}{A^*A} = \frac{|B|^2}{|A|^2} \tag{7.2}$$

Reflection coefficient for a barrier

In wave terminology, R is the fraction of wave intensity in the reflected beam; in particle language, R becomes the likelihood (probability) that a particle incident on the barrier from the left is reflected by it.

Similar arguments apply to the right of the barrier, where, again, the particle is free:

$$\Psi(x, t) = Fe^{i(kx-\omega t)} + Ge^{i(-kx-\omega t)} \tag{7.3}$$

This form for $\Psi(x, t)$ is valid in the range $x > L$, with the term proportional to F describing a wave traveling to the right, and that proportional to G a wave traveling to the left in this region. The latter has no physical interpretation for waves incident on the barrier from the left, and so is discarded by requiring $G = 0$. The former is that part of the incident wave that is transmitted through the barrier. The relative intensity of this transmitted wave is the **transmission coefficient** for the barrier T:

$$T = \frac{(\Psi^*\Psi)_{\text{transmitted}}}{(\Psi^*\Psi)_{\text{incident}}} = \frac{F^*F}{A^*A} = \frac{|F|^2}{|A|^2} \tag{7.4}$$

Transmission coefficient for a barrier

The transmission coefficient measures the likelihood (probability) that a particle incident on the barrier from the left penetrates to emerge on the other side. Since a particle incident on the barrier is either reflected or transmitted, the probabilities for these events must sum to unity:

$$R + T = 1 \tag{7.5}$$

Equation 7.5 expresses a kind of **sum rule** obeyed by the barrier coefficients. Further, the degree of transmission or reflection will depend on particle energy. In the classical case $T = 0$ (and $R = 1$) for $E < U$, but $T = 1$ (and $R = 0$) for $E > U$. The wave-mechanical predictions for the functions $T(E)$ and $R(E)$ are more complicated; to obtain them we must examine the matter wave within the barrier.

To find Ψ in the barrier, we must solve Schrödinger's equation. Let us consider stationary states $\psi(x)e^{-i\omega t}$ whose energy $E = \hbar\omega$ is below the top of the barrier. This is the case $E < U$ for which no barrier penetration is permitted classically. In the region of the barrier $(0 < x < L)$, $U(x) = U$ and the time-independent Schrödinger equation for $\psi(x)$ can be rearranged as

$$\frac{d^2\psi}{dx^2} = \left\{\frac{2m(U-E)}{\hbar^2}\right\}\psi(x)$$

With $E < U$, the term in braces is a positive constant, and solutions to this equation are the *real exponential* forms $e^{\pm\alpha x}$. Since $(d^2/dx^2)e^{\pm\alpha x} = (\alpha)^2 e^{\pm\alpha x}$, we should identify the term in braces with α^2 or, equivalently,

$$\alpha = \frac{\sqrt{2m(U-E)}}{\hbar} \tag{7.6}$$

For wide barriers, the probability of finding the particle should decrease steadily into the barrier; in such cases only the decaying exponential is important, and it is convenient to define a barrier **penetration depth** $\delta = 1/\alpha$. At a distance δ into the barrier, the wavefunction has fallen to $1/e$ of its value at the barrier edge; thus, the probability of finding the particle is appreciable only within about δ of the barrier edge.

The complete wavefunction in the barrier is, then,

$$\Psi(x, t) = \psi(x)e^{-i\omega t} = Ce^{-\alpha x - i\omega t} + De^{+\alpha x - i\omega t} \qquad \text{for } 0 < x < L \tag{7.7}$$

The coefficients C and D are fixed by requiring smooth joining of the wavefunction across the barrier edges; that is, both Ψ and $\partial\Psi/\partial x$ must be continuous at $x = 0$ and $x = L$. Writing out the joining conditions using Equations 7.1, 7.3, and 7.7 for Ψ in the regions to the left, to the right, and within the barrier, respectively, gives

Joining conditions at a square barrier

$$A + B = C + D \qquad \text{(continuity of } \Psi \text{ at } x = 0)$$

$$ikA - ikB = \alpha D - \alpha C \qquad \left(\text{continuity of } \frac{\partial\Psi}{\partial x} \text{ at } x = 0\right)$$

$$Ce^{-\alpha L} + De^{+\alpha L} = Fe^{ikL} \qquad \text{(continuity of } \Psi \text{ at } x = L) \tag{7.8}$$

$$(\alpha D)e^{+\alpha L} - (\alpha C)e^{-\alpha L} = ikFe^{ikL} \qquad \left(\text{continuity of } \frac{\partial\Psi}{\partial x} \text{ at } x = L\right)$$

In keeping with our previous remarks, we have set $G = 0$. Still, there is one more unknown than there are equations to find them. Actually this is as it should be, since the amplitude of the incident wave merely sets the scale for the other amplitudes. That is, doubling the incident wave amplitude simply doubles the amplitudes of the reflected and transmitted waves. Dividing Equations 7.8 through by A furnishes four equations for the four ratios B/A, C/A, D/A, and F/A. These equations may be solved by repeated substitution to find B/A and so on in terms of the barrier height U, the barrier width L, and the particle energy E. The result for the transmission coefficient T is (see Problem 7)

$$T(E) = \left\{ 1 + \frac{1}{4}\left[\frac{U^2}{E(U-E)}\right] \sinh^2 \alpha L \right\}^{-1} \qquad (7.9)$$

where sinh denotes the hyperbolic sine function: $\sinh x = (e^x - e^{-x})/2$.

A sketch of $T(E)$ for the square barrier is shown in Figure 7.3. Equation 7.9 holds only for energies E below the barrier height U. For $E > U$, α becomes imaginary and $\sinh(\alpha L)$ turns oscillatory. This leads to fluctuations in $T(E)$ and isolated energies for which transmission occurs with complete certainty, that is, $T(E) = 1$. Such *transmission resonances* arise from wave interference and constitute further evidence for the wave nature of matter (see Example 7.3).

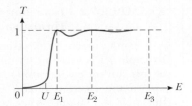

Figure 7.3 A sketch of the transmission coefficient $T(E)$ for a square barrier. Oscillation in $T(E)$ with E, and the transmission resonances at E_1, E_2, and E_3, are further evidence for the wave nature of matter.

EXAMPLE 7.1   Transmission Coefficient for an Oxide Layer

Two copper conducting wires are separated by an insulating oxide layer (CuO). Modeling the oxide layer as a square barrier of height 10.0 eV, estimate the transmission coefficient for penetration by 7.00-eV electrons (a) if the layer thickness is 5.00 nm and (b) if the layer thickness is 1.00 nm.

Solution   From Equation 7.6 we calculate α for this case, using $\hbar = 1.973$ keV·Å/c and $m_e = 511$ keV/c^2 for electrons to get

$$\alpha = \frac{\sqrt{2m_e(U-E)}}{\hbar}$$

$$= \frac{\sqrt{2(511 \text{ keV}/c^2)(3.00 \times 10^{-3} \text{ keV})}}{1.973 \text{ keV·Å}/c} = 0.8875 \text{ Å}^{-1}$$

The transmission coefficient from Equation 7.9 is then

$$T = \left\{ 1 + \frac{1}{4}\left[\frac{10^2}{7(3)}\right] \sinh^2 (0.8875 \text{ Å}^{-1})L \right\}^{-1}$$

Substituting $L = 50.0$ Å (5.00 nm) gives

$$T = 0.963 \times 10^{-38}$$

a fantastically small number on the order of 10^{-38}! With $L = 10.0$ Å (1.00 nm), however, we find

$$T = 0.657 \times 10^{-7}$$

We see that reducing the layer thickness by a factor of 5 enhances the likelihood of penetration by nearly 31 orders of magnitude!

Exercise 1   Go to our companion Web site (http://info.brookscole.com/mp3e) and select QMTools Simulations → Exercise 7.1. This particular Java applet shows the de Broglie wave (actually, just the real part) for an electron with energy 7.00 eV incident from the left on a square barrier 10.0 eV high and 1.0 Å wide. Compare this waveform with the illustration of Figure 7.2a. In fact, this wave is inherently complex valued, with a modulus and phase that varies from point to point. A more informative display plots the modulus in the usual way but uses color to represent the phase of the wave. Right-click on the waveform and select Properties . . . → Color-4-Phase → Apply to show the color-for-phase plotting style. Why does the transmitted wave (to the right of the barrier) now have a uniform height? What is the significance of this height? Follow the on-site instructions to display the incident component of this scattering wave and determine the transmission coefficient directly from the graphs. Compare your result with the prediction of Equation 7.9.

EXAMPLE 7.2   Tunneling Current Through an Oxide Layer

A 1.00-mA current of electrons in one of the wires of Example 7.1 is incident on the oxide layer. How much of this current passes through the layer to the adjacent wire if the electron energy is 7.00 eV and the layer thickness is 1.00 nm? What becomes of the remaining current?

Solution   Because each electron carries a charge equal to $e = 1.60 \times 10^{-19}$ C, an electron current of 1.00 mA represents $10^{-3}/(1.60 \times 10^{-19}) = 6.25 \times 10^{15}$ electrons per second impinging on the barrier. Of these, only the fraction T is transmitted, where $T = 0.657 \times 10^{-7}$ from Example 7.1. Thus, the number of electrons per second continuing on to the adjacent wire is

$$(6.25 \times 10^{15})(0.657 \times 10^{-7}) = 4.11 \times 10^{8} \text{ electrons/s}$$

This number represents a transmitted current of

$$(4.11 \times 10^{8}/\text{s})(1.60 \times 10^{-19} \text{ C}) = 6.57 \times 10^{-11} \text{ A}$$

$$= 65.7 \text{ pA (picoamperes)}$$

(Notice that the same transmitted current would be obtained had we simply multiplied the incident current by the transmission coefficient.) The remaining 1.00 mA $-$ 65.7 pA is reflected at the layer. It is important to note that the *measured* conduction current in the wire on the side of incidence is the net of the incident and reflected currents, or again 65.7 pA.

EXAMPLE 7.3   Transmission Resonances

Consider a particle incident from the left on a square barrier of width L in the case where the particle energy E exceeds the barrier height U. Write the necessary wavefunctions and impose the proper joining conditions to obtain a formula for the transmission coefficient for this case. Show that perfect transmission (resonance) results for special values of particle energy, and explain this phenomenon in terms of the interference of de Broglie waves.

Solution   To the left and right of the barrier, the wavefunctions are the free particle waves given by Equations 7.1 and 7.3 (again with $G = 0$ to describe a purely transmitted wave on the far side of the barrier):

$$\Psi(x, t) = Ae^{i(kx - \omega t)} + Be^{i(-kx - \omega t)} \qquad x < 0$$

$$\Psi(x, t) = Fe^{i(kx - \omega t)} \qquad x > L$$

The wavenumber k and frequency ω of these oscillations derive from the particle energy E in the manner characteristic of (nonrelativistic) de Broglie waves; that is, $E = (\hbar k)^2/2m = \hbar \omega$. Within the barrier, the wavefunction also is oscillatory. In effect, the decay constant α of Equation 7.6 has become imaginary, since $E > U$. Introducing a new wavenumber k' as $\alpha = ik'$, the barrier wavefunction

becomes

$$\Psi(x, t) = Ce^{i(-k'x - \omega t)} + De^{i(k'x - \omega t)} \qquad 0 < x < L$$

with $k' = [2m(E - U)/\hbar^2]^{1/2}$ a real number.

The barrier wavefunction will join smoothly to the exterior waveforms if the wavefunction and its slope are continuous at the barrier edges $x = 0$ and $x = L$. These continuity requirements are identical to Equations 7.8 with the replacement $\alpha = ik'$ everywhere. In particular, we now have

$$A + B = C + D \qquad \text{(continuity of } \Psi \text{ at } x = 0)$$

$$kA - kB = k'D - k'C \qquad \left(\text{continuity of } \frac{\partial \Psi}{\partial x} \text{ at } x = 0\right)$$

$$Ce^{-ik'L} + De^{ik'L} = Fe^{ikL} \qquad \text{(continuity of } \Psi \text{ at } x = L)$$

$$k'De^{ik'L} - k'Ce^{-ik'L} = kFe^{ikL} \qquad \left(\text{continuity of } \frac{\partial \Psi}{\partial x} \text{ at } x = L\right)$$

To isolate the transmission amplitude F/A, we must eliminate from these relations the unwanted coefficients B, C, and D. Dividing the second line by k and adding to the first eliminates B, leaving A in terms of C and D. In the same way, dividing the fourth line by k' and adding the result to the third line gives D (in terms of F), while subtracting the result from the third line gives C (in terms of F). Combining the previous results finally yields A in terms of F:

$$A = \frac{1}{4} Fe^{ikL} \left\{ \left[2 - \left(\frac{k'}{k} + \frac{k}{k'} \right) \right] e^{ik'L} \right.$$

$$\left. + \left[2 + \left(\frac{k'}{k} + \frac{k}{k'} \right) \right] e^{-ik'L} \right\}$$

The transmission probability is $T = |F/A|^2$. Writing $e^{\pm ik'L} = \cos k'L \pm i \sin k'L$ and simplifying, we obtain the final result

$$\frac{1}{T} = \left| \frac{A}{F} \right|^2 = \frac{1}{4} \left| 2 \cos k'L - i \left(\frac{k'}{k} + \frac{k}{k'} \right) \sin k'L \right|^2$$

$$= 1 + \frac{1}{4} \left[\frac{U^2}{E(E - U)} \right] \sin^2 k'L$$

We see that **transmission resonances** occur whenever $k'L$ is a multiple of π. Using $k' = [2m(E - U)/\hbar^2]^{1/2}$, we can express the resonance condition in terms of the particle energy E as

$$E = U + n^2 \frac{\pi^2 \hbar^2}{2mL^2} \qquad n = 1, 2, \ldots$$

Particles with these energies are transmitted perfectly ($T = 1$), with no chance of reflection ($R = 0$).

Resonances arise from the interference of the matter wave accompanying a particle. The wave reflected from the barrier can be regarded as the superposition of matter waves reflected from the leading and trailing edges of the barrier at $x = 0$ and $x = L$, respectively. If these reflected waves arrive phase shifted by odd multiples of 180° or π radians, they will interfere destructively, leaving no reflected wave ($R = 0$) and thus perfect transmission. Now the wave reflected from the rear of the barrier at $x = L$ must travel the extra distance $2L$ before recombining with the wave reflected at the front, leading to a phase difference of $2k'L$. But this wave also suffers an intrinsic phase shift of π radians, having been reflected from a medium with higher optical density.[1] Thus, the condition for destructive interference becomes $2k'L + \pi = (2n + 1)\pi$, or simply $k'L = n\pi$, where $n = 1, 2, \ldots$.

Perfect transmission also arises when particles are scattered by a potential well, a phenomenon known as the **Ramsauer–Townsend effect** (see Problem 11).

Exercise 2   Verify that for $E \gg U$, the transmission coefficient of Example 7.3 approaches unity. Why is this result expected? What happens to T in the limit as E approaches U?

EXAMPLE 7.4   Scattering by a Potential Step

The potential step shown in Figure 7.4 may be regarded as a square barrier in the special case where the barrier width L is infinite. Apply the ideas of this section to discuss the quantum scattering of particles incident from the left on a potential step, in the case where the step height U exceeds the total particle energy E.

Solution   The wavefunction everywhere to the right of the origin is the barrier wavefunction given by Equation 7.7. To keep Ψ from diverging for large x, we must take $D = 0$, leaving only the decaying wave

$$\Psi(x, t) = Ce^{-\alpha x - i\omega t} \qquad x > 0$$

This must be joined smoothly to the wavefunction on the left of the origin, given by Equation 7.1:

$$\Psi(x, t) = Ae^{ikx - i\omega t} + Be^{-ikx - i\omega t} \qquad x < 0$$

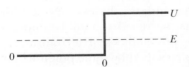

Figure 7.4   (Example 7.4) The potential step of height U may be thought of as a square barrier of the same height in the limit where the barrier width L becomes infinite. All particles incident on the barrier with energy $E < U$ are reflected.

The conditions for smooth joining at $x = 0$ yield

$$A + B = C \qquad \text{(continuity of } \Psi \text{)}$$

$$ikA - ikB = -\alpha C \qquad \left(\text{continuity of } \frac{\partial \Psi}{\partial x}\right)$$

Solving the second equation for C and substituting into the first (with $\delta = 1/\alpha$) gives $A + B = -ik\delta A + ik\delta B$, or

$$\frac{B}{A} = -\frac{(1 + ik\delta)}{(1 - ik\delta)}$$

The reflection coefficient is $R = |B/A|^2 = (B/A)(B/A)^*$, or

$$R = \left(\frac{(1 + ik\delta)}{(1 - ik\delta)}\right)\left(\frac{(1 - ik\delta)}{(1 + ik\delta)}\right) = 1$$

Thus, an *infinitely* wide barrier reflects all incoming particles with energies below the barrier height, in agreement with the classical prediction. Nevertheless, there is a nonzero wave in the step region since

$$\frac{C}{A} = 1 + \frac{B}{A} = \frac{-2ik\delta}{1 - ik\delta} \neq 0$$

But the wavefunction for $x > 0$, $\Psi(x, t) = Ce^{-\alpha x - i\omega t}$, is not a propagating wave at all; that is, there is no net transmission of particles to the right of the step. However, there will be quantum transmission through a barrier of *finite* width, no matter how wide (compare Eq. 7.9).

---

[1]This is familiar from the propagation of classical waves: A traveling wave arriving at the interface separating two media is partially transmitted and partially reflected. The reflected portion is phase shifted 180° only in the case where the wave speed is lower in the medium being penetrated. For matter waves, $p = h/\lambda$, and the wavelength (hence, wave speed) is largest in regions where the kinetic energy is smallest. Thus, the matter wave reflected from the front of a barrier suffers no change in phase, but that reflected from the rear is phase shifted 180°.

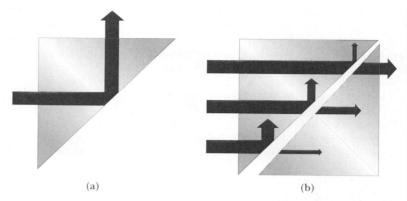

(a)   (b)

Figure 7.5   (a) Total internal reflection of light waves at a glass–air boundary. An evanescent wave penetrates into the space beyond the reflecting surface. (b) Frustrated total internal reflection. The evanescent wave is "picked up" by a neighboring surface, resulting in transmission across the gap. Notice that the light beam *does not* appear in the gap.

The existence of a barrier wave without propagation (as in Example 7.4) is familiar from the optical phenomenon of total internal reflection exploited in the construction of beam splitters (Figure 7.5): Light entering a right-angle prism is completely reflected at the hypotenuse face, even though an electromagnetic wave, the evanescent wave, penetrates into the space beyond. A second prism brought into near contact with the first can "pick up" this evanescent wave, thereby transmitting and redirecting the original beam (Fig. 7.5b). This phenomenon, known as **frustrated total internal reflection,** is the optical analog of tunneling: In effect, photons have tunneled across the gap separating the two prisms.

7.2   BARRIER PENETRATION: SOME APPLICATIONS

In actuality, few barriers can be modeled accurately using the square barrier discussed in the preceding section. Indeed, the extreme sensitivity to barrier constants found there suggests that barrier shape will be important in making reliable predictions of tunneling probabilities. The transmission coefficient for a barrier of arbitrary shape, as specified by some potential energy function $U(x)$, can be found from Schrödinger's equation. For high, wide barriers, where the likelihood of penetration is small, a lengthy treatment yields the approximate result

Approximate transmission coefficient of a barrier with arbitrary shape

$$T(E) \approx \exp\left(-\frac{2}{\hbar}\sqrt{2m}\int\sqrt{U(x)-E}\,dx\right) \tag{7.10}$$

The integral in Equation 7.10 is taken over the classically forbidden region where $E < U(x)$. A simple argument leading to this form follows by representing an arbitrary barrier as a succession of square barriers, all of which scatter independently, so that the transmitted wave intensity of one becomes the incident wave intensity for the next, and so forth (see Problem 15).

The use of Equation 7.10 is illustrated in the remainder of this section, where it is applied to several classic problems in contemporary physics.

Field Emission

In field emission, electrons bound to a metal are literally torn from the surface by the application of a strong electric field. In this way, the metal becomes a source that may be conveniently tapped to furnish electrons for many applications. In the past, such cold cathode emission, as it was known, was a popular way of generating electrons in vacuum tube circuits, producing less electrical "noise" than hot filament sources, where electrons were "boiled off" by heating the metal to a high temperature. Modern applications include the **field emission microscope** (Fig. 7.6) and a related device, the **scanning tunneling microscope** (see the essay at the end of this chapter), both of which use the escaping electrons to form an image of structural details at the emitting surface.

Field emission is a tunneling phenomenon. Figure 7.7a shows schematically how field emission can be obtained by placing a positively charged plate near the source metal to form, effectively, a parallel-plate capacitor. In the gap between the "plates" there is some electric field $\mathcal{E}$, but the electric field inside the metal remains zero due to the shielding by the mobile metal electrons attracted to the surface by the positively charged plate. Note that an electron in the bulk is virtually free, yet still bound to the metal by a potential well of depth U. The total electron energy E, which includes kinetic energy, is negative to indicate a bound electron; indeed, $|E|$ represents the energy needed to free this electron, a value at least equal to the work function of the metal.

Once beyond the surface $(x > 0)$, our electron is attracted by the electric force in the gap, $F = e\mathcal{E}$, represented by the potential energy $U(x) = -e\mathcal{E}x$. The potential energy diagram is shown in Figure 7.7b, together with the classically allowed and forbidden regions for an electron of energy E. The intersections of E with $U(x)$ at x_1 $(= 0)$ and x_2 $(= -E/e\mathcal{E})$ mark the classical turning points, where a classical particle with this energy would be turned around to keep it from entering the forbidden zone. Thus, from a classical viewpoint, an electron initially confined to the metal has insufficient energy to surmount the potential barrier at the surface and would remain in the bulk forever! It is only by virtue of its wave character that the electron can tunnel through this barrier to emerge on the other side. The probability of such an occurrence is measured by the transmission coefficient for the triangular barrier depicted in Figure 7.7b.

To calculate $T(E)$ we must evaluate the integral in Equation 7.10 over the classically forbidden region from x_1 to x_2. Since $U(x) = -e\mathcal{E}x$ in this region and $E = -e\mathcal{E}x_2$, we have

$$\int \sqrt{U(x) - E}\, dx = \sqrt{e\mathcal{E}} \int_0^{x_2} \sqrt{x_2 - x}\, dx$$

$$= -\frac{2}{3} \sqrt{e\mathcal{E}} \left\{x_2 - x\right\}^{3/2} \Big|_0^{x_2} = \frac{2}{3} \sqrt{e\mathcal{E}} \left(\frac{|E|}{e\mathcal{E}}\right)^{3/2}$$

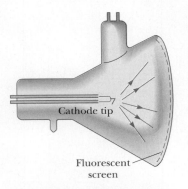

Cathode tip

Fluorescent screen

Figure 7.6 Schematic diagram of a field emission microscope. The intense electric field at the tip of the needle-shaped specimen allows electrons to tunnel through the work function barrier at the surface. Since the tunneling probability is sensitive to the exact details of the surface where the electron passes, the number of escaping electrons varies from point to point with the surface condition, thus providing a picture of the surface under study.

Tunneling model for field emission

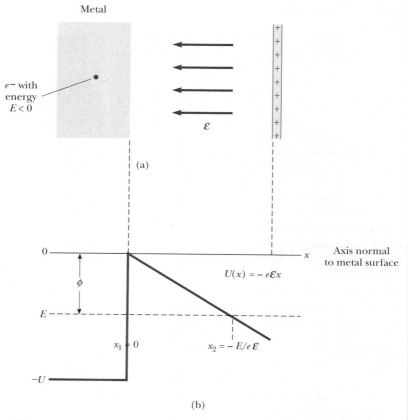

Figure 7.7 (a) Field emission from a metal surface. (b) The potential energy seen by an electron of the metal. The electric field produces the triangular potential barrier shown, through which electrons can tunnel to escape the metal. Turning points at $x_1 = 0$ and $x_2 = -E/e\mathcal{E}$ delineate the classically forbidden region. (Note that x_2 is positive since E is negative.) Tunneling is greatest for the most energetic electrons, for which $|E|$ is equal to the work function ϕ of the metal.

Using this result in Equation 7.10 gives the transmission coefficient for field emission as

Transmission coefficient for field emission

$$T(E) \approx \exp\left(\left\{-\frac{4\sqrt{2m}\,|E|^{3/2}}{3e\hbar}\right\}\frac{1}{\mathcal{E}}\right) \tag{7.11}$$

The strong dependence of T on electron energy E in the bulk is evident from this expression. It is also apparent that the quantity in curly brackets must have the dimensions of electric field and represents a characteristic field strength—say, $\mathcal{E}_c$—for field emission:

$$\mathcal{E}_c = \frac{4\sqrt{2m}\,|E|^{3/2}}{3e\hbar} \tag{7.12}$$

The escape probability is largest for the most energetic electrons; these are the ones most loosely bound and for which $|E| = \phi$, the work function of the

Figure 7.8   Field emission microscope image of the surface of a crystalline platinum alloy with a magnification of 3,000,000×. Individual atoms can be seen on surface layers using this technique. *(Manfred Kage/Peter Arnold, Inc.)*

metal. For $|E| = \phi = 4.0$ eV, a typical value for many metals, we calculate the characteristic field strength to be $\mathcal{E}_c = 5.5 \times 10^{10}$ V/m, a strong field by laboratory standards. Measurable emission occurs even with much weaker fields, however, since the emission *rate* depends on the product of the transmission coefficient and the number of electrons per second that collide with the barrier. This collision frequency is quite high for a bulk sample containing something like 10^{22} electrons per cubic centimeter, and values in excess of 10^{30} collisions per second per square centimeter are not uncommon (see Problem 18)! In this way field emission rates on the order of 10^{10} electrons per second (currents of about 1 nA) can be realized with applied fields as small as $\mathcal{E}_c/50$, or about 10^9 V/m.

EXAMPLE 7.5   Tunneling in a Parallel-Plate Capacitor

Estimate the leakage current due to tunneling that passes across a parallel-plate capacitor charged to a potential difference of 10 kV. Take the plate separation to be $d = 0.010$ mm and the plate area to be $A = 1.0$ cm^2.

Solution   The number of electrons per second impinging on the plate surface from the bulk is the collision frequency f, about 10^{30} per second per square centimeter for most metals. Of these, only the fraction given by the transmission coefficient T can tunnel through the potential barrier in the gap to register as a current through the device. Thus, the electron emission rate for a plate of area 1 cm^2 is

$$\lambda = fT(E) = 1.0 \times 10^{30} \exp(-\mathcal{E}_c/\mathcal{E})$$

The electric field $\mathcal{E}$ in the gap is 10 kV/0.010 mm = 1.0×10^9 V/m. Using this and $\mathcal{E}_c = 5.5 \times 10^{10}$ V/m gives for the exponential exp{−55} = 1.30×10^{-24} and an emission rate of $\lambda = 1.30 \times 10^6$ electrons per second. Since each electron carries a charge $e = 1.60 \times 10^{-19}$ C, the tunneling current is

$$I = 2.1 \times 10^{-13} \text{ A} = 0.21 \text{ pA}$$

α Decay

The decay of radioactive elements with the emission of α particles (helium nuclei composed of two protons and two neutrons) was among the long-standing puzzles to which the fledgling field of wave mechanics was first applied shortly after its inception in 1926. That α particles are a disintegration product of such species as radium, thorium, and uranium was well documented as early as 1900, but certain features of this decay remained a mystery, finally unraveled in 1928 in the now-classic works of George Gamow and R. W. Gurney and E. U. Condon. Their contribution was to recognize that the newly discovered tunnel effect lay behind the two most puzzling aspects of α decay:

- All α particles emitted from any one source have nearly the same energy and, for all known emitters, emerge with kinetic energies in the same narrow range, from about 4 to 9 MeV.
- In contrast to the uniformity of energies, the half-life of the emitter (time taken for half of the emitting substance to decay) varies over an enormous range—more than 20 orders of magnitude!—according to the emitting element (Table 7.1).

For instance, alphas emerge from the element thorium with kinetic energy equal to 4.05 MeV, only a little less than half as much as the alphas emitted from polonium (8.95 MeV). Yet the half-life of thorium is 1.4×10^{10} years, compared with only 3.0×10^{-7} *seconds* for the half-life of polonium!

Gamow attributed this striking behavior to a preformed α particle rattling around within the nucleus of the radioactive (parent) element, eventually tunneling through the potential barrier to escape as a detectable decay product (Fig. 7.9a). While inside the parent nucleus, the α is virtually free, but nonetheless confined to the nuclear potential well by the nuclear force. Once outside the nucleus, the α particle experiences only the Coulomb repulsion of the emitting (daughter) nucleus. (The nuclear force on the α outside the nucleus is insignificant due to its extremely short range, $\approx 10^{-15}$ m.) Figure 7.9b shows the potential-energy diagram for the α particle as a function of distance r from the emitting nucleus. The nuclear radius R is about 10^{-14} m, or 10 fm [note that 1 fm (fermi) = 10^{-15} m] for heavy nuclei[2]; beyond this there is only the energy of Coulomb repulsion, $U(r) = kq_1 q_2/r$, between the doubly charged α

Table 7.1   Characteristics of Some Common α Emitters

Element	α Energy	Half-Life*
$^{212}_{84}$Po	8.95 MeV	2.98×10^{-7} s
$^{240}_{96}$Cm	6.40 MeV	27 days
$^{226}_{88}$Ra	4.90 MeV	1.60×10^3 yr
$^{232}_{90}$Th	4.05 MeV	1.41×10^{10} yr

*Note that half-lives range over 24 orders of magnitude when α energy changes by a factor of 2.

[2]The fermi (fm) is a unit of distance commonly used in nuclear physics.

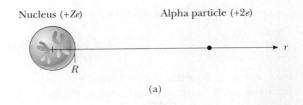

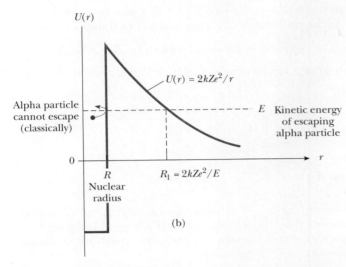

Figure 7.9 (a) α decay of a radioactive nucleus. (b) The potential energy seen by an α particle emitted with energy E. R is the nuclear radius, about 10^{-14} m, or 10 fm. α particles tunneling through the potential barrier between R and R_1 escape the nucleus to be detected as radioactive decay products.

($q_1 = +2e$) and a daughter nucleus with atomic number Z ($q_2 = +Ze$). Classically, even a 9-MeV α particle initially bound to the nucleus would have insufficient energy to overcome the Coulomb barrier (≈ 30 MeV high) and escape. But the α particle, with its wave attributes, may tunnel *through* the barrier to appear on the outside. The total α particle energy E inside the nucleus becomes the observed kinetic energy of the emerging α once it has escaped. **It is the sensitivity of the tunneling rate to small changes in particle energy that accounts for the wide range of half-lives observed for α emitters.**

The tunneling probability and associated decay rate are calculated in much the same way as for field emission, apart from the fact that the barrier shape now is Coulombic, rather than triangular. The details of this calculation are given in Example 13.9 (Chapter 13), with the result

> Tunneling through the
> Coulomb barrier

$$T(E) = \exp\left\{-4\pi Z\sqrt{\frac{E_0}{E}} + 8\sqrt{\frac{ZR}{r_0}}\right\} \qquad (7.13)$$

> Transmission coefficient for
> α particles of an unstable
> nucleus

In this expression, $r_0 = \hbar^2/m_\alpha ke^2$ is a kind of "Bohr" radius for the α particle. The mass of the α particle is $m_\alpha = 7295m_e$, so r_0 has the value $a_0/7295 = 7.25 \times 10^{-5}$ Å, or 7.25 fm. The length r_0, in turn, defines a convenient energy unit E_0 analogous to the Rydberg in atomic physics:

$$E_0 = \frac{ke^2}{2r_0} = \left(\frac{ke^2}{2a_0}\right)\left(\frac{a_0}{r_0}\right) = (13.6 \text{ eV})(7295) = 0.0993 \text{ MeV}$$

To obtain decay rates, $T(E)$ must be multiplied by the number of collisions per second that an α particle makes with the nuclear barrier. This collision frequency f is the reciprocal of the transit time for the α particle crossing the nucleus, or $f = v/2R$, where v is the speed of the α particle inside the nucleus. In most cases, f is about 10^{21} collisions per second (see Problem 17). The decay rate λ (the probability of α emission per unit time) is then

$$\lambda = fT(E) \approx 10^{21} \exp\{-4\pi Z\sqrt{E_0/E} + 8\sqrt{Z(R/r_0)}\}$$

The reciprocal of λ has dimensions of time and is related to the half-life of the emitter $t_{1/2}$ as

$$t_{1/2} = \frac{\ln 2}{\lambda} = \frac{0.693}{\lambda} \tag{7.14}$$

EXAMPLE 7.6   Estimating the Half-lives of Thorium and Polonium

Using the tunneling model just developed, estimate the half-lives for α decay of the radioactive elements thorium and polonium. The energy of the ejected alphas is 4.05 MeV and 8.95 MeV, respectively, and the nuclear size is about 9.00 fm in both cases.

Solution For thorium ($Z = 90$), the daughter nucleus has atomic number $Z = 88$, corresponding to the element radium. Using $E = 4.05$ MeV and $R = 9.00$ fm, we find for the transmission factor $T(E)$ in Equation 7.13

$$\exp\{-4\pi(88)\sqrt{(0.0993/4.05)} + 8\sqrt{88(9.00/7.25)}\}$$
$$= \exp\{-89.542\} = 1.3 \times 10^{-39}$$

Taking $f = 10^{21}$ Hz, we obtain for the decay rate the value $\lambda = 1.29 \times 10^{-18}$ alphas per second. The associated half-life is, from Equation 7.14,

$$t_{1/2} = \frac{0.693}{1.3 \times 10^{-18}} = 5.4 \times 10^{17} \text{ s} = 1.7 \times 10^{10} \text{ yr}$$

which compares favorably with the actual value for thorium, 1.4×10^{10} yr.

If polonium ($Z = 84$) is the radioactive species, the daughter element is lead, with $Z = 82$. Using for the disintegration energy $E = 8.95$ MeV, we obtain for the transmission factor

$$\exp\{-4\pi(82)\sqrt{(0.0993/8.95)} + 8\sqrt{82(9.00/7.25)}\}$$
$$= \exp\{-27.825\} = 8.2 \times 10^{-13}$$

Assuming f is unchanged at 10^{21} collisions per second, we get for this case $\lambda = 8.2 \times 10^{8}$ alphas per second and a half-life

$$t_{1/2} = \frac{0.693}{8.2 \times 10^8} = 8.4 \times 10^{-10} \text{ s}$$

The measured half-life of polonium is 3.0×10^{-7} s.

Given the crudeness of our method, both estimates should be considered satisfactory. Further, the calculations show clearly how a factor of only 2 in disintegration energy leads to half-lives differing by more than 26 orders of magnitude!

The radioactive decay process also can be understood in terms of the time evolution of a nonstationary state, in this case one representing the α particle initially confined to the parent nucleus. Solving the Schrödinger equation for the time-dependent waveform in this instance is complicated, making numerical studies the option of choice here. The interested reader is referred to our companion Web site for further details and a fully quantum-mechanical simulation of α decay from an unstable nucleus. Go to http://info.brookscole.com/mp3e, select QMTools Simulations → Leaky Wells (Tutorial) and follow the on-site instructions.

Ammonia Inversion

The "inversion" of the ammonia molecule is another example of tunneling, this time for an entire atom. The equilibrium configuration of the ammonia (NH_3) molecule is shown in Figure 7.10a: The nitrogen atom is situated at the apex of a pyramid whose base is the equilateral triangle formed by the three hydrogen atoms. But this equilibrium is not truly stable; indeed, there is a second equilibrium position for the nitrogen atom on the opposite side of the plane formed by the hydrogen atoms. With its two equilibrium locations, the nitrogen atom of the ammonia molecule constitutes a **double oscillator,** which can be modeled by using the potential shown in Figure 7.10b. A nitrogen atom initially located on one side of the symmetry plane will not remain there indefinitely, since there is some probability that it can tunnel through the oscillator barrier to emerge on the other side. When this occurs, the molecule becomes *inverted* (Fig. 7.10c). But the process does not stop there; the nitrogen atom, now on the opposite side of the symmetry plane, has a probability of tunneling back through the barrier to take up its original position! The molecule does not just undergo one inversion, but flip-flops repeatedly, alternating between the two classical equilibrium configurations. The "flopping" frequency is fixed by the tunneling rate and turns out to be quite high, on the order of 10^{10} Hz (microwave range of the electromagnetic spectrum)!

> **Double oscillator representation of the ammonia molecule**

We can estimate the tunneling probability for inversion using Equation 7.10. The double oscillator potential of Figure 7.10b is described by the potential energy function

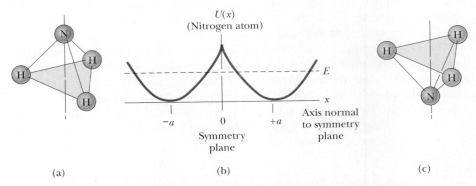

(a)   (b)   (c)

Figure 7.10   (a) The ammonia molecule NH_3. At equilibrium, the nitrogen atom is situated at the apex of a pyramid whose base is the equilateral triangle formed by the three hydrogen atoms. By symmetry, a second equilibrium configuration exists for the nitrogen atom on the opposite side of the plane formed by the hydrogen atoms. (b) The potential energy seen by the nitrogen atom along a line perpendicular to the symmetry plane. The two equilibrium points at $-a$ and $+a$ give rise to the double oscillator potential shown. A nitrogen atom with energy E can tunnel back and forth through the barrier from one equilibrium point to the other, with the result that the molecule alternates between the normal configuration in (a) and the inverted configuration shown in (c).

$$U(x) = \tfrac{1}{2} M\omega^2 (|x| - a)^2 \qquad (7.15)$$

with ω the classical frequency of vibration for the nitrogen atom around either of the equilibrium points $x = \pm a$. We suppose the nitrogen atom possesses the minimum energy of vibration, $E = \tfrac{1}{2}\hbar\omega$. There are four classical turning points for which $U(x) = E$ (see Fig. 7.10b); the limits for the tunneling integral are the pair closest to $x = 0$, and these are given by $x = \pm(a - A)$, where A is the vibration amplitude for the nitrogen atom in a single oscillator well. The vibration amplitude A is found from energy conservation by recognizing that here all the energy is in potential form: $\tfrac{1}{2}\hbar\omega = \tfrac{1}{2} M\omega^2 A^2$, or

$$A = \sqrt{\frac{\hbar}{M\omega}} \qquad (7.16)$$

Since the tunneling integral is symmetric about $x = 0$, we may write it as

$$\frac{2}{\hbar} \sqrt{2m} \int_{-(a-A)}^{a-A} \sqrt{U(x) - E}\, dx = \frac{4}{A^2} \int_0^{a-A} \sqrt{(x-a)^2 - A^2}\, dx$$

where $U(x)$ and E have been expressed in terms of A, using Equation 7.16. The integral on the right can be evaluated in terms of hyperbolic functions,[3] with the result

$$\frac{4}{A^2} \int_0^{a-A} \sqrt{(x-a)^2 - A^2}\, dx = \sinh(2y_0) - 2y_0$$

where y_0 is defined by the relation $\cosh(y_0) = a/A$. The transmission coefficient is then

$$T = e^{-[\sinh(2y_0) - 2y_0]} \qquad (7.17)$$

To get the tunneling rate λ (and its reciprocal—the tunneling time), we must multiply T by the frequency with which the nitrogen atom collides with the potential barrier. For an atom vibrating about its equilibrium position, this is the vibration frequency $f = \omega/2\pi$. From Equation 7.16, we see that f is related to the vibration amplitude as $f = \hbar/2\pi MA^2$.

The tunneling rate depends sensitively on the values chosen for a and A. For the equilibrium distance from the symmetry plane, we take $a = 0.38$ Å, an experimental value obtained from x-ray diffraction measurements.[4] The vibration amplitude A is not directly observable, but its value can be calculated from $U(0)$, the height of the potential barrier at $x = 0$, which is known to be 0.25690 eV. Using Equations 7.15 and 7.16, and taking $M = 14$ u for the mass of the nitrogen atom, we find

$$A = \left(\frac{\hbar^2 a^2}{2MU(0)} \right)^{1/4} = \left(\frac{(1.973 \text{ keV} \cdot \text{Å}/c)^2 (0.38\,\text{Å})^2}{2(14)(931.50 \times 10^3 \text{ keV}/c^2)(0.2569 \times 10^{-3} \text{ keV})} \right)^{1/4}$$

$$= 0.096\,\text{Å}$$

---

[3]Introduce a new integration variable y with the substitution $x - a = -A \cosh(y)$, and use the properties of the hyperbolic cosine and sine functions, $\cosh(y) = \tfrac{1}{2}(e^y + e^{-y})$, $\sinh(y) = \tfrac{1}{2}(e^y - e^{-y})$, to obtain the final form.

[4]B. H. Bransden and C. J. Joachain, *Physics of Atoms and Molecules*, New York, John Wiley and Sons, Inc., 1983, p. 456.

This underestimates the true value for A because we have (incorrectly) identified M as the mass of the nitrogen atom. In fact, M should be the *reduced mass* of the nitrogen–hydrogen group, about 2.47 u.[5] With this correction, we find $A = 0.148$ Å and a tunneling rate $\lambda = fT \approx 2.4 \times 10^{12}$ Hz. The observed tunneling rate, 2.4×10^{10} Hz, suggests a somewhat smaller value for A. By trial and error, we find the actual tunneling rate is reproduced with $A = 0.125$ Å, still a reasonable figure for the vibration amplitude of the nitrogen atom in the ammonia molecule.

Notice that because of tunneling, the nitrogen atom on one side of the symmetry plane or the other does *not* constitute a stationary state of the ammonia molecule, since the probability for finding it there changes over time. In fact, the flopping behavior stems from a simple combination of two stationary states of nearly equal energy for the nitrogen atom in this environment. Such superpositions of closely spaced (in energy) stationary states have applications that transcend this one example and are the subject of a computer-based tutorial available at our companion Web site. For more information, go to http://info.brookscole.com/mp3e, select QMTools Simulations → Two-Center Potentials (Tutorial), and follow the on-site instructions.

Since the flopping frequency is in the microwave range, the ammonia molecule can serve as an amplifier for microwave radiation. The **ammonia maser** operates on this principle. Because of the small energy difference between the ground and first excited states of the ammonia molecule, ammonia vapor at room temperature has roughly equal numbers of molecules in both states. Having opposite electric dipole moments, these states are easily separated by passing the vapor through a nonuniform electric field. In this way, ammonia vapor can be produced with the unusually large concentrations of excited molecules needed to create the population inversion necessary for maser operation. A spontaneous deexcitation to the ground state of one molecule releases a (microwave) photon, which, in turn, induces other molecules to deexcite. The result—much like a chain reaction—produces a photon cascade: an intense burst of coherent microwave radiation. The operation of masers and lasers is discussed in more detail in Chapter 12 and on our website at http://info.brookscole.com/mp3e.

Decay of Black Holes

Once inside the event horizon, nothing—not even light—can escape the gravitational pull of a black hole.[6] That was the view held until 1974, when the brilliant British astrophysicist Stephen Hawking proposed that black holes are indeed radiant objects, emitting a variety of particles by a mechanism involving tunneling through the (gravitational) potential barrier surrounding the black hole. The thickness of this barrier is proportional to the size of the black hole, so that the likelihood of a tunneling event initially may be extremely small. As the black hole emits particles, however, its mass and size steadily

---

[5]In this mode of vibration, all three hydrogen atoms move in unison as if they were a single object with mass 3 u. The reduced mass refers to the pair consisting of this total mass and the mass of the nitrogen atom (14 u).

[6]A brief introduction to black holes is found in Clifford Will's essay "The Renaissance of General Relativity" on our companion Web site.

Stephen W. Hawking (1942–   ), a British astrophysicist, has made major contributions to problems in cosmology, including the theory of black holes. (*MMP/Manni Mason's Pictures*)

decrease, making it easier for more particles to tunnel out. In this way emission continues at an ever-increasing rate, until eventually the black hole radiates itself out of existence in an explosive climax! Thus, Hawking's scenario leads inexorably to the decay and eventual demise of any black hole.

Calculations indicate that a black hole with the mass of our Sun would survive against decay by tunneling for about 10^{66} years. On the other hand, a black hole with the mass of only a billion tons and roughly the size of a proton (such mini black holes are believed to have been formed just after the Big Bang origin of the Universe) should have almost completely evaporated in the 10 billion years that have elapsed since the time of creation, and black holes a few times heavier should still be evaporating strongly. A large portion of the energy emitted by such holes would be in the form of gamma rays. Indeed, gamma rays from interstellar space have been observed, but in quantities and with properties that are readily explained in other ways. Currently there is no compelling observational evidence of black-hole evaporation in the Universe today.

SUMMARY

For potentials representing barriers, the stationary states are not localized, but extend throughout the entire space in a manner that describes particle scattering. When a matter wave encounters a potential barrier, part of the wave is reflected by the barrier and part is transmitted through the barrier. In particle language, an object colliding with the barrier does not predictably rebound or penetrate, but can only be assigned probabilities for reflection and transmission. These probabilities are given by the **reflection coefficient R** and **transmission coefficient T,** respectively. For any particle energy E, these two scattering coefficients must obey the **sum rule:**

$$R(E) + T(E) = 1 \qquad (7.5)$$

expressing the fact that the particle either is reflected or transmitted. Due to its wave nature, a particle has some nonzero probability of penetrating the barrier even when it has insufficient energy to do so on classical grounds. This process, called **tunneling,** is the basic mechanism underlying the phenomena of **field emission**, **α decay** of radioactive nuclei, **ammonia inversion, Hawking radiation** from black holes, and many more. The likelihood of a particle with mass m and energy E tunneling through a barrier of arbitrary shape $U(x)$ is given approximately as

$$T(E) \approx \exp\left(-\frac{2}{\hbar}\sqrt{2m}\int \sqrt{U(x) - E}\,dx\right) \qquad (7.10)$$

The integral in Equation 7.10 is taken over the range of x where $U(x) > E$, called the **classically forbidden region** because a classical particle in this interval would have to have a negative value of kinetic energy (an impossibility!).

SUGGESTIONS FOR FURTHER READING

1. Many of the topics in this chapter also are treated at about the same level by A. P. French and E. F. Taylor in *An Introduction to Quantum Physics*, New York, W. W. Norton and Company, Inc., 1978.
2. Experimental aspects of field emission microscopy and related analytical surface techniques are discussed by P. F. Kane in Chapter 6 of *Characterization of Solid Surfaces*, New York, Plenum Press, 1974, pp. 133–146.
3. For readable accounts of tunneling from black holes, see S. W. Hawking, "The Quantum Mechanics of Black Holes," *Sci. Am.*, January 1977, pp. 34–40; and J. D. Beckenstein, "Black-hole Thermodynamics," *Phys. Today*, January 1980, pp. 24–31. For a fascinating historical perspective of these discoveries in the making, written by one of the participants, see K. Thorne's *Black Holes and Time Warps*, New York, W. W. Norton & Company, Inc., 1994.

QUESTIONS

1. Consider a particle with energy E scattered from a potential barrier of height $U > E$. How does the amplitude of the reflected wave change as the barrier height is reduced? How does the amplitude of the incident wave change?
2. An electron and a proton of identical energy encounter the same potential barrier. For which is the probability of transmission greatest, and why?
3. In classical physics, only *differences* in energy have physical significance. Discuss how this carries over to the quantum scattering of particles by examining how adding a constant potential U_0 everywhere affects barrier reflection and transmission coefficients.
4. Suppose a particle with energy E is located within a barrier whose height U is greater than E. Will this particle be found to have a negative kinetic energy? Explain.
5. Explain how a barrier whose width is L might be considered wide for penetration by protons, yet at the same time narrow for penetration by electrons.
6. Discuss the suitability of portraying wavefunctions and potential barriers on the same graph, as in Figure 7.2a. Can a wave whose crest falls below the top of a square barrier ever penetrate the barrier? Explain.

PROBLEMS

7.1 The Square Barrier

1. A particle incident on the potential step of Example 7.4 with a certain energy $E < U$ is described by the wave

$$\psi(x) = \tfrac{1}{2}\{(1 + i)\,e^{ikx} + (1-i)\,e^{-ikx}\} \qquad \text{for } x \le 0$$

$$\psi(x) = e^{-kx} \qquad \text{for } x \ge 0$$

(a) Verify by direct calculation that the reflection coefficient is unity in this case. (b) How must k be related to E in order for $\psi(x)$ to solve Schrödinger's equation in the region to the left of the step ($x \le 0$)? to the right of the step ($x > 0$)? What does this say about the ratio E/U? (c) Evaluate the penetration depth $\delta = 1/k$ for 10 MeV protons incident on this step.

2. Consider the step potential of Example 7.4 in the case where $E > U$. (a) Examine the Schrödinger equation to the left of the step to find the form of the solution in the range $x < 0$. Do the same to the right of the step to obtain the solution form for $x > 0$. Complete the solution by enforcing whatever boundary and matching conditions may be necessary. (b) Obtain an expression for the reflection coefficient R in this case, and show that it can be written in the form

$$R = \frac{(k_1 - k_2)^2}{(k_1 + k_2)^2}$$

where k_1 and k_2 are wavenumbers for the incident and transmitted waves, respectively. Also write an expression for the transmission factor T using the sum rule obeyed by these coefficients. (c) Evaluate R and T in the limiting cases of $E \to U$ and $E \to \infty$. Are the results sensible? Explain. (This situation is analogous to the partial reflection and transmission of light striking an interface separating two different media.)

3. Use the results of the preceding problem to calculate the fraction of 25-MeV protons reflected and the fraction transmitted by a 20-MeV step. How do your answers change if the protons are replaced by electrons?

4. A 0.100-mA electron beam with kinetic energy 54.0 eV enters a sharply defined region of *lower* potential where the kinetic energy of the electrons is increased by 10.0 eV. What current is reflected at the boundary? (This simulates electron scattering at normal incidence from a metal surface, as in the Davisson–Germer experiment.)

5. (a) Tunneling of particles through barriers that are high or wide (or both) is very unlikely. Show that for a square barrier with

$$\frac{2mUL^2}{\hbar^2} \gg 1$$

and particle energies well below the top of the barrier ($E \ll U$) the probability for transmission is approximately

$$P \approx 16\,\frac{E}{U}\,e^{-2[\sqrt{2m(U-E)}/\hbar]L}$$

(The combination UL^2 is sometimes referred to as the barrier *strength*.) (b) Give numerical estimates for the exponential factor in P for each of the following cases: (1) an electron with $U - E = 0.01$ eV and $L = 0.1$ nm; (2) an electron with $U - E = 1$ eV and $L = 0.1$ nm; (3) an α particle ($m = 6.7 \times 10^{-27}$ kg) with $U - E = 10^6$ eV and $L = 10^{-15}$ m; and (4) a bowling ball ($m = 8$ kg) with $U - E = 1$ J and $L = 2$ cm (this corresponds to the ball's getting past a barrier 2 cm wide and too high for the ball to slide over).

6. A beam of electrons is incident on a barrier 5 eV high and 1 nm wide. Write a simple computer program to find what energy the electrons should have if 0.1% of them are to get through the barrier.

7. Starting from the joining conditions, Equations 7.8, obtain the result for the transmission coefficient of a square barrier given in Equation 7.9 (valid when the particle has insufficient energy to penetrate the barrier classically: $E < U$).

8. ◆ Use the Java applet available at our companion Web site (http://info.brookscole.com/ mp3e QMTools Simulations → Problem 7.8) to investigate the scattering of electrons from a square barrier 1.00 Å thick and 10.0 eV high, in the case where the electron energy is equal to the barrier height, $E = U$. What is the functional form of the wave in the barrier region? Determine the transmission coefficient at this energy, and compare your result with the prediction of Equation 7.9. What does classical physics predict for the probability of transmission in this case?

9. ◆ Use the Java applet referenced in the preceding problem to obtain transmission and reflection coefficients for a 5.00-eV electron incident on a square barrier that is 1.00 Å thick and 10.0 eV high. Verify the sum rule, Equation 7.5, and compare your result for $T(E)$ with the prediction of Equation 7.9. What must the barrier thickness be to transmit 5.00-eV *protons* with the same probability?

10. ◆ *Scattering resonances.* Use the Java applet available at our companion Web site (http://info. brookscole.com/mp3e QMTools Simulations → Problem 7.10) to locate the two lowest energies E giving rise to perfect transmission for electrons scattering from a square barrier of width 1.00 Å and height 10.0 eV (look for zero-amplitude reflections while varying E). For each energy use the "Trace" feature to estimate the electron wavelength λ in the barrier. (*Hint*: Zoom in for a close-up view of the barrier waveform and greater accuracy.) Compare λ with the barrier width L and dis-

cuss your findings in terms of the interference of waves reflected from the leading and trailing edges of the barrier (see Example 7.3).

11. *The Ramsauer–Townsend effect.* Consider the scattering of particles from the potential well shown in Figure P7.11. (a) Explain why the waves reflected from the well edges $x = 0$ and $x = L$ will cancel completely if $2L = \lambda_2$, where λ_2 is the de Broglie wavelength of the particle in region 2. (b) Write expressions for the wavefunctions in regions 1, 2, and 3. Impose the necessary continuity restrictions on Ψ and $\partial\Psi/\partial x$ to show explicitly that $2L = \lambda_2$ leads to no reflected wave in region 1. [This is a crude model for the Ramsauer–Townsend effect observed in the collisions of slow electrons with noble gas atoms like argon, krypton, and xenon. Electrons with just the right energy are diffracted around these atoms as if there were no obstacle in their path (perfect transmission). The effect is peculiar to the noble gases because their closed-shell configurations produce atoms with abrupt outer boundaries.]

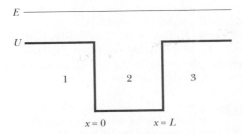

Figure P7.11

12. A potential model of interest for its simplicity is the *delta well*. The delta well may be thought of as a square well of width L and depth S/L in the limit $L \to 0$ (Fig. P7.12). The limit is such that S, the product of the well depth with its width, remains fixed at a finite value known as the well strength. The effect of a delta well is to introduce a discontinuity in the slope of the wavefunction at the well site, although the wave itself remains continuous here. In particular, it can be shown that

$$\left.\frac{d\psi}{dx}\right|_{0+} - \left.\frac{d\psi}{dx}\right|_{0-} = -\frac{2mS}{\hbar^2}\psi(0)$$

for a delta well of strength S situated at $x = 0$. (a) Solve Schrödinger's equation on both sides of the well ($x < 0$ and $x > 0$) for the case where particles are incident from the left with energy $E > 0$. Note that in these regions the particles are free, so that $U(x) = 0$. (b) Enforce the continuity of ψ and the slope condition at $x = 0$. Solve the resulting equations to obtain the transmission coefficient T as a function of particle energy E. Sketch $T(E)$ for $E \geq 0$. (c) If we allow E to be

negative, we find that $T(E)$ diverges for some particular energy E_0. Find this value E_0. (As it happens, E_0 is the energy of a *bound state* in the delta well. The calculation illustrates a general technique, in which bound states are sought among the singularities of the scattering coefficients for a potential well of arbitrary shape.) (d) What fraction of the particles incident on the well with energy $E = |E_0|$ is transmitted and what fraction is reflected?

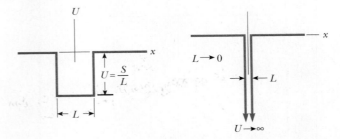

Figure P7.12

13. Obtain directly an expression for the reflection coefficient $R(E)$ for the delta well of Problem 12, and verify the sum rule

$$R(E) + T(E) = 1$$

for all particle energies $E > 0$.

14. Keeping constant speed 0.8 m/s, a marble rolls back and forth inside a shoebox. Make an order-of-magnitude estimate of the probability of its escaping through the wall of the box by quantum tunneling. State the quantities you take as data and the values you measure or estimate for them.

7.2   Barrier Penetration

15. A barrier of arbitrary shape can be approximated as a succession of square barriers, as shown in Figure P7.15. Write the transmission coefficient for this barrier using the result of Equation 7.9 for each of the individual barriers, assuming the transmitted wave intensity for one becomes the incident wave intensity for the barrier immediately following it in the series. Show that the form of Equation 7.10 is recovered in the case where $E < U$ and $\alpha L \gg 1$.

16. Consider an α particle confined to a thorium nucleus. Model the nuclear potential as a semi-infinite square well with an infinitely high wall at $r = 0$ and a wall of height 30.0 MeV at the nuclear radius $R = 9.00$ fm. Use the iterative method described in Example 6.8 to estimate the smallest values of energy and velocity permitted for the α particle. What conclusion can you draw from the fact that the *ejected* α is observed to have a kinetic energy of 4.05 MeV?

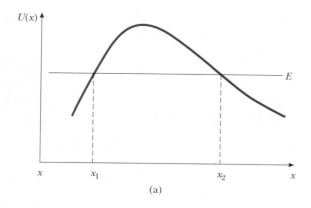

(a)

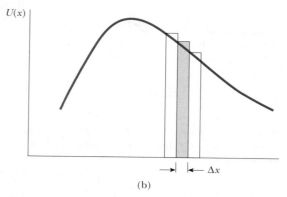

(b)

Figure P7.15

17. The attempt frequency of an α particle to escape the nucleus is the number of times per second it collides with the nuclear barrier. Estimate this collision frequency in the tunneling model for the α decay of thorium, assuming the α behaves like a true particle inside the nucleus with *total* energy equal to the observed kinetic energy of decay. The daughter nucleus for this case (radium) has $Z = 88$ and a radius of 9.00 fm. Take for the overall nuclear barrier 30.0 MeV, measured from the bottom of the nuclear well to the top of the Coulomb barrier (see Fig. 7.8).

18. Verify the claim of Section 7.2 that the electrons of a metal collide with the surface at a rate of about 10^{30} per second per square centimeter. Do this by estimating the collision frequency of electrons in a 1.00-cm cube of copper metal with one face of the cube surface. Assume that each copper atom contributes one conduction electron to the metal (the chemical valence of copper is $+1$) and that these conduction electrons move freely

with kinetic energy equal to 7.00 eV. In fact, not all the electrons have this energy; see Chapter 10.

19.   *Resonant tunneling.* Heterostructures formed from layered semiconductors have characteristics important to many modern electronic devices. Here, we use computer simulation to study tunneling in a three-layer gallium arsenide/gallium aluminum arsenide ($GaAs-Ga_{1-x}Al_xAs$) sandwich. The GaAs layer constitutes a potential well between two confining barriers formed by the $Ga_{1-x}Al_xAs$ layers. Unusually large transmission (resonant tunneling) through the device occurs when the energy of the incident electron coincides with that of a bound state in the central well. The Java applet simulating this device can be found at http://info.brookscole.com/mp3e QMTools Simulations → Problem 7.19. The barriers are 0.25 eV high and 5.0 nm wide, with a gap of equal width separating them. Note that electrons in these materials behave like free electrons with an effective mass $m^* = 0.067m_e$, only a fraction of the free electron value. Starting from $E = 0$, gradually increase the electron energy to find the lowest value for peak transmission. Investigate the width of the resonance by varying the electron energy further until $T(E)$ falls to half of its peak value. (In practice, the incident electron energy is fixed and the device is "tuned" to resonance by applying a suitable bias voltage that alters the bound-state energies of the central well.)

20.   *Ammonia inversion.* Inversion of the ammonia molecule can be simulated using the Java applet available at our companion Web site (http://info.brookscole.com/mp3e QMTools Simulations → Problem 7.20) The potential energy is the double oscillator of Equation 7.15 with parameter values chosen to model the nitrogen atom in NH_3 (as discussed in the text) and a reduced mass of 2.47 u for the atom in this environment. (a) Find and display the two lowest-lying stationary states of the nitrogen atom in the ammonia molecule. Describe the appearance of these waveforms (symmetry, number of nodes, and so on). (b) Construct an initial (nonstationary) state for the atom by mixing together these two stationary waves with equal amplitude. Describe this state. What does it imply for the location of the atom initially? (c) Explore the time evolution of the state constructed in (b). Verify that the atom flip-flops between the two equilibrium positions and determine the "flopping" frequency. Multiplying the flopping frequency by Planck's constant gives a characteristic energy for this process. How does this characteristic energy compare to the energy separation of the stationary states? Explain (see Problem 6.38).

Essay

THE SCANNING TUNNELING MICROSCOPE

The basic idea of quantum mechanics, that particles have properties of waves and vice versa, is among the strangest found anywhere in science. Because of this strangeness, and because quantum mechanics mostly deals with the very small, it might seem to have little practical application. As we will show in this essay, however, one of the basic phenomena of quantum mechanics—the tunneling of particles—is at the heart of a very practical device that is one of the most powerful microscopes ever built. This device, the scanning tunneling microscope, or STM, enables physicists to make highly detailed images of surfaces with resolution comparable to the size of a single atom. Such images promise to revolutionize our understanding of structures and processes on the atomic scale.

Before discussing how the STM works, we first look at a sample of what the STM can do. An image made by a scanning tunneling microscope of the surface of a piece of gold is shown in Figure 1. You can easily see that the surface is not uniformly flat, but is a series of terraces separated by steps that are only one atom high. Gentle corrugations can be seen in the terraces, caused by subtle rearrangements of the gold atoms.

Roger A. Freedman

Paul K. Hansma

Department of Physics,
University of California,
Santa Barbara

Figure 1 Scanning tunneling microscope image of the surface of crystalline gold. Successive scans are approximately 1.5 Å apart. The figure is from G. Binnig, H. Rohrer, Ch. Gerber, and E. Stoll, *Surface Sci.* 144:321, 1984.

What makes the STM so remarkable is the fineness of the detail that can be seen in images such as Figure 1. The *resolution* in this image—that is, the size of the smallest detail that can be discerned—is about 2 Å (2 × 10⁻¹⁰ m). For an ordinary microscope, the resolution is limited by the wavelength of the waves used to make the image. Thus an optical microscope has a resolution of no better than 2000 Å, about half the wavelength of visible light, and so could never show the detail displayed in Figure 1. Electron microscopes can have a resolution of 2 Å by using electron waves of wavelength 4 Å or shorter. From the de Broglie formula, $\lambda = h/p$, the electron momentum p required to give this wavelength is 3100 eV/c, corresponding to an electron speed $v = p/m_e = 1.8 \times 10^6$ m/s. Electrons traveling at this high speed would penetrate into the interior of the piece of gold in Figure 1 and so would give no information about individual surface atoms.

The image in Figure 1 was made by Gerd Binnig, Heinrich Rohrer, and collaborators at the IBM Research Laboratory in Zurich, Switzerland. Binnig and Rohrer invented the STM and shared the 1986 Nobel prize in physics for their work. Such is the importance of this device that unlike most Nobel prizes, which come decades after the original work, Binnig and Rohrer received their Nobel prize just six years after their first experiments with an STM.

One design for an STM is shown in Figure 2. The basic idea behind its operation is very simple, as shown in Figure 3. A conducting probe with a very sharp tip is brought near the surface to be studied. Because it is attracted to the positive ions in the surface, an electron in the surface has a lower total energy than would an electron in the empty space between surface and tip. The same thing is true for an electron in the tip. In clas-

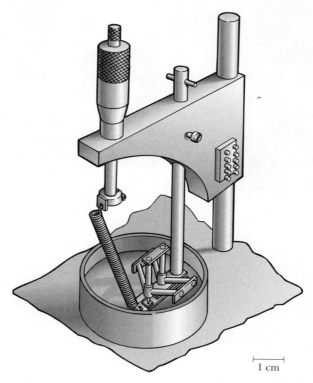

1 cm

Figure 2   One design for a scanning tunneling microscope (STM). The sample to be studied is mounted on a plate in the cylindrical dish. The probe extends beneath the left tripod. The micrometer attached to the spring is used to position the sample.

sical Newtonian mechanics, electrons could not move between the surface and tip because they would lack the energy to escape either material. But because the electrons obey quantum mechanics, they can "tunnel" across the barrier of empty space between the surface and the tip. Let us explore the operation of the STM in terms of the discussion of tunneling in Section 7.1.

For an electron in the apparatus of Figures 2 and 3, a plot of the energy as a function of position would look like Figure 7.2b. The horizontal coordinate in this figure represents electron position. Now L is to be interpreted as the distance between the surface and the tip, so that coordinates less than 0 refer to positions inside the surface material and coordinates greater than L refer to positions inside the tip. The barrier height $U = q\phi$ is the potential-energy difference between an electron outside the material and an electron in the material. That is, an electron in the surface or tip has potential energy $-U$ compared with one in vacuum. (We are assuming for the moment that the surface and tip are made of the same material. We will comment on this assumption shortly.) The kinetic energy of an electron in the surface is E, so that an amount of energy equal to $(U - E)$ must be given to an electron to remove it from the surface. Thus $(U - E)$ is the work function of an electron in the surface.

For the potential energy curve of Figure 7.2b, one could expect as much tunneling from the surface into the tip as in the opposite direction. In a STM, the direction in which electrons tend to cross the barrier is controlled by applying a voltage between the surface and the tip. With preferential tunneling from the surface into the tip, the tip samples the distribution of electrons in and above the surface. Because of this "bias" voltage, the work functions of surface and tip are different, giving a preferred direction of tunneling. This is also automatically the case if the surface and tip are made of different materials. In addition, the top of the barrier in Figure 7.2b will not be flat but will be tilted to reflect the electric field between the surface and the tip. If the barrier energy U is large compared with the difference between the surface and tip work functions, however, and if the bias voltage is small compared with $\phi = U/q$, we can ignore these complications in our calculations. Then all the results for a square barrier given in Section 7.1 may be applied to a STM. Detailed discussions of the effects that result when these complications are included can be found in the article by P. K. Hansma and J. Tersoff and in the Nobel prize lecture of Binnig and Rohrer (see Suggestions for Further Reading).

The characteristic scale of length for tunneling is set by the work function $(U - E)$. For a typical value $(U - E) = 4.0$ eV, this scale of length is

$$\delta = \frac{\hbar}{\sqrt{2m_e(U - E)}} = \frac{\hbar c}{\sqrt{2m_e c^2(U - E)}} = \frac{1.973 \text{ keV·Å}}{\sqrt{2(511 \text{ keV})(4.0 \times 10^{-3} \text{ keV})}}$$

$$= 0.98 \text{ Å} \approx 1.0 \text{ Å}$$

The probability that a given electron will tunnel across the barrier is just the transmission coefficient T (Equation 7.9). If the separation L between surface and tip is not small compared with δ, then the barrier is "wide" and we can use the approximate result of Problem 5 for T. The current of electrons tunneling across the barrier is simply proportional to T. The tunneling current density can be shown to be

$$j = \frac{e^2 V}{4\pi^2 L\delta\hbar} e^{2L/\delta}$$

In this expression e is the charge of the electron and V is the bias voltage between surface and tip.

We can see from this expression that the STM is very sensitive to the separation L between tip and surface. This is because of the exponential dependence of the tunneling current on L (this is much more important than the $1/L$ dependence). As we saw earlier, typically $\delta \approx 1.0$ Å. Hence, increasing the distance L by just 0.01 Å causes the tunneling current to be multiplied by a factor $e^{-2(0.01 \text{ Å})/(1.0 \text{ Å})} \approx 0.98$; that is, the

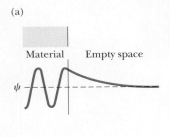

(a)

Material   Empty space

ψ

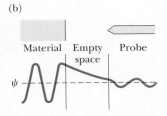

(b)

Material   Empty   Probe
space

ψ

Figure 3 (a) The wavefunction of an electron in the surface of the material to be studied. The wavefunction extends beyond the surface into the empty region. (b) The sharp tip of a conducting probe is brought close to the surface. The wavefunction of a surface electron penetrates into the tip, so that the electron can "tunnel" from surface to tip. Compare this figure to Figure 7.2a.

current decreases by 2%—a change that is measurable. For distance L greater than 10 Å (that is, beyond a few atomic diameters), essentially no tunneling takes place. This sensitivity to L is the basis of the operation of the STM: Monitoring the tunneling current as the tip is scanned over the surface gives a sensitive measure of the topography of the surface. In this way the STM can measure the height of surface features to within 0.01 Å, or approximately one one-hundredth of an atomic diameter.

The STM also has excellent lateral resolution, that is, resolution of features in the plane of the surface. This is because the tips used are *very* sharp indeed, typically only an atom or two wide at their extreme end. Thus the tip samples the surface electrons only in a very tiny region approximately 2 Å wide and so can "see" very fine detail. You might think that making such tips would be extremely difficult, but in fact it's relatively easy—sometimes just sharpening the tip on a fine grinding stone (or even with fine sandpaper) is enough to cause the tip atoms to rearrange by themselves into an atomically sharp configuration. (If you find this surprising, you're not alone. Binnig and Rohrer were no less surprised when they discovered this.)

There are two modes of operation for the STM, shown in Figure 4. In the **constant current mode** (Fig. 4a), a convenient operating voltage (typically between 2 mV and 2 V) is first established between surface and tip. The tip is then brought close enough to the surface to obtain measurable tunneling current. The tip is then scanned over the surface

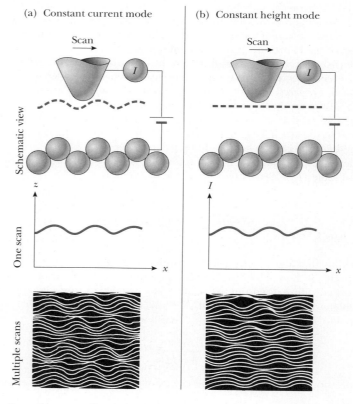

Figure 4   Scanning tunneling microscopes can be operated in either (a) the constant current mode or (b) the constant height mode. The images of the surface of graphite were made by Richard Sonnenfeld at the University of California at Santa Barbara. The constant height mode was first used by A. Bryant, D. P. E. Smith, and C. F. Quate, *Appl. Phys. Lett.* 48:832, 1986.

while the tunneling current I is measured. A feedback network changes the vertical position of the tip, z, to keep the tunneling current constant, thereby keeping the separation L between surface and tip constant. An image of the surface is made by plotting z versus lateral position (x, y). The simplest scheme for plotting the image is shown in the graph below the schematic view. The height z is plotted versus the scan position x. An image consists of multiple scans displaced laterally from each other in the y direction.

The constant current mode was historically the first to be used and has the advantage that it can be used to track surfaces that are not atomically flat (as in Fig. 1). The feedback network, however, requires that the scanning be done relatively slowly. As a result, the sample being scanned must be held fixed in place for relatively long times to prevent image distortion.

Alternatively, in the **constant height mode** (Fig. 4b), the tip is scanned across the surface at constant voltage and nearly constant height while the current is monitored. In this case the feedback network responds only rapidly enough to keep the average current constant, which means that the tip maintains the same average separation from the surface. The image is then a plot of current I versus lateral position (x, y), as shown in the graph below the schematic. Again, multiple scans along x are displayed laterally displaced in the y direction. The image shows the substantial variation of tunneling current as the tip passes over surface features such as individual atoms. The constant height mode allows much faster scanning of atomically flat surfaces (100 times faster than the constant current mode), since the tip does not have to be moved up and down over the surface "terrain." This fast scanning means that making an image of a surface requires only a short "exposure time." By making a sequence of such images, researchers may be able to study in real time processes in which the surfaces rearrange themselves—in effect making an STM "movie."

Individual atoms have been imaged on a variety of surfaces, including those of so-called *layered materials* in which atoms are naturally arranged into two-dimensional layers. Figure 5 shows an example of atoms on one of these layered materials. In this image it is fascinating not only to see individual atoms but also to note that some atoms are missing. Specifically, there are three atoms missing from Figure 5. Can you find the places where they belong?

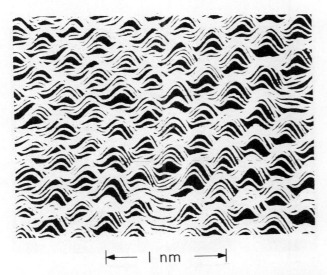

$\longmapsto$ 1 nm $\longmapsto$

Figure 5 Image of atoms on a surface of tantalum disulfide (TaS_2) immersed in liquid nitrogen; 1 nm = 10^{-9} m = 10 Å. The figure is from C. G. Slough, W. W. McNairy, R. V. Coleman, B. Drake, and P. K. Hansma, *Phys. Rev.* B34:994, 1986.

Another remarkable aspect of the STM image in Figure 5 is that it was obtained with the surface and tip immersed in liquid nitrogen. We assumed earlier in this essay that the space between the surface and tip must be empty, but in fact electron tunneling can take place not just through vacuum but also through gases and liquids—even water. This seems very surprising since we think of water, especially water with salts dissolved in it, as a conductor. But water is only an *ionic* conductor. For electrons, water behaves as an insulator just as vacuum behaves as an insulator. Thus electrons can flow through water only by tunneling, which makes scanning tunneling microscopy possible "under water."

As an example, Figure 6 shows individual carbon atoms on a graphite surface. It was obtained for a surface immersed in a silver-plating solution, which is highly conductive for ions but behaves as an insulator for electrons. (The sides of the conducting probe were sheathed with a nonconductor, so the predominant current into the probe comes from electrons tunneling into the exposed tip. The design of STM used to make this particular image is the one shown in Fig. 2.) Sonnenfeld and Schardt observed atoms on this graphite surface before plating it with silver, after "islands" of silver atoms were plated onto the surface, and after the silver was electrochemically stripped from the surface. Their work illustrates the promise of the scanning tunneling microscope for seeing processes that take place on an atomic scale.

The original STMs were one-of-a-kind laboratory devices, but commercial STMs have recently become available. Figure 7 is an image of a graphite surface in air made with such a commercial STM. Note the high quality of this image and the recognizable rings of carbon atoms. You may be able to see that three of the six carbon atoms in each ring *appear* lower than the other three. All six atoms are in fact at the same level, but the three that appear lower are bonded to carbon atoms lying directly beneath them in the underlying atomic layer. The atoms in the surface layer that appear higher do not lie directly over subsurface atoms and hence are not bonded to carbon atoms beneath them. For the higher-appearing atoms, some of the electron density that would have been involved in bonding to atoms beneath the surface instead extends into the space above the surface. This extra electron density makes these atoms appear higher in Figure 7, since what the STM maps is the topography of the electron distribution.

The availability of commercial instruments should speed the use of scanning tunneling microscopy in a variety of applications. These include characterizing electrodes for

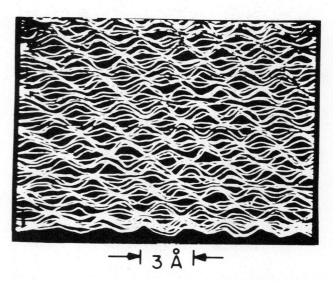

$$\vdash\!\!\dashv\ 3\ \overset{\circ}{A}\ \vdash\!\!\dashv$$

Figure 6   Image of a graphite electrode in an electrolyte used for silver plating. The figure is from R. Sonnenfeld and B. Schardt, *Appl. Phys. Lett.* 49:1172, 1986.

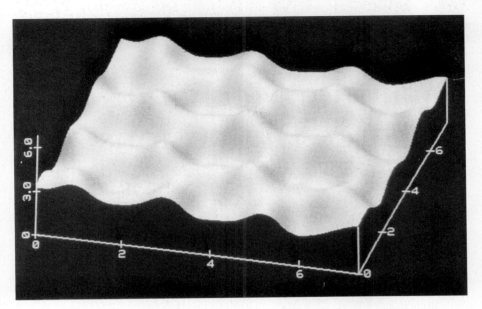

Figure 7 Image of a graphite surface in air obtained with a commercial STM, the Nanoscope II from Digital Instruments in Goleta, California.

electrochemistry (while the electrode is still in the electrolyte), characterizing the roughness of surfaces, measuring the quality of optical gratings, and even imaging replicas of biological structures.

Perhaps the most remarkable thing about the scanning tunneling microscope is that its operation is based on a quantum mechanical phenomenon—tunneling—that was well understood in the 1920s, yet the STM itself wasn't built until the 1980s. What other applications of quantum mechanics may yet be waiting to be discovered?

Suggestions for Further Reading

G. Binnig, H. Rohrer, Ch. Gerber, and E. Weibel, *Phys. Rev. Lett.* 49:57, 1982. The first description of the operation of a scanning tunneling microscope.

G. Binnig and H. Rohrer, *Sci. Am.*, August 1985, p. 50. A popular description of the STM and its applications.

C. F. Quate, *Phys. Today*, August 1986, p. 26. An overview of the field of scanning tunneling microscopy, including insights into how it came to be developed.

P. K. Hansma and J. Tersoff, *J. Appl. Phys.* 61:R1, 1987. A comprehensive review of the "state of the art" in scanning tunneling microscopy.

G. Binnig and H. Rohrer, *Rev. Mod. Phys.* 59:615, 1987. The text of the lecture given on the occasion of the presentation of the 1986 Nobel prize in physics.

8

Quantum Mechanics in Three Dimensions

Chapter Outline

So far we have shown how quantum mechanics can be used to describe motion in one dimension. Although the one-dimensional case illustrates such basic features of systems as the quantization of energy, we need a full three-dimensional treatment for the applications to atomic, solid-state, and nuclear physics that we will meet in later chapters. In this chapter we extend the concepts of quantum mechanics from one to three dimensions and explore the predictions of the theory for the simplest of real systems—the hydrogen atom.

With the introduction of new degrees of freedom (and the additional coordinates needed to describe them) comes a disproportionate increase in the level of mathematical difficulty. To guide our inquiry, we shall rely on classical insights to help us identify the observables that are likely candidates for quantization. These must come from the ranks of the so-called *sharp observables* and, with few exceptions, are the same as the constants of the classical motion.

8.1 PARTICLE IN A THREE-DIMENSIONAL BOX

Let us explore the workings of wave mechanics in three dimensions through the example of a particle confined to a cubic "box." The box has edge length L and occupies the region $0 < x, y, z < L$, as shown in Figure 8.1. We assume the walls of the box are *smooth*, so they exert forces only perpendicular to the surface, and that collisions with the walls are elastic. A classical particle would rattle around inside such a box, colliding with the walls. At each collision, the component of particle momentum normal to the wall is reversed (changes sign), while the other two components of momentum are unaffected (Fig. 8.2). Thus, the collisions preserve the magnitude of each momentum component, in addition to the total particle energy. These four quantities—$|p_x|$, $|p_y|$, $|p_z|$, and E—then, are constants of the classical motion, and it should be possible to find quantum states for which all of them are sharp.[1]

The wavefunction Ψ in three dimensions is a function of $\mathbf{r}$ and t. Again the magnitude of Ψ determines the probability density $P(\mathbf{r}, t) = |\Psi(\mathbf{r}, t)|^2$, which is now a *probability per unit volume.* Multiplication by the volume element $dV(=dx\,dy\,dz)$ gives the probability of finding the particle within the volume element dV at the point $\mathbf{r}$ at time t.

Since our particle is confined to the box, the wavefunction Ψ must be zero at the walls and outside. The wavefunction inside the box is found from Schrödinger's equation,

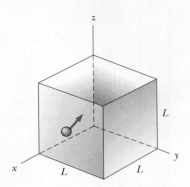

Figure 8.1 A particle confined to move in a cubic box of sides L. Inside the box $U = 0$. The potential energy is infinite at the walls and outside the box.

$$-\frac{\hbar^2}{2m}\nabla^2\Psi + U(\mathbf{r})\Psi = i\hbar\frac{\partial\Psi}{\partial t} \qquad (8.1)$$

We see that $\partial^2/\partial x^2$ in the one-dimensional case is replaced in three dimensions by the **Laplacian,**

$$\nabla^2 = \frac{\partial^2}{\partial x^2} + \frac{\partial^2}{\partial y^2} + \frac{\partial^2}{\partial z^2} \qquad (8.2)$$

Schrödinger equation in three dimensions

where $U(\mathbf{r})$ is still the potential energy but is now a function of all the space coordinates: $U(\mathbf{r}) = U(x, y, z)$. Indeed, the Laplacian together with its multiplying constant is just the kinetic energy operator of Table 6.2 extended to include the contributions to kinetic energy from motion in each of three mutually perpendicular directions:

$$-\frac{\hbar^2}{2m}\nabla^2 = \left(-\frac{\hbar^2}{2m}\frac{\partial^2}{\partial x^2}\right) + \left(-\frac{\hbar^2}{2m}\frac{\partial^2}{\partial y^2}\right) + \left(-\frac{\hbar^2}{2m}\frac{\partial^2}{\partial z^2}\right)$$
$$= [K_x] + [K_y] + [K_z] \qquad (8.3)$$

This form is consistent with our belief that the Cartesian axes identify independent but logically equivalent directions in space. With this identification of the kinetic energy operator in three dimensions, the left-hand side of Equation 8.1 is again the Hamiltonian operator $[H]$ applied to Ψ, and the right-hand side is the energy operator $[E]$ applied to Ψ (see Section 6.8). As it does in one dimension, Schrödinger's equation asserts the equivalence of these two operators when applied to the wavefunction of any physical system.

Stationary states are those for which all probabilities are constant in time, and are given by solutions to Schrödinger's equation in the separable form,

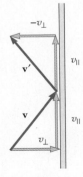

Figure 8.2 Change in particle velocity (or momentum) during collision with a wall of the box. For elastic collision with a smooth wall, the component normal to the wall is reversed, but the tangential components are unaffected.

[1]Recall from Section 6.7 that sharp observables are those for which there is no statistical distribution of measured values. Indeed, quantum wavefunctions typically are labeled by the sharp observables for that state. (For example, the oscillator states of Section 6.6 are indexed by the quantum number n, which specifies the sharp value of particle energy E. In this case, the sharp values of energy also are *quantized,* that is, limited to the discrete values $(n + \frac{1}{2})\hbar\omega$. It follows that any sharp observable is constant over time (unless the corresponding operator explicitly involves time). The converse—that quantum states exist for which all constants of the classical motion are sharp—is not always true but occurs frequently enough that it can serve as a useful rule of thumb.

$$\Psi(\mathbf{r}, t) = \psi(\mathbf{r}) e^{-i\omega t} \tag{8.4}$$

With this time dependence, the right-hand side of Equation 8.1 reduces to $\hbar\omega\Psi$, leaving $\psi(\mathbf{r})$ to satisfy the **time-independent Schrödinger equation** for a particle whose energy is $E = \hbar\omega$:

The time-independent Schrödinger equation

$$-\frac{\hbar^2}{2m} \nabla^2 \psi(\mathbf{r}) + U(\mathbf{r})\psi(\mathbf{r}) = E\psi(\mathbf{r}) \tag{8.5}$$

Since our particle is free inside the box, we take the potential energy $U(\mathbf{r}) = 0$ for $0 < x, y, z < L$. In this case the spatial wavefunction also is separable; that is, solutions to Equation 8.5 with $U(\mathbf{r}) = 0$ can be found in product form:

$$\psi(\mathbf{r}) = \psi(x, y, z) = \psi_1(x)\psi_2(y)\psi_3(z) \tag{8.6}$$

Substituting Equation 8.6 into Equation 8.5 and dividing every term by the function $\psi(x, y, z)$ gives (for $U(\mathbf{r}) = 0$)

$$-\frac{\hbar^2}{2m\psi_1}\frac{d^2\psi_1}{dx^2} - \frac{\hbar^2}{2m\psi_2}\frac{d^2\psi_2}{dy^2} - \frac{\hbar^2}{2m\psi_3}\frac{d^2\psi_3}{dz^2} = E$$

In this form the independent variables are isolated: the first term on the left depends only on x, the second only on y, and the third only on z. To satisfy the equation *everywhere* inside the cube, each of these terms must reduce to a constant:

$$-\frac{\hbar^2}{2m\psi_1}\frac{d^2\psi_1}{dx^2} = E_1$$

$$-\frac{\hbar^2}{2m\psi_2}\frac{d^2\psi_2}{dy^2} = E_2 \tag{8.7}$$

$$-\frac{\hbar^2}{2m\psi_3}\frac{d^2\psi_3}{dz^2} = E_3$$

The stationary states for a particle confined to a cube are obtained from these three separate equations. The energies E_1, E_2, and E_3 are **separation constants** and represent the energy of motion along the three Cartesian axes x, y, and z. Consistent with this identification, the Schrödinger equation requires that $E_1 + E_2 + E_3 = E$.

The first of Equations 8.7 is the same as that for the infinite square well in one dimension. Independent solutions to this equation are $\sin k_1 x$ and $\cos k_1 x$, where $k_1 = \sqrt{2mE_1/\hbar^2}$ is the wavenumber of oscillation. Only $\sin k_1 x$ satisfies the condition that the wavefunction must vanish at the wall $x = 0$, however. Requiring the wavefunction to vanish also at the opposite wall $x = L$ implies $k_1 L = n_1 \pi$, where n_1 is any positive integer. In other words, we must be able to fit an integral number of half-wavelengths into our box along the direction marked by x. It follows that the magnitude of particle momentum along this direction must be one of the special values

$$|p_x| = \hbar k_1 = n_1 \frac{\pi \hbar}{L} \qquad n_1 = 1, 2, \ldots$$

Identical considerations applied to the remaining two equations show that the magnitudes of particle momentum in all three directions are similarly quantized:

$$|p_x| = \hbar k_1 = n_1 \frac{\pi\hbar}{L} \qquad n_1 = 1, 2, \ldots$$

$$|p_y| = \hbar k_2 = n_2 \frac{\pi\hbar}{L} \qquad n_2 = 1, 2, \ldots \qquad (8.8)$$

$$|p_z| = \hbar k_3 = n_3 \frac{\pi\hbar}{L} \qquad n_3 = 1, 2, \ldots$$

Allowed values of momentum components for a particle in a box

Notice that $n_i = 0$ is not allowed, since that choice leads to a ψ_i that is also zero and a wavefunction $\psi(\mathbf{r})$ that vanishes everywhere. Since the momenta are restricted this way, the particle energy (all kinetic) is limited to the following discrete values:

$$E = \frac{1}{2m}(|p_x|^2 + |p_y|^2 + |p_z|^2) = \frac{\pi^2\hbar^2}{2mL^2}\{n_1^2 + n_2^2 + n_3^2\} \qquad (8.9)$$

Discrete energies allowed for a particle in a box

Thus, confining the particle to the cube serves to quantize its momentum and energy according to Equations 8.8 and 8.9. Note that **three quantum numbers are needed to specify the quantum condition, corresponding to the three independent degrees of freedom for a particle in space.** These quantum numbers specify the values taken by the sharp observables for this system.

Collecting the previous results, we see that the stationary states for this particle are

$$\Psi(x, y, z, t) = A \sin(k_1 x)\sin(k_2 y)\sin(k_3 z)e^{-i\omega t} \qquad \text{for } 0 < x, y, z < L$$
$$= 0 \qquad \text{otherwise} \qquad (8.10)$$

The multiplier A is chosen to satisfy the normalization requirement. Example 8.1 shows that $A = (2/L)^{3/2}$ for the ground state, and this result continues to hold for the excited states as well.

EXAMPLE 8.1   Normalizing the Box Wavefunctions

Find the value of the multiplier A that normalizes the wavefunction of Equation 8.10 having the lowest energy.

Solution   The state of lowest energy is described by $n_1 = n_2 = n_3 = 1$, or $k_1 = k_2 = k_3 = \pi/L$. Since Ψ is nonzero only for $0 < x, y, z < L$, the probability density integrated over the volume of this cube must be unity:

$$1 = \int_0^L dx \int_0^L dy \int_0^L dz \,|\Psi(x, y, z, t)|^2$$

$$= A^2 \left\{\int_0^L \sin^2(\pi x/L)\,dx\right\}\left\{\int_0^L \sin^2(\pi y/L)\,dy\right\}$$

$$\times \left\{\int_0^L \sin^2(\pi z/L)\,dz\right\}$$

Using $2\sin^2\theta = 1 - \cos 2\theta$ gives

$$\int_0^L \sin^2(\pi x/L)\,dx = \frac{L}{2} - \frac{L}{4\pi}\sin(2\pi x/L)\Big|_0^L = \frac{L}{2}$$

The same result is obtained for the integrations over y and z. Thus, normalization requires

$$1 = A^2\left(\frac{L}{2}\right)^3$$

or

$$A = \left(\frac{2}{L}\right)^{3/2}$$

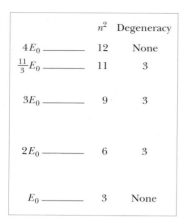

	n^2	Degeneracy
$4E_0$ ———	12	None
$\frac{11}{3}E_0$ ———	11	3
$3E_0$ ———	9	3
$2E_0$ ———	6	3
E_0 ———	3	None

Figure 8.3 An energy-level diagram for a particle confined to a cubic box. The ground-state energy is $E_0 = 3\pi^2\hbar^2/2mL^2$, and $n^2 = n_1^2 + n_2^2 + n_3^2$. Note that most of the levels are degenerate.

Exercise 1 With what probability will the particle described by the wavefunction of Example 8.1 be found in the volume $0 < x, y, z < L/4$?

Answer 0.040, or about 4%

Exercise 2 Modeling a defect trap in a crystal as a three-dimensional box with edge length 5.00 Å, find the values of momentum and energy for an electron bound to the defect site, assuming the electron is in the ground state.

Answer $|p_x| = |p_y| = |p_z| = 1.24$ keV/c; $E = 4.51$ eV

The ground state, for which $n_1 = n_2 = n_3 = 1$, has energy

$$E_{111} = \frac{3\pi^2\hbar^2}{2mL^2}$$

There are *three* first excited states, corresponding to the three different combinations of n_1, n_2, and n_3, whose squares sum to 6. That is, we obtain the same energy for the three combinations $n_1 = 2$, $n_2 = 1$, $n_3 = 1$, or $n_1 = 1$, $n_2 = 2$, $n_3 = 1$, or $n_1 = 1$, $n_2 = 1$, $n_3 = 2$. The first excited state has energy

$$E_{211} = E_{121} = E_{112} = \frac{6\pi^2\hbar^2}{2mL^2}$$

Note that each of the first excited states is characterized by a different wavefunction: ψ_{211} has wavelength L along the x-axis and wavelength $2L$ along the y- and z-axes, but for ψ_{121} and ψ_{112} the shortest wavelength is along the y-axis and the z-axis, respectively.

Whenever different states have the same energy, this energy level is said to be **degenerate.** In the example just described, the first excited level is three-fold (or triply) degenerate. This system has degenerate levels because of the high degree of symmetry associated with the cubic shape of the box. The degeneracy would be removed, or *lifted*, if the sides of the box were of unequal lengths (see Example 8.3). In fact, the extent of splitting of the originally degenerate levels increases with the degree of asymmetry.

Figure 8.3 is an energy-level diagram showing the first five levels of a particle in a cubic box; Table 8.1 lists the quantum numbers and degeneracies of the various levels. Computer-generated plots of the probability density $|\psi(x, y, z)|^2$ for the ground state and first excited states of a particle in a box are shown in Figure 8.4. Notice that the probabilities for the (degenerate) first excited states differ only in their orientation with respect to the coordinate axes, again a reflection of the cubic symmetry imposed by the box potential.

EXAMPLE 8.2   The Second Excited State

Determine the wavefunctions and energy for the second excited level of a particle in a cubic box of edge L. What is the degeneracy of this level?

Solution The second excited level corresponds to the three combinations of quantum numbers $n_1 = 2$, $n_2 = 2$, $n_3 = 1$ or $n_1 = 2$, $n_2 = 1$, $n_3 = 2$ or $n_1 = 1$, $n_2 = 2$, $n_3 = 2$. The corresponding wavefunctions inside the box are

$$\Psi_{221} = A \sin\left(\frac{2\pi x}{L}\right)\sin\left(\frac{2\pi y}{L}\right)\sin\left(\frac{\pi z}{L}\right)e^{-iE_{221}t/\hbar}$$

$$\Psi_{212} = A \sin\left(\frac{2\pi x}{L}\right)\sin\left(\frac{\pi y}{L}\right)\sin\left(\frac{2\pi z}{L}\right)e^{-iE_{212}t/\hbar}$$

$$\Psi_{122} = A \sin\left(\frac{\pi x}{L}\right)\sin\left(\frac{2\pi y}{L}\right)\sin\left(\frac{2\pi z}{L}\right)e^{-iE_{122}t/\hbar}$$

The level is threefold degenerate, since each of these wavefunctions has the same energy,

$$E_{221} = E_{212} = E_{122} = \frac{9\pi^2\hbar^2}{2mL^2}$$

Table 8.1   Quantum Numbers and Degeneracies of the Energy Levels for a Particle Confined to a Cubic Box*

n_1	n_2	n_3	n^2	Degeneracy
1	1	1	3	None
1	1	2	6	
1	2	1	6	Threefold
2	1	1	6	
1	2	2	9	
2	1	2	9	Threefold
2	2	1	9	
1	1	3	11	
1	3	1	11	Threefold
3	1	1	11	
2	2	2	12	None

*Note: $n^2 = n_1^2 + n_2^2 + n_3^2$.

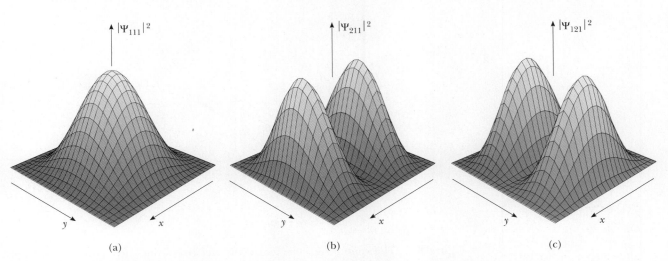

(a)   (b)   (c)

Figure 8.4 Probability density (unnormalized) for a particle in a box: (a) ground state, $|\Psi_{111}|^2$; (b) and (c) first excited states, $|\Psi_{211}|^2$ and $|\Psi_{121}|^2$. Plots are for $|\Psi|^2$ in the plane $z = \frac{1}{2}L$. In this plane, $|\Psi_{112}|^2$ (not shown) is indistinguishable from $|\Psi_{111}|^2$.

EXAMPLE 8.3   Quantization in a Rectangular Box

Obtain a formula for the allowed energies of a particle confined to a rectangular box with edge lengths L_1, L_2, and L_3. What is the degeneracy of the first excited state?

Solution   For a box having edge length L_1 in the x direction, ψ will be zero at the walls if L_1 is an integral number of half-wavelengths. Thus, the magnitude of particle momentum in this direction is quantized as

$$|p_x| = \hbar k_1 = n_1 \frac{\pi \hbar}{L_1} \qquad n_1 = 1, 2, \ldots$$

Likewise, for the other two directions, we have

$$|p_y| = \hbar k_2 = n_2 \frac{\pi \hbar}{L_2} \qquad n_2 = 1, 2, \ldots$$

$$|p_z| = \hbar k_3 = n_3 \frac{\pi \hbar}{L_3} \qquad n_3 = 1, 2, \ldots$$

The allowed energies are

$$E = (|p_x|^2 + |p_y|^2 + |p_z|^2)/2m$$
$$= \frac{\pi^2 \hbar^2}{2m} \left\{ \left(\frac{n_1}{L_1} \right)^2 + \left(\frac{n_2}{L_2} \right)^2 + \left(\frac{n_3}{L_3} \right)^2 \right\}$$

The lowest energy occurs again for $n_1 = n_2 = n_3 = 1$. Increasing one of the integers by 1 gives the next-lowest, or first, excited level. If L_1 is the largest dimension, then $n_1 = 2$, $n_2 = 1$, $n_3 = 1$ produces the smallest energy increment and describes the first excited state. Further, so long as both L_2 and L_3 are not equal to L_1, the first excited level is nondegenerate, that is, there is no other state with this energy. If L_2 or L_3 equals L_1, the level is doubly degenerate; if all three are equal, the level will be triply degenerate. Thus, the higher the symmetry, the more degeneracy we find.

8.2   CENTRAL FORCES AND ANGULAR MOMENTUM

The formulation of quantum mechanics in Cartesian coordinates is the natural way to generalize from one to higher dimensions, but often it is not the best suited to a given application. For instance, an atomic electron is attracted to the nucleus of the atom by the Coulomb force between opposite charges. This is an example of a **central force,** that is, one directed toward a fixed point. The nucleus is the center of force, and the coordinates of choice here are spherical coordinates r, θ, ϕ centered on the nucleus (Fig. 8.5). If the central force is conservative, the particle energy (kinetic plus potential) stays

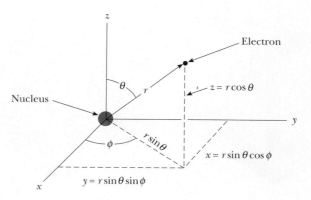

Figure 8.5   The central force on an atomic electron is one directed toward a fixed point, the nucleus. The coordinates of choice here are the spherical coordinates r, θ, ϕ centered on the nucleus.

constant and E becomes a candidate for quantization. In that case, the quantum states are stationary waves $\psi(\mathbf{r})e^{-i\omega t}$, with $\omega = E/\hbar$, where E is the sharp value of particle energy.

But for central forces, angular momentum $\mathbf{L}$ about the force center also is constant (a central force exerts no torque about the force center), and we might expect that wavefunctions can be found for which all three angular momentum components take sharp values. This imposes severe constraints on the form of the wavefunction. In fact, these constraints are so severe that it is impossible to find a wavefunction satisfying all of them at once; that is, not all components of angular momentum can be known simultaneously!

The dilemma is reminiscent of our inability to specify simultaneously the position and momentum of a particle. Indeed, if the direction of $\mathbf{L}$ were known precisely, the particle would be confined to the orbital plane (the plane perpendicular to the vector $\mathbf{L}$), and its coordinate in the direction normal to this plane would be accurately known (and unchanging) (Fig. 8.6). In that case, however, the particle could have no momentum out of the orbital plane, so that its linear momentum perpendicular to this plane also would be known (to be zero), in violation of the uncertainty principle. The argument just given may be refined to establish an uncertainty principle for angular momentum: **it is impossible to specify simultaneously any two components of angular momentum.** Alternatively, if one component of $\mathbf{L}$ is sharp, then the remaining two must be "fuzzy."[2]

Along with E and one component of $\mathbf{L}$, then, what else might be quantized, or sharp, for central forces? The answer is contained in the following observation: With only one component of $\mathbf{L}$ sharp, there is no redundancy in having the magnitude $|\mathbf{L}|$ sharp also. In this way, E, $|\mathbf{L}|$, and one component of $\mathbf{L}$, say L_z, become the sharp observables subject to quantization in the central force problem.

Wavefunctions for which $|\mathbf{L}|$ and L_z are both sharp follow directly from separating the variables in Schrödinger's equation for a central force. We take the time-independent wavefunction in spherical coordinates r, θ, ϕ to be the product

$$\psi(\mathbf{r}) = \psi(r, \theta, \phi) = R(r)\Theta(\theta)\Phi(\phi) \tag{8.11}$$

and write Schrödinger's time-independent equation (Eq. 8.5) in these coordinates using the spherical coordinate form for the Laplacian[3]:

$$\nabla^2 = \frac{\partial^2}{\partial r^2} + \left(\frac{2}{r}\right)\frac{\partial}{\partial r} + \frac{1}{r^2}\left[\frac{\partial^2}{\partial \theta^2} + \cot\theta\frac{\partial}{\partial \theta} + \csc^2\theta\frac{\partial^2}{\partial \phi^2}\right]$$

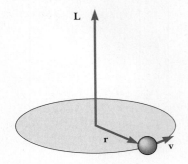

Figure 8.6 The angular momentum $\mathbf{L}$ of an orbiting particle is perpendicular to the plane of the orbit. If the direction of $\mathbf{L}$ were known precisely, both the coordinate and momentum in the direction perpendicular to the orbit would be known, in violation of the uncertainty principle.

Uncertainty principle for angular momentum

Separation of variables for the stationary state wavefunction

---

[2]Angular momentum is a notable exception to the argument that constants of the classical motion correspond to sharp observables in quantum mechanics. The failure is rooted in another maxim—the uncertainty principle—that takes precedence. The only instance in which two or more angular momentum components may be known exactly is the trivial case $L_x = L_y = L_z = 0$.

[3]The Laplacian in spherical coordinates is given in any more advanced scientific text or may be constructed from the Cartesian form by following the arguments of Section 8.4.

After dividing each term in Equation 8.5 by $R\Theta\Phi$, we are left with

$$\frac{-\hbar^2}{2mR}\left[\frac{d^2R}{dr^2} + \frac{2}{r}\frac{dR}{dr}\right] - \frac{\hbar^2}{2m\Theta}\frac{1}{r^2}\left[\frac{d^2\Theta}{d\theta^2} + \cot\theta\frac{d\Theta}{d\theta}\right]$$
$$- \frac{\hbar^2}{2m\Phi}\frac{1}{r^2\sin^2\theta}\frac{d^2\Phi}{d\phi^2} + U(r) = E$$

Notice that ordinary derivatives now replace the partials and that $U(\mathbf{r})$ becomes simply $U(r)$ for a central force seated at the origin. We can isolate the dependence on ϕ by multiplying every term by $r^2\sin^2\theta$ to get, after some rearrangement,

$$\frac{1}{\Phi}\frac{d^2\Phi}{d\phi^2} = -\sin^2\theta\left\{\frac{r^2}{R}\left[\frac{d^2R}{dr^2} + \frac{2}{r}\frac{dR}{dr}\right] + \frac{1}{\Theta}\left[\frac{d^2\Theta}{d\theta^2} + \cot\theta\frac{d\Theta}{d\theta}\right]\right.$$
$$\left. + \frac{2mr^2}{\hbar^2}[E - U(r)]\right\} \quad (8.12)$$

In this form the left side is a function only of ϕ while the right side depends only on r and θ. Because these are independent variables, equality of the two sides requires that each side reduce to a constant. Following convention, we write this separation constant as $-m_\ell^2$, where m_ℓ is the **magnetic quantum number.**[4]

Equating the left side of Equation 8.12 to $-m_\ell^2$ gives an equation for $\Phi(\phi)$:

$$\frac{d^2\Phi}{d\phi^2} = -m_\ell^2\Phi(\phi) \quad (8.13)$$

A solution to Equation 8.13 is $\Phi(\phi) = \exp(im_\ell\phi)$; this is periodic with period 2π if m_ℓ is restricted to integer values. Periodicity is necessary here because all physical properties that derive from the wavefunction should be unaffected by the replacement $\phi \to \phi + 2\pi$, both of which describe the same point in space (see Fig. 8.5).

Equating the right side of Equation 8.12 to $-m_\ell^2$ gives, after some further rearrangement,

$$\frac{r^2}{R}\left[\frac{d^2R}{dr^2} + \frac{2}{r}\frac{dR}{dr}\right] + \frac{2mr^2}{\hbar^2}[E - U(r)] = -\frac{1}{\Theta}\left[\frac{d^2\Theta}{d\theta^2} + \cot\theta\frac{d\Theta}{d\theta}\right]$$
$$+ \frac{m_\ell^2}{\sin^2\theta} \quad (8.14)$$

In this form the variables r and θ are separated, the left side being a function only of r and the right side a function only of θ. Again, each side must reduce to a constant. This furnishes two more equations, one for each of the remaining functions $R(r)$ and $\Theta(\theta)$, and introduces a second separation

---

[4]This seemingly peculiar way of writing the constant multiplier is based on the physical significance of the function $\Phi(\phi)$ and is discussed at length in (optional) Section 8.4. The student is referred there for a concise treatment of the quantum central force problem based on the operator methods of Section 6.8.

constant, which we write as $\ell(\ell + 1)$. Equating the right side of Equation 8.14 to $\ell(\ell + 1)$ requires $\Theta(\theta)$ to satisfy

$$\frac{d^2\Theta}{d\theta^2} + \cot\theta\,\frac{d\Theta}{d\theta} - m_\ell^2\,\csc^2\theta\,\Theta(\theta) = -\ell(\ell + 1)\Theta(\theta) \qquad (8.15)$$

ℓ is called the **orbital quantum number.** For $\psi(\mathbf{r})$ to be an acceptable wavefunction, $\Theta(\theta)$ also must be bounded and single-valued, conditions which are met for functions satisfying Equation 8.15 only if ℓ is a nonnegative integer, and then only if m_ℓ is limited to absolute values not larger than ℓ. The resulting solutions $\Theta(\theta)$ are polynomials in $\cos\theta$ known as **associated Legendre polynomials.** A few of these polynomials are listed in Table 8.2 for easy reference. The products $\Theta(\theta)\Phi(\phi)$ specify the full angular dependence of the central force wavefunction and are known as **spherical harmonics,** denoted by $Y_\ell^{m_\ell}(\theta, \phi)$. Some spherical harmonics are given in Table 8.3. The constant prefactors are chosen to normalize these functions.[5]

Table 8.2   Some Associated Legendre Polynomials $P_\ell^{m_\ell}(\cos\theta)$

$P_0^0 = 1$

$P_1^0 = 2\cos\theta$

$P_1^1 = \sin\theta$

$P_2^0 = 4(3\cos^2\theta - 1)$

$P_2^1 = 4\sin\theta\cos\theta$

$P_2^2 = \sin^2\theta$

$P_3^0 = 24(5\cos^3\theta - 3\cos\theta)$

$P_3^1 = 6\sin\theta(5\cos^2\theta - 1)$

$P_3^2 = 6\sin^2\theta\cos\theta$

$P_3^3 = \sin^3\theta$

Table 8.3   The Spherical Harmonics $Y_\ell^{m_\ell}(\theta, \phi)$

$$Y_0^0 = \frac{1}{2\sqrt{\pi}}$$

$$Y_1^0 = \frac{1}{2}\sqrt{\frac{3}{\pi}}\cdot\cos\theta$$

$$Y_1^{\pm 1} = \mp\frac{1}{2}\sqrt{\frac{3}{2\pi}}\cdot\sin\theta\cdot e^{\pm i\phi}$$

$$Y_2^0 = \frac{1}{4}\sqrt{\frac{5}{\pi}}\cdot(3\cos^2\theta - 1)$$

$$Y_2^{\pm 1} = \mp\frac{1}{2}\sqrt{\frac{15}{2\pi}}\cdot\sin\theta\cdot\cos\theta\cdot e^{\pm i\phi}$$

$$Y_2^{\pm 2} = \frac{1}{4}\sqrt{\frac{15}{2\pi}}\cdot\sin^2\theta\cdot e^{\pm 2i\phi}$$

$$Y_3^0 = \frac{1}{4}\sqrt{\frac{7}{\pi}}\cdot(5\cos^3\theta - 3\cos\theta)$$

$$Y_3^{\pm 1} = \mp\frac{1}{8}\sqrt{\frac{21}{\pi}}\cdot\sin\theta\cdot(5\cos^2\theta - 1)\cdot e^{\pm i\phi}$$

$$Y_3^{\pm 2} = \frac{1}{4}\sqrt{\frac{105}{2\pi}}\cdot\sin^2\theta\cdot\cos\theta\cdot e^{\pm 2i\phi}$$

$$Y_3^{\pm 3} = \mp\frac{1}{8}\sqrt{\frac{35}{\pi}}\cdot\sin^3\theta\cdot e^{\pm 3i\phi}$$

[5]The normalization is such that the integral of $|Y_\ell^{m_\ell}|^2$ over the surface of a sphere with unit radius is 1.

In keeping with our earlier remarks, the separation constants $\ell(\ell + 1)$ and m_ℓ should relate to the sharp observables $|\mathbf{L}|$, L_z, and E for central forces. The connection is established by the more detailed arguments of Section 8.4, with the result

Angular momentum and its z component are quantized

$$|\mathbf{L}| = \sqrt{\ell(\ell + 1)}\hbar \qquad \ell = 0, 1, 2, \ldots$$
$$L_z = m_\ell \hbar \qquad m_\ell = 0, \pm 1, \pm 2, \ldots, \pm \ell \qquad (8.16)$$

We see that the limitation on the magnetic quantum number m_ℓ to values between $-\ell$ and $+\ell$, obtained on purely mathematical grounds from separating variables, has an obvious physical interpretation: the z component of angular momentum, L_z, must never exceed the magnitude of the vector, $|\mathbf{L}|$! Notice, too, that ℓ and m_ℓ are quantum numbers for angular momentum only; their connection with particle energy E must depend on the potential energy function $U(r)$ and is prescribed along with the radial wavefunction $R(r)$ in the final stage of the separation procedure.

To obtain $R(r)$, we return to Equation 8.14 and equate the left side to $\ell(\ell + 1)$. Rearranging terms once more, we find that $R(r)$ must satisfy

Radial wave equation

$$-\frac{\hbar^2}{2m}\left[\frac{d^2R}{dr^2} + \frac{2}{r}\frac{dR}{dr}\right] + \frac{\ell(\ell + 1)\hbar^2}{2mr^2}R(r) + U(r)R(r) = ER(r) \qquad (8.17)$$

This is the **radial wave equation;** it determines the radial part of the wavefunction ψ and the allowed particle energies E. As the equation contains the orbital quantum number ℓ, each angular momentum orbital is expected to give rise to a different radial wave and a distinct energy. By contrast, the magnetic quantum number m_ℓ appears nowhere in this equation. Thus, the radial wave and particle energy remain the same for different m_ℓ values consistent with a given value of ℓ. In particular, **for a fixed ℓ the particle energy E is independent of m_ℓ, and so is at least $2\ell + 1$-fold degenerate.** Such degeneracy—a property of all central forces—stems from the spherical symmetry of the potential and illustrates once again the deep-seated connection between symmetry and the degeneracy of quantum states.

The reduction from Schrödinger's equation to the radial wave equation represents enormous progress and is valid for any central force. Still, the task of solving this equation for a specified potential $U(r)$ is a difficult one, often requiring methods of considerable sophistication. In Section 8.5 we tackle this task for the important case of the electron in the hydrogen atom.

EXAMPLE 8.4   Orbital Quantum Number for a Stone

A stone with mass 1.00 kg is whirled in a horizontal circle of radius 1.00 m with a period of revolution equal to 1.00 s. What value of orbital quantum number ℓ describes this motion?

Solution   The speed of the stone in its orbit is

$$v = \frac{2\pi R}{T} = \frac{2\pi(1.00 \text{ m})}{1.00 \text{ s}} = 6.28 \text{ m/s}$$

The corresponding angular momentum has magnitude

$$|\mathbf{L}| = mvR = (1.00 \text{ kg})(6.28 \text{ m/s})(1.00 \text{ m})$$
$$= 6.28 \text{ kg} \cdot \text{m}^2/\text{s}$$

But angular momentum is quantized as $\sqrt{\ell(\ell + 1)}\hbar$, which is approximately $\ell\hbar$ when ℓ is large. Then

$$\ell \approx \frac{|\mathbf{L}|}{\hbar} = \frac{6.28 \text{ kg} \cdot \text{m}^2/\text{s}}{1.055 \times 10^{-34} \text{ kg} \cdot \text{m}^2/\text{s}} = 5.96 \times 10^{34}$$

Again, we see that macroscopic objects are described by enormous quantum numbers, so that quantization is not evident on this scale.

EXAMPLE 8.5   The Bohr Atom Revisited

Discuss angular momentum quantization in the Bohr model. What orbital quantum number describes the electron in the first Bohr orbit of hydrogen?

Solution   Angular momentum in the Bohr atom is quantized according to the Bohr postulate

$$|\mathbf{L}| = mvr = n\hbar$$

with $n = 1$ for the first Bohr orbit. Comparing this with the quantum mechanical result, Equation 8.14, we see that the two rules are incompatible! The magnitude $|\mathbf{L}|$ can never be an integral multiple of $\hbar$—the smallest nonzero value consistent with Equation 8.16 is $|\mathbf{L}| = \sqrt{2}\hbar$ for $\ell = 1$.

In fairness to Bohr, we should point out that the Bohr model makes no distinction between the quantization of $\mathbf{L}$ and quantization of its components along the coordinate axes. From the classical viewpoint, the coordinate system can always be oriented to align one of the axes, say the z-axis, along the direction of $\mathbf{L}$. In that case, $\mathbf{L}$ may be identified with $|L_z|$. The Bohr postulate in this form agrees with the quantization of L_z in Equation 8.16 and indicates $\ell = 1$ for the first Bohr orbit! This conflicting result derives from a false assertion, namely, that we may orient a coordinate axis along the direction of $\mathbf{L}$. The freedom to do so must be abandoned if the quantization rules of Equation 8.16 are correct! This stunning conclusion is one of the great mysteries of quantum physics and is implicit in the notion of space quantization that we discuss in the next section.

8.3   SPACE QUANTIZATION

For wavefunctions satisfying Equations 8.13 and 8.15, the orbital angular momentum magnitude $|\mathbf{L}|$, and L_z, the projection of $\mathbf{L}$ along the z-axis, are both sharp and quantized according to the restrictions imposed by the orbital and magnetic quantum numbers, respectively. Together, ℓ and m_ℓ specify the orientation of the angular momentum vector $\mathbf{L}$. The fact that the direction of $\mathbf{L}$ is quantized with respect to an arbitrary axis (the z-axis) is referred to as **space quantization.**

Let us look at the possible orientations of $\mathbf{L}$ for a given value of orbital quantum number ℓ. Recall that m_ℓ can have values ranging from $-\ell$ to $+\ell$. If $\ell = 0$, then $m_\ell = 0$ and $L_z = 0$. In fact, for this case $|\mathbf{L}| = 0$, so that all components of $\mathbf{L}$ are 0. If $\ell = 1$, then the possible values for m_ℓ are -1, 0, and $+1$, so that L_z may be $-\hbar$, 0, or $+\hbar$. If $\ell = 2$, m_ℓ can be -2, -1, 0, $+1$, or $+2$, corresponding to L_z values of $-2\hbar$, $-\hbar$, 0, $+\hbar$, or $+2\hbar$, and so on. A classical visualization of the algebra describing space quantization for the case $\ell = 2$ is shown in Figure 8.7a. Note that $\mathbf{L}$ can never be aligned with the z-axis, since L_z must be smaller than the total angular momentum L. From a three-dimensional perspective, $\mathbf{L}$ must lie on the surface of a cone that makes an angle θ with the z-axis, as shown in Figure 8.7b. From the figure, we see that θ also is quantized and that its values are specified by the relation

$$\cos \theta = \frac{L_z}{|\mathbf{L}|} = \frac{m_\ell}{\sqrt{\ell(\ell + 1)}} \tag{8.18}$$

The orientations of L are restricted (quantized)

Classically, θ can take any value; that is, the angular momentum vector $\mathbf{L}$ can point in any direction whatsoever. According to quantum mechanics, the possible orientations for $\mathbf{L}$ are those consistent with Equation 8.18. Furthermore, these special directions have nothing to do with the forces acting on

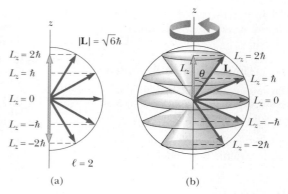

Figure 8.7 (a) The allowed projections of the orbital angular momentum for the case $\ell = 2$. (b) From a three-dimensional perspective, the orbital angular momentum vector **L** lies on the surface of a cone. The fuzzy character of L_x and L_y is depicted by allowing **L** to precess about the z-axis, so that L_x and L_y change continually while L_z maintains the fixed value $m_\ell \hbar$.

the particle, provided only that these forces are central. Thus, **the rule of Equation 8.18 does not originate with the law of force but derives from the structure of space itself,** hence the name *space quantization*.

Figure 8.7 is misleading in showing **L** with a specific direction, for which all three components L_x, L_y, and L_z are known exactly. As we have mentioned, there is no quantum state for which this condition is true. If L_z is sharp, then L_x and L_y must be fuzzy. Accordingly, it is more proper to visualize the vector **L** of Figure 8.7b as precessing around the z-axis so as to trace out a cone in space. This allows the components L_x and L_y to change continually, while L_z maintains the fixed value $m_\ell \hbar$.[6]

EXAMPLE 8.6   Space Quantization for an Atomic Electron

Consider an atomic electron in the $\ell = 3$ state. Calculate the magnitude $|\mathbf{L}|$ of the total angular momentum and the allowed values of L_z and θ.

Solution   With $\ell = 3$, Equation 8.16 gives

$$|\mathbf{L}| = \sqrt{3(3+1)}\, \hbar = 2\sqrt{3}\hbar$$

The allowed values of L_z are $m_\ell \hbar$, with $m_\ell = 0, \pm 1, \pm 2,$ and ± 3. This gives

$$L_z = -3\hbar, -2\hbar, -\hbar, 0, \hbar, 2\hbar, 3\hbar$$

Finally, we obtain the allowed values of θ from

$$\cos \theta = \frac{L_z}{|\mathbf{L}|} = \frac{m_\ell}{2\sqrt{3}}$$

Substituting the values for m_ℓ gives

$$\cos \theta = \pm 0.866, \quad \pm 0.577, \quad \pm 0.289, \quad \text{and} \quad 0$$

or

$$\theta = \pm 30°, \quad \pm 54.8°, \quad \pm 73.2°, \quad \text{and} \quad 90°$$

---

[6]This precession of the classical vector to portray the inherent fuzziness in L_x and L_y is meant to be suggestive only. In effect, we have identified the quantum averages $\langle L_x \rangle$ and $\langle L_y \rangle$ with their averages over *time* in a classical picture, but the two kinds of averaging in fact are quite distinct.

Exercise 3   Compare the minimum angles between **L** and the z-axis for the electron of Example 8.6 and for the 1.00-kg stone of Example 8.4.

Answer   30° for the electron but only 2.3×10^{-16} degrees for the stone

8.4   QUANTIZATION OF ANGULAR MOMENTUM AND ENERGY

O P T I O N A L

We saw in Section 8.2 that angular momentum plays an essential role in the quantization of systems with spherical symmetry. Here we develop further the properties of angular momentum from the viewpoint of quantum mechanics and show in more detail how angular momentum considerations facilitate the solution to Schrödinger's equation for central forces.

Our treatment is based on the operator methods introduced in Section 6.8, which the reader should review at this time. In particular, the eigenvalue condition

$$[Q]\Psi = q\Psi \tag{8.19}$$

used there as a *test* for sharp observables becomes here a tool for discovering the form of the unknown wavefunction Ψ. (Recall that $[Q]$ denotes the operator for some observable Q and q is the sharp value of that observable when measured for a system described by the wavefunction Ψ.) We look on Equation 8.19 now as a *constraint* imposed on the wavefunction Ψ that guarantees that the observable Q will take the sharp value q in that state. The more sharp observables we can identify for some system, the more we can learn in advance about the wavefunction describing that system. **With few exceptions, the sharp observables are just those that are constants of the classical motion.** (See footnote 1.)

Consider the particle energy. If total energy is constant, we should be able to find wavefunctions Ψ for which E is a sharp observable. Otherwise, repeated measurements made on identical systems would reveal a statistical distribution of values for the particle energy, inconsistent with the idea of a quantity not changing over time. Thus, energy conservation suggests $\Delta E = 0$, which, in turn, requires Ψ to be an eigenfunction of the energy operator

$$[E]\Psi = E\Psi$$

Because $[E] = i\hbar\,\partial/\partial t$ this eigenvalue condition is met by the stationary waves $\psi(\mathbf{r})e^{-i\omega t}$, with $E = \hbar\omega$ the sharp value of particle energy.

The argument for energy applies equally well to other constants of the classical motion.[7] If the only forces are central, angular momentum about the force center is a constant of the motion. This is a vector quantity $\mathbf{L} = \mathbf{r} \times \mathbf{p}$, whose rectangular components are expressed in terms of position and momentum components as

$$L_z = (\mathbf{r} \times \mathbf{p})_z = xp_y - yp_x$$

and so on. In the same way, the *operators* for angular momentum are found from the coordinate and momentum *operators*. From Chapter 6, the operator for x is just the

---

[7] Exceptions to this rule do exist. For instance, an atomic electron cannot have all three angular momentum components sharp at once, even though in this case all components are constant classically. For such *incompatible* observables, quantum mechanics adopts the broader interpretation of a conserved quantity as one whose *average value does not change over time*, no matter what may be the initial state of the system. With this definition, all components of angular momentum for an atomic electron remain constant, but no more than one can be sharp in a given state.

coordinate itself, and the operator for momentum in this direction is $(\hbar/i)\,\partial/\partial x$. Similar relations should apply to the directions labeled y and z,[8] so that the operator for L_z becomes

$$[L_z] = [x][p_y] - [y][p_x] = \frac{\hbar}{i}\left(x\frac{\partial}{\partial y} - y\frac{\partial}{\partial x}\right) \qquad (8.20)$$

Angular momentum finds its natural expression in the spherical coordinates of Figure 8.5. A little geometry applied to Figure 8.5 shows that the spherical-to-Cartesian coordinate transformation equations are

$$z = r\cos\theta$$
$$x = r\sin\theta\cos\phi \qquad (8.21)$$
$$y = r\sin\theta\sin\phi$$

The inverse transformations are

$$r = \{x^2 + y^2 + z^2\}^{1/2}$$
$$\cos\theta = \frac{z}{r} = z\{x^2 + y^2 + z^2\}^{-1/2} \qquad (8.22)$$
$$\tan\phi = \frac{y}{x}$$

The procedure for transcribing operators such as $[L_z]$ from Cartesian to spherical form is straightforward, but tedious. For instance, an application of the chain rule gives

$$\frac{\partial f}{\partial z} = \frac{\partial r}{\partial z}\frac{\partial f}{\partial r} + \frac{\partial\theta}{\partial z}\frac{\partial f}{\partial\theta} + \frac{\partial\phi}{\partial z}\frac{\partial f}{\partial\phi}$$

for any function f. On the left we think of f expressed as a function of x, y, z, but on the right the same function is expressed in terms of r, θ, ϕ. The partial derivatives $\partial r/\partial z$, and so on, are to be taken with the aid of the transformation equations, holding x and y fixed. The simplest of these is $\partial\phi/\partial z$. From the inverse transformations in Equations 8.22, we see that ϕ is independent of z, so that $\partial\phi/\partial z = 0$. From the same equations we find that

$$\frac{\partial r}{\partial z} = \frac{1}{2}\{x^2 + y^2 + z^2\}^{-1/2}(2z) = \frac{z}{r} = \cos\theta$$

To obtain $\partial\theta/\partial z$ we differentiate the second of Equations 8.22 implicitly to get

$$-\sin\theta\,\frac{\partial\theta}{\partial z} = \frac{\partial\{\cos\theta\}}{\partial z} = z(-\tfrac{1}{2})\{x^2 + y^2 + z^2\}^{-3/2}(2z) + \{x^2 + y^2 + z^2\}^{-1/2}$$

Converting the right-hand side back to spherical coordinates gives

$$-\sin\theta\,\frac{\partial\theta}{\partial z} = -\frac{(r\cos\theta)^2}{r^3} + \frac{1}{r} = \frac{1 - \cos^2\theta}{r} = \frac{\sin^2\theta}{r}$$

or $\partial\theta/\partial z = -\sin\theta/r$. Collecting the previous results, we have

$$\frac{\partial}{\partial z} = \frac{\partial r}{\partial z}\frac{\partial}{\partial r} + \frac{\partial\theta}{\partial z}\frac{\partial}{\partial\theta} = \cos\theta\,\frac{\partial}{\partial r} - \frac{\sin\theta}{r}\frac{\partial}{\partial\theta}$$

---

[8]For the momentum operator along each axis we use the one-dimensional form from Chapter 6, ($[p_x] = (\hbar/i)\partial/\partial x$, etc.), consistent with the belief that Cartesian axes identify independent but logically equivalent directions in space.

In like fashion, we can obtain spherical representations for $\partial/\partial x$, $\partial/\partial y$ and ultimately for the angular momentum operators themselves:

$$[L_x] = i\hbar \left\{ \sin\phi \, \frac{\partial}{\partial\theta} + \cot\theta \, \cos\phi \, \frac{\partial}{\partial\phi} \right\}$$

$$[L_y] = -i\hbar \left\{ \cos\phi \, \frac{\partial}{\partial\theta} - \cot\theta \, \sin\phi \, \frac{\partial}{\partial\phi} \right\} \qquad (8.23)$$

$$[L_z] = -i\hbar \, \frac{\partial}{\partial\phi}$$

Notice that the operators for angular momentum in spherical form do not contain the radial coordinate r—our reward for selecting the right system of coordinates! By insisting that angular momentum be sharp in the quantum state Ψ, and using the differential forms (Eqs. 8.23) for the angular momentum operators, we can discover how the wavefunction depends on the spherical angles θ and ϕ *without knowing the details of the central force.*

L_z Is Sharp: The Magnetic Quantum Number

For L_z to be sharp, our wavefunction $\Psi = \psi(r) e^{-i\omega t}$ must be an eigenfunction of $[L_z]$, or

$$-i\hbar \, \frac{\partial\psi}{\partial\phi} = L_z\psi \qquad (8.24)$$

with L_z the eigenvalue (a number). Equation 8.24 prescribes the functional dependence of ψ on the azimuthal angle ϕ to be

$$\psi(\mathbf{r}) = C e^{iL_z\phi/\hbar} \qquad (8.25)$$

with C still any function of r and θ. The values taken by L_z must be restricted, however. Since increasing ϕ by 2π radians returns us to the same place in space (see Fig. 8.5), the wavefunction also should return to its initial value, that is, the solutions represented by Equation 8.25 must be periodic in ϕ with period 2π. This will be true if $L_z/\hbar$ is any integer, say, m_ℓ or $L_z = m_\ell\hbar$. The magnetic quantum number m_ℓ indicates the (sharp) value for the z component of angular momentum in the state described by ψ.

Because all components of **L** are constant for central forces, we should continue by requiring the wavefunction of Equation 8.25 also to be an eigenfunction of the operator for a second angular momentum component, say $[L_x]$, so that L_z and L_x might both be sharp in the state ψ. For ψ to be an eigenfunction of $[L_x]$ with eigenvalue L_x requires

$$i\hbar \left\{ \sin\phi \, \frac{\partial\psi}{\partial\theta} + \cot\theta \, \cos\phi \, \frac{\partial\psi}{\partial\phi} \right\} = L_x\psi$$

This relation must hold for all values of θ and ϕ. In particular, for $\phi = 0$ the requirement is

$$i\hbar \cot\theta \, \frac{\partial\psi}{\partial\phi} = L_x\psi$$

Because ψ varies exponentially with ϕ according to Equation 8.25, the indicated derivative may be taken and evaluated at $\phi = 0$ to get

$$(-L_z \cot\theta) \, \psi = L_x\psi$$

which can be satisfied for *all* θ only if $L_x = L_z = 0$ or if ψ vanishes identically. A similar difficulty arises if we attempt to make L_y sharp together with L_z. Therefore,

unless the angular momentum is exactly zero (all components), there is no wavefunction for which two or more components are simultaneously sharp!

$|\mathbf{L}|$ Is Sharp: The Orbital Quantum Number

To make further progress, we must look to other constants of the classical motion. In addition to each component of angular momentum, the magnitude $|\mathbf{L}|$ of the vector also is constant and becomes a candidate for an observable that can be made sharp together with L_z. Consider simply the squared-magnitude $L^2 = \mathbf{L} \cdot \mathbf{L}$. The operator for L^2 is

$$[L]^2 = [L_x]^2 + [L_y]^2 + [L_z]^2$$

Using Equation 8.23, we find the spherical coordinate form for $[L^2]$:

$$[L^2] = -\hbar^2 \left\{ \frac{\partial^2}{\partial \theta^2} + \cot \theta \frac{\partial}{\partial \theta} + \csc^2 \theta \frac{\partial^2}{\partial \phi^2} \right\} \tag{8.26}$$

For L^2 to be sharp requires $[L^2]\psi = |\mathbf{L}|^2 \psi$ or

$$-\hbar^2 \left\{ \frac{\partial^2 \psi}{\partial \theta^2} + \cot \theta \frac{\partial \psi}{\partial \theta} + \csc^2 \theta \frac{\partial^2 \psi}{\partial \phi^2} \right\} = |\mathbf{L}|^2 \psi$$

But from Equation 8.24, we see that differentiating ψ with respect to ϕ is equivalent to multiplication by $iL_z/\hbar = im_\ell$, so that the last equation reduces to

$$-\hbar^2 \left\{ \frac{\partial^2 \psi}{\partial \theta^2} + \cot \theta \frac{\partial \psi}{\partial \theta} - m_\ell^2 \csc^2 \theta \psi \right\} = |\mathbf{L}|^2 \psi \tag{8.27}$$

For wavefunctions satisfying Equation 8.27, the magnitude of angular momentum will be sharp at the value $|\mathbf{L}|$. The equation prescribes the dependence of ψ on the polar angle θ. The solutions are not elementary functions but can be investigated with the help of more advanced techniques. The results of such an investigation are reported in Section 8.2 and repeated here for completeness: Physically acceptable solutions to Equation 8.27 can be found provided $|\mathbf{L}| = \sqrt{\ell(\ell + 1)}\,\hbar$, where ℓ, the orbital quantum number, must be a nonnegative integer. Furthermore, the magnetic quantum number, also appearing in Equation 8.27, must be limited as $|m_\ell| \leq \ell$. With these restrictions, the solutions to Equation 8.27 are the associated Legendre polynomials in $\cos \theta$, denoted $P_\ell^{m_\ell}(\cos \theta)$. Several of these polynomials are listed in Table 8.2 for easy reference; you may verify by direct substitution that they satisfy Equation 8.27 for the appropriate values of ℓ and m_ℓ.

The associated Legendre polynomials may be multiplied by $\exp(im_\ell \phi)$ and still satisfy the orbital equation Equation 8.27. Indeed, the products $P_\ell^{m_\ell}(\cos \theta) \exp(im_\ell \phi)$ satisfy both Equations 8.24 and 8.27; thus, they represent waves for which $|\mathbf{L}|$ and L_z are simultaneously sharp. Except for a multiplicative constant, these are just the spherical harmonics $Y_\ell^{m_\ell}(\theta, \phi)$ introduced in Section 8.2.

E Is Sharp: The Radial Wave Equation

For energy to be constant, E should be a sharp observable. The stationary state form $\Psi(\mathbf{r}, t) = \psi(\mathbf{r})e^{-i\omega t}$ results from imposing the eigenvalue condition for the energy operator $[E]$—but what of the other energy operator $[H] = [K] + [U]$ (the Hamiltonian)? In fact, requiring $[H]\Psi = E\Psi$ is equivalent to writing the time-independent Schrödinger equation for $\psi(\mathbf{r})$

$$-\frac{\hbar^2}{2m} \nabla^2 \psi + U(\mathbf{r})\psi = E\psi \tag{8.28}$$

because $[U] = U(\mathbf{r})$ and the kinetic energy operator $[K]$ in three dimensions is none other than

$$[K] = \frac{[p_x]^2 + [p_y]^2 + [p_z]^2}{2m} = \frac{(\hbar/i)^2\{(\partial/\partial x)^2 + (\partial/\partial y)^2 + (\partial/\partial z)^2\}}{2m} = -\frac{\hbar^2}{2m}\nabla^2$$

The spherical form of the Laplacian given in Section 8.2 may be used to write $[K]$ in spherical coordinates:

$$[K] = -\frac{\hbar^2}{2m}\left\{\frac{\partial^2}{\partial r^2} + \frac{2}{r}\frac{\partial}{\partial r} + \frac{1}{r^2}\left[\frac{\partial^2}{\partial\theta^2} + \cot\theta\frac{\partial}{\partial\theta} + \csc^2\theta\frac{\partial^2}{\partial\phi^2}\right]\right\}$$

Comparing this with the spherical representation for $[L^2]$ from Equation 8.26 shows that $[K]$ is the sum of two terms,

$$[K] = [K_{\text{rad}}] + [K_{\text{orb}}] = [K_{\text{rad}}] + \frac{1}{2mr^2}[L^2] \qquad (8.29)$$

representing the separate contributions to the kinetic energy from the orbital and radial components of motion. The comparison also furnishes the spherical form for $[K_{\text{rad}}]$:

$$[K_{\text{rad}}] = -\frac{\hbar^2}{2m}\left\{\frac{\partial^2}{\partial r^2} + \frac{2}{r}\frac{\partial}{\partial r}\right\} \qquad (8.30)$$

These expressions for $[K]$ are completely general. When applied to waves for which $|\mathbf{L}|$ is sharp, however, the *operator* $[L^2]$ may be replaced with the *number* $|\mathbf{L}|^2 = \ell(\ell+1)\hbar^2$. Therefore, the Schrödinger equation (Eq. 8.28) becomes

$$[K_{\text{rad}}]\psi(\mathbf{r}) + \frac{\ell(\ell+1)\hbar^2}{2mr^2}\psi(\mathbf{r}) + U(r)\psi(\mathbf{r}) = E\psi(\mathbf{r}) \qquad (8.31)$$

For central forces, all terms in this equation, including $U(r)$, involve only the spherical coordinate r: the angle variables θ and ϕ have been eliminated from further consideration by the requirement that ψ be an eigenfunction of $[L^2]$! It follows that the solutions to Equation 8.31 take the form of a radial wave $R(r)$ multiplied by a spherical harmonic:

$$\psi(\mathbf{r}) = R(r)Y_\ell^{m_\ell}(\theta, \phi) \qquad (8.32)$$

The spherical harmonic may be divided out of each term in Equation 8.31, in effect replacing $\psi(\mathbf{r})$ by $R(r)$. The result is just the radial wave equation of Section 8.2; it determines the radial wavefunction $R(r)$ and the allowed particle energies E, once the potential energy function $U(r)$ is specified.

8.5   ATOMIC HYDROGEN AND HYDROGEN-LIKE IONS

In this section we study the hydrogen atom from the viewpoint of wave mechanics. Its simplicity makes atomic hydrogen the ideal testing ground for comparing theory with experiment. Furthermore, the hydrogen atom is the prototype system for the many complex ions and atoms of the heavier elements. Indeed, our study of the hydrogen atom ultimately will enable us to understand the periodic table of the elements, one of the greatest triumphs of quantum physics.

The object of interest is the orbiting electron, with its mass m and charge $-e$, bound by the force of electrostatic attraction to the nucleus, with its much larger mass M and charge $+Ze$, where Z is the atomic number. The choice $Z = 1$ describes the hydrogen atom, while for singly ionized helium (He^+) and doubly ionized lithium (Li^{2+}), we take $Z = 2$ and $Z = 3$, respectively, and so on. Ions with only one electron, like He^+ and Li^{2+}, are called **hydrogen-like.** Because $M \gg m$, we will assume that the nucleus does not move but simply exerts the attractive force that binds the electron. This force is the coulomb force, with its associated potential energy

$$U(r) = \frac{k(+Ze)(-e)}{r} = -\frac{kZe^2}{r} \tag{8.33}$$

where k is the coulomb constant.

The hydrogen atom constitutes a central force problem; according to Section 8.2, the stationary states for any central force are

$$\Psi(r, \theta, \phi, t) = R(r) Y_\ell^{m_\ell}(\theta, \phi) e^{-i\omega t} \tag{8.34}$$

where $Y_\ell^{m_\ell}(\theta, \phi)$ is a spherical harmonic from Table 8.3. The radial wavefunction $R(r)$ is found from the radial wave equation of Section 8.2,

$$-\frac{\hbar^2}{2m}\left[\frac{d^2R}{dr^2} + \frac{2}{r}\frac{dR}{dr}\right] + \frac{\ell(\ell+1)\hbar^2}{2mr^2}R(r) + U(r)R(r) = ER(r) \tag{8.17}$$

$U(r)$ is the potential energy of Equation 8.33; the remaining terms on the left of Equation 8.17 are associated with the kinetic energy of the electron. The term proportional to $\ell(\ell+1)\hbar^2 = (|\mathbf{L}^2|)$ is the orbital contribution to kinetic energy, K_{orb}. To see this, consider a particle in circular orbit, in which all the kinetic energy is in orbital form (since the distance from the center remains fixed). For such a particle $K_{\text{orb}} = \frac{1}{2}mv^2$ and $|\mathbf{L}| = mvr$. Eliminating the particle speed v, we get

$$K_{\text{orb}} = \frac{m}{2}\left(\frac{|\mathbf{L}|}{mr}\right)^2 = \frac{|\mathbf{L}|^2}{2mr^2} \tag{8.35}$$

Although it was derived for circular orbits, this result correctly represents the orbital contribution to kinetic energy of a mass m in general motion with angular momentum $\mathbf{L}$.[9]

The derivative terms in Equation 8.17 are the radial contribution to the kinetic energy, that is, they represent the contribution from electron motion toward or away from the nucleus. The leftmost term is just what we should write for the kinetic energy of a matter wave $\psi = R(r)$ associated with motion along the coordinate line marked by r. But what significance should we attach to the first derivative term dR/dr? In fact, the presence of this term is evidence that *the effective one-dimensional matter wave is not $R(r)$, but $rR(r)$.* In support of this claim we note that

$$\frac{d^2(rR)}{dr^2} = r\left[\frac{d^2R}{dr^2} + \frac{2}{r}\frac{dR}{dr}\right]$$

Then the radial wave equation written for the effective one-dimensional matter wave $g(r) = rR(r)$ takes the same form as Schrödinger's equation in

---

[9]For any planar orbit we may write $|\mathbf{L}| = rp_\perp$, where $p_\perp$ is the component of momentum in the orbital plane that is normal to the radius vector $\mathbf{r}$. The orbital part of the kinetic energy is then $K_{\text{orb}} = p_\perp^2/2m = |\mathbf{L}|^2/2mr^2$.

one dimension,

$$-\frac{\hbar^2}{2m}\frac{d^2g}{dr^2} + U_{\text{eff}}(r)g(r) = Eg(r) \tag{8.36}$$

but with an *effective potential energy*

$$U_{\text{eff}} = \frac{|\mathbf{L}|^2}{2mr^2} + U(r) = \frac{\ell(\ell+1)\hbar^2}{2mr^2} + U(r) \tag{8.37}$$

The magnitude of $g(r)$ also furnishes probabilities, as described later in this section.

The solution of Equation 8.36 with $U(r) = -kZe^2/r$ (for one-electron atoms or ions) requires methods beyond the scope of this text. Here, as before, we shall present the results and leave their verification to the interested reader. Acceptable wavefunctions can be found only if the energy E is restricted to be one of the following special values:

$$E_n = -\frac{ke^2}{2a_0}\left\{\frac{Z^2}{n^2}\right\} \qquad n = 1, 2, 3, \ldots \tag{8.38}$$

Allowed energies for hydrogen-like atoms

This result is in exact agreement with that found from the simple Bohr theory (see Chapter 4): $a_0 = \hbar^2/m_e ke^2$ is the Bohr radius, 0.529 Å, and $ke^2/2a_0$ is the Rydberg energy, 13.6 eV. The integer n is called the **principal quantum number.** Although n can be any positive integer, the orbital quantum number now is limited to values less than n; that is, ℓ can have only the values

$$\ell = 0, 1, 2, \ldots, (n-1) \tag{8.39}$$

Allowed values of ℓ

The cutoff at $(n-1)$ is consistent with the physical significance of these quantum numbers: The magnitude of orbital angular momentum (fixed by ℓ) cannot become arbitrarily large for a given energy (fixed by n). A semiclassical argument expressing this idea leads to $\ell_{\max} = n - 1$ (see Problem 20). In the same spirit, the restriction on the magnetic quantum number m_ℓ to values between $-\ell$ and $+\ell$ guarantees that the z component of angular momentum never exceeds the magnitude of the vector.

Principal quantum number n
$$n = 1, 2, 3, \ldots$$
Orbital quantum number ℓ
$$\ell = 0, 1, 2, \ldots, (n-1)$$
Magnetic quantum number m_ℓ
$$m_\ell = 0, \pm1, \pm2, \ldots, \pm\ell$$

The radial waves $R(r)$ for hydrogen-like atoms are products of exponentials with polynomials in r/a_0. These radial wavefunctions are tabulated as $R_{n\ell}(r)$ in Table 8.4 for principal quantum numbers up to and including $n = 3$.

For hydrogen-like atoms, then, the quantum numbers are n, ℓ, and m_ℓ, associated with the sharp observables E, $|\mathbf{L}|$, and L_z, respectively. Notice that the energy E depends on n, but not at all on ℓ or m_ℓ. The energy is independent of m_ℓ because of the spherical symmetry of the atom, and this will be true for any central force that varies only with distance r. The fact that E also is independent of ℓ is a special consequence of the coulomb force, however, and is not to be expected in heavier atoms, say, where the force on any one electron includes the electrostatic repulsion of the remaining electrons in the atom, as well as the coulombic attraction of the nucleus.

For historical reasons, all states with the same principal quantum number n are said to form a **shell.** These shells are identified by the letters K, L, M, $\ldots$, which designate the states for which $n = 1, 2, 3, \ldots$. Likewise, states

Table 8.4   **The Radial Wavefunctions $R_{n\ell}(r)$ of Hydrogen-like Atoms for $n = 1, 2,$ and 3**

n	ℓ	$R_{n\ell}(r)$
1	0	$\left(\dfrac{Z}{a_0}\right)^{3/2} 2e^{-Zr/a_0}$
2	0	$\left(\dfrac{Z}{2a_0}\right)^{3/2} \left(2 - \dfrac{Zr}{a_0}\right) e^{-Zr/2a_0}$
2	1	$\left(\dfrac{Z}{2a_0}\right)^{3/2} \dfrac{Zr}{\sqrt{3}\,a_0} e^{-Zr/2a_0}$
3	0	$\left(\dfrac{Z}{3a_0}\right)^{3/2} 2\left[1 - \dfrac{2Zr}{3a_0} + \dfrac{2}{27}\left(\dfrac{Zr}{a_0}\right)^2\right] e^{-Zr/3a_0}$
3	1	$\left(\dfrac{Z}{3a_0}\right)^{3/2} \dfrac{4\sqrt{2}}{3} \dfrac{Zr}{a_0}\left(1 - \dfrac{Zr}{6a_0}\right) e^{-Zr/3a_0}$
3	2	$\left(\dfrac{Z}{3a_0}\right)^{3/2} \dfrac{2\sqrt{2}}{27\sqrt{5}}\left(\dfrac{Zr}{a_0}\right)^2 e^{-Zr/3a_0}$

having the same value of both n and ℓ are said to form a **subshell.** The letters $s, p, d, f, \ldots$ are used to designate the states for which $\ell = 0, 1, 2, 3, \ldots$.[10] This **spectroscopic notation** is summarized in Table 8.5.

The shell (and subshell) energies for several of the lowest-lying states of the hydrogen atom are illustrated in the energy-level diagram of Figure 8.8. The figure also portrays a few of the many electronic transitions possible within the atom. Each such transition represents a change of energy for the atom and must be compensated for by emission (or absorption) of energy in some other form. For **optical transitions,** photons carry off the surplus energy, but not all energy-conserving optical transitions may occur. As it happens, photons also carry angular momentum. To conserve *total* angular momentum (atom + photon) in optical transitions, the angular momentum of the electron in the

Table 8.5   **Spectroscopic Notation for Atomic Shells and Subhells**

n	Shell Symbol	ℓ	Shell Symbol
1	K	0	s
2	L	1	p
3	M	2	d
4	N	3	f
5	O	4	g
6	P	5	h
$\ldots$		$\ldots$	

[10]s, p, d, f are one-letter abbreviations for *sharp, principal, diffuse,* and *fundamental.* The nomenclature is a throwback to the early days of spectroscopic observations, when these terms were used to characterize the appearance of spectral lines.

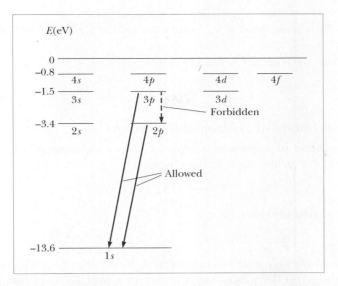

Figure 8.8   Energy-level diagram of atomic hydrogen. Allowed photon transitions are those obeying the selection rule $\Delta\ell = \pm 1$. The $3p \rightarrow 2p$ transition ($\Delta\ell = 0$) is said to be forbidden, though it may still occur (but only rarely).

initial and final states must differ by exactly one unit, that is

$$|\ell_f - \ell_i| = 1 \qquad \text{or} \qquad \Delta\ell = \pm 1 \qquad (8.40)$$

Selection rule for allowed transitions

Equation 8.40 expresses a **selection rule** that must be obeyed in optical transitions.[11] As Figure 8.8 indicates, the transitions $3p \rightarrow 1s$ and $2p \rightarrow 1s$ are **allowed** by the rule ($\Delta\ell = -1$), but the $3p \rightarrow 2p$ transition is said to be **forbidden** ($\Delta\ell = 0$). (Such transitions can occur, but with negligible probability compared with that of allowed transitions.) Clearly, selection rules play a vital role in the interpretation of atomic spectra.

EXAMPLE 8.7   The $n = 2$ Level of Hydrogen

Enumerate all states of the hydrogen atom corresponding to the principal quantum number $n = 2$, giving the spectroscopic designation for each. Calculate the energies of these states.

Solution   When $n = 2$, ℓ can have the values 0 and 1. If $\ell = 0$, m_ℓ can only be 0. If $\ell = 1$, m_ℓ can be -1, 0, or $+1$. Hence, we have a $2s$ state with quantum numbers

$$n = 2, \qquad \ell = 0, \qquad m_\ell = 0$$

and three $2p$ states for which the quantum numbers are

$$n = 2, \qquad \ell = 1, \qquad m_\ell = -1$$
$$n = 2, \qquad \ell = 1, \qquad m_\ell = 0$$
$$n = 2, \qquad \ell = 1, \qquad m_\ell = +1$$

Because all of these states have the same principal quantum number, $n = 2$, they also have the same energy, which can be calculated from Equation 8.38. For $Z = 1$ and $n = 2$, this gives

$$E_2 = -(13.6 \text{ eV})\{1^2/2^2\} = -3.4 \text{ eV}$$

---

[11]$\Delta\ell = 0$ also is allowed by angular momentum considerations but forbidden by parity conservation. Further, since the angular momentum of a photon is just $\pm\hbar$, the z component of the atom's angular momentum cannot change by more than $\pm\hbar$, giving rise to a *second selection rule*, $\Delta m_\ell = 0, \pm 1$.

Exercise 4   How many possible states are there for the $n = 3$ level of hydrogen? For the $n = 4$ level?

Answers   Nine states for $n = 3$, and 16 states for $n = 4$.

The Ground State of Hydrogen-like Atoms

The ground state of a one-electron atom or ion with atomic number Z, for which $n = 1$, $\ell = 0$, and $m_\ell = 0$, has energy

$$E_1 = -(13.6 \text{ eV})Z^2 \tag{8.41}$$

The wavefunction for this state is

$$\psi_{100} = R_{10}(r)Y_0^0 = \pi^{-1/2}(Z/a_0)^{3/2}e^{-Zr/a_0} \tag{8.42}$$

The constants are such that ψ is normalized. Notice that ψ_{100} does not depend on angle, since it is the product of a radial wave with $Y_0^0 = 1/\sqrt{4\pi}$. In fact, all the $\ell = 0$ waves share this feature; that is, **all *s*-state waves are spherically symmetric.**

The electron described by the wavefunction of Equation 8.42 is found with a probability per unit volume given by

$$|\psi_{100}|^2 = \frac{Z^3}{\pi a_0^3} e^{-2Zr/a_0}$$

(Three-dimensional renditions of the probability per unit volume $|\psi(\mathbf{r})|^2$—often called electron "clouds"—are constructed by making the shading at every point proportional to $|\psi(\mathbf{r})|^2$.) The probability distribution also is spherically symmetric, as it would be for any *s*-state wave; that is, the likelihood for finding the electron in the atom is the same at all points equidistant from the center (nucleus). Thus, it is convenient to define another probability function, called the **radial probability distribution,** with its associated density $P(r)$, such that ***P(r) dr* is the probability of finding the electron anywhere in the spherical shell of radius *r* and thickness *dr*** (Fig. 8.9). The shell volume is its surface area, $4\pi r^2$, multiplied by the shell thickness, dr. Since the probability density $|\psi_{100}|^2$ is the same everywhere in the shell, we have

$$P(r)\,dr = |\psi|^2\,4\pi r^2\,dr$$

or, for the hydrogen-like 1*s* state,

$$P_{1s}(r) = \frac{4Z^3}{a_0^3}r^2e^{-2Zr/a_0} \tag{8.43}$$

The same result is obtained for $P_{1s}(r)$ from the intensity of the effective one-dimensional matter wave $g(r) = rR(r)$:

$$P(r) = |g(r)|^2 = r^2|R(r)|^2 \tag{8.44}$$

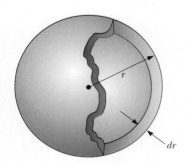

Figure 8.9 $P(r)\,dr$ is the probability that the electron will be found in the volume of a spherical shell with radius r and thickness dr. The shell volume is just $4\pi r^2\,dr$.

The radial probability density for any state

In fact, *Equation 8.44 gives the correct radial probability density for any state;* for the spherically symmetric *s*-states this is the same as $4\pi r^2|\psi|^2$, since then $\psi(r) = (1/\sqrt{4\pi})R(r)$.[12]

A plot of the function $P_{1s}(r)$ is presented in Figure 8.10a; Figure 8.10b shows the 1*s* electron "cloud" or probability per unit volume $|\psi_{100}|^2$ from which $P_{1s}(r)$ derives. $P(r)$ may be loosely interpreted as the probability of finding the electron at distance *r* from the nucleus, irrespective of its angular position. Thus, the peak of the curve in Figure 8.10a represents the most probable distance of the 1*s* electron from the nucleus. Furthermore, the normalization condition becomes

$$1 = \int_0^\infty P(r)\ dr \tag{8.45}$$

where the integral is taken over all *possible* values of *r*. The average distance of the electron from the nucleus is found by weighting each possible distance with the probability that the electron will be found at that distance:

$$\langle r \rangle = \int_0^\infty rP(r)\ dr \tag{8.46}$$

The average distance of an electron from the nucleus

In fact, the average value of *any* function of distance $f(r)$ is obtained

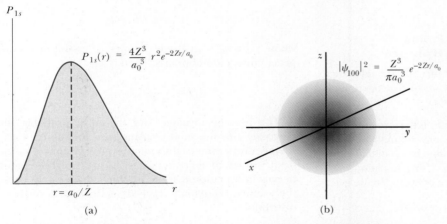

Figure 8.10 (a) The curve $P_{1s}(r)$ representing the probability of finding the electron as a function of distance from the nucleus in a 1*s* hydrogen-like state. Note that the probability takes its maximum value when *r* equals a_0/Z. (b) The spherical electron "cloud" for a hydrogen-like 1*s* state. The shading at every point is proportional to the probability density $|\psi_{1s}(\mathbf{r})|^2$.

[12]From its definition, $P(r)\ dr$ always may be found by integrating $|\psi|^2 = |R(r)|^2|Y_\ell^{m_\ell}|^2$ over the volume of a spherical shell having radius *r* and thickness *dr*. Since the volume element in spherical coordinates is $dV = r^2 dr\sin\theta\ d\theta\ d\phi$, and the integral of $|Y_\ell^{m_\ell}|^2$ over angle is unity (see footnote 5), this leaves $P(r) = r^2|R(r)|^2$.

by weighting the function value at every distance with the probability at that distance:

$$\langle f \rangle = \int_0^\infty f(r) P(r) \; dr \qquad (8.47)$$

EXAMPLE 8.8   Probabilities for the Electron in Hydrogen

Calculate the probability that the electron in the ground state of hydrogen will be found outside the first Bohr radius.

Solution   The probability is found by integrating the radial probability density for this state, $P_{1s}(r)$, from the Bohr radius a_0 to ∞. Using Equation 8.43 with $Z = 1$ for hydrogen gives

$$P = \frac{4}{a_0^3} \int_{a_0}^\infty r^2 e^{-2r/a_0} \; dr$$

We can put the integral in dimensionless form by changing variables from r to $z = 2r/a_0$. Noting that $z = 2$ when $r = a_0$, and that $dr = (a_0/2) \; dz$, we get

$$P = \frac{1}{2} \int_2^\infty z^2 e^{-z} dz = -\frac{1}{2} \{z^2 + 2z + 2\} e^{-z} \Big|_2^\infty = 5e^{-2}$$

This is about 0.677, or 67.7%.

EXAMPLE 8.9   The Electron–Proton Separation in Hydrogen

Calculate the most probable distance of the electron from the nucleus in the ground state of hydrogen, and compare this with the average distance.

Solution   The most probable distance is the value of r that makes the radial probability $P(r)$ a maximum. The slope here is zero, so the most probable value of r is obtained by setting $dP/dr = 0$ and solving for r. Using Equation 8.43 with $Z = 1$ for the $1s$, or ground, state of hydrogen, we get

$$0 = \left(\frac{4}{a_0^3}\right) \frac{d}{dr} \{r^2 e^{-2r/a_0}\} = \left(\frac{4}{a_0^3}\right) e^{-2r/a_0} \left\{-\frac{2r^2}{a_0} + 2r\right\}$$

The right-hand side is zero for $r = 0$ and for $r = a_0$. Since $P(0) = 0$, $r = 0$ is a minimum of $P(r)$, not a maximum. Thus, the most probable distance is

$$r = a_0$$

The average distance is obtained from Equation 8.46, which in this case becomes

$$\langle r \rangle = \frac{4}{a_0^3} \int_0^\infty r^3 e^{-2r/a_0} \; dr$$

Again introducing $z = 2r/a_0$, we obtain

$$\langle r \rangle = \frac{a_0}{4} \int_0^\infty z^3 e^{-z} \; dz$$

The definite integral on the right is one of a broader class,

$$\int_0^\infty z^n e^{-z} dz = n!$$

whose value $n! = n(n-1) \; \ldots \; (1)$ is established by repeated integration by parts. Then

$$\langle r \rangle = \frac{a_0}{4} (3!) = \frac{3}{2} a_0$$

The average distance and the most probable distance are not the same, because the probability curve $P(r)$ is not symmetric about the peak distance a_0. Indeed, values of r greater than a_0 are weighted more heavily in Equation 8.46 than values smaller than a_0, so the average $\langle r \rangle$ actually exceeds a_0 for this probability distribution.

Excited States of Hydrogen-like Atoms

There are four first excited states for hydrogen-like atoms: ψ_{200}, ψ_{210}, ψ_{211}, and ψ_{21-1}. All have the same principal quantum number $n = 2$, hence the same total energy

$$E_2 = -(13.6 \text{ eV}) \frac{Z^2}{4}$$

Accordingly, the first excited level, E_2, is fourfold degenerate.

The 2s state, ψ_{200}, is again spherically symmetric. Plots of the radial probability density for this and several other hydrogen-like states are shown in Figure 8.11. Note that the plot for the 2s state has two peaks. In this case, the most probable distance ($\sim 5a_0/Z$) is marked by the highest peak. An electron in the 2s state would be much farther from the nucleus (on the average) than an electron in the 1s state. Likewise, the most probable distances are even greater for an electron in any of the $n = 3$ states (3s, 3p, or 3d). Observations such as these continue to support the old idea of a shell structure for the atom, even in the face of the uncertainties inherent in the wave nature of matter.

The remaining three first excited states, ψ_{211}, ψ_{210}, and ψ_{21-1}, have $\ell = 1$ and make up the 2p subshell. These states are not spherically symmetric. All of them have the same radial wavefunction $R_{21}(r)$, but they are multiplied by different spherical harmonics and thus depend differently on the angles θ and ϕ. For example, the wavefunction ψ_{211} is

$$\psi_{211} = R_{21}(r)\, Y_1^1 = \pi^{-1/2}\left(\frac{Z}{a_0}\right)^{3/2}\left(\frac{Zr}{8a_0}\right) e^{-Zr/2a_0} \sin\theta\, e^{i\phi} \qquad (8.48)$$

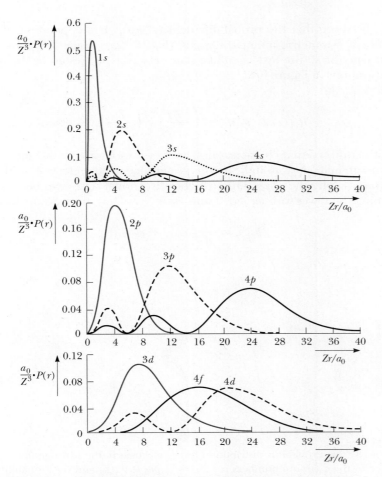

Figure 8.11 The radial probability density function for several states of hydrogen-like atoms.

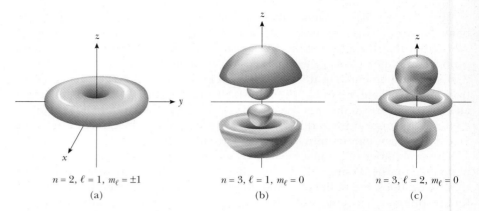

$n = 2, \ell = 1, m_\ell = \pm 1$
(a)

$n = 3, \ell = 1, m_\ell = 0$
(b)

$n = 3, \ell = 2, m_\ell = 0$
(c)

Figure 8.12 (a) The probability density $|\psi_{211}|^2$ for a hydrogen-like $2p$ state. Note the axial symmetry about the z-axis. (b) and (c) The probability densities $|\psi(\mathbf{r})|^2$ for several other hydrogen-like states. The electron "cloud" is axially symmetric about the z-axis for all the hydrogen-like states $\psi_{n\ell m_\ell}(\mathbf{r})$.

Notice, however, that the probability density $|\psi_{211}|^2$ is independent of ϕ and therefore is symmetric about the z-axis (Fig. 8.12a). Since $|e^{im_l\phi}|^2 = 1$, the same is true for all the hydrogen-like states $\psi_{n\ell m_\ell}$, as suggested by the remaining illustrations in Figure 8.12.

The ψ_{210} state

$$\psi_{210} = R_{21}(r) Y_1^0 = \pi^{-1/2} \left(\frac{Z}{2a_0} \right)^{3/2} \left(\frac{Zr}{2a_0} \right) e^{-Zr/2a_0} \cos\theta \qquad (8.49)$$

has distinct directional characteristics, as shown in Figure 8.13a, and is sometimes designated $[\psi_{2p}]_z$ to indicate the preference for the electron in this state to be found along the z-axis. Other highly directional states can be formed by combining the waves with $m_\ell = +1$ and $m_\ell = -1$. Thus, the wavefunctions

$$[\psi_{2p}]_x = \frac{1}{\sqrt{2}} \{\psi_{211} + \psi_{21-1}\} \qquad (8.50a)$$

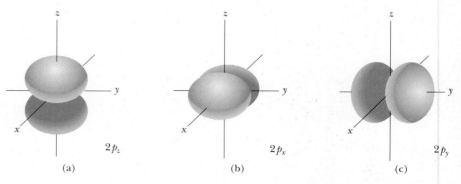

$2p_z$
(a)

$2p_x$
(b)

$2p_y$
(c)

Figure 8.13 (a) Probability distribution for an electron in the hydrogenlike $2p_z$ state, described by the quantum numbers $n = 2, \ell = 1, m_\ell = 0$. (b) and (c) Probability distributions for the $2p_x$ and $2p_y$ states. The three distributions $2p_x$, $2p_y$, and $2p_z$ have the same structure, but differ in their spatial orientation.

and

$$[\psi_{2p}]_y = \frac{1}{\sqrt{2}} \{\psi_{211} - \psi_{21-1}\} \qquad (8.50b)$$

have large probability densities along the x- and y-axes, respectively, as shown in Figures 8.13b and 8.13c. The wavefunctions of Equations 8.50, formed as superpositions of first excited state waves with *identical* energies E_2, are themselves stationary states with this same energy; indeed, the three wavefunctions $[\psi_{2p}]_z$, $[\psi_{2p}]_x$, and $[\psi_{2p}]_y$ together constitute an equally good description of the $2p$ states for hydrogenlike atoms. Wavefunctions with a highly directional character, such as these, play an important role in chemical bonding, the formation of molecules, and chemical properties.

The next excited level, E_3, is ninefold degenerate; that is, there are nine different wavefunctions with this energy, corresponding to all possible choices for ℓ and m_ℓ consistent with $n = 3$. Together, these nine states constitute the third shell, with subshells $3s$, $3p$, and $3d$ composed of one, three, and five states, respectively. The wavefunctions for these states may be constructed from the entries in Tables 8.3 and 8.4. Generally, they are more complicated than their second-shell counterparts because of the increasing number of nodes in the radial part of the wavefunction.

The progression to higher-lying states leads to still more degeneracy and wavefunctions of ever-increasing complexity. The nth shell has energy

$$E_n = -(13.6 \text{ eV}) \frac{Z^2}{n^2}$$

and contains exactly n subshells, corresponding to $\ell = 0, 1, \ldots, n - 1$. Within the ℓth subshell there are $2\ell + 1$ orbitals. Altogether, the nth shell contains a total of n^2 states, all with the same energy, E_n; that is, **the energy level E_n is n^2 degenerate.** Equivalently, we say the nth shell can hold as many as n^2 electrons, assuming no more than one electron can occupy a given orbital. This argument underestimates the actual shell capacity by a factor of 2, owing to the existence of electron spin, as discussed in the next chapter.

8.6 ANTIHYDROGEN

The constituents of hydrogen atoms, protons and electrons, are abundant in the universe and among the elementary particles that make up all matter around us. But each of these elementary particles has a partner, its *antiparticle*, identical to the original in all respects other than carrying charge of the opposite sign.[13] The anti-electron, or positron, was discovered in 1932 by Carl Anderson.[14] The positron has the same mass as the electron but carries charge

---

[13]Strictly speaking, the antiparticle also has a magnetic moment opposite that of its companion particle. Magnetic moments are discussed in Chapter 9, and are intimately related to a new particle property called *spin*. The spin is the same for particle and antiparticle.

[14]The idea of antiparticles received a solid theoretical underpinning in 1928 with P. A. M. Dirac's relativistic theory of the electron. While furnishing an accurate quantum description of electrons with relativistic energies, Dirac's theory also included mysterious "negative energy" states. Eventually Dirac realized that these "negative energy" states actually describe antiparticles with positive energy. Dirac's conjecture was subsequently confirmed with the discovery of the positron in 1932. See Section 15.2.

$+e$. Because it cannot be produced without a powerful particle accelerator, the much more massive antiproton was not observed until 1955. Again, the antiproton has the same mass as the proton but is oppositely charged. Just as atoms and ordinary matter are made up of particles, it is easy to conceive of anti-atoms and all forms of antimatter built out of antiparticles. Indeed, on the face of it there would be no way to tell if we lived in an antimatter world and were ourselves composed entirely of antimatter!

The simplest such anti-atom is **antihydrogen,** the most fundamental neutral unit of antimatter. Antihydrogen consists of a positron bound electrically to an antiproton. Many physicists believe that the study of antihydrogen can answer the question of whether there is some fundamental, heretofore unknown difference between matter and antimatter and why our Universe seems to be composed almost exclusively of ordinary matter. For the reasons outlined here, the production of antihydrogen is fraught with difficulties, and it was not until the mid 1990s that two groups, one at CERN and the other at Fermilab, reported success in producing antihydrogen at high energies. In 2002, the CERN group reported antihydrogen production at the very low energies required for precision comparison measurements with ordinary hydrogen. Trapping the anti-atoms long enough so that experiments can be performed on them is much more difficult and has not yet been achieved.

When a particle and its antiparticle collide, both disappear in a burst of electromagnetic energy. This is **pair annihilation,** the direct conversion of mass into energy in accord with Einstein's famous relation $E = mc^2$. An electron and a positron combine to produce two (sometimes three) gamma-ray photons (one photon alone cannot conserve both energy *and* momentum). The collision of a proton and antiproton produces three or four other elementary particles called *pions*. The problem that experimentalists face is that their laboratories and measuring instruments are made of ordinary matter and antiparticles will self-destruct on first contact with the apparatus. A similar problem arises in the containment of plasmas, which are tamed using a magnetic trap, that is, a configuration of magnetic fields that exert forces on charged particles of the plasma to keep them confined. But neutral anti-atoms experience only weak magnetic forces and will quickly escape the trap unless they are moving very slowly. Thus, the antihydrogen atom must be *cold* if it is to survive long enough to be useful for precision experiments, and this presents yet another challenge. While positrons are readily available as decay products of naturally occurring radioactive species like ^{22}Na, antiprotons must be created artificially in particle accelerators by bombarding heavy targets (Be) with ultra-energetic ($\sim$GeV) protons. The positrons and antiprotons so produced are very energetic and must be slowed down enormously to form cold antihydrogen. The slowdown is achieved at the cost of lost particles in what is essentially an accelerator run backwards. The CERN experiment yields about 50,000 antihydrogen atoms starting from some 1.5 million antiprotons. And of these, only a small fraction is actually detected.

Detecting antihydrogen is a challenge in its own right. The existence of an antiproton is confirmed by the decay products (pions) it produces on annihilation with its antimatter counterpart, the proton. These decay products leave directional traces in the detectors that surround the anti-atom sample. From the directional traces, physicists are able to reconstruct the precise location of the annihilation event. To confirm the existence of antihydrogen, however, one must also record electron–positron annihilation in the same place at the same time. The tell-tale gamma-ray photons produced in that annihilation can also be traced back to a point of origination. Thus, the signature of antihydrogen is the coincidence of multiple distinct detection events, as illustrated in Figure 8.14.

According to the theory presented in this chapter, the energy spectrum of antihydrogen should be identical to that of ordinary hydrogen. The hydrogen atom is the best known of all physical systems, and extremely precise measurements of its spectrum have been made, the best of which is accurate to about 2 parts in 10^{14}. Thus, comparing spectra of antihydrogen with ordinary hydrogen would allow a stringent test of the symmetry expected between matter and antimatter in atomic interactions. This symmetry is rooted in so-called CPT invariance, which states that if one were to take any lump of matter, reverse the sign of all the elementary charges (C), the direction of time's flow (T), and another property of particles called parity (P), the specimen would obey the same laws of physics. CPT invariance is a very general consequence of quantum theory and the covariance of quantum laws under Lorentz transformation demanded by special relativity and is the cornerstone for every modern theory of matter. If CPT invariance is violated, the whole of physical theory at the fundamental level will have to be rewritten. While nearly all physicists agree that is very unlikely, the prospect of CPT violation holds tantalizing possibilities and may shed light on one of the most perplexing problems of modern cosmology: why there is now a preponderance of matter in the Universe, when the Big Bang theory predicts that matter and antimatter should have been created in equal amounts. Furthermore, if antihydrogen responds differently to gravity, the theory of relativity in its present form would be compromised, a development that could point the way to the long-sought unification of relativity and quantum theory.

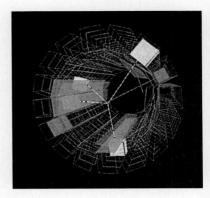

Figure 8.14 Antihydrogen is detected through its destruction in collisions with matter particles. The annihilation of the antiproton produces pions that, picked up in the surrounding detectors (light colored), can be traced back (four light colored dashed lines) to the annihilation point. Similarly, the annihilation of the positron produces a distinctive back-to-back two-photon signature (two dashed tracks at 180° to one another). Overlap of the two annihilation points signifies that the positron and antiproton were bound together in an atom of antihydrogen. (Adapted from *Nature*, 419, 456–459, October 3, 2002.)

SUMMARY

In three dimensions, the matter wave intensity $|\Psi(\mathbf{r}, t)|^2$ represents the probability per unit volume for finding the particle at $\mathbf{r}$ at time t. Probabilities are found by integrating this probability density over the volume of interest.

The wavefunction itself must satisfy the Schrödinger equation

$$-\frac{\hbar^2}{2m}\nabla^2\Psi + U(\mathbf{r})\Psi = i\hbar\frac{\partial\Psi}{\partial t} \tag{8.1}$$

Stationary states are solutions to this equation in separable form: $\Psi(\mathbf{r}, t) = \psi(\mathbf{r})e^{-i\omega t}$ with $\psi(\mathbf{r})$ satisfying

$$-\frac{\hbar^2}{2m}\nabla^2\psi(\mathbf{r}) + U(\mathbf{r})\psi(\mathbf{r}) = E\psi(\mathbf{r}) \tag{8.5}$$

This is the **time-independent Schrödinger equation,** from which we obtain the time-independent wavefunction $\psi(\mathbf{r})$ and the allowed values of particle energy E.

For a particle confined to a cubic box whose sides are L, the magnitudes of the components of particle momentum normal to the walls of the box can be made sharp, as can the particle energy. The sharp momentum values are quantized as

$$|p_x| = n_1 \frac{\pi\hbar}{L}$$

$$|p_y| = n_2 \frac{\pi\hbar}{L} \tag{8.8}$$

$$|p_z| = n_3 \frac{\pi\hbar}{L}$$

and the allowed energies are found to be

$$E = \frac{\pi^2\hbar^2}{2mL^2}\{n_1^2 + n_2^2 + n_3^2\} \tag{8.9}$$

The three **quantum numbers** n_1, n_2, and n_3 are all positive integers, one for each degree of freedom of the particle. Many levels of this system are **degenerate;** that is, there is more than one wavefunction with the same energy.

For particles acted on by a **central force,** the angular momentum **L** is a constant of the classical motion and is quantized along with particle energy. Wavefunctions for which the z component L_z and magnitude $|\mathbf{L}|$ of angular momentum are simultaneously sharp are the **spherical harmonics** $Y_\ell^{m_\ell}(\theta, \phi)$ in the spherical coordinate angles θ and ϕ. For any central force, angular momentum is quantized by the rules

$$|\mathbf{L}| = \sqrt{\ell(\ell + 1)}\,\hbar$$

and

$$L_z = m_\ell\hbar \tag{8.16}$$

The **orbital quantum number** ℓ must be a nonnegative integer. For a fixed value of ℓ, the **magnetic quantum number** m_ℓ is restricted to integer values lying between $-\ell$ and $+\ell$. Since $|\mathbf{L}|$ and L_z are quantized differently, the classical freedom to orient the z-axis in the direction of **L** must be abandoned. This stunning conclusion is the essence of **space quantization.** Furthermore, no two components of **L**, such as L_z and L_y, can be sharp simultaneously. This implies a lower limit to the uncertainty product $\Delta L_z \Delta L_y$ and gives rise to an uncertainty principle for the components of angular momentum.

A central force of considerable importance is the force on the electron in a one-electron atom or ion. This is the coulomb force, described by the potential energy $U(r) = -kZe^2/r$, where Z is the atomic number of the nucleus. The allowed energies for this case are given by

$$E_n = -\frac{ke^2}{2a_0}\left\{\frac{Z^2}{n^2}\right\} \qquad n = 1, 2, \ldots \tag{8.38}$$

This coincides exactly with the results obtained from the Bohr theory. The energy depends only on the **principal quantum number** n. For a fixed value of n, the orbital and magnetic quantum numbers are restricted as

$$\ell = 0, 1, 2, \ldots, n - 1$$

$$m_\ell = 0, \pm1, \pm2, \ldots, \pm\ell \tag{8.39}$$

All states with the same principal quantum number n form a **shell,** identified by the letters K, L, M, ... (corresponding to $n = 1, 2, 3, ...$). All states with the same values of n and ℓ form a **subshell,** designated by the letters s, p, d, f, ... (corresponding to $\ell = 0, 1, 2, ...$).

The wavefunctions for an electron in hydrogen or a hydrogen-like ion,

$$\psi(r, \theta, \phi) = R_{n\ell}(r) Y_\ell^{m_\ell}(\theta, \phi)$$

depend on the three quantum numbers n, ℓ, and m_ℓ and are products of spherical harmonics multiplied by radial wavefunctions $R_{n\ell}(r)$. The **effective one-dimensional wavefunction** $g(r) = rR_{n\ell}(r)$ is analogous to the wavefunction $\psi(x)$ in one dimension; the intensity of $g(r)$ gives the **radial probability density,**

$$P(r) = |g(r)|^2 = r^2 |R_{n\ell}(r)|^2 \qquad (8.44)$$

$P(r) \, dr$ is the probability that the electron will be found at a distance between r and $r + dr$ from the nucleus. The most probable distance is the one that maximizes $P(r)$ and generally differs from the average distance $\langle r \rangle$, calculated as

$$\langle r \rangle = \int_0^\infty rP(r) \, dr \qquad (8.46)$$

The most probable values are found to coincide with the radii of the allowed orbits in the Bohr theory.

SUGGESTIONS FOR FURTHER READING

1. For more on the use of angular momentum to simplify the three-dimensional Schrödinger equation in central force applications, see Chapter 11 of *An Introduction to Quantum Physics*, by A. P. French and Edwin F. Taylor, New York, W. W. Norton and Company, Inc., 1978.
2. A discussion of the radial probability density and its use for the lowest states of hydrogen is contained in Chapter 7 of *Modern Physics*, 2nd ed., by Kenneth Krane, New York, John Wiley and Sons, Inc., 1996.
3. Illustrations in perspective of the electron "cloud" $|\psi(\mathbf{r})|^2$ for various hydrogen-like states may be found in *Quantum Physics of Atoms, Molecules, Solids, Nuclei, and Particles*, 2nd ed., by R. Eisberg and R. Resnick, New York, John Wiley and Sons, Inc., 1985.

QUESTIONS

1. Why are three quantum numbers needed to describe the state of a one-electron atom?
2. Compare the Bohr theory with the Schrödinger treatment of the hydrogen atom. Comment specifically on the total energy and orbital angular momentum.
3. How do we know whether a given $2p$ electron in an atom has $m_\ell = 0, +1$, or -1? What value of m_ℓ characterizes a directed orbital such as $[\psi_{2p}]_x$ of Equation 8.50?
4. For atomic s states, the probability density $|\psi|^2$ is largest at the origin, yet the probability for finding the electron a distance r from the nucleus, given by $P(r)$, goes to zero with r. Explain.
5. For the electron in the ground state of hydrogen— as with many other quantum systems—the kinetic and potential energies are fuzzy observables, but their sum, the total energy, is sharp. Explain how this can be so.
6. Discuss the relationship between space quantization and Schrödinger's equation. If the latter were invalid, would space quantization still hold?

PROBLEMS

8.1   Particle in a Three-Dimensional Box

1. A particle of mass m moves in a three-dimensional box with edge lengths L_1, L_2, and L_3. Find the energies of the six lowest states if $L_1 = L$, $L_2 = 2L$, and $L_3 = 2L$. Which of these energies are degenerate?

2. An electron moves in a cube whose sides have a length of 0.2 nm. Find values for the energy of (a) the ground state and (b) the first excited state of the electron.

3. A particle of mass m moves in a three-dimensional box with sides L. If the particle is in the third excited level, corresponding to $n^2 = 11$, find (a) the energy of the particle, (b) the combinations of n_1, n_2, and n_3 that would give this energy, and (c) the wavefunctions for these different states.

4. A particle of mass m moves in a two-dimensional box of sides L. (a) Write expressions for the wavefunctions and energies as a function of the quantum numbers n_1 and n_2 (assuming the box is in the xy plane). (b) Find the energies of the ground state and first excited state. Is either of these states degenerate? Explain.

5. Assume that the nucleus of an atom can be regarded as a three-dimensional box of width 2×10^{-14} m. If a proton moves as a particle in this box, find (a) the ground-state energy of the proton in MeV and (b) the energies of the first and second excited states. (c) What are the degeneracies of these states?

6. Obtain the stationary states for a *free* particle in three dimensions by separating the variables in Schrödinger's equation. Do this by substituting the separable form $\Psi(\mathbf{r}, t) = \psi_1(x)\psi_2(y)\psi_3(z)\phi(t)$ into the time-dependent Schrödinger equation and dividing each term by $\Psi(\mathbf{r}, t)$. Isolate all terms depending only on x from those depending only on y, and so on, and argue that four separate equations must result, one for each of the unknown functions ψ_1, ψ_2, ψ_3, and ϕ. Solve the resulting equations. What dynamical quantities are sharp for the states you have found?

7. For a particle confined to a cubic box of dimension L, show that the normalizing factor is $A = (2/L)^{3/2}$, the *same* value for *all* the stationary states. How is this result changed if the box has edge lengths L_1, L_2, and L_3, all of which are different?

8. Consider a particle of mass m confined to a three-dimensional cube of length L so small that the particle motion is *relativistic*. Obtain an expression for the allowed particle energies in this case. Compute the ground-state energy for an electron if $L = 10$ fm (10^{-5} nm), a typical nuclear dimension.

8.2   Central Forces and Angular Momentum

9. If an electron has an orbital angular momentum of 4.714×10^{-34} J·s, what is the orbital quantum number for this state of the electron?

10. Consider an electron for which $n = 4$, $\ell = 3$, and $m_\ell = 3$. Calculate the numerical value of (a) the orbital angular momentum and (b) the z component of the orbital angular momentum.

11. The orbital angular momentum of the Earth in its motion about the Sun is 4.83×10^{31} kg·m²/s. Assuming it is quantized according to Equation 8.16, find (a) the value of ℓ corresponding to this angular momentum and (b) the fractional change in $|\mathbf{L}|$ as ℓ changes from ℓ to $\ell + 1$.

8.5   Atomic Hydrogen and Hydrogen-like Ions

12. The normalized ground-state wavefunction for the electron in the hydrogen atom is

$$\psi(r, \theta, \phi) = \frac{1}{\sqrt{\pi}}\left(\frac{1}{a_0}\right)^{3/2} e^{-r/a_0}$$

where r is the radial coordinate of the electron and a_0 is the Bohr radius. (a) Sketch the wavefunction versus r. (b) Show that the probability of finding the electron between r and $r + dr$ is given by $|\psi(r)|^2 4\pi r^2\, dr$. (c) Sketch the probability versus r and from your sketch find the radius at which the electron is most likely to be found. (d) Show that the wavefunction as given is normalized. (e) Find the probability of locating the electron between $r_1 = a_0/2$ and $r_2 = 3a_0/2$.

13. (a) Determine the quantum numbers ℓ and m_ℓ for the He$^+$ ion in the state corresponding to $n = 3$. (b) What is the energy of this state?

14. (a) Determine the quantum numbers ℓ and m_ℓ for the Li^{2+} ion in the states for which $n = 1$ and $n = 2$. (b) What are the energies of these states?

15. In obtaining the results for hydrogen-like atoms in Section 8.5, the atomic nucleus was assumed to be immobile due to its much larger mass compared with that of the electron. If this assumption is relaxed, the results remain valid if the electron mass m is replaced everywhere by the *reduced mass* μ of the electron–nucleus combination:

$$\mu = \frac{mM}{m + M}$$

Here M is the nuclear mass. (a) Making this replacement in Equation 8.38, show that a more general expression for the allowed energies of a one-electron atom with atomic number Z is

$$E_n = -\frac{\mu k^2 e^4}{2\hbar^2}\left\{\frac{Z^2}{n^2}\right\}$$

(b) The wavelength for the $n = 3$ to $n = 2$ transition of the hydrogen atom is 656.3 nm (visible red light). What is the wavelength of this same transition in singly ionized helium? In positronium? (*Note:* Positronium is an "atom" consisting of a bound positron–electron pair. A positron is a positively charged electron.)

16. Calculate the possible values of the z component of angular momentum for an electron in a d subshell.

17. Calculate the angular momentum for an electron in (a) the 4d state and (b) the 6f state of hydrogen.

18. A hydrogen atom is in the 6g state. (a) What is the principal quantum number? (b) What is the energy of the atom? (c) What are the values for the orbital quantum number and the magnitude of the electron's orbital angular momentum? (d) What are the possible values for the magnetic quantum number? For each value, find the corresponding z component of the electron's orbital angular momentum and the angle that the orbital angular momentum vector makes with the z-axis.

19. Prove that the nth energy level of an atom has degeneracy equal to n^2.

20. For fixed electron energy, the orbital quantum number ℓ is limited to $n - 1$. We can obtain this result from a semiclassical argument using the fact that the largest angular momentum describes circular orbits, where all the kinetic energy is in orbital form. For hydrogen-like atoms, $U(r) = -Zke^2/r$, and the energy in circular orbits becomes

$$E = \frac{|\mathbf{L}|^2}{2mr^2} - \frac{Zke^2}{r}$$

Quantize this relation using the rules of Equations 8.16 and 8.38, together with the Bohr result for the allowed values of r, to show that the largest integer value of ℓ consistent with the total energy is $\ell_{max} = n - 1$.

21. Suppose that a hydrogen atom is in the 2s state. Taking $r = a_0$, calculate values for (a) $\psi_{2s}(a_0)$, (b) $|\psi_{2s}(a_0)|^2$, and (c) $P_{2s}(a_0)$.

22. The radial part of the wavefunction for the hydrogen atom in the 2p state is given by

$$R_{2p}(r) = Are^{-r/2a_0}$$

where A is a constant and a_0 is the Bohr radius. Using this expression, calculate the average value of r for an electron in this state.

23. A dimensionless number that often appears in atomic physics is the *fine-structure constant* α, given by

$$\alpha = \frac{ke^2}{\hbar c}$$

where k is the Coulomb constant. (a) Obtain a numerical value for $1/\alpha$. (b) In scattering experiments, the "size" of the electron is the *classical electron radius*, $r_0 = ke^2/m_ec^2$. In terms of α, what is the ratio of the Compton wavelength, $\lambda = h/m_ec$, to the classical electron radius? (c) In terms of α, what is the ratio of the Bohr radius, a_0, to the Compton wavelength? (d) In terms of α, what is the ratio of the *Rydberg wavelength*, $1/R$, to the Bohr radius?

24. Calculate the average potential and kinetic energies for the electron in the ground state of hydrogen.

25. Compare the most probable distances of the electron from the proton in the hydrogen 2s and 2p states with the radius of the second Bohr orbit in hydrogen, $4a_0$.

26. Compute the probability that a 2s electron of hydrogen will be found inside the Bohr radius for this state, $4a_0$. Compare this with the probability of finding a 2p electron in the same region.

27. Use the Java applet available at our companion Web site (http://info.brookscole.com/mp3e → QMTools Simulations → Problem 8.27) to display the radial waveforms for the $n = 3$ level of atomic hydrogen. Locate the most probable distance from the nucleus for an electron in the 3s state. Do the same for an electron in the 3p and 3d states. What does the simple Bohr theory predict for this case?

28. *Angular Variation of Hydrogen Wavefunctions.* Use the Java applet of the preceding problem to display the electron clouds for the $n = 4$ states of atomic hydrogen. Observe the distinctly different symmetries of the s, p, d, and f orbitals in the case $m_\ell = 0$. Which of these orbitals is most extended, that is, in which orbital is the electron likely to be found furthest away from the nucleus? Explore the effect of the magnetic quantum number m_ℓ on the overall appearance and properties of the $n = 4$ orbitals. Can you identify any trends?

29. As shown in Example 8.9, the average distance of the electron from the proton in the hydrogen ground state is 1.5 bohrs. For this case, calculate Δr, the uncertainty in distance about the average value, and compare it with the average itself. Comment on the significance of your result.

30. Calculate the uncertainty product $\Delta r \Delta p$ for the 1s electron of a hydrogen-like atom with atomic number Z. (*Hint:* Use $\langle p \rangle = 0$ by symmetry and deduce $\langle p^2 \rangle$ from the average kinetic energy, calculated as in Problem 24.)

ADDITIONAL PROBLEMS

31. An electron outside a dielectric is attracted to the surface by a force $F = -A/x^2$, where x is the perpendicular distance from the electron to the surface and A is a constant. Electrons are prevented from crossing the surface, since there are no quantum states in the dielectric for them to occupy. Assume that the surface is infinite in extent, so that the problem is effectively one-dimensional. Write the Schrödinger equation for an electron outside the surface $x > 0$. What is the appropriate boundary condition at $x = 0$? Obtain a formula for the allowed energy

levels in this case. (*Hint:* Compare the equation for $\psi(x)$ with that satisfied by the effective one-dimensional wave-function $g(r) = rR(r)$ for hydrogen-like atoms.)

32.   *The Spherical Well.* The three-dimensional analog of the square well in one dimension, the spherical well is commonly used to model the potential energy of nucleons (protons, neutrons) in an atomic nucleus. It is defined by a potential $U(r)$ that is zero everywhere inside a sphere and takes a large (possibly infinite) positive value outside this sphere. Use the Java applet available at our companion Web site (http://info.brookscole.com/mp3e → QMTools Simulations → Problem 8.32) to find the ground-state energy for a proton bound to a spherical well of radius 9.00 fm and height 30.0 MeV. Is the ground state an *s* state? Explain. Also report the most probable distance from the center of the well for this nucleon.

33.   Use the Java applet of Problem 32 to find the first three excited-state energy levels for the spherical well described there. What orbital quantum numbers ℓ describe these states? Determine the degeneracy of each excited level and display the probability clouds for the degenerate wavefunctions.

34. Example 8.9 found the most probable value and the average value for the distance of the electron from the proton in the ground state of a hydrogen atom. For comparison, find the *median* value as follows. (a) Derive an expression for the probability, as a function of r, that the electron in the ground state of hydrogen will be found outside a sphere of radius r centered on the nucleus. (b) Find the value of r for which the probability of finding the electron outside a sphere of radius r is equal to the probability of finding the electron inside this sphere. (You will need to solve a transcendental equation numerically.)

9
Atomic Structure

Chapter Outline

Much of what we have learned about the hydrogen atom with its single electron can be used directly to describe such single-electron ions as He^+ and Li^{2+}, which are hydrogen-like in their electronic structure. Multielectron atoms, however, such as neutral helium and lithium, introduce extra complications that stem from the interactions among the atomic electrons. Thus, the study of the atom inevitably involves us in the complexities of systems consisting of many interacting electrons. In this chapter we will learn some of the basic principles needed to treat such systems effectively and apply these principles to describe the physics of electrons in atoms.

Being of like charge and confined to a small space, the electrons of an atom repel one another strongly through the Coulomb force. In addition, we shall discover that the atomic electrons behave like tiny bar magnets, interacting magnetically with one another as well as with any external magnetic field applied to the atom. These magnetic properties derive in part from a new concept—electron spin—which will be explored at some length in this chapter.

Another new physical idea, known as the exclusion principle, is also presented in this chapter. This principle is extremely important in understanding the properties of multielectron atoms and the periodic table. In fact, the implications of the exclusion principle are almost as far-reaching as those of the Schrödinger equation itself.

9.1 ORBITAL MAGNETISM AND THE NORMAL ZEEMAN EFFECT

An electron orbiting the nucleus of an atom should give rise to magnetic effects, much like those arising from an electric current circulating in a wire loop. In particular, the motion of charge generates a magnetic field within the atom, and the atom as a whole is subject to forces and torques when it is placed in an external magnetic field. These magnetic interactions can all be described in terms of a single property of the atom—the magnetic dipole moment.

To calculate the magnetic moment of an orbiting charge, we reason by analogy with a current-carrying loop of wire. The moment $\boldsymbol{\mu}$ of such a loop has magnitude $|\boldsymbol{\mu}| = iA$, where i is the current and A is the area bounded by the loop. The direction of this moment is perpendicular to the plane of the loop, and its sense is given by a right-hand rule, as shown in Figure 9.1a. This characterization of a current loop as a magnetic dipole implies that its magnetic behavior is similar to that of a bar magnet with its north-south axis directed along $\boldsymbol{\mu}$ (Fig. 9.1b).

For a circulating charge q, the (time-averaged) current is simply q/T, where T is the orbital period. Furthermore, A/T is just the area swept out per unit time and equals the magnitude of the angular momentum $|\mathbf{L}|$ of the orbiting charge divided by twice the particle mass m.[1] This relation is

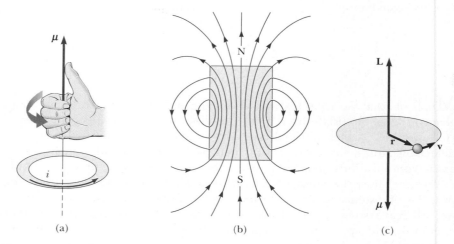

(a)   (b)   (c)

Figure 9.1   (a) The magnetic field in the space surrounding a current-carrying wire loop is that of a magnetic dipole with moment $\boldsymbol{\mu}$ perpendicular to the plane of the loop. The vector $\boldsymbol{\mu}$ points in the direction of the thumb if the fingers of the right hand are curled in the sense of the current i (right-hand rule). (b) The magnetic field in the space surrounding a bar magnet is also that of a magnetic dipole. The dipole moment vector $\boldsymbol{\mu}$ points from the south to the north pole of the magnet. (c) The magnetic moment $\boldsymbol{\mu}$ of an orbiting electron with angular momentum $\mathbf{L}$. Since the electron is negatively charged, $\boldsymbol{\mu}$ and $\mathbf{L}$ point in opposite directions.

---

[1]This is one of Kepler's laws of planetary motion, later shown by Newton to be a consequence of any central force.

easily verified for circular orbits, where $|\mathbf{L}| = mvr$, $v = 2\pi r/T$, and $A = \pi r^2$, so that

$$|\mathbf{L}| = m\left(\frac{2\pi r}{T}\right)r = 2m\left(\frac{\pi r^2}{T}\right) = 2m\left(\frac{A}{T}\right) \quad \text{or} \quad \frac{A}{T} = \frac{|\mathbf{L}|}{2m}$$

The same result holds for orbital motion of any kind (see Problem 2), so that $|\boldsymbol{\mu}| = iA$ becomes

$$\boldsymbol{\mu} = \frac{q}{2m}\mathbf{L} \tag{9.1}$$

Magnetic moment of an orbiting charge

for the magnetic moment of an orbiting charge q. Since $\mathbf{L}$ is perpendicular to the orbital plane, so too is $\boldsymbol{\mu}$. You may verify that the sense of the vector described by Equation 9.1 is consistent with that expected from the right-hand rule. Thus, the magnetic moment vector is directed along the angular momentum vector, and its magnitude is fixed by the proportionality constant $q/2m$, called the *gyromagnetic ratio*. For electrons, $q = -e$ so the direction of $\boldsymbol{\mu}$ is *opposite* the direction of $\mathbf{L}$ (Fig. 9.1c).

On the atomic scale, the elemental unit of angular momentum is $\hbar$. It follows that the natural unit for atomic moments is the quantity $e\hbar/2m_e$, called the **Bohr magneton** and designated by the symbol μ_B. Its value in SI units (joules/tesla) is

$$\mu_B = \frac{e\hbar}{2m_e} = 9.274 \times 10^{-24}\,\text{J/T} \tag{9.2}$$

Bohr magneton

Because $\boldsymbol{\mu}$ is proportional to $\mathbf{L}$, the orbital magnetic moment is subject to space quantization, as illustrated in Figure 9.2. In particular, the z component of the orbital magnetic moment is fixed by the value of the magnetic quantum number m_ℓ as

$$\mu_z = -\frac{e}{2m_e}L_z = -\frac{e\hbar}{2m_e}m_\ell = -\mu_B m_\ell \tag{9.3}$$

Just as with angular momentum, the magnetic moment vector can be visualized as precessing about the z-axis, thereby preserving this sharp value of μ_z while depicting the remaining components μ_x and μ_y as fuzzy.

The interaction of an atom with an applied magnetic field depends on the size and orientation of the atom's magnetic moment. Suppose an external field $\mathbf{B}$ is applied along the z-axis of an atom. According to classical electromagnetism, the atom experiences a torque

$$\boldsymbol{\tau} = \boldsymbol{\mu} \times \mathbf{B} \tag{9.4}$$

A magnetic moment precesses in a magnetic field

that tends to align its moment with the applied field. Instead of aligning itself with $\mathbf{B}$, however, the moment actually precesses around the field direction! This unexpected precession arises because $\boldsymbol{\mu}$ is proportional to the angular momentum $\mathbf{L}$. The motion is analogous to that of a spinning top precessing in the Earth's gravitational field. The gravitational torque acting to tip it over instead results in precession because of the angular

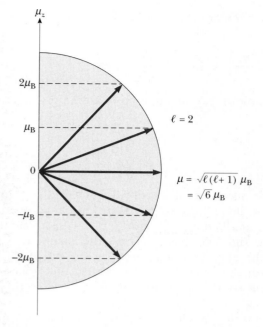

Figure 9.2 The orientations in space and z components of the orbital magnetic moment for the case $\ell = 2$. There are $2\ell + 1 = 5$ different possible orientations.

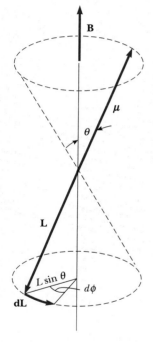

Figure 9.3 Larmor precession of the orbital moment $\boldsymbol{\mu}$ in an applied magnetic field **B**. Since $\boldsymbol{\mu}$ is proportional to **L**, the torque of the applied field causes the moment vector $\boldsymbol{\mu}$ to precess around the direction of **B** with frequency $\omega_L = eB/2m_e$.

momentum possessed by the spinning top. Returning to the atomic case, because $\boldsymbol{\tau} = d\mathbf{L}/dt$, we see from Equation 9.4 that the change in angular momentum, $d\mathbf{L}$, is always perpendicular to both **L** and **B**. Figure 9.3 depicts the motion (precession) that results. For atoms in a magnetic field this is known as **Larmor precession.**

From the geometry of Figure 9.3, we see that in a time dt the precession angle increases by $d\phi$, where

$$L \sin \theta \cdot d\phi = |d\mathbf{L}|$$

But Equations 9.1 and 9.4 can be combined to give

$$|d\mathbf{L}| = |\boldsymbol{\tau}| \, dt = \left| \frac{q}{2m_e} LB \sin\theta \right| dt$$

For electrons we take $q = -e$ and the frequency of precession, or **Larmor frequency** ω_L, becomes

$$\omega_L = \frac{d\phi}{dt} = \frac{1}{L \sin\theta} \frac{|d\mathbf{L}|}{dt} = \frac{e}{2m_e} B \qquad (9.5)$$

It is useful to introduce the quantum of energy $\hbar\omega_L$ associated with the Larmor frequency ω_L. This energy is related to the work required to reorient the atomic moment against the torque of the applied field. Remembering that the work of a torque τ to produce an angular displacement $d\theta$ is $dW = \tau \, d\theta$, we have from Equation 9.4

$$dW = -\mu B \sin \theta \, d\theta = d(\mu B \cos \theta) = d(\boldsymbol{\mu} \cdot \mathbf{B})$$

The minus sign signifies that the external torque must *oppose* that produced by the magnetic field **B**. The work done is stored as orientational potential energy of the dipole in the field. Writing $dW = -dU$, we identify the **magnetic potential energy** U as

$$U = -\boldsymbol{\mu} \cdot \mathbf{B} \qquad (9.6)$$

The energy of a magnetic moment depends on its orientation in a magnetic field

Equation 9.6 expresses the fact that the energy of a magnetic dipole in an external magnetic field **B** depends on its orientation in this field. The magnetic energy is minimal when $\boldsymbol{\mu}$ and **B** are aligned; therefore, this alignment is the preferred orientation. Because the possible orientations for $\boldsymbol{\mu}$ are restricted by space quantization, the magnetic energy is quantized accordingly. Taking the z-axis along **B**, and combining Equations 9.1, 9.3, and 9.6, we find

$$U = \frac{e}{2m_e} \mathbf{L} \cdot \mathbf{B} = \frac{eB}{2m_e} L_z = \hbar \omega_L m_\ell \qquad (9.7)$$

From Equation 9.7 we see that the magnetic energy of an atomic electron depends on the magnetic quantum number m_ℓ (so named for this dependence!) and, therefore, is quantized. The total energy of this electron is the sum of its magnetic energy U plus whatever energy it had in the absence of an applied field—say, E_0. Therefore,

$$E = E_0 + \hbar \omega_L m_\ell \qquad (9.8)$$

For atomic hydrogen, E_0 depends only on the principal quantum number n; in more complex atoms, the atomic energy also varies according to the subshell label ℓ, as discussed further in Section 9.5.

Unlike energies, the wavefunctions of atomic electrons are unaffected by the application of a magnetic field. This somewhat surprising result can be partly understood by recognizing that according to classical physics, the only effect of the field is to cause (Larmor) precession around the direction of **B**. For atomic electrons, this translates into precession of **L** about the z-axis. However, such a precession is already implicit in our semiclassical picture of electron orbits in the absence of external fields, as required by the sharpness of L_z while L_x and L_y remain fuzzy. From this viewpoint, the introduction of an applied magnetic field merely transforms this *virtual precession*[2] into a real one at the Larmor frequency!

---

[2]In zero magnetic field, the precession of the classical vector may be termed virtual (not real) since even though the same value may not be obtained for L_x (or L_y) in successive measurements, the average value $\langle L_x \rangle$ (or $\langle L_y \rangle$) does not change over time. With **B** nonzero, however, it can be shown that $(d^2/dt^2) \langle L_x \rangle = -\omega_L^2 \langle L_x \rangle$ (and similarly for $\langle L_y \rangle$), indicating that $\langle L_x \rangle$ (and $\langle L_y \rangle$) oscillates at the Larmor frequency ω_L.

EXAMPLE 9.1   Magnetic Energy of the Electron in Hydrogen

Calculate the magnetic energy and Larmor frequency for an electron in the $n = 2$ state of hydrogen, assuming the atom is in a magnetic field of strength $B = 1.00$ T.

Solution   Taking the z-axis along **B**, we calculate the magnetic energy from Equation 9.7 as

$$U = \frac{eB}{2m_e} L_z = \frac{e\hbar}{2m_e} Bm_\ell = \hbar\omega_L m_\ell$$

For a 1.00 T field, the energy quantum $\hbar\omega_L$ has the value

$$\hbar\omega_L = \frac{e\hbar}{2m_e} B = \mu_B B = (9.27 \times 10^{-24} \, \text{J/T})(1.00 \, \text{T})$$

$$= 9.27 \times 10^{-24} \, \text{J} = 5.79 \times 10^{-5} \, \text{eV}$$

With $n = 2$, ℓ can be 0 or 1, and m_ℓ is 0(twice) and ± 1. Thus, the magnetic energy U can be 0, $+\hbar\omega_L$, or $-\hbar\omega_L$. In such applications, the energy quantum $\hbar\omega_L$ is called the **Zeeman energy.** This Zeeman energy divided by $\hbar$ is the Larmor frequency:

$$\omega_L = \frac{5.79 \times 10^{-5} \, \text{eV}}{6.58 \times 10^{-16} \, \text{eV} \cdot \text{s}} = 8.80 \times 10^{10} \, \text{rad/s}$$

Evidence for the existence of atomic moments is the appearance of extra lines in the spectrum of an atom that is placed in a magnetic field. Consider a hydrogen atom in its first excited ($n = 2$) state. For $n = 2$, ℓ can have values 0 and 1. The magnetic field has no effect on the state for which $\ell = 0$, since then $m_\ell = 0$. For $\ell = 1$, however, m_ℓ can take values of 1, 0, and -1, and the first excited level is split into three levels by the magnetic field (Figure 9.4).

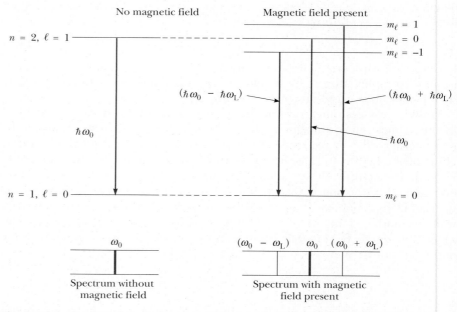

Figure 9.4   Level splittings for the ground and first excited states of a hydrogen atom immersed in a magnetic field **B**. An electron in one of the excited states decays to the ground state with the emission of a photon, giving rise to emission lines at ω_0, $\omega_0 + \omega_L$, and $\omega_0 - \omega_L$. This is the normal Zeeman effect. When **B** = 0, only the line at ω_0 is observed.

The original (Lyman) emission line is replaced by the three lines depicted in Figure 9.4. The central line appears at the same frequency ω_0 as it would without a magnetic field. This is flanked on both sides by new lines at frequencies $\omega_0 \pm \omega_L$. Therefore, the magnetic field splits the original emission line into three lines. Because ω_L is proportional to B, the amount of splitting increases linearly with the strength of the applied field. This effect of **spectral line splitting by a magnetic field is known as the normal Zeeman effect** after its discoverer, Pieter Zeeman.

The normal Zeeman effect

Zeeman spectra of atoms excited to higher states should be more complex, because many more level splittings are involved. For electrons excited to the $n = 3$ state of hydrogen, the expected Zeeman lines and the atomic transitions that give rise to them are shown in Figure 9.5. Accompanying each hydrogen line are anywhere from two to six satellites at frequencies removed from the original by multiples of the Larmor frequency. But the observed Zeeman spectrum is not this complicated, owing to selection rules that limit the transitions

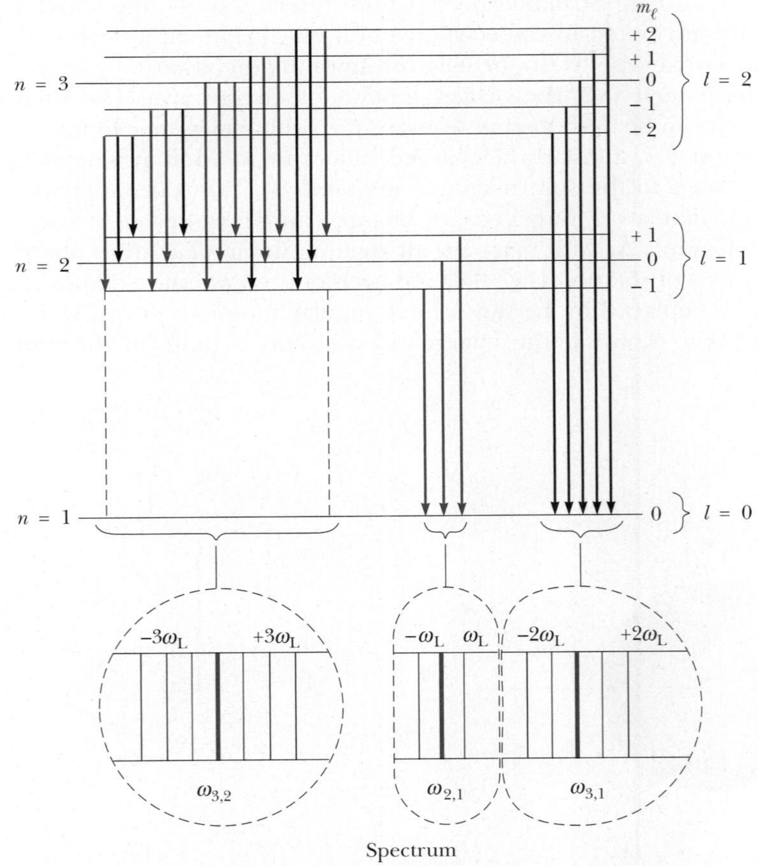

Figure 9.5 Zeeman spectral lines and the underlying atomic transitions that give rise to them for an electron excited to the $n = 3$ state of hydrogen. Because of selection rules, only the transitions drawn in color actually occur. Transitions from the $n = 3$, $l = 1$ orbitals (not shown) to the $n = 1$ state give rise to the colored lines in the illustration at the bottom right.

to those for which ℓ changes by 1 and m_ℓ changes by 0, +1, or −1. The result is that satellites appear at the Larmor frequency *only* and not at multiples of this frequency. The selection rules express conservation of angular momentum for the system, taking into account the angular momentum of the emitted photon. (See Section 8.5.)

Finally, even the splitting of an emission line into a triplet of equally spaced lines as predicted here, called the normal Zeeman effect, frequently is not observed. More commonly, splittings into four, six, or even more unequally spaced lines are seen. This is the anomalous Zeeman effect, which has its roots in the existence of electron spin.

9.2   THE SPINNING ELECTRON

The anomalous Zeeman splittings are only one of several phenomena not explained by the magnetic interactions discussed thus far. Another is the observed doubling of many spectral lines referred to as fine structure. Both effects are attributed to the existence of a new magnetic moment—the spin moment—that arises from the electron spinning on its axis.

We have seen that the orbital motion of charge gives rise to magnetic effects that can be described in terms of the orbital magnetic moment $\boldsymbol{\mu}$ given by Equation 9.1. Similarly, a charged object in rotation produces magnetic effects related to the spin magnetic moment $\boldsymbol{\mu}_s$. The spin moment is found by noting that a rotating body of charge can be viewed as a collection of charge elements Δq with mass Δm all rotating in circular orbits about a fixed line, the axis of rotation (Fig. 9.6). To each of these we should apply Equation 9.1 with $\mathbf{L}$ replaced by $\mathbf{L}_i$, the orbital angular momentum of the ith charge element (Fig. 9.6b). If the charge-to-mass ratio is uniform throughout the

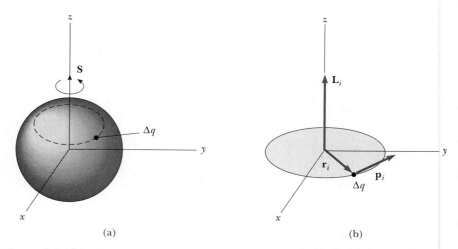

(a)   (b)

Figure 9.6   (a) A spinning charge q may be viewed as a collection of charge elements Δq orbiting a fixed line, the axis of rotation. (b) The circular path followed by one such charge element. The angular momentum of this charge element $\mathbf{L}_i = \mathbf{r}_i \times \mathbf{p}_i$ lies along the axis of rotation. The magnetic moments accompanying these orbiting charge elements are summed to give the total magnetic moment of rotation, or spin magnetic moment, of the charge q.

Otto Stern was one of the finest experimental physicists of the 20th century. Born and educated in Germany (Ph.D. in physical chemistry in 1912), he at first worked with Einstein on theoretical issues in molecular theory, in particular applying the new quantum ideas to theories of the specific heat of solids. From about 1920, Stern devoted himself to his real life's work, the development of the molecular beam method, which enabled him to investigate the properties of free or isolated atoms and culminated in a Nobel prize in 1945. In this method, a thin stream of atoms is introduced into a high-vacuum chamber where the atoms are free, and hence the properties of individual atoms may be investigated by applying external fields or by some other technique.

Stern first used this method to confirm that silver atoms obey the Maxwell speed distribution. Shortly after, in a series of elegant and difficult experiments with Walter Gerlach, Stern showed that silver atoms obey space quantization and succeeded in measuring the magnetic moment of the silver atom. In the period from 1923 to 1933, Stern directed a remarkably productive molecular beam laboratory at the University of Hamburg. With his students and coworkers he directly demonstrated the wave

BIOGRAPHY

OTTO STERN
(1888–1969)

nature of helium atoms and measured the magnetic moments of many atoms. Finally, with a great deal of effort, he succeeded in measuring the very small magnetic moments of the proton and deuteron. For these last important fundamental measurements he was awarded the Nobel prize. In connection with the measurement of the proton's magnetic moment, an interesting story is told by Victor Weisskopf, which should gladden the hearts of experimentalists everywhere:

"There was a seminar held by the theoretical group in Göttingen, and Stern came down and gave a talk on the measurements he was about to finish of the magnetic moment of the proton. He explained his apparatus, but he did not tell us the result. He took a piece of paper and went to each of us saying, 'What is your prediction of the magnetic moment of the proton?' Every theoretician from Max Born down to Victor Weisskopf said, 'Well, of course, the great thing about the Dirac equation is that it predicts a magnetic moment of one Bohr magneton for a particle of spin one-half!' Then he asked us to write down the prediction; everybody wrote 'one magneton.' Then, two months later, he came to give again a talk about the finished experiment, which showed that the value was 2.8 magnetons. He then projected the paper with our predictions on the screen. It was a sobering experience."*

In protest over Nazi dismissals of some of his closest coworkers, Stern resigned his post in Hamburg and came to the Carnegie Institute of Technology in the United States in 1933. Here he worked on molecular beam research until his retirement in 1946.

*From Victor F. Weisskopf, *Physics in the Twentieth Century; Selected Essays: My Life as a Physicist*, Cambridge, MA, The MIT Press, 1972.

(Photo courtesy of University of Hamburg, Germany)

body, then $\Delta q/\Delta m$ is the ratio of total charge q to total mass m_e and we get for the spin moment

$$\boldsymbol{\mu}_s = \frac{q}{2m_e} \sum \mathbf{L}_i = \frac{q}{2m_e} \mathbf{S} \qquad (9.9)$$

A rotating charge gives rise to a spin magnetic moment

where $\mathbf{S}$, the *spin* angular momentum, is the total angular momentum of rotation. The spin angular momentum $\mathbf{S}$ points along the axis of rotation according to a right-hand rule, as shown in Figure 9.6; its magnitude depends on the size and shape of the object, as well as its speed of rotation. If the charge-to-mass ratio is not uniform, the gyromagnetic ratio in Equation 9.9, $q/2m_e$, must be multiplied by a dimensionless constant, the **g factor,** whose value reflects the detailed charge-to-mass distribution within the body. Note that g factors different from unity imply a distribution of charge that is not tightly linked to the distribution of mass, an unusual circumstance but one that cannot be excluded.[3]

The existence of a spin magnetic moment for the electron was first demonstrated in 1921 in a classic experiment performed by Otto Stern and Walter Gerlach. Electron spin was unknown at that time; the Stern–Gerlach experiment was originally conceived to demonstrate the space quantization associated with orbiting electrons in atoms. In their experiment, a beam of silver atoms was passed through a *nonuniform* magnetic field created in the gap between the pole faces of a large magnet. The beam was then detected by being deposited on a glass collector plate (Fig. 9.7). A nonuniform field exerts a force on any magnetic moment, so that each atom is deflected in the gap by an amount governed by the orientation of its moment with respect to the direction of inhomogeneity (the z-axis), as illustrated in Figure 9.7b. If the moment directions are restricted by space quantization as in Figure 9.2, so too are the deflections. Thus, the atomic beam should split into a number of discrete components, one for each distinct moment orientation present in the beam. This is contrary to the classical expectation that any moment orientation (and hence any beam deflection) is possible, and all would combine to produce a continuous fanning of the atomic beam (Fig. 9.7c).

The Stern–Gerlach experiment produced a staggering result: The silver atomic beam was clearly split—but into only *two* components, not the odd number $(2\ell + 1)$ expected from the space quantization of orbital moments! This is all the more remarkable when we realize that silver atoms in their ground state have no orbital angular momentum $(\ell = 0)$, because the outermost electron in silver normally would be in an *s* state. The result was so surprising that the experiment was repeated in 1927 by T. E. Phipps and J. B. Taylor with a beam of hydrogen atoms replacing silver, thereby eliminating any uncertainties arising from the use of the more complex silver atoms. The results, however, were unchanged. From these experiments, we are forced to conclude that there is some contribution to the atomic magnetic moment other than the orbital motion of electrons and that this moment is subject to space quantization.

Our present understanding of the situation dates to the 1925 paper of Samuel Goudsmit and George Uhlenbeck, then graduate students at the University of Leiden. Goudsmit and Uhlenbeck believed that the unknown moment had its origin in the spinning motion of atomic electrons, with the spin angular momentum obeying the same quantization rules as orbital angular momentum. The magnetic moment seen in the Stern–Gerlach experiment is attributed to the spin of the outermost electron in silver. Because all allowed orientations of the spin moment should be represented in the atomic beam, the observed splitting presents a dramatic confirmation of space quantization as applied to electron spin, with the number of components $(2s + 1)$ indicating the value of the spin quantum number *s*.

The spin magnetic moment suggests that the electron can be viewed as a charge in rotation, although the classical picture of a spinning body of

---

[3]It is only fair to caution the reader at this point not to take the classical view of an electron as a tiny charged ball spinning on its axis too literally. Although such a picture is useful in first introducing and visualizing electron spin, it is not technically correct. Several shortcomings of the classical picture are discussed in detail on pp. 306 and 307.

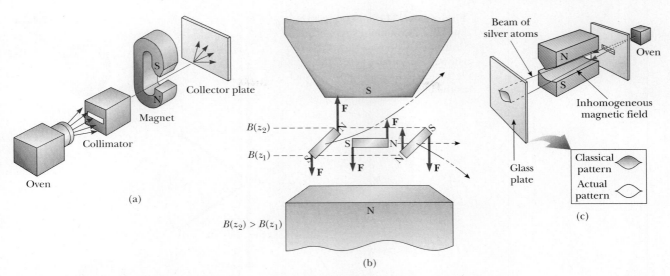

Figure 9.7 The Stern–Gerlach experiment to detect space quantization. (a) A beam of silver atoms is passed through a nonuniform magnetic field and detected on a collector plate. (b) The atoms, with their magnetic moment, are equivalent to tiny bar magnets. In a nonuniform field, each atomic magnet experiences a net force that depends on its orientation. (c) If any moment orientation were possible, a continuous fanning of the beam would be seen at the collector. For space quantization, the fanning is replaced by a set of discrete lines, one for each distinct moment orientation present in the beam.

charge must be adjusted to accommodate the wave properties of matter. The resulting semiclassical model of electron spin can be summarized as follows:

- The **spin quantum number** s for the electron is $\frac{1}{2}$! This value is dictated by the observation that an atomic beam passing through the Stern–Gerlach magnet is split into just two components ($= 2s + 1$). Accordingly, there are exactly two orientations possible for the spin axis, described as the "spin-up" and "spin-down" states of the electron. This is space quantization again, according to the quantization rules for angular momentum[4] as applied to a spin of $\frac{1}{2}$:

$$S_z = m_s \hbar \qquad \text{where } m_s = \tfrac{1}{2} \text{ or } -\tfrac{1}{2} \qquad (9.10)$$

The two values $\pm\hbar/2$ for S_z correspond to the two possible orientations for **S** shown in Figure 9.8. The value $m_s = +\frac{1}{2}$ refers to the spin-up case, sometimes designated with an up arrow ($\uparrow$) or simply a plus sign ($+$). Likewise, $m_s = -\frac{1}{2}$ is the spin-down case, ($\downarrow$) or ($-$). The fact that s has a nonintegral value suggests that spin is not merely another manifestation of orbital motion, as the classical picture implies.

Properties of electron spin

[4]For integer angular momentum quantum numbers, the z component is quantized as $m_s = 0, \pm 1, \ldots$ $\pm s$, which can also be written as $m_s = s, s-1, \ldots, -s$. For $s = \frac{1}{2}$, the latter implies $m_s = \frac{1}{2}$ or $-\frac{1}{2}$.

Figure 9.8   The spin angular momentum also exhibits space quantization. This figure shows the two allowed orientations of the spin vector **S** for a spin $\frac{1}{2}$ particle, such as the electron.

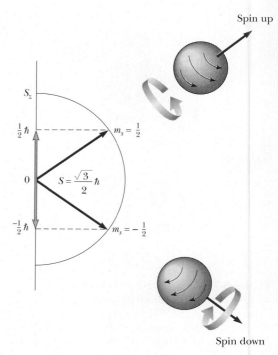

- The magnitude of the spin angular momentum is

**The spin angular momentum
of an electron**

$$|\mathbf{S}| = \sqrt{s(s + 1)}\,\hbar = \frac{\sqrt{3}}{2}\,\hbar \qquad (9.11)$$

and *never changes!* **This angular momentum of rotation cannot be changed in any way, but is an intrinsic property of the electron,** like its mass or charge. The notion that $|\mathbf{S}|$ is fixed contradicts classical laws, where a rotating charge would be slowed down by the application of a magnetic field owing to the Faraday emf that accompanies the changing magnetic field (the diamagnetic effect). Furthermore, if the electron were viewed as a spinning ball with angular momentum $\hbar\sqrt{3}/2$ subject to classical laws, parts of the ball near its surface would be rotating with velocities in excess of the speed of light![5] All of this is taken to mean that the classical picture of the electron as a charge in rotation must not be pressed too far; ultimately, the spinning electron is a quantum entity defying any simple classical description.

- The spin magnetic moment is given by Equation 9.9 with a *g factor of 2;* that is, the moment is twice as large as would be expected for a body with

---

[5]This follows from the extremely small size of the electron. The exact size of the electron is unknown, but an upper limit of 10^{-6} Å is deduced from experiments in which electrons are scattered from other electrons. According to some current theories, the electron may be a true point object, that is, a particle with zero size!

spin angular momentum given by Equation 9.10. The value $g = 2$ is required by the amount of beam deflection produced by the Stern–Gerlach magnet; the larger the magnetic moment, the greater will be the deflection of the atomic beam. As already mentioned, any g factor other than unity implies a nonuniform charge-to-mass ratio in the classical picture. The g factor of 2 can be realized classically but suggests a bizarre picture that cannot be taken seriously (see Problem 8). The correct g factor of 2 is predicted by the relativistic quantum theory of the electron put forth by Paul Dirac in 1929.[6]

With the recognition of electron spin we see that an additional quantum number, m_s, is needed to specify the internal, or spin, state of an electron. Therefore, the state of an electron in hydrogen must be described by the four quantum numbers n, ℓ, m_ℓ, and m_s. Furthermore, the total magnetic moment now has orbital and spin contributions:

$$\boldsymbol{\mu} = \boldsymbol{\mu}_0 + \boldsymbol{\mu}_s = \frac{-e}{2m_e}\{\mathbf{L} + g\mathbf{S}\} \qquad (9.12)$$

The total magnetic moment of an electron

Because of the electron g factor, the total moment $\boldsymbol{\mu}$ is no longer in the same direction as the total (orbital plus spin) angular momentum $\mathbf{J} = \mathbf{L} + \mathbf{S}$. The component of the total moment $\boldsymbol{\mu}$ along $\mathbf{J}$ is sometimes referred to as the **effective moment.** When the magnetic field $\mathbf{B}$ applied to an atom is weak, the effective moment determines the magnetic energy of atomic electrons according to Equation 9.6. As we shall discover in Section 9.3, the number of possible orientations for $\mathbf{J}$ (and, hence, for the effective moment) is even, leading to the even number of spectral lines seen in the anomalous Zeeman effect.

EXAMPLE 9.2   Semiclassical Model for Electron Spin

Calculate the angles between the z-axis and the spin angular momentum $\mathbf{S}$ of the electron in the up and down spin states. How should we portray the fuzziness inherent in the x and y components of the spin angular momentum?

Solution   For the electron, the magnitude of the spin angular momentum is $|\mathbf{S}| = \hbar\sqrt{3}/2$, and the z component of spin is $S_z = \pm\hbar/2$. Thus, the spin vector $\mathbf{S}$ is inclined from the z-axis at angles given by

$$\cos\theta = \frac{S_z}{|\mathbf{S}|} = \pm\frac{1}{\sqrt{3}}$$

For the up spin state, we take the plus sign and get $\cos\theta = 0.577$, or $\theta = 54.7°$. The down spin orientation is described by the minus sign and gives $\cos\theta = -0.577$, or $\theta = 125.3°$. Because the axis of rotation coincides with the direction of the spin vector, these are the angles the rotation axis makes with the z-axis.

While S_z is sharp in either the up or down spin orientation, both S_x and S_y are fuzzy. This fuzziness may be depicted by allowing the spin vector to precess about the z-axis, as we did for the orbital angular momentum in Chapter 8.

---

[6]The g factor for the electron is not exactly 2. The best value to date is $g = 2.00232$. The discrepancy between Dirac's predicted value and the observed value is attributed to the electron interacting with the "vacuum." Such effects are the subject of quantum electrodynamics, developed by Richard Feynman in the early 1950s.

Exercise 1   The photon is a spin 1 particle, that is, $s = 1$ for the photon. Calculate the possible angles between the z-axis and the spin vector of the photon.

Answer   45°, 90°, and 135°

EXAMPLE 9.3   Zeeman Spectrum of Hydrogen Including Spin

Examine the Zeeman spectrum produced by hydrogen atoms initially in the $n = 2$ state when electron spin is taken into account, assuming the atoms to be in a magnetic field of magnitude $B = 1.00$ T.

Solution   The electron energies now have a magnetic contribution from both the orbital and spin motions. Choosing the z-axis along the direction of **B**, we calculate the magnetic energy from Equations 9.6 and 9.12:

$$U = -\boldsymbol{\mu} \cdot \mathbf{B} = \frac{e}{2m_e} B\{L_z + gS_z\} = \frac{e\hbar}{2m_e} B(m_\ell + gm_s)$$

The energy $(e\hbar/2m_e)B$ is the Zeeman energy $\mu_B B$ or $\hbar\omega_L$; its value in this example is

$$\mu_B B = (9.27 \times 10^{-24}\,\text{J/T})(1.00\,\text{T}) = 9.27 \times 10^{-24}\,\text{J}$$
$$= 5.79 \times 10^{-5}\,\text{eV}$$

For the $n = 2$ state of hydrogen, the shell energy is $E_2 = -(13.6\,\text{eV})/2^2 = -3.40\,\text{eV}$. Because m_ℓ takes the values 0 (twice) and ±1, there is an orbital contribution to the magnetic energy $U_0 = m_\ell \hbar \omega_L$ that introduces new levels at $E_2 \pm \hbar\omega_L$, as discussed in Example 9.1. The presence of electron spin splits each of these into a pair of levels, the additional (spin) contribution to the energy being $U_s = (gm_s)\hbar\omega_L$ (Fig. 9.9). Because $g = 2$ and m_s is $\pm\frac{1}{2}$ for the electron, the spin energy in the field $|U_s|$ is again the Zeeman energy $\hbar\omega_L$. Therefore, an electron in this shell can have any one of the energies

$$E_2, \qquad E_2 \pm \hbar\omega_L, \qquad E_2 \pm 2\hbar\omega_L$$

In making a downward transition to the $n = 1$ shell with energy $E_1 = -13.6$ eV, the final state of the electron may have energy $E_1 + \hbar\omega_L$ or $E_1 - \hbar\omega_L$, depending on the orientation of its spin in the applied field. Therefore, the energy of transition may be any one of the following possibilities:

$$\Delta E_{2,1}, \quad \Delta E_{2,1} \pm \hbar\omega_L, \quad \Delta E_{2,1} \pm 2\hbar\omega_L, \quad \Delta E_{2,1} \pm 3\hbar\omega_L$$

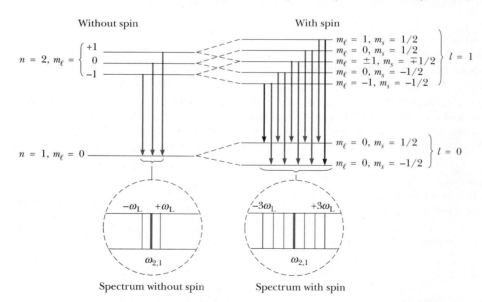

Figure 9.9   (Example 9.3) Predicted Zeeman pattern and underlying atomic transitions for an electron excited to the $n = 2$ state of hydrogen, when electron spin is taken into account. Again, selection rules prohibit all but the colored transitions. Because of the neglect of the spin–orbit interaction, the effect shown here (called the Paschen–Back effect) is observed only in very intense applied magnetic fields.

Photons emitted with these energies have frequencies

$$\omega_{2,1}, \qquad \omega_{2,1} \pm \omega_L, \qquad \omega_{2,1} \pm 2\omega_L, \qquad \omega_{2,1} \pm 3\omega_L$$

Therefore the spectrum should consist of the original line at $\omega_{2,1}$ flanked on both sides by satellite lines separated from the original by the Larmor frequency, twice the Larmor frequency, and three times this frequency. Notice that the lines at $\omega_{2,1} \pm 2\omega_L$ and $\omega_{2,1} \pm 3\omega_L$ appear solely because of electron spin.

Again, however, the observed pattern is not the predicted one. Selection rules inhibit transitions unless $m_\ell + m_s$ changes by 0, +1, or −1. This has the effect of eliminating the satellites at $\omega_{2,1} \pm 3\omega_L$. Furthermore, the spin moment and the orbital moment of the electron interact with *each other*, a circumstance not recognized in our calculation. Only when this spin–orbit interaction energy is small compared with the Zeeman energy, $\hbar\omega_L$, do we observe the spectral lines predicted here. This is the case for the **Paschen–Back effect,** in which the magnetic field applied to the atom is intense enough to make $\hbar\omega_L$ the dominant energy. Typically, to observe the Paschen–Back effect requires magnetic fields in excess of several tesla.

9.3   THE SPIN–ORBIT INTERACTION AND OTHER MAGNETIC EFFECTS

The existence of both spin and orbital magnetic moments for the electron inevitably leads to their mutual interaction. This so-called **spin–orbit interaction** is best understood from the vantage point of the orbiting electron, which "sees" the atomic nucleus circling it (Fig. 9.10). The apparent orbital motion of the nucleus generates a magnetic field at the electron site, and the electron spin moment acquires magnetic energy in this field according to Equation 9.6. This can be thought of as an internal Zeeman effect, with **B** arising from the orbital motion of the electron itself. The electron has a higher energy when its spin is up, or aligned with **B**, than when its spin is down, or aligned opposite to **B** (Fig. 9.10b).

The energy difference between the two spin orientations is responsible for the **fine structure doubling** of many atomic spectral lines. For example, the $2p \rightarrow 1s$ transition in hydrogen is split into two lines because the $2p$ level is actually a spin doublet with a level spacing of about 5×10^{-5} eV (Fig. 9.11), while the $1s$ level remains unsplit (there is no orbital field in a state with zero orbital angular momentum). Similarly, the spin–orbit doubling of the sodium $3p$ level gives rise to the well-known sodium D lines to be discussed in Example 9.4.

The coupling of spin and orbital moments implies that **neither orbital angular momentum nor spin angular momentum is conserved separately.** But total angular momentum **J** = **L** + **S** is conserved, so long as no external torques are present. Consequently, quantum states exist for which $|\mathbf{J}|$ and J_z are sharp observables quantized in the manner we have come to expect for angular momentum:

$$|\mathbf{J}| = \sqrt{j(j+1)}\,\hbar \tag{9.13}$$

$$J_z = m_j\hbar \qquad \text{with } m_j = j, j-1, \ldots, -j$$

Permissible values for the total angular momentum quantum number j are

$$j = \ell + s, \qquad \ell + s - 1, \ldots, |\ell - s| \tag{9.14}$$

in terms of the orbital (ℓ) and spin (s) quantum numbers. For an atomic electron $s = \frac{1}{2}$ and $\ell = 0, 1, 2, \ldots$, so $j = \frac{1}{2}$ (for $\ell = 0$) and $j = \ell \pm \frac{1}{2}$ (for $\ell > 0$).

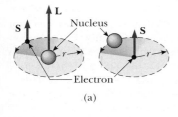

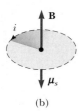

(b)

Figure 9.10 (a) *Left:* An electron with angular momentum **L** orbiting the nucleus of an atom. In the spin-up orientation shown here, the spin angular momentum **S** of the electron is "aligned" with **L**. *Right:* From the viewpoint of the orbiting electron, the nucleus circulates as shown. (b) The apparently circulating nuclear charge is represented by the current i and causes a magnetic field **B** at the site of the electron. In the presence of **B**, the electron spin moment $\boldsymbol{\mu}_s$ acquires magnetic energy $U = -\boldsymbol{\mu}_s \cdot \mathbf{B}$. The spin moment $\boldsymbol{\mu}_s$ is opposite the spin vector **S** for the negatively charged electron. The direction of **B** is given by a right-hand rule: With the thumb of the right hand pointing in the direction of the current i, the fingers give the sense in which the **B** field circulates about the orbit path. The magnetic energy is highest for the case shown, where **S** and **L** are "aligned."

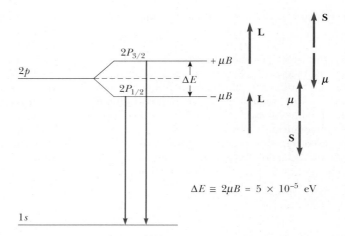

Figure 9.11 The $2p$ level of hydrogen is split by the spin–orbit effect into a doublet separated by the spin–orbit energy $\Delta E = 5 \times 10^{-5}$ eV. The higher energy state is the one for which the spin angular momentum of the electron is "aligned" with its orbital angular momentum. The $1s$ level is unaffected, since no magnetic field arises for orbital motion with zero angular momentum.

These results can be deduced from the vector addition model shown in Figure 9.12a. With $j = \frac{1}{2}$, there are only two possibilities for m_j, namely $m_j = \pm\frac{1}{2}$. For $j = \ell \pm \frac{1}{2}$, the number of possibilities $(2j + 1)$ for m_j becomes either 2ℓ or $2\ell + 2$. Notice that the number of m_j values is always *even* for a single electron, leading to an even number of orientations in the semiclassical model for **J**

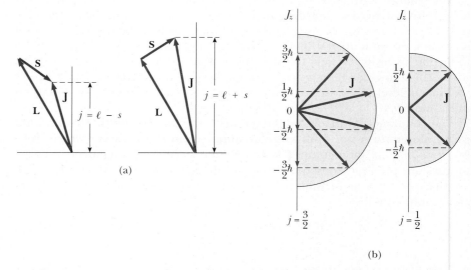

Figure 9.12 (a) A vector model for determining the total angular momentum $\mathbf{J} = \mathbf{L} + \mathbf{S}$ of a single electron. (b) The allowed orientations of the total angular momentum **J** for the states $j = \frac{3}{2}$ and $j = \frac{1}{2}$. Notice that there are now an even number of orientations possible, not the odd number familiar from the space quantization of **L** alone.

(Fig. 9.12b), rather than the odd number predicted for the orbital angular momentum **L** alone.

A common spectroscopic notation is to use a subscript after a letter to designate the total angular momentum of an atomic electron, where the letter itself (now uppercase) describes its orbital angular momentum. For example, the notation $1S_{1/2}$ describes the ground state of hydrogen, where the 1 indicates $n = 1$, the S tells us that $\ell = 0$, and the subscript $\frac{1}{2}$ denotes $j = \frac{1}{2}$. Likewise, the spectroscopic notations for the $n = 2$ states of hydrogen are $2S_{1/2}(\ell = 0, j = \frac{1}{2})$, $2P_{3/2}(\ell = 1, j = \frac{3}{2})$, and $2P_{1/2}(\ell = 1, j = \frac{1}{2})$. Again, the spin–orbit interaction splits the latter two states in energy by about 5×10^{-5} eV.

Spectroscopic notation extended to include spin

EXAMPLE 9.4   The Sodium Doublet

The famed sodium doublet arises from the spin–orbit splitting of the sodium $3p$ level, and consists of the closely spaced pair of spectral lines at wavelengths of 588.995 nm and 589.592 nm. Show on an energy-level diagram the electronic transitions giving rise to these lines, labeling the participating atomic states with their proper spectroscopic designations. From the doublet spacing, determine the magnitude of the spin–orbit energy.

Solution   The outer electron in sodium is the first electron to occupy the $n = 3$ shell, and it would go into the lowest-energy subshell, the $3s$ or $3S_{1/2}$ level. The next-highest levels belong to the $3p$ subshell. The $2(2\ell + 1) = 6$ states of this subshell are grouped into the $3P_{1/2}$ level with two states, and the $3P_{3/2}$ level with four states. The spin–orbit effect splits these levels by the spin–orbit energy. The outer electron, once it is excited to either of these levels by some means (such as an electric discharge in the sodium vapor lamp), returns to the $3S_{1/2}$ level with the emission of a photon. The two possible transitions $3P_{3/2} \rightarrow 3S_{1/2}$ and $3P_{1/2} \rightarrow 3S_{1/2}$ are shown in Figure 9.13. The emitted photons have nearly the same energy but differ by the small amount ΔE representing the spin–orbit splitting of the initial levels. Since $E = hc/\lambda$ for photons, ΔE is found as

$$\Delta E = \frac{hc}{\lambda_1} - \frac{hc}{\lambda_2} = \frac{hc(\lambda_2 - \lambda_1)}{\lambda_1 \lambda_2}$$

For the sodium doublet, the observed wavelength difference is

$$\lambda_2 - \lambda_1 = 589.592 \text{ nm} - 588.995 \text{ nm} = 0.597 \text{ nm}$$

Using this with $hc = 1240$ eV·nm gives

$$\Delta E = \frac{(1240 \text{ eV·nm})(0.597 \text{ nm})}{(589.592 \text{ nm})(588.995 \text{ nm})} = 2.13 \times 10^{-3} \text{ eV}$$

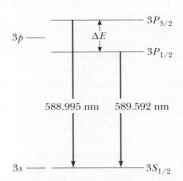

Figure 9.13   (Example 9.4). The transitions $3P_{3/2} \rightarrow 3S_{1/2}$ and $3P_{1/2} \rightarrow 3S_{1/2}$ that give rise to the sodium doublet. The $3p$ level of sodium is split by the spin–orbit effect, but the $3s$ level is unaffected. In the sodium vapor lamp, electrons normally in the $3s$ level are excited to the $3p$ levels by an electric discharge.

Exercise 2   Using the spin–orbit interaction energy calculated in Example 9.4, calculate the magnitude of the magnetic field at the site of the orbiting $3p$ electron in sodium.

Answer $B = 18.38$ T, a large field by laboratory standards.

9.4 EXCHANGE SYMMETRY AND THE EXCLUSION PRINCIPLE

As mentioned earlier, the existence of spin requires that the state of an atomic electron be specified with four quantum numbers. In the absence of spin–orbit effects these could be n, ℓ, m_ℓ, and m_s; if the spin–orbit interaction is taken into account, m_ℓ and m_s are replaced by j and m_j. In either case, four quantum numbers are required, one for each of the four degrees of freedom possessed by a single electron.

In those systems where two or more electrons are present, we might expect to describe each electronic state by giving the appropriate set of four quantum numbers. In this connection an interesting question arises, namely, "How many electrons in an atom can have the same four quantum numbers, that is, be in the same state?" This important question was answered by Wolfgang Pauli in 1925 in a powerful statement known as the exclusion principle. **The exclusion principle states that no two**

Wolfgang Pauli was an extremely talented Austrian theoretical physicist who made important contributions in many areas of modern physics. At the age of 21, Pauli gained public recognition with a masterful review article on relativity, which is still considered to be one of the finest and most comprehensive introductions to the subject. Other major contributions were the discovery of the exclusion principle, the explanation of the connection between particle spin and statistics, theories of relativistic quantum electrodynamics, the neutrino hypothesis, and the hypothesis of nuclear spin. An article entitled "The Fundamental Principles of Quantum Mechanics," written by Pauli in 1933 for the *Handbuch der Physik*, is widely acknowledged to be one of the best treatments of quantum physics ever written. Pauli was a forceful and colorful character, well known for his witty and often caustic remarks directed at those who presented new theories in a less than perfectly clear manner. Pauli exerted great influence on his students and colleagues by forcing them with his sharp criticism to a deeper and clearer under-

B I O G R A P H Y

———

WOLFGANG PAULI
(1900–1958)

standing. Victor Weisskopf, one of Pauli's famous students, has aptly described him as "the conscience of theoretical physics." Pauli's sharp sense of humor was also nicely captured by Weisskopf in the following anecdote:

"In a few weeks, Pauli asked me to come to Zurich. I came to the big door of his office, I knocked, and no answer. I knocked again and no answer. After about five minutes he said, rather roughly, "Who is it? Come in!"

I opened the door, and here was Pauli—it was a very big office—at the other side of the room, at his desk, writing and writing. He said, "Who is this? First I must finish calculating." Again he let me wait for about five minutes and then: "Who is that?" "I am Weisskopf." "Uhh, Weisskopf, ja, you are my new assistant." Then he looked at me and said, "Now, you see I wanted to take Bethe, but Bethe works now on the solid state. Solid state I don't like, although I started it. This is why I took you." Then I said, "What can I do for you, sir?" and he said "I shall give you right away a problem." He gave me a problem, some calculation, and then he said, "Go and work." So I went, and after 10 days or so, he came and said, "Well, show me what you have done." And I showed him. He looked at it and exclaimed: "I should have taken Bethe!"*

*From Victor F. Weisskopf, *Physics in the Twentieth Century: Selected Essays: My Life as a Physicist*. Cambridge, MA, The MIT Press, 1972, p. 10.

(Photo taken by S. A. Goudsmit, AIP Niels Bohr Library)

electrons in an atom can have the same set of quantum numbers. We should point out that if this principle were not valid, every electron would occupy the $1s$ atomic state (this being the state of lowest energy), the chemical behavior of the elements would be drastically different, and nature as we know it would not exist!

The exclusion principle follows from our belief that electrons are identical particles—that it is impossible to distinguish one electron from another. This seemingly innocuous statement takes on added importance in view of the wave nature of matter, and has far-reaching consequences. To gain an appreciation for this point, let us consider a collision between two electrons, as shown in Figure 9.14. Figures 9.14a and 9.14b depict two distinct events, the scattering effect being much stronger in the latter where the electrons are turned through a larger angle. Each event, however, arises from the same initial condition and leads to the same outcome—both electrons are scattered and emerge at angles θ relative to the axis of incidence. Had we not followed their paths, we could not decide which of the two collisions actually occurred, and the separate identities of the electrons would have been lost in the process of collision.

But paths are classical concepts, blurred by the wave properties of matter according to the uncertainty principle. That is, there is an inherent fuzziness to these paths, which blends them inextricably in the collision region, where the electrons may be separated by only a few de Broglie wavelengths. The quantum viewpoint is better portrayed in Figure 9.14c, where the two distinct possibilities (from a classical standpoint) merge into a single quantum event—the scattering of two electrons through an angle θ. Note that indistinguishability plays no role in classical physics: All particles, even identical ones, are distinguishable classically through their *paths!* With our acceptance of matter waves, we must conclude that **identical particles cannot be told**

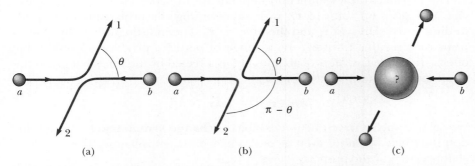

(a)   (b)   (c)

Figure 9.14   The scattering of two electrons as a result of their mutual repulsion. The events depicted in (a) and (b) produce the same outcome for identical electrons but are nonetheless distinguishable classically because the path taken by each electron is different in the two cases. In this way, the electrons retain their separate identities during collision. (c) According to quantum mechanics, the paths taken by the electrons are blurred by the wave properties of matter. In consequence, once they have interacted, the electrons cannot be told apart in any way!

**Electrons are truly
indistinguishable**

apart in any way—they are truly indistinguishable. Incorporating this remarkable fact into the quantum theory leads to the exclusion principle discovered by Pauli.

Let us see how indistinguishability affects our mathematical description of a two-electron system, say, the helium atom. Each electron has kinetic energy and the atom has electrostatic potential energy associated with the interaction of the two electrons with the doubly charged helium nucleus. These contributions are represented in Schrödinger's equation for one electron by terms

$$-\frac{\hbar^2}{2m_e}\nabla_1^2\psi + \frac{k(2e)(-e)}{r_1}\psi$$

where ∇_1^2 is the Laplacian in this electron's coordinate, $\mathbf{r}_1$. For brevity, let us write the sum of both terms simply as $h(1)\psi$, with the label 1 referring to $\mathbf{r}_1$. For the second electron, we write the same expression, except that $\mathbf{r}_1$ must be replaced everywhere by $\mathbf{r}_2$, the coordinate of the second electron. The stationary states for our two-electron system satisfy Schrödinger's time-independent equation,

$$h(1)\psi + h(2)\psi = E\psi \tag{9.15}$$

The fact that $h(1)$ and $h(2)$ are the *same* but for their arguments reflects the indistinguishability of the two electrons.

Equation 9.15 accounts for the electrons' kinetic energy and the atom's potential energy, but ignores the interaction between the two electrons. In fact, the electrons repel each other through the Coulomb force, leading to an interaction energy that must be added to the left-hand side of Equation 9.15. For simplicity, we shall ignore this interaction and treat the electrons as independent objects, each unaffected by the other's presence. In Section 9.5 we show how this *independent particle approximation* can be improved to give a better description of reality.

The two-electron wavefunction depends on the coordinates of both particles, $\psi = \psi(\mathbf{r}_1, \mathbf{r}_2)$, with $|\psi(\mathbf{r}_1, \mathbf{r}_2)|^2$ representing the probability density for finding one electron at $\mathbf{r}_1$ and the other at $\mathbf{r}_2$. The indistinguishability of electrons requires that a formal interchange of particles produce no observable effects. In particular, all probabilities are unaffected by the interchange, so the wavefunction ψ must be one for which

$$|\psi(\mathbf{r}_1, \mathbf{r}_2)|^2 = |\psi(\mathbf{r}_2, \mathbf{r}_1)|^2$$

We say that such a wavefunction exhibits **exchange symmetry.** The wavefunction itself may be either even or odd under particle exchange. The former is characterized by the property

**Exchange symmetry
for bosons**

$$\psi(\mathbf{r}_1, \mathbf{r}_2) = \psi(\mathbf{r}_2, \mathbf{r}_1) \tag{9.16}$$

and describes a class of particles called **bosons.** Photons belong to this class, as do some more exotic particles such as pions. Electrons, as well as protons and neutrons, are examples of **fermions,** for which

$$\psi(\mathbf{r}_1, \mathbf{r}_2) = -\psi(\mathbf{r}_2, \mathbf{r}_1) \qquad (9.17)$$

Exchange symmetry for fermions

Therefore, our two-electron helium wavefunction must obey Equation 9.17 to account for the indistinguishability of electrons.[7]

To recover the Pauli principle, we must examine the wavefunction more closely. For independent electrons, solutions to Equation 9.15 are easily found. Because each electron "sees" only the helium nucleus, the wavefunction in each coordinate must be an atomic function of the type discussed in Chapter 8. We denote these atomic functions by ψ_a, where a is a collective label for the four quantum numbers n, ℓ, m_ℓ, and m_s (or n, ℓ, j, and m_j if spin–orbit effects are included). The products $\psi_a(\mathbf{r}_1)\psi_b(\mathbf{r}_2)$ satisfy our equation, because

$$h(1)\psi_a(\mathbf{r}_1)\psi_b(\mathbf{r}_2) = E_a\psi_a(\mathbf{r}_1)\psi_b(\mathbf{r}_2)$$

$$h(2)\psi_a(\mathbf{r}_1)\psi_b(\mathbf{r}_2) = E_b\psi_a(\mathbf{r}_1)\psi_b(\mathbf{r}_2)$$

E_a and E_b are hydrogen-like energies for the states labeled a and b (see Eq. 8.38). Therefore,

$$[h(1) + h(2)]\psi_a(\mathbf{r}_1)\psi_b(\mathbf{r}_2) = (E_a + E_b)\psi_a(\mathbf{r}_1)\psi_b(\mathbf{r}_2) \qquad (9.18)$$

and $E = E_a + E_b$ is the total energy of this two-electron state.

Notice that the one-electron energies are simply additive, as we might have anticipated for independent particles. Furthermore, the solution $\psi_a(\mathbf{r}_1)\psi_b(\mathbf{r}_2)$ describes one electron occupying the atomic state labeled a and the other the state labeled b. But this product is *not* odd under particle exchange, as required for identical fermions. However, you can verify that $\psi_a(\mathbf{r}_2)\psi_b(\mathbf{r}_1)$ also is a solution to Equation 9.15 with energy $E = E_a + E_b$, corresponding to our two electrons having exchanged states. The antisymmetric combination of these two

$$\psi_{ab}(\mathbf{r}_1, \mathbf{r}_2) = \psi_a(\mathbf{r}_1)\psi_b(\mathbf{r}_2) - \psi_a(\mathbf{r}_2)\psi_b(\mathbf{r}_1) \qquad (9.19)$$

does display the correct exchange symmetry, that is,

$$\psi_{ab}(\mathbf{r}_2, \mathbf{r}_1) = \psi_a(\mathbf{r}_2)\psi_b(\mathbf{r}_1) - \psi_a(\mathbf{r}_1)\psi_b(\mathbf{r}_2)$$
$$= -\psi_{ab}(\mathbf{r}_1, \mathbf{r}_2)$$

Therefore, Equation 9.19 furnishes an acceptable description of the system. Notice, however, that it is now impossible to decide which electron occupies which state—as it should be for identical electrons! Finally, we see that when a and b label the same state ($a = b$), ψ_{ab} is identically zero—the theory allows no solution (description) in this case, in agreement with the familiar statement of the exclusion principle.

---

[7]It is an experimental fact that integer spin particles are bosons, but half-integer spin particles are fermions. This connection between spin and symmetry under particle exchange can be shown to have a theoretical basis when the quantum theory is formulated so as to conform to the requirements of special relativity.

EXAMPLE 9.5   Ground State of the Helium Atom

Construct explicitly the two-electron ground-state wavefunction for the helium atom in the independent particle approximation, using the prescription of Equation 9.19. Compare the predicted energy of this state with the measured value, and account in a qualitative way for any discrepancy.

Solution   In the independent-particle approximation, each helium electron "sees" only the doubly charged helium nucleus. Accordingly, the ground-state wavefunction of the helium atom is constructed from the lowest-energy hydrogen-like wavefunctions, with atomic number $Z = 2$ for helium. These are states for which $n = 1$, $\ell = 0$, and $m_\ell = 0$. Referring to Equation 8.42 of Chapter 8, we find (with $Z = 2$)

$$\psi_{100}(\mathbf{r}) = \pi^{-1/2}(2/a_0)^{3/2}e^{-2r/a_0}$$

To this orbital function we must attach a spin label ($\pm$) indicating the direction of electron spin. Thus, the one-electron state labels a and b in this example are given by $a = (1, 0, 0, +)$, $b = (1, 0, 0, -)$. Because there is no orbital field to interact with the electron spin, the energies of these two states are identical and are just the hydrogen-like levels of Equation 8.38 with $n = 1$ and $Z = 2$:

$$E_a = E_b = -(2^2/1^2)(13.6 \text{ eV}) = -54.4 \text{ eV}$$

The antisymmetric two-electron wavefunction for the ground state of helium is then

$$\psi(\mathbf{r}_1, \mathbf{r}_2) = \psi_{100+}(\mathbf{r}_1)\psi_{100-}(\mathbf{r}_2) - \psi_{100-}(\mathbf{r}_1)\psi_{100+}(\mathbf{r}_2)$$

Both terms have the same spatial dependence but differ as to their spin. The first term of the antisymmetric wave describes electron 1 as having spin up and electron 2 as having spin down. These spin directions are reversed in the second term. If we introduce the notation $|+-\rangle$ to describe the two electron spins in the first term, then the second term becomes $|-+\rangle$, and the total two-electron wavefunction for the helium ground state can be written

$$\psi(\mathbf{r}_1, \mathbf{r}_2) = \pi^{-1}(2/a_0)^3 e^{-2(r_1 + r_2)/a_0}\{|+-\rangle - |-+\rangle\}$$

The equal admixture of the spin states $|+-\rangle$ and $|-+\rangle$ means the spin of any one of the helium electrons is just as likely to be up as it is to be down. Notice, however, that the spin of the remaining electron is *always* opposite the first. Such spin–spin *correlations* are a direct consequence of the exclusion principle. (The valence electrons in different orbitals of many higher-Z atoms tend to align their spins. This tendency—known as **Hund's rule**—is another example of spin–spin correlations induced by the exclusion principle.)

The total electronic energy of the helium atom in this approximation is the sum of the one-electron energies E_a and E_b:

$$E = E_a + E_b = -54.4 \text{ eV} - 54.4 \text{ eV} = -108.8 \text{ eV}$$

The magnitude of this number, 108.8 eV, represents the energy (work) required to remove both electrons from the helium atom in the independent particle model. The measured value is substantially lower, about 79.0 eV, because of the mutual repulsion of the two electrons. Specifically, it requires less energy—only about 24.6 eV—to remove the first electron from the atom, because the electron left behind *screens* the nuclear charge, making it appear less positive than a bare helium nucleus.

O P T I O N A L

9.5   ELECTRON INTERACTIONS AND SCREENING EFFECTS

The preceding discussion of the helium atom exposes an issue that arises whenever we treat a system with two or more electrons, namely, how to handle the effects of electron–electron repulsion. Electrons confined to the small space of an atom are expected to exert strong repulsive electrical forces on one another. To ignore these altogether, as in the independent-particle model, is simply too crude; to include them exactly is unmanageable, since precise descriptions even for the classical motion in this case are unknown except through numerical computation. Accordingly, some workable approximation scheme is needed. A most fruitful approach to this problem begins with the notion of an effective field.

Any one atomic electron is subject to the Coulomb attraction of the nucleus as well as the Coulomb repulsion of every other electron in the atom. These influences largely cancel each other, leaving a net effective field with potential energy $U_{\text{eff}}(\mathbf{r})$.

U_{eff} may not be Coulombic—or even spherically symmetric—and may be different for each atomic electron. The success of this approach hinges on how simply and accurately we can model the effective potential. A few of the more obvious possibilities are outlined here.

The outermost, or valence, electrons of an atom "see" not the bare nucleus, but one shielded, or *screened*, by the intervening electrons. The attraction is more like that arising from a nucleus with an effective atomic number Z_{eff} somewhat less than the actual number Z and would be described by

$$U_{eff}(\mathbf{r}) = \frac{k(Z_{eff}e)(-e)}{r} \tag{9.20}$$

For a Z-electron atom, $Z_{eff} = Z$ would represent no screening whatever; at the opposite extreme is perfect screening by the $Z - 1$ other electrons, giving $Z_{eff} = Z - (Z - 1) = 1$. The best choice for Z_{eff} need not even be integral, and useful values may be deduced from measurements of atomic ionization potentials (see Example 9.6). Furthermore, the degree of screening depends on how much time an electron spends near the nucleus, and we should expect Z_{eff} to vary with the shell and subshell labels of the electron in question. In particular, a $4s$ electron is screened more effectively than a $3s$ electron, since its average distance from the nucleus is greater. Similarly, a $3d$ electron is better screened than a $3s$, or even a $3p$ electron (lower angular momentum implies more eccentric classical orbits, with greater penetration into the nuclear region). The use of a Z_{eff} for valence electrons is appropriate whenever a clear distinction exists between these and inner (core) electrons of the atom, as in the alkali metals.

EXAMPLE 9.6 Z_{eff} for the 3s Electron in Sodium

The outer electron of the sodium atom occupies the $3s$ atomic level. The observed value for the ionization energy of this electron is 5.14 eV. From this information, deduce a value of Z_{eff} for the $3s$ electron in sodium.

Solution   Since the ionization energy, 5.14 eV, represents the amount of energy that must be expended to remove the $3s$ electron from the atom, we infer that the energy of the $3s$ electron in sodium is $E = -5.14$ eV. This should be compared with the energy of a $3s$ electron in a hydrogen-like atom with atomic number Z_{eff}, or

$$E = -\frac{Z_{eff}^2}{3^2}(13.6 \text{ eV})$$

Equating this to -5.14 eV and solving for Z_{eff} gives

$$Z_{eff} = 3\sqrt{\frac{5.14}{13.6}} = 1.84$$

In principle, nuclear shielding can be better described by allowing Z_{eff} to vary continuously throughout the atom in a way that mimics the tighter binding accompanying electron penetration into the core. Two functional forms commonly are used for this purpose. For **Thomas–Fermi screening** we write

$$Z_{eff}(r) = Z e^{-r/a_{TF}} \tag{9.21}$$

The Thomas–Fermi atom

where a_{TF} is the Thomas–Fermi screening length. According to Equation 9.21, Z_{eff} is very nearly Z close to the nucleus ($r \approx 0$) but drops off quickly in the outer region, becoming essentially zero for $r \gg a_{TF}$. In this way, a_{TF} becomes an indicator of atomic size. The Thomas–Fermi model prescribes a_{TF} proportional to $Z^{-1/3}$; the weak variation with Z suggests that all atoms are essentially the same size, regardless of how many electrons they may have. Because the Thomas–Fermi potential is not Coulombic, the one-electron energies that result from the use of Equation 9.21 vary within a given shell; that is, they depend on the principal (n) *and* orbital (ℓ) quantum numbers. The study of these energies and their associated wavefunctions requires numerical methods, or further approximation. The Thomas–Fermi approximation improves with larger values of Z and so is especially well suited to describe the outer electronic structure of the heavier elements.

In another approach, called the quantum-defect method, nuclear shielding is described by

$$Z_{eff}(r) = 1 + \frac{b}{r} \tag{9.22}$$

where b is again a kind of screening length. This form is appropriate to the alkali metals, where a lone outer electron is responsible for the chemical properties of the atom. From Equation 9.22, this electron "sees" $Z_{eff} \approx 1$ for $r \gg b$ and larger values in the core. The special virtue of Equation 9.22 is that it leads to one-electron energies and wavefunctions that can be found without further approximation. In particular, the energy levels that follow from Equation 9.22 can be shown to be

Quantum defects

$$E_n = \frac{ke^2}{2a_0} \{n - D(\ell)\}^{-2} \tag{9.23}$$

where $D(\ell)$ is termed the **quantum defect,** since it measures the departure from the simple hydrogen-atom level structure. As the notation suggests, the quantum defect for an s electron differs from that for a p or d electron, but all s electrons have the same quantum defect, regardless of their shell label. Table 9.1 lists some quantum defects deduced experimentally for the sodium atom. Taking $b = 0$ in Equation 9.22 causes all quantum defects to vanish, returning us to the hydrogen-like level structure discussed in Chapter 8.

The use of a simple Z_{eff}, or the more complicated forms of the Thomas–Fermi or quantum-defect method, still results in a U_{eff} with spherical symmetry; that is, the electrons move in a central field. The **Hartree theory** discards even this feature in order to achieve more accurate results. According to Hartree, the electron "cloud" in the atom should be treated as a classical body of charge distributed with some volume charge density $\rho(\mathbf{r})$. The potential energy of any one atomic electron is then

$$U_{eff}(\mathbf{r}) = \frac{kZe^2}{r} - \int ke \left[\frac{\rho(\mathbf{r'})}{|\mathbf{r} - \mathbf{r'}|} \right] dV' \tag{9.24}$$

The first term is the attractive energy of the nucleus, and the second term is the repulsive energy of all other atomic electrons. This U_{eff} gives rise to a one-electron Schrödinger equation for the energies E_i and wavefunctions ψ_i of this, say the ith, atomic electron.

Table 9.1   Some Quantum Defects for the Sodium Atom

Subshell	s	p	d	f
$D(\ell)$	1.35	0.86	0.01	~0

But the Hartree theory is *self-consistent*. That is, the charge density $\rho(\mathbf{r})$ due to the other atomic electrons is itself calculated from the electron wavefunctions as

$$\rho(\mathbf{r}) = -e \sum |\psi_j(\mathbf{r})|^2 \qquad (9.25)$$

Hartree's self-consistent fields

The sum in Equation 9.25 includes all occupied electron states ψ_j except the ith state. In this way the mathematical problem posed by U_{eff} is turned back on itself: We must solve not one Schrödinger equation, but N of them in a single stroke, one for each of the N electrons in the atom! This is accomplished using numerical methods in an iterative solution scheme. An educated guess is made initially for each of the N ground-state electron waves. Starting with this guess, the ρ and U_{eff} for every electron can be computed and all N Schrödinger equations solved. The resulting wavefunctions are compared with the initial guesses; if discrepancies appear, the calculation is repeated with the new set of electron wavefunctions replacing the old ones. After several such iterations, agreement is attained between the starting and calculated wavefunctions. The resulting N electron wavefunctions are said to be fully self-consistent. Implementation of the Hartree method is laborious and demands considerable skill, but the results for atomic electrons are among the best available. Indeed, the Hartree and closely related **Hartree–Fock** methods are the ones frequently used today when accurate atomic energy levels and wavefunctions are required.

9.6   THE PERIODIC TABLE

In principle, it is possible to predict the properties of all the elements by applying the procedures of wave mechanics to each one. Because of the large number of interactions possible in multielectron atoms, however, approximations must be used for all atoms except hydrogen. Nevertheless, the electronic structure of even the most complex atoms can be viewed as a succession of filled levels increasing in energy, with the outermost electrons primarily responsible for the chemical properties of the element.

In the central field approximation, the atomic levels can be labeled by the quantum numbers n and ℓ. From the exclusion principle, the maximum number of electrons in one such subshell level is $2(2\ell + 1)$. The energy of an electron in this level depends primarily on the quantum number n, and to a lesser extent on ℓ. The levels can be grouped according to the value of n (the shell label), and all those within a group have energies that increase with increasing ℓ. The order of filling the subshell levels with electrons is as follows: Once a subshell is filled, the next electron goes into the vacant level that is lowest in energy. This **minimum energy principle** can be understood by noting that if the electron were to occupy a higher level, it would spontaneously decay to a lower one with the emission of energy.

The chemical properties of atoms are determined predominantly by the least tightly bound, or valence, electrons, which are in the subshell of highest energy. The most important factors are the occupancy of this subshell and the energy separation between this and the next-higher (empty) subshell. For example, an atom tends to be chemically inert if its highest subshell is full and there is an appreciable energy gap to the next-higher subshell, since then electrons are not readily shared with other atoms to form a molecule. The quasi-periodic recurrence of similar highest-shell structures as Z increases is responsible for the periodic system of the chemical elements.

The specification of n and ℓ for each atomic electron is called the **electron configuration** of that atom. We are now in a position to describe the electron configuration of any atom in its ground state:

Hydrogen has only one electron, which, in its ground-state, is described by the quantum numbers $n = 1$, $\ell = 0$. Hence, its electron configuration is designated as $1s^1$.

Helium, with its two electrons, has a ground-state electron configuration of $1s^2$. That is, both electrons are in the same (lowest-energy) $1s$ subshell. Since two is the maximum occupancy for an s subshell, the subshell (and in this case also the shell) is said to be *closed*, and helium is inert.

Lithium has three electrons. Two of these are assigned to the $1s$ subshell, and the third must be assigned to the $2s$ subshell, because this subshell has slightly lower energy than the $2p$ subshell. Hence, the electron configuration of lithium is $1s^22s^1$.

With the addition of another electron to make *beryllium*, the $2s$ subshell is closed. The electron configuration of beryllium, with four electrons altogether, is $1s^22s^2$. (Beryllium is not inert, however, because the energy gap separating the $2s$ level from the next available level—the $2p$—is not very large.)

Boron has a configuration of $1s^22s^22p^1$. (With spin–orbit doubling, the $2p$ electron in boron actually occupies the $2P_{1/2}$ sublevel, corresponding to $n = 2$, $\ell = 1$, and $j = \frac{1}{2}$.)

Carbon has six electrons, and a question arises of how to assign the two $2p$ electrons. Do they go into the same orbital with paired spins (↑ ↓), or do they occupy different orbitals with unpaired spins (↑ ↑)? Experiments show that the energetically preferred configuration is the latter, in which the spins are aligned. This is one illustration of **Hund's rule,** which states that electrons usually fill different orbitals with unpaired spins, rather than the same orbital with paired spins. Hund's rule can be partly understood by noting that electrons in the same orbital tend to be closer together, where their mutual repulsion contributes to a higher energy than if they were separated in different orbitals. Some exceptions to this rule do occur in those elements with subshells that are nearly filled or half-filled. The progressive filling of the $2p$ subshell illustrating Hund's rule is shown schematically in Figure 9.15. With *neon*, the $2p$ subshell is also closed. The neon atom has ten electrons in the configuration $1s^22s^22p^6$. Because the energy gap separating the $2p$ level from the next available level—the $3s$—is quite large, the neon configuration is exceptionally stable and the atom is chemically inert.

A complete list of electron configurations for all the known elements is given in Table 9.2. Note that beginning with *potassium* ($Z = 19$), the $4s$ subshell starts to fill while the $3d$ level remains empty. Only after the $4s$ subshell is closed to form *calcium* does the $3d$ subshell begin to fill. We infer that the $3d$ level has a higher energy than the $4s$ level, even though it belongs to a lower-indexed shell. This should come as no surprise, because the energy

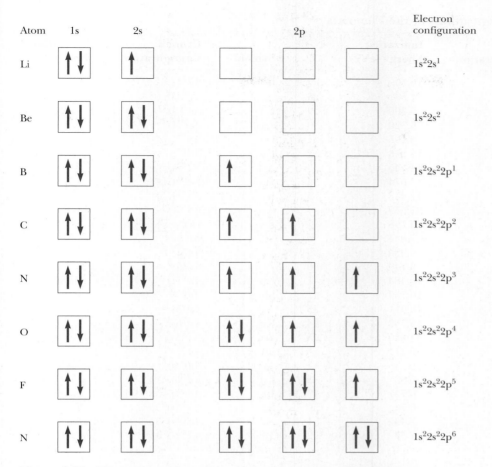

Atom	1s	2s	2p			Electron configuration
Li	↑↓	↑				$1s^2 2s^1$
Be	↑↓	↑↓				$1s^2 2s^2$
B	↑↓	↑↓	↑			$1s^2 2s^2 2p^1$
C	↑↓	↑↓	↑	↑		$1s^2 2s^2 2p^2$
N	↑↓	↑↓	↑	↑	↑	$1s^2 2s^2 2p^3$
O	↑↓	↑↓	↑↓	↑	↑	$1s^2 2s^2 2p^4$
F	↑↓	↑↓	↑↓	↑↓	↑	$1s^2 2s^2 2p^5$
N	↑↓	↑↓	↑↓	↑↓	↑↓	$1s^2 2s^2 2p^6$

Figure 9.15 Electronic configurations of successive elements from lithium to neon. The filling of electronic states must obey the Pauli exclusion principle and Hund's rule.

separating consecutive shells becomes smaller with increasing n (see the hydrogen-like spectrum), but the energy separating subshells is more nearly constant because of the screening discussed in Section 9.5. (In fact, the energy separating the $3d$ and $4s$ levels is very small, as evidenced by the electron configuration of *chromium*, in which the $3d$ subshell temporarily regains an electron from the $4s$.) The same phenomenon occurs again with *rubidium* ($Z = 37$), in which the $5s$ subshell begins to fill at the expense of the $4d$ and $4f$ subshells. Energetically, the electron configurations shown in the table imply the following ordering of subshells with respect to energy:

$$1s < 2s < 2p < 3s < 3p < 4s \sim 3d < 4p < 5s < 4d < 5p < 6s < 4f \sim 5d$$
$$< 6p < 7s < 6d \sim 5f \ldots$$

Ordering of subshells by energy

The elements from *scandium* ($Z = 21$) to zinc ($Z = 30$) form the first **transition series.** These transition elements are characterized by progres-

Table 9.2   Electronic Configurations of the Elements

Z	Symbol	Ground Configuration	Ionization Energy (eV)	Z	Symbol	Ground Configuration	Ionization Energy (eV)
1	H	$1s^1$	13.595	27	Co	$3d^74s^2$	7.86
2	He	$1s^2$	24.581	28	Ni	$3d^84s^2$	7.633
3	Li	$[He]\,2s^1$	5.390	29	Cu	$3d^{10}4s^1$	7.724
4	Be	$2s^2$	9.320	30	Zn	$3d^{10}4s^2$	9.391
5	B	$2s^22p^1$	8.296	31	Ga	$3d^{10}4s^24p^1$	6.00
6	C	$2s^22p^2$	11.256	32	Ge	$3d^{10}4s^24p^2$	7.88
7	N	$2s^22p^3$	14.545	33	As	$3d^{10}4s^24p^3$	9.81
8	O	$2s^22p^4$	13.614	34	Se	$3d^{10}4s^24p^4$	9.75
9	F	$2s^22p^5$	17.418	35	Br	$3d^{10}4s^24p^5$	11.84
10	Ne	$2s^22p^6$	21.559	36	Kr	$3d^{10}4s^24p^6$	13.996
11	Na	$[Ne]\,3s^1$	5.138	37	Rb	$[Kr]\,5s^1$	4.176
12	Mg	$3s^2$	7.644	38	Sr	$5s^2$	5.692
13	Al	$3s^23p^1$	5.984	39	Y	$4d5s^2$	6.377
14	Si	$3s^23p^2$	8.149	40	Zr	$4d^25s^2$	6.835
15	P	$3s^23p^3$	10.484	41	Nb	$4d^45s^1$	6.881
16	S	$3s^23p^4$	10.357	42	Mo	$4d^55s^1$	7.10
17	Cl	$3s^23p^5$	13.01	43	Tc	$4d^55s^2$	7.228
18	Ar	$3s^23p^6$	15.755	44	Ru	$4d^75s^1$	7.365
19	K	$[Ar]\,4s^1$	4.339	45	Rh	$4d^85s^1$	7.461
20	Ca	$4s^2$	6.111	46	Pd	$4d^{10}$	8.33
21	Sc	$3d4s^2$	6.54	47	Ag	$4d^{10}5s^1$	7.574
22	Ti	$3d^24s^2$	6.83	48	Cd	$4d^{10}5s^2$	8.991
23	V	$3d^34s^2$	6.74	49	In	$4d^{10}5s^25p^1$	5.785
24	Cr	$3d^54s$	6.76	50	Sn	$4d^{10}5s^25p^2$	7.342
25	Mn	$3d^54s^2$	7.432	51	Sb	$4d^{10}5s^25p^3$	8.639
26	Fe	$3d^64s^2$	7.87	52	Te	$4d^{10}5s^25p^4$	9.01

sive filling of the $3d$ subshell while the outer electron configuration is unchanged at $4s^2$ (except in the case of copper). Consequently, all the transition elements exhibit similar chemical properties. This belated occupancy of inner d subshells is encountered again in the second and third transition series, marked by the progressive filling of the $4d$ and $5d$ subshells, respectively. The second transition series includes the elements *yttrium* ($Z = 39$) to *cadmium* ($Z = 48$); the third contains the elements *lutetium* ($Z = 71$) to *mercury* ($Z = 80$).

Related behavior is also seen as the $4f$ and $5f$ subshells are filled. The **lanthanide series,** stretching from *lanthanum* ($Z = 57$) to *ytterbium* ($Z = 70$), is marked by a common $6s^2$ valence configuration, with the added electrons completing the $4f$ subshell (the nearby $5d$ levels also are occupied in some instances). The lanthanide elements, or lanthanides, also are known as the **rare earths** because of their low natural abundance. *Cerium* ($Z = 58$), which forms 0.00031% by weight of the Earth's crust, is the most abundant of the lanthanides.

Table 9.2   Electronic Configurations of the Elements

Z	Symbol	Ground Configuration	Ionization Energy (eV)	Z	Symbol	Ground Configuration	Ionization Energy (eV)
53	I	$4d^{10}5s^25p^5$	10.454	79	Au	$[\text{Xe}, 4f^{14}5d^{10}]\,6s^1$	9.22
54	Xe	$4d^{10}5s^25p^6$	12.127	80	Hg	$6s^2$	10.434
55	Cs	$[\text{Xe}]\,6s^1$	3.893	81	Tl	$6s^26p^1$	6.106
56	Ba	$6s^2$	5.210	82	Pb	$6s^26p^2$	7.415
57	La	$5d6s^2$	5.61	83	Bi	$6s^26p^3$	7.287
58	Ce	$4f5d6s^2$	6.54	84	Po	$6s^26p^4$	8.43
59	Pr	$4f^36s^2$	5.48	85	At	$6s^26p^5$	9.54
60	Nd	$4f^46s^2$	5.51	86	Rn	$6s^26p^6$	10.745
61	Pm	$4f^56s^2$	5.60	87	Fr	$[\text{Rn}]\,7s^1$	3.94
62	Fm	$4f^66s^2$	5.644	88	Ra	$7s^2$	5.277
63	Eu	$4f^76s^2$	5.67	89	Ac	$6d7s^2$	5.17
64	Gd	$4f^75d6s^2$	6.16	90	Th	$6d^27s^2$	6.08
65	Tb	$4f^96s^2$	6.74	91	Pa	$5f^26d7s^2$	5.89
66	Dy	$4f^{10}6s^2$	6.82	92	U	$5f^36d7s^2$	6.194
67	Ho	$4f^{11}6s^2$	6.022	93	Np	$5f^46d7s^2$	6.266
68	Er	$4f^{12}6s^2$	6.108	94	Pu	$5f^67s^2$	6.061
69	Tm	$4f^{13}6s^2$	6.185	95	Am	$5f^77s^2$	5.99
70	Yb	$4f^{14}6s^2$	6.22	96	Cm	$5f^76d7s^2$	6.02
71	Lu	$4f^{14}5d6s^2$	6.15	97	Bk	$5f^86d7s^2$	6.23
72	Hf	$4f^{14}5d^26s^2$	6.83	98	Cf	$5f^{10}7s^2$	6.30
73	Ta	$4f^{14}5d^36s^2$	7.88	99	Es	$5f^{11}7s^2$	6.42
74	W	$4f^{14}5d^46s^2$	7.98	100	Fm	$5f^{12}7s^1$	6.50
75	Re	$4f^{14}5d^56s^2$	7.87	101	Mv	$5f^{13}7s^2$	6.58
76	Os	$4f^{14}5d^66s^2$	8.71	102	No	$5f^{14}7s^2$	6.65
77	Ir	$4f^{14}5d^76s^2$	9.12	103	Lw	$5f^{14}6d7s^2$	
78	Pt	$4f^{14}5d^86s^2$	8.88	104	Ku	$5f^{14}6d^27s^2$	

Note: The bracket notation is used as a shorthand method to avoid repetition in indicating inner-shell electrons. Thus, [He] represents $1s^2$, [Ne] represents $1s^22s^22p^6$, [Ar] represents $1s^22s^22p^63s^23p^6$, and so on.

In the **actinide series** from *actinium* ($Z = 89$) to *nobelium* ($Z = 102$), the valence configuration remains $7s^2$, as the $5f$ subshell progressively fills (along with occasional occupancy of the nearby $6d$ level).

Table 9.2 also lists the ionization energies of the elements. The ionization energy for each element is plotted against its atomic number Z in Figure 9.16a. This plot shows that the ionization energy tends to increase within a shell, then drops dramatically as the filling of a new shell begins. The behavior repeats, and it is from this recurring pattern that the periodic table gets its name. A similar repetitive pattern is observed in a plot of the atomic volume per atom versus atomic number (see Fig. 9.16b).

The primary features of these plots can be understood from simple arguments. First, the larger nuclear charge that accompanies higher values of Z tends to pull the electrons closer to the nucleus and binds them more tightly. Were this the only effect, the ionization energy would increase and the atomic volume would decrease steadily with increasing Z. But the innermost, or core, electrons screen the nuclear charge, making it less effective in binding the outer electrons. The screening effect varies in a

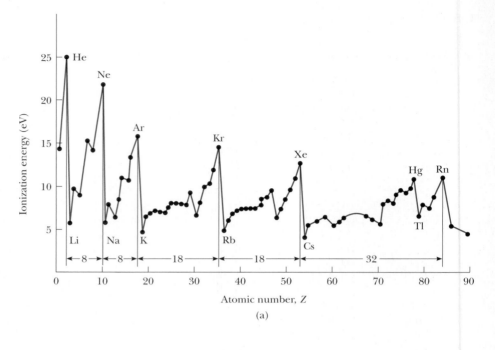

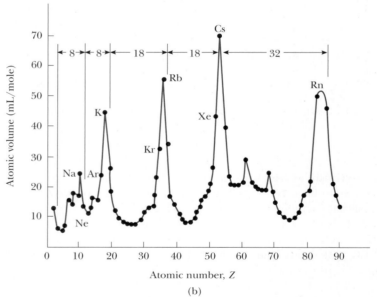

Figure 9.16   (a) Ionization energy of the elements versus atomic number Z. (b) Atomic volume of the elements versus atomic number Z. The recurring pattern with increasing atomic number exemplifies the behavior from which the periodic table gets its name.

complicated way from one element to the next, but it is most pronounced for a lone electron outside a closed shell, as in the alkali metals (Li, Na, K, Rb, Cs, and Fr). For these configurations the ionization energy drops sharply, only to rise again as the nuclear charge intensifies at higher Z. The variation in ionization energy is mirrored by the behavior of atomic volume,

Associated with the Larmor frequency is the energy quantum $\hbar\omega_L = \mu_B B$. The ℓth subshell level of an atom placed in a magnetic field is split by the field into $2\ell + 1$ sublevels separated by the Larmor energy $\hbar\omega_L$. This is the **Zeeman effect,** and $\hbar\omega_L$ is also known as the Zeeman energy. The magnetic contribution to the energy of the atom is

$$U = \hbar\omega_L m_\ell \qquad (9.7)$$

where m_ℓ is the same magnetic quantum number discussed in Chapter 8. Equation 9.7 is a special case of the more general result

$$U = -\boldsymbol{\mu} \cdot \mathbf{B} \qquad (9.6)$$

for the magnetic potential energy U of any magnetic moment $\boldsymbol{\mu}$ in an applied field $\mathbf{B}$.

In addition to any orbital magnetic moment, the electron possesses an intrinsic magnetic moment called the **spin moment, $\boldsymbol{\mu}_s$**. In a classical picture, the spin moment arises from the rotation of the electron on its axis and is proportional to the angular momentum of rotation, or spin, $\mathbf{S}$. The magnitude of the spin angular momentum is

$$|\mathbf{S}| = \frac{\sqrt{3}}{2}\hbar \qquad (9.11)$$

corresponding to a **spin quantum number** $s = \frac{1}{2}$, analogous to the orbital quantum number ℓ. The z component of $\mathbf{S}$ is quantized as

$$S_z = m_s \hbar \qquad (9.10)$$

where the **spin magnetic quantum number** m_s is the analog of the orbital magnetic quantum number m_ℓ. For the electron, m_s can be either $+\frac{1}{2}$ or $-\frac{1}{2}$, which describes the spin-up and spin-down states, respectively. Equations 9.10 and 9.11 imply that electron spin also is subject to space quantization. This was confirmed experimentally by Stern and Gerlach, who observed that a beam of silver atoms passed through a nonuniform magnetic field was split into two distinct components. The same experiment shows that the spin moment is related to the spin angular momentum by

$$\boldsymbol{\mu}_s = -\frac{e}{m_e}\mathbf{S}$$

This is *twice* as large as the orbital moment for the same angular momentum. The anomalous factor of 2 is called the **g factor** of the electron. With the recognition of spin, four quantum numbers—n, ℓ, m_ℓ, and m_s—are needed to specify the state of an atomic electron.

The spin moment of the electron interacts with the magnetic field arising from its orbital motion. The energy difference between the two spin orientations in this orbital field is responsible for the **fine structure doubling** of atomic spectral lines. With the spin–orbit interaction, atomic states are labeled by a quantum number j for the *total* angular momentum $\mathbf{J} = \mathbf{L} + \mathbf{S}$. The value of j is included as a subscript in the spectroscopic notation of atomic states. For example, $3P_{1/2}$ specifies a state for which $n = 3$, $\ell = 1$, and $j = \frac{1}{2}$. For each value of j, there are $2j + 1$ possibilities for the total magnetic quantum number m_j.

The **exclusion principle** states that no two electrons can be in the same quantum state; that is, no two electrons can have the same four quantum numbers. The exclusion principle derives from the notion that electrons are identical particles called **fermions.** Fermions are described by wavefunctions that are antisymmetric in the electron coordinates. Wavefunctions that are symmetric in the particle coordinates describe another class of objects called **bosons,** to which no exclusion principle applies. All known particles are either fermions or bosons. An example of a boson is the photon.

Using the exclusion principle and the principle of minimum energy, one can determine the electronic configurations of the elements. This serves as a basis for understanding atomic structure and the physical and chemical properties of the elements.

One can catalog the discrete x-ray line spectra emitted by different metals in terms of electronic transitions within inner shells. When electron bombardment creates a vacancy in an inner K or L shell, a higher-lying atomic electron quickly fills the vacancy, giving up its excess energy as an x-ray photon. According to **Moseley's law,** the square root of this photon frequency should be proportional to the atomic number of the emitting atom.

SUGGESTIONS FOR FURTHER READING

1. A classic work on the physics of atoms is H. E. White, *Introduction to Atomic Spectra*, New York, McGraw-Hill, 1934.

The following sources contain more extensive discussions of the topics found in this chapter:

2. A. P. French and E. F. Taylor, *An Introduction to Quantum Physics*, New York, W. W. Norton and Company, Inc., 1978.

3. R. Eisberg and R. Resnick, *Quantum Physics of Atoms, Molecules, Solids, Nuclei, and Particles*, 2nd ed., New York, John Wiley and Sons, Inc., 1985.

4. B. H. Bransden and C. J. Joachain, *Physics of Atoms and Molecules*, New York, John Wiley and Sons, Inc., 1990.

QUESTIONS

1. Why is the direction of the orbital angular momentum of an electron opposite that of its magnetic moment?

2. Why is an inhomogeneous magnetic field used in the Stern–Gerlach experiment?

3. Could the Stern–Gerlach experiment be performed with ions rather than neutral atoms? Explain.

4. Describe some experiments that would support the conclusion that the spin quantum number for electrons can have only the values $\pm\frac{1}{2}$.

5. Discuss some of the consequences of the exclusion principle.

6. Why do lithium, potassium, and sodium exhibit similar chemical properties?

7. From Table 9.2, we find that the ionization energies for Li, Na, K, Rb, and Cs are 5.390, 5.138, 4.339, 4.176, and 3.893 eV, respectively. Explain why these values are to be expected in terms of the atomic structures.

8. Although electrons, protons, and neutrons obey the exclusion principle, some particles that have integral spin, such as photons (spin = 1), do not. Explain.

9. How do we know that a photon has a spin of 1?

10. An energy of about 21 eV is required to excite an electron in a helium atom from the 1s state to the 2s state. The same transition for the He$^+$ ion requires about twice as much energy. Explain why this is so.

11. Discuss degeneracy as it applies to a multielectron atom. Can a one-electron atom have degeneracy? Explain.

12. The absorption or emission spectrum of a gas consists of lines that broaden as the density of gas molecules increases. Why do you suppose this occurs?

13. For a one-electron atom or ion, spin–orbit coupling splits all states except s states into doublets. Why are s states exceptions to this rule?

14. Why is it approximately correct to neglect the screening effect of outer-shell electrons (for example, electrons in the M and N shells) on an electron in the L shell?

PROBLEMS

9.1   Orbital Magnetism and the Normal Zeeman Effect

1. In the technique known as electron spin resonance (ESR), a sample containing unpaired electrons is placed in a magnetic field. Consider the simplest situation, that in which there is only one electron and therefore only two possible energy states, corresponding to $m_s = \pm\frac{1}{2}$. In ESR, the electron's spin magnetic moment is "flipped" from a lower energy state to a higher energy state by the absorption of a photon. (The lower energy state corresponds to the case in which the magnetic moment $\boldsymbol{\mu}_s$ is aligned with the magnetic field, and the higher energy state corresponds to the case where $\boldsymbol{\mu}_s$ is aligned against the field.) What is the photon frequency required to excite an ESR transition in a magnetic field of 0.35 T?

2. Show that for a mass m in orbit with angular momentum $\mathbf{L}$ the rate at which area is swept out by the orbiting particle is

$$\frac{dA}{dt} = \frac{|\mathbf{L}|}{2m}$$

(*Hint:* First show that in its displacement, $\mathbf{dr}$, *along the path*, the particle sweeps out an area $dA = \frac{1}{2}|\mathbf{r} \times \mathbf{dr}|$, where $\mathbf{r}$ is the position vector of the particle drawn from some origin.)

9.2   The Spinning Electron

3. How many different sets of quantum numbers are possible for an electron for which (a) $n = 1$, (b) $n = 2$, (c) $n = 3$, (d) $n = 4$, and (e) $n = 5$? Check your results to show that they agree with the general rule that the number of different sets of quantum numbers is equal to $2n^2$.

4. List the possible sets of quantum numbers for an electron in (a) the $3d$ subshell and (b) the $3p$ subshell.

5. The force on a magnetic moment $\boldsymbol{\mu}_z$ in a nonuniform magnetic field B_z is given by

$$F_z = \mu_z \frac{dB_z}{dz}$$

If a beam of silver atoms travels a horizontal distance of 1 m through such a field and each atom has a speed of 100 m/s, how strong must the field gradient dB_z / dz be in order to deflect the beam 1 mm?

6. Consider the original Stern–Gerlach experiment employing an atomic beam of silver, for which the magnetic moment is due entirely to the spin of the single valence electron of the silver atom. Assuming the magnetic field $\mathbf{B}$ has magnitude 0.500 T, compute the energy difference in electron volts of the silver atoms in the two exiting beams.

7. When the idea of electron spin was introduced, the electron was thought to be a tiny charged sphere (today it is considered a point object with no extension in space). Find the equatorial speed under the assumption that the electron is a uniform sphere of radius 3×10^{-6} nm, as early theorists believed, and compare your result to the speed of light, c.

8. Consider a right circular cylinder of radius R, with mass M uniformly distributed throughout the cylinder volume. The cylinder is set into rotation with angular speed ω about its longitudinal axis. (a) Obtain an expression for the angular momentum $\mathbf{L}$ of the rotating cylinder. (b) If charge Q is distributed uniformly over the *curved* surface only, find the magnetic moment $\boldsymbol{\mu}$ of the rotating cylinder. Compare your expressions for $\boldsymbol{\mu}$ and $\mathbf{L}$ to deduce the g factor for this object.

9. An exotic elementary particle called the *omega minus* (symbol Ω^-) has spin $\frac{3}{2}$. Calculate the magnitude of the spin angular momentum for this particle and the possible angles the spin angular momentum vector makes with the z-axis. Does the Ω^- obey the Pauli exclusion principle? Explain.

9.3   The Spin–Orbit Interaction and Other Magnetic Effects

10. Consider a single-electron atom in the $n = 2$ state. Find all possible values for j and m_j for this state.

11. Find all possible values of j and m_j for a d electron.

12. Give the spectroscopic notation for the following states: (a) $n = 7$, $\ell = 4$, $j = \frac{9}{2}$; (b) all the possible states of an electron with $n = 6$ and $\ell = 5$.

13. An electron in an atom is in the $4F_{5/2}$ state. (a) Find the values of the quantum numbers n, ℓ, and j. (b) What is the magnitude of the electron's total angular momentum? (c) What are the possible values for the z component of the electron's total angular momentum?

14. (a) Starting with the expression $\mathbf{J} = \mathbf{L} + \mathbf{S}$ for the total angular momentum of an electron, derive an expression for the scalar product $\mathbf{L} \cdot \mathbf{S}$ in terms of the quantum numbers j, ℓ, and s. (b) Using $\mathbf{L} \cdot \mathbf{S} = |\mathbf{L}|\,|\mathbf{S}|\cos\theta$, where θ is the angle between $\mathbf{L}$ and $\mathbf{S}$, find the angle between the electron's orbital angular momentum and spin angular momentum for the following states: (1) $P_{1/2}$, $P_{3/2}$ and (2) $H_{9/2}$, $H_{11/2}$.

15. *Spin–Orbit energy in an atom.* Estimate the magnitude of the spin–orbit energy for an atomic electron in the hydrogen $2p$ state. (*Hint:* From the vantage point of the moving electron, the nucleus circles it in an orbit with radius equal to the Bohr radius for this state. Treat the orbiting nucleus as a current in a circular wire loop and use the result from classical electromagnetism,

$$B = \frac{2k_m\mu}{r^3}$$

for the **B** field at the center of loop with radius r and magnetic moment μ. Here, $k_m = 10^{-7}\,\text{N/A}^2$ is the magnetic constant in SI units.)

9.4 Exchange Symmetry and the Exclusion Principle

16. Show that the symmetric combination of two single particle wavefunctions

$$\psi_{ab}(\mathbf{r}_1, \mathbf{r}_2) = \psi_a(\mathbf{r}_1)\psi_b(\mathbf{r}_2) + \psi_a(\mathbf{r}_2)\psi_b(\mathbf{r}_1)$$

displays the exchange symmetry characteristic of bosons, Equation 9.16. Is it possible for two bosons to occupy the same quantum state? Explain.

17. Eight identical, noninteracting particles are placed in a cubical box of sides $L = 0.200$ nm. Find the lowest energy of the system (in electron volts) and list the quantum numbers of all occupied states if (a) the particles are electrons and (b) the particles have the same mass as the electron but do not obey the exclusion principle.

9.5 Electron Interactions and Screening Effects (Optional)

18. The claim is made in Section 9.5 that a d electron is screened more effectively from the nuclear charge in an atom than is a p electron or an s electron. Give a classical argument based on the definition of angular momentum $\mathbf{L} = \mathbf{r} \times \mathbf{p}$ that indicates that smaller values of angular momentum are associated with orbits of larger eccentricity. Verify this quantum mechanically by calculating the probability that a $2p$ electron of hydrogen will be found inside the $n = 1$ atomic shell and comparing this with the probability of finding a hydrogen $2s$ electron in this same region. For which is the probability largest, and what effect does this have on the degree of screening? The relevant wavefunctions may be found in Table 8.4 of Chapter 8.

19. *Multielectron atoms.* For atoms containing many electrons, the potential seen by the outer, or valence, electrons is often described by the Thomas–Fermi form (see Equation 9.21)

$$U(r) = -\frac{Zke^2}{r}\,e^{-r/a}$$

where Z is the atomic number and a is the Thomas–Fermi screening length. Use the Java applet available at our companion Web site (http://info.brookscole.com/mp3e QMTools Simulations → Problem 9.19) to find the lowest valence energy and wavefunction for gold (Au), taking $Z = 79$ and $a = 0.39\,a_0$. (According to Table 9.2, gold has a valence electron configuration of $6s^1$.) How many nodes does this wavefunction exhibit? Use the results of this study to esti-mate the ionization energy of gold, and compare with the experimental value given in Table 9.2. Also report the most probable distance from the nucleus for the $6s$ electron in gold, according to your findings. What size would you assign to the gold atom? How does this size compare with that of the hydrogen atom?

20. *Quantum defects.* According to Table 9.1, the p-wave quantum defect for sodium is 0.86. What is the energy of the $2p$ level in sodium? the $3p$ level? Use the Java applet available at our companion Web site (http://info.brookscole.com/mp3e QMTools Simulations → Problem 9.20) to determine the screening length b in Equation 9.22 that reproduces the observed p-state energies for the sodium atom. Based on your findings, report the most probable distance from the nucleus for the $2p$ electrons in sodium.

9.6 The Periodic Table

21. (a) Write out the electronic configuration for oxygen ($Z = 8$). (b) Write out the values for the set of quantum numbers n, ℓ, m_ℓ, and m_s for each of the electrons in oxygen.

22. Which electronic configuration has a lower energy: $[\text{Ar}]3d^44s^2$ or $[\text{Ar}]3d^54s^1$? Identify this element and discuss Hund's rule in this case. (*Note:* The notation $[\text{Ar}]$ represents the filled configuration for Ar.)

23. Which electronic configuration has the lesser energy and the greater number of unpaired spins: $[\text{Kr}]4d^95s^1$ or $[\text{Kr}]4d^{10}$? Identify this element and discuss Hund's rule in this case. (*Note:* The notation $[\text{Kr}]$ represents the filled configuration for Kr.)

24. Devise a table similar to that shown in Figure 9.15 for atoms with 11 through 19 electrons. Use Hund's rule and educated guesswork.

25. The states of matter are solid, liquid, gas, and plasma. Plasma can be described as a gas of charged particles, or a gas of ionized atoms. Most of the matter in the Solar System is plasma (throughout the interior of the Sun). In fact, most of the matter in the Universe is plasma; so is a candle flame. Use the information in Figure 9.16 to make an order-of-magnitude estimate for the temperature to which a typical chemical element must be raised to turn into plasma by ionizing most of the atoms in a sample. Explain your reasoning.

9.7 X-ray Spectra and Moseley's Law

26. Show that Moseley's law for K_α radiation may be expressed as

$$\sqrt{f} = \sqrt{\frac{3}{4}\left(\frac{13.6\,\text{eV}}{h}\right)}\,(Z - 1)$$

where f is the x-ray frequency and Z is the atomic number. (b) Check the agreement of the original 1914 data shown in Figure 9.18 with Moseley's law. Do this by comparing the least-squares slope and intercept of the K_α line in Figure 9.18 to the theoretical slope and intercept predicted by Moseley's law. (c) Is the screened charge seen by the L shell electron equal to $Z - 1$?

27. (a) Derive an equation similar to that in Problem 26, but for L_α x rays. Assume, as in the case of K_α x rays, that electrons in the shell of origin (in this case M) produce no screening and that all screening is attributed to electrons in the inner shells (in this case L and K). (b) Test your equation by comparing its slope and intercept with that of the experimental L_α line in Figure 9.18. (c) From the intercept of the experimental L_α line, deduce the average screened charge seen by the M shell electron.

10

Statistical Physics

Chapter Outline

Thermodynamics is based on macroscopic or bulk properties, such as temperature and pressure of a gas. In this chapter we explain thermodynamic properties in terms of the motion of individual atoms. The goal of this microscopic approach, known as statistical physics, or statistical mechanics, is to explain the relationships between thermodynamic bulk properties using a more fundamental atomic picture. It is possible in principle to calculate the detailed motion of individual atoms from Newton's laws or the Schrödinger equation. The number of atoms in the average size sample ($\approx 10^{22}$ atoms/cm^3), however, makes such calculations impractical, and we must rely on a statistical approach.

In this chapter we introduce the laws of statistical physics and discuss systems of particles that obey either classical or quantum mechanics. We will show how a fixed amount of energy may be shared or distributed among a large number of particles in thermal equilibrium at temperature T. We investigate this energy distribution by calculating the average number of particles with a specific energy or, what is essentially the same thing, by finding the probability that a single particle has a certain energy.

10.1 THE MAXWELL–BOLTZMANN DISTRIBUTION

The satisfying explanation of thermodynamics in terms of averages over atomic properties was given in the second half of the 1800s by three physicists: James Clerk Maxwell, Ludwig Boltzmann, and Josiah Willard Gibbs. Maxwell, a Scottish professor at Cambridge, was extremely impressed by the work of Rudolf Clausius in explaining the apparent contradiction between the high speed of gas molecules at room temperature (about 400 m/s) and the slow diffusion rate of a gas. Clausius had explained this riddle by reasoning that gas molecules do not all travel at a single high speed, but that there is a well-defined distribution of molecular speeds in a gas that depends on the gas temperature; furthermore, the gas molecules collide and hence follow long zigzag paths from one spot to another. Building on this idea, Maxwell was able to derive the functional form of the equilibrium speed distribution, which is the number of gas molecules per unit volume having speeds between v and $v + dv$ at a specific temperature. Applying the theory of statistics to this distribution, Maxwell was able to calculate the temperature dependence of quantities such as the average molecular speed, the most probable speed, and the dispersion, or width, of the speed distribution.

In 1872, Boltzmann, an Austrian professor at the University of Vienna, profoundly impressed with Darwin's ideas on evolution, took Maxwell's work a step further. He not only wanted to establish the properties of the equilibrium or most probable distribution but he also wished to describe the evolution in time of a gas toward the Maxwellian distribution—the so-called approach-to-equilibrium problem. With the use of a time-dependent speed distribution function and his kinetic equation, Boltzmann was able to show that a system of particles that starts off with a non-Maxwellian speed distribution steadily approaches and eventually achieves an equilibrium Maxwellian speed distribu-

James Clerk Maxwell (1831–1879) is generally regarded as the greatest theoretical physicist of the 19th century. (*Courtesy AIP Emilio Segrè Visual Archives*)

Is the universe a gambling casino? (*Courtesy of Tropicana Casino And Resort*)

Ludwig Boltzmann (1844–1908), an Austrian theoretical physicist. (*Courtesy AIP Niels Bohr Library, Lande Collection*)

Josiah Willard Gibbs (1839–1903), an American theoretical physicist who made significant contributions to statistical mechanics. (*Burndy Library/Courtesy AIP Emilio Segrè Visual Archives*)

Assumptions of the Maxwell–Boltzmann distribution

tion. Boltzmann, a staunch advocate of the reality of molecules, was subjected to personal attacks at the hands of critics who rejected the molecular theory of matter in the late 1800s. Depressed over the lack of universal acceptance of his theories, he committed suicide in 1908.

Gibbs, in contrast to Boltzmann, led a rather sheltered and secluded life as a professor at Yale. The son of a Yale professor, he lived his adult life in the same house in New Haven in which he had grown up, quietly establishing statistical mechanics and the kinetic theory of gases on a rigorous mathematical basis. Gibbs published his work in the obscure *Transactions of the Connecticut Academy of Arts and Sciences,* and his work remained relatively unknown during his lifetime.

Having briefly discussed the contributions of Maxwell, Boltzmann, and Gibbs to statistical mechanics, let us examine the underlying assumptions and explicit form of the Maxwell–Boltzmann distribution for a system of particles. The basic assumptions are:

- The particles are identical in terms of physical properties but distinguishable in terms of position, path, or trajectory. It will be demonstrated later in this chapter that this assumption is equivalent to the statement that the particle size is small compared with the average distance between particles.
- The equilibrium distribution is the most probable way of distributing the particles among various allowed energy states subject to the constraints of a fixed number of particles and fixed total energy.
- There is no theoretical limit on the number of particles in a given energy state, but the density of particles is sufficiently low and the temperature sufficiently high that no more than one particle is likely to be in a given state at the same time.

To make these assumptions concrete, let us consider the analysis of a manageable-sized system of distinguishable particles. In particular, consider the distribution of a total energy of 8E among six particles where E is an indivisible unit of energy. To work with a diagram of reasonable size, Figure 10.1a enumerates the 20 possible ways of sharing an energy of 8E among six *indistinguishable* particles. Since we are actually interested in *distinguishable* particles, each of the 20 arrangements can be decomposed into many distinguishable substates, or **microstates,** as shown explicitly for one arrangement in Figure 10.1b. The number of microstates for each of the 20 arrangements is given in parentheses in Figure 10.1a and may be computed from the relation

$$N_{MB} = \frac{N!}{n_1!n_2!n_3!\ \cdots}$$ (10.1)

where N_{MB} is the Maxwell–Boltzmann number of microstates, N is the total number of particles, and n_1, n_2, n_3, . . . are the numbers of particles in *occupied* states of a certain energy. This result may be understood by arguing that the first energy level may be assigned in N ways, the second in $N-1$ ways, and so on, giving $N!$ in the numerator. The factor in the denominator of Equation 10.1 corrects for indistinguishable order arrangements when more than one particle occupies the same energy level. As an example of the use of Equation 10.1, consider the energy distribution of six particles, with two having energy 1E, one having energy 6E, and three having energy 0. This energy distribution is shown in the fourth diagram from the left in the top row of Figure 10.1a. In

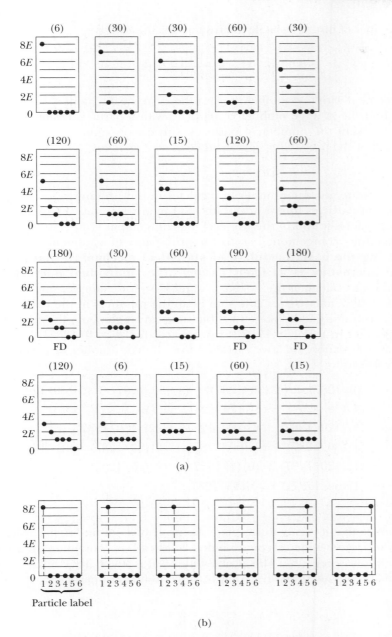

Figure 10.1 (a) The 20 arrangements of six *indistinguishable* particles with a total energy of 8*E*. (b) The decomposition of the upper left-hand arrangement of part (a) into six distinguishable states for *distinguishable* particles.

this case, $N = 6$, $n_1(0E) = 3$ (that is, the number of particles in the 0 energy state is 3), $n_2(1E) = 2$, $n_3(2E) = 0$, $n_4(3E) = 0$, $n_5(4E) = 0$, $n_6(5E) = 0$, $n_7(6E) = 1$, $n_8(7E) = 0$, and $n_9(8E) = 0$. Since only the numbers of particles in occupied levels appear in the denominator of Equation 10.1, we find that the number of distinguishable microstates for this case is

$$N_{MB} = \frac{6!}{3!2!1!} = 60$$

in agreement with the number in parentheses in the diagram.

To find the average number of particles with a particular value of energy, say E_j, we sum the number of particles with energy E_j in each arrangement weighted by the probability of realizing that arrangement:

$$\overline{n_j} = n_{j_1}p_1 + n_{j_2}p_2 + \cdots \tag{10.2}$$

Here $\overline{n_j}$ is the average number of particles in the jth energy level, n_{j_1} is the number of particles found in the jth level in arrangement 1, n_{j_2} is the number of particles found in the jth level in arrangement 2, p_1 is the probability of observing arrangement 1, p_2 is the probability of arrangement 2, and so on. **Using the basic postulate of statistical mechanics, that any individual microstate is as likely as any other individual microstate,** we may go on to calculate the various p's and $\overline{n_j}$'s. For example, since there are a total of 1287 microstates (the sum of all the numbers in the parentheses), and 6 distinguishable ways of obtaining arrangement 1 (the leftmost arrangement in row 1 in Fig. 10.1a), we see that $p_1 = 6/1287$. Using these ideas and Equation 10.2, we calculate the average number of particles with energy 0 as follows:

$$
\begin{aligned}
\overline{n_0} = \ & (5)(6/1287) + (4)(30/1287) + (4)(30/1287) + (3)(60/1287) \\
& + (4)(30/1287) + (3)(120/1287) + (2)(60/1287) + (4)(15/1287) \\
& + (3)(120/1287) + (3)(60/1287) + (2)(180/1287) \\
& + (1)(30/1287) + (3)(60/1287) + (2)(90/1287) + (2)(180/1287) \\
& + (1)(120/1287) + (0)(6/1287) + (2)(15/1287) \\
& + (1)(60/1287) + (0)(15/1287) \\
= \ & 2.307
\end{aligned}
$$

Now notice that it is easy to calculate the probability of finding a particle with energy 0 if we imagine reaching randomly into a box containing the six particles with total energy $8E$. This probability, $p(0)$, is simply the average number of particles with energy 0 divided by the total number of particles:

$$p(0) = \frac{\overline{n_0}}{6} = \frac{2.307}{6} = 0.385$$

It is left as a problem (Problem 1) to show that the probabilities of finding a particle with energies from $1E$ through $8E$ are as follows:

$$p(1E) = 0.256$$

$$p(2E) = 0.167$$

$$p(3E) = 0.0978$$

$$p(4E) = 0.0543$$

$$p(5E) = 0.0272$$

$$p(6E) = 0.0117$$

$$p(7E) = 0.00388$$

$$p(8E) = 0.000777$$

These results, which are plotted in Figure 10.2, show that this simple system follows an approximately exponential decrease in probability with energy. (See Problem 5.) The rapid decrease in probability with increasing energy shown in Figure 10.2 indicates that we are more likely to find the energy uniformly distributed among many particles of the system rather than concentrated in a few particles.

One may rigorously derive the Maxwell–Boltzmann distribution for a system in thermal equilibrium at the absolute temperature T containing a *large* number of particles by using calculus (see reference 1 in Suggestions for Further Reading at the end of this chapter). The expression for the number of ways of distributing the particles among the allowed energy states is maximized subject to two constraints. These constraints are (1) that the total number of particles is constant at any temperature and (2) that the total system energy is fixed at a given temperature. One finds an exponential form

$$f_{MB} = Ae^{-E_i/k_B T} \qquad (10.3)$$

**Maxwell–Boltzmann
distribution**

where f_{MB} is the Maxwell–Boltzmann probability of finding a particle with energy E_i, or in the language of statistical mechanics, the probability that a state with energy E_i is occupied at the absolute temperature T. If the number of states with the same energy E_i is denoted by g_i (g_i is called the **degeneracy** or **statistical weight**), then the number of particles, n_i, with energy E_i is equal to the product of the statistical weight and the probability that the state E_i is occupied, or

$$n_i = g_i f_{MB} \qquad (10.4)$$

The parameter A in Equation 10.3 is a normalization coefficient, which is similar to the normalization constant in quantum physics. A is determined by requiring the number of particles in the system to be constant, or

$$N = \sum n_i \qquad (10.5)$$

where N is the total number of particles in the system.

When the allowed energy states are numerous and closely spaced, the discrete quantities are replaced by continuous functions as follows:

$$g_i \longrightarrow g(E)$$

$$f_{MB} \longrightarrow Ae^{-E/k_B T}$$

where $g(E)$ is the **density of states** or the number of energy states per unit volume in the interval dE. In a similar manner, Equations 10.4 and 10.5 may be replaced as follows:

$$n_i = g_i f_{MB} \longrightarrow n(E)\,dE = g(E)\,f_{MB}(E)\,dE \qquad (10.6)$$

$$N = \sum n_i \longrightarrow \frac{N}{V} = \int_0^\infty n(E)\,dE = \int_0^\infty g(E)f_{MB}(E)\,dE \qquad (10.7)$$

where $n(E)\,dE$ is the number of particles per unit volume with energies between E and $E + dE$. Note that Equations 10.6 and 10.7 may also be used for a system of quantum particles, provided that $g(E)$ and $f_{MB}(E)$ are replaced with the appropriate density of states and quantum distribution functions.

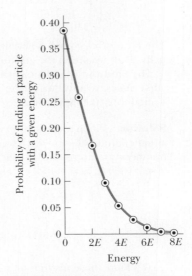

Figure 10.2 The distribution function for an assembly of six distinguishable particles with a total energy of 8E.

**Number of particles per unit
volume with energy between
E and $E + dE$**

**EXAMPLE 10.1   Emission Lines from
Stellar Hydrogen**

(a) Find the populations of the first and second excited states *relative* to the ground state for atomic hydrogen at room temperature, assuming that hydrogen obeys Maxwell–Boltzmann statistics.

Solution   For a gas at ordinary pressures, the atoms maintain the discrete quantum levels of isolated atoms. Recall that the discrete energy levels of atomic hydrogen are given by $E_n = (-13.6/n^2)$ eV and the degeneracy by $g_n = 2n^2$. Thus we have

$$\text{Ground state:} \quad E_1 = -13.6 \text{ eV} \quad g_1 = 2$$

$$\text{First excited state:} \quad E_2 = -3.40 \text{ eV} \quad g_2 = 8$$

$$\text{Second excited state:} \quad E_3 = -1.51 \text{ eV} \quad g_3 = 18$$

Using Equation 10.4 gives

$$\frac{n_2}{n_1} = \frac{g_2 A e^{-E_2/k_B T}}{g_1 A e^{-E_1/k_B T}} = \frac{g_2}{g_1} e^{(E_1 - E_2)/k_B T}$$

$$= \frac{8}{2} \exp\{(-10.2 \text{ eV})/(8.617 \times 10^{-5} \text{ eV/K})(300 \text{ K})\}$$

$$= 4e^{-395} \approx 0$$

The ratio of n_3/n_1 will be even smaller. Therefore, essentially all atoms are in the ground state at 300 K.

(b) Find the populations of the first and second excited states relative to the ground state for hydrogen heated to 20,000 K in a star.

Solution   When a gas is at very high temperatures (as in a flame, under electric discharge, or in a star), detectable numbers of atoms are in excited states. In this case, $T = 20,000$ K, $k_B T = 1.72$ eV, and we find

$$\frac{n_2}{n_1} = \frac{g_2}{g_1} e^{(E_1 - E_2)/k_B T} = 4e^{-10.2/1.72} = 0.0107$$

$$\frac{n_3}{n_1} = \frac{g_3}{g_1} e^{(E_1 - E_3)/k_B T} = 9e^{-12.1/1.72} = 0.0807$$

(c) Find the emission strengths of the spectral lines corresponding to the transitions $E_3 \to E_1$ and $E_3 \to E_2$ relative to $E_2 \to E_1$ at 20,000 K, assuming equal probability of transition for $E_3 \to E_1$, $E_3 \to E_2$, and $E_2 \to E_1$.

Solution   The strength of an emission or absorption line is proportional to the number of atomic transitions per unit time. For particles obeying Maxwell–Boltzmann statistics, the number of transitions per unit time from some initial state (i) to some final state (f) equals the product of the population of the initial state and the probability for the transition i → f. Note that *the transition rate for particles obeying MB statistics*

depends only on the initial population, since there are no restrictions on the number of particles in the final state. Returning to the calculation of the emission strength, S, we find the relative values:

$$\frac{S(3 \to 1)}{S(2 \to 1)} = \frac{n_3 P(3 \to 1)}{n_2 P(2 \to 1)}$$

$$\frac{S(3 \to 2)}{S(2 \to 1)} = \frac{n_3 P(3 \to 2)}{n_2 P(2 \to 1)}$$

In reality, the transition probabilities depend on the wavefunctions of the states involved, but to simplify matters we assume equal probabilities of transition; that is, $P(2 \to 1) = P(3 \to 1) = P(3 \to 2)$. This yields

$$\frac{S(3 \to 1)}{S(2 \to 1)} = \frac{n_3}{n_2} = \frac{g_3}{g_2} e^{(E_2 - E_3)/k_B T} = \frac{18}{8} e^{-1.89/1.73}$$

$$= 0.75$$

$$\frac{S(3 \to 2)}{S(2 \to 1)} = 0.75$$

If the emission lines are narrow, the measured heights of the $3 \to 1$ and $3 \to 2$ lines will be 75% of the height of the $2 \to 1$ line, as shown in Figure 10.3. For broader lines, the area under the peaks must be used as the experimental measure of emission strength.

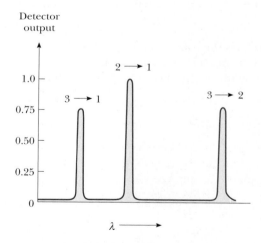

Figure 10.3   The predicted emission spectrum for the $2 \to 1$, $3 \to 1$, and $3 \to 2$ transitions for atomic hydrogen at 20,000 K.

The Maxwell Speed Distribution for Gas Molecules in Thermal Equilibrium at Temperature T

Maxwell's important formula for the equilibrium speed distribution, or the number of molecules with *speeds* between v and $v + dv$ in a gas at temperature T, may be found by using the Maxwell–Boltzmann distribution in its continuous form (Eqs. 10.6 and 10.7). In particular, we shall show that

$$n(v)\,dv = \frac{4\pi N}{V}\left(\frac{m}{2\pi k_\mathrm{B} T}\right)^{3/2} v^2 e^{-mv^2/2k_\mathrm{B}T}\,dv \qquad (10.8)$$

where $n(v)\,dv$ is the number of gas molecules per unit volume with speeds between v and $v + dv$, N/V is the total number of molecules per unit volume, m is the mass of a gas molecule, k_B is Boltzmann's constant, and T is the absolute temperature. This speed distribution function is sketched in Figure 10.4. The v^2 term determines the behavior of the distribution as $v \to 0$, and the exponential term determines what happens as $v \to \infty$.

For an ideal gas of point particles (no internal structure and no interactions between particles), the energy of each molecule consists only of translational kinetic energy and we have

$$E = \tfrac{1}{2}mv^2$$

for each molecule. Since the gas molecules have speeds that are continuously distributed from 0 to ∞, the energy distribution of molecules is also continuous and we may write the number of molecules per unit volume with energy between E and $E + dE$ as

$$n(E)\,dE = g(E)f_\mathrm{MB}(E)\,dE = g(E)Ae^{-mv^2/2k_\mathrm{B}T}\,dE$$

To find the density of states, $g(E)$, we introduce the concept of **velocity space.** According to this idea, the velocity of each molecule may be represented by a velocity vector with components v_x, v_y, and v_z or by a *point* in velocity space with coordinate axes v_x, v_y, and v_z (Fig. 10.5). From Figure 10.5 we note that the number of states $f(v)\,dv$ with speeds between v and $v + dv$ is proportional to the volume of the spherical shell between v and $v + dv$:

$$f(v)\,dv = C4\pi v^2\,dv \qquad (10.9)$$

where C is some constant. Because $E = \tfrac{1}{2}mv^2$, each speed v corresponds to a single energy E, and the number of energy states, $g(E)\,dE$, with energies between E and $E + dE$ is the same as the number of states with speeds between v and $v + dv$. Thus,

$$g(E)\,dE = f(v)\,dv = C4\pi v^2\,dv$$

Substituting this expression for $g(E)\,dE$ into our expression for $n(E)\,dE$, we obtain

$$n(E)\,dE = A4\pi v^2 e^{-mv^2/2k_\mathrm{B}T}\,dv$$

where the constant C has been absorbed into the normalization coefficient A. Since the number of molecules with energy between E and $E + dE$ equals the number of molecules with speed between v and $v + dv$, we may write

$$n(E)\,dE = n(v)\,dv = A4\pi v^2 e^{-mv^2/2k_\mathrm{B}T}\,dv \qquad (10.10)$$

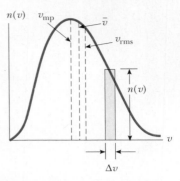

Figure 10.4 The speed distribution of gas molecules at some temperature. The number of molecules in the range Δv is equal to the area of the shaded rectangle, $n(v)\Delta v$. The most probable speed, v_mp, the average speed, $\bar{v}$, and the root mean square speed, v_rms, are indicated.

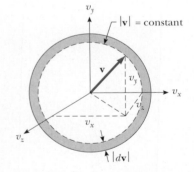

Figure 10.5 Velocity space. The number of states with speeds between v and $v + dv$ is proportional to the volume of a spherical shell with radius v and thickness dv.

To find A we use the fact that the total number of particles per unit volume is N/V:

$$\frac{N}{V} = \int_0^\infty n(v)\,dv = \int_0^\infty 4\pi A v^2 e^{-mv^2/2k_B T}\,dv \qquad (10.11)$$

Because

$$\int_0^\infty z^{2j} e^{-az^2}\,dz = \frac{1\cdot 3\cdot 5\,\cdots\,(2j-1)}{2^{j+1}a^j}\sqrt{\frac{\pi}{a}} \qquad j = 1, 2, 3, \ldots$$

we find with $j = 1$ and $a = m/2k_B T$

$$\frac{N}{V} = \frac{(4\pi A)}{2^2(m/2k_B T)}\sqrt{\frac{\pi 2k_B T}{m}} = A\left(\frac{2\pi k_B T}{m}\right)^{3/2}$$

or

$$A = \frac{N}{V}\left(\frac{m}{2\pi k_B T}\right)^{3/2}$$

Therefore, the normalization coefficient A depends on the number of particles per unit volume, the particle mass, and the temperature. Substituting this value for A into Equation 10.10, we finally obtain Maxwell's famous 1859 result:

$$n(v)\,dv = \frac{4\pi N}{V}\left(\frac{m}{2\pi k_B T}\right)^{3/2} v^2 e^{-mv^2/2k_B T}\,dv \qquad (10.8)$$

To find the average speed, $\overline{v}$, indicated in Figure 10.4, we multiply $n(v)\,dv$ by v, integrate over all speeds from 0 to ∞, and divide the result by the total number of molecules per unit volume:

$$\overline{v} = \frac{\int_0^\infty v n(v)\,dv}{N/V} = \frac{4\pi(N/V)(m/2\pi k_B T)^{3/2}\int_0^\infty v^3 e^{-mv^2/2k_B T}\,dv}{N/V}$$

Using the definite integral formula

$$\int_0^\infty z^3 e^{-az^2}\,dz = \frac{1}{2a^2}$$

gives

$$\overline{v} = 4\pi\left(\frac{m}{2\pi k_B T}\right)^{3/2}\left(\frac{1}{2}\right)\left(\frac{2k_B T}{m}\right)^2 = \sqrt{\frac{8k_B T}{\pi m}} \qquad (10.12)$$

This important result, first proved by Maxwell, shows that the average speed of the molecules in a gas is proportional to the square root of the temperature and inversely proportional to the square root of the molecular mass. The root mean square speed may be found by finding the average of v^2, denoted $\overline{v^2}$, and then taking its square root. Consequently, we have

$$\overline{v^2} = \frac{\int_0^\infty v^2 n(v)\,dv}{N/V} = 4\pi\left(\frac{m}{2\pi k_B T}\right)^{3/2}\int_0^\infty v^4 e^{-mv^2/2k_B T}\,dv$$

Using the definite integral formula

$$\int_0^\infty z^4 e^{-az^2}\, dz = \frac{3}{8a^2}\sqrt{\frac{\pi}{a}}$$

gives

$$\overline{v^2} = 4\pi\left(\frac{m}{2\pi k_B T}\right)^{3/2}\left[\frac{3}{8(m/2k_B T)^2}\right]\left(\frac{2\pi k_B T}{m}\right)^{1/2} = \frac{3k_B T}{m} \tag{10.13}$$

Since the root mean square speed, v_{rms}, is defined as $v_{rms} = \sqrt{\overline{v^2}}$, we have

$$v_{rms} = \sqrt{\frac{3k_B T}{m}} \tag{10.14}$$

Note that v_{rms} is *not* the same as the average speed, $\overline{v}$, but is about 10% greater, as indicated in Figure 10.4. The derivation of the most probable speed, v_{mp}, is left to Problem 2.

The Equipartition of Energy

As a final remark to this section, we observe that Equation 10.13 may be rewritten as

$$\tfrac{1}{2}m\overline{v^2} = \overline{K} = \tfrac{3}{2}k_B T$$

In this form, Equation 10.13 is consistent with the result known as the **equipartition of energy,** or the **equipartition theorem.** According to this theorem, **a classical molecule in thermal equilibrium at temperature T has an average energy of $k_B T/2$ for each independent mode of motion or so-called degree of freedom.** In this case there are 3 degrees of freedom corresponding to translational motion of the molecule along the independent x, y, and z directions in space; hence the average kinetic energy in each independent direction is $k_B T/2$:

$$\tfrac{1}{2}m\overline{v_x^2} = \tfrac{1}{2}m\overline{v_y^2} = \tfrac{1}{2}m\overline{v_z^2} = \tfrac{1}{2}k_B T$$

Equipartition of energy

The total average kinetic energy consequently equals 3 times $k_B T/2$, in agreement with Equation 10.13:

$$\tfrac{1}{2}m\overline{v^2} = \tfrac{1}{2}m\overline{v_x^2} + \tfrac{1}{2}m\overline{v_y^2} + \tfrac{1}{2}m\overline{v_z^2} = \tfrac{3}{2}k_B T$$

Note that degrees of freedom are not only associated with translational velocities. A degree of freedom is also associated with each rotational velocity as well so that $\tfrac{1}{2}I_1\overline{\omega_1^2} = \tfrac{1}{2}k_B T$ for a molecule with moment of inertia I_1 rotating about an axis with angular velocity ω_1. In fact, **each variable that occurs squared in the formula for the energy of a particular system represents a degree of freedom subject to the equipartition of energy.** For example, a one-dimensional harmonic oscillator with $E = \tfrac{1}{2}mv_x^2 + \tfrac{1}{2}kx^2$ has 2 degrees of freedom, one associated with its kinetic energy and the variable v_x^2 and the other with its potential energy and the variable x^2. Thus, each oscillator in a group in thermal equilibrium at T has $\overline{K} = \tfrac{1}{2}m\overline{v_x^2} = \tfrac{1}{2}k_B T$ and $\overline{U} = \tfrac{1}{2}k\overline{x^2} = \tfrac{1}{2}k_B T$. The average total energy of each one-dimensional harmonic oscillator is then $\overline{E}_{total} = \overline{K} + \overline{U} = k_B T/2 + k_B T/2 = k_B T$. This result will be of use to us shortly when we model the atoms of a solid as a system of vibrating harmonic oscillators.

Exercise 1   (a) Show that the formula for the number of molecules with energies between E and $E + dE$ in an ideal gas at temperature T is given explicitly in terms of E by

$$n(E)\,dE = \frac{2\pi(N/V)}{(\pi k_{\mathrm{B}}T)^{3/2}}\,E^{1/2}e^{-E/k_{\mathrm{B}}T}\,dE$$

(b) Use this result to show that the total energy per unit volume of the gas is given by

$$E_{\mathrm{total}} = \frac{3}{2}\frac{NkT}{V}$$

in agreement with the equipartition theorem.

Exercise 2   Confirm Maxwell's 1859 result that the "spread-outness" of the speed distribution increases as $\sqrt{T}$. Do this by showing that the standard deviation of the molecular speeds is given by

$$\sigma_v = \sqrt{3 - \frac{8}{\pi}} \cdot \sqrt{\frac{k_{\mathrm{B}}T}{m}}$$

10.2 UNDER WHAT PHYSICAL CONDITIONS ARE MAXWELL–BOLTZMANN STATISTICS APPLICABLE?

If we reexamine the assumptions that led to the Maxwell–Boltzmann distribution for classical particles, keeping the quantum mechanical wave nature of particles in mind, we immediately find a problem with the assumption of distinguishability. Since particles exhibit wave-like behavior, they are necessarily fuzzy and are not distinguishable when they are close together because their wavefunctions overlap. (See Section 9.4, "Exchange Symmetry and the Exclusion Principle," for a review of this issue.) If trading molecule A for molecule B no longer counts as a different configuration, then the number of ways a given energy distribution can be realized changes, as does the equilibrium or most probable distribution. Thus the classical Maxwell–Boltzmann distribution must be replaced by a quantum distribution when there is wavefunction overlap or when the particle concentration is high. The MB distribution is a valid approximation to the correct quantum distribution, however, in the common case of gases at ordinary conditions. Quantum statistics are required for cases involving high particle concentrations, such as electrons in a metal[1] or photons in a blackbody cavity.

It is useful to develop a criterion to determine when the classical distribution is valid. We may say that the **Maxwell–Boltzmann distribution is valid when the average distance between particles, *d*, is large compared with the quantum uncertainty in particle position, Δx, or**

$$\Delta x \ll d \tag{10.15}$$

To find Δx we use the uncertainty principle and evaluate Δp_x for a particle of mass *m*. For such a particle that is part of a system of particles in thermal

[1] The density of conduction electrons in a metal is several thousand times the density of molecules in a gas at standard temperature and pressure.

equilibrium at temperature T, $\overline{p_x} = 0$ and $\overline{p_x^2}/2m = k_B T/2$ from the equipartition theorem. Thus

$$\Delta p_x = \sqrt{\overline{p_x^2} - (\overline{p_x})^2} = \sqrt{mk_B T} \qquad (10.16)$$

Substituting this expression for Δp_x into $\Delta p_x \, \Delta x \geq \hbar/2$, we find

$$\Delta x \geq \frac{\hbar}{2\sqrt{mk_B T}} \qquad (10.17)$$

As mentioned before, the uncertainty in particle position, Δx, must be much less than the average distance, d, between particles if the particles are to be distinguishable and the Maxwell–Boltzmann distribution is to hold. Substituting $d = (V/N)^{1/3}$ and $\Delta x = \hbar/2\sqrt{mk_B T}$ into the relation $\Delta x \ll d$ gives

$$\frac{\hbar}{2\sqrt{mk_B T}} \ll \left(\frac{V}{N}\right)^{1/3}$$

or cubing both sides,

$$\left(\frac{N}{V}\right) \frac{\hbar^3}{8(mk_B T)^{3/2}} \ll 1 \qquad (10.18)$$

Criterion for the validity of Maxwell–Boltzmann statistics

Equation 10.18 shows that the Maxwell–Boltzmann distribution holds for low particle concentration and for high particle mass and temperature.

EXAMPLE 10.2   When Can We Use Maxwell–Boltzmann Statistics?

(a) Are Maxwell–Boltzmann statistics valid for hydrogen gas at standard temperature and pressure (STP)?

Solution   Under standard conditions of 273 K and 1 atmosphere, 1 mol of H_2 gas (6.02×10^{23} molecules) occupies a volume of 22.4×10^{-3} m³. Using $m_{H_2} = 3.34 \times 10^{-27}$ kg, and $k_B T = 3.77 \times 10^{-21}$ J, we find

$$\left(\frac{N}{V}\right) \frac{\hbar^3}{8(mk_B T)^{3/2}} = \left(\frac{6.02 \times 10^{23}}{22.4 \times 10^{-3} \text{ m}^3}\right)$$

$$\times \frac{(1.055 \times 10^{-34})^3 \text{ (J·s)}^3}{8[(3.34 \times 10^{-27} \text{ kg})(3.77 \times 10^{-21} \text{ J})]^{3/2}}$$

$$= 8.83 \times 10^{-8}$$

This is much less than 1, and from the condition given by Equation 10.18, we conclude that even hydrogen, the lightest gas, is described by Maxwell–Boltzmann statistics.

(b) Are Maxwell–Boltzmann statistics valid for conduction electrons in silver at 300 K?

Solution   Silver has a density of 10.5 g/cm³ and a molar weight of 107.9 g. Assuming one free electron per silver atom, the density of free electrons in silver is found to be

$$\frac{10.5 \text{ g/cm}^3}{107.9 \text{ g/mol}} (6.02 \times 10^{23} \text{ electrons/mol})$$

$$= 5.86 \times 10^{22} \text{ electrons/cm}^3$$

$$= 5.86 \times 10^{28} \text{ electrons/m}^3$$

Note that the density of free electrons in silver is about 2000 times greater than the density of hydrogen gas molecules at STP; that is,

$$\frac{(N/V)\text{electrons in Ag}}{(N/V)_{H_2 \text{ at STP}}} = \frac{5.86 \times 10^{28} \text{ m}^{-3}}{2.69 \times 10^{25} \text{ m}^{-3}} = 2180$$

Using $\hbar^3 = 1.174 \times 10^{-102}$ (J·s)³, $m_e = 9.109 \times 10^{-31}$ kg, and $k_B T = 4.14 \times 10^{-21}$ J (at $T = 300$ K), we find

$$\left(\frac{N}{V}\right) \frac{\hbar^3}{8(m_e k_B T)^{3/2}} = 4.64$$

Comparing this result to the condition given by Equation 10.18, we conclude that the Maxwell–Boltzmann distribution *does not hold* for electrons in silver because of the small mass of the electron and the high free electron density. We shall see that the correct quantum distribution for electrons is the Fermi–Dirac distribution.

10.3   QUANTUM STATISTICS

Wavefunctions and the Bose–Einstein Condensation and Pauli Exclusion Principle

Maxwell–Boltzmann statistics apply to systems of *identical, distinguishable* particles. As mentioned in the previous section, in quantum terms this means the wavefunctions of the particles do not overlap. If the individual particle wavefunctions do overlap, then the particles become indistinguishable or interchangeable, and this forces the system wavefunction to be either even or odd under particle exchange (see Section 9.4). In order to understand the important connection between wavefunctions and distribution functions, as well as the origin of the Bose–Einstein condensation for a system of particles with no actual attractive physical forces between particles, we look at a simple system of two particles with two possible quantum states to expose the essential features.

Consider two independent particles—particle 1 located at the position $\mathbf{r}_1$ and particle 2 located at $\mathbf{r}_2$—and two quantum states—state a and state b. For distinguishable particles there exist two possible system wavefunctions, which are simple products of normalized single particle wavefunctions:

$$\psi_A = \psi_a(\mathbf{r}_1)\psi_b(\mathbf{r}_2)$$

$$\psi_B = \psi_a(\mathbf{r}_2)\psi_b(\mathbf{r}_1)$$

Now we ask for the probability that both particles are in the same state, say a. In this case, both ψ_A and ψ_B are the same, and we find the probability of two distinguishable particles described by the Maxwell–Boltzmann distribution to be in the same state to be given by

$$\psi_{MB}^* \, \psi_{MB} = \psi_a^*(\mathbf{r}_1)\psi_a^*(\mathbf{r}_2)\psi_a(\mathbf{r}_1)\psi_a(\mathbf{r}_2) = |\psi_a(\mathbf{r}_1)|^2 \, |\psi_a(\mathbf{r}_2)|^2$$

If the particles are *indistinguishable*, we can't tell if a given particle is in state a or b, and to reflect this fact, the system wavefunction must be a *combination* of the distinguishable wavefunctions ψ_A and ψ_B. As mentioned in Section 9.4 and Problem 9.16, bosons have a symmetric wavefunction, ψ_B, given by

$$\psi_B = \frac{1}{\sqrt{2}} \left[\psi_a(\mathbf{r}_1)\psi_b(\mathbf{r}_2) + \psi_a(\mathbf{r}_2)\psi_b(\mathbf{r}_1) \right]$$

where we have added $\dfrac{1}{\sqrt{2}}$ as the normalization constant. Fermions have an antisymmetric wavefunction ψ_F, where

$$\psi_F = \frac{1}{\sqrt{2}} \left[\psi_a(\mathbf{r}_1)\psi_b(\mathbf{r}_2) - \psi_a(\mathbf{r}_2)\psi_b(\mathbf{r}_1) \right]$$

For comparison to the case of distinguishable particles, we now recalculate the probability that two bosons or fermions occupy the *same* state. For bosons the wavefunction becomes

$$\psi_B = \frac{1}{\sqrt{2}} \left[\psi_a(\mathbf{r}_1)\psi_a(\mathbf{r}_2) + \psi_a(\mathbf{r}_2)\psi_a(\mathbf{r}_1) \right] = \sqrt{2}\psi_a(\mathbf{r}_1)\psi_a(\mathbf{r}_2)$$

and the probability for two bosons to be in the same state is

$$\psi_B^* \, \psi_B = 2|\psi_a(\mathbf{r}_1)|^2 \, |\psi_a(\mathbf{r}_2)|^2 = 2\psi_{MB}^* \, \psi_{MB}$$

Thus we have the amazing result that bosons are twice as probable to occupy the same state as distinguishable particles! This is an entirely quantum mechanical effect and it is *as if* there were a force attracting additional bosons once a boson occupies a state, even though there are no actual attractive physical forces, such as electromagnetic intermolecular forces, present. Einstein was the first to point out that an *ideal* gas of bosons (with no attractive intermolecular forces!) could still undergo a strange kind of condensation at low-enough temperatures called a **Bose–Einstein condensation (BEC).** A Bose–Einstein condensation is a single cooperative state with all individual particle wavefunctions in phase in the ground state. In 1995 Einstein's prediction was directly confirmed by a group at the University of Colorado led by Eric Cornell and Carl Wieman who observed a BEC in a cloud of rubidium atoms cooled to less than 100 nK (see the guest essay by Steven Chu at the end of this chapter for details).

For fermions, as in Chapter 9, we find a probability of zero for two fermions to be in the same state since the wavefunction is zero:

$$\psi_F = \frac{1}{\sqrt{2}} \left[\psi_a(\mathbf{r}_1)\psi_a(\mathbf{r}_2) - \psi_a(\mathbf{r}_2)\psi_a(\mathbf{r}_1) \right] = 0$$

This is just the Pauli exclusion principle again.

Bose–Einstein and Fermi–Dirac Distributions

As we have seen there are two distributions for indistinguishable particles that flow from parity requirements on the system wavefunctions, the **Bose–Einstein distribution** and the **Fermi–Dirac distribution.** To obtain the Bose–Einstein distribution, we retain the MB assumption of no theoretical limit on the number of particles per state. Particles that obey the Bose–Einstein distribution are called **bosons** and are observed to have **integral spin.** Some examples of bosons are the alpha particle ($S = 0$), the photon ($S = 1$), and the deuteron ($S = 1$). To obtain the Fermi–Dirac distribution we stipulate that only one particle can occupy a given quantum state. Particles that obey the Fermi–Dirac distribution are called **fermions** and are observed to have **half integral spin.** Some important examples of fermions are the electron, the proton, and the neutron, all with spin $\frac{1}{2}$.

To see the essential changes in the distribution function introduced by quantum statistics, let us return to our simple system of six particles with a total energy of $8E$. First we consider the Bose–Einstein case; the particles are indistinguishable and there is no limit on the number of particles in a particular energy state. Figure 10.1a was drawn to represent this situation. Since the particles are indistinguishable, each of the 20 arrangements shown in Figure 10.1a is equally likely, so the probability of each arrangement is $1/20$. The average number of particles in a particular energy level may be calculated by again using Equation 10.2. The average number of particles in the zero energy level is found to be

$$\begin{aligned}
\overline{n_0} = {} & (5)(1/20) + (4)(1/20) + (4)(1/20) + (3)(1/20) + (4)(1/20) \\
& + (3)(1/20) + (2)(1/20) + (4)(1/20) + (3)(1/20) + (3)(1/20) \\
& + (2)(1/20) + (1)(1/20) + (3)(1/20) + (2)(1/20) + (2)(1/20) \\
& + (1)(1/20) + (0)(1/20) + (2)(1/20) + (1)(1/20) + (0)(1/20) \\
= {} & 49/20 = 2.45
\end{aligned}$$

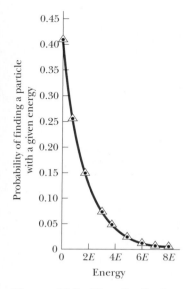

Figure 10.6 The distribution function for six indistinguishable particles with total energy 8E (bosons).

Similarly, we find $\overline{n_1} = 31/20 = 1.55$, $\overline{n_2} = 18/20 = 0.90$, $\overline{n_3} = 9/20 = 0.45$, $\overline{n_4} = 6/20 = 0.30$, $\overline{n_5} = 3/20 = 0.15$, $\overline{n_6} = 2/20 = 0.10$, $\overline{n_7} = 1/20 = 0.05$, and $\overline{n_8} = 0.05$. Once again using the idea that the probability of finding a particle with a given energy, $p(E)$, is simply the average number of particles with that energy divided by the total number of particles, we find

$$p(0) = \frac{\overline{n_0}}{6} = \frac{2.45}{6} = 0.408$$

In like manner we find: $p(1E) = 0.258$, $p(2E) = 0.150$, $p(3E) = 0.0750$, $p(4E) = 0.0500$, $p(5E) = 0.0250$, $p(6E) = 0.0167$, $p(7E) = 0.00833$, and $p(8E) = 0.00833$. A plot of these values in Figure 10.6 shows that the Bose–Einstein distribution gives results similar, but not identical, to the Maxwell–Boltzmann distribution. In general, the Bose–Einstein distribution tends to have more particles in the lowest energy levels. At higher energies, the curves come together and both exhibit a rapid decrease in probability with increasing energy.

To illustrate the distinctive shape of the Fermi–Dirac distribution, again consider our simple example of six indistinguishable particles with energy 8E. Since the particles are fermions, we impose the constraint that no more than two particles can be assigned to a given energy state (corresponding to electrons with spin up and down). There are only three arrangements (denoted by FD) out of the 20 shown in Figure 10.1a that meet this additional constraint imposed by the Pauli exclusion principle. Since each of these arrangements is equally likely, each has a probability of occurrence of 1/3, and we again use Equation 10.2 to calculate the average number of fermions in the zero energy level, as follows:

$$\overline{n_0} = (2)(1/3) + (2)(1/3) + (2)(1/3) = 2.00$$

Similarly, we find for the average number of fermions with energies of 1E through 8E the following:

$$\overline{n_1} = 5/3 = 1.67, \overline{n_2} = 3/3 = 1, \overline{n_3} = 3/3 = 1, \overline{n_4} = 1/3 = 0.33,$$
$$\overline{n_5} = \overline{n_6} = \overline{n_7} = \overline{n_8} = 0$$

Finally, we obtain the probabilities of finding a fermion with energies 0 through 8E:

$$p(0) = 2.00/6 = 0.333, p(1E) = 0.278, p(2E) = 0.167, p(3E) = 0.167,$$
$$p(4E) = 0.0550, \text{ and } p(5E) = p(6E) = p(7E) = p(8E) = 0$$

When this distribution is plotted, we discover a distinctly different shape from the Maxwell–Boltzmann or Bose–Einstein curves (Fig. 10.7). The results show a leveling off of probability at both low and high energies. Although it is not entirely clear that the points plotted in Figure 10.7 conform to the smooth curve drawn, consideration of systems with more than six particles proves this to be the case (see Problem 10.11).

When large numbers of quantum particles are considered, *continuous* distribution functions may be rigorously derived for both the Bose–Einstein (BE) and Fermi–Dirac (FD) cases. By maximizing the number of ways of distributing the indistinguishable quantum particles among the allowed energy states, again subject to the two constraints of a fixed number of particles and a fixed

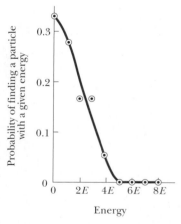

Figure 10.7 The distribution function for six indistinguishable particles with total energy 8E constrained so that no more than two particles occupy the same energy state (fermions).

total energy, we find the distribution functions to have the explicit forms:

$$f_{BE}(E) = \frac{1}{Be^{E/k_B T} - 1} \tag{10.19}$$

$$f_{FD}(E) = \frac{1}{He^{E/k_B T} + 1} \tag{10.20}$$

where $f(E)$ is the probability of finding a particle in a particular state of energy E at a given absolute temperature T. As noted earlier, the number of particles per unit volume with energy between E and $E + dE$ is given by

$$n(E)\ dE = g(E)f_{BE}(E)\ dE \tag{10.21}$$

or

$$n(E)\ dE = g(E)f_{FD}(E)\ dE \tag{10.22}$$

Thus the parameters B and H in Equations 10.19 and 10.20 may be determined from the total number of particles, N, since integrating Equations 10.21 and 10.22 yields

$$\left(\frac{N}{V}\right)_{bosons} = \int_0^\infty \frac{g(E)}{Be^{E/k_B T} - 1}\ dE \tag{10.23}$$

and

$$\left(\frac{N}{V}\right)_{fermions} = \int_0^\infty \frac{g(E)}{He^{E/k_B T} + 1}\ dE \tag{10.24}$$

In general we find that B and H depend on the system temperature and particle density as shown by Equations 10.23 and 10.24. For a system of bosons that are not fixed in number with temperature, Equation 10.23 no longer serves to determine B. By maximizing the ways of distributing the bosons among allowed states subject to the single constraint of fixed energy, it can be shown that the coefficient B in Equation 10.19 is equal to 1. This is a particularly important case since both photons in a blackbody cavity and phonons in a solid are bosons whose numbers per unit volume increase with increasing temperature. (We define phonons shortly.) Thus,

$$f(E) = \frac{1}{e^{E/k_B T} - 1} \qquad \text{(for photons or phonons)}$$

For the case of the Fermi–Dirac distribution, H depends strongly on temperature and is often written in an explicitly temperature-dependent form as $H = e^{-E_F/k_B T}$, where E_F is called the **Fermi energy**.[2] With this substitution, Equation 10.20 changes to the more common form

$$f_{FD}(E) = \frac{1}{e^{(E - E_F)/k_B T} + 1} \tag{10.25}$$

Fermi–Dirac distribution

---

[2]If we force the functional form of H to be $H = e^{-E_F/k_B T}$, E_F will itself have a weak dependence on T. Fortunately, this dependence of E_F on T is so weak that we can ignore it here.

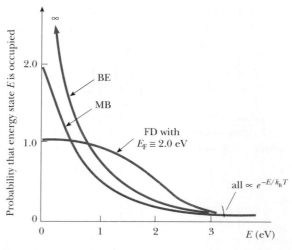

Figure 10.8   A comparison of Maxwell–Boltzmann, Bose–Einstein, and Fermi–Dirac distribution functions at 5000 K.

Satyendranath Bose (1894–1974). (*Indian National Council of Science Museums, Courtesy of AIP Emilio Segré Visual Archives*)

Enrico Fermi (1901–1954). (*Fermi National Accelerator Laboratory*)

This expression shows the meaning of the Fermi energy: The probability of finding an electron with an energy equal to the Fermi energy is exactly 1/2 at any temperature.

A plot comparing the Maxwell–Boltzmann, Bose–Einstein, and Fermi–Dirac distributions as functions of energy at a common temperature of 5000 K is shown in Figure 10.8. Note that for large E, all occupation probabilities decrease to zero as $e^{-E/k_B T}$. For small values of E, the FD probability saturates at 1 as required by the exclusion principle, the MB probability constantly increases but remains finite, and the BE probability tends to infinity. This very high probability for bosons to have low energies means that at low temperatures most of the particles drop into the ground state. When this happens, a new phase of matter with different physical properties can occur. This change in phase for a system of bosons is called a **Bose–Einstein condensation (BEC),** and it occurs in liquid helium at a temperature of 2.18 K. Below 2.18 K liquid helium becomes a mixture of the normal liquid and a phase with all molecules in the ground state. The ground-state phase, called liquid helium II, exhibits many interesting properties, one being zero viscosity. For more on Bose condensation and applications of this remarkable state of matter, see the essay by Steven Chu at the end of this chapter.

The history of the discovery of the quantum distributions is interesting. The first quantum distribution to be discovered was the Bose–Einstein function introduced in 1924 by Satyendranath Bose (Indian physicist, 1894–1974), working in isolation. He sent his paper, which contained a new proof of the Planck formula for blackbody radiation, to Einstein. In this paper, Bose applied the normal methods of statistical mechanics to light quanta but treated the quanta as absolutely indistinguishable. Einstein was impressed by Bose's work and proceeded to translate the paper into German for publication in the *Zeitschrift für Physik*.[3] To obtain the quantum theory of the ideal gas,

---

[3]S. N. Bose, *Z. Phys.*, 26:178, 1924.

Einstein extended the method to molecules in several papers published in 1924 and 1925.

In 1925, Wolfgang Pauli, after an exhaustive study of the quantum numbers assigned to atomic levels split by the Zeeman effect, announced his new and fundamental principle of quantum theory, the exclusion principle: Two electrons in an atom cannot have the same set of quantum numbers. In 1926, Enrico Fermi obtained the second type of quantum statistics that occurs in nature by combining Pauli's exclusion principle with the requirement of indistinguishability. Paul Dirac is also credited for this work, since he performed a more rigorous quantum mechanical treatment of these statistics in 1926. The empirical observation that particles with integral spin obey BE statistics and particles with half-integral spin obey FD statistics was explained much later (in 1940) by Pauli, using relativity and causality arguments.

The exclusion principle

Paul Adrien Maurice Dirac (1902–1984). (*Courtesy AIP Emilio Segré Visual Archives*)

10.4 APPLICATIONS OF BOSE–EINSTEIN STATISTICS

Blackbody Radiation

In this section we apply BE statistics to the problem of determining the energy density (energy per unit volume) of electromagnetic radiation in an enclosure heated to temperature T, now treating the radiation as a gas of photons. (In Chapter 3 we discussed the importance of this blackbody problem for quantum physics.) Since photons have spin 1, they are bosons and follow Bose–Einstein statistics. The number of photons per unit volume with energy between E and $E + dE$ is given by $n(E)\ dE = g(E)f_{BE}(E)\ dE$. The *energy density* of photons in the range from E to $E + dE$ is

$$u(E)\ dE = En(E)\ dE = \frac{g(E)E\ dE}{e^{E/k_B T} - 1} \tag{10.26}$$

To complete our calculation, we need the factor $g(E)$, the density of states for photons in an enclosure. This important calculation, given in Web Appendix 1 on our Web site, shows that the number of photon states per unit volume with frequencies between f and $f + df$ is

$$N(f)\ df = \frac{8\pi f^2\ df}{c^3} = \frac{8\pi(hf)^2\ d(hf)}{(hc)^3} = \frac{8\pi E^2\ dE}{(hc)^3}$$

using $E = hf$ for photons. Since the number of photon states per unit volume with frequencies between f and $f + df$ is equal to the number of photon states with energies between E and $E + dE$, we have

$$N(f)\ df = \frac{8\pi E^2\ dE}{(hc)^3} = g(E)\ dE$$

Thus we find that the density of states for photons is

$$g(E) = \frac{8\pi E^2}{(hc)^3} \tag{10.27}$$

Substituting Equation 10.27 into Equation 10.26 gives the expression for the energy density:

$$u(E) \, dE = \frac{8\pi}{(hc)^3} \frac{E^3 \, dE}{e^{E/k_{\mathrm{B}}T} - 1} \tag{10.28}$$

Converting from photon energy to frequency using $E = hf$ in Equation 10.28, we immediately retrieve the Planck blackbody formula:

$$u(f, T) = \frac{8\pi h}{c^3} \frac{f^3}{e^{hf/k_{\mathrm{B}}T} - 1} \tag{3.9}$$

Thus the Planck formula for a blackbody follows directly and simply from Bose–Einstein statistics.

EXAMPLE 10.3   Photons in a Box

(a) Find an expression for the number of photons per unit volume with energies between E and $E + dE$ in a cavity at temperature T.

Solution

$$n(E) \, dE = g(E) f(E) \, dE = \frac{8\pi E^2 \, dE}{(hc)^3 (e^{E/k_{\mathrm{B}}T} - 1)}$$

(b) Find an expression for the total number of photons per unit volume (all energies).

Solution

$$\frac{N}{V} = \int_0^\infty n(E) \, dE = \frac{8\pi (k_{\mathrm{B}}T)^3}{(hc)^3} \int_0^\infty \frac{(E/k_{\mathrm{B}}T)^2 (dE/k_{\mathrm{B}}T)}{e^{E/k_{\mathrm{B}}T} - 1}$$

or

$$\frac{N}{V} = 8\pi \left(\frac{k_{\mathrm{B}}T}{hc} \right)^3 \int_0^\infty \frac{z^2 \, dz}{e^z - 1}$$

(c) Calculate the number of photons/cm^3 inside a cavity whose walls are heated to 3000 K. Compare this with a cavity whose walls are at 3.00 K.

Solution   From standard tables,

$$\int_0^\infty \frac{z^2 \, dz}{e^z - 1} \approx 2.40$$

Therefore,

$$\frac{N}{V} \text{ (at 3000 K)} = (8\pi) \left[\frac{(8.62 \times 10^{-5} \text{ eV/K})(3000 \text{ K})}{1.24 \times 10^{-4} \text{ eV} \cdot \text{cm}} \right]^3 \cdot (2.40)$$

$$= 5.47 \times 10^{11} \text{ photons/cm}^3$$

Likewise, N/V (at 3.00 K) = 5.47×10^2 photons/cm^3. Therefore, the photon density decreases by a factor of 10^9 when the temperature drops from 3000 K to 3.00 K.

Einstein's Theory of Specific Heat

Recall that the molar specific heat of a substance, C, is the ratio of the differential thermal energy, dU, added to a mole of substance divided by the resulting differential increase in temperature, dT, or

$$C = \frac{dU}{dT} \tag{10.29}$$

Thus C has units of calories per mole per kelvin (cal/mol·K). To develop a theoretical expression for comparison to the experimental curves of C versus T measured for different elemental solids, we need an expression for U, the internal thermal energy of the solid, as a function of the solid's temperature, T. Differentiation of this expression will then yield the specific heat as a function of temperature.

To find an expression for U, let us model the solid as a collection of atoms vibrating *independently* on springs with equal force constants in the x,

y, and z directions, each atom being represented by three identical one-dimensional harmonic oscillators. The internal energy of each atom may then be calculated from the classical equipartition theorem. A one-dimensional harmonic oscillator has 2 degrees of freedom: one for its kinetic energy and one for its potential energy. (Physically this means that thermal energy added to the atoms in a solid may go into atomic vibration or into work done to stretch the springs holding the atoms in place.) Because the equipartition theorem states that the average thermal energy per degree of freedom should be $k_B T/2$, the internal energy per atom of a solid should be ($k_B T/2$ per degree of freedom) $\times$ (2 degrees of freedom per one-dimensional oscillator) $\times$ (three oscillators per atom) $= 3k_B T$. As a mole contains Avogadro's number of atoms, N_A, the total internal energy per mole, U, is predicted to be

$$U = 3N_A k_B T = 3RT \qquad (10.30)$$

where R is the universal gas constant given by $R = N_A k_B = 8.31 \ \text{J/mol} \cdot \text{K} = 1.99 \ \text{cal/mol} \cdot \text{K}$. Using $C = dU/dT$, we immediately see that C should be constant with temperature:

$$C = \frac{d}{dT}(3RT) = 3R = 5.97 \ \text{cal/mol} \cdot \text{K} \qquad (10.31)$$

The specific heat of many solids is indeed constant with temperature, especially at higher temperatures, as can be seen in Figure 10.9, showing good agreement with the classical idea that the average thermal energy is $k_B T/2$ per degree of freedom. However, as can also be seen in Figure 10.9, the specific heat of all solids drops sharply at some temperature and approaches zero as the temperature approaches 0 K.

The explanation of why classical physics failed to give the correct value of specific heat at all temperatures was given by Einstein in 1907. He realized that the quantized energies of vibrating atoms in a solid must be explicitly considered at low temperatures to secure agreement with experimental measurements of specific heat. Einstein assumed that the atoms of the solid

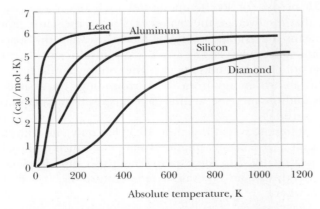

Figure 10.9   The dependence of specific heat on temperature for several solid elements.

could be modeled as a system of *independent* or *uncoupled* three-dimensional *quantum* harmonic oscillators with equal spring constants in the *x*, *y*, and *z* directions. He then showed that the average energy of a *one-dimensional* oscillator with frequency ω at temperature T was given by the Bose–Einstein distribution, or

$$\overline{E} = \frac{\hbar\omega}{e^{\hbar\omega/k_BT} - 1}$$

Because the atoms are considered to be independent, he gave the internal energy of a mole of atoms, or N_A atoms, as

$$U = 3N_A\overline{E} = 3N_A\,\frac{\hbar\omega}{e^{\hbar\omega/k_BT} - 1}$$

Finally, he obtained the molar specific heat:

$$C = \frac{dU}{dT} = 3R\left(\frac{\hbar\omega}{k_BT}\right)^2 \frac{e^{\hbar\omega/k_BT}}{(e^{\hbar\omega/k_BT} - 1)^2} \tag{10.32}$$

It is left as an exercise to show that Equation 10.32 predicts that C approaches zero for small T as $e^{-\hbar\omega/k_BT}$, and that C approaches $3R$ for large T.

To understand Equation 10.32 qualitatively, consider the quantity $\overline{E}$, the average one-dimensional quantum oscillator energy at temperature, T:

$$\overline{E} = \frac{\hbar\omega}{e^{\hbar\omega/k_BT} - 1} \tag{10.33}$$

Recall that the vibrating atoms of the solid have quantized energy levels spaced $\hbar\omega$ apart. For high temperatures such that $\hbar\omega \ll k_BT$, the energy level spacing $\hbar\omega$ is small relative to the average thermal energy per atom, and we can expect many atoms to be in excited energy levels. In fact, we can expand the exponential in the denominator of Equation 10.33 as $\exp(\hbar\omega/k_BT) = 1 + \hbar\omega/k_BT + \cdots$ to get

$$\overline{E} = \frac{\hbar\omega}{e^{\hbar\omega/k_BT} - 1} \approx k_BT$$

In this case the atomic energies appear to be continuous and the classical result $C = 3R$ holds. For low temperatures such that $\hbar\omega \gg k_BT$, Equation 10.33 shows that the average thermal energy of an oscillator rapidly tends to zero. This means the average energy is much less than the spacing between adjacent atomic energy levels, $\hbar\omega$, and there is insufficient thermal energy to raise an atom out of its ground state to higher energy levels. In this case atoms are unable to absorb energy from the surroundings for a small increase in temperature, and the increase in internal energy with temperature or specific heat tends to zero.

A final point to note is that Equation 10.32 has only one adjustable parameter, ω, the harmonic oscillator vibration frequency, which is chosen to give the best fit of Equation 10.32 to the experimental heat capacity data. Frequently, ω is given in terms of an equivalent temperature T_E, called the **Einstein temperature,** where

$$\hbar\omega = k_B T_E \tag{10.34}$$

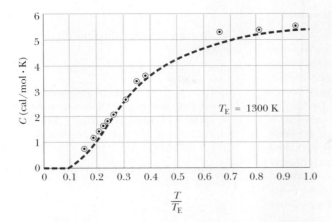

Figure 10.10   Einstein's specific heat formula fitted to Weber's experimental data for diamond. This figure is adapted from A. Einstein, *Ann. Physik.,* 4(22):180, 1907.

In his pioneering 1907 paper, Einstein found good agreement between his formula and Heinrich Weber's data on diamond, with $T_E = 1300$ K. This agreement is shown in Figure 10.10.

The too rapid falloff of the Einstein formula at low temperatures hinted at in Figure 10.10 was confirmed in 1911 by Hermann Nernst. Although it was generally felt that the problem with Einstein's result was the assumption that each atom vibrated independently of its neighbors at a single fixed frequency, no one really knew how to treat a band or spread of frequencies corresponding to groups of neighboring atoms interacting and moving together. In 1912, however, Peter Debye obtained the experimentally observed temperature dependence of $C \propto T^3$ for low temperatures by modeling a solid as a continuous elastic object whose internal energy was made up of the energy in standing sound (elastic) waves. These sound waves are both transverse and longitudinal in a solid and possess a range of frequencies from zero to some maximum value determined by the dependence of the minimum wavelength on the interatomic spacing. Furthermore, these elastic waves or lattice vibrations are *quantized*, like electromagnetic waves or photons. **A quantized elastic vibration of frequency ω, called a phonon, travels at the speed of sound in a solid, and carries a quantum of elastic energy $\hbar\omega$.** Debye was able to show that a "phonon gas" with a distribution of allowed frequencies was a better model of a solid at low temperatures than a system of independent harmonic oscillators all having the same frequency. Since the introduction of the idea of phonons by Debye, the concept has found many applications in condensed matter physics, including the electron–phonon interaction in superconductivity and the coupling of phonons to the motion of impurity atoms and molecules in a lattice.

Exercise 3   Show that Equation 10.32 predicts that C approaches zero for small T as $e^{-\hbar\omega/k_B T}$ and that C approaches $3R$ for large T.

EXAMPLE 10.4   The Specific Heat of Diamond

As we have seen, a solid at temperature T can be viewed as a system of quantized harmonic oscillators with discrete energy levels separated by $\hbar\omega$. The oscillators can only absorb thermal energy, however, if the temperature is high enough that the average thermal energy of the oscillator, $\overline{E}$, is approximately equal to the oscillator energy-level spacing, $\hbar\omega$. For low temperatures such that $\overline{E} \ll \hbar\omega$, there is so little thermal energy available that the atoms cannot even be raised to the first excited state and the specific heat tends to zero. In the following example we show that the carbon atoms in diamond are effectively decoupled from thermal energy at room temperature but can absorb energy at a temperature of 1500 K.

(a) Calculate the vibration frequency of the carbon atoms in diamond if the Einstein temperature is 1300 K. Also find the energy-level spacing for the carbon atoms.

Solution   Since $\hbar\omega = k_B T_E$, the frequency of vibration of carbon atoms in diamond is

$$\omega = \frac{k_B T_E}{\hbar} = \frac{(8.62 \times 10^{-5} \text{ eV/K})(1300 \text{ K})}{6.58 \times 10^{-16} \text{ eV·s}}$$

$$= 1.70 \times 10^{14} \text{ Hz}$$

The spacing between adjacent oscillator energy levels in carbon is

$$\hbar\omega = (6.58 \times 10^{-16} \text{ eV·s})(1.70 \times 10^{14} \text{ Hz}) = 0.112 \text{ eV}$$

(b) Calculate the average oscillator energy $\overline{E}$ at room temperature and at 1500 K and compare this energy with the carbon energy-level spacing $\hbar\omega$. Is there sufficient thermal energy on average to excite carbon atoms at 300 K? at 1500 K?

Solution   The average oscillator energy at room temperature (300 K) is

$$\overline{E} = \frac{\hbar\omega}{e^{\hbar\omega/k_B T} - 1}$$

$$= \frac{0.112 \text{ eV}}{e^{0.112 \text{ eV}/(8.62 \times 10^{-5} \text{ eV/K})(300 \text{ K})} - 1} = 0.00149 \text{ eV}$$

while the average oscillator energy at 1500 K is

$$\overline{E} = \frac{0.112 \text{ eV}}{e^{0.112 \text{ eV}/(8.62 \times 10^{-5} \text{ eV/K})(1500 \text{ K})} - 1} = 0.0813 \text{ eV}$$

Comparing, we see that $\overline{E}$ at 300 K is about $0.01\hbar\omega$, and $\overline{E}$ at 1500 K is approximately equal to $\hbar\omega$. This means that at 300 K most carbon atoms are frozen into the oscillator ground state and the specific heat tends to zero.

Exercise 4   (a) Calculate the vibration frequency of lead atoms and their energy-level spacing if the Einstein temperature of lead is 70 K. (b) Explain the low Einstein temperature of lead relative to that for diamond in terms of the physical properties of lead. (c) Calculate the average one-dimensional oscillator energy in lead at room temperature. Is there enough energy to raise lead atoms out of the ground state at 300 K?

10.5   AN APPLICATION OF FERMI–DIRAC STATISTICS: THE FREE ELECTRON GAS THEORY OF METALS

Because the outer electrons are weakly bound to individual atoms in a metal, we can treat these outer conduction electrons as a gas of fermions trapped within a cavity formed by the metallic surface. Many interesting physical quantities, such as the average energy, Fermi energy, specific heat, and thermionic emission rate, may be derived from the expression for the concentration of electrons with energies between E and $E + dE$:

$$n(E) \, dE = g(E) f_{FD}(E) \, dE \tag{10.22}$$

Recall that the probability of finding an electron in a particular energy state E is given by the Fermi–Dirac distribution function,

$$f_{FD}(E) = \frac{1}{e^{(E - E_F)/k_B T} + 1}$$

Plots of this function versus energy are shown in Figure 10.11 for the cases $T = 0$ K and $T > 0$. Note that at $T = 0$ K, $f_{FD} = 1$ for $E < E_F$, and $f_{FD} = 0$ for $E > E_F$. Thus all states with energies less than E_F are completely filled and all states with energies greater than E_F are empty. This is in sharp contrast to the predictions of MB and BE statistics, in which all particles condense to a state of zero energy at absolute zero. In fact, far from having zero speed, a conduction electron in a metal with the cutoff energy E_F has a speed v_F which satisfies the relation

$$\tfrac{1}{2} m_e v_F^2 = E_F \qquad (10.35)$$

where v_F is called the **Fermi speed.** Substituting a typical value of 5 eV for the Fermi energy yields the remarkable result that electrons at the Fermi level possess speeds of the order of 10^6 m/s at 0 K!

Figure 10.11b shows that as T increases, the distribution rounds off slightly, with states between E and $E - k_B T$ losing population and states between E and $E + k_B T$ gaining population. In general, E_F also depends on temperature, but the dependence is weak in metals, and we may say that $E_F(T) \approx E_F(0)$ up to several thousand kelvin.

Let us now turn to the calculation of the density of states, $g(E)$, for conduction electrons in a metal. Since the electrons may be viewed as a system of matter waves whose wavefunctions vanish at the boundaries of the metal, we obtain the same result for electrons as for electromagnetic waves confined to a cavity. In the latter case, we found (see our Web site at http://info. brookscole.com/mp3e) that the number of states per unit volume with wavenumber between k and $k + dk$ is

$$g(k)\ dk = \frac{k^2\ dk}{2\pi^2} \qquad (3.44)$$

To apply this expression to electrons in a metal, we must multiply it by a factor of 2 to account for the two allowed spin states of an electron with a given momentum or energy:

$$g(k)\ dk = \frac{k^2\ dk}{\pi^2} \qquad (10.36)$$

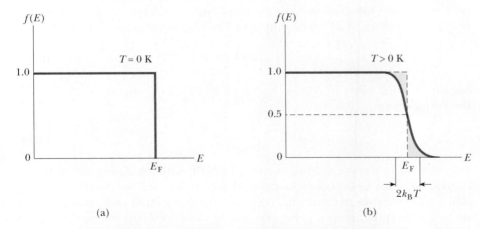

(a)   (b)

Figure 10.11   A comparison of the Fermi–Dirac distribution functions at (a) absolute zero and (b) finite temperature.

To obtain $g(E)$ from $g(k)$, we assume nonrelativistic free electrons. Thus

$$E = \frac{p^2}{2m_e} = \frac{\hbar^2 k^2}{2m_e}$$

or

$$k = \left(\frac{2m_e E}{\hbar^2}\right)^{1/2} \tag{10.37}$$

and

$$dk = \frac{1}{2}\left(\frac{2m_e}{\hbar^2}\right)^{1/2} E^{-1/2}\, dE \tag{10.38}$$

Substituting Equations 10.37 and 10.38 into Equation 10.36 yields

Density of states for conduction electrons

$$g(E)\, dE = DE^{1/2}\, dE \tag{10.39}$$

where

$$D = \frac{8\sqrt{2}\,\pi m_e^{3/2}}{h^3} \tag{10.40}$$

Thus the key expression for the number of electrons per unit volume with energy between E and $E + dE$ becomes

Number of electrons per unit volume with energy between E and $E + dE$

$$n(E)\, dE = \frac{DE^{1/2}\, dE}{e^{(E-E_F)/k_B T} + 1} \tag{10.41}$$

Figure 10.12 is a plot of $n(E)$ versus E, showing the product of an increasing density of states and the decreasing FD distribution. Because

$$\frac{N}{V} = \int_0^\infty n(E)\, dE = D\int_0^\infty \frac{E^{1/2}\, dE}{e^{(E-E_F)/k_B T} + 1} \tag{10.42}$$

we can determine the Fermi energy as a function of the electron concentration, N/V. For arbitrary T, Equation 10.42 must be integrated numerically. At $T = 0$ K, the integration is simple since $f_{FD}(E) = 1$ for $E < E_F$ and is 0 for $E > E_F$. Therefore, at $T = 0$ K, Equation 10.42 becomes

$$\frac{N}{V} = D\int_0^{E_F} E^{1/2}\, dE = \tfrac{2}{3}DE_F^{3/2} \tag{10.43}$$

Substituting the value of D from Equation 10.40 into Equation 10.43 gives for the Fermi energy at 0 K, $E_F(0)$,

$$E_F(0) = \frac{h^2}{2m_e}\left(\frac{3N}{8\pi V}\right)^{2/3} \tag{10.44}$$

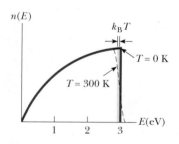

Figure 10.12 The number of electrons per unit volume with energy between E and $E + dE$. Note that $n(E) = g(E)f_{FD}(E)$.

Equation 10.44 shows a gradual increase in $E_F(0)$ with increasing electron concentration. This is expected, because the electrons fill the available energy states, two electrons per state, in accordance with the Pauli exclusion principle up to a maximum energy E_F. Representative values of $E_F(0)$ for various metals calculated from Equation 10.44 are given in Table 10.1. This table also lists values of the Fermi speed and the **Fermi temperature,** T_F, defined by

Table 10.1   Calculated Values of Various Parameters for Metals Based on the Free Electron Theory

Metal	Electron Concentration (m^{-3})	Fermi Energy (eV)	Fermi Speed (m/s)	Fermi Temperature (K)
Li	4.70×10^{28}	4.72	1.29×10^6	5.48×10^4
Na	2.65×10^{28}	3.23	1.07×10^6	3.75×10^4
K	1.40×10^{28}	2.12	0.86×10^6	2.46×10^4
Cu	8.49×10^{28}	7.05	1.57×10^6	8.12×10^4
Ag	5.85×10^{28}	5.48	1.39×10^6	6.36×10^4
Au	5.90×10^{28}	5.53	1.39×10^6	6.41×10^4

$$T_{\mathrm{F}} \equiv \frac{E_{\mathrm{F}}}{k_{\mathrm{B}}} \tag{10.45}$$

As a final note, it is interesting that a long-standing puzzle concerning the anomalously small contribution of the conduction electron "gas" to the heat capacity of a solid has a qualitative solution in terms of the Fermi–Dirac distribution. If conduction electrons behaved classically, warming a gas of N electrons from 0 to 300 K should result in an average energy increase of $3k_{\mathrm{B}}T/2$ for *each* particle, or a total thermal energy per mole, U, given by

$$U = N_{\mathrm{A}} \left(\tfrac{3}{2} k_{\mathrm{B}} T \right) = \tfrac{3}{2} RT$$

Thus the electronic heat capacity per mole should be given by

$$C_{\mathrm{el}} = \frac{dU}{dT} = \tfrac{3}{2} R$$

assuming one free electron per atom. An examination of Figure 10.12, however, shows that on heating from 0 K, very few electrons become excited and gain an energy $k_{\mathrm{B}}T$. Only a small fraction f within $k_{\mathrm{B}}T$ of E_{F} can be excited thermally. The fraction f may be approximated by the ratio of the area of a thin rectangle of width $k_{\mathrm{B}}T$ and height $n(E_{\mathrm{F}})$ to the total area under $n(E)$. Thus

$$f \approx \frac{\text{area of shaded rectangle in Figure 10.12}}{\text{total area under } n(E)}$$

$$= \frac{(k_{\mathrm{B}}T) g(E_{\mathrm{F}})}{D \displaystyle\int_0^{E_{\mathrm{F}}} E^{1/2}\, dE} = \frac{(k_{\mathrm{B}}T) D (E_{\mathrm{F}})^{1/2}}{\tfrac{2}{3} D E_{\mathrm{F}}^{3/2}} = \frac{3}{2} \frac{k_{\mathrm{B}}T}{E_{\mathrm{F}}} = \frac{3}{2} \frac{T}{T_{\mathrm{F}}}$$

Since only fN of the electrons gain an energy of the order of $k_{\mathrm{B}}T$, the actual total thermal energy gained per mole is

$$U = \left(\frac{3}{2} \frac{T}{T_{\mathrm{F}}} \right) (N_{\mathrm{A}} k_{\mathrm{B}} T) = \frac{3}{2} \frac{RT^2}{T_{\mathrm{F}}}$$

From this result, we find that the electronic heat capacity is

$$C_{\mathrm{el}} = \frac{dU}{dT} \approx 3R \frac{T}{T_{\mathrm{F}}}$$

Substituting $T = 300$ K and $T_F = 5 \times 10^4$ K, we find a very small value for the electronic heat capacity at ordinary temperatures:

$$C_{el} \approx 3R \left(\frac{300 \text{ K}}{50,000 \text{ K}} \right) = 0.018R$$

Thus, the electrons contribute only $0.018R/1.5R$, or about 1% of the classically expected amount, to the heat capacity.

EXAMPLE 10.5   The Fermi Energy of Gold

(a) Calculate the Fermi energy of gold at 0 K.

Solution   The density of gold is 19.32 g/cm³, and its molar weight is 197 g/mol. Assuming each gold atom contributes one free electron to the Fermi gas, we can calculate the electron concentration as follows:

$$\frac{N}{V} = (19.32 \text{ g/cm}^3)\left(\frac{1}{197 \text{ g/mol}} \right) \times (6.02 \times 10^{23} \text{ electrons/mol})$$

$$= 5.90 \times 10^{22} \text{ electrons/cm}^3$$

$$= 5.90 \times 10^{28} \text{ electrons/m}^3$$

Using Equation 10.44, we find

$$E_F(0) = \frac{h^2}{2m_e} \left(\frac{3N}{8\pi V} \right)^{2/3}$$

$$= \frac{(6.625 \times 10^{-34} \text{ J} \cdot \text{s})^2}{2(9.11 \times 10^{-31} \text{ kg})} \left(\frac{3 \times 5.90 \times 10^{28} \text{ m}^{-3}}{8\pi} \right)^{2/3}$$

$$= 8.85 \times 10^{-19} \text{ J} = 5.53 \text{ eV}$$

(b) Calculate the Fermi speed for gold at 0 K.

Solution   Since $\frac{1}{2}m_e v_F^2 = E_F$,

$$v_F = \left(\frac{2E_F}{m_e} \right)^{1/2} = \left(\frac{2 \times 5.85 \times 10^{-19} \text{ J}}{9.11 \times 10^{-31} \text{ kg}} \right)^{1/2}$$

$$= 1.39 \times 10^6 \text{ m/s}$$

(c) Calculate the Fermi temperature for gold at 0 K.

Solution   The Fermi temperature is given by

$$T_F = \frac{E_F}{k_B} = \frac{5.53 \text{ eV}}{8.62 \times 10^{-5} \text{ eV/K}}$$

$$= 64,000 \text{ K}$$

Thus a gas of classical particles would have to be heated to about 64,000 K to have an average energy per particle equal to the Fermi energy at 0 K!

SUMMARY

Statistical physics deals with the distribution of a fixed amount of energy among a number of particles that are identical and indistinguishable in any way (quantum particles) or identical particles that are distinguishable in the classical limit of narrow particle wave packets and low particle density. In most situations, one is not interested in the energies of all the particles at a given instant, but rather in the time average of the number of particles in a particular energy level. The average number of particles in a given energy level is of special interest in spectroscopy because the intensity of radiation emitted or absorbed is proportional to the number of particles in a particular energy state.

For a system described by a continuous distribution of energy levels, the number of particles per unit volume with energy between E and $E + dE$ is given by

$$n(E) \ dE = g(E)f(E) \ dE \tag{10.6}$$

where $g(E)$ is the density of states or the number of energy states per unit volume in the interval dE and $f(E)$ is the probability that a particle is in the energy state E. The function $f(E)$ is called the **distribution function.**

Three distinct distribution functions are used, depending on whether the particles are distinguishable and whether there is a restriction on the number of particles in a given energy state:

- **Maxwell–Boltzmann Distribution (Classical).** The particles are distinguishable, and there is no limit on the number of particles in a given energy state.

$$f_{MB}(E) = Ae^{-E/k_B T} \qquad (10.3)$$

- **Bose–Einstein Distribution (Quantum).** The particles are indistinguishable, and there is no limit on the number of particles in a given energy state.

$$f_{BE}(E) = \frac{1}{Be^{E/k_B T} - 1} \qquad (10.19)$$

- **Fermi–Dirac Distribution (Quantum).** The particles are indistinguishable, and there can be no more than one particle per quantum state.

$$f_{FD}(E) = \frac{1}{e^{(E - E_F)/k_B T} + 1} \qquad (10.25)$$

where E_F is the **Fermi energy.** At $T = 0$ K, all levels below E_F are filled and all levels above E_F are empty.

At low particle concentrations and high temperature, most systems are well described by Maxwell–Boltzmann statistics. The criterion that determines when the classical Maxwell–Boltzmann distribution is valid is

$$\left(\frac{N}{V}\right) \frac{h^3}{(8mk_B T)^{3/2}} \ll 1 \qquad (10.18)$$

where N/V is the particle concentration, m is the particle mass, and T is the absolute temperature. For high particle concentration, low particle mass, and modest temperature, there is considerable overlap between the particles' wavefunctions, and quantum distributions must be used to describe these systems of indistinguishable particles.

A system of photons in thermal equilibrium at temperature T is described by the Bose–Einstein distribution with $B = 1$ and a density of states given by

$$g(E) = \frac{8\pi E^2}{(hc)^3} \qquad (10.27)$$

Thus, the concentration of photons with energies between E and $E + dE$ is

$$n(E) \, dE = \frac{8\pi E^2}{(hc)^3} \left(\frac{1}{e^{E/k_B T} - 1}\right) dE$$

Phonons, which are quantized lattice vibrations of a solid, are also described by the Bose–Einstein distribution with $B = 1$.

Free (conduction) electrons in metals obey the Pauli exclusion principle, and we must use the Fermi–Dirac distribution to treat such a system. The density of states for electrons in a metal is

$$g(E) = \frac{8\sqrt{2}\,\pi m_e^{3/2}}{h^3} E^{1/2} \qquad (10.39)$$

hence the number of electrons per unit volume with energy between E and $E + dE$ is

$$n(E)\ dE = \frac{8\sqrt{2}\pi m_e^{3/2}\ E^{1/2}}{h^3(e^{(E-E_F)/k_B T} + 1)}\ dE \qquad (10.41)$$

An expression for the Fermi energy at 0 K as a function of electron concentration may be obtained by integrating Equation 10.41. One finds

$$E_F(0) = \frac{h^2}{2m_e}\left(\frac{3N}{8\pi V}\right)^{2/3} \qquad (10.44)$$

The small electronic contribution to the heat capacity of a metal can be explained by noting that only a small fraction of the electrons near E_F gain $k_B T$ in thermal energy when the metal is heated from 0 K to T K.

SUGGESTIONS FOR FURTHER READING

1. A. Beiser, *Concepts of Modern Physics*, 5th ed., New York, McGraw-Hill Book Co., 1995.

More advanced treatments of statistical physics may be found in the following books:

2. P. M. Morse, *Thermal Physics*, New York, Benjamin, 1965.
3. C. Kittel, *Thermal Physics*, New York, Wiley, 1969.
4. D. Griffiths, *Introduction to Quantum Mechanics*, Englewood Cliffs, NJ, Prentice-Hall, 1994.

QUESTIONS

1. Discuss the basic assumptions of Maxwell–Boltzmann, Fermi–Dirac, and Bose–Einstein statistics. How do they differ, and what are their similarities?

2. Explain the role of the Pauli exclusion principle in describing the electrical properties of metals.

PROBLEMS

10.1   The Maxwell–Boltzmann Distribution

1. Verify that for a system of six distinguishable particles with total energy $8E$, the probabilities of finding a particle with energies $1E$ through $8E$ are: 0.256, 0.167, 0.0978, 0.0543, 0.0272, 0.0117, 0.00388, 0.000777.

2. Show that the most probable speed of a gas molecule is

$$v_{mp} = \sqrt{\frac{2k_B T}{m}}$$

Note that the most probable speed corresponds to the point where the Maxwellian speed distribution curve, $n(v)$, has a maximum.

3. Figure P10.3 shows an apparatus similar to that used by Otto Stern in 1920 to verify the Maxwell speed distribution. A collimated beam of gas molecules from an oven, O, is allowed to enter a rapidly rotating cylinder when slit S is coincident with the beam. The pulse of molecules created by the rapid rotation of S then strikes and adheres to a glass plate detector, D. The velocity of a molecule may be determined from its position on the glass plate (fastest molecules to the right). The number of molecules arriving with a given velocity may be determined by measuring the density of molecules deposited on D at a given position. Suppose that

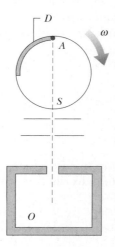

Figure P10.3   A schematic drawing of an apparatus used to verify the Maxwell speed distribution.

the oven contains a gas of bismuth molecules (Bi_2) at 850 K, and that the cylinder is 10 cm in diameter and rotates at 6250 rpm. (a) Find the distance from A of the impact points of molecules traveling at $\bar{v}$, v_{rms}, and v_{mp}. (b) Why do you suppose that measurements were originally made with Bi_2 instead of O_2 or N_2?

4. Energy levels known as tunneling levels have been observed from CN^- ions incorporated into KCl crystals. These levels arise from the rotational motion of CN^- ions as they tunnel, at low temperatures, through barriers separating crystalline potential minima. According to one model, the tunneling levels should consist of four equally spaced levels in the far IR with spacing of 12.41×10^{-5} eV (1 cm^{-1}). Assuming that the CN^- ions obey Maxwell–Boltzmann statistics and using Figure P10.4, calculate and sketch the expected appearance of the absorption spectra (strength of absorption vs. energy) at 4 K and 1 K. Assume equal transition probabilities for all allowed transitions, and make the strength of absorption proportional to the peak height. Use $k_B = 8.62 \times 10^{-5}$ eV/K.

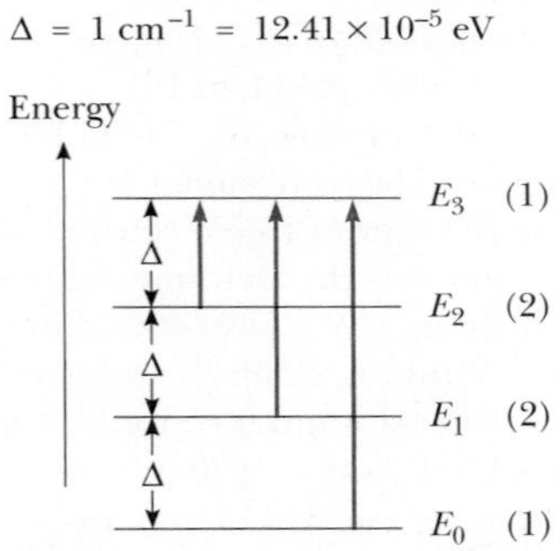

$$\Delta = 1 \text{ cm}^{-1} = 12.41 \times 10^{-5} \text{ eV}$$

Figure P10.4 Tunneling energy levels. Allowed transitions are indicated by vertical arrows. The degeneracy of each level is indicated in parentheses.

5. Fit an exponential curve $P(E) = Ae^{-BE}$ to Figure 10.2 to see how closely the system of six distinguishable particles comes to an exponential distribution. Use the values at energies of 0 and $1E$ to determine A and B.

6. The energy difference between the first excited state of mercury and the ground state is 4.86 eV. (a) If a sample of mercury vaporized in a flame contains 10^{20} atoms in thermal equilibrium at 1600 K, calculate the number of atoms in the $n = 1$ (ground) and $n = 2$ (first excited) states. (Assume the Maxwell–Boltzmann distribution applies and that the $n = 1$ and $n = 2$ states have equal statistical weights.) (b) If the mean lifetime of the $n = 2$ state is τ seconds, the transition probability is $1/\tau$ and the number of photons emitted per second by the $n = 2$ state is n_2/τ, where n_2 is the number of atoms in state 2.

If the lifetime of the $n = 2$ state is 100 ns, calculate the power (in watts) emitted by the hot atoms.

7. Consider a molecule with a permanent electric dipole moment **p** placed in an electric field $\mathcal{E}$, with **p** aligned either parallel or antiparallel to $\mathcal{E}$. (a) Recall that the energy of a dipole in an electric field is given by $E = -\mathbf{p} \cdot \mathcal{E}$, and show that this system has two allowed energy states separated by $2p\mathcal{E}$. (b) Assume a ground-state energy of 0 and an excited-state energy of $2p\mathcal{E}$ and degeneracies in the ratio $g(2p\mathcal{E})/g(0) = 2/1$. For a collection of N molecules obeying Maxwell–Boltzmann statistics, calculate the ratio of the number of molecules in the excited state to the number in the ground state at temperature T. (c) For high T such that $k_B T \gg 2p\mathcal{E}$, the ratio of the number of molecules in the upper state to the number in the lower is 2 to 1. Taking reasonable estimates of $p = 1.0 \times 10^{-30}$ C·m and $\mathcal{E} = 1.0 \times 10^6$ V/m, find the temperature at which the ratio has fallen by a measurable 10% to 1.9 to 1. (d) Calculate the average energy $\bar{E}$ at T and show that $\bar{E} \rightarrow 0$ as $T \rightarrow 0$ and $\bar{E} \rightarrow 4p\mathcal{E}/3$ as $T \rightarrow \infty$. (e) Find E_{total} from $\bar{E}$, and show that the heat capacity for this two-level system is

$$C = \left(\frac{Nk_B}{2}\right)\left(\frac{2p\mathcal{E}}{k_B T}\right)^2\left(\frac{e^{2p\mathcal{E}/k_B T}}{(1 + \frac{1}{2}e^{2p\mathcal{E}/k_B T})^2}\right)$$

(f) Sketch C as a function of $2p\mathcal{E}/k_B T$. Find the value of $2p\mathcal{E}/k_B T$ at which C is a maximum, and explain, in physical terms, the dependence of C on T.

8. Use the distribution function given in Exercise 10.1,

$$n(E) \, dE = \frac{2\pi(N/V)}{(\pi k_B T)^{3/2}} E^{1/2} e^{-E/k_B T} \, dE$$

to find (a) the most probable kinetic energy of gas molecules at temperature T, (b) the mean kinetic energy at T, and (c) the root mean square kinetic energy at T.

9. The light from a heated atomic gas is shifted in frequency because of the random thermal motion of light-emitting atoms toward or away from an observer. Estimate the fractional Doppler shift ($\Delta f/f_0$), assuming that light of frequency f_0 is emitted in the rest frame of each atom, that the light-emitting atoms are iron atoms in a star at temperature 6000 K, and that the atoms are moving relative to an observer with the mean speed

$$\bar{v} = \pm\sqrt{\frac{8k_B T}{\pi m}}$$

Must we use the relativistic Doppler shift formulas

$$f = f_0 \frac{\sqrt{1 \pm v/c}}{\sqrt{1 \mp v/c}}$$

for this calculation? Such thermal Doppler shifts are measurable and are used to determine stellar surface temperatures.

10.2 Under What Physical Conditions Do Maxwell–Boltzmann Statistics Apply?

10. Helium atoms have spin 0 and are therefore bosons. (a) Must we use the Bose–Einstein distribution at standard temperature and pressure to describe helium gas, or will the Maxwell–Boltzmann distribution suffice? (b) Helium becomes a liquid with a density of 0.145 g/cm^3 at 4.2 K and atmospheric pressure. Must the Bose–Einstein distribution be used in this case? Explain.

10.3 Quantum Statistics

11. To obtain a more clearly defined picture of the Fermi–Dirac distribution, consider a system of 20 Fermi–Dirac particles sharing 94 units of energy. By drawing diagrams like Figure P10.11, show that there are nine different microstates. Using Equation 10.2, calculate and plot the average number of particles in each energy level from 0 to 14E. Locate the Fermi energy at 0 K on your plot from the fact that electrons at 0 K fill all the levels consecutively up to the Fermi energy. (At 0 K the system no longer has 94 units of energy, but has the *minimum* amount of 90E.)

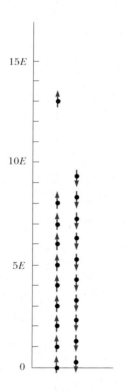

1 Microstate...8 others?

Figure P10.11 One of the nine equally probable microstates for 20 FD particles with a total energy of 94E.

10.4 Bose–Einstein Statistics

12. (a) Find the average energy per photon for photons in thermal equilibrium with a cavity at temperature T. (b) Calculate the average photon energy in electron volts at $T = 6000$ K. *Hint:* Two useful integrals are

$$\int_0^\infty \frac{z^2\,dz}{e^z - 1} \approx 2.41 \quad \text{and} \quad \int_0^\infty \frac{z^3\,dz}{e^z - 1} = \frac{\pi^4}{15}$$

13. (a) Show that the specific heat of any substance in the Einstein model equals 5.48 cal/mol·K at $T = T_E$. (b) Using this result, estimate the Einstein temperatures of lead, aluminum, and silicon from Figure 10.9. (c) Using the temperatures found in part (b), calculate the Einstein specific heats of each element at 50-K intervals and compare with the experimental results shown in Figure 10.9. You may wish to photocopy Figure 10.9 and plot your calculated values on this figure for easy comparison.

10.5 Fermi–Dirac Statistics

14. The Fermi energy of copper at 0 K is 7.05 eV. (a) What is the average energy of a conduction electron in copper at 0 K? (b) At what temperature would the average energy of a molecule of an ideal gas equal the energy obtained in (a)? (See Problem 16)

15. The Fermi energy of aluminum is 11.63 eV. (a) Assuming that the free electron model applies to aluminum, calculate the number of free electrons per unit volume at low temperatures. (b) Determine the valence of aluminum by dividing the answer found in part (a) by the number of aluminum atoms per unit volume as calculated from the density and the atomic weight. Note that aluminum has a density of 2.70 g/cm^3.

16. Show that the average kinetic energy of a conduction electron in a metal at 0 K is given by $\overline{E} = 3E_F/5$. By way of contrast, note that *all* of the molecules in an ideal gas at 0 K have *zero energy!* *Hint:* Use the standard definition of an average given by

$$\overline{E} = \frac{\int_0^\infty E g(E) f_{FD}(E)\,dE}{N/V}$$

where E_F is in electron volts when n is in electrons per cubic meter.

17. Although we usually apply Fermi–Dirac statistics to free electrons in a conductor, Fermi–Dirac statistics apply to any system of spin $\frac{1}{2}$ particles, including protons and neutrons in a nucleus. Since protons are distinguishable from neutrons, assume that each set of nucleons independently obeys the Fermi–Dirac distribution and that the number of protons equals the number of neutrons. Using these ideas, estimate E_F and $\overline{E}$ for the nucleons in Zn. (Zn has 30 protons, 34 neutrons, and a radius of 4.8×10^{-15} m.) Are your answers reasonable? Explain.

18. Show that Equation 10.44 can be expressed as

$$E_F = (3.65 \times 10^{-19}) n^{2/3} \text{ eV}$$

where E_F is in electron volts when n is in electrons per cubic meter.

19. Calculate the probability that a conduction electron in copper at 300 K has an energy equal to 99% of the Fermi energy.

20. Find the probability that a conduction electron in a metal has an energy equal to the Fermi energy at the temperature 300 K.

21. Sodium is a monovalent metal having a density of 0.971 g/cm^3 and a molar mass of 23.0 g/mol. Use this information to calculate (a) the density of charge carriers, (b) the Fermi energy, and (c) the Fermi speed for sodium.

22. Calculate the energy of a conduction electron in silver at 800 K if the probability of finding the electron in that state is 0.95. Assume that the Fermi energy for silver is 5.48 eV at this temperature.

23. Consider a cube of gold 1 mm on an edge. Calculate the approximate number of conduction electrons in this cube whose energies lie in the range from 4.000 to 4.025 eV at 300 K. Assume $E_F(300 \text{ K}) = E_F(0)$.

24. (a) Consider a system of electrons confined to a three-dimensional box. Calculate the ratio of the number of allowed energy levels at 8.5 eV to the number of allowed energy levels at 7.0 eV. (b) Copper has a Fermi energy of 7.0 eV at 300 K. Calculate the ratio of the number of occupied levels at an energy of 8.5 eV to the number of occupied levels at the Fermi energy. Compare your answer with that obtained in part (a).

Calculator/Computer Problems

25. Consider a system of 10^4 oxygen molecules per cubic centimeter at a temperature T. Calculate and plot a graph of the Maxwell distribution, $n(v)$, as a function of v and T. Use your program to evaluate $n(v)$ for speeds ranging from $v = 0$ to $v = 2000$ m/s (in intervals of 100 m/s) at temperatures of (a) 300 K and (b) 1000 K. (c) Make graphs of $n(v)$ versus v and use the graph at $T = 1000$ K to estimate the number of molecules per cubic centimeter having speeds between 800 m/s and 1000 m/s at $T = 1000$ K. (d) Calculate and indicate on each graph the root mean square speed, the average speed, and the most probable speed (see Problem 2).

26. Graph the Fermi–Dirac distribution function, Equation 10.25, versus energy. Plot $f(E)$ versus E for (a) $T = 0.2T_F$ and (b) $T = 0.5T_F$, where T_F is the Fermi temperature, defined by Equation 10.45.

27. Copper has a Fermi energy of 7.05 eV at 300 K and a conduction electron concentration of 8.49×10^{28} m^{-3}. Calculate and plot (a) the density of states, $g(E)$; versus E; (b) the particle distribution function, $n(E)$, as a function of energy at $T = 0$ K; and (c) the particle distribution function versus E at $T = 1000$ K. Your energy scales should range from $E = 0$ to $E = 10$ eV.

Ludwig Boltzmann spent the summer of 1905 teaching at the University of California at Berkeley. Some of his fellow Viennese, including the cartoonist who drew this image, were amused at the thought of Boltzmann lecturing on theoretical physics in the "Wild West." You can read more about Boltzmann's experience in L. Boltzmann, "A German Professor's Trip to El Dorado," *Physics Today*, January 1992, p. 44. (*University of Vienna/Courtesy AIP Emilio Segrè Visual Archives*)

LASER MANIPULATION OF ATOMS

by Steven Chu

Stanford University

The ability to control charged particles at a distance with electric and magnetic fields has led to an enormous number of advances in science and technology. Applications include particle accelerators, x-ray sources, television, and vacuum tubes. Until recently, our ability to control the motion of neutral particles has been much more limited. Usually, the device we use to hold an object such as the tips of your fingers or the wall of a bottle do not exert any force on an object until the device is brought within a few atomic diameters of the object. Once the atoms at the surface of your fingers are atomically close to an object, the electrons on your finger are repelled by the electrons of the object. The electric forces generated by this repulsion allow us to grasp neutral objects.

In the last two decades, scientists have developed the ability to manipulate neutral particles, such as atoms, molecules, and micron-sized particles. This new ability has quickly led to a wide number of applications, including the study of atom–atom interactions and light–matter interactions in an entirely new regime, the production of new quantum states of matter, and the construction of exquisitely accurate atomic clocks and accelerometers.

LASER COOLING

The revolution in the control of neutral atoms is based on the ability to laser-cool atoms to extremely low temperatures. How cold? The surface of the Sun (5000 K) is 18 times hotter than the freezing point of water and 1250 times hotter than the temperature at which helium liquefies (4 K). By comparison, atoms have been cooled by lasers to less than 10^{-9} K, a billion times colder than liquid helium temperatures.

The first methods of laser cooling were simple. Since photons have momentum, light can exert forces on atoms. For a photon of frequency f, its momentum is $p_\gamma = E/c = hf/c$. As an example, a sodium atom absorbing one quantum of yellow light at 589 nm will recoil with a velocity change $\Delta v = p_\gamma/m_{Na} = 3$ cm/s. Although typical atomic speeds are on the order of 10^5 cm/s, laser light directed against a beam of atoms can be used to slow the atoms down since a laser can induce over 10^7 photon absorptions per second. Typically, atoms in a thermal beam can be stopped in a few milliseconds.

Atoms absorb light only if the laser is tuned to a resonance frequency. An atom moving toward a laser beam with velocity v will experience an upward Doppler fractional frequency shift given by $\Delta f/f = v/c$. Thus, an atom moving toward a laser beam at 10^5 cm/s will be in resonance with the laser beam only if the light is tuned 1.7 GHz below the atomic resonance. As the atom slows down, the laser frequency has to be continuously increased in order to keep it in resonance.

Once the atoms are fairly cold, further laser cooling is accomplished by surrounding the atom with counterpropagating laser beams tuned slightly to the low-frequency side of an atomic resonance as in Figure 1. The atoms absorb more photons from the beam opposing its motion because of the Doppler shift, independent of the direction of motion. By surrounding the atom with six laser beams, cooling in all three dimensions is accomplished. This method of laser cooling was first proposed in 1975 by Theodore Hänsch and Arthur Schawlow at Stanford University and first demonstrated by Steve Chu and his colleagues at AT&T Bell Labs in 1985. At that time, sodium atoms were cooled from over 500 K to 240 μK. The sea of photons surrounding the atoms was dubbed "optical molasses" because the light field serves as a viscous damping medium that slows the motion of the atoms.

As the atoms cool to temperatures at which the average Doppler shift is a small fraction of the resonance linewidth, the differential absorption, and hence the cooling

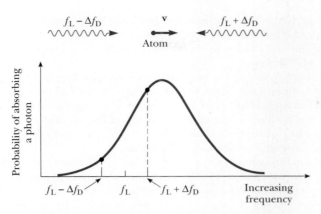

Figure 1 An atom irradiated by two counterpropagating laser beams tuned to frequency f_L will see frequencies $f_L \pm \Delta f_D$. If f_L is tuned below the atomic resonance, the atom will receive more photon momentum kicks from the beam opposing the motion.

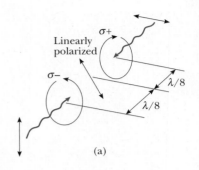

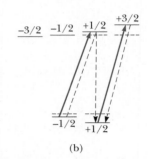

Figure 2 (a) Polarization gradients that result from the superposition of orthogonal linearly polarized light. The local polarization of the light changes from positive helicity (σ^+) to linear to negative helicity polarization (σ^-) in a distance $\lambda/4$. (b) The energy levels of an atom with an $F = 1/2$ ground state and an $F = 3/2$ excited state irradiated with positive-helicity light. The dashed lines show the allowed spontaneous emission paths, and the dotted lines denote the energy levels in the absence of the light shifts. Because the excitation light is always trying to add $+\hbar$ unit of angular momentum to the atom, repeated excitation and spontaneous emission will optically pump the atom into the $+1/2$ ground state. The light shifts are greatly exaggerated with respect to the separation of the energy levels. (*From S. Chu, Science, 253:861–866, 1991*).

force, decreases. An equilibrium temperature is reached when the cooling rate of optical molasses balances the heating rate due to the random absorption and reemission of photons. For a simple atom consisting of one ground and one excited state separated by $\hbar\omega$, the minimum equilibrium temperature was calculated to be $k_B T = \hbar\omega/2$. In 1988, a NIST[1] group at Gaithersburg, Maryland, led by William Phillips discovered that atoms could be cooled in optical molasses to temperatures far below this theoretical minimum temperature.

Jean Dalibard and Claude Cohen-Tannoudji at the École Normale Supérieure, and Steve Chu, now at Stanford, realized that the low temperatures were due to the interplay between several physical effects. (1) Atoms with nonzero angular momentum have several Zeeman sublevels in the ground and excited states. (2) These Zeeman energy levels shift with the presence of light by different amounts, depending on the strength of the coupling of the atom to the light field. If the laser is tuned below the resonance frequency, the states are shifted to lower energy. (3) For a given polarization of the light field, the atom is preferentially driven into the quantum states with the lowest energy. (The population of specific Zeeman states due to polarized light is known as "optical pumping." Alfred Kastler was awarded a Nobel prize in 1966 for his studies of this effect.)

Two counterpropagating laser beams with orthogonal linear polarization generate a laser field having spatially dependent polarization as in Figure 2. An atom in a region in space where the light has positive helicity (σ^+) will optically pump into a low-energy ground state $m_F = +F$. If it then moves into a region of space where the light has negative helicity (σ^-), the $+F$ state becomes the highest-energy ground state. The kinetic energy of the atom is thus converted into internal energy. This internal energy is dissipated by the optical pumping effect, which drives the atom into the $m_F = -F$ state, the low-energy state for σ^- light. In effect, the atom slows down as it rolls up a potential hill created by the light shift of the atomic energy states. Near the top of the hill, the atom optically pumps into the new low-energy state; that is, it finds itself at the bottom of a new hill. Optical molasses with polarization gradients is sometimes referred to as Sisyphus cooling, after the mythical condemned man who faced the perpetual torment of rolling a rock up a hill only to have it roll back down again. This cooling process has cooled sodium atoms to 35 μK and cesium atoms to 3 μK.

Other laser cooling techniques that use a number of quantum coherence tricks and that are given exotic names, such as "coherent population trapping" and "stimulated

[1]National Institute of Standards and Technology.

Raman cooling," have cooled atoms to temperatures in the nanokelvin range at which the average speed is less than the speed change an atom experiences if it recoils from a single photon. An explanation of these cooling methods can be found in the articles listed at the end of this essay.

Once atoms are cooled to low temperatures corresponding to speeds on the order of 1 cm/s, they can be easily manipulated with light, as well as static magnetic or electric fields. For example, atoms cooled to microkelvin energies have been tossed upward with light to form an atomic fountain. As they follow the ballistic trajectory dictated by gravity, they are in a nearly perturbation-free environment, and very accurate measurements of atomic energy level splittings are possible.

A fundamental limitation to the accuracy of any measurement is the Heisenberg uncertainty principle, which states that the quantum measurement time Δt and the uncertainty in the energy measurement ΔE must be greater than $\hbar/2$: $\Delta E \, \Delta t \geq \hbar/2$. For an atom in an atomic beam the measurement time is limited by the transit time of the atom through the measuring device. With an atomic fountain, the measurement time has been increased from a few milliseconds to roughly one second. This thousandfold increase in the measurement time will undoubtedly lead to more accurate atomic clocks. The current time standard, defined by the energy level splitting between two ground states of the cesium atom, is accurate to one part in 10^{14}.

Atomic fountains have also been used to construct atom interferometers. Similar to an optical interferometer, the atom interferometer splits the atom into a superposition of two coherent states that separate spatially and recombine to form interference fringes. Atom interferometers make extremely sensitive inertial sensors because of the long transit time of the atoms through the device. The Stanford group has measured g, the acceleration of an atom due to gravity, with a resolution of one part in 10^8 with an atom interferometer, and it is likely that the uncertainty will be reduced to less than one part in 10^{11}. A portable version of this device could replace the mechanical "g" meters now used in oil exploration. A low value of g could signify the presence of porous, oil-laden rock, which has a lower density than solid rock.

ATOM TRAPPING

Atoms cooled below a millikelvin can also be held in space with either static magnetic fields, laser fields, or a combination of a weak magnetic field and circularly polarized light. Atom traps have been used to accumulate a large number of laser-cooled atoms, confine them for further cooling, and use them for studies of atomic collisions at very low temperatures.

A magnetic trap exerts forces on an atom via the atom's magnetic moment. The potential energy of a magnetic dipole in a magnetic field **B** is given by $-\boldsymbol{\mu} \cdot \mathbf{B}$, where μ is the magnetic moment, typically on the order of 1 $\mu_{\rm B}$. Atoms were first magnetically trapped in 1985 by the NIST group with a "spherical quadrupole field" as shown in Figure 3. At the center of the trapping coils, the **B**-field is 0 and increases linearly as one moves radially outward. An atom with its magnetic dipole aligned antiparallel to the **B**-field minimizes its energy by seeking regions where the field magnitude is smallest; that is, the atom experiences a force that drives it to the center of the trap. If the atom moves slowly in the trap, it can remain aligned antiparallel to the magnetic field, even if the field changes direction. Classically, if the magnetic moment precesses rapidly around a slowly changing magnetic field, it continues to spin around a changing field axis. Quantum mechanically, the state of the atom "adiabatically" follows external field conditions that define that quantum state.

Atoms can also be held with a laser beam by using intense light tuned far from an atomic resonance to polarize the atom. The induced dipole moment **p** on the atom points in the same direction as the driving electric field as long as the frequency of the light is well below the atomic resonance frequency. Thus, the potential energy of the

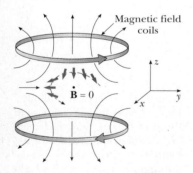

Figure 3 A spherical quadrupole trap showing some of the magnetic field lines and the direction of the currents. The dashed line indicates an atom moving in the trap with its magnetic moment (bright colored arrows) remaining antiparallel to the field lines.

atom in the light field, $-\mathbf{p} \cdot \boldsymbol{\mathcal{E}}$, causes the atom to seek regions of highest laser intensity. The first optical trap, demonstrated by the Bell Labs group in 1986, was based on a single focused laser beam. Since the focal point of a laser beam can be easily moved with mirrors or lenses, this trap was dubbed "optical tweezers." If the light is tuned above the atomic resonance, an atom is repelled by the light. A box of light formed from focused sheets of light has also been used to trap atoms by the Stanford group. By using an optical microscope to focus the laser light and simultaneously observe samples, this type of trap has also been used to manipulate particles, such as micron-sized spheres, bacteria, and viruses and individual molecules of DNA in aqueous solution.

The most widely used atom trap, first demonstrated at Bell Labs in 1987, uses a combination of a weak spherical magnetic quadrupole field and circularly polarized light. The magnetic field shifts the Zeeman energy levels of the atom as in Figure 4. When illuminated with circularly polarized light, an atom to the right of the trap center will predominately scatter σ^+ light, and atoms left of center will scatter more σ^- light. Thus, the atoms are pushed to the trap center by the same scattering force used in the first laser cooling experiments. The advantage of this trap is that it combines laser trapping with laser cooling and needs only low-intensity laser light and weak magnetic fields. Another advantage is that atoms can be loaded directly inside a low-pressure vapor cell without the initial slowing from an atomic beam. Over 10^{10} atoms can be collected in a magneto-optic/cell trap in a few seconds.

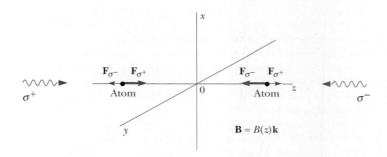

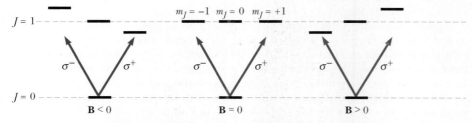

Figure 4   In a spherical quadrupole magnetic field, the energy levels of an atom with a $J = 0$ ground state and a $J = 1$ excited state will be shifted in energy in a spatially dependent way. For example, along the z-axis of the field given in Figure 3, the $\mathbf{B}$ field is positive and increases linearly along the $+z$-axis. It reverses direction along the $-z$-axis. An atom irradiated by two counterpropagating beams tuned below the transition frequency as in optical molasses will cool. If the light is circularly polarized as shown, the momentum exchange will also trap the atom by tending to force it to the $\mathbf{B} = 0$ region. Because σ^- light excites the transition $m_J = 0 \rightarrow m_J = -1$, and σ^+ light excites $m_J = 0 \rightarrow m_J = +1$, the σ^- beam is slightly more in resonance with an atom located in the positive z direction than the σ^+ light. Consequently, more σ^- photons than σ^+ photons are scattered by an atom in the $+z$ position.

BOSE–EINSTEIN CONDENSATION

Laser-cooled atoms in traps allow one to study atomic collisions in the new domain of ultracold temperatures. At very low temperatures, the atoms will no longer exhibit a Maxwell–Boltzmann distribution of energies. Depending on whether a gas of atoms has total angular momentum $0\hbar$, $1\hbar$, $2\hbar$, . . . or $\hbar/2$, $3\hbar/2$, $5\hbar/2$, . . . the atoms obey either Bose–Einstein or Fermi–Dirac statistics. One of the most dramatic quantum gas effects predicted by Einstein should occur when the de Broglie wavelength $\lambda = h/p$ of Bose atoms is comparable to the interatomic spacing. At the critical phase space density of $n\lambda^3 = 2.612$, where n is the density of atoms in the gas, a sizable fraction of the atoms should begin to condense into a single quantum state.

In 1995, a group at the University of Colorado at Boulder led by Eric Cornell and Carl Wieman verified this prediction for a gas of rubidium atoms by combining many of the cooling and trapping techniques developed in the previous ten years. They started by cooling and trapping atoms in a magneto-optic trap. Further cooling was accomplished using optical molasses and Sisyphus cooling in the absence of a magnetic field. At this point, the phase space density of the atoms was increased above that of an intense thermal atomic beam by 13 orders of magnitude. The atoms were then optically pumped into a particular atomic Zeeman state and loaded into a magnetic trap.

The spherical quadrupole magnetic trap they used was modified to correct a known flaw of the NIST trap. If an atom in the trap ventures close to the zero magnetic field point, the precession frequency of the atom goes to zero and its spin can no longer follow the direction of the field line. The spin could then flip to become parallel to the field and hence antitrapped: the trap has a "hole" at its center. The Boulder group solved the problem by rotating the zero point of the magnetic field with additional magnetic field coils. If the rotation frequency is made much faster than the atomic motion in the trap, the atoms experience a time-averaged potential centered on the point of rotation. Since the hole rotates outside of the ensemble of atoms, the atoms never see the hole.

The final cooling to about 20 nK was accomplished by "evaporization" from the magnetic trap, a cooling technique first demonstrated by Harold Hess, Tom Greytak, Dan Kleppner, and colleagues at MIT in 1989. Evaporization allows the very hottest atoms to leave the sample, thereby reducing the average energy of the remaining atoms. Collisions rethermalize the sample to produce more hot atoms, and as the evaporization proceeds, the evaporization barrier is lowered in order to continue the cooling process. Using this technique, they were able to increase the phase space density by another 5 orders of magnitude.

Figure 5 shows the velocity distribution of a cloud of about 2000 rubidium atoms cooled in the vicinity of the Bose condensation threshold. The real experimental difficulty is to both cool and maintain sufficiently low density in the cloud (interatomic distance about 10^{-4} cm) to preserve the ideal gas character of the cloud, minimize interatomic forces, and prevent liquification. Just above the required phase space density, the atomic energy is distributed in all directions equally among the many quantum states, in accord with the equipartition theorem of statistical physics. As the threshold condition is crossed, a central peak at $v = 0$ begins to form, signifying the onset of a condensation. With further cooling, most of the atoms condense into the ground quantum state of the system. Since the trap potential is not spherically symmetrical, the ground quantum state is also asymmetrical.

Once the atoms are in a Bose condensate, their energy is defined in terms of the localization energy $\Delta x \Delta p \geq \hbar/2$. Since the atoms remain in the Bose condensate as the confining forces of the trap are relaxed, the effective temperature $k_{B}T/2 = 3\Delta p^2/2m$ can be brought into the picokelvin range.

The activity in laser cooling and atom trapping has exploded into many areas of physics, chemistry, biology, and medicine. One of the joys of science is that many of these applications were not foreseen by the inventors of the field in the early days.

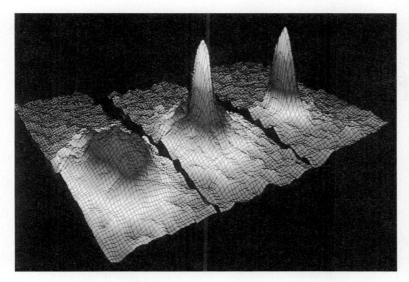

Figure 5 Computer-generated images of the velocity distribution of a cloud of rubidium atoms. Before the Bose–Einstein threshold is reached, the velocity distribution is uniform. The leftmost image was obtained just before the appearance of the Bose–Einstein condensate. Immediately after the appearance of the condensate (middle image), an asymmetrical velocity spike centered at $v = 0$ appears above the broader uniform distribution. At still higher-phase space densities (rightmost image), most of the remaining atoms have condensed into a ground quantum state. The field of view of each frame is $200 \times 270 \ \mu$m, corresponding to the distance the atoms have moved in about 1/20 of a second. (*Images courtesy of Eric Cornell and Mike Matthews, JILA Research Team, University of Colorado at Boulder*)

Similarly, the applications of a Bose–Einstein condensate of atoms cannot be predicted. Since the condensate is analogous to in-phase photons in a laser cavity, one can, in principle, construct an exquisitely intense beam of atoms or atom laser. Another unforeseen application is the remarkable slowing and stopping of light pulses in a Bose–Einstein condensate of sodium atoms first observed in 2000 by the Danish-American physicist Lene Vestergaard Hau.[2] Since all of the information content in a light pulse slowed in a BEC is recoverable at later times, this effect can probably be used in opto-electronic components such as switches, memories, and delay lines.

This is indeed still an exciting time for physics.

Suggestions for Further Reading

For general reviews of cooling and trapping, see S. Chu, *Sci. Am.*, February, 1992; and S. Chu, *Science*, 253:861, 1991.

For a description of newer methods of laser cooling, see C. Cohen-Tannoudji and W. D. Phillips, *Phys. Today*, 43:33, 1990. Raman cooling is described in M. Kasevich and S. Chu, *Phys. Rev. Lett.*, 69:1741, 1992.

The first Bose–Einstein condensation of a gas is described in M. H. Anderson, J. R. Ensher, M. R. Matthews, C. E. Wieman, and E. A. Cornell, *Science*, 269:198, 1995.

[2]Hau et al., *Nature*, 18 February, 1999.

11

Molecular Structure

Chapter Outline

Except for the inert gases, elements generally combine to form chemical compounds whose basic unit is the *molecule*, an aggregate of individual atoms joined by chemical bonds. The physical and chemical properties of molecules derive from their constituent atoms—their arrangement, the manner and degree to which they interact, and their individual electronic structures.

The properties of molecules can be studied experimentally by examining their spectra. As with atoms, a molecule can emit or absorb photons, with accompanying electronic transitions among the allowed energy levels of the molecule. The resulting emission or absorption spectrum is different for each molecule and acts as a sort of fingerprint of its electronic structure.

But molecules also emit or absorb energy in ways not found in atoms. Molecules can rotate, storing energy in the form of kinetic energy of rotation, and they can vibrate, and so possess energy of vibration. As we shall discover, both the rotational and vibrational energies are quantized and so give rise to their own unique spectra. It follows that molecular spectra are vastly more complicated than atomic spectra, but also carry a good deal more information. In particular, the vibration–rotation spectrum tells us how the individual atoms that form the molecule are arranged and the strength of their interaction.

In this chapter we shall describe the bonding mechanisms in molecules, the various modes of molecular excitation, and the radiation emitted or absorbed by molecules. In the course of our study we shall encounter the quantum origins of the chemical bond and discover why some atoms bond to form a molecule and others do not. Central to this inquiry are the roles played by the exclusion principle and tunneling, nonclassical ideas that underscore the importance of wave mechanics to the study of molecular structure.

11.1 BONDING MECHANISMS: A SURVEY

Two atoms combine to form a molecule because of a net attractive force between them. Furthermore, the total energy of the bound molecule is *less* than the total energy of the separated atoms; the energy difference is the energy that must be supplied to break apart the molecule into its constituent atoms.

Fundamentally, the bonding mechanisms in a molecule are primarily due to electrostatic forces between atoms (or ions). When two atoms are separated by an infinite distance, the force between them is zero, as is their electrostatic energy. As the atoms are brought closer together, both attractive and repulsive forces come into play. At very large separations, the dominant forces are attractive in nature. For small separations, repulsive forces between like charges begin to dominate. The potential energy of the pair can be positive or negative, depending on the separation between the atoms.

The total potential energy U of a system of two atoms often is approximated by the expression

$$U = -\frac{A}{r^n} + \frac{B}{r^m}$$

where r is the internuclear separation distance between the two atoms, A and B are constants associated with the attractive and repulsive forces, and n and m are small integers. Figure 11.1 presents a sketch of the total potential energy

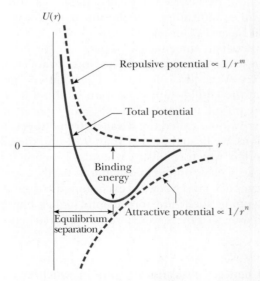

Figure 11.1 The total particle energy as a function of the internuclear separation for a system of two atoms.

versus internuclear separation for a two-atom system. Note that the potential energy for large separations is negative, corresponding to a net attractive force. At the equilibrium separation, the attractive and repulsive forces just balance. At this point, the potential energy has its minimum value and the slope of the curve is zero.

A complete description of the binding mechanisms in molecules is a highly complex problem because bonding involves the mutual interactions of many particles. In this section we shall discuss some simplified models in order of decreasing bond strength: the ionic bond, the covalent bond, the van der Waals bond, and the hydrogen bond.

Ionic Bonds

When two atoms combine in such a way that one or more electrons are transferred from one atom to the other, the bond formed is called an **ionic bond.** Ionic bonds are fundamentally caused by the Coulomb attraction between oppositely charged ions. A familiar example of an ionically bonded molecule is sodium chloride, NaCl, or common table salt. Sodium, which has an electronic configuration $1s^2 2s^2 2p^6 3s$, gives up its $3s$ valence electron to form a Na^+ ion. The energy required to ionize the atom to form Na^+ is 5.1 eV. Chlorine, which has an electronic configuration $1s^2 2s^2 2p^5$, is one electron short of the closed-shell structure of argon. Because closed-shell configurations are energetically more favorable, the Cl^- ion is more stable than the neutral Cl atom. The energy released when an atom takes on an electron is the **electron affinity** of the atom. For chlorine, the electron affinity is 3.7 eV. Therefore, the energy required to form Na^+ and Cl^- from isolated atoms is $5.1 - 3.7 = 1.4$ eV. It costs 5.1 eV to remove the electron from the Na atom but you gain 3.7 eV of it back when that electron joins the Cl atom. The difference, in this case 1.4 eV, is called the **activation energy** of the molecule. As the ions are brought closer together, their mutual energy decreases due to electrostatic attraction. At sufficiently small separations, the energy of formation becomes negative, indicating that the ion pair now is energetically preferred over neutral Na and Cl atoms.

The total energy versus internuclear separation for Na^+ and Cl^- ions is sketched in Figure 11.2. At very large separation distances, the energy of the system of ions is 1.4 eV, as just calculated. The total energy has a minimum value of -4.2 eV at the equilibrium separation of about 0.24 nm. This means that the energy required to break the $Na^+ - Cl^-$ bond and form neutral sodium and chlorine atoms, called the **dissociation energy,** is 4.2 eV.

When the two ions are brought closer than 0.24 nm, the electrons in closed shells begin to overlap, which results in a repulsion between the closed shells. This repulsion is partly electrostatic in origin and partly a result of the identity of electrons. Because they must obey the exclusion principle (Chapter 9), some electrons in overlapping shells are forced into higher energy states and the system energy increases, as if there were a repulsive force between the ions.

Covalent Bonds

A **covalent bond** between two atoms is one in which electrons supplied by either one or both atoms are shared by the two atoms. Many diatomic molecules, such as H_2, F_2, and CO, owe their stability to covalent bonds. In the case

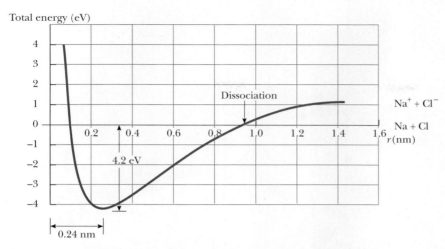

Figure 11.2 Total energy versus the internuclear separation for Na^+ and Cl^- ions. The energy required to separate the NaCl molecule into neutral atoms of Na and Cl is the dissociation energy, 4.2 eV.

of the H_2 molecule, the two electrons are equally shared between the nuclei and form a so-called molecular orbital. The two electrons are more likely to be found between the two nuclei, hence the electron density is large in this region. The formation of the molecular orbital from the s orbitals of the two hydrogen atoms is represented in Figure 11.3. Because of the exclusion principle, the two electrons in the ground state of H_2 must have antiparallel spins. If a third H atom is brought near the H_2 molecule, the third electron would have to occupy a higher-energy quantum state because of the exclusion principle, which is an energetically unfavorable situation. Hence, the H_3 molecule is not stable and does not form. The stability of H_2 and related species is examined in detail in Section 11.4.

More complex stable molecules, such as H_2O, CO_2, and CH_4, are also formed by covalent bonds. Consider methane, CH_4, a typical organic molecule shown schematically in the electron-sharing diagram of Figure 11.4a. Note that covalent bonds are formed between the carbon atom and each of the four hydrogen atoms. The spatial electron distribution of the four covalent bonds is shown in Figure 11.4b. The four hydrogen nuclei are at the corners of a regular tetrahedron, with the carbon nucleus at the center.

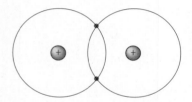

Figure 11.3 Classical orbit model for the covalent bond formed by the two $1s$ electrons of the H_2 molecule.

van der Waals Bonds

Ionic and covalent bonds occur between atoms to form molecules or ionic solids, so that they can be described as bonds *within* molecules. Two additional types of bonds, van der Waals bonds and hydrogen bonds, can occur *between* molecules.

We might expect that two neutral molecules would not interact by means of the electric force because they each have zero net charge. We find, however, that they are attracted to each other by weak electrostatic forces called van der Waals forces. Likewise, atoms that do not form ionic or covalent bonds are attracted to each other by van der Waals forces. Inert gases, for example,

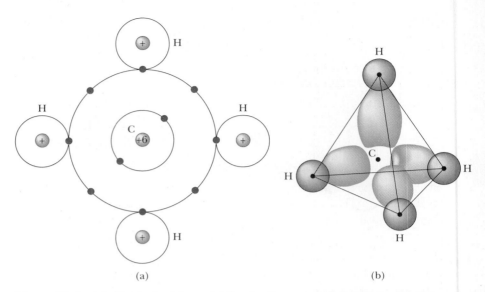

(a)   (b)

Figure 11.4 (a) Classical orbit model for the four covalent bonds in the CH_4 molecule. (b) Quantum-mechanical picture of the spatial arrangement of the four covalent bonds of the CH_4 molecule. The carbon atom is at the center of a tetrahedron with hydrogen atoms at its corners. The orbitals supplied by carbon are actually sp^3 hybrid orbitals as explained at the end of Section 11.5. These orbitals have two lobes, and only the larger lobes, which are shown here greatly narrowed for ease of depiction, participate in the bonding. Each C—H bond consists of an overlapping $1s$ orbital from hydrogen and an sp^3 hybrid orbital from carbon. (*Adapted from D. Ebbing*, General Chemistry, 5th ed., *Boston, Houghton Mifflin Co., 1996*)

because of their filled shell structure, do not generally form molecules. Because of van der Waals forces, however, at sufficiently low temperatures at which thermal excitations are negligible, inert gases first condense to liquids and then solidify (with the exception of helium, which does not solidify at atmospheric pressure). The van der Waals force arises when an electrically neutral molecule has centers of positive and negative charge that do not coincide. As a result, the molecule constitutes an electric dipole. The interaction between electric dipoles causes two molecules to attract one another.

There are three types of van der Waals forces, which we shall briefly describe. The first type, called the **dipole–dipole force,** is an interaction between two molecules, each having a permanent electric dipole moment. For example, polar molecules such as HCl and H_2O have permanent electric dipole moments and attract other polar molecules. In effect, one molecule interacts with the electric field produced by another molecule.

The second type, the **dipole-induced force,** results when a polar molecule having a permanent electric dipole moment induces a dipole moment in a nonpolar molecule. In this case, the electric field of the polar molecule creates the dipole moment in the nonpolar molecule, which then results in an attractive force between the molecules.

The third type, called the **dispersion force,** is an attractive force that occurs between two nonpolar molecules. Although the average dipole moment of a nonpolar molecule is zero, charge fluctuations can cause two

nonpolar molecules near each other to have dipole moments that are corre-
lated in time so as to produce an attractive van der Waals force.

Because all three types of van der Waals forces are dipolar in origin, they all
fall off with separation distance r as $1/r^7$; however, the proportionality constant is different for each type.

The Hydrogen Bond

Because hydrogen has only one electron, it is expected to form a covalent bond
with only one other atom within a molecule. A hydrogen atom in a given molecule can also form a second type of bond between molecules called a **hydrogen
bond.** One example of a hydrogen bond is the hydrogen difluoride ion, $(HF_2)^-$,
shown in Figure 11.5. The two negative fluorine ions are bound by the positively
charged proton between them. This is a relatively weak chemical bond, with a
binding energy of only about 0.1 eV. The water molecule, H_2O, is another example of a system that contains hydrogen bonds. In the two covalent bonds in this
molecule, the electrons from the hydrogen atoms are more likely to be found
near the oxygen atom than near the hydrogen atoms. This leaves essentially bare
protons at the positions of the hydrogen atoms. This unshielded positive charge
can be attracted to the negative end of another polar molecule. Because the proton is unshielded by electrons, the negative end of the other molecule can come
very close to the proton to form a bond that is strong enough to form a solid crystalline structure, such as that of ice. The bonds within a water molecule are covalent, but the bonds between water molecules in ice are hydrogen bonds.

The hydrogen bond is relatively weak compared with other chemical
bonds—it can be broken with an input energy of about 0.1 eV. Because of
this, ice melts at the low temperature of 0°C. Despite the fact that this bond is
very weak, hydrogen bonding is a critical mechanism responsible for the link-
ing of biological molecules and polymers. For example, in the case of the
DNA (deoxyribonucleic acid) molecule, which has a double-helix structure,
hydrogen bonds formed by the sharing of a proton between two atoms create
linkages between the turns of the helix.

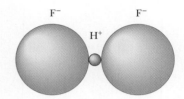

Figure 11.5 Hydrogen bonding in the $(HF_2)^-$ molecular
ion. The two negative fluorine
ions are bound by the positively
charged proton between them.

11.2   MOLECULAR ROTATION AND VIBRATION

As is the case with atoms, we can study the structure and properties of molecules by examining the radiation they emit or absorb. Before we describe
these processes, it is important to first understand the various ways of exciting
a molecule.

Consider an individual molecule in the gaseous phase of a substance. The
energy of the molecule can be divided into four categories: (1) electronic
energy, due to the interactions between the molecule's electrons and nuclei;
(2) translational energy, due to the motion of the molecule's center of mass
through space; (3) rotational energy, due to the rotation of the molecule
about its center of mass; and (4) vibrational energy, due to the vibration of the
molecule's constituent atoms.

$$E = E_{el} + E_{trans} + E_{rot} + E_{vib}$$

Because the translational energy is unrelated to internal structure, this component is unimportant in interpreting molecular spectra. The electronic energy

of a molecule is very complex because it involves the interaction of many charged particles. Electronic energies are the subject of Section 11.4. Here we concentrate on the significant information about a molecule that can be deduced by analyzing its rotational and vibrational energy states, which give spectral lines in the infrared region of the electromagnetic spectrum.

Molecular Rotation

Let us consider the rotation of a molecule about its center of mass. We confine our discussion to a diatomic molecule, although the same ideas can be extended to polyatomic molecules. As shown in Figure 11.6a, the diatomic molecule has only 2 rotational degrees of freedom, corresponding to rotations about the y- and z-axes, that is, the axes perpendicular to the molecular axis.[1]

The energy of a rigid rotating molecule is all kinetic. Let m_1 and m_2 denote the atomic masses, with speeds v_1 and v_2. For a molecule in rotation, the speeds v_1 and v_2 are interrelated. In terms of the angular velocity of rotation ω, we have

$$v_1 = \omega r_1 \quad \text{and} \quad v_2 = \omega r_2$$

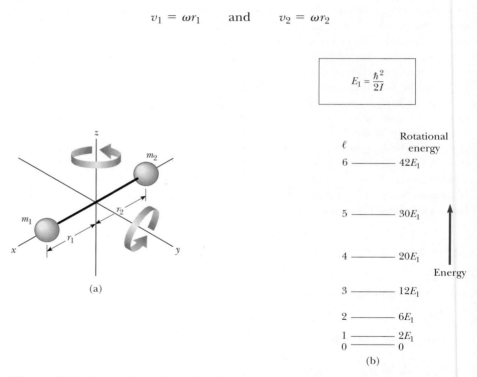

$$E_1 = \frac{\hbar^2}{2I}$$

Figure 11.6 (a) A diatomic molecule oriented along the x-axis has 2 rotational degrees of freedom, corresponding to rotation about the y- and z-axes. (b) Allowed rotational energies of a diatomic molecule as calculated using Equation 11.5.

---

[1]The excitation energy for rotations about the molecular axis is so large that such modes are not observable. This follows because nearly all the molecular mass is concentrated within nuclear dimensions of the rotation axis, giving a negligibly small moment of inertia about the internuclear line (see Problem 12).

where r_1 is the distance of m_1 from the axis of rotation and similarly for r_2 (see Fig. 11.6a). The angular momentum of rotation about the z-axis is

$$L = m_1 v_1 r_1 + m_2 v_2 r_2 = \{m_1 r_1^2 + m_2 r_2^2\}\omega = I\omega$$

The quantity in curly brackets {. . .} is the *moment of inertia,* denoted I. With this identification, the energy of rotation is

$$E_{rot} = \tfrac{1}{2}m_1 v_1^2 + \tfrac{1}{2}m_2 v_2^2 = \tfrac{1}{2}I\omega^2$$

Eliminating ω from the preceding two equations gives the simple result

$$E_{rot} = \frac{L^2}{2I} \qquad (11.1)$$

Rotational energy and angular momentum

Comparing Equation 11.1 with the kinetic energy of a particle in translation, $p^2/2m$, we see that the moment of inertia I of the molecule measures its resistance to changes in rotation in the same way that the mass m of a single particle measures its resistance to changes in translation. The value for I depends on the rotation axis, however. For the important case where the axis of rotation passes through the center of mass,[2] we have $m_1 r_1 = m_2 r_2$, and r_1 and r_2 can be written in terms of the atomic separation $R_0 = r_1 + r_2$ as

$$R_0 = \left(\frac{m_2}{m_1} + 1\right) r_2 = \left(1 + \frac{m_1}{m_2}\right) r_1$$

Then the moment of inertia about the center of mass, I_{CM}, becomes

$$I_{CM} = \left(\frac{m_1 m_2}{m_1 + m_2}\right) R_0^2 \equiv \mu R_0^2 \qquad (11.2)$$

where μ is the **reduced mass** of the molecule,

$$\mu = \frac{m_1 m_2}{m_1 + m_2} \qquad (11.3)$$

Unlike moment of inertia, which is a property of the molecule, angular momentum L is a dynamical variable; in the transition to quantum mechanics, L^2 becomes quantized. As shown in Chapter 8, the correct quantization rule is

$$L^2 = \ell(\ell + 1)\hbar^2 \qquad \ell = 0, 1, 2, \ldots \qquad (11.4)$$

which, in turn, restricts the energy of rotation to be one of the discrete values

$$E_{rot} = \frac{\hbar^2}{2I_{CM}} \ell(\ell + 1) \qquad (11.5)$$

Allowed energies for rotation

In the context of molecular rotation, the integer ℓ is called the **rotational quantum number.** Thus, we see that **the rotational energy of the molecule is quantized and depends on the moment of inertia of the molecule.** The allowed rotational energies of a diatomic molecule are sketched in

---

[2]Since there is no axis about which the molecule is constrained to rotate, the correct axis for calculating rotational energy is one passing through the center of mass. Energy associated with rotation about any other axis would include some energy of translation, as well as energy of rotation.

Table 11.1   Microwave Absorption Lines for Several Rotational Transitions of the CO Molecule

Rotational Transition	Wavelength of Absorption Line (m)	Frequency of Absorption Line (Hz)
$\ell = 0 \rightarrow \ell = 1$	2.60×10^{-3}	1.15×10^{11}
$\ell = 1 \rightarrow \ell = 2$	1.30×10^{-3}	2.30×10^{11}
$\ell = 2 \rightarrow \ell = 3$	8.77×10^{-4}	3.46×10^{11}
$\ell = 3 \rightarrow \ell = 4$	6.50×10^{-4}	4.61×10^{11}

Figure 11.6b. These results apply also to polyatomic molecules, provided the appropriate generalization of I_{CM} is used.

The spacing between adjacent rotational levels can be calculated from Equation 11.5:

$$\Delta E = E_\ell - E_{\ell-1} = \frac{\hbar^2}{2I_{CM}} \{\ell(\ell+1) - (\ell-1)\ell\} = \frac{\hbar^2}{I_{CM}} \ell \quad (11.6)$$

where ℓ is the quantum number of the higher energy state. In going from one rotational state to the next, the molecule loses (or gains) energy, ΔE. The loss (or gain) typically is accompanied by photon emission (or absorption) at the frequency $\omega = \Delta E/\hbar$. Thus, photons should be observed at the frequencies $\omega_0 = \hbar/I_{CM}$, $2\omega_0$, $3\omega_0$, ... These predictions are in excellent agreement with experiment.[3] The wavelengths and frequencies for the absorption spectrum of the CO molecule are given in Table 11.1. The lowest frequencies lie in the *microwave range* of the electromagnetic spectrum, as is typical of the rotational spectra of all molecules. This indicates that the energy required to excite a molecule into rotation is quite small, on the order of 10^{-4} eV. From the data, one can deduce the moment of inertia and the bond length of the molecule, as shown in the following example.

EXAMPLE 11.1   Rotation of the CO Molecule

The $\ell = 0$ to $\ell = 1$ rotational transition of the CO molecule occurs at a frequency of 1.15×10^{11} Hz. (a) Use this information to calculate the moment of inertia of the molecule about its center of mass.

Solution   From Equation 11.6, we see that the energy difference between the $\ell = 0$ and $\ell = 1$ rotational levels is $\hbar^2/I_{CM}$. Equating this to the energy of the absorbed photon, we get

$$\frac{\hbar^2}{I_{CM}} = \hbar\omega$$

The angular frequency ω of the absorbed radiation is

$$\omega = 2\pi f = 2\pi(1.15 \times 10^{11} \text{ Hz}) = 7.23 \times 10^{11} \text{ rad/s}$$

so that I_{CM} becomes

$$I_{CM} = \frac{\hbar}{\omega} = \frac{1.055 \times 10^{-34} \text{ J} \cdot \text{s}}{7.23 \times 10^{11} \text{ rad/s}} = 1.46 \times 10^{-46} \text{ kg} \cdot \text{m}^2$$

(b) Calculate the bond length of the molecule.

Solution   Equation 11.2 can be used to calculate the bond length once the reduced mass of the molecule is found. Since the carbon and oxygen atomic masses are 12.0 u and 16.0 u, respectively, the reduced mass of the CO molecule is, from Equation 11.3,

$$\mu = \frac{(12.0 \text{ u})(16.0 \text{ u})}{12.0 \text{ u} + 16.0 \text{ u}} = 6.857 \text{ u} = 1.14 \times 10^{-26} \text{ kg}$$

---

[3]These simple statements must be refined when molecular vibration is taken into account, as discussed in Section 11.3.

where the conversion $1\,u = 1.66 \times 10^{-27}$ kg has been used. Then

$$R_0 = \sqrt{\frac{I_{CM}}{\mu}} = \sqrt{\frac{1.46 \times 10^{-46}\ \text{kg}\cdot\text{m}^2}{1.14 \times 10^{-26}\ \text{kg}}}$$

$$= 1.13 \times 10^{-10}\ \text{m} = 0.113\ \text{nm}$$

This example hints at the immense power of spectroscopic measurements to determine molecular properties!

Molecular Vibration

A molecule is a flexible structure whose atoms are bonded together by what can be considered "effective springs." If disturbed, the molecule can vibrate, taking on vibrational energy. This energy of vibration may be altered if the molecule is exposed to radiation of the proper frequency.

Consider again a diatomic molecule. The potential energy $U(r)$ versus atomic separation r for such a molecule is sketched in Figure 11.7a. The equilibrium separation of the atoms is denoted there by R_0; for small displacements from equilibrium, the atoms vibrate, as if connected by a spring with unstretched length R_0 and force constant K (Fig. 11.7b).[4] Atomic displacements in the direction of the molecular axis give rise to oscillations along the line joining the atoms. For these *longitudinal* vibrations, the system is effectively one-dimensional, with the coordinates of each atom measured along the molecular axis.

We denote by ξ_1 and ξ_2 the displacements from equilibrium of m_1 and m_2, respectively. In terms of these displacements, the effective spring is stretched a net amount $\xi_1 - \xi_2$, and the elastic energy of the two-atom pair is

$$U = \tfrac{1}{2}K(\xi_1 - \xi_2)^2$$

Harmonic approximation to molecular vibration

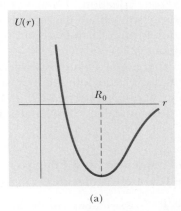

(a)

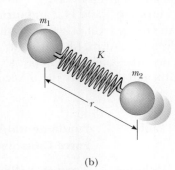

(b)

Figure 11.7 (a) A plot of the potential energy of a diatomic molecule versus atomic separation. The parameter R_0 is the equilibrium separation of the atoms. (b) Model of a diatomic molecule whose atoms are bonded by an effective spring of force constant K. The fundamental vibration is along the molecular axis.

[4]In Section 6.6 of Chapter 6 we showed that the effective force constant K is given by the curvature of $U(r)$ evaluated at the equilibrium separation R_0; that is, $K = \partial^2 U/\partial r^2|_{R_0}$.

The kinetic energy of the pair is $(p_1^2/2m_1) + (p_2^2/2m_2)$, but some of this represents translational energy for the molecule as a whole. To isolate the vibrational component, we examine the problem in the center-of-mass (CM) frame, where the total momentum of the molecule is zero. In the CM, the atomic momenta are equal in magnitude but oppositely directed—that is, $p_2 = -p_1$—and the kinetic energy (now of vibration only) is simply

$$KE_{vib} = p_1^2 \left(\frac{1}{2m_1} + \frac{1}{2m_2} \right) = \frac{p_1^2}{2\mu}$$

where μ is again the reduced mass defined in Equation 11.3.

The preceding relations describe a one-dimensional oscillator with vibration coordinate $\xi = \xi_1 - \xi_2$. The Schrödinger equation for this case is

$$-\frac{\hbar^2}{2\mu} \frac{d^2\psi}{d\xi^2} + \tfrac{1}{2}K\xi^2 \psi(\xi) = E_{vib}\psi(\xi) \tag{11.7}$$

which is that for the quantum oscillator studied in Chapter 6, with x there replaced by ξ. The allowed energies of vibration are the oscillator levels (compare Equation 6.29),

Allowed energies for vibration

$$E_{vib} = (v + \tfrac{1}{2})\hbar\omega \qquad v = 0, 1, 2, \ldots \tag{11.8}$$

where v is an integer called the **vibrational quantum number.** The angular frequency ω is the classical frequency of vibration and is related to the force constant K by

$$K = \mu\omega^2 \tag{11.9}$$

In the lowest vibrational state, $v = 0$, we find $E_{vib} = \tfrac{1}{2}\hbar\omega$, the so-called zero-point energy. The accompanying vibration—the zero-point motion—is present even when the molecule is not excited. From Equation 11.8 we see also that the energy difference between any two successive vibrational levels is the same and is given by $\Delta E_{vib} = \hbar\omega$. A typical value for ΔE_{vib} can be found from the entries in Table 11.2 and turns out to be about 0.3 eV.

An energy level diagram for the vibrational energies of a diatomic molecule is given in Figure 11.8. Normally, most molecules are in the lowest energy state because the thermal energy at ordinary temperatures ($k_B T \approx 0.025$ eV) is insufficient to excite the molecule to the next available vibrational state.

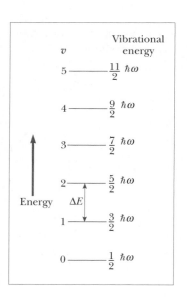

Figure 11.8 Allowed vibrational energies of a diatomic molecule, where ω is the fundamental frequency of vibration given by $\omega = \sqrt{K/\mu}$. Note that the spacings between adjacent vibrational levels are equal.

Table 11.2   Fundamental Vibrational Frequencies and Effective Force Constants for Some Diatomic Molecules

Molecule	Frequency (Hz), $v = 0$ to $v = 1$	Force Constant (N/m)
HF	8.72×10^{13}	970
HCl	8.66×10^{13}	480
HBr	7.68×10^{13}	410
HI	6.69×10^{13}	320
CO	6.42×10^{13}	1860
NO	5.63×10^{13}	1530

From G. M. Barrows, *The Structure of Molecules,* New York, W. A. Benjamin, 1963.

Electromagnetic radiation, however, can stimulate transitions to the first excited level. Such a transition would be accompanied by the absorption of a photon to conserve energy. Once excited, the molecule can return to the lower vibrational state by emitting a photon of the same energy. For molecular vibrations, photon absorption and emission occur in the infrared region of the spectrum. Absorption frequencies for the $v = 0$ to $v = 1$ transitions of several diatomic molecules are listed in Table 11.2, together with the effective force constants K calculated from Equation 11.9. Since larger force constants describe stiffer springs, K indicates the strength of the molecular bond. Notice that the CO molecule, which is bonded by several electrons, is much more rigid than such single-bonded molecules as HCl.

EXAMPLE 11.2   Vibration of the CO Molecule

The CO molecule shows a strong absorption line at the frequency 6.42×10^{13} Hz. (a) Calculate the effective force constant for this molecule.

Solution   The absorption process is accompanied by a molecular transition from the $v = 0$ to the $v = 1$ vibrational level. Since the energy difference between these levels is $\Delta E_{\text{vib}} = \hbar\omega$, the absorbed photon must have carried this much energy. It follows that the photon frequency is just $\Delta E_{\text{vib}}/\hbar = \omega$, the frequency of the CO oscillator! From the information given, we calculate

$$\omega = 2\pi f = 2\pi(6.42 \times 10^{13}) = 4.03 \times 10^{14} \text{ rad/s}$$

Using 12.0 u and 16.0 u for the carbon and oxygen atomic masses, we find $\mu = 6.857$ u $= 1.14 \times 10^{-26}$ kg, as in Example 11.1. Then

$$K = \mu\omega^2 = (1.14 \times 10^{-26})(4.03 \times 10^{14})^2$$
$$= 1.86 \times 10^3 \text{ N/m}$$

Thus, infrared spectroscopy furnishes useful information on the elastic properties (bond strengths) of molecules.

(b) What is the classical amplitude of vibration for a CO molecule in the $v = 0$ vibrational state?

Solution   The total vibrational energy for the molecule is $E_{\text{vib}} = \frac{1}{2}\hbar\omega$. At maximum displacement, the CO molecule has transformed all this into elastic energy of the spring,

$$\tfrac{1}{2}\hbar\omega = \tfrac{1}{2}KA^2$$

where A is the vibration amplitude. Using $K = \mu\omega^2$ and $\mu = 1.14 \times 10^{-26}$ kg, we get

$$A = \left(\frac{\hbar}{\mu\omega}\right)^{1/2} = \left\{\frac{1.055 \times 10^{-34}}{(1.14 \times 10^{-26})(4.03 \times 10^{14})}\right\}^{1/2}$$
$$= 4.79 \times 10^{-12} \text{ m} = 0.00479 \text{ nm}$$

Comparing this with the bond length of 0.113 nm, we see that the vibration amplitude is only about 4% of the bond length.

Exercise 1   Compare the effective force constant for the CO molecule deduced here with that of an ordinary laboratory spring that stretches 0.5 m when a 1.0 kg mass is suspended from it.

Answer 1.86×10^3 N/m for the molecule versus only 19.6 N/m for the laboratory spring.

A proper treatment of molecular rotation and vibration should begin with Schrödinger's wave equation in three dimensions. However, because such problems involve many particles, approximation methods are necessary. For diatomic molecules, the particle mass m is replaced by the reduced mass μ of the system, and the potential $U(\mathbf{r})$ describes the interaction between atoms. Since the force between atoms depends only on their separation, U is spherically symmetric, and the wavefunctions are just spherical harmonics, $Y_\ell^{m_\ell}(\theta, \phi)$, multiplied by solutions, $R(r)$, to the radial wave equation (Section 8.2). In this context, the angular momentum quantum numbers ℓ and m_ℓ specify the rotational state of the molecule. Molecular vibration is described by the radial wave $R(r)$ and reflects the choice of interatomic potential $U(r)$. Near equilibrium, $U(r)$ closely resembles the potential of a spring, leading to vibrational energies characteristic of the quantum oscillator.

In actuality, the atoms of a molecule exert complicated forces on one another; these forces are harmonic only when the atoms are close to their equilibrium positions. More vibrational energy implies larger vibration amplitude, and with it a breakdown of the harmonic approximation—the effective spring must give way to a more accurate representation of the true interatomic force. One way to do this consists of replacing the harmonic potential $\frac{1}{2}K\xi^2$ in Equation 11.7 with the **Morse potential** shown in Figure 11.9a. Near equilibrium ($r = R_0$) the Morse potential is harmonic; further away, $U(r)$ mimics the asymmetry characteristic of interatomic forces, becoming zero when the atoms are widely separated ($r \rightarrow \infty$) but rising sharply as the atoms come close together. In consequence, oscillations take place about an average position $\langle r \rangle$ that *increases* with the energy of vibration. This is the origin of thermal expansion, in which the increased vibrational energy results from raising the temperature of the sample. Unlike the harmonic oscillator, the interval separating successive levels of the Morse oscillator diminishes steadily at higher energies, as illustrated in Figure 11.9b. The Morse oscillator is explored further in Problems 14–17.

Anharmonic effects

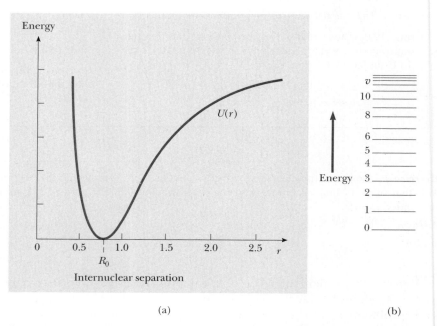

(a)   (b)

Figure 11.9   (a) Potential energy $U(r)$ for the Morse oscillator. In the equilibrium region around R_0, $U(r)$ is harmonic, but rises sharply as the atoms are brought closer together. (b) Allowed energies of vibration for the Morse oscillator. Notice that the separation between adjacent levels *decreases* with increasing energy.

11.3   MOLECULAR SPECTRA

In general, a molecule rotates and vibrates simultaneously. To a first approximation, these motions are independent of each other and the total rotational and vibrational energy of the molecule is given by the sum of Equations 11.5 and 11.8:

$$E_{\text{rot-vib}} = \frac{\hbar^2}{2I_{\text{CM}}}\,\ell(\ell+1) + (v+\tfrac{1}{2})\hbar\omega \qquad (11.10)$$

The levels prescribed by Equation 11.10 constitute the simplest approximation to the **rotation–vibration spectrum** of any molecule. Each level is indexed by the *two* quantum numbers ℓ and v, specifying the state of rotation and vibration, respectively. For each allowed value of the vibrational quantum number v, there is a complete set of rotational levels, corresponding to $\ell = 0$, 1, 2, ... Since successive rotational levels are separated by energies much smaller than the vibrational energy $\hbar\omega$, the rotation–vibration levels of a typical molecule appear as shown in Figure 11.10. Normally, the molecule would take on the configuration with lowest energy, in this case the one designated by $v = 0$ and $\ell = 0$. External influences, such as temperature or the presence of electromagnetic radiation, can change the molecular condition, resulting in a transition from one rotation–vibration level to another. The attendant change in molecular energy must be compensated by absorption or emission of energy in some other form. When electromagnetic radiation is involved, the transitions—referred to as **optical transitions**—are subject to other conservation laws as well, since photons carry both momentum and energy.

Any optical transition between molecular levels with energies E_1 and E_2 must be accompanied by photon emission or absorption at the frequency

$$f = \frac{|E_2 - E_1|}{h} \qquad \text{or} \qquad \Delta E = \pm hf \qquad (11.11)$$

Because hf is the photon energy, this is energy conservation for the system of molecule plus photon. Equation 11.11 expresses a kind of *resonance* between the two; unless photons of the correct frequency (energy) are available, no transition is possible. Similar restrictions apply to other quantities that we know must be conserved in the process of transition. In particular, the initial and final states for an optical transition also must differ by exactly one angular momentum unit:

$$|\ell_2 - \ell_1| = 1 \qquad \text{or} \qquad \Delta\ell = \pm 1 \qquad (11.12)$$

The inference to be drawn from Equation 11.12 is that the photon carries angular momentum in the amount of $\hbar$, that is, the photon is a spin 1 particle with spin quantum number $s = 1$. Equation 11.12 then expresses angular momentum conservation for the system molecule plus photon. Equations 11.11 and 11.12 are the **selection rules for optical transitions.**

For the lower vibrational levels, there is also a restriction on the vibrational quantum number v:

$$|v_2 - v_1| = 1 \qquad \text{or} \qquad \Delta v = \pm 1 \qquad (11.13)$$

Equation 11.13 reflects the harmonic character of the interatomic force rather than any photon property; indeed, for higher vibrational energies the concept

Rotation–vibration spectrum of a diatomic molecule

Figure 11.10 The rotation–vibration levels for a typical molecule. Note that the vibrational levels are separated by much larger energies so that a complete rotational spectrum can be associated with each vibrational level.

of an effective spring joining the atoms is inaccurate, and Equation 11.13 ceases to be valid.

Selection rules greatly restrict the number of photon frequencies or wavelengths observed in molecular spectra, since transitions are prohibited unless all rules are obeyed simultaneously.[5] For instance, a pure rotational transition normally would not be observed, since this requires $\Delta v = 0$ in violation of Equation 11.13. In the same way, a pure vibrational transition ($\Delta \ell = 0$) is forbidden, and we conclude that optical transitions usually involve both molecular vibration and rotation.

The spectrum of a particular molecule can be predicted by considering a collection of such molecules, initially undisturbed. At ordinary temperatures there is insufficient thermal energy to excite any but the $v = 0$ vibrational mode, although the molecules will be in various states of rotation. Since a pure rotational transition is forbidden, optical absorption must result from transitions in which v increases by one unit but ℓ either increases or decreases, also by one unit (Figure 11.11a). Therefore, the molecular absorption spectrum consists of two sequences of lines, represented by $\Delta \ell = \pm 1$, with $\Delta v = +1$ for both cases. The energies of the absorbed photons are readily calculated from Equation 11.14:

$$\Delta E = \hbar \omega + \frac{\hbar^2}{I_{CM}} (\ell + 1) \qquad \ell = 0, 1, \ldots \qquad (\Delta \ell = +1)$$

$$\Delta E = \hbar \omega - \frac{\hbar^2}{I_{CM}} \ell \qquad \ell = 1, 2, \ldots \qquad (\Delta \ell = -1)$$

$$(11.14)$$

Here ℓ is the rotational quantum number of the initial state. The first of Equations 11.14 generates a series of equally spaced lines at frequencies *above* the characteristic vibration frequency ω, and the second generates a series *below* this frequency. Adjacent lines are separated in (angular) frequency by the fundamental unit $\hbar / I_{CM}$. Notice that ω itself is excluded, since ℓ cannot be zero if the transition is one for which ℓ decreases ($\Delta \ell = -1$). Figure 11.11b shows the expected frequencies in the absorption spectrum for the molecule; these same frequencies appear in the emission spectrum.

The absorption spectrum of the HCl molecule shown in Figure 11.12 follows this pattern very well and reinforces our model. However, one peculiarity is apparent: Each line in the HCl spectrum is split into a doublet. This doubling occurs because the sample is a mixture of two chlorine isotopes, ^{35}Cl and ^{37}Cl, whose different masses give rise to two distinct values for I_{CM}. Notice, too, that not all spectral lines appear with the same intensity, because even the allowed transitions occur at different *rates* (number of photons absorbed per second). Transition rates are governed chiefly by the populations of the initial and final states, and these depend on the degeneracy of the levels as well as the temperature of the system.

---

[5]The selection rules given are for harmonic oscillations and rotations of a rigid rotor. In practice, violations of these rules are observed for predictable reasons, such as anharmonicity or rotation–vibration coupling.

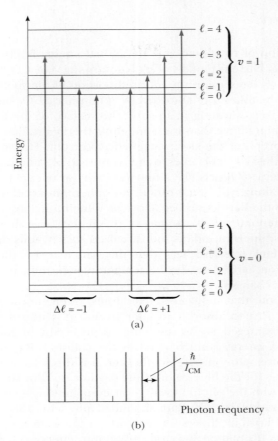

(a)

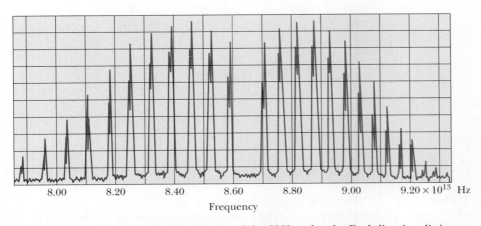

(b)

Figure 11.11 (a) Absorptive transitions between the $v = 0$ and $v = 1$ vibrational states of a diatomic molecule. The transitions obey the selection rule $\Delta\ell = \pm 1$ and fall into two sequences: those for which $\Delta\ell = +1$ and those for which $\Delta\ell = -1$. The transition energies are given by Equation 11.14. (b) Expected lines in the optical absorption spectrum of a molecule. The lines on the right side of center correspond to transitions in which ℓ changes by $+1$, and the lines to the left of center correspond to transitions for which ℓ changes by -1. These same lines appear in the emission spectrum.

Figure 11.12 The absorption spectrum of the HCl molecule. Each line is split into a doublet because chlorine has two isotopes, ^{35}Cl and ^{37}Cl, which have different nuclear masses. (This is an adaptation of data taken by T. Faulkner and T. Nestrick at Oakland University, Rochester, MI.)

387

The excitation of rotational and vibrational energy levels is an important consideration in current models of global warming. For CO_2 molecules, most of the absorption lines are in the infrared portion of the spectrum. Thus, visible light from the Sun is not absorbed by atmospheric CO_2 but instead strikes the Earth's surface, warming it. In turn, the surface of the Earth, being at a much lower temperature than the Sun, emits thermal radiation that peaks in the infrared portion of the electromagnetic spectrum. This infrared radiation is absorbed by the CO_2 molecules in the air instead of radiating out into space. Thus, atmospheric CO_2 acts like a one-way valve for energy from the Sun and is responsible, along with some other atmospheric molecules, for raising the temperature of the Earth's surface above its value in the absence of an atmosphere. This phenomenon is commonly called the "greenhouse effect." The burning of fossil fuels in today's industrialized society adds more CO_2 to the atmosphere. Many scientists fear that this will increase the absorption of infrared radiation, raising the Earth's temperature further, and may cause substantial climatic changes.

Finally, we note that molecules, like atoms, often simply *scatter* radiation, without having first to absorb and later re-emit it. (Indeed, photons of any energy can be scattered, so no resonance is involved.) In **Rayleigh scattering,** the photon energy is unchanged by the collision. Rayleigh scattering is stronger at shorter photon wavelengths, and it is this selectivity in the scattering that accounts for the blue color of the daytime sky. But **Raman scattering** also can occur, with the photon losing—or even gaining—energy in the collision. (Sir Chandrasekhara V. Raman, Indian physicist, 1888–1970, received the 1930 Nobel Prize in Physics for his work on the scattering of light and the effect that bears his name.) Because the photon energy changes, the Raman effect is an example of an inelastic process. For such processes, energy is still conserved overall, with the photon energy loss or gain compensated by a suitable change in the rotational and/or vibrational state of the molecule. Where rotation is involved, the energy exchanged in Raman scattering is consistent with the selection rule $\Delta \ell = \pm 2$. Figure 11.13 depicts a typical Raman process

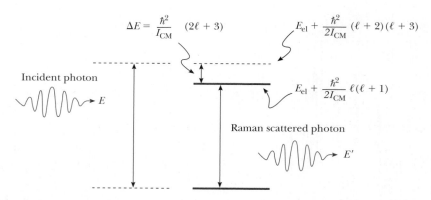

$$\Delta E = \frac{\hbar^2}{I_{CM}} \ (2\ell + 3) \qquad E_{el} + \frac{\hbar^2}{2I_{CM}} \ (\ell + 2)(\ell + 3)$$

Incident photon

E

$$E_{el} + \frac{\hbar^2}{2I_{CM}} \ \ell(\ell + 1)$$

Raman scattered photon

E'

Figure 11.13   An illustration of Raman scattering. In this case an incoming photon with energy E scatters from a molecule and emerges with reduced energy E'. The energy lost by the photon increases the rotational energy of the molecule in accordance with the selection rule $\Delta \ell = 2$. The energy loss translates into a change in photon frequency, the Raman shift, that can be used to probe molecular structure.

where an incoming photon with energy E is scattered and emerges with a reduced energy E', the difference being expended to excite a higher rotational state of the molecule. The excitation energy ΔE is found from Equation 11.10 using ℓ, v for the quantum numbers of the initial state and $\ell + 2$, v for those of the final state ($\Delta \ell = 2$, $\Delta v = 0$). The scattered photon has lower frequency f' compared to the original; the **Raman shift** $f - f'$ is just the excitation energy ΔE divided by h, or

$$f - f' = \frac{\hbar}{2\pi I_{\text{CM}}} (2\ell + 3) \qquad (11.15)$$

Raman shift

Measurements of the Raman shift can be used to determine the moment of inertia of the molecule, which furnishes important clues about the molecular structure. Indeed, Raman spectra serve as a kind of "fingerprint" for molecules, and have been used successfully to identify minerals in lunar soil samples.

Raman scattering is relatively weak, and observable only if the incident radiation is sufficiently intense. With the advent of powerful monochromatic laser sources, Raman spectroscopy has found application in the remote monitoring of pollutants. For example, the scattering produced by a laser beam directed on the plume from an industrial smokestack can be used to monitor the effluent for levels of those molecules that produce recognizable Raman lines.

In the Raman process, the electronic state of the molecule is unchanged. Molecular spectra for which changes occur in the electronic state as well as in the vibrational and/or rotational states of the molecule are called **electronic spectra.** Because the electronic energy levels of a molecule are separated by much larger energies (~ 1 eV) than vibrational (or rotational) levels, electronic transitions give rise to spectral lines that lie in the visible or ultraviolet range. For the same reason, a complete set of vibrational levels may be associated with each electronic level, just as a complete rotational spectrum accompanies each vibrational level.

In cataloging the electronic spectra of molecules, we find that other photon processes can occur that, like Raman scattering, have no counterpart in atomic spectra. Molecules that absorb electromagnetic energy in the visible or near-ultraviolet range may re-emit it at a longer wavelength in a process called **fluorescence.** Fluorescence follows the three-step sequence shown schematically in Figure 11.14. Step 1 is photon absorption that excites the molecule to a more energetic electronic-vibrational state. In step 2, the molecule rids itself of some vibrational energy in collisions with neighboring molecules. This is followed by deexcitation to the original level in step 3 with the emission of a photon having less energy (longer wavelength) than the absorbed photon. The difference in energy, called the **Stokes shift,** is fundamental to the sensitivity of fluorescence techniques because it allows the emitted photons to be detected against a background isolated from the absorbed photons. In a related process called **phosphorescence,** step 2 proceeds through a different mechanism that leaves the molecule in a metastable state. Step 3 then must proceed via a forbidden transition, that is, one that violates the selection rules for optical transitions and so on average takes a much longer time to occur. In some phosphorescent materials, emission is delayed by as much as minutes or even *hours* following absorption. The afterglow associated with phosphorescence is exploited in assorted "glow-in-the-dark" items, such as watch faces and numerous novelty items.

Fluorescence and phosphorescence

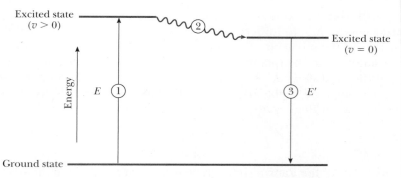

Figure 11.14 Fluorescence is a three-step process. In step 1 a photon with energy E is absorbed, in the process exciting a higher vibrational state of the molecule. This excess energy of vibration is lost in step 2 to collisions with neighboring molecules. The molecule returns to its original state in step 3 by emitting a photon with reduced energy E'. The difference between the absorbed and emitted photon energies is known as the Stoke's shift. In phosphorescence, the final transition is forbidden by selection rules, resulting in photon emission that is delayed by minutes—or even hours—after the initial absorption.

Fluorescence plays a prominent role in our everyday lives, and is increasingly becoming a vital tool for biomedical research. The aptly named fluorescent lamp is a tube coated on the inside surface with a phosphor that fluoresces when exposed to the ultraviolet light produced by passing an electric current through the mercury vapor that fills the tube. And most laundry detergents contain a dye that fluoresces strongly in the blue when exposed to sunlight—this blue cast makes clothes appear cleaner than they really are! Modern fluorescence techniques can accurately measure ion concentrations in living cells down to femtomolar (10^{-15} molar) levels with no adverse effects on cell behavior and can monitor changes in concentrations on a time scale of picoseconds (10^{-12} s). As a result, cell membrane structure and function are actively being studied using fluorescent probes.

11.4 ELECTRON SHARING AND THE COVALENT BOND

In this section we examine the allowed energies and wavefunctions for electrons in a molecule. There are two aspects to the problem: One deals with the complexity of electron motion in the field of several nuclei; the other with the effect of all other electrons on the motion of any one. The two may be divided by first treating a one-electron molecular ion such as H_2^+ and subsequently examining the effects of adding one more electron to give the neutral hydrogen molecule H_2.

The Hydrogen Molecular Ion

The hydrogen molecular ion H_2^+ consists of one electron in the field of two protons (Fig. 11.15). The electron is drawn to each proton by the force of Coulomb attraction, resulting in a total potential energy of

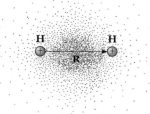

Figure 11.15 The hydrogen molecular ion H_2^+. The lone electron is attracted to both protons by the electrostatic force between opposite charges. The equilibrium separation $|\mathbf{R}|$ of the protons in H_2^+ is about 0.1 nm. (*Adapted from D. Ebbing, General Chemistry 5th ed., Boston, Houghton Mifflin Co., 1996.*)

$$U(\mathbf{r}) = -\frac{ke^2}{|\mathbf{r}|} - \frac{ke^2}{|\mathbf{r} - \mathbf{R}|} \qquad (11.16)$$

In this expression, one proton is assumed to be at the origin of coordinates $\mathbf{r} = 0$, and the other at $\mathbf{R}$. The equilibrium separation $R = |\mathbf{R}|$ of the protons in H_2^+ is about 0.1 nm. The stationary state wavefunctions and energies for the electron in H_2^+ are solutions to the time-independent Schrödinger equation with the potential $U(\mathbf{r})$ of Equation 11.16.

As a guide, it is useful to consider the problem in the two limiting cases $R \rightarrow 0$ and $R \rightarrow \infty$, for which the results are already known. As $R \rightarrow 0$, the two protons coalesce to form a nucleus of helium; consequently, the wavefunctions and energies in this **united atom limit** are those of He^+, given by the formulas of Chapter 8 with atomic number $Z = 2$. In particular, the ground and first excited states are both spherically symmetric waves with energies of -54.4 eV and -13.6 eV, respectively. In the **separated atom limit** $R \rightarrow \infty$, the electron sees only the field of a single proton and the levels are those of atomic hydrogen, -13.6 eV for the ground state and -3.4 eV for the first excited state. Energies for these extreme cases can be marked on the **correlation diagram** of Figure 11.16. The degeneracy of these levels (excluding spin) is important and is given by the numbers in parentheses. Notice especially that the ground state of the united atom is nondegenerate, but the ground state for the separated atoms is doubly degenerate. The extra degeneracy stems from the presence of the second proton in H_2^+: In the separated atom limit, the electron can occupy a hydrogen-like orbital around *either* proton. The same argument shows that all of the atomic levels in the separated atom limit have an extra, twofold degeneracy, corresponding to hydrogen-like wavefunctions centered at the site of either proton. The completed correlation diagram for H_2^+ includes the energy levels at any proton separation R.

To find out what happens at intermediate values of R, consider first the case where R is finite but large, so that the influence of the other proton, though weak, cannot be ignored. This appearance of a second attractive force center means that the electron does not remain with its parent proton indefinitely. Given sufficient time, the electron can tunnel through the intervening potential barrier to occupy an atomic orbital around the far proton. Some time later, the electron tunnels back to the original proton. In principle, the situation is no different from the inversion of the ammonia molecule discussed in Section 7.2: In effect, H_2^+ is a double oscillator, for which the stationary states are not wavefunctions concentrated at a single force center but divided equally between them. The tunneling time increases with proton separation, and may be very long indeed—something like 1 s for protons 1 nm apart (an eternity by atomic standards). But the time itself is inconsequential, since we are interested here in the stationary states that have, so to speak, "all the time in the world" to form.

From the symmetry of H_2^+, the tunneling probability from one proton to the other must be the same in either direction, and so we expect the electron to spend equal amounts of time in the vicinity of each. It follows that the stationary state wave is an equal mixture of atomic-like wavefunctions centered on each proton as, for example,

$$\psi_+(\mathbf{r}) = \psi_a(\mathbf{r}) + \psi_a(\mathbf{r} - \mathbf{R}) \qquad (11.17)$$

Molecular orbitals from atomic wavefunctions

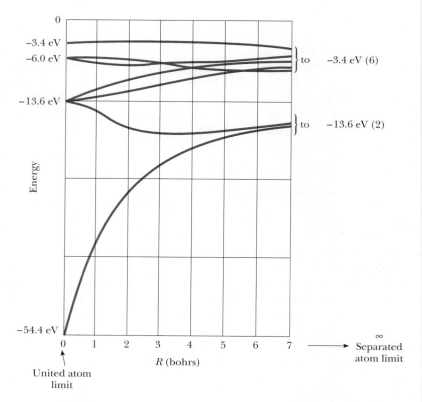

Figure 11.16 Correlation diagram for H_2^+, showing the two lowest electron energy levels as a function of proton separation, in bohrs. [Recall that 1 bohr (a_0) = 0.529 Å.] At $R = 0$, the levels are those of the united atom (ion) He^+, and at $R = \infty$ the levels are those of neutral H. The degeneracy of the various levels (excluding spin) is given by the numbers in parentheses.

The subscript a on the right side of Eq. 11.17 is a collective label for all the quantum numbers needed to specify an atomic state; for the ground state of atomic hydrogen, a is (1, 0, 0), meaning $n = 1$, $\ell = 0$, and $m_\ell = 0$. Equation 11.17 describes the *symmetric* combination of two atomic wavefunctions centered at the proton sites $\mathbf{r} = 0$ and $\mathbf{r} = \mathbf{R}$. The wavefunction and probability density given by $\psi_+(\mathbf{r})$ are sketched in Figure 11.17a and b; both are symmetric about the molecular center located at $\mathbf{r} = \mathbf{R}/2$, or halfway between the two protons.

However, we could also have written the *antisymmetric* form

$$\psi_-(\mathbf{r}) = \psi_a(\mathbf{r}) - \psi_a(\mathbf{r} - \mathbf{R}) \tag{11.18}$$

This again leads to a symmetric probability distribution, but one with a node at the molecular center (Fig. 11.17c and d). In fact, Equations 11.17 and 11.18 both approximate true molecular wavefunctions when R is large, but they describe states having distinctly different energies. In particular, ψ_- has a somewhat higher energy than ψ_+, since the electron in the antisymmetric state is more confined, being relegated largely to the vicinity of one or the other nucleus. As with the particle in a box, this greater degree of confinement

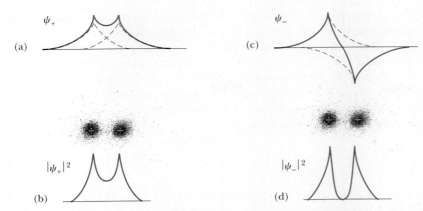

Figure 11.17   The wavefunction (a) and probability density (b) for the approximate molecular wave ψ_+ formed from the symmetric combination of atomic orbitals centered at $\mathbf{r} = 0$ and $\mathbf{r} = \mathbf{R}$. (c) and (d): Wavefunction and probability density for the approximate molecular wave ψ_- formed from the antisymmetric combination of these same orbitals.

comes at the expense of increased kinetic energy. Only in the limit $R \rightarrow \infty$ do the two energies merge to form the (doubly degenerate) atomic level of the separate atoms. As $R \rightarrow 0$, continuity requires the lower of the two energies to approach -54.4 eV, the ground-state energy of the united atom He^+; similarly, the higher energy goes over to the first excited-state energy of He^+, -13.6 eV.

We have come upon the quantum origins of the covalent bond. The energetically preferred state ψ_+ is referred to as the **bonding orbital,** since the electron in this state spends much of its time in the space between the two protons, shuttling to and fro and acting as a kind of "glue" that holds the molecule together. The more energetic state ψ_- is the **antibonding orbital** and decreases the molecular stability. Notice that a bonding–antibonding orbital pair is associated with every orbital of the separated atoms, not just with the ground state.

The energy of the bonding orbitals in H_2^+ can be estimated from Schrödinger's equation. For the electron in H_2^+ this is

$$-\frac{\hbar^2}{2m}\nabla^2\psi + U(\mathbf{r})\psi = E\psi$$

with $U(\mathbf{r})$ the potential energy of the electron in the field of the two protons, as in Equation 11.16. Let us multiply every term by ψ^* and integrate over the whole space. Then the right-hand side becomes just E (assuming ψ is normalized),

$$E\int|\psi|^2\,dV = E$$

leaving for the left-hand side an expression that we can use to compute the value of E:

$$E = \int\psi^*\left[-\frac{\hbar^2}{2m}\nabla^2 - \left\{\frac{ke^2}{|\mathbf{r}|} + \frac{ke^2}{|\mathbf{r}-\mathbf{R}|}\right\}\right]\psi\,dV \qquad (11.19)$$

If ψ is the true molecular wavefunction, this equation furnishes the exact value of particle energy in the state ψ; if ψ is an approximate wavefunction, Equation 11.19 yields an approximation to E.

Substituting for ψ the symmetric combination ψ_+ (suitably normalized—see the Chapter 11 Web Appendix titled "Overlap Integrals of Atomic Wavefunctions") gives a first approximation to the energy of the bonding orbitals in H_2^+. For the case where ψ_a refers to the hydrogen-atom ground state, the molecular energy E_+ can be shown to be

Bonding orbital of H_2^+

$$E_+ = -1 - \frac{2}{1 + \Delta}\left\{\frac{1}{R} - \frac{1}{R}(1 + R)e^{-2R} + (1 + R)e^{-R}\right\} \quad (11.20)$$

with Δ the overlap integral of atomic wavefunctions found in the Chapter 11 Web Appendix:

$$\Delta = \left(1 + R + \frac{R^2}{3}\right)e^{-R}$$

These expressions for E_+ and Δ are given in atomic units, where the rydberg (1 Ry = -13.6 eV) is adopted as the unit of energy to go along with the bohr unit of length.

In the limit of large R, $\Delta \rightarrow 0$ and E_+ approaches the energy of an isolated hydrogen atom, $E_a = -1$ Ry, as it should. As R decreases, E_+ becomes more and more negative, finally reaching -3 Ry at $R = 0$, where the nuclei coalesce. (The correct value of -4 Ry for the ground state of He^+ is not reproduced owing to the approximation inherent in our use of Equation 11.17 for the molecular wavefunction, which fails completely as $R \rightarrow 0$.) Since E_+ decreases steadily with R, the molecule appears to be unstable and should collapse to $R = 0$ under the bonding tendency of the orbiting electron. But this overlooks the Coulomb energy of the protons, which must be included to obtain the total molecular energy. Two protons separated by a distance R repel each other with energy ke^2/R, or $2/R$ Ry. The Coulomb repulsion of the protons offsets the bonding attraction of the electron to stabilize the molecule at that separation for which the total energy is minimal. The total molecular energy given by our model is sketched as a function of R in Figure 11.18. The minimum comes at $R_0 = 2.49$ bohrs, or 0.132 nm, and agrees reasonably well with the observed bond length for H_2^+, 0.106 nm.

At the equilibrium separation of 2.49 bohrs, we find $E_+ = -2.13$ Ry, and a total molecular energy of $E_+ + 2/R_0 = -1.13$ Ry. The negative of this, 1.13 Ry or 15.37 eV, represents the work required to separate the molecule into its constituents and is the dissociation energy introduced in Section 1. The measured dissociation energy for H_+ is about 16.3 eV. The **bond energy,** or the work required to separate H_2^+ into H and H^+, is the difference between the molecular energy at $R = \infty$ and at equilibrium. Our model predicts a bond energy of 15.37 eV $-$ 13.6 eV = 1.77 eV, which is somewhat less than the actual value of 2.65 eV. The difference can be attributed to our use of Equation 11.17 for the molecular wavefunction, an approximation that is best for larger values of R.

To obtain the energy of the antibonding orbitals for H_2^+, we replace ψ in Equation 11.19 with the antisymmetric wave ψ_- of Equation 11.18 (again,

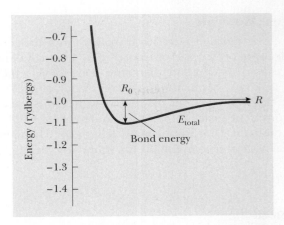

Figure 11.18   The total molecular energy for the bonding orbital of H_2^+, as given by the approximate wavefunction ψ_+ of Equation 11.17. The predicted bond length occurs at the point of stable equilibrium, around $R = 2.5$ bohrs. The predicted bond energy is about 1.77 eV.

suitably normalized). With ψ_a the hydrogen ground state, we find for the energy of the lowest antibonding orbital in H_2^+

$$E_- = -1 - \frac{2}{1 - \Delta} \left\{ \frac{1}{R} - \frac{1}{R} (1 + R)e^{-2R} - (1 + R)e^{-R} \right\} \quad (11.21)$$

Antibonding orbital of H_2^+

For large separations, E_- coincides with E_+ at -1 Ry. With decreasing R, however, E_- steadily increases and always lies above E_+. In particular, E_- for any finite value of R is higher than the energy of $H + H^+$ in isolation, and no stable molecular ion will be formed in this state. Inclusion of the nuclear repulsion only enhances the instability. This is the antibonding effect for H_2^+.

Although Eq. 11.21 seriously overestimates the true molecular energy for small values of R, the conclusion reached about the instability of this state is, nonetheless, correct. The true antibonding orbital energy approaches -1 Ry in the united atom limit and exhibits a broad, shallow minimum near $R = 3$ bohrs (see Fig. 11.16). Equilibrium cannot be sustained, however, when the Coulomb repulsion of the nuclei is included; that is, the curve of total molecular energy shows no minimum for the antibonding state.

The previous discussion of bonding and antibonding orbitals for H_2^+ exemplifies the complexity of covalent bonding in real systems. To illustrate the fundamental ideas without incurring all the mathematical "baggage" that accompanies real-world applications, simplified models are often employed along with numerical methods of solution. Common to all such models is a potential energy having two (or more) points of stable equilibrium (attractive force centers). To explore the allowed energies and wavefunctions in these cases, go to http://info.brookscole.com/mp3e, select QMTools Simulations → Two-Center Potentials (Tutorial), and follow the on-site instructions. You will also find there specific applications to the covalent bond in diatomic molecules composed of like atoms.

The Hydrogen Molecule

The addition of one more electron to H_2^+ gives the hydrogen molecule, H_2. The second electron provides even more "glue" for covalent bonding and should produce a stronger bond than that seen in H_2^+. Indeed, the bond energy for H_2 is 4.5 eV, compared with 2.65 eV for H_2^+. (Naively, we might have expected the bond energy to double with the number of electrons, but the electrons repel each other—an antibonding effect.) The second electron also results in a measured bond length for H_2 equal to 0.074 nm, which is noticeably shorter than the 0.1 nm found for the single-bonded molecular ion H_2^+.

EXAMPLE 11.3   The Hydrogen Molecular Bond

Estimate the bond length and the bond energy of H_2, assuming that each electron moves independently of the other and is described by the approximate wavefunction of Equation 11.17.

Solution   If the electrons do not interact, the energy of each must be the bonding energy E_+ of Equation 11.20. The total energy of the molecule in this *independent particle model* is then

$$E_{\text{total}} = \frac{2}{R} + 2E_+$$

$$= \frac{2}{R} - 2 - \frac{4}{1 + \Delta}\left\{\frac{1}{R} - \frac{1}{R}(1 + R)e^{-2R} + (1 + R)e^{-R}\right\}$$

By trial and error, we discover that the minimum energy occurs at about $R_0 = 1.44$ bohrs, or 0.076 nm, which becomes the bond length predicted by our model. The molecular energy at this separation is $E_0 = -3.315$ Ry. Comparing this with the total energy of the separated hydrogen atoms, -2 Ry, leaves for the bond energy 1.315 Ry, or 17.88 eV. Due to our neglect of electron repulsion, this model grossly exaggerates the bond energy, although it does get the bond length about right. Even the latter agreement is probably fortuitous, however, since Equation 11.17 is not expected to be accurate even for noninteracting electrons at nuclear separations as small as 1.5 bohrs.

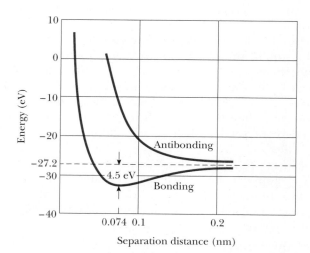

Figure 11.19   Total molecular energy for the bonding and antibonding orbitals of H_2. For both electrons to be in the bonding orbital, their spins must be opposite. The bond energy for H_2 is 4.5 eV and the bond length is 0.074 nm. Since the energy of the antibonding orbital exceeds that of the isolated H atoms, no stable molecule can be formed in this state.

So far we have not recognized that the two electrons in H_2, being identical fermions, are subject to the exclusion principle. If both electrons occupy the bonding orbital of H_2—as we have been assuming—*their spins must be opposite.* If their spins were parallel, one electron would be forced into the antibonding orbital, where it has higher energy. Figure 11.19 shows the electronic energy of the molecule for the two cases. These curves were obtained from laborious calculations that account in an approximate way for the effect of electron–electron repulsion. The results imply that no stable molecule is formed in the parallel spin case.

11.5   BONDING IN COMPLEX MOLECULES

O P T I O N A L

The bonds we found in H_2^+ and H_2 are known as **sigma-type** molecular bonds, denoted σ. They arise from the overlap of atomic s states and are characterized by an electron density that is axially symmetric about the line joining the two atoms. Bonding between atoms having more electrons often involves the overlap of p states or other atomic orbitals and leads to other types of molecular bonds with their own distinctive characteristics.

Consider the nitrogen molecule, N_2, the simplest molecule to exhibit bonds other than the σ type. The electron configuration of the nitrogen atom is $1s^2 2s^2 2p^3$. The molecular bond in N_2 is due primarily to the three valence electrons in the $2p$ subshell; the inner s electrons are too tightly bound to their parent atoms to participate in the sharing necessary for bond formation. The $2p$ subshell is made up of three atomic orbitals ($m_\ell = 1, 0, -1$), and each of these gives rise to a bonding and an antibonding orbital for the molecule. The N_2 molecule has six valence electrons to accommodate, three from each N atom. With two electrons in each of the three bonding orbitals, the N_2 molecule is especially stable, with a bond energy of 9.8 eV and a bond length of 0.11 nm.

Of the three bonding–antibonding orbital pairs in N_2, however, only one is a sigma bond. Recall that atomic p states are lobed structures with highly directional characteristics. The electron density in the p_z orbital ($m_\ell = 0$) is concentrated along the z-axis; similarly, the p_x and p_y orbitals are marked by electron densities along the x- and y-axes, respectively (Fig. 8.13). (The p_x and p_y orbitals are not eigenstates of L_z, but are formed from the $m_\ell = \pm 1$ states according to the prescription of Equation 8.50 in Chapter 8). When two N atoms are brought together, one of the three axes—say the z-axis—will become the molecular axis for N_2 in order to maximize atomic overlaps and produce the strongest bond. This is the sigma bond, because it has axial symmetry. The p_x orbitals, however, also overlap to form a different kind of bond—the **pi bond** (π)—which has a plane of symmetry in the nodal plane of the p orbitals. The same is true of the p_y orbitals. Figure 11.20 illustrates the different kinds of bonds in the N_2 molecule. The pi bonds are weaker than the sigma bond because they involve less electron overlap.

Sigma and pi bonds for N_2

So far we have been discussing only molecules formed from like atoms. These *homonuclear* molecules exemplify the pure covalent bond. The joining of two different atomic species to form a *heteronuclear* molecule produces polar covalent bonds. The hydrogen fluoride molecule, HF, is a good example. The F atom has nine electrons in the configuration $1s^2 2s^2 2p^5$. Of the five $2p$ electrons, four completely fill two of the $2p$ orbitals, leaving one $2p$ electron to be shared with the H atom (the filled orbitals are especially stable and do not significantly affect the molecular bonding in HF). When the two atoms are brought together, the $2p$ orbital of F overlaps the $1s$ orbital of H to form a bonding–antibonding orbital pair for the HF

Bonding in heteronuclear molecules

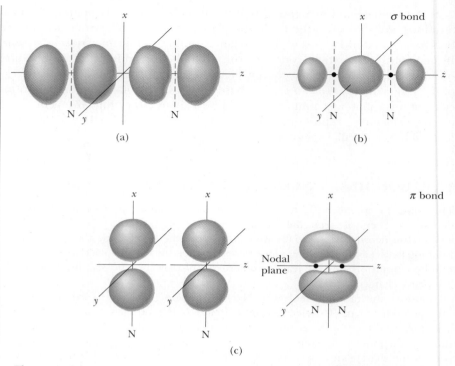

Figure 11.20 (a) and (b): Formation of a sigma bond in N_2 from the overlap of the $2p_z$ orbitals on adjacent N atoms. (c) Formation of a pi bond by overlap of the $2p_x$ orbitals on adjacent N atoms. A similar bond is formed by overlap of the $2p_y$ orbitals.

molecule. The result is an **s–p bond,** with both electrons occupying the more stable bonding orbital. The s–p bond in HF is polar since the shared electrons spend more time in the vicinity of the F atom due to its high electronegativity. Equivalently, the bond is partly ionic, and the HF molecule shows a permanent electric dipole moment. The bond energy for HF is 5.90 eV, and the bond length is 0.092 nm.

The case of HF suggests correctly that purely covalent bonds are found only in homonuclear molecules; heteronuclear molecules always form bonds with some degree of ionicity, as measured by a dipole moment. The water molecule, H_2O, is another example of an s–p-bonded structure with a dipole moment, as is the ammonia molecule, NH_3. In NH_3, each of the three $2p$ electrons in the N atom forms an s–p bond with an H atom. Since the p_x, p_y, and p_z atomic orbitals are directed along mutually perpendicular axes, we would expect the three s–p bonds to be at right angles. In fact, the measured bond angle in NH_3 is somewhat larger—about 107°—because of the electrostatic repulsion of the H nuclei.

In closing, we mention briefly the bonds formed by the carbon atom. These bonds result from **s–p hybridization,** a concept that accounts for the almost endless variety of organic compounds. The C atom has six electrons in the configuration $1s^2 2s^2 2p^2$. We might conclude that only the two $2p$ electrons are prominent in molecular bonding and that carbon is divalent. The existence of hydrocarbons like CH_4 (methane), however, shows that the C atom shares all four of its second-shell electrons, suggesting that the binding energy of the $2s$ electrons in the carbon atom is not much different from that of the $2p$ electrons. But most surprising is the fact that all four bonds are equivalent! There are not two s–p bonds and two s–s bonds as we might have anticipated, but four structurally identical molecular bonds. These bonds arise from the overlap of the $1s$ orbital in H with atomic orbitals in C formed from a *mixture* of carbon $2s$ and $2p$ orbitals. In CH_4 these mixed, or hybrid, orbitals are represented by wavefunctions such as

$$\psi = \psi_{2s} + [\psi_{2p}]_x + [\psi_{2p}]_y + [\psi_{2p}]_z \qquad (11.22)$$

Other combinations arise by subtracting, rather than adding, the individual wavefunctions, but all mix one $2s$ and three $2p$ atomic orbitals to give four sp^3 *hybrids*. In other carbon compounds, hybridization may involve only one or two of the $2p$ orbitals; these are described as sp and sp^2 hybrids, respectively. It is this complexity that gives rise to the rich variety of organic materials.

SUMMARY

Two or more atoms may combine to form molecules because of a net attractive force between them. The resulting molecular bonds are classified according to the bonding mechanisms and are of the following types:

1. **Ionic bonds.** Certain molecules form ionic bonds because of the Coulomb attraction between oppositely charged ions. Sodium chloride (NaCl) is one example of an ionically bonded molecule.
2. **Covalent bonds.** The covalent bond in a molecule is formed by the sharing of valence electrons of its constituent atoms. For example, the two electrons of the H_2 molecule are equally shared between the nuclei.
3. **van der Waals bonds.** This is a weak electrostatic bond between molecules or atoms that do not form ionic or covalent bonds. It is responsible for the condensation of inert gas atoms and nonpolar molecules into the liquid phase.
4. **Hydrogen bonds.** This type of bonding corresponds to the attraction of two negative ions by an intermediate hydrogen atom (a proton).

The energy of a molecule in a gas consists of contributions from the electronic interactions, the translation of the molecule, rotations, and vibrations. The allowed values of the rotational energy of a diatomic molecule are given by

$$E_{\text{rot}} = \frac{\hbar^2}{2I_{\text{CM}}} \ell(\ell + 1) \qquad \ell = 0, 1, 2, \ldots \qquad (11.5)$$

where I_{CM} is the moment of inertia of the molecule about its center of mass and ℓ is an integer called the **rotational quantum number.**

The allowed values of the vibrational energy of a diatomic molecule are given by

$$E_{\text{vib}} = (v + \tfrac{1}{2})\hbar\omega \qquad v = 0, 1, 2, \ldots \qquad (11.8)$$

where v is the **vibrational quantum number.** The quantity ω is the classical frequency of vibration and is related to μ, the reduced mass of the molecule, and K, the force constant of the effective spring bonding the molecule, by the relation $\omega = \sqrt{K/\mu}$.

The internal state of motion of a molecule is some combination of rotation and vibration. Any change in the molecular condition is described as a transition from one rotation–vibration level to another. When accompanied by the emission or absorption of photons, these are called **optical transitions.** Besides conserving energy, optical transitions must conform to the **selection rules**

$$\Delta\ell = \pm 1 \qquad \text{and} \qquad \Delta v = \pm 1 \qquad (11.12, 11.13)$$

In the most common case, the absorption spectrum of a diatomic molecule consists of *two* sequences of lines, corresponding to $\Delta\ell = \pm 1$, with $\Delta v = +1$ for both sequences. Measurements made on such spectra can be used to determine the length and strength of the molecular bonds.

In **Raman scattering,** the incident and emergent photons have different energies, with the discrepancy attributed to a change in the rotation–vibration state of the scattering molecule. The colliding photon can gain or lose energy in this process, depending on whether energy is delivered to or extracted from the rotation–vibration state of the molecule.

Other inelastic photon processes also can occur with molecules. Molecules that absorb electromagnetic energy in the visible or near-ultraviolet range may re-emit it at a longer wavelength in a process called **fluorescence.** In this process, the energy absorbed in vibrational form is dissipated through collisions with other molecules, leaving less energy for the emitted photon. In a related process called **phosphorescence,** the transition giving rise to the emitted photon has an unusually long lifetime, with the result that the emitted radiation is delayed minutes—or even hours—following the initial absorption.

The electronic states of a molecule are classified as **bonding** or **antibonding.** A bonding state is one for which the electron density is large in the space between the nuclei. An electron in a bonding orbital can be thought of as shuttling rapidly from one nucleus to the other, drawing them together as a kind of "glue." The bonding state is energetically preferred over the antibonding state, where the electron spends more of its time outside the bonding region. In a diatomic molecule composed of like atoms (homonuclear molecule), each atomic orbital gives rise to a bonding–antibonding orbital pair for the molecule.

SUGGESTIONS FOR FURTHER READING

1. Molecular bonding is discussed extensively at an introductory level in Chapter 24 of J. Brackenridge and R. Rosenberg, *The Principles of Physics and Chemistry*, New York, McGraw-Hill Book Company, Inc., 1970. This work also includes a number of nice illustrations of molecular bonds in various compounds.
2. For an in-depth treatment of the bonding in H_2^+ and the H_2 molecule, see J. C. Slater, *Quantum Theory of Molecules and Solids*, Vol. 1, New York, McGraw-Hill Book Company, Inc., 1963.
3. An elaborate discussion of chemical bonding in organic compounds may be found in Chapter 9 of K. Krane, *Modern Physics*, 2nd ed., New York, John Wiley and Sons, Inc., 1996.

QUESTIONS

1. List three ways (modes) a diatomic molecule can store energy internally. Which of the three modes is easiest to excite, that is, requires the least energy? Which requires the most energy?
2. How do the effective force constants for the molecules listed in Table 11.2 compare with those found for typical laboratory springs? Comment on the significance of your findings.
3. Describe hybridization, why it occurs, and what it has to do with molecular bonding. Give an explicit example of where it occurs.
4. Discuss the mechanisms responsible for the different types of bonds that can occur to form stable molecules.
5. Explain the role of the Pauli exclusion principle in determining the stability of molecules, using H_2 and the (hypothetical) species H_3 as examples.

6. Discuss the relationship between tunneling and the covalent bond. From the viewpoint of possible chemical species, what would be the similarities and differences between the world in which we live and one devoid of tunneling?

7. Distinguish between the dissociation energy of a molecule and its bond energy. Which of the two is expected to be greater?

8. In treating the rotational levels of a diatomic molecule, why do we ignore rotation about the internuclear line? Can you think of a case involving a triatomic molecule in which similar considerations might apply?

9. Explain why the noble gases tend to be monatomic rather than diatomic.

PROBLEMS

11.1 Bonding Mechanisms: A Survey

1. Potassium iodide can be taken as a medicine to reduce radiation dosage to the thyroid gland, before or after exposure to radioactive iodine. In the potassium iodide molecule, assume that the atoms K and I bond ionically by the transfer of one electron from K to I. (a) The ionization energy of K is 4.34 eV, and the electron affinity of I is 3.06 eV. What energy is needed to transfer an electron from K to I, to form K^+ and I^- ions from neutral atoms? (b) A model potential energy function for the KI molecule is the Lennard–Jones (12, 6) potential:

$$U(r) = 4\mathcal{E} \left[\left(\frac{\sigma}{r} \right)^{12} - \left(\frac{\sigma}{r} \right)^{6} \right] + E_a$$

where r is the internuclear separation distance, and σ and $\mathcal{E}$ are adjustable constants. E_a is added to ensure the correct asymptotic behavior at large r and is the activation energy calculated in (a). At the equilibrium separation distance $r = r_0 = 0.305$ nm, $U(r)$ is a minimum, and $dU/dr = 0$. $U(r_0)$ is the negative of the dissociation energy: $U(r_0) = -3.37$ eV. Use the experimental values for the equilibrium separation and dissociation energy of KI to determine σ and $\mathcal{E}$. (c) Calculate the force needed to rupture the molecule.

10.2 Molecular Rotation and Vibration

2. Use the data in Table 11.2 to calculate the reduced mass of the NO molecule; then compute a value for μ using Equation 11.3. Compare the two results.

3. The CO molecule undergoes a rotational transition from the $\ell = 1$ level to the $\ell = 2$ level. Using Table 11.1, calculate the values of the reduced mass and the bond length of the molecule. Compare your results with those of Example 11.1.

4. Use the data in Table 11.2 to calculate the maximum amplitude of vibration for (a) the HI molecule and (b) the HF molecule. Which molecule has the weaker bond?

5. The $\ell = 5$ to $\ell = 6$ rotational absorption line of a diatomic molecule occurs at a wavelength of 1.35 cm (in the vapor phase). (a) Calculate the wavelength and frequency of the $\ell = 0$ to $\ell = 1$ rotational absorption line. (b) Calculate the moment of inertia of the molecule.

6. The HF molecule has a bond length of 0.092 nm. (a) Calculate the reduced mass of the molecule. (b) Sketch the potential energy versus internuclear separation in the vicinity of $r = 0.092$ nm.

7. The HCl molecule is excited to its first rotational energy level, corresponding to $\ell = 1$. If the distance between its nuclei is 0.1275 nm, what is the angular velocity of the molecule about its center of mass?

8. The $v = 0$ to $v = 1$ vibrational transition of the HI molecule occurs at a frequency of 6.69×10^{13} Hz. The same transition for the NO molecule occurs at a frequency of 5.63×10^{13} Hz. Calculate (a) the effective force constant and (b) the amplitude of vibration for each molecule. (c) Explain why the force constant of the NO molecule is so much larger than that of the HI molecule.

9. Consider the HCl molecule, which consists of a hydrogen atom of mass 1 u bound to a chlorine atom of mass 35 u. The equilibrium separation between the atoms is 0.128 nm, and it requires 0.15 eV of work to increase or decrease this separation by 0.01 nm. (a) Calculate the four lowest rotational energies (in eV) that are possible, assuming the molecule rotates rigidly. (b) Find the molecule's "spring constant" and its classical frequency of vibration. (*Hint:* Recall that $U = \frac{1}{2}Kx^2$.) (c) Find the two lowest vibrational energies and the classical amplitude of oscillation corresponding to each of these energies. (d) Determine the longest wavelength radiation that the molecule can emit in a pure rotational transition and in a pure vibrational transition.

10. The hydrogen molecule comes apart (dissociates) when it is excited internally by 4.5 eV. Assuming that this molecule behaves exactly like a harmonic oscillator with classical frequency $\omega = 8.277 \times 10^{14}$ rad/s, find the vibrational quantum number corresponding to its 4.5-eV dissociation energy.

11. As a model for a diatomic molecule, consider two identical point masses m connected by a rigid massless rod of length R_0. Suppose that this molecule rotates about an axis perpendicular to the rod through its midpoint. Use the Bohr quantization rule for angular momentum to obtain the allowed rotational energies in this approximation. Compare your result with the correct quantum mechanical treatment of this model.

12. For the model of a diatomic molecule described in Problem 11, derive an expression for the minimum energy required to excite the molecule into rotation about the *internuclear line*, that is, about the axis joining the two masses. Assume that the masses are uniform spheres of radius r. Apply your result to the hydrogen molecule, H_2, by taking m equal to the proton mass and r equal to the nuclear size, about 10 fm.

13. The rotational motion of molecules has an effect on the equilibrium separation of the nuclei, a phenomenon known as *bond stretching*. To model this effect, consider a diatomic molecule with reduced mass μ, oscillator frequency ω_0, and internuclear separation R_0 when the angular momentum is zero. The effective potential energy for nonzero values of ℓ is then (see Section 8.5)

$$ U_{\text{eff}} = \tfrac{1}{2}\mu\omega_0^2(r - R_0)^2 + \ell(\ell + 1)\,\frac{\hbar^2}{2\mu r^2} $$

(a) Minimize the effective potential $U_{\text{eff}}(r)$ to find an equation for the equilibrium separation of the nuclei, R_ℓ, when the angular momentum is ℓ. Solve this equation approximately, assuming $\ell \ll \mu\omega_0 R_0^2/\hbar$. (b) Near the corrected equilibrium point, R_ℓ, the effective potential again is nearly harmonic and can be written approximately as

$$ U_{\text{eff}} \approx \tfrac{1}{2}\mu\omega^2(r - R_\ell)^2 + U_0 $$

Find expressions for the corrected oscillator frequency ω and the energy offset U_0 by matching U_{eff} and its first two derivatives at the equilibrium point R_ℓ. Show that the fractional change in frequency is given by

$$ \frac{\Delta\omega}{\omega_0} \approx \frac{3\ell(\ell + 1)\hbar^2}{2\mu^2\omega_0^2 R_0^4} $$

14. As an alternative to harmonic interactions, the *Morse potential*,

$$ U(r) = U_0\{1 - e^{-\beta(r - R_0)}\}^2 $$

can be used to describe the vibrations of a diatomic molecule. The parameters R_0, U_0, and β are chosen to fit the data for a particular atom pair. (a) Show that R_0 is the equilibrium separation and that the potential energy far from equilibrium approaches U_0. (b) Show that near equilibrium ($r \approx R_0$) the Morse potential is harmonic, with force constant $K = m\omega^2 = 2U_0\beta^2$. (c) The lowest vibrational energy for the Morse oscillator can be shown to be

$$ E_{\text{vib}} = \tfrac{1}{2}\hbar\omega - \frac{(\hbar\omega)^2}{16U_0} $$

Obtain from this an expression for the dissociation energy of the molecule. (d) Apply the results of parts (b) and (c) to deduce the Morse parameters U_0 and β for the hydrogen molecule. Use the experimental values 573 N/m and 4.52 eV for the effective spring constant and dissociation energy, respectively. (The measured value of R_0 for H_2 is 0.074 nm.)

15. The allowed energies of vibration for the Morse oscillator can be shown to be

$$ E_{\text{vib}} = (v + \tfrac{1}{2})\hbar\omega - (v + \tfrac{1}{2})^2\,\frac{(\hbar\omega)^2}{4U_0} $$

where $v = 0, 1, \ldots$. Obtain from this an expression for the interval separating successive levels of the Morse oscillator, and show that this interval diminishes steadily at higher energies, as illustrated in Figure 11.9b. From your result, deduce an upper limit for the vibrational quantum number v. What is the largest vibrational energy permitted for the Morse oscillator?

16. *The Morse Oscillator Spectrum.* Use the Java applet available at our companion Web site (http://info.brookscole.com/mp3e → QMTools Simulations → Problem 11.16) to display the first three pure vibrational states ($\ell = 0$, $v = 0, 1, 2$) of the H_2 molecule in the Morse oscillator approximation to the interatomic potential. The number of Morse vibrational states is limited. See if you can find the highest-lying pure vibrational state for this case. What is the vibrational quantum number v for this state?

17. Consider higher rotational states of the Morse oscillator described in Problem 16. Use the Java applet referenced there to find the energies of the two lowest rotational levels ($\ell = 1$ and $\ell = 2$) associated with the vibrational ground state of the H_2 molecule in this model. Compare your results with the predictions of Equation 11.10 for this case. Are the rotations and vibrations of the H_2 molecule really independent?

18. An H_2 molecule is in its vibrational and rotational ground states. It absorbs a photon of wavelength 2.2112 μm and jumps to the $v = 1$, $\ell = 1$ energy level. It then drops to the $v = 0$, $\ell = 2$ energy level, while emitting a photon of wavelength 2.4054 μm. Calculate (a) the moment of inertia of the H_2 molecule about an axis through its center of mass, (b) the vibrational frequency of the H_2 molecule, and (c) the equilibrium separation distance for this molecule.

11.4 Electron Sharing and the Covalent Bond

19. A one-dimensional model for the electronic energy of a diatomic homonuclear molecule is described by the potential well and barrier shown in Figure P11.19. In the simplest case, the barrier width is shrunk to zero ($w \to 0$) while the barrier height increases without limit ($U \to \infty$) in such a way that the product Uw approaches a finite value S called the barrier strength. Such a barrier—known as a delta function barrier—can be shown to produce a discontinuous slope in the

wavefunction at the barrier site $L/2$ given by

$$\frac{d\psi}{dx}\bigg|_{L/2^+} - \frac{d\psi}{dx}\bigg|_{L/2^-} = \frac{2mS}{\hbar^2}\,\psi(L/2)$$

Solve the wave equation in the well subject to this condition to obtain expressions for the energies of the ground state and first excited state as functions of the barrier strength S. Examine carefully the limits $S \rightarrow 0$ and $S \rightarrow \infty$ and comment on your findings.

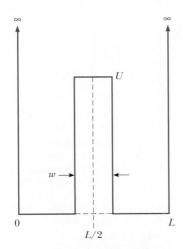

Figure P11.19

20. In Section 11.4 it is stated that the approximate electronic energy for the bonding state of H_2^+ as a function of the internuclear separation R is (in atomic units)

$$E_+ = -1 - \frac{2}{1+\Delta}\left\{\frac{1}{R} - \frac{1}{R}(1+R)e^{-2R} + (1+R)e^{-R}\right\}$$

To this we add the Coulomb energy of the two protons, $2/R$, to get the total energy of this molecular ion, E_{total}. (a) Write a simple computer program to evaluate E_{total} for any given value of R. Use your program to verify that the equilibrium separation of the protons in H_2^+ is $R_0 = 2.49$ bohrs, according to this model. (b) Use this model to predict a value for the effective spring constant that governs the vibrations of the H_2^+ molecular ion, and compare your result with the values reported in Table 11.2. (*Hint:* Use the well-known finite difference approximation to the second derivative,

$$\frac{d^2f}{dx^2} \approx (1/\Delta x)^2\{f(x+\Delta x) - 2f(x) + f(x-\Delta x)\}$$

where Δx is a small increment.)

21. Repeat the calculations of Problem 20 for the case of the neutral hydrogen molecule, H_2. Take for the molecular energy of H_2

$$E_{\text{total}} = \frac{2}{R} + 2E_+$$

where E_+ is the bonding energy of H_2^+ given in Problem 20. Compare your result for the effective spring constant with the experimental value for H_2, about 573 N/m.

22. *Modeling a Heteronuclear Molecule.* Before attempting this problem, review the on-line tutorial at http://info.brookscole.com/mp3e $\rightarrow$ QMTools Simulations $\rightarrow$ Two-Center Potentials (Tutorial) and the application to covalent bonding presented there. The application references a Java applet that uses a divided square well to model the potential energy of an electron in a diatomic molecule. The defaults portray "atomic" wells 100 eV high and 2.00 Å wide, separated by a barrier 0.500 Å wide and 100 eV high. Moving the divider off-center destroys the identity of the atomic wells to either side and transforms the model to one of a heteronuclear molecule. Without changing the barrier width or height, reposition the divider to leave atomic wells with widths 1.75 Å and 2.25 Å. Find the bonding–antibonding orbital pair that derives from the atomic ground states for this heteronuclear molecule. Contrast the energy-splitting and molecular wavefunctions with those found for a homonuclear molecule, specifically noting any similarities or differences.

23. The Java applet referenced in Problem 22 can be used to examine the behavior of an electron initially confined to one side of a divided square well. Here we use the applet defaults to model a homonuclear molecule. Take for the initial state $\Psi(x, 0)$ a sum of the bonding and antibonding orbitals that derive from the atomic ground state. How much time, in seconds, elapses before the electron can be said to have crossed the divider? The reciprocal of this time is the crossing frequency. Compare this crossing frequency with the frequency $\Delta E/h$ representing the energy splitting of the bonding and antibonding states in the mixture. How would your results change if the width and/or height of the central divider were increased?

12

The Solid State

Chapter Outline

Matter in the solid state has been a subject of enormous fascination since the beginnings of civilization. Primitive people were attracted to the solid state by its beauty, as in radiant, symmetric gemstones, and by its utility, as in metal tools. These two attributes, *utility* and *beauty*, are just as important in physics today. Industrial applications have made solid-state physics, or condensed-matter physics, the largest subfield of physics, as evidenced by the number of pages allocated in physics journals and the number of physicists employed in this field. The beautiful symmetry and regularity of crystalline solids have both allowed and stimulated rapid progress in the physics of crystalline solids in the 20th century. Interestingly, although rapid theoretical progress has occurred with the most random (gases) and the most regular (crystalline solids) atomic arrangements, much less has been done with liquids and amorphous (irregular) solids until quite recently. Applications such as solar cells, memory elements, fiber-optic waveguides, and xerography have driven the relatively recent rush of interest in low-cost amorphous materials.

In this chapter we first describe how molecules combine to form crystalline and amorphous solids. We then introduce one of the simplest classical models of conductors, the free electron gas model, to gain physical insight into the processes of electrical and thermal conduction. Next we consider the quantum theory of metals to explain the deficiencies of the classical model. We describe the band theory of solids to explain the differences between insulators, conductors, and semiconductors and include a brief discussion of *p-n* junctions, semiconductor devices, and superconductivity. Finally we discuss the general principles of lasers and some of the specifics of gas and semiconductor lasers.

12.1   BONDING IN SOLIDS

A crystalline solid consists of a large number of atoms arranged in a regular array, forming a periodic structure. The bonding schemes for molecules discussed in Chapter 11 are also appropriate for describing the bonding mechanisms in solids. For example, the ions in the NaCl crystal are ionically bonded, while the carbon atoms in the diamond structure form covalent bonds. Another type of bonding mechanism is the metallic bond, which is responsible for the cohesion of copper, silver, sodium, and other metals. Finally, the weakest type of bonding, van der Waals bonding, is responsible for the cohesion of organic solids and rare gas crystals. Refer to Section 11.1 for a survey of molecular bonding mechanisms.

Ionic Solids

Many crystals form by ionic bonding, where the dominant effect is the Coulomb interaction between the ions. Consider the NaCl crystal shown in Figure 12.1, where each Na$^+$ ion has six nearest-neighbor Cl$^-$ ions and each

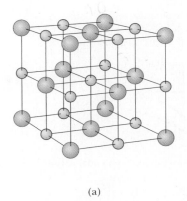

(a)

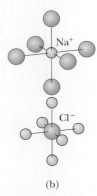

(b)

Figure 12.1   (a) The crystal structure of NaCl. The unit cell shown here represents four NaCl units and is the smallest repeating unit having the symmetry of the crystal. (b) In the NaCl structure, each positive sodium ion is surrounded by six negative chlorine ions, and each chlorine ion is surrounded by six sodium ions.

Cl^- ion has six nearest-neighbor Na^+ ions. Each Na^+ ion is attracted to the six Cl^- ions. The corresponding attractive potential energy is $-6ke^2/r$, where r is the Na^+–Cl^- separation and k is the Coulomb constant $k = 1/4\pi\epsilon_0$. In addition, there are 12 Na^+ ions at a distance of $\sqrt{2}r$ from the Na^+, which produce a weaker repulsive force on the Na^+ ion. Furthermore, beyond these 12 Na^+ ions one finds more Cl^- ions that produce an attractive force, and so on. The net effect of all these interactions is an attractive electrostatic potential energy,

$$U_{\text{attractive}} = -\alpha k \frac{e^2}{r} \tag{12.1}$$

where α is a dimensionless number called the **Madelung constant.** The value of α depends only on the particular type of crystal lattice or structure. For example, $\alpha = 1.7476$ for the NaCl structure. When the constituent ions of a crystal are brought close together, the electrons in closed shells begin to overlap, which results in a repulsion between the closed shells that is partly electrostatic in origin and partly a result of the Pauli exclusion principle. This introduces a repulsive potential energy term into the expression for the total potential energy of the crystal, which we write as B/r^m, where m is an integer on the order of 10. The total potential energy per ion pair of the crystal is therefore of the form

$$U_{\text{total}} = -\alpha k \frac{e^2}{r} + \frac{B}{r^m} \tag{12.2}$$

(We do not sum over neighbors in the case of the repulsive force because this force is negligible when ions are separated by a distance larger than the equilibrium separation, r_0.) Figure 12.2 is a plot of the total potential energy per ion pair of the crystal versus ion separation. The potential energy has its

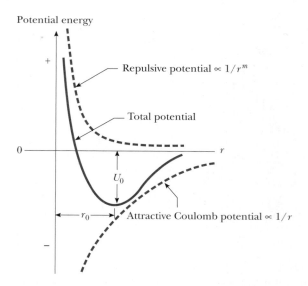

Figure 12.2 Potential energy per ion pair versus ion separation for an ionic solid. U_0 is the ionic cohesive energy, and r_0 is the equilibrium separation between ions.

minimum value U_0 at the equilibrium separation where $r = r_0$. It is left to an end-of-chapter problem (Problem 1) to show that B may be expressed in terms of r_0 and m and that the minimum energy U_0 is

$$U_0 = -\alpha k \frac{e^2}{r_0} \left(1 - \frac{1}{m}\right) \qquad (12.3)$$

Atomic cohesive energy

The absolute value of U_0, called the **ionic cohesive energy** of the solid, is the energy required to pull the solid apart into a collection of infinitely separated positive and negative ions. The measured ionic cohesive energy of NaCl is 7.84 eV per ion pair or about 760 kJ/mol. In calculating the **atomic cohesive energy,** which is the energy needed to pull the solid apart into a collection of infinitely separated Na and Cl neutral atoms, we must keep in mind that one gains 5.14 eV in going from Na^+ to Na and one must supply 3.61 eV in order to convert Cl^- to Cl. Thus the atomic cohesive energy of NaCl per atom pair may be computed as

$$+7.84 \text{ eV} - 5.14 \text{ eV} + 3.61 \text{ eV} = +6.31 \text{ eV}$$

The atomic cohesive energy is an important comparative measure of the strengths of differently bonded solids. Table 12.1 lists measured atomic cohesive energies per atom pair and melting points for a few ionic compounds.

Ionic crystals have the following general properties:

- They form relatively stable and hard crystals.
- They are poor electrical conductors because there are no available free electrons.
- They have fairly high melting and boiling points since appreciable thermal energy must be added to the crystal to overcome the large cohesive energy.
- They are transparent to visible radiation but absorb strongly in the infrared region. This occurs because the electrons form such tightly bound shells in ionic solids that visible radiation does not contain sufficient energy to promote electrons to the next allowed shell and so is not absorbed. The strong infrared absorption (at 20 to 150 μm) occurs because the vibrations of the rather massive ions have a low natural frequency and experience resonant absorption in the low-energy infrared region.
- They are generally quite soluble in polar liquids such as water. The water molecule, which has a permanent electric dipole moment, exerts an attractive force on the charged ions, which breaks the ionic bonds and dissolves the solid.

Table 12.1   Properties of Some Ionic Crystals

Crystal	Equilibrium Separation (Å)	Atomic Cohesive Energy (eV per atom pair)	Melting Point (K)
LiF	2.01	8.32	1143
NaCl	2.82	6.31	1074
RbF	2.82	7.10	1068
KCl	3.15	6.48	1043
CsI	3.95	5.36	621

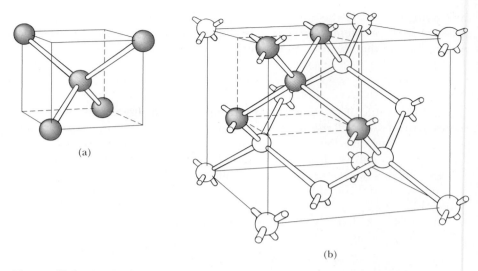

Figure 12.3   (a) Each carbon atom in diamond is covalently bonded to four other carbons and the four outer carbons form a tetrahedron. (b) The crystal structure of diamond, showing the tetrahedral bond arrangement. (*After W. Shockley*, Electrons and Holes in Semiconductors, *New York, Van Nostrand, 1950*)

Covalent Solids

As we found in Chapter 11, the covalent bond is very strong. Solid carbon, in the form of diamond, is a crystal whose atoms are covalently bonded. Because atomic carbon has the electron configuration $1s^2 2s^2 2p^2$, it lacks four electrons with respect to a filled shell ($2p^6$). Hence each carbon atom in diamond bonds covalently to four other carbon atoms to form a stable closed-shell structure.

In the diamond structure, each carbon atom is covalently bonded to four other carbon atoms at the corners of a cube, as illustrated in Figure 12.3a. Figure 12.3b shows the crystal structure of diamond. Note that each carbon atom forms covalent bonds with four nearest-neighbor atoms. The basic structure of diamond is called *tetrahedral* (each carbon atom is at the center of a regular tetrahedron), and the angle between the bonds is 109.5°. Other covalent crystals, such as silicon and germanium, have similar structures.

Table 12.2 gives the properties of some covalent solids. Note that the atomic cohesive energies are greater than for ionic solids, which accounts for the

Table 12.2   Properties of Some Covalent Crystals

Crystal	Atomic Cohesive Energy (eV/atom)*	Melting Point (K)
C (diamond)	7.37	~4000
SiC	6.15	2870
Si	4.63	1687
Ge	3.85	1211

*Since covalent atomic cohesive energies are given in electron volts per atom, they should be multiplied by 2 for a proper comparison to Table 12.1.

hardness of covalent solids. Diamond is particularly hard and has an extremely high melting point (about 4000 K). In general, covalently bonded solids are very hard, have large bond energies and high melting points, and are good insulators, since the electrons are tightly bound in localized bonds. Because electrons are so tightly bound in diamond and other covalent solids, there is insufficient energy in visible light to raise electrons to excited states. Consequently, many covalent solids do not absorb visible light and so appear transparent.

Metallic Solids

Metallic bonds are generally weaker than ionic or covalent bonds. The valence electrons in a metal are relatively free to move throughout the material. There is a large number of such mobile electrons in a metal, typically one or two electrons per atom. The metal structure can be viewed as a lattice of positive ions surrounded by a "gas" of nearly free electrons (Fig. 12.4). The binding mechanism in a metal is the attractive force between the positive ions and the electron gas.

Metals have an atomic cohesive energy in the range of 1 to 4 eV, which is smaller than the cohesive energies of covalent solids but still large enough to produce strong solids (see Table 12.3). Visible light interacts strongly with the free electrons in metals because these conduction electrons can move with large amplitude in the oscillating electric field of the light wave, both strongly absorbing the light wave and reradiating it. Hence visible light is absorbed and re-emitted quite close to the surface of a metal, which accounts for both the nontransparency to visible light and the shiny nature of metallic surfaces. In addition to the high electrical conductivity of metals produced by the free electrons, the nondirectional nature of the metallic bond allows many different types of metallic atoms to be dissolved in a host metal in varying amounts. The resulting solid solutions, or **alloys,** may be designed to have particular properties, such as high strength, low density, ductility, resistance to corrosion, and so on.

Molecular Crystals

A fourth class of binding can occur even when electrons are not available to participate in bond formation, as in the cases of saturated organic molecules (CH_4) and inert gas atoms with closed electron shells. The weak electric forces at work in this type of bonding include **van der Waals forces** and arise from

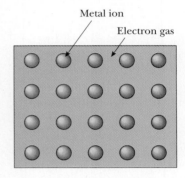

Figure 12.4   The free electron model of a metal. The negative electron gas acts as a kind of glue to hold the positive metal ions together.

Table 12.3   Properties of Some Metallic Crystals

Crystal	Atomic Cohesive Energy (eV/atom)*	Melting Point (K)
Fe	4.28	2082
Cu	3.49	1631
Au	3.81	1338
Ag	2.95	1235
Pb	2.04	874
Zn	1.35	693

*Since metallic atomic cohesive energies are given in electron volts per atom, they should be multiplied by 2 for a proper comparison to Table 12.1.

the attractive force between electric dipoles. (See the discussion of van der Waals bonds in Section 11.1.) Some simple molecules containing hydrogen, such as water, have permanent dipole moments (due to the built-in separation of positive and negative centers of charge) and form relatively strong bonds known as **hydrogen bonds.** Ice, for example, is hydrogen bonded with a cohesive energy of about 0.52 eV/molecule or 12 kcal/mol. (Note that 1 eV/molecule = 23 kcal/mol.) Actually, hydrogen bonding involves more than dipole attractive forces and may be considered a sort of covalent bond where protons are shared rather than electrons. Two molecules without permanent dipole moments also experience fluctuation-induced dipole–dipole attractions, but these van der Waals forces are much weaker. Such weak forces hold together many organic solids with comparatively low melting and boiling points, as well as inert gas crystals. Two typical cases are solid methane, with a cohesive energy of about 0.10 eV/molecule and a melting point of 91 K, and solid argon, with a cohesive energy of 0.078 eV/molecule and a melting point of 84 K.

Amorphous Solids

The apparent simplicity and regularity of crystalline solids suggest that it will be most profitable to focus attention on perfect crystals. This is the direction we shall take in most of this chapter. However, real crystals are far from perfect, since they contain irregularities and impurities that can have profound effects on their strength, conductivity, and other properties. This gives us some motivation for a brief discussion of crystal defects and the nature of amorphous solids.

It is well known that crystals of a given material form when the liquid state of that material is cooled sufficiently. For example, water forms crystals when cooled to 0°C. In general, more perfect crystals form if a liquid is cooled slowly, allowing the molecules to relax gradually to states having minimum potential energy. Conversely, rapid cooling of a liquid causes many dislocations in the resulting lattice structure. If the cooling is rapid enough, almost any liquid can form an **amorphous solid** (or glass) without long-range order, as shown in Figure 12.5. Although an amorphous solid has no long-range order, there is extensive short-range order in that bond lengths and bond

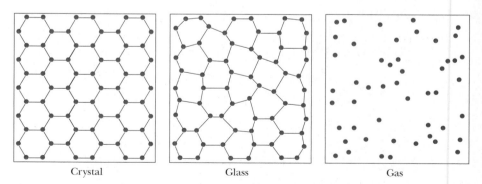

Crystal   Glass   Gas

Figure 12.5 Two-dimensional sketches of the atoms in a crystal, a glass, and a gas. Note that the crystal has equal bond lengths and angles, the glass has a distribution of nearly equal bond lengths and angles, and the gas has a completely random spatial distribution of atoms. (*After R. Zallen*, The Physics of Amorphous Solids, *New York, John Wiley and Sons, 1983*)

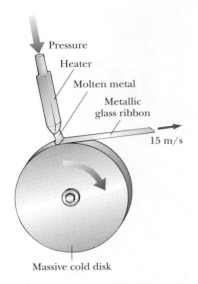

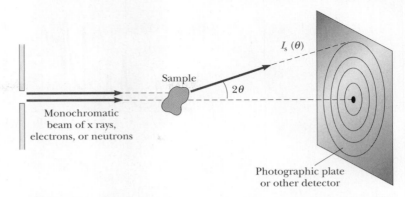

Figure 12.7 A basic experimental setup for determining the distribution of atoms in a powder sample. Rings of constant intensity on the photographic plate arise from the tiny, randomly oriented crystals comprising the powder. (*After R. Zallen*, The Physics of Amorphous Solids, *New York, John Wiley and Sons, 1983*)

Figure 12.6 Preparing an amorphous metal ribbon by melt spinning.

angles for nearest neighbors are nearly the same throughout the solid. Even metals can be made to solidify in an amorphous form (called a **metallic glass**) if the metal in the liquid state can be cooled by 1000 K in about 1 millisecond. Figure 12.6 illustrates one clever technique for achieving cooling rates of 10^6 K/s, known as **melt spinning.**

The main experimental techniques for determining the distribution of atoms in crystalline and amorphous solids are x-ray diffraction, electron diffraction, and neutron diffraction. The primary experimental data are measurements of the scattered intensity $I_s(\theta)$, as sketched in Figure 12.7. Figure 12.8 shows actual photographic records of the electron intensity scattered from amorphous and crystalline iron samples. Note that the regular spacing of atoms over many atomic separations in the crystalline sample leads to many sharp diffraction lines. This is in contrast to the metallic glass sample, in which the distribution of atomic separations and bond angles over one or two atomic separations leads to fewer, more diffuse lines.

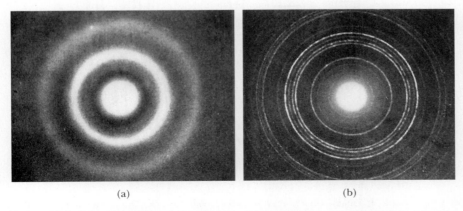

(a)

(b)

Figure 12.8 Electron diffraction patterns of (a) amorphous iron and (b) crystalline iron. (*From T. Ichikawa*, Phys. Status, Solidi A, *19:707, 1973*)

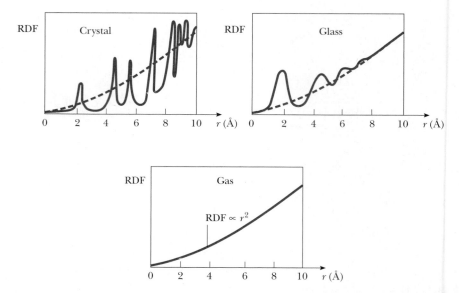

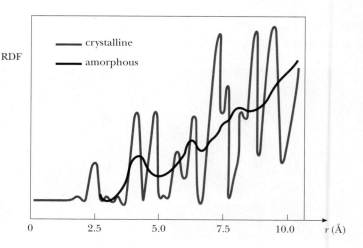

Figure 12.9   Representative radial distribution functions (RDF) for a substance in crystalline, amorphous, and gas phases. (*After R. Zallen*, The Physics of Amorphous Solids, *New York, John Wiley and Sons, 1983*)

The distribution of atoms in space, which is described by the **radial distribution function** (RDF), $\rho(r)$, may be obtained by taking the Fourier transform of $I_s(\theta)$.[1] Note that $\rho(r)\, dr$ is the probability of finding neighboring atoms between r and $r + dr$ from any atom chosen as the origin. Thus we expect $\rho(r)$ to consist of many sharp maxima for crystals, because their structures consist of extremely regular shells of nearest-neighbor atoms, next-nearest-neighbor atoms, and so forth. On the other hand, we expect fewer and broader maxima for glasses because of their short-range order and distribution of bond lengths. Figure 12.9 shows schematic illus-

Figure 12.10   RDFs for amorphous and crystalline germanium, calculated from x-ray scattering. (*From R. J. Temkin, W. Paul, and G. A. N. Connell*, Adv. Phys., *22:581, 1973*)

[1]See, for example, R. Zallen, *The Physics of Amorphous Solids*, New York, John Wiley and Sons, 1983, Chapter 2.

**Table 12.4   Examples of Technological Applications
of Amorphous Solids**

Type	Material	Use
Oxide glass	$(SiO_2)_{0.8}(Na_2O)_{0.2}$	Transparent window glass
Oxide glass	$(SiO_2)_{0.9}(GeO_2)_{0.1}$	Ultratransparent optical fibers
Organic polymer	Polystyrene	Strong, low-density plastics
Chalcogenide glass	Se, As_2Se_3	Photoconductive films used for xerography
Amorphous silicon	$Si_{0.9}H_{0.1}$	Solar cells
Metallic glass	$Fe_{0.8}B_{0.2}$	Ferromagnetic low-loss ribbons used as transformer cores

trations of these differences in the radial distribution function for the crystal, glass, and gas phases. Note that the RDFs for the crystal and glass phases are superimposed on a background equivalent to the gas RDF, $\rho_{gas}(r) = Ar^2$, where A is a constant. The RDFs for both crystal and glass approach the gas RDF for large r values because the distribution of atoms becomes effectively continuous over large distances and smoothes out to an average value of $\overline{n}$ atoms per unit volume, as in the gas phase. Because the volume of a spherical shell is $4\pi r^2\,dr$, the number of atoms between r and $r + dr$ in the gas phase is given by $\rho_{gas}(r)\,dr = \overline{n}(4\pi r^2\,dr)$. Thus $\rho_{gas}(r) = Ar^2$, where $A = 4\pi\overline{n}$.

Figure 12.10 shows actual RDFs for crystalline and amorphous germanium obtained from x-ray diffraction measurements. Note that the amorphous RDF merges with the smeared background after about four oscillations, indicating only short-range order. However, the crystalline RDF still varies sharply after 14 oscillations. Furthermore, note that the crystalline RDF peaks are not as sharp as might be expected but are broadened by thermal and zero-point vibrations of the atoms about their lattice positions.

Amorphous solids have many useful and interesting physical properties, such as extremely high optical transparency, extremely high strength, and low density. You may wish to read Zallen (cited in Suggestions for Further Reading at the end of this chapter) for more information about amorphous solids. We conclude our brief introduction to this rapidly expanding field with Table 12.4, which lists some of the technological applications of amorphous solids.

12.2   CLASSICAL FREE ELECTRON MODEL OF METALS

Shortly after Thomson's discovery of the electron, Drude and Thomson proposed the free electron theory of metals. According to this theory, the physical properties of a metal may be explained by modeling the metal as a classical gas of conduction electrons moving through a fixed lattice of positive ion cores. Thomson, Drude, and Lorentz used this picture of a highly

Table 12.5   Thermal Conductivity, K, and Electrical Conductivity, σ, of Selected Substances at Room Temperature

Substance	K in $W \cdot m^{-1} K^{-1}$	σ in $(\Omega \cdot m)^{-1}$
Silver	427	62×10^6
Copper	390	59×10^6
Gold	314	41×10^6
Aluminum	210	35×10^6
Iron	63	10×10^6
Steel	50	1.4×10^6
Nichrome	14	0.9×10^6
Quartz	13	
NaCl	7.0	$<10^{-4}$

mobile electron "fluid" to explain the high electrical and thermal conductivities of metals, shown in Table 12.5, with considerable success. In particular, the model predicts the correct functional form of Ohm's law and the remarkably simple empirical connection between the electrical and thermal conductivities of a metal known as the Wiedemann–Franz relation.[2] However, the model does not accurately predict the experimental values of electrical and thermal conductivities when classical electronic mean free paths are used in the calculations. In fact, we shall see that the shortcomings of the classical model can be remedied only by taking into account the wave nature of the electron. This involves replacing the Maxwell–Boltzmann rms velocity for electrons with the Fermi velocity (see Section 10.5) as well as replacing the classical mean free path of electrons with the much longer quantum mean free path, which may be hundreds to thousands of times the interatomic distance. Such long mean free paths occur because the electron is a wave and, as such, is able to pass freely through a nearly perfect lattice, scattering only when it encounters impurity atoms or other deviations from crystalline regularity.

Ohm's Law

Ohm's law was first established as an experimental result applicable to a wide range of metals and semiconductors. It states that the current density in a material is directly proportional to the applied electric field. That is,

$$\mathbf{J} = \sigma \mathbf{E} \tag{12.4}$$

where $\mathbf{J}$ is the current density (A/m^2), σ is the electrical conductivity of the material $(\Omega^{-1} \, m^{-1})$, and $\mathbf{E}$ is the electric field (V/m). Rather than viewing Equation 12.4 simply as a proportionality between applied field and resulting current, it is more instructive to interpret this equation as a definition of σ. Interpreted in this way, Equation 12.4 tells us that a single constant, σ, which

---

[2]The Wiedemann–Franz relation states that the ratio of thermal conductivity to electrical conductivity for metals is proportional to the temperature and that the value of the proportionality constant is independent of the metal considered.

depends on the material and temperature *but not on applied field or voltage*, completely characterizes the electrical conduction in an ohmic solid.

We can derive Ohm's law by considering a conductor to consist of a gas of classical particles (conduction electrons) moving through a background of immobile, heavy ions. The electrons in a metal move randomly along straight-line trajectories, which are constantly interrupted by collisions with lattice ions (see Fig. 12.11a). The root mean square speed at room temperature is fairly high ($\sim 10^5$ m/s) and may be calculated from the classical equipartition theorem:

$$\tfrac{1}{2} m_e \overline{v^2} = \tfrac{3}{2} k_B T$$

or

$$v_{\text{rms}} = \sqrt{\overline{v^2}} = \left(\frac{3k_B T}{m_e} \right)^{1/2} \tag{12.5}$$

Each step of the motion (the path between collisions) shown in Figure 12.11a is a "free path"; the average free path, or **mean free path,** L, is related to the **mean free time,** τ, the average time between collisions, and to v_{rms} by

$$L = v_{\text{rms}} \tau \tag{12.6}$$

In Drude's original model, L was taken to be several angstroms, consistent with the view that an electron generally travels one interatomic spacing before bumping into a large ion.

When an electric field is applied to the sample, an electric force is exerted on an electron during each interval between collisions, resulting in a displacement that is small compared to the mean free path. The cumulative effect of these displacements may be viewed in terms of a small average **drift speed,** v_d, superimposed on the rather high random thermal speed, as shown in Figure 12.11b.

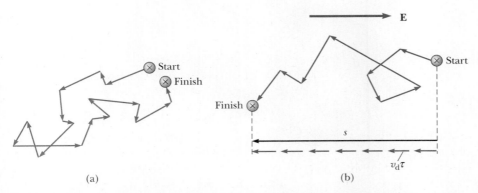

(a)   (b)

Figure 12.11   (a) Random successive displacements of an electron in a metal without an applied electric field. (b) A combination of random displacements and displacements produced by an external electric field. The net effect of the electric field is to add together multiple displacements of length $v_d \tau$ opposite the field direction. For purposes of illustration, this figure greatly exaggerates the size of v_d compared with v_{rms}.

We may derive an expression for v_d by assuming that the total displacements, s, of an electron after b collisions is

$$s = \frac{a}{2}\left(t_1^2 + t_2^2 + t_3^2 + \cdots + t_b^2\right) \qquad (12.7)$$

where t_1, t_2, and so on are the successive times between collisions and a is the acceleration ($a = eE/m_e$) produced by the electric field. Note that we have ignored the random initial velocities of the electron in Equation 12.7, since these average to zero. We may write Equation 12.7 in terms of averages as

$$s = \frac{a}{2}\,(b)\,(\overline{t^2}) \qquad (12.8)$$

Because the average value of t^2 is $\overline{t^2} = 2\tau^2$ (see Problem 9), Equation 12.8 becomes

$$s = ab\tau^2$$

or

$$s = \frac{eE}{m_e}\,b\tau^2 \qquad (12.9)$$

Comparing Equation 12.9 to the expression for the displacement in terms of the drift speed, $s = bv_d\tau$, we find

$$v_d = \frac{eE\tau}{m_e} \qquad (12.10)$$

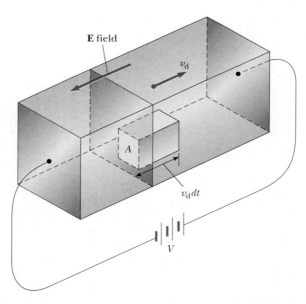

Figure 12.12   The connection between current density, J, and drift velocity, v_d. The charge that passes through A in time dt is the charge contained in the small parallelepiped, $neAv_d\,dt$.

To find the magnitude of the current density, J, when n electrons per unit volume all move with speed v_d, we note that in a time dt the electrons move a distance $v_d dt$, so that $nAv_d dt$ electrons cross an area A perpendicular to the direction of electron flow (Fig. 12.12). Since each electron has a charge e, the charge crossing the area A in the time dt is $neAv_d dt$, and the current density is

$$J = \frac{neAv_d \, dt}{A dt} = nev_d \tag{12.11}$$

Substituting $v_d = eE\tau/m_e$ into Equation 12.11 yields

$$J = nev_d = \frac{ne^2\tau}{m_e}E \tag{12.12}$$

The proportionality of J to E given by Equation 12.12 shows that the classical free electron model predicts the observed Ohm's law dependence of J on E. For this case, the conductivity, which is *independent of E*, is

$$\sigma = \frac{ne^2\tau}{m_e} \tag{12.13}$$

or, using $\tau = L/v_{rms}$,

$$\sigma = \frac{ne^2 L}{m_e v_{rms}} \tag{12.14}$$

Substituting the Maxwell–Boltzmann rms thermal speed into Equation 12.14 yields

$$\sigma = \frac{ne^2 L}{(3k_B Tm_e)^{1/2}} \tag{12.15}$$

Table 12.6 lists measured values of σ for various metals, for comparison with calculations using Equation 12.15. Since the resistivity, ρ, is the reciprocal of the conductivity, Equation 12.15 may also be written

$$\rho = \frac{(3k_B Tm_e)^{1/2}}{ne^2 L} \tag{12.16}$$

Classical expressions for conductivity and resistivity

Table 12.6   Electrical Conductivity of Metals at 300 K

Substance	Measured σ in $(\Omega \cdot m)^{-1}$
Copper	59×10^6
Aluminum	35×10^6
Sodium	22×10^6
Iron	10×10^6
Mercury	1.0×10^6

EXAMPLE 12.1   The Classical Free Electron Model of Conductivity in Solids

(a) Show that the rms thermal speed of electrons at 300 K is many orders of magnitude higher than the drift speed, v_d. To find v_d, assume that a copper wire with a cross section of 2 mm × 2 mm carries a current of 10 A, and that each copper atom contributes 1 free electron. The density of copper at room temperature is 8.96 g/cm^3. (b) Estimate τ, the average time between collisions for copper at room temperature, assuming that the mean free path is the interatomic distance, 2.6 Å. (c) Calculate the conductivity of copper at room temperature as a test of the classical free electron gas theory, and compare this to the measured value found in Table 12.6.

Solution   (a) By Equation 12.5,

$$v_{rms} = \left(\frac{3k_B T}{m_e}\right)^{1/2}$$

$$= \left[\frac{3(1.38 \times 10^{-23}\,\text{J/K})(300\,\text{K})}{9.11 \times 10^{-31}\text{kg}}\right]^{1/2}$$

$$= 1.17 \times 10^5\,\text{m/s}$$

To calculate v_d we use $J = nev_d$, or $v_d = J/ne$. The number of free (conduction) electrons per cubic centimeter, n, in copper is

$$n = \left(\frac{1\,\text{free electron}}{\text{atom}}\right)\left(\frac{6.02 \times 10^{23}\,\text{atoms}}{\text{mol}}\right)$$

$$\times \left(\frac{8.96\,\text{g}}{\text{cm}^3}\right)\left(\frac{1\,\text{mol}}{63.5\,\text{g}}\right)$$

$$= 8.49 \times 10^{22}\,\text{electrons/cm}^3$$

Thus,

$$v_d = \frac{J}{ne} = \frac{(10\,\text{A})/(4 \times 10^{-6}\,\text{m}^2)}{(8.49 \times 10^{28}\,\text{m}^{-3})(1.6 \times 10^{-19}\,\text{C})}$$

$$= 1.8 \times 10^{-4}\,\text{m/s}$$

The ratio of drift speed to rms speed is

$$\frac{v_d}{v_{rms}} = 1.5 \times 10^{-9}$$

(b) Using Equation 12.6, we find for the average time between collisions

$$\tau = \frac{L}{v_{rms}} = \frac{2.6 \times 10^{-10}\,\text{m}}{1.2 \times 10^5\,\text{m/s}} = 2.2 \times 10^{-15}\,\text{m/s}$$

Thus, in this model, the electrons experience several hundred trillion collisions per second!

(c) Equation 12.15 gives for $T = 300$ K

$$\sigma = \frac{ne^2 L}{(3k_B T m_e)^{1/2}} = 5.3 \times 10^6\,(\Omega\cdot\text{m})^{-1}$$

From Table 12.6 we see that this value of the conductivity is about 10 times smaller than the measured value!

Although the classical electron gas model does predict Ohm's law, we see from Example 12.1 that it results in conductivity values that differ from measured values by an order of magnitude. Worse yet, the *measured resistivity* of most metals is found to be proportional to the absolute temperature (see Fig. 12.13), yet the classical electron gas model predicts a much weaker dependence of ρ on T. From Equation 12.16 we see that the classical model incorrectly predicts that resistivity should be proportional to the square root of the absolute temperature. In Section 12.3 we will give a different model of electron scattering, which explains the observed linear dependence of ρ on T at high temperatures.

Classical Free Electron Theory of Heat Conduction

The thermal conductivity of a substance, K, is defined in a way similar to electrical conductivity, σ. In the case of electrical conductivity we found current density equal to conductivity times voltage gradient, or

$$J = -\sigma\frac{\Delta V}{\Delta x} \tag{12.17}$$

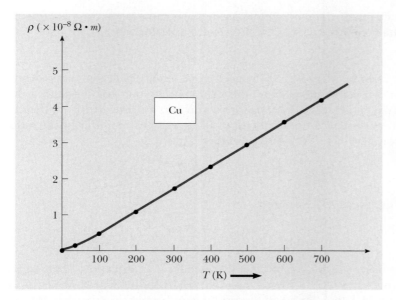

$\rho\ (\times 10^{-8}\ \Omega \cdot m)$

Figure 12.13   The resistivity of pure copper as a function of temperature.

Likewise, we now have thermal current density (W/m^2) equal to thermal conductivity times temperature gradient, or

$$\frac{\Delta Q}{A\ \Delta t} = -K\ \frac{\Delta T}{\Delta x} \tag{12.18}$$

Here ΔQ is the thermal energy conducted through a cross-sectional area A in time Δt between two planes with a temperature gradient of $\Delta T/\Delta x$. Since good electrical conductors are also good conductors of thermal energy, it is natural to assume that the highly mobile electron gas is responsible for transporting charge as well as thermal energy through the metal via random collision processes. A remarkable triumph of the classical free electron model was to show that if free electrons were responsible for both electrical and thermal conduction in metals, then the ratio of K/σ should be a universal constant, *the same for all metals*, and *dependent only on the absolute temperature*. Since such an unusually simple connection between K and σ had already been observed experimentally (the Wiedemann–Franz law), this prediction confirmed that the motion of the electron gas was basically responsible for both electrical and thermal conductivity.

Since the Wiedemann–Franz law is such a simple result, it bears consideration in more detail. The kinetic theory of gases may be applied to the free electron gas to calculate the flux of thermal energy (W/m^2) carried by electrons moving from a region at temperature $T + \Delta T$ to a region at lower temperature T.[3] Using this result, one immediately finds that the thermal

---

[3]See C. Kittel, *Introduction to Solid State Physics*, 6th ed., New York, John Wiley and Sons, 1986, Chapter 5.

Table 12.7 Experimental Lorentz Numbers $K/\sigma T$ in Units of 10^{-8} W·Ω/K²*

Metal	273 K	373 K
Ag	2.31	2.37
Au	2.35	2.40
Cd	2.42	2.43
Cu	2.23	2.33
Ir	2.49	2.49
Mo	2.61	2.79
Pb	2.47	2.56
Pt	2.51	2.60
Sn	2.52	2.49
W	3.04	3.20
Zn	2.31	2.33

*From C. Kittel, *Introduction to Solid State Physics*, 2nd ed., New York, John Wiley and Sons, 1965.

conductivity, K, depends simply on the heat capacity per unit volume, C_v, v_{rms}, and the mean free path, L, of the electrons, as follows:

$$K = \tfrac{1}{3} C_v v_{rms} L \tag{12.19}$$

If we assume that the electrons behave as a classical ideal gas and obey Maxwell–Boltzmann statistics, the heat capacity per *mole* is found to be $3R/2$ or $3N_A k_B/2$, where N_A is Avogadro's number. To convert this to heat capacity per unit volume for use in Equation 12.19, we must multiply by the ratio of the electron density to Avogadro's number. Thus,

$$C_v = (\tfrac{3}{2} N_A k_B) \left(\frac{n}{N_A} \right) = \tfrac{3}{2} k_B n$$

and Equation 12.19 becomes

$$K = \frac{k_B n v_{rms} L}{2} \tag{12.20}$$

If we simply use the interatomic distance for L in Equation 12.20, we will find incorrect values, as in the case of electrical conductivity. However, if we form the ratio of K to σ, we avoid the problem of having to assign values to n and L, as these quantities cancel. Thus,

$$\frac{K}{\sigma} = \frac{k_B n v_{rms} L/2}{ne^2 L / m_e v_{rms}} = \frac{k_B (v_{rms})^2 m_e}{2e^2}$$

Substituting $v_{rms}^2 = 3k_B T / m_e$ into this expression gives the desired classical result, known as the **Wiedemann–Franz law:**

Wiedemann–Franz law

$$\frac{K}{\sigma} = \frac{3k_B^2}{2e^2} T \tag{12.21}$$

Equation 12.21 shows that the ratio of K/σ for any metal is proportional to T and depends only on the universal constants k_B and e. This result agrees with **the empirical law of Wiedemann and Franz** (1853) which states that $K/\sigma T$ has the same value for all metals. The ratio $K/\sigma T$ is known as the **Lorentz number,** and according to our classical theory, it has the same value, $3k_B^2/2e^2 = 1.1 \times 10^{-8}$ W·Ω/K², *for any metal at any temperature.* Table 12.7 shows that, indeed, the ratio of $K/\sigma T$ is nearly constant from metal to metal and with varying temperatures, but that the value of $3k_B^2/2e^2$ does not agree precisely with the measured Lorentz numbers. Overall, however, the agreement of the simple classical theory with experiment is good and is taken to constitute strong evidence that the free electron gas accounts for both electrical and thermal conductivity in metals.

12.3 QUANTUM THEORY OF METALS

Although the classical electron gas model broadly describes the electrical and thermal properties of metals, there are notable deficiencies in the predictions of this model—in the numerical values of K and σ, in the temperature dependence of σ, and in the prediction of an excess heat capacity for metals (see Section 10.5.) These deficiencies can be rectified by replacing the Maxwell–Boltzmann distribution with the Fermi–Dirac distribution for the conduction electrons in the metal and by calculating the electron mean free path while explicitly taking the wave nature of the electron into account. Since quantum mechanical calculations of electron mean free path are complicated, we shall rely on qualitative arguments and inferences from measured quantities to give physical insight into the surprising transparency of metals to conduction electrons.

Replacement of v_{rms} with v_F

As we have seen in Section 10.3, the electrons in a metal must be described by Fermi–Dirac statistics, and essentially all energy levels are filled up to the Fermi energy, E_F. Because of restrictions imposed by the Pauli principle on the quantum states of scattered electrons, one might anticipate a "bottleneck" in electron transport through a metal because of a lack of empty final states. That this is not the case is illustrated in Figure 12.14, which shows the velocity distribution in three dimensions of conduction electrons in a metal. Essentially all the electrons have velocities within a radius of $\sqrt{2E_F/m_e}$ in velocity space [$\frac{1}{2}m_e(v_x^2 + v_y^2 + v_z^2) \leq E_F$]. Figure 12.14b shows that the application of an electric field causes all the electron velocities to increase by an increment v_d. Since the shift of the electron originally at A creates an empty final state for the electron originally at B, there is no bottleneck. Furthermore, the net effect of the E field is to leave an intact core of states for electrons with $E < E_F$ and to produce a displacement of those electrons near the Fermi surface having $v \approx v_F$. Thus only those electrons with $v \approx v_F$ are free to move and participate in electrical and thermal conduction, and we can presumably use the classical expressions for σ and K,

$$\sigma = \frac{ne^2L}{m_e v_{rms}} \quad \text{and} \quad K = \frac{1}{3}C_v v_{rms}L$$

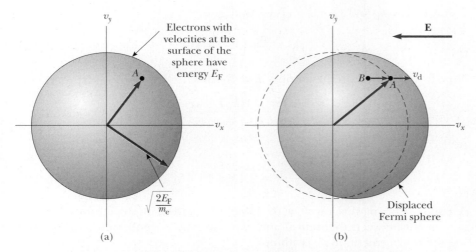

(a)   (b)

Figure 12.14   (a) Allowed velocity vectors, or positions in velocity space, of conduction electrons in a metal without an applied electric field. (b) The net effect of an applied electric field is a small displacement of the Fermi sphere, the displaced electrons having a speed of approximately v_F.

with the Maxwell–Boltzmann rms speed replaced by the Fermi speed; thus,

$$\sigma = \frac{ne^2 L}{m_e v_F} \tag{12.22}$$

and

$$K = \tfrac{1}{3} C_v v_F L \tag{12.23}$$

Wiedemann–Franz Law Revisited

Let us test the validity of the quantum replacement of v_{rms} with v_F by recalculating $K/\sigma T$ and checking for improved agreement with experiment. First we note that using a Maxwell–Boltzmann molar heat capacity of $3R/2$ for electrons in a metal is incorrect, since conduction electrons obey Fermi–Dirac statistics. As already mentioned in Section 10.5, only a fraction of the electrons within $k_B T$ of E_F change energy as the temperature changes, leading to a molar heat capacity that is a small fraction of $3R/2$:

$$C \approx 2 \left(\frac{k_B T}{E_F} \right) (\tfrac{3}{2} R)$$

An exact calculation is rather complicated[4] but leads to a similar result:

$$C = \frac{\pi^2}{3} \left(\frac{k_B T}{E_F} \right) (\tfrac{3}{2} R) \tag{12.24}$$

Changing C to a heat capacity per unit volume and substituting into $K = \tfrac{1}{3} C v_F L$ yields

$$K = \tfrac{1}{3} \left(\frac{\pi^2}{3} \right) \left(\frac{k_B T}{E_F} \right) (\tfrac{3}{2} R) \left(\frac{n}{N_A} \right) v_F L$$

Using $R = N_A k_B$ and $E_F = \tfrac{1}{2} m_e v_F^2$, we find

$$K = \frac{\pi^2}{3} \left(\frac{k_B^2 T}{m_e v_F} \right) n L \tag{12.25}$$

Now, forming the Lorentz number $K/\sigma T$ yields

$$\frac{K}{\sigma T} = \frac{(\pi^2/3)(k_B^2 T / m_e v_F)(nL)}{(ne^2 L / m_e v_F) T}$$

or

Quantum form of the Wiedemann–Franz law

$$\frac{K}{\sigma T} = \frac{\pi^2 k_B^2}{3 e^2} = 2.45 \times 10^{-8} \, \text{W} \cdot \Omega / \text{K}^2 \tag{12.26}$$

A comparison of this result with Table 12.7 shows improved and excellent agreement between theoretical and experimental Lorentz values, thus justifying the replacement of v_{rms} with v_F.

---

[4]See Kittel, op. cit., Chapter 6.

Quantum Mean Free Path of Electrons

Changes from the classical value of electron mean free path arise from the wave properties of the electron. Consider, once again, our familiar copper sample. If we simply use the interatomic distance between copper atoms as the mean free path L for electrons, and v_F for v_{rms} in $\sigma = ne^2L/m_e v_{rms}$, we find a value for σ that is about 200 times *smaller* than the measured value of the conductivity. This discrepancy implies that we are using the wrong value for L and that the scattering sites for electrons are not adjacent ion cores but more widely separated scattering centers. These scattering centers consist of departures from perfect lattice regularity, such as thermal displacements of ions from equilibrium lattice sites, dislocations, and impurity atoms. Because quantum calculations of L that include these effects are complicated, let us make a rough estimate of L for copper at room temperature by using the measured value of σ. Solving $\sigma = ne^2L/m_e v_F$ for L yields

$$L = \frac{m_e v_F \sigma}{ne^2} \tag{12.27}$$

Using a Fermi energy of 7.05 eV for copper (see Table 10.1) gives

$$v_F = \left(\frac{2E_F}{m_e}\right)^{1/2} = \left(\frac{(2)(7.05 \times 1.6 \times 10^{-19} \text{J})}{9.11 \times 10^{-31} \text{kg}}\right)^{1/2} = 1.57 \times 10^6 \text{ m/s}$$

Substituting this value into Equation 12.27 gives

$$L = \frac{(9.11 \times 10^{-31} \text{ kg})(1.57 \times 10^6 \text{ m/s})(5.9 \times 10^7 \, \Omega^{-1} \text{ m}^{-1})}{(8.49 \times 10^{28} \text{ electrons/m}^3)(1.60 \times 10^{-19} \text{ C})^2}$$
$$= 3.9 \times 10^{-8} \text{ m} = 390 \text{ Å}$$

This is about 150 times the distance between copper atoms. Using $\tau = L/v_F$, we can also estimate the mean free time between electronic collisions in copper at room temperature:

$$\tau = \frac{3.90 \times 10^{-8} \text{ m}}{1.57 \times 10^6 \text{ m/s}} = 2.50 \times 10^{-14} \text{ s}$$

We can account for the unexpectedly long electron mean free path by taking the wave nature of the electron into account. Quantum mechanical calculations show that electron waves with a broad range of energies can pass through a *perfect* lattice of ion cores unscattered, without resistance, and with an infinite mean free path. The actual resistance of a metal is due to the random thermal displacements (thermal vibrations) of ions about lattice points and to other deviations from a perfect lattice, such as impurity atoms and defects that scatter electron waves. The lack of electron scattering by a perfect lattice can be understood by noting that the electron wave generally travels through the metal unattenuated, just as does light through a transparent crystal. Strong reflections of electron waves are set up only for specific electron energies, and when this occurs, the electron wave cannot travel freely through the crystal. As we shall see in the next section, these strong reflections occur when the lattice spacing is equal to an integral number of electronic wavelengths, resulting in a discrete set of forbidden energy bands for electrons.

Causes of resistance in a metal

From a classical viewpoint, the observed proportionality of resistivity to absolute temperature at high temperature is the result of the scattering of electrons by lattice ions vibrating with larger amplitude the higher the temperature. In the quantum view, as we saw in Chapter 10, lattice vibrations have a quantized energy $\hbar\omega$, where ω is the angular frequency of vibration of the lattice ion. These quantized lattice vibrations are called **phonons,** and for purposes of calculation, the vibrating lattice ions are replaced by phonons. The number of phonons with energy $\hbar\omega$ that are available at a given temperature T is denoted by n_{p} and is proportional to the Bose–Einstein distribution function:

$$n_{\mathrm{p}} \propto \frac{1}{e^{\hbar\omega/k_{\mathrm{B}}T} - 1} \tag{10.33}$$

At high temperatures, $k_{\mathrm{B}}T$ is much greater than $\hbar\omega$, and this becomes

$$n_{\mathrm{p}} \propto \frac{k_{\mathrm{B}}T}{\hbar\omega}$$

Thus the number of phonons available to scatter electrons is directly proportional to T. Finally, since the number of electron scatterers is proportional to the temperature, so is the resistivity, ρ.

In addition to the temperature-dependent part of the resistivity, there is also a temperature-independent contribution to the resistivity of the metal, which clearly manifests itself for temperatures less than about 10 K. This residual resistivity ρ_i, which remains as $T \to 0$, is produced by electron waves scattering from impurities and structural imperfections in a given sample (see Figure 12.15). The parallel nature of the curves shown in Figure 12.15 implies that the resistivity caused by thermal motion of the lattice, ρ_L, is independent of the impurity concentration and that ρ_i is independent of temperature. This result is formalized in a result known as **Matthiessen's rule,** which states that

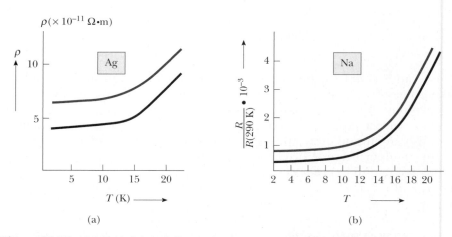

Figure 12.15 (a) Resistivity of silver at low temperature. The colored curve is for a silver sample with a higher concentration of impurity atoms. (*From W. J. de Haas, G. J. van den Berg, J. de Boer*, Physica, *2:453, 1935*) (b) Resistance of sodium as a function of temperature. The colored curve is for a sodium sample with a higher concentration of imperfections. (*From D. K. C. MacDonald and K. Mendelssohn*, Proc. Roy. Soc. (London), *A202:103, 1950*)

the resistivity of a metal may be written in the form

$$\rho = \rho_i + \rho_L \qquad (12.28)$$

where ρ_i depends only on the concentration of crystal imperfections and ρ_L depends only on T.

12.4   BAND THEORY OF SOLIDS

In Sections 12.2 and 12.3 we discussed the classical and quantum models of metals, making the simplifying assumption that the metal consists of a gas of free electrons. Since good insulators have enormous resistivity compared to metals (a factor of 10^{24} greater), their electronic configurations must be quite different. In fact, most outer-shell electrons in an insulator are not free but are involved in ionic or covalent bonds, as discussed earlier. Furthermore, in order to more fully understand the electronic properties of solids, one must consider the effect of the lattice ions.

In the general case, there are two approaches to the determination of electronic energies in a solid. One is to follow the behavior of energy levels of isolated atoms as they are brought closer and closer to form a solid. The other is to show that energy bands arise when the Schrödinger equation is solved for electrons subject to a periodic potential representing the lattice ions. We shall follow the isolated-atom approach here first, since it is simpler and immediately leads to an explanation of the differences between conductors, insulators, and semiconductors. Later in this section we shall explicitly consider the effect of a periodic potential on electron waves, to show how energy bands arise from another viewpoint.

Isolated-Atom Approach to Band Theory

If two identical atoms are very far apart, they do not interact, and their electronic energy levels can be considered to be those of isolated atoms. Suppose the two atoms are sodium, each having an outermost $3s$ electron with a specific energy. As the two sodium atoms are brought closer together, their wavefunctions overlap, and the two degenerate, isolated $3s$ energy levels are split into two different levels, as shown in Figure 12.16a. We can understand this splitting by considering the appropriate electronic wavefunctions for the case of widely separated atoms and the case of neighboring atoms. (For a more complete and careful treatment of this splitting, see the discussion of covalent bonding in Section 11.4.) Figure 12.17 shows idealized isolated atom wavefunctions ψ_1 and ψ_2 as well as the linear combinations $\psi_1 + \psi_2$ and $\psi_1 - \psi_2$ that represent approximate electronic wavefunctions for the two atoms close together. Note that an electron in the state $\psi_1 + \psi_2$ has a substantial probability of being found midway between the ion cores, while in the state $\psi_1 - \psi_2$ the probability density vanishes at the midpoint. Since the electron spends part of its time midway between the two attractive ion cores in the state $\psi_1 + \psi_2$, the electron is more tightly bound (has lower energy) in the state $\psi_1 + \psi_2$ than in $\psi_1 - \psi_2$. This leads to the two different $3s$ energy levels shown in Figure 12.16a.[5]

---

[5]The width of the energy band ΔE depends on the amount of charge distribution overlap between adjacent atoms and hence on the interatomic separation, r_0.

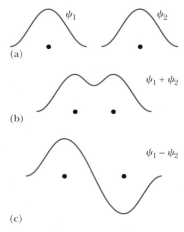

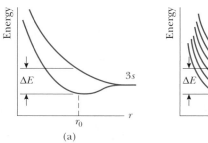

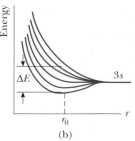

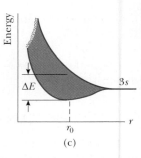

Figure 12.16 (a) The splitting of the $3s$ levels when two sodium atoms are brought together. (b) The splitting when six sodium atoms are brought together. (c) The formation of a $3s$ band when a large number of sodium atoms are assembled to form a solid. Note that r_0 is the actual lattice constant.

Figure 12.17 Idealized sodium wavefunctions in various stages of combination. (a) Wavefunctions of electrons bound to two ion cores at large separation. (b) One linear combination of wavefunctions that is appropriate for electrons bound to ion cores at small separation is $\psi_1 + \psi_2$. (c) The other linear combination is $\psi_1 - \psi_2$.

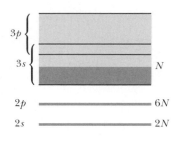

Figure 12.18 The energy bands of sodium are shaded gray in this figure. The solid contains N atoms. Note the energy gaps between allowed bands and that the $3s$ and $3p$ bands overlap in sodium. The number of electrons in each energy band is indicated to the right of the band.

When a large number of atoms are brought together to form a solid, a similar phenomenon occurs. As the atoms are brought close together, the various isolated-atom energy levels begin to split. This splitting is shown in Figure 12.16b for six atoms in proximity. In this case there are six energy levels corresponding to six different linear combinations of isolated-atom wavefunctions. The width of an energy band (designated ΔE in Fig. 12.16) depends only on the number of atoms close enough to interact strongly, which is always a small number. If we consider the total number of atoms in a solid ($N \approx 10^{23}$ atoms/cm^3), we find a very large number of levels (determined by N) spaced within the width ΔE, so the levels may be regarded as a **continuous band** of energy levels (see Fig. 12.16c). In the case of sodium, it is common to refer to the continuous distribution of allowed energy levels as the **3s band,** because it originates from the $3s$ levels of individual sodium atoms. In general, a crystalline solid has numerous allowed energy bands, one band arising from each atomic energy level. Figure 12.18 shows the allowed energy bands of sodium. Note that energy gaps, or forbidden energy regions, separate the allowed energy bands. Forbidden energy regions arise from the separation between different atomic levels and will always be present unless individual atomic levels broaden so much that they overlap, as do the $3s$ and $3p$ bands in sodium.

If the solid contains N atoms, each energy band has N energy levels. The $1s$, $2s$, and $2p$ bands of sodium are each full of electrons, as indicated by the dark gray-shaded areas in Figure 12.18. A level whose orbital angular momentum is ℓ can hold $2(2\ell + 1)$ electrons. The factor of 2 arises from the two possible electron-spin orientations, while the factor $2\ell + 1$ corresponds to the number of possible orientations of the orbital angular momentum. The capacity of each band for a system of N atoms is $2(2\ell + 1)N$ electrons. Hence the $1s$ and $2s$ bands each contain $2N$ electrons ($\ell = 0$), while the $2p$ band contains $6N$ electrons ($\ell = 1$). Because sodium has only one $3s$ electron and there is a total of N atoms in the solid, the $3s$ band contains only N electrons and is only half full. The $3p$ band, which is above the $3s$ band, is completely empty.

Conduction in Metals, Insulators, and Semiconductors

The enormous variation in electrical conductivity of metals, insulators, and semiconductors may be explained qualitatively in terms of energy bands. We

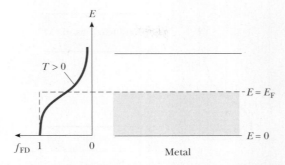

Figure 12.19 A half-filled band of a conductor such as the $3s$ band of sodium. At $T = 0$ K, the Fermi energy lies in the middle of the band. The Fermi–Dirac probability that an energy state E is occupied at $T > 0$ K is shown at the left.

shall see that the position and the electronic occupation of the highest band or, at most, of the highest two bands determine the conductivity of a solid.

Metals.   We can understand electrical conduction in metals by considering the half-filled $3s$ band of sodium. Figure 12.19 shows this typical half-filled metallic band at $T = 0$ K, where the shaded region represents levels that are filled with electrons. Since electrons obey Fermi–Dirac statistics, all levels below the Fermi energy, E_F, are filled with electrons, while all levels above E_F are empty. In the case of sodium, the Fermi energy lies in the middle of the band. At temperatures greater than 0 K, some electrons are thermally excited to levels above E_F (as shown by the Fermi–Dirac distribution to the left of Fig. 12.19), but overall there is little change from the 0-K case. **If an electric field is applied to the metal, electrons with energies near the Fermi energy can gain a small amount of additional energy from the field and reach nearby empty energy states.** Thus electrons are free to move with only a small applied field in a metal because there are many unoccupied energy states very close to occupied energy states.

Insulators.   Now consider the two highest-energy bands of a material having the lower band completely filled with electrons and the higher completely empty at 0 K (see Fig. 12.20). It is common to refer to the separation between the outermost filled and empty bands as the **energy gap,** E_g, of the material. The lower band filled with electrons is called the **valence band,** and the upper, empty band is the **conduction band.** The Fermi energy is at the midpoint of the energy gap, as shown in Figure 12.20. Since the energy gap for an insulator is large (~ 10 eV) compared to $k_B T$ at room temperature ($k_B T = 0.025$ eV at 300 K), the Fermi–Dirac distribution predicts that very few electrons will be thermally excited into the upper band at normal temperatures, as can be seen by the small value of f_{FD} at the bottom of the conduction band in Figure 12.20. **Although an insulator has many vacant states in the conduction band that can accept electrons, there are so few electrons actually occupying conduction-band states at room temperature that the overall contribution to electrical conductivity is very small, resulting in a high resistivity for insulators.**

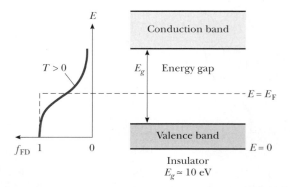

Figure 12.20   An insulator at $T = 0$ K has a filled valence band and an empty conduction band. The Fermi level lies midway between these bands. The Fermi–Dirac probability that an energy state E is occupied at $T > 0$ K is shown to the left.

Table 12.8   Energy-Gap Values for Some Semiconductors*

Crystal	E_g(eV) 0 K	300 K
Si	1.17	1.14
Ge	0.744	0.67
InP	1.42	1.35
GaP	2.32	2.26
GaAs	1.52	1.43
CdS	2.582	2.42
CdTe	1.607	1.45
ZnO	3.436	3.2
ZnS	3.91	3.6

*From C. Kittel, *Introduction to Solid State Physics*, 6th ed., New York, John Wiley and Sons, 1986.

Semiconductors.   Now consider a material that has a much smaller energy gap, on the order of 1 eV. Such materials are called **semiconductors;** Table 12.8 shows the energy gaps for some representative semiconductors. At $T = 0$ K, all electrons are in the valence band, and there are no electrons in the conduction band. Thus semiconductors are *poor* conductors at low temperatures. At ordinary temperatures, however, the situation is quite different. As shown in Figure 12.21, the populations of the valence and conduction bands are altered. Because the Fermi level, E_F, is located at about the middle of the gap for a semiconductor and E_g is small, appreciable numbers of electrons are thermally excited from the valence band to the conduction band. Since there are many empty nearby states in the conduction band, a small applied potential can easily raise the energy of the electrons in the conduction band, resulting in a moderate current. Because thermal excitation across the narrow gap is more probable at higher temperatures, the conductivity of semiconductors depends strongly on temperature and *increases* rapidly with temperature. This contrasts sharply with the conductivity of a metal, which *decreases slowly* with temperature (see Section 12.3).

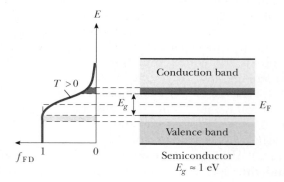

Figure 12.21   The band structure of a semiconductor at ordinary temperatures ($T \approx 300$ K). Note that the energy gap is much smaller than in an insulator and that many electrons occupy states in the conduction band.

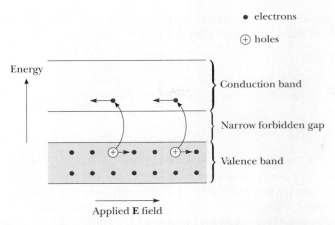

Figure 12.22   An intrinsic semiconductor. The electrons move toward the left and the holes move toward the right when the applied electric field is to the right as shown.

It is important to point out that there are both negative and positive charge carriers in a semiconductor. When an electron moves from the valence band into the conduction band, it leaves behind a vacant valence electron site, or so-called **hole,** in the otherwise filled valence band. This hole (electron-deficient site) appears as a positive charge, $+e$. The hole acts as a charge carrier in the sense that a valence electron from a nearby site can transfer into the hole, thereby filling it and leaving a hole behind in the electron's original place. Thus the hole migrates through the valence band. In a pure crystal containing only one element or compound, there are equal numbers of conduction electrons and holes. Such combinations of charges are called electron–hole pairs, and a pure semiconductor that contains such pairs is called an **intrinsic semiconductor** (see Fig. 12.22). In the presence of an electric field, the holes move in the direction of the field and the conduction electrons move opposite the field.

EXAMPLE 12.2

Estimate the electric field strength required to produce conduction in diamond, an excellent insulator at room temperature. Assume a mean free path of 5×10^{-8} m and an energy gap of 7 eV in diamond.

Solution   If an electron in diamond is to conduct, it must be supplied with 7 eV of energy from the electric field. Since the electron generally loses most of its excess energy in each collision, the field must supply 7 eV in a *single* mean free path. Thus, for an electron to gain 7 eV in a distance L, we require an electric field of

$$E = \frac{V}{L} = \frac{7\,\text{V}}{5 \times 10^{-8}\,\text{m}} = 1.4 \times 10^{8}\ \text{V/m}$$

This is an enormous field compared to the field required to produce conduction in metals.

Energy Bands from Electron Wave Reflections

We now take a completely different approach to understanding the origin of energy bands in solids. This approach involves modifying the free electron wavefunctions to take into account the scattering of electron waves by the periodic crystal lattice.

Recall that a completely free electron moving in the $+x$ direction is represented by a traveling wave with wavenumber $k = 2\pi/\lambda$, described by

$$\Psi_{\text{free}} = Ae^{i(kx-\omega t)}$$

According to de Broglie, the wave carries momentum $p = \hbar k$ and energy $E = \hbar\omega$. Furthermore, the energy of the free electron as a function of k is

$$E = \frac{p^2}{2m_e} = \frac{\hbar^2 k^2}{2m_e}$$

and a plot of E versus k yields a parabola (see Fig. 12.23). From Fig. 12.23 we see that allowed energy values are distributed continuously from zero to infinity, and there are no breaks or gaps in energy at particular k values.

Now consider what happens to a traveling electron wave when it passes through a one-dimensional crystal lattice with atomic spacing a. We start with incident electron waves of very long wavelength, or small k, and low energy, traveling to the right as in Figure 12.24a. In this case, waves reflected from successive atoms and traveling to the left are all slightly *out-of-phase* and cancel out on average. Thus the electron wave does not get reflected, and the electron moves through the lattice like a free particle. If, however, we make the electron wavelength shorter and shorter, when we reach the condition $\lambda = 2a$,

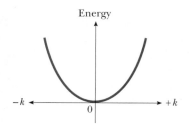

Figure 12.23 Energy versus wavenumber k for a free electron, where $k = 2\pi/\lambda$.

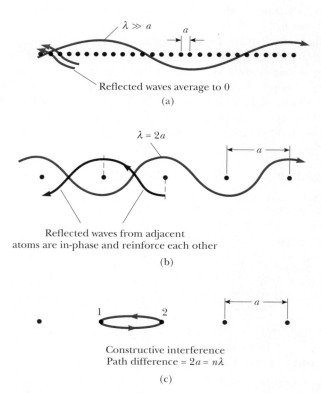

Figure 12.24 (a) Reflection of electron waves when $\lambda \gg a$. (b) Reflection of electron waves when $\lambda = 2a$. (c) Constructive interference of waves reflected from atoms 1 and 2.

waves reflected from adjacent atoms will be *in-phase*, as shown in Figure 12.24b, and there will be a strong reflected wave. This occurs because the path length difference is $2a$ for a wave reflected to the left directly from atom 1 compared to a wave that travels from 1 to 2 and then is reflected back to 1, as shown in Figure 12.24c. The longest wavelength at which constructive interference of reflected waves occurs is thus

$$2a = \lambda$$

In general, constructive interference also occurs at other, shorter wavelengths,

$$2a = \pm n\lambda \qquad n = 1, 2, 3, \ldots \tag{12.29}$$

where the negative sign arises from reflection from the atom to the left of atom 1. In terms of wavenumber k, Equation 12.29 predicts strong reflected electron waves when

$$k = \pm \frac{n\pi}{a} \qquad n = 1, 2, 3, \ldots \tag{12.30}$$

Thus, for $k = \pm n\pi/a$ the electron wavefunctions are not just waves traveling to the right but are composed of equal parts of waves traveling to the right (incident) and to the left (reflected). Since the waves traveling left and right can be added or subtracted, we have two different possible standing wave types, denoted by Ψ_- and Ψ_+:

$$\Psi_- = Be^{i(kx-\omega t)} - Be^{i(-kx-\omega t)} \tag{12.31}$$

$$\Psi_+ = Be^{i(kx-\omega t)} + Be^{i(-kx-\omega t)} \tag{12.32}$$

where $Be^{i(kx-\omega t)}$ describes a wave traveling to the right and $Be^{i(-kx-\omega t)}$ a wave traveling to the left. Using Euler's identity, $e^{i\theta} = \cos\theta + i\sin\theta$, one can show that for $k = \pm \pi/a$, Ψ_- and Ψ_+ take the more useful forms

$$\Psi_- = \pm 2Bie^{-i\omega t}\sin\frac{\pi x}{a} \tag{12.33}$$

$$\Psi_+ = 2Be^{-i\omega t}\cos\frac{\pi x}{a} \tag{12.34}$$

The key point is that Ψ_- has a slightly *higher* energy than Ψ_{free}, and Ψ_+ has a slightly *lower* energy than Ψ_{free}, at $k = \pi/a$. This leads to a discontinuity in energy, or a band gap, at $k = \pi/a$ as shown in Figure 12.25. To show that the Ψ^- state has a higher energy than the Ψ^+ state, we must consider both kinetic and potential energies of the electron in these states. The kinetic energy of the electron is the same in both states, since both have the same magnitude of momentum, $p = \hbar k = \hbar\pi/a$. Thus the average kinetic energy of both Ψ_+ and Ψ_- is

$$\langle K \rangle = \left\langle \frac{p^2}{2m_e} \right\rangle = \frac{\hbar^2 k^2}{2m_e} = \frac{\hbar^2 \pi^2}{2m_e a^2}$$

The potential energies of the two states are different, however, since Ψ_+ and Ψ_- distribute the electrons differently with respect to the positions of the positive ion cores. Figure 12.26 shows the distributions of electronic charge for the case of a traveling wave and the two standing wave cases, Ψ_+ and Ψ_-. They are

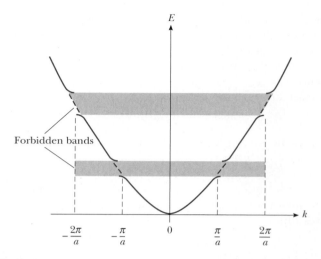

Figure 12.25 Energy versus wavenumber for a one-dimensional lattice with atomic separation a. The E versus k curve for the free electron case is shown dashed. Note the forbidden bands corresponding to impossible energy states for the electron.

calculated from the probability densities as follows:

$$\Psi^*_{free}\Psi_{free} = [Ae^{i(kx-\omega t)}]^*[Ae^{i(kx-\omega t)}] = |A|^2$$

$$\Psi^*_+\Psi_+ = \left[2Be^{-i\omega t}\cos\frac{\pi x}{a}\right]^*\left[2Be^{-i\omega t}\cos\frac{\pi x}{a}\right] = 4|B|^2\cos^2\frac{\pi x}{a}$$

$$\Psi^*_-\Psi_- = \left[\pm 2Bie^{-i\omega t}\sin\frac{\pi x}{a}\right]^*\left[\pm 2Bie^{-i\omega t}\sin\frac{\pi x}{a}\right] = 4|B|^2\sin^2\frac{\pi x}{a}$$

Note that the traveling wave solution distributes electronic charge uniformly along the lattice. On the other hand, Ψ_- concentrates electronic charge midway

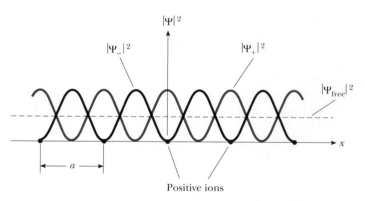

Figure 12.26 Probability densities of standing waves and traveling (free) electron waves in a one-dimensional lattice. Note that the state Ψ_+ places the electron over the positive ion and consequently has the lowest potential energy.

between the ions, while Ψ_+ concentrates electronic charge right over the positive ion cores, minimizing the electron's potential energy.[6] Thus we would expect the average potential energy of the electron in the state Ψ_+ to be somewhat below the free electron energy, and the average potential energy in the state Ψ_- to be a bit above the free electron energy. This difference in potential energy between Ψ_+ and Ψ_- leads to a difference in total energy, since the kinetic energies of Ψ_+ and Ψ_- are the same. Thus the total energy of the electron in the state Ψ_+ at $k = \pi/a$ shown in Figure 12.25 lies slightly *below* the free electron energy parabola, and the total energy of the state Ψ_- lies slightly *above* the free electron energy. Also, Figure 12.25 shows that the overall effect of the lattice is to introduce forbidden gaps in the free electron E versus k plot at values of $k = \pm n\pi/a$.

12.5   SEMICONDUCTOR DEVICES

The *p-n* Junction

In order to make devices, one must be able to fabricate semiconductors with well-defined regions of different conductivity. Both the type (positive or negative) and number of carriers in a semiconductor may be tailored to the needs of a particular device by the addition of specific impurities in a process called **doping.** For example, when an atom with five valence electrons, such as arsenic, is added to a semiconductor from Group IV of the periodic table, four valence electrons participate in the covalent bonds and one electron is left over (see Fig. 12.27a). This extra electron is nearly free and has an energy level that lies within the energy gap, just below the conduction band, as shown in Figure 12.27b. Such a pentavalent atom in effect donates an electron to the

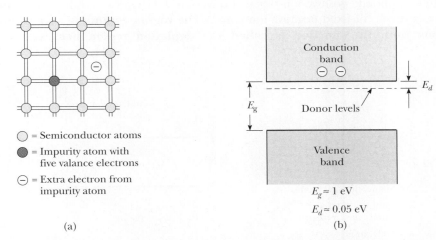

○ = Semiconductor atoms

● = Impurity atom with
     five valance electrons

⊖ = Extra electron from
     impurity atom

(a)

$E_g \approx 1$ eV

$E_d \approx 0.05$ eV

(b)

Figure 12.27   (a) A two-dimensional representation of a semiconductor containing a donor atom (colored spot). (b) An energy-band diagram of a semiconductor in which the donor levels lie within the forbidden gap, just below the bottom of the conduction band.

---

[6]Recall that the electron's electrical potential energy is the work done by an external force to move the electron to a particular point. To move the electron to a *positive* ion requires the minimum external work. Thus the electron has the minimum potential energy in this position.

semiconductor and is referred to as a **donor atom.** Since the energy spacing between the donor levels and the bottom of the conduction band is very small (typically about 0.05 eV), a small amount of thermal energy causes an electron in these levels to move into the conduction band. (Recall that the average thermal energy of an electron at room temperature is $k_BT \approx 0.025$ eV.) Semiconductors doped with donor atoms are called **n-type semiconductors** because the charge carriers are negatively charged electrons.

If the semiconductor is doped with atoms having three valence electrons, such as indium and aluminum, the three electrons form covalent bonds with neighboring atoms, leaving an electron deficiency, or hole, in the fourth bond (see Fig. 12.28a). The energy levels of such impurities also lie within the energy gap, just above the valence band, as indicated in Figure 12.28b. Electrons from the valence band have enough thermal energy at room temperature to fill these impurity levels, leaving behind a hole in the valence band. Because a trivalent atom in effect accepts an electron from the valence band, such impurities are referred to as **acceptors.** A semiconductor doped with trivalent (acceptor) impurities is known as **a p-type semiconductor** because the charge carriers are positively charged holes. When conduction is dominated by acceptor or donor impurities, the material is called an **extrinsic semiconductor** or **impurity semiconductor.** The typical range of doping densities for n- or p-type semiconductors is 10^{13} to 10^{19} cm^{-3}.

Now consider what happens when a p-type semiconductor is joined to an n-type semiconductor to form a **p-n junction.** We find that the completed junction consists of three distinct semiconductor regions, as shown in Figure 12.29a: a p-type region, a **depletion region,** and an n-type region. The depletion region may be visualized as arising when the two halves of the junction are brought together and mobile donor electrons diffuse to the p side of the junction, leaving behind immobile positive ion cores. (Conversely, holes diffuse to the n side and leave a region of fixed negative ion cores.) **The region extending several microns from the junction is called the depletion region because it is**

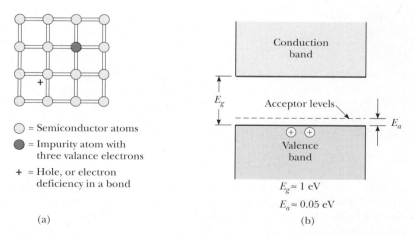

= Semiconductor atoms

= Impurity atom with three valance electrons

+ = Hole, or electron deficiency in a bond

$E_g \approx 1$ eV

$E_a \approx 0.05$ eV

(a)   (b)

Figure 12.28   (a) A two-dimensional representation of a semiconductor containing an acceptor atom (colored spot). (b) An energy-band diagram of a semiconductor in which the acceptor levels lie within the forbidden gap, just above the top of the valence band.

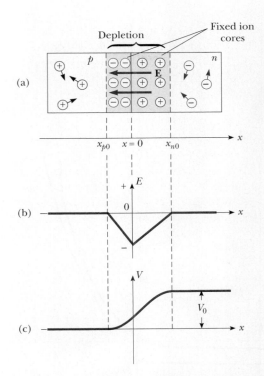

Figure 12.29   (a) The physical arrangement of a *p-n* junction. (b) Built-in electric field versus *x* for the *p-n* junction. (c) Built-in potential versus *x* for the *p-n* junction.

depleted of mobile charge carriers. It also contains a built-in electric field on the order of 10^3 to 10^5 V/cm, which serves to sweep mobile charge out of this region and keep it truly depleted. This electric field creates a potential barrier V_0 that prevents the further diffusion of holes and electrons across the junction and ensures zero current through the junction when no external voltage is applied.

Perhaps the most notable feature of the *p-n* junction is its ability to pass current in only one direction. Such *diode* action is easiest to understand in terms of the potential diagram in Figure 12.29c. If a positive external voltage is applied to the *p* side of the junction, the overall barrier is decreased, resulting in a current that increases exponentially with increasing forward voltage or bias. For reverse bias (a positive external voltage applied to the *n* side of the junction), the potential barrier is increased, resulting in a very small reverse current that quickly reaches a saturation value, I_0, with increasing reverse bias. The current–voltage relation for an ideal diode is

$$I = I_0(e^{qV/k_BT} - 1) \tag{12.35}$$

where q is the electronic charge, k_B is Boltzmann's constant, and T is the temperature in kelvins. Figure 12.30 shows a plot of an *I–V* characteristic for a real diode, along with a schematic of a diode under forward bias. Region 1 shows reverse-bias operation, region 2 ordinary forward bias, and region 3 the extreme forward bias needed for a *p-n* junction laser. We treat *p-n* junction lasers in Section 12.7.

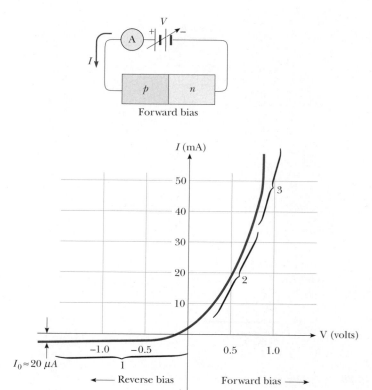

Figure 12.30   The characteristic curve for a real diode.

EXAMPLE 12.3   Forward and Reverse Currents in a Diode

Calculate the ratio of the forward current to the reverse current for a diode described by Equation 12.35 for an applied voltage across the diode of +1.0 V (forward bias) and −1.0 V (reverse bias).

Solution

$$\frac{I_{\text{forward}}}{I_{\text{reverse}}} = \frac{I_0(e^{+qV/k_BT} - 1)}{I_0(e^{-qV/k_BT} - 1)}$$

Taking $k_B T = 0.025$ eV at room temperature gives

$$\frac{I_f}{I_r} = \frac{I_0(e^{(1.0\text{ eV})/(0.025\text{ eV})} - 1)}{I_0(e^{-(1.0\text{ eV})/(0.025\text{ eV})} - 1)} = \frac{(e^{40} - 1)}{(e^{-40} - 1)}$$

$$= -2.4 \times 10^{17}$$

Light-Emitting and -Absorbing Diodes—LEDs and Solar Cells

Light emission and absorption in semiconductors is similar to light emission and absorption by gaseous atoms, except that discrete atomic energy levels must be replaced by bands in semiconductors. As shown in Figure 12.31a, an electron excited electrically into the conduction band can easily recombine with a hole (especially if the electron is injected into a p-type region), emitting a photon of band-gap energy. Examples of p-n junctions that convert electrical input to light output are light-emitting diodes (LEDs) and injection lasers.

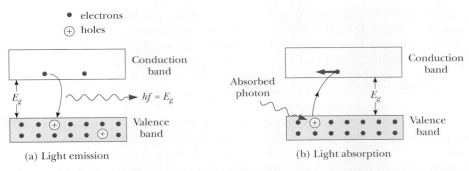

• electrons
⊕ holes

(a) Light emission

E_g

$hf = E_g$

Conduction band

Valence band

(b) Light absorption

Absorbed photon

E_g

Conduction band

Valence band

Figure 12.31   (a) Light emission from a semiconductor. (b) Light absorption by a semiconductor.

Conversely, an electron in the valence band may absorb a photon and be promoted to the conduction band, leaving a hole behind (see Fig. 12.31b). Devices in which light-generated electrons and holes are separated by a junction field and collected as current are called **solar cells** or **photovoltaic devices.** They are described in detail in an essay by John Meakin on our Web site http://info.brookscole.com/mp3e.

The Junction Transistor

The invention of the transistor by American physicists John Bardeen (1908–1991), Walter Brattain (1902–1987), and William Shockley (1910–1989) in 1948 totally revolutionized the world of electronics and led to a Nobel prize, shared by the three men, in 1956. By 1960 the transistor had replaced the vacuum tube in many electronic applications, giving rise to a multibillion-dollar

The Nobel prize was awarded in 1956 to Americans (L–R) John Bardeen, William Shockley, and William Brattain. (*Lucent Technologies' Bell Laboratory, courtesy of AIP Emilio Segrè Visual Archives*)

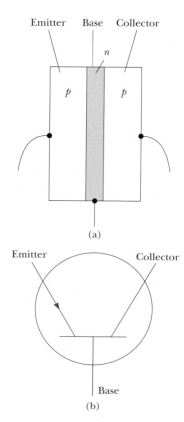

Emitter   Base   Collector

(a)

Emitter   Collector

Base

(b)

Figure 12.32   (a) The *pnp* transistor consists of an *n* region (base) sandwiched between two *p* regions (the emitter and the collector). (b) The circuit symbol for the *pnp* transistor.

industry that produces such popular devices as pocket radios, handheld organizers, computers, CD and DVD players, cell phones, and electronic games.

The junction transistor consists of a semiconducting material with a very narrow *n* region sandwiched between two *p* regions, called the ***pnp* transistor,** or a *p* region sandwiched between two *n* regions, called the ***npn* transistor.** Because the operations of the two transistors are essentially the same, we shall describe only the *pnp* transistor.

Figure 12.32 shows the structure of the *pnp* transistor together with its circuit symbol. The outer regions of the transistor are called the *emitter* and the *collector*, and the narrow central region is called the *base*. Note that the configuration contains two junctions. One junction is the interface between the emitter and the base, and the other is that between the base and the collector.

Suppose a voltage is applied to the transistor such that the emitter is at a higher potential than the collector. (This is accomplished with battery V_{ec} in Figure 12.33.) If we think of the transistor as two diodes back to back, we see that the emitter–base junction is forward-biased and the base–collector junction is reverse-biased. Because the *p*-type emitter is heavily doped relative to the base, nearly all of the current consists of holes moving across the emitter–base junction. Most of these holes do not recombine in the base because it is very narrow. The holes are finally accelerated across the reverse-biased base–collector junction, producing the current I_c in Figure 12.33.

Although only a small percentage of holes recombine in the base, those that do limit the emitter current to a small value, because positive charge carriers accumulate in the base and prevent holes from flowing into this region. To prevent this limitation of current, some of the positive charge on the base must be drawn off; this is accomplished by connecting the base to a second battery, V_{eb} in Figure 12.33. Those positive charges that are not swept across the collector–base junction leave the base through this added pathway. This base current, I_b, is very small, but a small change in it can significantly change the larger collector current, I_c. If the transistor is properly biased, the collector (output) current is directly proportional to the base (input) current, and the transistor acts as a current amplifier. This condition may be written

$$I_c = \beta I_b \qquad (12.36)$$

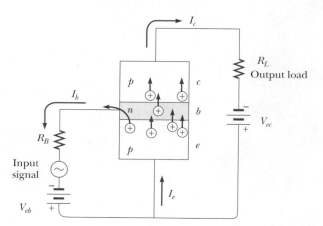

Figure 12.33   A small base current, I_b, controls a large collector current, I_c. Note that $I_e = I_b + I_c$, $I_b \ll I_c$, and $I_c = \beta I_b$, where $\beta \approx 10$ to 100.

(a)   (b)

(a) Jack Kilby, co-inventor of the integrated circuit, is surrounded by devices containing integrated circuits at the Texas Instruments headquarters. (*Roger Ressmeyer/CORBIS*) (b) Robert Noyce, former Sematech chief executive, holding up a semiconductor wafer at Sematech in Austin, Texas in 1989. (*AP photo/David Breslaner*)

where β, the current gain, is typically in the range from 10 to 100. Thus the transistor may be used to amplify a small time-varying signal. The small voltage to be amplified is placed in series with the battery V_{eb} as shown in Figure 12.33. The time-dependent input signal produces a small variation in the base current. This variation results in a large change in collector current and hence a large change in voltage across the output resistor.

The Field-Effect Transistor (FET)

During the 1960s, the electronics industry converted many electronic applications from the junction transistor to the **field-effect transistor (FET),** which is easier to manufacture and uses much less power than the junction transistor. Figure 12.34 shows the structure of a very common device, the **MOSFET,** or **metal-oxide-semiconductor field-effect transistor.** You are likely using millions of MOSFET devices as switches when you are working on your new laptop.

There are three metal connections to the transistor—the *source, drain,* and *gate.* The source and drain are connected to *n*-type semiconductor regions at either end of the structure. These regions are connected by a narrow channel of additional *n*-type material, the *n* channel. The source and drain regions and the *n* channel are embedded in a *p*-type substrate material. This forms a depletion region, as in the junction diode, along the bottom of the *n* channel. (Depletion regions also exist at the junctions underneath the source and drain regions, but we will ignore these because the operation of the device depends primarily on the behavior in the channel.) The gate is separated from the *n* channel by a layer of insulating silicon dioxide. Thus, it does not make electrical contact with the rest of the semiconducting material.

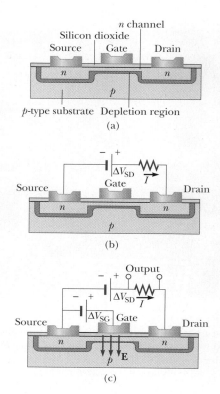

Figure 12.34 (a) The structure of a metal-oxide-semiconductor field-effect transistor (MOSFET). (b) A source–drain voltage is applied, with the result that current exists in the circuit. (c) A gate voltage is applied. The gate voltage can be used to control the source–drain current, so that the MOSFET acts as an amplifier.

Imagine that a voltage source ΔV_{SD} is applied across the source and drain as shown in Figure 12.34b. With this situation, electrons will flow through the upper region of the n channel. Electrons cannot flow through the depletion region in the lower part of the n channel because this region is depleted of charge carriers. Now a second voltage ΔV_{SG} is applied across the source and gate as in Figure 12.34c. The positive potential on the gate electrode results in an electric field below the gate that is directed downward in the n channel. (This is the field in "field-effect.") This electric field exerts upward forces on electrons in the region below the gate, causing them to move into the n channel. This causes the depletion region to become smaller, widening the area through which there is current between the top of the n channel and the depletion region. As the area widens, the current increases.

If a varying voltage is applied to the gate, the area through which the source–drain current exists varies in size according to the varying gate voltage. A small variation in gate voltage results in a large variation in current and a correspondingly large voltage across the resistor in Figure 12.34c. Therefore, the MOSFET acts as a voltage amplifier. A circuit consisting of a chain of such transistors can result in a very small initial signal from a microphone being amplified enough to drive powerful speakers at an outdoor concert.

The MOSFET can be used as a switch by reversing the potential difference ΔV_{SG} in Figure 12.34c. In this case, increasing the voltage causes the n channel to decrease in size. If this voltage is large enough to completely block the area through which electrons pass, the current falls to zero. Thus the MOSFET can be viewed as a low-power switch: A very low input current in the source–gate circuit essentially opens and closes the source–drain circuit by changing the resistance of the source–drain by several orders of magnitude. This very low power switching behavior is especially useful in integrated circuits with millions of transistors, where power dissipation is a major problem.

The Integrated Circuit

The integrated circuit (IC), invented independently by Americans Jack Kilby (b. 1923; Nobel prize, 2000) at Texas Instruments in late 1958 and Robert Noyce (1927–1992) at Fairchild Camera and Instrument in early 1959, has justly been called the most remarkable technology ever to hit mankind. ICs have started a second industrial revolution and are found at the hearts of computers, watches, cameras, automobiles, aircraft, robots, space vehicles, and all sorts of communication and switching networks. In simplest terms, an **integrated circuit** is a collection of interconnected transistors, diodes, resistors, and capacitors fabricated on a single piece of silicon, affectionately known as a chip. State-of-the-art chips easily contain several million components in a 1-cm^2 area (Fig. 12.35).

Interestingly, integrated circuits were invented partly in an effort to achieve circuit miniaturization and partly to solve the interconnection problem spawned by the transistor. In the era of vacuum tubes, the relatively great power and size of individual components severely limited the number of components that could be interconnected in a given circuit. With the advent of the tiny, low-power, highly reliable transistor, design limits on the number of components disappeared and were replaced by the challenge of wiring together hundreds of thousands of components. The magnitude of this problem can be appreciated when one considers that second-generation computers (consisting of discrete transistors) contained several hundred thousand components requiring more than a million hand-soldered joints to be made and tested.

Figure 12.35   This powerful single-chip 64-bit microprocessor (the R10000 RISC) is designed for a broad range of computer applications, from personal computers to supercomputers. The processor has an area of 298 mm^2 and contains approximately 6 million transistors. (*Courtesy of MIPS Technologies, Inc.*)

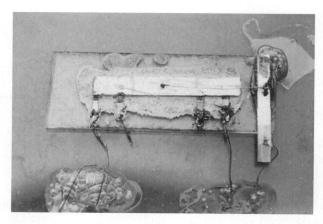

Figure 12.36 Jack Kilby's first integrated circuit, tested on September 12, 1958. (*Courtesy of Texas Instruments, Inc.*)

Figures 12.36 and 12.37 show pictures of the first ICs that contained multiple electronic devices on a single piece of silicon. Noyce's IC also shows interconnections between devices *fabricated as part of the chip manufacturing process.* In addition to solving the interconnection problem, ICs possess the advantages of miniaturization and fast response, two attributes that are critical for high-speed computers. The fast response is actually also a product of the miniaturization and close packing of components, because the response time of a circuit depends on the time it takes for electrical signals traveling at about 1 ft/ns to pass from one component to another. This time is reduced by the closer packing of components. Additional information on the history, theory of operation, and fabrication of chips may be found in the Suggestions for Further Reading at the end of this chapter.

Figure 12.38a illustrates the dramatic advances made in chip technology since Intel introduced the first microprocessor in 1971. Figure 12.38b is a graph of the logarithm of the number of transistors in a chip as a function of the year in which the chip was introduced. Because this growth follows an

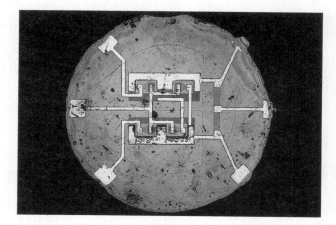

Figure 12.37 One of Robert Noyce's first integrated circuits. Note that the leads connecting different components are built in as part of the chip. (*Courtesy of Fairchild Semiconductor Corporation*)

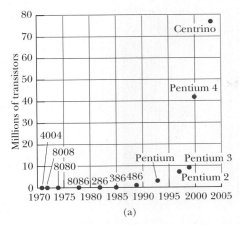

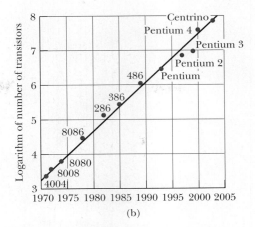

(a)   (b)

Figure 12.38   Dramatic advances in chip technology related to computer microchips manufactured by Intel. (a) A plot of the number of transistors on a single computer chip versus year of manufacture. (b) A plot of the logarithm of the number of transistors in part (a). The fact that this plot is approximately a straight line shows that the growth is exponential.

approximately straight line, we conclude that that the growth is exponential. "Moore's law," proposed in 1965 by Gordon Moore, a cofounder of Intel with Noyce, claims that the number of transistors per square inch on integrated circuits should double every 18 months. The doubling time in Figure 12.38b is longer than 18 months because this graph shows only the total number of transistors and does not take into account the shrinking size of the integrated circuit over the years. When the density of transistors per square centimeter is graphed, the results are similar to those predicted by Moore's law.

12.6 SUPERCONDUCTIVITY

There is a class of metals and compounds known as **superconductors** whose electrical resistance R decreases to virtually zero below a certain temperature T_c, called the *critical temperature* (Table 12.9). Let us now look at these amazing materials with infinite conductivity in greater detail, using what we have just learned about the properties of solids to help us understand the behavior of superconductors.

We start by examining the **Meissner effect,** a phenomenon in which magnetic fields are expelled from the interior of superconductors. Simple arguments based on the laws of electricity and magnetism can be used to show that the magnetic field inside a superconductor cannot change with time. According to Ohm's law, the resistance of a conductor is given by $R = \Delta V / I$, where ΔV is the potential difference across the conductor and I is the current in the conductor. Because the potential difference ΔV is proportional to the electric field inside the conductor, the electric field is proportional to the resistance of the conductor. Therefore, because $R = 0$ for a superconductor at or below its critical temperature, **the electric field in its interior must be zero.** Now recall that Faraday's law of induction can be expressed as

$$\oint \mathbf{E} \cdot d\mathbf{s} = -\frac{d\Phi_B}{dt}$$

Table 12.9   Critical Temperatures for Various Superconductors

Material	T_c(K)
Zn	0.88
Al	1.19
Sn	3.72
Hg	4.15
Pb	7.18
Nb	9.46
Nb_3Sn	18.05
Nb_3Ge	23.2
$YBa_2Cu_3O_7$	92
Bi–Sr–Ca–Cu–O	105
Tl–Ba–Ca–Cu–O	125
$HgBa_2Ca_2Cu_3O_8$	134

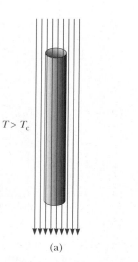
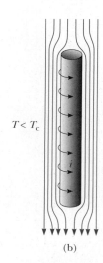

$T > T_c$ $T < T_c$

(a)   (b)

Figure 12.39 A superconductor in the form of a long cylinder in the presence of an external magnetic field. (a) At temperature above T_c, the field lines penetrate the cylinder because it is in its normal state. (b) When the cylinder is cooled to $T < T_c$ and becomes superconducting, magnetic flux is excluded from its interior by the induction of surface currents.

That is, the line integral of the electric field around any closed loop is equal to the negative rate of change in the magnetic flux Φ_B through the loop. Because **E** is zero everywhere inside the superconductor, the integral over any closed path inside the superconductor is zero. Therefore, $d\Phi_B/dt = 0$; this tells us that **the magnetic flux in the superconductor cannot change.** From this information, we conclude that $B = \Phi_B/A$ must remain constant inside the superconductor.

Before 1933, it was assumed that superconductivity was a manifestation of perfect conductivity. If a perfect conductor is cooled below its critical temperature in the presence of an applied magnetic field, the field should be trapped in the interior of the conductor even after the external field is removed. In addition, the final state of the perfect conductor should depend on which occurs first, the application of the field or the cooling to below T_c. If the field is applied after the material has been cooled, the field should be expelled from the superconductor. If the field is applied before the material is cooled, the field should not be expelled once the material has been cooled. In 1933, however, W. Hans Meissner and Robert Ochsenfeld discovered that when a metal becomes superconducting in the presence of a weak magnetic field, the field is expelled. Thus, the same final state **B** = 0 is achieved whether the field is applied before or after the material is cooled below its critical temperature.

The Meissner effect is illustrated in Figure 12.39 for a superconducting material in the shape of a long cylinder. Note that the field penetrates the cylinder when its temperature is greater than T_c (Fig. 12.39a). As the temperature is lowered to below T_c, however, the field lines are spontaneously expelled from the interior of the superconductor (Fig. 12.39b). Thus, a superconductor is more than a perfect conductor (resistivity $\rho = 0$); it is also a perfect diamagnet (**B** = 0). The property that **B** = 0 in the interior of a superconductor is as fundamental as the property of zero resistance. If the mag-

Figure 12.40   A small permanent magnet levitated above a pellet of the $YBa_2CU_3O_{7-\delta}$ superconductor cooled to the temperature of liquid nitrogen, 77 K. (*Courtesy of IBM Research*)

nitude of the applied magnetic field exceeds a critical value B_c, defined as the value of B that destroys a material's superconducting properties, the field again penetrates the sample.

Because a superconductor is a perfect diamagnet having a *negative* magnetic susceptibility, it repels a permanent magnet. In fact, one can perform a demonstration of the Meissner effect by floating a small permanent magnet above a superconductor and achieving magnetic levitation, as seen in Figure 12.40. Recall from your study of electricity that a good conductor expels static electric fields by moving charges to its surface. In effect, the surface charges produce an electric field that exactly cancels the externally applied field inside the conductor. In a similar manner, a superconductor expels magnetic fields by forming surface currents. To see why this happens, consider again the superconductor shown in Figure 12.39. Let us assume that the sample is initially at a temperature $T > T_c$ as illustrated in Figure 12.39a, so that the magnetic field penetrates the cylinder. As the cylinder is cooled to a temperature $T < T_c$, the field is expelled, as shown in Figure 12.39b. Surface currents induced on the superconductor's surface produce a magnetic field that exactly cancels the externally applied field inside the superconductor. As you would expect, the surface currents disappear when the external magnetic field is removed.

A successful theory of superconductivity in metals was published in 1957 by American physicists John Bardeen, Leon N. Cooper (b. 1930), and J. Robert Schreiffer (b. 1931) and is generally called BCS theory, based on the first letters of their last names. This theory led to a Nobel Prize in Physics for the three scientists in 1972 (the second Nobel prize for Bardeen!). In this theory, two electrons can interact via distortions in the array of lattice ions so that there is a net attractive force between the electrons.[7] As a result, the two electrons are bound into an entity called a **Cooper pair.** The Cooper pair behaves like a particle with integral spin. Recall that particles with integral spin are *bosons.* (Particles with half-integer spins are *fermions.*) An important feature of bosons is that they *do not* obey the Pauli exclusion principle. Consequently, at very low temperatures, it is possible for all bosons in a collection of such particles to be in the lowest quantum state. As a result, **the entire collection of Cooper pairs in the metal is described by a single wavefunction.** Above the energy level associated with this wavefunction is an energy gap equal to the binding energy of a Cooper pair. Under the action of an applied electric field, the Cooper pairs experience an electric force that causes them to move through the metal. A random-scattering event of a Cooper pair from a lattice ion represents resistance to the electric current. Such a collision changes the energy of the Cooper pair because some energy is transferred to the lattice ion. But there are no available energy levels below that of the Cooper pair (it is already in the lowest energy level), and none available above, because of the energy gap. As a result, collisions are forbidden and there is no resistance to the movement of Cooper pairs.

An important development in physics that elicited much excitement in the scientific community was the discovery of high-temperature copper oxide–based superconductors. The excitement began with a 1986 publication by J. Georg Bednorz (b. 1950) and K. Alex Müller (b. 1927), scientists at the IBM Zurich Research Laboratory in Switzerland. In their seminal paper,[8] Bednorz and Müller reported strong evidence for superconductivity at 30 K in an oxide of barium, lanthanum, and copper. They were awarded the Nobel Prize for Physics in 1987 for their remarkable discovery. Shortly thereafter, a new family of compounds was open for investigation, and research activity in the field of superconductivity proceeded vigorously. In early 1987, groups at the University of Alabama at Huntsville and the University of Houston announced superconductivity at about 92 K in an oxide of yttrium, barium, and copper ($YBa_2Cu_3O_7$). Later that year, teams of scientists from Japan and the United States reported superconductivity at 105 K in an oxide of bismuth, strontium, calcium, and copper. More recently, scientists have reported superconductivity at temperatures as high as 150 K in an oxide containing mercury. Today, one cannot rule out the possibility of room-temperature superconductivity, and the mechanisms responsible for the behavior of high-temperature superconductors are still under investigation. The search for novel superconducting materials continues both for scientific reasons and because practical applications become more probable and widespread as the critical temperature is raised.

While BCS theory was very successful in explaining superconductivity in metals, there is currently no widely accepted theory for high-temperature superconductivity. This remains an area of active research. The interested reader should see our companion Web site http://info.brookscole.com/mp3e for a

[7]A highly simplified explanation of this attraction between electrons is as follows. Around one electron, the attractive Coulomb force causes surrounding positively charged lattice ions to move slightly inward toward the electron. As a result, there is a higher concentration of positive charge in this region than elsewhere in the lattice. A second electron is attracted to the higher concentration of positive charge.

[8]J. G. Bednorz and K. A. Müller, *Z. Phys. B,* 64:189, 1986.

detailed treatment of superconductivity, including discussions of magnetism in matter, Type I and II superconductors, BCS theory, Josephson tunneling, high-temperature superconductivity, and superconducting devices.

12.7 LASERS

Lasers provide an interesting example of the basic principles of the interaction of radiation with atoms. Because of this and because of the large number of engineering applications of lasers—such as CD-ROMs, precision surveying and length measurement, precision cutting and shaping of materials, and communication by optical fibers—it is worthwhile to discuss the principles of operation of lasers in some detail.

Absorption, Spontaneous Emission, and Stimulated Emission

To understand the operation of a laser, we must become familiar with the processes that describe the emission and absorption of radiation by atoms. Although it would be stretching the truth to say that Einstein anticipated the laser, he laid the foundations for the device with his explanation in 1917 of the processes of atomic absorption, spontaneous emission, and stimulated emission.[9] Einstein showed that an atom can absorb a photon of energy, hf, from a radiation field and make a transition from a lower energy state, E_1, to a higher state, E_2, where $E_2 - E_1 = hf$. Let us set the probability of absorption per unit time per atom equal to $B_{12}u(f, T)$, where B_{12} is Einstein's coefficient of absorption and $u(f, T)$ is the energy density per unit frequency in the radiation field. The probability of absorption of radiation is thus assumed to be proportional to the density of radiation. Once excited, an atom in state 2 is observed to have a definite probability per second of making a **spontaneous transition** (that is, one not induced by radiation) back to level 1. The spontaneous transition rate may be characterized by the coefficient A_{21} (transitions per unit time) or by the spontaneous emission lifetime, $t_s = 1/A_{21}$ (time per transition).

    In addition to spontaneous emission, which is independent of the radiation density, another kind of emission can occur that is dependent on the radiation density. If a photon of energy hf interacts with an atom when it is in level 2, the electric field associated with this photon can stimulate or induce atomic emission such that the emitted electromagnetic wave (photon) vibrates in-phase with the stimulating wave (photon) and travels in the same direction. Two such photons are said to be **coherent.** We may set the stimulated transition rate per atom equal to $B_{21}u(f, T)$, where B_{21} is Einstein's coefficient of stimulated emission. Figure 12.41 summarizes the three processes of absorption, spontaneous emission, and stimulated emission.[10]

Spontaneous transition

[9]A. Einstein, "Zur Quanten Theorie der Strahlung (On the Quantum Theory of Radiation)," *Phys. Zeit.*, 18:121–128, 1917.

[10]Although spontaneous emission has no classical analog and is a truly quantum effect (arising from quantization of the radiation field), absorption and stimulated emission do have classical analogs. If the atom–photon interaction is modeled as a charged harmonic oscillator driven by an electromagnetic wave, the classical oscillator can either absorb energy from the wave or give up energy, depending on the relative phases of oscillator and wave.

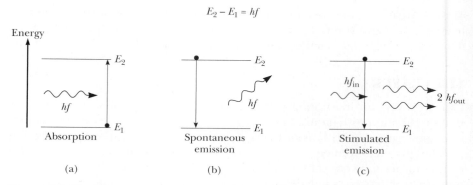

Figure 12.41 The processes of (a) absorption and (b) spontaneous emission. The lifetime of the upper state is t_s, and the photon is emitted in a random direction. (c) Stimulated emission. In this process, the emitted photons are in phase with the stimulating photon, and all have the same direction of travel.

Next, following Einstein, consider a mixture of atoms and radiation in thermal equilibrium at temperature T. The populations N_1 and N_2 of the energy levels E_1 and E_2 obey the Boltzmann relation (discussed in Chapter 10).

$$\frac{N_2}{N_1} = e^{-(E_2-E_1)/k_{\mathrm{B}}T} = e^{-hf/k_{\mathrm{B}}T} \tag{12.37}$$

This equilibrium ratio of populations represents a case of *dynamic equilibrium* in which the number of atomic transitions per second up (from E_1 to E_2) is balanced by the number of spontaneous *and* stimulated transitions per second down (from E_2 to E_1). Since the number of atoms making transitions is proportional to the population of the starting level as well as to the transition probability, we may write

Number of atoms going from 1 to 2 per unit time $= N_1 u(f, T)B_{12}$

where $u(f, T)$ is the radiation energy density per unit frequency ($\mathrm{J/m^3 \cdot Hz}$).

Number of atoms going from 2 to 1 per unit time by stimulated emission
$$= N_2 u(f, T)B_{21}$$

For spontaneous emission there is no dependence on the radiation field, and we merely have

Number of atoms going from 2 to 1 per unit time spontaneously $= N_2 A_{21}$

Since the number of upward transitions per unit time must equal the number of downward transitions per unit time, we have

$$N_1 u(f, T)B_{12} = N_2[B_{21}u(f, T) + A_{21}] \tag{12.38}$$

Using Equation 12.37, this becomes

$$u(f, T) = \frac{A_{21}}{B_{12}e^{hf/k_{\mathrm{B}}T} - B_{21}} \tag{12.39}$$

Equation 12.39 is an expression for the radiation density per unit frequency of a system of atoms and radiation in equilibrium at temperature T expressed in terms of Einstein coefficients A_{21}, B_{12}, and B_{21}. But Planck's blackbody radiation law (Chapter 3) provides us with an expression for a system of radiation and oscillators in thermal equilibrium:

$$u(f,\, T) = \frac{8\pi h f^3}{c^3} \frac{1}{(e^{hf/k_{\mathrm{B}}T} - 1)} \tag{3.9}$$

Comparing Equations 12.39 and 3.9, we find the interesting results

$$B_{21} = B_{12} = B \tag{12.40}$$

and

$$\frac{A_{21}}{B} = \frac{8\pi h f^3}{c^3} \tag{12.41}$$

Relation of Einstein coefficients of stimulated emission, absorption, and spontaneous emission

Equation 12.40 states that any atom that has a finite probability per unit time of absorption has an *equal* probability of stimulated emission. Equation 12.41 shows that the process of spontaneous emission dominates the process of stimulated emission at higher frequency.

Population Inversion and Laser Action

Since $B_{12} = B_{21}$, there is an equal probability that a photon will cause an upward (absorption) or downward (stimulated emission) atomic transition. When light is incident on a system of atoms in thermal equilibrium, there is usually a net absorption of energy since, according to the Boltzmann distribution, there are more atoms in the ground state than in excited states. However, if one can invert the situation so that there are more atoms in an excited state than in a lower state—a condition called **population inversion**—amplification of photons can result. Under the proper conditions, a single input photon can result in a cascade of stimulated photons, all of which are in-phase, traveling in the same direction and of the same frequency as the input photon. Since the device that does this is a light amplifier, it is called a **laser,** an acronym for **l**ight **a**mplification by **s**timulated **e**mission of **r**adiation.

Population inversion

Although there are many different types of lasers, most lasers have certain essential features:

- An energy source capable of producing either pulsed or continuous population inversions. In the case of the helium–neon gas laser, the energy source is an electrical discharge that imparts energy by electron–atom collisions. In the case of the ruby crystal laser, the population inversion is produced by intense flashes of broadband illumination from flash lamps. The process of excitation with intense illumination is called **optical pumping.**
- A lasing medium with at least *three* energy levels: a ground state; an intermediate (metastable) state with a relatively long lifetime, t_s; and a high-energy pump state (Fig. 12.42). To obtain population inversion, t_s must be greater than t_2, the lifetime of the pump state E_2. Note, too, that amplification cannot be obtained with only two levels, because such a

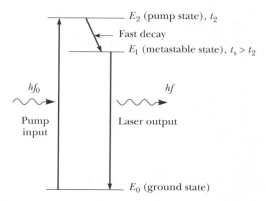

Figure 12.42 A three-level laser system. t_2 is the lifetime of the state E_2, and t_s is the lifetime of the state E_1.

system cannot support a population inversion. At most, with extremely intense optical pumping we can increase the population of the upper state in a two-level system until it equals the population of the lower state. Further pumping serves only to excite as many downward transitions as upward transitions, since the probability of absorption is equal to the probability of stimulated emission for a *given* transition. For a population inversion to be produced, energy absorption must occur for a transition *different* from the transition undergoing stimulated emission—thus the need for at least a three-level system.

- A method of containing the initially emitted photons within the laser so that they can stimulate further emission from other excited atoms. In practice this is usually achieved by placing mirrors at the ends of the lasing medium so that photons make multiple passes through the laser. Thus the laser may be thought of as an optical resonator or oscillator with two opposing reflectors at right angles to the laser beam. The oscillation consists of a plane wave bouncing back and forth between the reflectors. The oppositely traveling plane waves, in turn, generate a highly monochromatic standing wave, which is strongest at resonance when an integral number of half-wavelengths just fits between the reflectors (Fig. 12.43). To extract a highly collimated beam from the laser, one of the parallel mirrors is made slightly transmitting so that a small amount of energy leaks out of the cavity.

These three features of lasers lead to several unique characteristics of laser light that make it a much more powerful technological tool than light from ordinary sources. These unique characteristics are high monochromaticity, strong intensity, spatial coherence, and low beam divergence. Table 12.10 shows typical values of these quantities for a low-power gas laser. In comparison with a single emission line in a strong conventional light source, the He–Ne laser is about 100 times more monochromatic, generates about 100 times more beam power, and has about 1000 times more exit irradiance (power per unit area). The beam divergence, or spread, is measured in milliradians (mrad). A 1.0-mrad laser would have a beam diameter of 1 cm at a distance of 10 m.

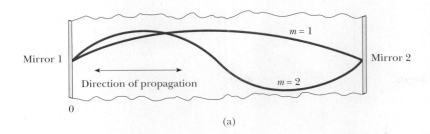

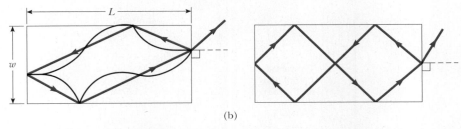

Figure 12.43   Cavity modes in a laser. (a) Longitudinal modes have $L = m\lambda/2$, where m is an integer, λ is the laser wavelength in the laser material, and L is the distance between the two mirrors. (b) Modes with transverse components can exist if both ends and sides of the laser are made reflective. Arrows show the direction of propagation of light rays.

Semiconductor Lasers

One of the most important light sources for fiber-optic communication is the semiconductor laser. This laser is eminently suited to fiber-optic communication because of its small size (maximum dimension several tenths of a millimeter), high efficiency (25 to 30%), simplicity, long lifetime (p-n junctions have estimated working lifetimes of about 100 years), ease of modulation (through control of the junction current), and fast response (extending well into the GHz range). Figure 12.44 shows a typical gallium arsenide (GaAs) p-n junction laser along with nominal operating values. The first semiconductor laser was made of GaAs and operated at 0.842 μm in the IR.[11,12] Shortly thereafter,

Table 12.10   Typical He-Ne Gas Laser Characteristics

Monochromaticity	Beam Power	Beam Diameter at Laser Exit	Beam Divergence
632.81 ± 0.002 nm	10 mW	0.50 mm	1.0 mrad

Irradiance at Laser Exit	Irradiance for a Beam Focused to a 10-μm Diameter
5100 mW/cm^2	1.27×10^7 mW/cm^2

[11]R. N. Hall, G. E. Fenner, J. D. Kingsley, T. J. Soltys, and R. O. Carlson, "Coherent Light Emission from GaAs Junctions," *Phys. Rev. Lett.,* 9:336, 1962.

[12]M. I. Nathan, W. P. Dumke, G. Burns, F. H. Dill, and G. Lasher, "Stimulated Emission of Radiation from GaAs p-n Junctions," *Appl. Phys. Lett.,* 1:62, 1962.

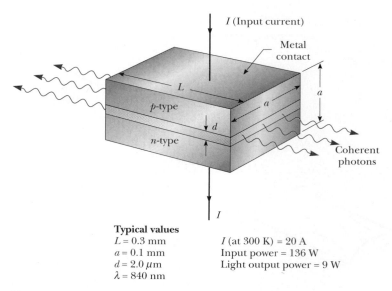

Typical values
$L = 0.3$ mm I (at 300 K) = 20 A
$a = 0.1$ mm   Input power = 136 W
$d = 2.0$ μm   Light output power = 9 W
$\lambda = 840$ nm

Figure 12.44   A gallium arsenide p-n junction laser.

Holonyak and Bevacqua reported the operation of a visible GaAs p-n junction laser,[13] and the laser was on its way to becoming an important part of semiconductor device technology.

Semiconductor lasers are similar to other lasers in that they require a population inversion between quantum levels in order to produce laser action. They are notably different from gas lasers in that (a) discrete energy levels in gas lasers become *energy bands* in semiconductors, and (b) optical pumping and electrical discharge in conventional lasers are replaced by **injection pumping** in semiconductors. In injection pumping, large forward currents are passed through the laser diode, and electrons and holes are injected into p and n regions, respectively, where they recombine and emit radiation. Let us briefly sketch the main points needed to understand semiconductor lasers.

In semiconductors, the discrete energy levels found in isolated atoms widen into bands because of the perturbations caused by neighboring atoms. We are basically interested in the two outermost energy bands, known as the valence (v) and conduction (c) bands. In a semiconductor such as silicon in thermal equilibrium at room temperature, the valence band is filled with electrons and the conduction band, where electrons can move freely, contains no electrons. Since the two levels are separated by a forbidden energy region or gap, $E_g = 1.1$ eV, very few electrons possess enough thermal energy at room temperature to cross the gap and carry a current in the material. Hence silicon, like other semiconductors, is normally a poor conductor of electricity. However, light, heat, or electrical energy added to a semiconductor can promote electrons into the conduction band (c band) and increase the conductivity of a semiconductor. The hole (vacancy) left by the electron in the valence band (v band) plays an equally important role in conduction in a semiconductor, since it behaves as a positive charge moving opposite in direction to the conduction electron. In fact, the hole actually completes the current path

[13]N. Holonyak, Jr., and S. F. Bevacqua, *Appl. Phys. Lett.*, 1:82, 1962.

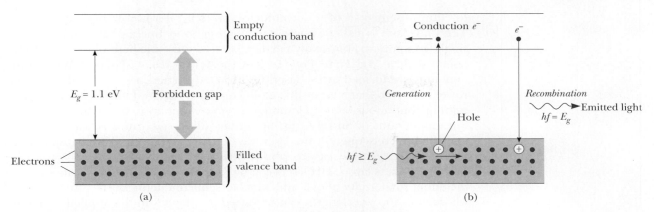

Figure 12.45   (a) Energy bands in pure silicon at 0 K. (b) Generation and recombination of electron–hole pairs in silicon.

through the semiconductor. This process of the creation of an electron–hole pair is called **pair generation.** An electron in the c band can also "fall into" a hole and recombine with it, emitting light of band-gap energy in the process. Both of these important processes are indicated schematically in Figure 12.45. If the thermal equilibrium situation of empty c band and full v band is disturbed by intense optical or electrical excitation, a semiconductor population inversion can be set up, with the v band empty of electrons (full of holes) to a depth X and the c band full of electrons to a height Y. As can be seen in Figure 12.46, once a few spontaneously emitted photons with energy E_g induce stimulated emission at E_g, stimulated emission eventually far exceeds absorption since photons with energy E_g lack sufficient energy to be absorbed. Ultimately, the dominance of the rate of stimulated emission over the rate of absorption leads to line narrowing and laser action.

Other Types of Semiconductor Lasers.   For simplicity, we have described only the basic GaAs injection laser. To leave the reader with the impression that this is the most important laser would be a serious distortion of the truth,

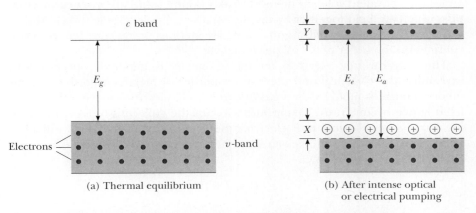

Figure 12.46   A population inversion in a semiconductor. Note that E_e, the emission energy, is equal to E_g and that E_a, the absorption energy, is equal to $E_g + X + Y$. The absorption energy is $E_g + X + Y$, and not just $E_g + X$, because electrons obey the Pauli exclusion principle. Thus electrons must absorb enough energy to occupy states above Y.

for the present field of semiconductor lasers is enormously rich and varied. The GaAs laser is a single-homojunction laser (one junction fabricated from a single compound), but many current lasers are double-heterojunction (DH) lasers, such as the PbTe–PbSnTe and the GaAs–AlGaAs lasers.[14] When two junctions are formed in the laser, the injected carriers are confined to a narrower region so that a lower threshold current is required for laser action, permitting continuous laser operation at room temperature. In addition, by varying the amount of Sn or Al in the ternary alloy, one can vary the band-gap energy and, consequently, the laser wavelength. Also, GaAs–AlGaAs multilayer junctions are relatively easy to fabricate since AlGaAs layers may be grown epitaxially on GaAs layers. DH lasers can be formed with many other compounds, including quaternary alloys, and tunable semiconductor lasers can be made whose emission wavelengths can be varied by changes in the applied pressure or magnetic field. These techniques have resulted in semiconductor lasers that span the broad range from 0.3 to 40 μm. These and other developments in the field of semiconductor lasers, which is a truly fascinating combination of optics, material science, and device physics, are treated more extensively in some of the Suggestions for Further Reading cited at the end of this chapter.

SUMMARY

Bonding mechanisms in solids can be classified in a manner similar to that used to describe bonding of atoms to form molecules. **Ionic solids,** such as NaCl, are formed by the net attractive coulomb force between positive and negative ion neighbors. Ionic crystals are fairly hard, have high melting points, and have strong cohesive energies of several electron volts. **Covalent solids,** such as diamond, form when valence electrons are shared by adjacent atoms, producing bonds of great stability. Covalent solids are generally very hard, have high melting points, and possess the highest cohesive energies, ranging up to 8 eV per atom. **Metallic solids** are held together by the metallic bond, which arises from the attractive force between the positive ion cores and the electron "gas." This electron gas acts as a kind of "cement" to hold the ion cores together. Generally, metallic solids are softer and have lower cohesive energies than covalent solids. The most weakly bound solids are **molecular crystals,** which are held together by van der Waals or dipole–dipole forces. Molecular crystals, such as solid argon, have low melting points and low cohesive energies on the order of 0.1 eV per molecule.

The classical free electron model of metals is generally successful in explaining the electrical and thermal properties of metals. According to this theory, a metal is modeled by a gas of conduction electrons moving through a fixed lattice of ion cores. Taking into account the collisions of electrons with the lattice, the classical free electron model predicts that metals will obey Ohm's law, $\mathbf{J} = \sigma\mathbf{E}$, with the electrical conductivity σ given by

$$\sigma = \frac{ne^2 L}{(3k_B T m_e)^{1/2}} \tag{12.15}$$

---

[14]PbSnTe and AlGaAs are called *ternary* (three-component) *alloys.* They are crystalline solid solutions with Sn and Al randomly replacing some Pb and Ga atoms, respectively.

where n is the concentration of conduction electrons, L is the electronic mean free path, T is the absolute temperature, and m_e is the mass of the electron. The classical model may also be applied to heat conduction by electrons in metals to show that K/σ, where K is the thermal conductivity, is a universal constant for all metals and depends only on the absolute temperature **(Wiedemann–Franz law.)** The classical model yields for this ratio

$$\frac{K}{\sigma} = \frac{3k_B^2}{2e^2}\, T \qquad (12.21)$$

Although the classical electron gas model broadly describes the electrical and thermal properties of metals, there are notable deficiencies in its predictions—namely, in the numerical values of σ and K and in the temperature dependence of σ. These deficiencies may be rectified by taking into account the wave nature and indistinguishability of electrons, by replacing the Maxwell–Boltzmann distribution with the Fermi–Dirac distribution. When this is done, Equation 12.21 for K/σ takes the new form

$$\frac{K}{\sigma} = \frac{\pi^2 k_B^2}{3e^2}\, T = 2.45 \times 10^{-8}\, T \qquad (12.26)$$

and shows excellent agreement with experimental values. The correct quantum expression for the conductivity of a metal is

$$\sigma = \frac{ne^2 L}{m_e v_F}$$

where v_F is the Fermi speed and L is the quantum mean free path, *which is usually several hundred times the metal's interatomic distance.* This unusually long mean free path occurs because electron waves can pass through a lattice of ions without appreciable scattering, since they scatter only from widely spaced crystal imperfections, such as thermal displacements of ions and impurity atoms. The observed temperature dependence of σ can also be explained as the result of electron scattering from thermal lattice vibrations, or phonons.

When many atoms are brought together, the individual atomic levels form a set of allowed bands separated by forbidden energy regions, or gaps. One can understand the properties of metals, insulators, and semiconductors in terms of the **band theory of solids.** The highest occupied band of a *metal* is partially filled with electrons. Therefore, many electrons are free to move throughout the metal and contribute to the conduction current when raised in energy a small amount by a small applied electric field. In an *insulator* at $T = 0$ K, the valence band is completely filled with electrons and the conduction band is empty. The region between the valence band and the conduction band is called the **energy gap** of the material. The energy gap of an insulator is of the order of 10 eV. Because this gap is large compared to $k_B T$ at ordinary temperatures, very few electrons are thermally excited into the conduction band, which explains the small electrical conductivity of an insulator.

A **semiconductor** is a material with a small energy gap, of the order of 1 eV, and a valence band that is filled at 0 K. Because of this small energy gap, a significant number of electrons can be thermally excited from the valence band into the conduction band as temperature increases, thereby producing a semiconductor conductivity that *increases* with temperature. The electrical properties of a semiconductor can be modified by adding donor atoms with five valence electrons (such as arsenic) or by adding acceptor atoms with three valence electrons (such as boron). A semiconductor doped with donor atoms is called an ***n*-type semiconductor,** and one doped with acceptor atoms is called ***p*-type.** The energy levels of these donor and acceptor atoms fall within the energy gap of the material, the donor levels falling just below the conduction band and the acceptor levels just above the valence band.

When a *p*-type material is joined to an *n*-type material, a ***p-n* junction** is formed with the current–voltage characteristics of a diode. If a p-n junction is operated under strong forward bias, electrons can be injected into the *p*-type material and holes into the *n*-type material, where they recombine with holes and electrons, respectively, emitting photons with band-gap energy. This is the principle of operation of the **light-emitting diode** (LED) and the **injection laser.** If a *p-n* junction absorbs a photon with energy greater than or equal to the band gap, an electron may be promoted to the conduction band (leaving a hole in the valence band), thereby producing a current. Devices in which light-generated electrons and holes are separated by a built-in junction field and collected as current are called **solar cells.** Other semiconductor devices that have revolutionized life in the latter half of the 20th century are the **transistor** and the **integrated circuit (IC).** The transistor is a small, low-power, highly reliable amplifying device that basically consists of two *p-n* junctions placed back to back. The integrated circuit, which is the heart of the small computer, consists of millions of interconnected transistors, diodes, resistors, and capacitors fabricated on a single chip of silicon. Interestingly, integrated circuits were invented partly in an effort to achieve greater miniaturization and partly to solve the interconnection problem spawned by the transistor.

Superconductivity occurs when the resistance of a sample falls suddenly to zero at some critical temperature T_c. At T_c magnetic field lines are also completely expelled from the interior to give $B = 0$ inside the superconductor, a phenomenon known as the Meissner effect. BCS theory explains superconductivity in terms of the formation of pairs of free electrons, called Cooper pairs. Cooper pairs can move unhindered through the lattice without scattering from lattice ions.

The processes of atomic absorption, spontaneous emission, and stimulated emission are crucial to an understanding of laser action. By considering a group of atoms and radiation in thermal equilibrium and balancing the rate of upward transitions against the rate of downward transitions, one can show that

$$B_{21} = B_{12} = B \tag{12.40}$$

and

$$\frac{A_{21}}{B} = \frac{8\pi h f^3}{c^3} \tag{12.41}$$

where B_{12} is Einstein's coefficient of absorption, B_{21} is Einstein's coefficient of stimulated emission, and A_{21} is the probability of spontaneous emission per second per atom.

Amplification of the light from a particular atomic transition and generation of an intense, monochromatic, coherent, and highly collimated beam (laser action) may be achieved if certain conditions are met:

- The lasing medium must contain at least three energy levels: a ground state, an intermediate state with a long lifetime, and a high-energy pump state.
- There must be an electrical or optical energy source capable of "pumping" atoms into excited states faster than they leave, so that a population inversion is produced.
- There must be a method of confining the first wave of emitted photons within the laser so that they can stimulate further emission.

Actual lasers include solid-state crystalline lasers such as the ruby laser, gas lasers such as the He–Ne laser, and semiconductor lasers such as the GaAs laser. The most important light source for optical signal processing is the semiconductor laser. This laser is essentially a *p-n* junction operated under strong forward bias. Typical advantages of these lasers are small size, high efficiency, simplicity, ease of modulation, and fast response.

SUGGESTIONS FOR FURTHER READING

1. R. Zallen, *The Physics of Amorphous Solids*, New York, John Wiley and Sons, 1983. A comprehensive treatment of amorphous solids at a senior undergraduate level. Great care is exercised in introducing new concepts verbally before launching into mathematical arguments.
2. T. R. Reid, *The Chip: How Two Americans Invented the Microchip and Launched a Revolution*, New York, Simon and Schuster, 1984. An informal, nonmathematical, extremely interesting chip history.
3. B. G. Streetman, *Solid State Electronic Devices*, 2nd ed., Englewood Cliffs, NJ, Prentice-Hall, 1980. A comprehensive treatment of devices at the junior-senior undergraduate level.
4. W. Shockley, *Electrons and Holes in Semiconductors*, Melbourne, FL, Krieger Publishing Co., 1950. The classic "granddaddy" text on semiconductors. Although somewhat dated, the book is a marvelous compendium of Shockley's insights in the field of semiconductor physics. It starts out at a very clear introductory level and

about halfway through reaches advanced undergraduate and graduate levels.
5. C. Kittel, *Introduction to Solid State Physics*, 7th ed., New York, John Wiley and Sons, 1996. A detailed, comprehensive, up-to-date treatment of solid-state physics at the senior or graduate level. It contains many useful tables of the properties of solids.

Readings on Lasers

6. C. K. N. Patel, "High-Power Carbon Dioxide Lasers," *Sci. Amer.*, August 1968, pp. 22–23.
7. A. L. Schawlow, "Laser Light," *Sci. Amer.*, September 1968, pp. 120–126.

More advanced treatments of lasers may be found in

8. J. I. Pankove, *Optical Processes in Semiconductors*, New York, Dover Publications, 1971.
9. S. M. Sze, *Physics of Semiconductor Devices*, 2nd ed., New York, John Wiley and Sons, 1981.

QUESTIONS

1. Explain how the energy bands of metals, semiconductors, and insulators account for the following general optical properties: (a) Metals are opaque to visible light. (b) Semiconductors are opaque to visible light but transparent to infrared (many IR lenses are made of germanium). (c) Many insulators, such as diamond, are transparent to visible light.
2. Table 12.8 shows that the energy gaps for semiconductors decrease with increasing temperature. What accounts for this behavior?
3. The resistivity of most metals increases with increasing temperature, whereas the resistivity of a semiconductor decreases with increasing temperature. What explains this difference in behavior?
4. Discuss the differences among the band structures of metals, insulators, and semiconductors. How does the band structure model enable you to better understand the electrical properties of these materials?
5. Discuss the electrical, physical, and optical properties of ionically bonded solids.

6. Discuss the electrical and physical properties of covalently bonded solids.

7. When a photon is absorbed by a semiconductor, an electron–hole pair is said to be created. Give a physical explanation of this concept, using the energy band model as the basis for your description.

8. In a semiconductor such as silicon, pentavalent atoms such as arsenic are donor atoms, and trivalent atoms such as indium are acceptors. Inspect the periodic table shown on the endpapers, and determine which other elements would be considered either donors or acceptors.

9. Explain how a *p-n* junction diode operates as a rectifier, as a solar cell, and as a LED.

10. What are the basic assumptions made in the classical free electron theory of metals? How does the energy band model differ from the classical free electron theory in describing the properties of metals?

11. Explain the similarities of electrical conduction and heat conduction in metals.

12. Discuss the differences between crystalline solids, amorphous solids, and gases.

13. Discuss the physical sources of electrical resistance in a metal. Does the resistance depend upon the strong reflections set up when the lattice spacing is equal to an integral number of electronic wavelengths?

14. Does it seem reasonable that the average of the square of a quantity is greater than the square of the average? (Recall from Section 12.2 that $\bar{l} = \tau$ and $\overline{t^2} = 2\tau^2$.)

15. How are the equations governing heat flow and charge flow similar for metals?

16. Radiative emission by an atom can be spontaneous or stimulated. Distinguish between these two processes, identifying the characteristic features of each. Describe the role(s) each one plays in the operation of a laser.

17. Why must a lasing medium possess at least three energy levels?

18. Since the light from semiconductor lasers can be modulated easily, these lasers are of great use in light-wave communication systems. Show that laser communication systems possess a great advantage over microwave communication systems in terms of the amount of information that can be carried. (*Hint:* If each speech channel requires a bandwidth of 5 kHz, calculate the number of channels carried by a laser beam of 10^{16} Hz and by a microwave beam of 10^{10} Hz.)

19. Discuss the main criteria that must be met to achieve laser action in a three-level system.

20. What are the three fundamental ways in which light (photons) interacts with matter (atoms)? Briefly explain each.

21. Assuming that K is a positive constant in Equation 12.18, explain why there is a negative sign in this equation.

22. Explain why Cooper pairs do not scatter from positive lattice ions in a metal.

PROBLEMS

12.1   Bonding in Solids

1. Show that the minimum potential energy of an ion pair in an *ionic* solid is

$$U_0 = -\alpha k \frac{e^2}{r_0}\left(1 - \frac{1}{m}\right)$$

where r_0 is the equilibrium separation between ions, α is the Madelung constant, and the repulsive potential energy between ions is B/r^m. $|U_0|$ is called the *ionic cohesive energy* and is the energy required to separate to infinity an ion pair originally at a separation of r_0 in the crystal. *Hint:*

$$\left.\frac{dU}{dr}\right|_{r=r_0} = 0$$

2. Show that the angle between carbon bonds shown in Figure 12.3a is 109.5°.

3. Calculate the ionic cohesive energy for NaCl, using Equation 12.3. Take $\alpha = 1.7476$, $r_0 = 0.281$ nm, and $m = 8$.

4. LiCl, which has the same crystal structure as NaCl, has $r_0 = 0.257$ nm and has a measured ionic cohesive energy of 199 kcal/mol. (a) Find the integral value of m

that best agrees with the measured value 199 kcal/mol. (b) Is the value of U_0 very sensitive to a change in *m*? Compute the percent change in U_0 when *m* increases by 1.

5. Consider a one-dimensional chain of alternating positive and negative ions. Show that the potential energy of an ion in this hypothetical crystal is

$$U(r) = -k\alpha \frac{e^2}{r}$$

where $\alpha = 2 \ln 2$ (the Madelung constant), and *r* is the interionic spacing. [*Hint:* Make use of the series expansion for $\ln(1 + x)$.]

6. (a) Show that the force on an ion in an ionic solid can be written

$$F = -k\alpha \frac{e^2}{r^2}\left[1 - \left(\frac{r_0}{r}\right)^{m-1}\right]$$

where r_0 is the equilibrium separation and α is the Madelung constant. (*Hint:* Make use of the value of B found in Problem 1.) (b) Imagine that an ion in the solid is displaced a small distance *x* from r_0, and show that the ion experiences a restoring force $F = -Kx$, where

$$K = \frac{k\alpha e^2}{r_0^3}(m - 1)$$

(c) Use the result of (b) to estimate the frequency of vibration of an Na^+ ion in NaCl. Take $m = 8$ and use the value $\alpha = 1.7476$. (d) At what wavelength would the Na^+ ion absorb incident radiation? Is this wavelength in the UV, visible, or IR part of the spectrum?

7. (a) Calculate the ionic cohesive energy for KCl, which has the same crystal structure as NaCl. Take $r_0 = 0.314$ nm and $m = 9$. (b) Calculate the atomic cohesive energy of KCl by using the facts that the ionization energy of potassium is 4.34 eV (that is, $K + 4.34$ eV $\rightarrow K^+ + e$) and that the electron affinity of chlorine is 3.61 eV ($Cl^- + 3.61$ eV $\rightarrow Cl + e$).

8. The Madelung constant for the NaCl structure may be found by summing an infinite alternating series of terms giving the electrostatic potential energy between an Na^+ ion and its 6 nearest Cl^- neighbors, its 12 next-nearest Na^+ neighbors, and so on (see Fig. 12.1a). (a) From this expression, show that the first *three* terms of the infinite series for the Madelung constant for the NaCl structure yield $\alpha = 2.13$. (b) Does this infinite series converge rapidly? Calculate the fourth term as a check.

12.2 Classical Free Electron Model of Metals

9. Consider a group of N electrons, all of which experience a collision with lattice ions at $t = 0$. One can show that the number of electrons that suffer their next collision between t and $t + dt$ follows the exponentially decreasing distribution

$$n(t)\,dt = \frac{Ne^{-t/\tau}}{\tau}\,dt$$

where τ is the mean free time. (a) Show that $\int_0^\infty n(t)\,dt = N$, as expected. (b) Show that this distribution leads to $\bar{t} = \tau$. *Hint:*

$$\bar{t} = \frac{n_1 t_1 + n_2 t_2 + \cdots + n_f t_f}{N} = \frac{1}{N}\int_0^\infty t \cdot n(t)\,dt$$

(c) Show, similarly, that $\overline{t^2} = 2\tau^2$, as stated in the derivation of Equation 12.9.

10. Silver has a density of 10.5×10^3 kg/m³ and a resistivity of 1.60×10^{-8} $\Omega \cdot$m at room temperature. (a) On the basis of the classical free electron gas model, and assuming that each silver atom contributes one electron to the electron gas, calculate the average time between collisions of the electrons. (b) Calculate the mean free path from τ and the electron's thermal velocity. (c) How does the mean free path compare to the lattice spacing?

11. The contribution of a single electron or hole to the electric conductivity of a semiconductor can be expressed by an important property called the *mobility*. The mobility, μ, is defined as the particle drift speed per unit electric field or, in terms of a formula, $\mu = v_d/E$. Note that mobility describes the ease with which charge carriers drift in an electric field and that the mobility of an electron, μ_n, may be different from the mobility of a hole, μ_p, in the same material. (a) Show that the current density may be written in terms of mobility as $J = ne\mu_n E$, where $\mu_n = e\tau/m_e$. (b) Show that when both electrons and holes are present, the conductivity may be expressed in terms of μ_p and μ_n as $\sigma = ne\mu_n + pe\mu_p$, where n is the electron concentration and p is the hole concentration. (c) If a germanium sample has $\mu_n = 3900$ cm²/V·s, calculate the drift speed of an electron when a field of 100 V/cm is applied. (d) A pure (intrinsic) sample of germanium has $\mu_n = 3900$ cm²/V·s and $\mu_p = 1900$ cm²/V·s. If the hole concentration is equal to 3.0×10^{13} cm⁻³, calculate the resistivity of the sample.

12. Gallium arsenide (GaAs) is a semiconductor material of great interest for its high power-handling capabilities and fast response time. (a) Calculate the drift speed of electrons in GaAs for a field of 10 V/cm if the electron mobility is $\mu_n = 8500$ cm²/V·s. (See Problem 11 for the definition of μ.) (b) What percent of the electron's thermal speed at 300 K is this drift speed? (c) Assuming an effective electron mass equal to the free electron mass, calculate the average time between electron collisions. (d) Calculate the electronic mean free path.

12.3 Quantum Theory of Metals

13. Assuming that conduction electrons in silver are described by the Fermi free electron gas model with $E_F = 5.48$ eV, repeat the calculations of Problem 10.

14. Sodium is a monovalent metal having a density of 0.971 g/cm³, an atomic weight of 23.0 g/mol, and a resistivity of 4.20 $\mu\Omega \cdot$cm at 300 K. Use this information to calculate (a) the free electron density; (b) the Fermi energy, E_F, at 0 K; (c) the Fermi velocity, v_F; (d) the average time between electronic collisions; (e) the mean free path of the electrons, assuming that E_F at 0 K is the same as E_F at 300 K; and (f) the thermal conductivity. For comparison to (e), the nearest-neighbor distance in sodium is 0.372 nm.

12.4 Band Theory of Solids

15. The energy gap for Si at 300 K is 1.14 eV. (a) Find the lowest-frequency photon that will promote an electron from the valence band to the conduction band of silicon. (b) What is the wavelength of this photon?

16. From the optical absorption spectrum of a certain semiconductor, one finds that the longest wavelength

of radiation absorbed is 1.85 μm. Calculate the energy gap for this semiconductor.

17. The simplest way to model the energy-level splitting that is produced when two originally isolated atoms are brought close together is with two finite wells. In this model, the Coulomb potential experienced by the outermost electron in each atom is approximated by a one-dimensional finite square well of depth U and width a. The energy levels for two atoms close together may be found by solving the time-independent Schrödinger equation for a potential consisting of two finite wells separated by a distance b. (a) Start this problem by "warming up" with a solution to the single finite well shown in Figure P12.17a. Justify the solutions listed for regions I, II, and III, and apply the standard boundary conditions (ψ and $d\psi/dx$ continuous at $x = 0$ and $x = a$) to obtain a transcendental equation for the bound-state energies. (b) Solve numerically for the bound-state energies when an electron is confined to a well with $U = 100$ eV and $a = 1$ Å. You should find two bound states at approximately 19 and 70 eV. (c) Now consider the finite wells separated by a distance b, as shown in Figure P12.17b. Impose the conditions of continuity in ψ and $d\psi/dx$ at $x = 0$ and $x = a$ to obtain

$$\frac{D}{E'} = \frac{2e^{-2Ka}[\cos ka + \frac{1}{2}(K/k - k/K)\sin ka]}{(k/K + K/k)\sin ka}$$

Show that the boundary conditions at $x = a + b$ yield

$$\frac{F}{G} = \frac{(D/E')e^{\beta}[\cos \alpha - (K/k)\sin \alpha] + e^{-\beta}[\cos \alpha + (K/k)\sin \alpha]}{(D/E')e^{\beta}[\sin \alpha + (K/k)\cos \alpha] + e^{-\beta}[\sin \alpha - (K/k)\cos \alpha]}$$

where $\alpha = k(a + b)$ and $\beta = K(a + b)$. The boundary conditions at $x = 2a + b$ yield

$$\frac{F}{G} = \frac{\cos k(2a + b) + (K/k)\sin k(2a + b)}{\sin k(2a + b) - (K/k)\cos k(2a + b)}$$

Thus, the two expressions for F/G may be set equal and the expression for D/E' used to obtain a transcendental equation for the energy E. (d) Solve for the bound-state energies when $U = 100$ eV and $a = b = 1$ Å. You should observe twofold splitting of the bound-state energies found at 19 and 70 eV for the single well.

12.5   Semiconductor Devices

18. One can roughly calculate the weak binding energy of a donor electron as well as its orbital radius in a semiconductor on the basis of the Bohr theory of the atom. Recall that for a single electron bound to a nucleus of charge Z, the binding energy of the ground state is given by

$$\frac{ke^2 Z^2}{2a_0} = 13.6 Z^2 \text{ eV} \qquad (4.36)$$

and the ground-state radius by

$$r_1 = a_0/Z \qquad (4.35)$$

For the case of a phosphorus donor atom in silicon, the outermost donor electron is attracted by a nuclear charge of $Z = 1$. However, because the phosphorus nucleus is embedded in the polarizable silicon, the effective nuclear charge seen by the electron is reduced to Z/κ, where κ is the dielectric constant. (a) Calculate the binding energy of a donor electron in Si($\kappa = 12$) and Ge($\kappa = 16$), and compare to the thermal energy available at room temperature. (b) Calculate the radius of the first Bohr orbit of a donor electron in Si

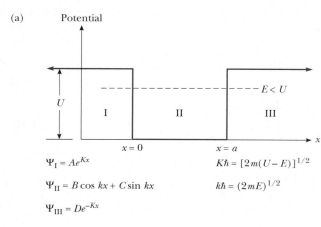

(a)   Potential

$\Psi_I = Ae^{Kx}$ $K\hbar = [2m(U - E)]^{1/2}$

$\Psi_{II} = B\cos kx + C\sin kx$ $k\hbar = (2mE)^{1/2}$

$\Psi_{III} = De^{-Kx}$

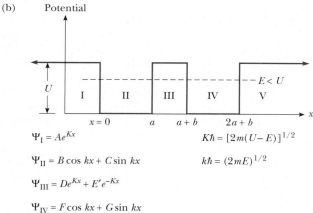

(b)   Potential

$\Psi_I = Ae^{Kx}$ $K\hbar = [2m(U - E)]^{1/2}$

$\Psi_{II} = B\cos kx + C\sin kx$ $k\hbar = (2mE)^{1/2}$

$\Psi_{III} = De^{Kx} + E'e^{-Kx}$

$\Psi_{IV} = F\cos kx + G\sin kx$

$\Psi_V = He^{-Kx}$

Figure P12.17 (a) Potential and eigenfunctions for a single finite well of depth U, where $E < U$. (b) Potential and eigenfunctions for two finite wells. The width of each well is a, and the wells are separated by a distance b.

and Ge. How does the Bohr orbit radius compare to the nearest-neighbor distance in Si(2.34 Å) and in Ge(2.43 Å)?

12.7 Lasers

19. *Cavity modes in lasers.* Under high resolution and at threshold current, the emission spectrum from a GaAs laser is seen to consist of many sharp lines, as shown in Figure P12.19. Although one line generally dominates at higher currents, let us consider the origin and spacing of the multiple lines, or modes, shown in this figure. Recall that standing waves or resonant modes corresponding to the sharp emission lines are formed when an integral number m of half-wavelengths fits between the cleaved GaAs surfaces. If L is the distance between cleaved faces, n is the index of refraction of the semiconductor, and λ is the wavelength in air, we have

$$m = \frac{L}{\lambda/2n}$$

(a) Show that the wavelength separation between adjacent modes (the change in wavelength for the case $\Delta m = +1$) is

$$|\Delta\lambda| = \frac{\lambda^2}{2L\left(n - \lambda\dfrac{dn}{d\lambda}\right)}$$

(*Hint:* Since m is large and we want the small change in wavelength that corresponds to $\Delta m = 1$, we can consider m to be a continuous function of λ and differentiate $m = 2Ln/\lambda$). (b) Using the following typical values for GaAs, calculate the wavelength separation between adjacent modes. Compare your results to Figure P12.19.

$$\lambda = 837 \text{ nm}$$
$$2L = 0.60 \text{ nm}$$
$$n = 3.58$$
$$dn/d\lambda = 3.8 \times 10^{-4} \text{ (nm)}^{-1} \bigg\} \text{ at } \lambda = 837 \text{ nm}$$

(c) Estimate the mode separation for the He–Ne gas laser using $\lambda = 632.8$ nm, $2L = 0.6$ m, $n \approx 1$, and $dn/d\lambda = 0$. On the basis of this calculation and the result of part (b), what is the controlling factor in mode separation in both solids and gases?

20. *How monochromatic is a laser?* The natural line width emitted by a collection of Cr^{3+} ions in ruby in thermal equilibrium at 290 K is about 0.4 nm for the 694.4-nm line. This is the *thermal line width.* Show that the emission from a ruby laser is much more monochromatic than 694.4 ± 0.2 nm if laser emission is confined to a single-cavity mode. Take the length of the ruby between polished faces to be 10 cm, $n = 1.8$, and $dn/d\lambda \approx 0$. (See Problem 19.)

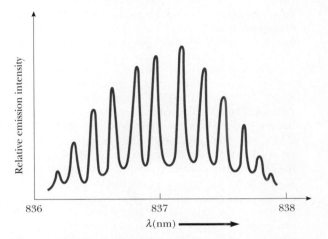

Figure P12.19 A high-resolution emission spectrum of the GaAs laser, operated just below the laser threshold.

ADDITIONAL PROBLEMS

21. A thin rod of superconducting material 2.50 cm long is placed into a 0.540-T magnetic field with its cylindrical axis along the magnetic field lines. (a) Sketch the directions of the applied field and the induced surface current. (b) Find the magnitude of the surface current on the curved surface of the rod.

22. Determine the current generated in a superconducting ring of niobium metal 2.00 cm in diameter when a 0.020 0-T magnetic field directed perpendicular to the ring is suddenly decreased to zero. The inductance of the ring is 3.10×10^{-8} H.

23. *A convincing demonstration of zero resistance.* A direct and relatively simple demonstration of zero dc resistance can be carried out using the four-point probe method. The probe shown in Figure P12.23 consists of a disk of $YBa_2Cu_3O_7$ (a high-T_c superconductor) to which four wires are attached by indium solder. Current is maintained through the sample by applying a dc voltage between points a and b, and it is measured with a dc ammeter. The current can be varied with the variable resistance R. The potential difference ΔV_{cd} between c and d is measured with a digital voltmeter. When the

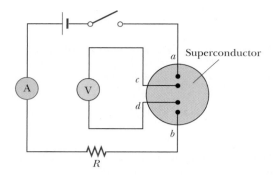

Figure P12.23   Circuit diagram used in the four-point probe measurement of the dc resistance of a sample. A dc digital ammeter is used to measure the current, and the potential difference between c and d is measured with a dc digital voltmeter. Note that there is no voltage source in the inner loop circuit where ΔV_{cd} is measured.

Table P12.23   Current Versus Potential Difference ΔV_{cd} Measured in a Bulk Ceramic Sample of $YBa_2Cu_3O_{7-\delta}$ at Room Temperature

I (mA)	ΔV_{cd} (mV)
57.8	1.356
61.5	1.441
68.3	1.602
76.8	1.802
87.5	2.053
102.2	2.398
123.7	2.904
155	3.61

probe is immersed in liquid nitrogen, the sample quickly cools to 77 K, below the critical temperature of the material, 92 K. The current remains approximately constant, but ΔV_{cd} *drops abruptly to zero*. (a) Explain this observation on the basis of what you know about super-conductors. (b) The data in Table P12.23 represent actual values of ΔV_{cd} for different values of I taken on the sample at room temperature. A 6-V battery in series with a variable resistor R supplied the current. The values of R ranged from 10 Ω to 100 Ω. The data are from the author's [RAS] laboratory. Make an I–ΔV plot of the data, and determine whether the sample behaves in a linear manner. From the data obtain a value for the dc resistance of the sample at room temperature. (c) At room temperature it is found that $\Delta V_{cd} = 2.234$ mV for $I = 100.3$ mA, but after the sample is cooled to 77 K, $\Delta V_{cd} = 0$ and $I = 98.1$ mA. What do you think might cause the slight decrease in current?

24. Under pressure, liquid helium can solidify as each atom bonds with four others, and each bond has an average energy of 1.74×10^{-23} J. Find the latent heat of fusion for helium in joules per gram. (The molar mass of He is 4.00 g/mol.)

25. A diode, a resistor, and a battery are connected in a series circuit. The diode is at a temperature for which $k_B T = 25.0$ meV and has a saturation current of 1.00 μA. The resistance of the resistor is 745 Ω, and the battery maintains a constant potential difference of 2.42 V. (a) Find graphically the current in the loop. Proceed as follows: On the same axes, draw graphs of the current in the diode I_D and the current in the wire I_W versus the voltage across the diode ΔV. Choose values of ΔV ranging from 0 to 0.250 V in steps of 0.025 V. Determine the value of ΔV at the intersection of the two graph lines, and calculate the corresponding currents I_D and I_W. Do they agree? (b) Find the ohmic resistance of the diode, which is defined as the ratio $\Delta V/I_D$. (c) Find the dynamic resistance of the diode, which is defined as the derivative $d(\Delta V)/dI_D$.

13
Nuclear Structure

Chapter Outline

In 1896, the year that marks the birth of nuclear physics, the French physicist Henri Becquerel (1852–1908) discovered radioactivity in uranium compounds. Following his discovery, scientists did a great deal of research in attempts to understand the nature of the radiation emitted by radioactive nuclei. Pioneering work by Ernest Rutherford showed that the emitted radiation was of three types, which he called alpha, beta, and gamma rays. Rutherford classified the rays according to the nature of the electric charges they possessed and their abilities to penetrate matter and ionize air. Later experiments showed that alpha rays are helium nuclei, beta rays are electrons, and gamma rays are high-energy photons.

In 1911 Rutherford and his students Geiger and Marsden performed a number of important scattering experiments involving alpha particles. The experiments established that the nucleus of an atom can be regarded as essentially a point mass and point charge and that most of the atomic mass is contained in the nucleus. Subsequent studies revealed the presence of a previously unknown short-range type of force, the nuclear force, which is predominant at distances of less than approximately 10^{-14} m and is zero at great distances.

Ernest Rutherford (1871–1937), a physicist from New Zealand, was awarded a Nobel prize in 1908 for discovering that atoms can be broken apart by alpha rays and for studying radioactivity. "On consideration, I realized that this scattering backward must be the result of a single collision, and when I made calculations I saw that it was impossible to get anything of that order of magnitude unless you took a system in which the greater part of the mass of the atom was concentrated in a minute nucleus. It was then that I had the idea of an atom with a minute massive center carrying a charge." (*Photo courtesy of AIP Niels Bohr Library*)

Other milestones in the development of nuclear physics include:

- The observation of nuclear reactions in 1930 by Cockroft and Walton, using *artificially* accelerated particles
- The discovery of the neutron in 1932 by Chadwick and the conclusion that neutrons make up about half of the nucleus
- The discovery of artificial radioactivity in 1933 by Joliot and Irene Curie
- The discovery of nuclear fission in 1938 by Meitner, Hahn, and Strassmann
- The development of the first controlled fission reactor in 1942 by Fermi and his collaborators

In this chapter we discuss the properties and structure of the atomic nucleus. We first describe the basic properties of nuclei, then discuss nuclear forces and binding energy, nuclear models, and the phenomenon of radioactivity. Finally, we examine the processes by which nuclei decay.

13.1   SOME PROPERTIES OF NUCLEI

All nuclei are composed of two types of particles: protons and neutrons. The only exception is the ordinary hydrogen nucleus, which is a single proton. In describing the atomic nucleus, we must use the following quantities:

- The **atomic number,** Z (sometimes called the *charge number*), which equals the number of protons in the nucleus
- The **neutron number,** N, which equals the number of neutrons in the nucleus.
- The **mass number,** A, which equals the number of nucleons (neutrons plus protons) in the nucleus.

In representing nuclei, it is convenient to have a system of symbols to show how many protons and neutrons are present. The symbol used is $^{A}_{Z}X$, where X represents the chemical symbol for the element. For example, $^{56}_{26}Fe$(iron) has a mass number of 56 and an atomic number of 26; it therefore contains 26 protons and 30 neutrons. When no confusion is likely to arise, we omit the subscript Z because the chemical symbol can always be used to determine Z.

The nuclei of all atoms of a particular element contain the same number of protons but often contain different numbers of neutrons. Nuclei that are related in this way are called **isotopes.**

Isotopes

The isotopes of an element have the same Z value but different N and A values.

The natural abundances of isotopes can differ substantially. For example, $^{11}_{6}C$, $^{12}_{6}C$, $^{13}_{6}C$, and $^{14}_{6}C$ are four isotopes of carbon. The natural abundance of the $^{12}_{6}C$ isotope is about 98.9%, whereas that of the $^{13}_{6}C$ isotope is only about 1.1%. Some isotopes do not occur naturally but can be produced in the laboratory through nuclear reactions. Even the simplest element, hydrogen, has isotopes: $^{1}_{1}H$, the ordinary hydrogen nucleus; $^{2}_{1}H$, deuterium; and $^{3}_{1}H$, tritium.

Charge and Mass

The proton carries a single positive charge, equal in magnitude to the electron charge (where $e = 1.602\,177\,3 \times 10^{-19}$ C). The neutron is electrically neutral, as its name implies. Because the neutron has no charge, it is more difficult to detect.

Atomic mass (the mass of an atom containing a nucleus and Z electrons) can be measured with great precision with the mass spectrometer. The proton is approximately 1836 times as massive as the electron, and the masses of the proton and the neutron are almost equal. It is convenient to define, for atomic masses, the **atomic mass unit,** u, in such a way that the mass of the isotope ^{12}C is exactly 12 u. That is, the mass of an atom is measured relative to the mass of an atom of the neutral carbon-12 isotope (the nucleus plus six electrons). Thus the mass of ^{12}C is exactly 12 u, where $1\,u = 1.660\,540 \times 10^{-27}$ kg. The proton and neutron each have a mass of approximately 1 u, and the electron has a mass that is only a small fraction of an atomic mass unit:

$$\text{mass of proton} = 1.007\,276\,5 \text{ u}$$

$$\text{mass of neutron} = 1.008\,664\,9 \text{ u}$$

$$\text{mass of electron} = 0.000\,548\,579\,90 \text{ u}$$

Because the rest energy of a particle is given by $E = mc^2$, it is often convenient to express the atomic mass unit in terms of its rest-energy equivalent. For one atomic mass unit, we have

$$E = mc^2 = (1.660\,540 \times 10^{-27} \text{ kg}) \frac{(2.997\,924\,6 \times 10^8 \text{ m/s})^2}{(1.602\,177\,3 \times 10^{-19} \text{ J/eV})}$$

$$= 931.494\,3 \text{ MeV}$$

As noted in Chapter 2, physicists often express mass in terms of the unit MeV/c^2, so here the mass of 1 u is

$$1\,u \equiv 931.494\,3 \,\frac{\text{MeV}}{c^2}$$

Table 13.1 gives the masses of the proton, neutron, and electron in different units. The masses and some other properties of selected isotopes are provided in Appendix B.

Table 13.1   Masses of the Proton, Neutron, and Electron in Various Units

Particle	kg	u	MeV/c^2
		Mass	
Proton	$1.672\,623 \times 10^{-27}$	1.007 276	938.272 3
Neutron	$1.674\,929 \times 10^{-27}$	1.008 665	939.565 6
Electron	$9.109\,390 \times 10^{-31}$	$5.48\,579\,9 \times 10^{-4}$	0.510 999 1

EXAMPLE 13.1   The Atomic Mass Unit

Use Avogadro's number to show that the atomic mass unit is $1 \text{ u} = 1.66 \times 10^{-27}$ kg.

Solution   We know that exactly 12 g of ^{12}C contains Avogadro's number of atoms. Avogadro's number, N_A, has the value 6.02×10^{23} atoms/mol. Thus the mass of one carbon atom is

$$\text{mass of one } {}^{12}\text{C atom} = \frac{0.012 \text{ kg}}{6.02 \times 10^{23} \text{ atoms}}$$

$$= 1.99 \times 10^{-26} \text{ kg}$$

Since one atom of ^{12}C is defined to have a mass of 12 u, we find that

$$1 \text{ u} = \frac{1.99 \times 10^{-26} \text{ kg}}{12} = 1.66 \times 10^{-27} \text{ kg}$$

Size and Structure of Nuclei

The size and structure of nuclei were first investigated in Rutherford's scattering experiments, discussed in Section 4.2. In those experiments, Rutherford directed positively charged nuclei of helium atoms (alpha particles) at a thin piece of metal foil. As the α particles moved through the foil, they often passed near a metal nucleus. Because of the positive charge on both the incident particles and the nuclei, particles were deflected from their straight-line paths by the Coulomb repulsive force. In fact, some particles were even deflected backward, through an angle of 180° from the incident direction. Those particles were apparently moving directly toward a nucleus in a head-on collision course.

Rutherford employed an energy calculation and found an expression for the distance, d, at which a particle approaching a nucleus is turned around by Coulomb repulsion. In such a head-on collision, the kinetic energy of the incoming alpha particle must be converted completely to electrical potential energy when the particle stops at the point of closest approach and turns around (Fig. 13.1). If we equate the initial kinetic energy of the alpha particle to the electrical potential energy of the system (α particle plus target nucleus), we have

$$\tfrac{1}{2} m v^2 = k \frac{q_1 q_2}{r} = k \frac{(2e)(Ze)}{d}$$

Solving for d, the distance of closest approach, we get

$$d = \frac{4kZe^2}{mv^2}$$

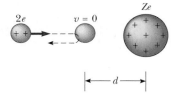

Figure 13.1   An alpha particle on a head-on collision course with a nucleus of charge Ze. Because of the Coulomb repulsion between the like charges, the alpha particle approaches to a distance d from the nucleus, called the *distance of closest approach*.

From this expression, Rutherford found that 7-MeV α particles approached nuclei to within 3.2×10^{-14} m when the foil was made of gold. Thus the radius of the gold nucleus must be less than this value. For silver atoms, the distance of closest approach was found to be 2×10^{-14} m. From these results, Rutherford concluded that the positive charge in an atom is concentrated in a small sphere with a radius of approximately 10^{-14} m, which he called the nucleus. Because such small dimensions are common in nuclear physics, a convenient length unit is the femtometer (fm), sometimes called the **fermi,** defined as

$$1 \text{ fm} \equiv 10^{-15} \text{ m}$$

In the early 1920s it was known from the work of Rutherford, Bohr, and Moseley that the nucleus contained Z protons with a charge $+Ze$ but had a mass approximately equal to that of A protons, where $A \cong 2Z$. To account for the nuclear mass, Rutherford proposed that the nucleus contained $(A - Z)$ neutral proton–electron combination particles in addition to the Z protons. He called the hypothetical combination particle a neutron. This hypothesis prompted an experimental search to confirm the existence of the neutron in a free state outside the nucleus—a search that culminated in 1932. In that year, the important experimental work of the British physicist James Chadwick (1891–1974) clearly showed the existence of a particle with a charge of zero and a mass of approximately 1 u. Chadwick (Fig. 13.2) was awarded the 1935 Nobel prize for this basic discovery.

Although Rutherford's idea that the neutron was a neutral proton–electron combination led to the important search for the neutron, both it and the idea that electrons exist inside the nucleus and can leak out in beta decay have since been abandoned. The current view is that the neutron is a fundamental particle, not a proton–electron combination, and that the β particles (electrons) emitted in radioactive nuclear decay are created at the moment of decay. As mentioned in Chapter 5 (see Example 5.9), if electrons were confined in the nucleus, the uncertainty principle would require them to possess unrealistically large kinetic energies compared to the energies actually observed in beta decay.

Another argument against electrons in the nucleus concerns the magnetic moments of nuclei, which are observed to be approximately 2000 times smaller than the magnetic moment of the electron. As we saw in Chapter 9, the magnetic moment of the electron is of the order of a Bohr magneton, $\mu_B = e\hbar/2m_e$, where m_e is the mass of the electron. Thus, if the same kind of theory accounts for the magnetic moment of a nucleus and there are no electrons in the nucleus, we expect the nuclear magnetic moment to be of the order of a nuclear magneton, $\mu_n = e\hbar/2m_p$, where m_p is the mass of a proton. Since m_p is about 2000 times m_e, this model accounts for the observed size difference between nuclear and electronic magnetic moments. If there were electrons in the nucleus, their much larger moments would dominate and nuclei would have magnetic moments of the order of $e\hbar/2m_e$, in disagreement with actual observations.

Table 13.2 lists important properties of the electron, proton, and neutron. This table presents yet another reason for discarding the idea that a neutron is a bound electron–proton pair: The spin of the neutron is $\frac{1}{2}$ and cannot be composed from the electron and proton spins, both of which are $\frac{1}{2}$.

Since the time of Rutherford's scattering experiments, a multitude of other experiments have shown that most nuclei are approximately spherical and have an average radius of

$$r = r_0 A^{1/3} \tag{13.1}$$

Nuclear radius

Figure 13.2 James Chadwick (1891–1974). Chadwick, a student of Rutherford, was one of the greatest experimental nuclear physicists. He was awarded the 1935 Nobel prize for the discovery of the neutron. (*Photo by Bortzells Esselte, courtesy AIP Emilo Segrè Visual Archives*)

Table 13.2   Masses, Spins, and Magnetic Moments of the Proton, Neutron, and Electron

Particle	Mass (MeV/c^2)	Spin	Magnetic Moment
Proton	938.28	$\frac{1}{2}$	$2.7928\mu_n$
Neutron	939.57	$\frac{1}{2}$	$-1.9135\mu_n$
Electron	0.510 99	$\frac{1}{2}$	$-1.0012\mu_B$

Figure 13.3 A nucleus can be modeled as a cluster of tightly packed spheres, each of which is a nucleon.

where A is the mass number and r_0 is a constant equal to 1.2×10^{-15} m. Because the volume of a sphere is proportional to the cube of its radius, it follows from Equation 13.1 that the volume of a nucleus (assumed to be spherical) is directly proportional to A, the total number of nucleons. This suggests that **all nuclei have nearly the same density.** When the nucleons combine to form a nucleus, they combine as though they were tightly packed spheres (Fig. 13.3). This fact has led to an analogy between the nucleus and a drop of liquid, in which the density of the drop is independent of its size. We shall discuss the liquid-drop model in Section 13.3.

EXAMPLE 13.2   The Volume and Density of a Nucleus

Find (a) an approximate expression for the mass of a nucleus of mass number A, (b) an expression for the volume of this nucleus in terms of the mass number, and (c) a numerical value for its density.

Solution (a) The mass of the proton is approximately equal to that of the neutron. Thus, if the mass of one of these particles is m, the mass of the nucleus is approximately Am.

(b) Assuming that the nucleus is spherical and using Equation 13.1, we find that the volume is

$$V = \tfrac{4}{3}\pi r^3 = \tfrac{4}{3}\pi r_0^3 A$$

(c) The nuclear density can be found as follows:

$$\rho_n = \frac{\text{mass}}{\text{volume}} = \frac{Am}{\tfrac{4}{3}\pi r_0^3 A} = \frac{3m}{4\pi r_0^3}$$

Taking $r_0 = 1.2 \times 10^{-15}$ m and $m = 1.67 \times 10^{-27}$ kg, we find that

$$\rho_n = \frac{3(1.67 \times 10^{-27}\ \text{kg})}{4\pi(1.2 \times 10^{-15}\ \text{m})^3} = 2.3 \times 10^{17}\ \frac{\text{kg}}{\text{m}^3}$$

The nuclear density is approximately 2.3×10^{14} times as great as the density of water ($\rho_{\text{water}} = 1.0 \times 10^3$ kg/m^3)!

Nuclear Stability

The nucleus consists of a closely packed collection of protons and neutrons. Like charges (the protons) in proximity exert very large repulsive electrostatic forces on each other, which should cause the nucleus to fly apart. However, nuclei are stable because of the presence of the **nuclear force.** This is an attractive force, with a very short range (about 2 fm), that acts between *all* nuclear particles. The protons attract each other via the nuclear force, and at the same time they repel each other through the Coulomb force. The attractive nuclear force also acts between pairs of neutrons and between neutrons and protons.

There are approximately 260 stable nuclei; hundreds of other nuclei have been observed but these are unstable. Figure 13.4 is a plot of N versus Z for some stable nuclei. Note that light nuclei are most stable if they contain equal numbers of protons and neutrons—that is, if $N = Z$. Furthermore, heavy nuclei are more stable if the number of neutrons exceeds the number of protons. We can understand this by recognizing that as the number of protons increases, the strength of the Coulomb force increases, which tends to break the nucleus apart. As a result, more neutrons are needed to keep the nucleus stable, since neutrons experience only attractive nuclear forces. Eventually, the repulsive forces between protons cannot be compensated for by the addition of more neutrons; this occurs when $Z = 83$. Elements that contain more than 83 protons do not have stable nuclei.

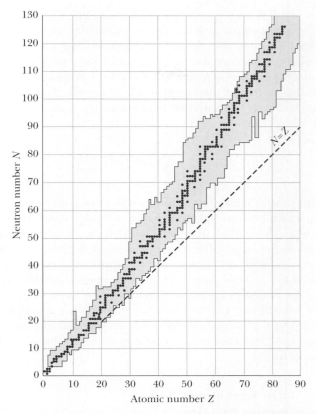

Figure 13.4   A plot of neutron number N versus atomic number Z for the stable nuclei (solid points). The dashed line, corresponding to the condition $N = Z$, is called the *line of stability*. The shaded area shows radioactive nuclei.

Interestingly, most stable nuclei have even values of A. In fact, certain values of Z and N correspond to unusually high stability in nuclei. These values of N and Z, called **magic numbers,** are

$$Z \text{ or } N = 2, 8, 20, 28, 50, 82, 126 \qquad (13.2)$$

For example, the helium nucleus (two protons and two neutrons), with $Z = 2$ and $N = 2$, is very stable. The unusual stability of nuclei with progressively larger magic numbers suggests a shell structure for the nucleus that is similar to atomic shell structure. In Section 13.3 we briefly treat the shell model of the nucleus, which explains magic numbers.

Nuclear Spin and Magnetic Moment

In Chapter 9 we discussed the fact that an electron has an intrinsic angular momentum associated with its spin. A nucleus, like an electron, has an intrinsic angular momentum that arises from relativistic properties. The magnitude of the nuclear angular momentum is $\sqrt{I(I + 1)}\,\hbar$, where I is a quantum number called the **nuclear spin** and may be an integer or a half-integer. Nuclear angular momentum is the total angular momentum of all the nucleons,

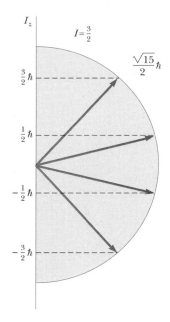

Figure 13.5 The possible orientations of the nuclear angular momentum and its projections along the z-axis for the case $I = \frac{3}{2}$.

including both orbital and spin angular momentum of each nucleon. The maximum component of the angular momentum projected along the z-axis is $I\hbar$. Figure 13.5 illustrates the possible orientations of the nuclear angular momentum and its projections along the z-axis for the case where $I = \frac{3}{2}$.

The nuclear angular momentum has a nuclear magnetic moment associated with it, similar to that of the electron. The magnetic moment of a nucleus is measured in terms of the **nuclear magneton** μ_n, a unit of magnetic moment defined as

$$\mu_n \equiv \frac{e\hbar}{2m_p} = 5.05 \times 10^{-27}\,\text{J/T} \tag{13.3}$$

where m_p is the proton mass. This definition is analogous to that of the Bohr magneton, μ_B, which corresponds to the spin magnetic moment of a free electron (Section 9.2). Note that μ_n is smaller than μ_B by a factor of approximately 2000 due to the large difference in masses of the proton and electron.

The magnetic moment of a free proton is approximately $2.7928\mu_n$. Unfortunately, there is no general theory of nuclear magnetism that explains this value. Another surprising point is that a neutron also has a magnetic moment, with a value of $-1.9135\mu_n$. The fact that the neutron has a magnetic moment is evidence that the uncharged neutron has an internal charge distribution. The minus sign indicates that this moment is opposite the spin angular momentum of the neutron.

Nuclear Magnetic Resonance and Magnetic Resonance Imaging

Nuclear magnetic moments (as well as electronic magnetic moments) precess when placed in an external magnetic field. The frequency at which they precess, called the **Larmor precessional frequency,** ω_L, is directly proportional to the magnetic field. This precession is sketched in Figure 13.6a, where the magnetic field is along the z-axis. For example, the Larmor frequency of a proton in a 1-T magnetic field is 42.577 MHz. The potential energy of a magnetic dipole moment $\boldsymbol{\mu}$ in an external magnetic field $\mathbf{B}$ is given by $-\boldsymbol{\mu} \cdot \mathbf{B}$. When the magnetic moment $\boldsymbol{\mu}$ is lined up with the field as closely as quantum physics allows, the potential energy of the dipole moment in the field has its minimum value, $E_{\min}$. When $\boldsymbol{\mu}$ is as antiparallel as possible, the potential energy has its maximum value, $E_{\max}$. Figure 13.6b shows these two energy states for a nucleus with a spin of $\frac{1}{2}$.

It is possible to observe transitions between these two spin states through a technique called **nuclear magnetic resonance (NMR).** A constant magnetic field ($\mathbf{B}$ in Fig. 13.6a) is introduced to align magnetic moments, along with a second, weak, oscillating magnetic field oriented perpendicular to $\mathbf{B}$. When the frequency of the oscillating field is adjusted to match the Larmor precessional frequency, a torque acting on the precessing moments causes them to "flip" between the two spin states. These transitions result in a net absorption of energy by the spin system, an absorption that can be detected electronically. Figure 13.7 is a sketch of the apparatus used in NMR. The absorbed energy is supplied by the generator producing the oscillating magnetic field. Nuclear magnetic resonance and a related technique called electron spin resonance are extremely important methods of studying nuclear and atomic systems and

Larmor frequency

Nuclear magnetic resonance

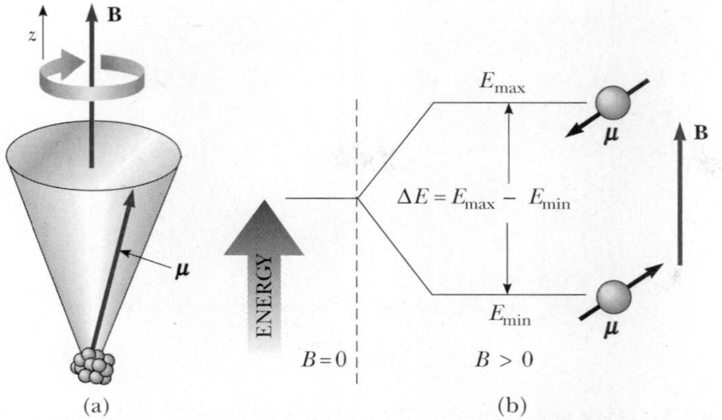

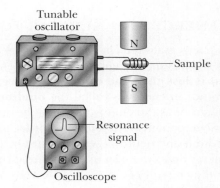

Figure 13.6 (a) When a nucleus is placed in an external magnetic field, **B**, the magnetic moment precesses about the magnetic field with a frequency proportional to the field. (b) A nucleus with spin $\frac{1}{2}$ can occupy one of two energy states when placed in an external magnetic field. The lower energy state E_{min} corresponds to the case where the spin is aligned with the field as much as possible according to quantum mechanics, and the higher energy state E_{max} corresponds to the case where the spin is opposite the field as much as possible.

how those systems interact with their surroundings. Figure 13.8 shows a typical NMR spectrum.

A widely used diagnostic technique called **magnetic resonance imaging (MRI)** is based on nuclear magnetic resonance. Because about two-thirds of the atoms in the human body are hydrogen, which gives a strong NMR signal, MRI works exceptionally well for viewing internal tissues. In MRI, the patient is placed inside a large solenoid that supplies a spatially varying magnetic field. Because of the gradient in the magnetic field, protons in different parts of the

Figure 13.7 An experimental arrangement for nuclear magnetic resonance. The radio-frequency magnetic field of the coil, provided by the variable-frequency oscillator, must be perpendicular to the dc magnetic field. When the nuclei in the sample meet the resonance condition, the spins absorb energy from the field of the coil, and this absorption changes the Q of the circuit in which the coil is included. Most modern NMR spectrometers use superconducting magnets at fixed field strengths and operate at frequencies of approximately 200 MHz.

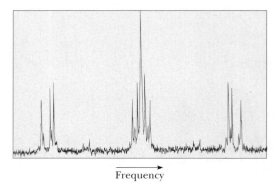

Figure 13.8   An NMR spectrum of ^{31}P in a bridged metallic complex containing platinum. The lines that flank the central strong peak are due to the interaction between ^{31}P and other distant ^{31}P nuclei. The outermost set of lines is due to the interaction between ^{31}P and neighboring platinum nuclei. The spectrum was recorded at a fixed magnetic field of about 4 T, and the mean frequency was 200 MHz.

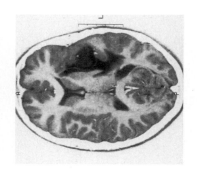

Figure 13.9 A computer-digitized MRI scan of a brain with a glioma tumor. (© *1996 Scott Camazine*)

body precess at different frequencies, and so the resonance signal can provide information on the positions of the protons. A computer is used to analyze the position information to provide data for the construction of a final image. Figure 13.9 shows an MRI scan of a human brain.

Another advantage of MRI over other imaging techniques is that it causes minimal damage to cellular structures. Photons associated with the radio-frequency signals used in MRI have energies of only about 10^{-7} eV. Because molecular bond strengths are much greater (approximately 1 eV), the radio-frequency radiation causes little cellular damage. In comparison, x rays and gamma rays have energies ranging from 10^4 to 10^6 eV and can cause considerable cellular damage.

13.2   BINDING ENERGY AND NUCLEAR FORCES

The total mass of a nucleus is always less than the sum of the masses of its nucleons. Because mass is a measure of energy, **the total energy of the bound system (the nucleus) is less than the combined energy of the separated nucleons.** This difference in energy is called the **binding energy** of the nucleus and can be thought of as the energy that must be added to a nucleus to break it apart into its components. Therefore, in order to separate a nucleus into its constituent protons and neutrons, energy must be put into the system.

Conservation of energy and the Einstein mass–energy equivalence relationship show that the binding energy of any nucleus of mass M_A is

Binding energy of a nucleus

$$E_b(\text{MeV}) = [ZM(\text{H}) + Nm_\text{n} - M_A] \times 931.494 \, \frac{\text{MeV}}{\text{u}} \qquad (13.4)$$

where $M(\text{H})$ is the atomic mass of hydrogen, M_A represents the atomic mass of the element ^A_ZX, m_n is the mass of the neutron, and the masses are all expressed in atomic mass units. Note that the mass of the Z electrons included in the first term in Equation 13.4 cancels with the mass of the Z electrons included in the term M_A, within a small difference associated with the atomic binding energy of the electrons. Since atomic binding energies are typically several electron volts and nuclear binding energies are several MeV, this difference is negligible.

**EXAMPLE 13.3   The Binding Energy
of the Deuteron**

Calculate the binding energy of the deuteron, which consists of a proton and a neutron, given that the atomic mass of the deuteron is $M_2 = 2.014102$ u.

Solution   We know that the atomic mass of hydrogen and the mass of the neutron are

$$M(\text{H}) = 1.007\ 825\ \text{u}$$

$$m_{\text{n}} = 1.008\ 665\ \text{u}$$

Using Eq. 13.4, we find for the deuteron binding energy

$$E_b = [M(\text{H}) + m_{\text{n}} - M_2] \times 931.494\ \text{MeV/u}$$

$$= [2.016\ 490\ \text{u} - 2.014\ 102\ \text{u}] \times 931.494\ \text{MeV/u}$$

$$E_b = [0.002\ 388\ \text{u}] \times 931.494\ \text{MeV/u}$$

$$= 2.224\ \text{MeV}$$

This result tells us that in order to separate a deuteron into its constituent parts at rest (a proton and a neutron), it is necessary to add 2.224 MeV of energy to the deuteron. One way of supplying the deuteron with this energy is by bombarding it with energetic particles.

If the binding energy of a nucleus were zero, the nucleus would separate into its constituent protons and neutrons without the addition of any energy; that is, it would spontaneously break apart.

A plot of **binding energy per nucleon,** E_b/A, as a function of mass number for various stable nuclei is shown in Figure 13.10. Except for the lighter nuclei, the average binding energy per nucleon is about 8 MeV. For the deuteron, the average binding energy per nucleon is $E_b/A = 2.224/2$ MeV $= 1.112$ MeV. Note that the curve in Figure 13.10 peaks in the vicinity of $A = 60$. That is, nuclei with mass numbers greater or less than 60 are not as tightly bound as

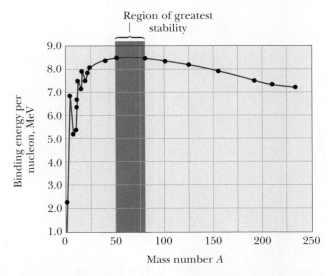

Figure 13.10   A plot of binding energy per nucleon versus mass number for the stable nuclei shown in Figure 13.4.

those near the middle of the periodic table. The higher values of binding energy near $A = 60$ imply that energy is released when a heavy nucleus with $A \approx 200$ splits or fissions into several lighter nuclei that lie near $A = 60$. Energy is released in fission because the final state, consisting of two lighter fragments, is more tightly bound, or lower in energy, than the original nucleus. Similarly, energy can be released when two light nuclei with $A \leq 20$ combine or fuse to form one heavier nucleus. These two important processes, *fission* and *fusion*, are considered in detail in Chapter 14.

Another important feature of Figure 13.10 is that the binding energy per nucleon is approximately constant for $A > 20$. In this case, the nuclear forces between a particular nucleon and all the other nucleons in the nucleus are said to be saturated; that is, a particular nucleon forms attractive bonds with only a limited number of other nucleons. Because of the short-range character of the nuclear force, these other nucleons can be viewed as being the nearest neighbors in the close-packed structure shown in Figure 13.3. If every nucleon could interact with every other nucleon, each nucleon would form $(A - 1)$ bonds, and the binding energy per nucleon would be proportional to $(A - 1)$ rather than constant as observed.

Saturation of nuclear forces

The general features of the nuclear binding force have been revealed in a wide variety of experiments. We summarize them as follows.

- The attractive nuclear force is a different kind of force from the common forces of electromagnetism and gravitation, and since it dominates the repulsive Coulomb force between protons in the nucleus, it is stronger than the electromagnetic force.
- The nuclear force is a short-range force that rapidly falls to zero when the separation between nucleons exceeds several fermis. Evidence for the limited range of nuclear forces comes from scattering experiments and from the saturation of nuclear forces already mentioned. The neutron–proton (n–p) potential energy plot of Figure 13.11a, obtained

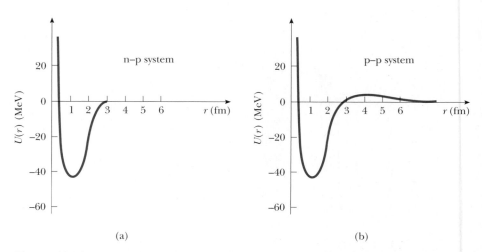

Figure 13.11   (a) Potential energy versus separation for the neutron–proton system. (b) Potential energy versus separation for the proton–proton system. The difference in the two curves is due mainly to the Coulomb repulsion in the case of the proton–proton interaction.

by scattering neutrons from a target containing hydrogen, shows the short range of the nuclear force. The depth of the n–p potential energy is 40 to 50 MeV and contains a strong repulsive component that prevents the nucleons from approaching much closer than 0.4 fm. Another interesting feature of the nuclear force is that its size depends on the relative spin orientations of the nucleons, as shown by scattering experiments performed with spin-polarized beams and targets.

- From n–n, n–p, and p–p scattering experiments and other indirect evidence, it is found that the nuclear force is independent of the electric charge of the interacting nucleons. As might be expected from this "charge-blind" character of the nuclear force, the nuclear force does not affect electrons, and so it enables energetic electrons to serve as point-like probes of the charge density of nuclei. The charge independence of the nuclear force also means that the main difference between the n–p and p–p interactions is that the p–p potential energy consists of a *superposition* of nuclear and coulomb interactions, as shown in Figure 13.11b. At distances less than 2 fm, p–p and n–p potential energies are nearly identical, but at distances greater than this, the p–p potential has a positive energy barrier, with a maximum of about 1 MeV at 4 fm.

Charge independence of the nuclear force

Although the preceding "laundry list" of observed properties of the nuclear force may be interesting, it leaves unanswered the deeper question of what mechanism accounts for the powerful nuclear force. One model that successfully explains the nuclear force is the **exchange force model.** According to this model, two nucleons—a proton and a neutron, for example—experience an attractive force when one spontaneously emits a particle and the other absorbs that particle. From a classical point of view, the emitting and absorbing nucleons are both strongly attracted to the exchanged particle as if connected to it by elastic bands or springs. Since both nucleons are strongly attracted to the same particle, they appear to be strongly attracted to each other.

Particle exchange model

You may wonder how the exchange of a particle of mass m between the proton and neutron can occur without violating conservation of energy and conservation of momentum. A violation of energy conservation must occur if a proton at rest with energy $m_p c^2$ emits a particle of mass m with energy mc^2 and yet remains a proton with energy $m_p c^2$. Also, conservation of momentum would cause a paired proton and neutron to recoil away from each other as the proton emitted a particle and the neutron absorbed it, instead of producing motion of the proton toward the neutron as required for an attractive nuclear force.

The answer to both dilemmas seems to be a trick: The exchanged particle exists for such a short time and is confined to such a small region of space that it is allowed to violate the conservation of energy and momentum. It is a so-called **virtual particle.** A virtual particle cannot be directly detected or measured during the momentum- and energy-violating processes taking place inside the nucleus. Perhaps the best way to understand this is through the uncertainty principles $\Delta E \Delta t \geq \hbar/2$ and $\Delta p_x \Delta x \geq \hbar/2$. These relationships imply that energy conservation can be violated by an amount ΔE for a short time interval Δt and that momentum conservation can be violated by an amount Δp_x over a small nuclear region of size Δx. Although it may appear that a virtual particle can never be detected, if an incident particle strikes the nucleus in just the right way, it can supply the missing momentum and energy required to make a virtual particle real, thereby permitting the exchanged particle to be measured in the laboratory.

A simple application of the energy–time uncertainty principle enables us to estimate the mass of the exchanged particle that should be responsible for the nuclear force. Since the amount by which energy conservation is violated in the exchange process is $\Delta E = mc^2$, where m is the mass of the exchanged particle, we have

$$\Delta t \cong \frac{\hbar}{\Delta E} = \frac{\hbar}{mc^2} \qquad (13.5)$$

Realizing that Δt is the time it takes the exchanged particle to travel a distance d between nucleons, and that the maximum value of d is $d = c\Delta t$ for an exchanged particle traveling at the speed of light, we find that

$$d_{max} = c\Delta t = c\left(\frac{\hbar}{mc^2}\right) = \frac{\hbar}{mc} \qquad (13.6)$$

This expression shows that **the range of the nuclear force, d_{max}, is inversely proportional to the mass of the exchanged particle,** a general feature of exchange-force models. Finally, because the range of the nuclear force is approximately 2 fm, we can estimate the rest energy of the exchanged particle that is responsible for the nuclear force:

$$mc^2 \cong \frac{\hbar c}{d_{max}} = \frac{(1.05 \times 10^{-34}\,\text{J·s})(3.00 \times 10^8\,\text{m/s})}{2.0 \times 10^{-15}\,\text{m}}$$
$$= 1.6 \times 10^{-11}\,\text{J} = 100\,\text{MeV}$$

Twelve years after the Japanese physicist Hideki-Yukawa (1907–1981) proposed the exchange theory of the nuclear force just described, a strongly interacting particle with a mass of about 140 MeV/c^2, called the pi meson, was found in cosmic-ray interactions, closely confirming his predictions. We discuss the pi meson and other force-carrying exchange particles more fully in Chapter 15.

13.3 NUCLEAR MODELS

Although the details of nuclear forces are still not completely understood, several nuclear models with adjustable parameters have been proposed, and they help us understand various features of nuclear experimental data and the mechanisms responsible for binding energy. The models we shall discuss are (1) the liquid-drop model, which accounts for the nuclear binding energy; (2) the independent-particle model, which accounts for more detailed features of nuclei, including large differences in stability between nuclei with similar Z and A values; and (3) the collective model.

Liquid-Drop Model

The **liquid-drop model,** proposed by C. F. von Weizsächer in 1935, treats the nucleons as though they were molecules in a drop of liquid. The nucleons interact strongly with each other and undergo frequent collisions as they jiggle around within the nucleus. This jiggling motion is analogous to the thermally agitated motions of molecules in a drop of liquid.

Three major effects influence the binding energy of the nucleus in the liquid-drop model:

- **The volume effect.** Earlier we showed that the binding energy per nucleon is approximately constant, indicating that the nuclear force saturates (Fig. 13.10). Therefore, the total binding energy of the nucleus is proportional to A and to the nuclear volume. The contribution to the binding energy of this volume effect is $C_1 A$, where C_1 is an adjustable constant.

- **The surface effect.** Because nucleons on the surface of the drop have fewer neighbors than those in the interior, surface nucleons reduce the binding energy by an amount proportional to the number of surface nucleons. Because the number of surface nucleons is proportional to the surface area of the nucleus, $4\pi r^2$, and $r^2 \propto A^{2/3}$ (Eq. 13.1), the surface term can be expressed as $-C_2 A^{2/3}$, where C_2 is a constant.

- **The Coulomb repulsion effect.** Each proton repels every other proton in the nucleus. The corresponding potential energy per pair of interacting particles is ke^2/r, where k is the Coulomb constant. The total Coulomb energy represents the work required to assemble Z protons from infinity to a sphere of volume V. This energy is proportional to the number of proton pairs $Z(Z-1)/2$ and inversely proportional to the nuclear radius. Consequently, the reduction in energy that results from the Coulomb effect is $-C_3 Z(Z-1)/A^{1/3}$, where C_3 is a constant.

Another effect that decreases the binding energy is significant for heavy nuclei with large excesses of neutrons. Since it is observed that for a given A value, nuclei with $Z = N$ have the largest binding energy, the binding energy must be corrected by a symmetry term that favors $Z = N$ and decreases the binding energy symmetrically for $N > Z$ or $N < Z$. The symmetry term is generally written as $-C_4(N-Z)^2/A$, where C_4 is an adjustable constant.

Adding these contributions, we get as the total binding energy

$$E_b = C_1 A - C_2 A^{2/3} - C_3 \frac{Z(Z-1)}{A^{1/3}} - C_4 \frac{(N-Z)^2}{A} \qquad (13.7)$$

Semiempirical binding energy formula

This equation is often referred to as the **Weizsächer semiempirical binding energy formula,** because it has some theoretical justification but contains four constants that are adjusted to fit this expression to experimental data. For nuclei with $A \geq 15$, the constants have the values

$$C_1 = 15.7 \text{ MeV} \qquad C_2 = 17.8 \text{ MeV}$$

$$C_3 = 0.71 \text{ MeV} \qquad C_4 = 23.6 \text{ MeV}$$

Equation 13.7, together with these constants, fits the known nuclear binding energy values very well. However, the liquid-drop model does not account for some finer details of nuclear structure, such as certain stability rules and angular momentum. On the other hand, it does provide a qualitative description of nuclear fission, shown schematically in Figure 13.12. If the drop vibrates with a large amplitude (which may be initiated by collision with another particle), it distorts, and under the right conditions, it breaks apart. We shall discuss the process of fission further in Chapter 14.

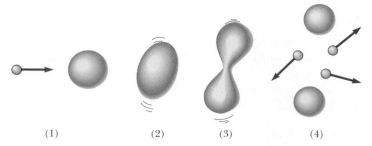

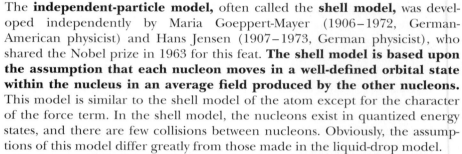

(1)   (2)   (3)   (4)

Figure 13.12   Steps leading to fission according to the liquid-drop model of the nucleus.

Independent-Particle Model

The **independent-particle model,** often called the **shell model,** was developed independently by Maria Goeppert-Mayer (1906–1972, German-American physicist) and Hans Jensen (1907–1973, German physicist), who shared the Nobel prize in 1963 for this feat. **The shell model is based upon the assumption that each nucleon moves in a well-defined orbital state within the nucleus in an average field produced by the other nucleons.** This model is similar to the shell model of the atom except for the character of the force term. In the shell model, the nucleons exist in quantized energy states, and there are few collisions between nucleons. Obviously, the assumptions of this model differ greatly from those made in the liquid-drop model.

Each quantized orbital state for a proton or neutron is analogous to one of the orbital states of an electron in an atomic subshell, with the exception that the ordering of nuclear states in energy is more closely modeled by a spherical-well potential than by the Coulomb potential that is appropriate for atomic electrons. (See Problem 32 in Chapter 8 for a solution of the spherical well.) The quantized states occupied by the nucleons can be described by a set of quantum numbers. Because both the proton and the neutron have spin $\frac{1}{2}$, we can apply the Pauli exclusion principle to describe the allowed states (as we did for electrons in Chapter 9). That is, each orbital state can contain only two protons (or two neutrons) having *opposite* spin (Fig. 13.13). The protons have a set of allowed states, which differ slightly from those of the neutrons because they move in different potential wells. The proton levels are higher in energy than the neutron levels because the protons experience a superposition of Coulomb potential and nuclear potential, while the neutrons experience only a nuclear potential. These two different wells are sketched in Figure 13.13.

Using only the fact that nuclear energy levels exist and that they fill according to the Pauli exclusion principle, it is possible to explain the tendency for N to equal Z in stable, light nuclei. To achieve maximum stability for a given A, all the nucleons must be packed sequentially into the lowest energy levels, leaving no gaps in filled levels. Since any energy state is filled when it contains two protons (or two neutrons), another proton (or neutron) added to the nucleus produces an increase in energy and a decrease in stability of the nucleus. Thus, comparing $^{12}_{5}B$ and $^{12}_{6}C$, the seventh neutron in boron must occupy a higher energy level than the sixth neutron in carbon (two nucleons per energy state), and consequently $^{12}_{5}B$ has higher energy and is less stable than $^{12}_{6}C$.

Maria Goeppert-Mayer (1906–1972) was born and educated in Germany. She is best known for her development of the shell model (independent-particle model) of the nucleus, published in 1950. A similar model was simultaneously developed by Hans Jensen, another German scientist. Goeppert-Mayer and Jensen were awarded the Nobel Prize in Physics in 1963 for their extraordinary work in understanding the structure of the nucleus. (*Courtesy of Louise Barker/AIP Niels Bohr Library*)

To understand the observed characteristics of nuclear ground states, such as angular momentum and the high stability of magic-number nuclei, it is necessary to include *nuclear spin-orbit effects.* Unlike the spin-orbit interaction for an electron in an atom, which is magnetic in origin, the spin-orbit effect for nucleons in a nucleus is due to the nuclear force. It is much stronger than in the atomic case and has opposite sign, so that in nuclei, spin-orbit split states with higher angular momentum lie lower in energy. When the nuclear spin-orbit effect is added to a spherical finite potential, the magic numbers are predicted, because the spin-orbit potential produces especially large jumps between certain energy levels. In effect, the spin-orbit interaction substantially raises the energy levels containing 9, 21, 29, 51, 83, and 127 nucleons, thereby making the levels that contain 8, 20, 28, 50, 82, and 126 nucleons unusually stable.

Finally, it is possible to understand how individual nucleons can be considered to be moving in well-defined orbitals within the cramped confines of the nucleus, which is literally swarming with other nucleons. Under these circumstances, it would seem that a given nucleon would undergo many collisions and not move in a well-defined orbital. However, for the low-energy nuclear ground state, the exclusion principle inhibits energy-changing collisions by preventing colliding nucleons from occupying already filled low-lying energy states. In effect, the exclusion principle prevents nucleon collisions within the densely packed nucleus at low energy and justifies the shell-model approach.

Collective Model

A third model of nuclear structure, known as the **collective model,** combines some features of the liquid-drop model and the independent-particle model. The nucleus is considered to have some "extra" nucleons moving in quantized orbits in addition to the filled core of nucleons. The extra nucleons are subject to the field produced by the core, as in the independent-particle scheme. Deformations can be set up in the core as a result of a strong interaction between the core and the extra nucleons, thereby initiating vibrational and rotational motions, as in the liquid-drop model. The collective model has been very successful in explaining many nuclear phenomena.

13.4   RADIOACTIVITY

In 1896 Henri Becquerel (1852–1908, French physicist) accidentally discovered that uranyl potassium sulfate crystals emitted an invisible radiation that could darken a photographic plate when the plate was covered to exclude light. After a series of experiments, he concluded that the radiation emitted by the crystals was of a new type, one that required no external stimulation and was so penetrating that it could darken protected photographic plates and ionize gases. This process of spontaneous emission of radiation by uranium was soon to be called **radioactivity.** Subsequent experiments by other scientists showed that other substances were even more powerfully radioactive. Marie (1867–1934) and Pierre Curie (1859–1906) conducted the most significant investigations of this type. After several years of careful and laborious chemical separation processes on tons of pitchblende, a radioactive ore, the Curies reported the discovery of two previously unknown elements, both radioactive, which they named polonium and radium. Subsequent experiments, including Rutherford's famous

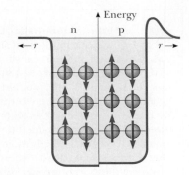

Figure 13.13   A square-well potential containing 12 nucleons. The gray circles represent protons, and the colored circles represent neutrons. The energy levels for the protons are slightly higher than those for the neutrons because of the Coulomb potential in the case of the protons. The difference in the levels increases as Z increases. Note that only two nucleons with opposite spin can occupy a given level, as required by the Pauli exclusion principle.

BIOGRAPHY

MARIE SKLODOWSKA CURIE
(1867–1934)

Marie Sklodowska Curie was born in Poland shortly after the unsuccessful Polish revolt against Russia in 1863. After high school, she worked diligently to help meet the educational expenses of her older brother and sister in Paris. At the same time, she managed to save enough money for her own move to Paris, where she entered the Sorbonne in 1891. Although she lived very frugally during this period (she once fainted from hunger in the classroom), she graduated at the top of her class.

In 1895 Marie Sklodowska married the French physicist Pierre Curie (1859–1906), who was already known for the discovery of piezoelectricity. (A piezoelectric crystal exhibits a potential difference under pressure.) Using piezoelectric materials to measure the activity of radioactive substances, Marie Curie demonstrated the radioactive nature of the elements uranium and thorium. In 1898 she and her husband discovered a new radioactive element contained in uranium ore, which they called polonium after her native land. By the end of 1898, the Curies succeeded in isolating trace amounts of an even more radioactive element, which they named radium. In an effort to produce weighable quantities of radium, they took on the painstaking job of isolating radium from pitchblende, an ore rich in uranium. After four years of purifying and repurifying tons of ore, and using their own life savings to finance their work, the Curies succeeded in preparing about 0.1 g of radium. In 1903, along with Henri Becquerel, they received the Nobel Prize in Physics for their studies of radioactive substances.

After her husband's death in an accident in 1906, Marie Curie assumed his professorship at the Sorbonne. Unfortunately, she experienced prejudice in the scientific community because she was a woman. For example, after being nominated to the French Academy of Sciences, she was refused membership after losing by one vote.

In 1911 Marie Curie was awarded a second Nobel prize, this one in chemistry, for the discovery of radium and polonium. She spent the last few decades of her life supervising the Paris Institute of Radium.

(Courtesy of AIP Niels Bohr Library / W. F. Meggers Collection)

work on alpha-particle scattering, suggested that radioactivity was the result of the decay, or disintegration, of unstable nuclei.

Three types of radiation can be emitted by a radioactive substance: alpha (α), in which the emitted particles are ^{4}He nuclei; beta (β), in which the emitted particles are either electrons or positrons; and gamma (γ), in which the emitted "rays" are high-energy photons. A **positron** is a particle like the electron in all respects except that the positron has a charge of $+e$. In this book, the symbol e^- is used to designate an electron, and e^+ designates a positron.

It is possible to distinguish these three forms of radiation using the scheme shown in Figure 13.14. The radiation from a radioactive sample is directed into a region in which there is a magnetic field. The beam splits into three components, two bending in opposite directions and the third experiencing no change in direction. From this simple observation, we can conclude that the radiation of the undeflected beam carries no charge (the gamma ray), the component deflected upward consists of positively charged particles (α particles), and the component deflected downward consists of negatively charged particles (e^-). If the beam includes a positron (e^+), it is deflected upward.

The three types of radiation have quite different penetrating powers. Alpha particles barely penetrate a sheet of paper, beta particles can penetrate a few

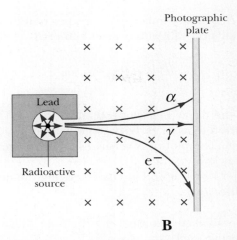

B

Figure 13.14 The radiation from a radioactive source can be separated into three components through the use of a magnetic field to deflect the charged particles. The photographic plate at the right records the events. The gamma ray is not deflected by the magnetic field.

millimeters of aluminum, and gamma rays can penetrate several centimeters of lead.

The rate at which a particular decay process occurs in a radioactive sample is proportional to the number of radioactive nuclei present (that is, those nuclei that have not yet decayed). If N is the number of radioactive nuclei present at some instant, the rate of change of N is

$$\frac{dN}{dt} = -\lambda N \tag{13.8}$$

where λ, called the **decay constant,** is the probability per unit time that a nucleus will decay. The minus sign indicates that dN/dt is negative because λ and N are both positive; that is, N is *decreasing* in time.

If we write Equation 13.8 in the form

$$\frac{dN}{N} = -\lambda \, dt$$

we can integrate the expression to give

$$\int_{N_0}^{N} \frac{dN}{N} = -\lambda \int_{0}^{t} dt$$

$$\ln\left(\frac{N}{N_0}\right) = -\lambda t$$

or

$$N = N_0 e^{-\lambda t} \tag{13.9}$$ **Exponential decay**

where the constant N_0 represents the number of radioactive nuclei at $t = 0$. Equation 13.9 shows that **the number of radioactive nuclei in a sample decreases exponentially with time.**

The **decay rate** R, or the number of decays per unit time, can be obtained by differentiating Equation 13.9 with respect to time:

Decay rate

$$R = \left| \frac{dN}{dt} \right| = N_0 \lambda e^{-\lambda t} = R_0 e^{-\lambda t} \qquad (13.10)$$

where $R_0 = N_0 \lambda$ is the decay rate at $t = 0$ and $R = \lambda N$. The decay rate of a sample is often referred to as its **activity.** Note that both N and R decrease exponentially with time. The plot of N versus t shown in Figure 13.15 illustrates the exponential decay law.

Another parameter that is useful in characterizing the decay of a particular nucleus is the **half-life,** $T_{1/2}$:

> The **half-life** of a radioactive substance is the time it takes half of a given number of radioactive nuclei to decay.

Setting $N = N_0/2$ and $t = T_{1/2}$ in Equation 13.9 gives

$$\frac{N_0}{2} = N_0 e^{-\lambda T_{1/2}}$$

Writing this in the form $e^{\lambda T_{1/2}} = 2$ and taking the natural logarithm of both sides, we get

$$T_{1/2} = \frac{\ln 2}{\lambda} = \frac{0.693}{\lambda} \qquad (13.11)$$

This is a convenient expression for relating half-life to the decay constant. Note that after an elapsed time of one half-life, $N_0/2$ radioactive nuclei remain (by definition); after two half-lives, half of these have decayed and $N_0/4$ radioactive nuclei remain; after three half-lives, $N_0/8$ remain; and so on. In general, after n half-lives, the number of radioactive nuclei remaining is $N_0/2^n$. Thus we see that nuclear decay is independent of the past history of a sample.

A frequently used unit of activity is the **curie** (Ci), defined as

The curie

$$1 \text{ Ci} \equiv 3.7 \times 10^{10} \text{ decays/s}$$

This value was originally selected because it is the approximate activity of 1 g of radium. The SI unit of activity is the **becquerel** (Bq):

The becquerel

$$1 \text{ Bq} \equiv 1 \text{ decay/s}$$

Therefore, $1 \text{ Ci} = 3.7 \times 10^{10}$ Bq. The curie is a rather large unit, and the more frequently used activity units are the millicurie mCi (10^{-3} Ci) and the microcurie, μCi (10^{-6} Ci).

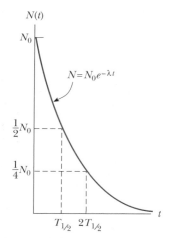

Figure 13.15 A plot of the exponential decay law for radioactive nuclei. The vertical axis represents the number of radioactive nuclei present at any time t, and the horizontal axis is time. The time $T_{1/2}$ is the half-life of the sample.

EXAMPLE 13.4   How Many Nuclei Are Left?

The isotope carbon-14, $^{14}_{6}$C, is radioactive and has a half-life of 5730 years (yr). If you start with a sample of 1000 carbon-14 nuclei, how many will still be around in 22,920 yr?

Solution   In 5730 yr half the sample will have decayed, leaving 500 carbon-14 nuclei. In another 5730 yr (for a total elapsed time of 11,460 yr), the number will be reduced to 250 nuclei. After another 5730 yr (total time

17,190 yr), 125 will remain. Finally, after four half-lives (22,920 yr), only about 62 will remain.

These numbers represent ideal circumstances. Calculation of radioactive decay is an averaging process conducted with a very large number of atoms, and the actual outcome depends on statistics. Our original sample in this example contained only 1000 nuclei, certainly not a very large number. Thus, if we actually counted the number remaining from this small sample after one half-life, it probably would not be 500.

EXAMPLE 13.5   The Activity of Radium

The half-life of the radioactive nucleus $^{226}_{88}Ra$ is about 1.6×10^3 yr. (a) What is the decay constant of $^{226}_{88}Ra$?

Solution   (a) We can calculate the decay constant λ by using Equation 13.11 and the fact that

$$T_{1/2} = 1.6 \times 10^3 \text{ yr}$$
$$= (1.6 \times 10^3 \text{ yr})(3.16 \times 10^7 \text{ s/yr})$$
$$= 5.0 \times 10^{10} \text{ s}$$

Therefore,

$$\lambda = \frac{0.693}{T_{1/2}} = \frac{0.693}{5.0 \times 10^{10} \text{ s}} = 1.4 \times 10^{-11} \text{ s}^{-1}$$

Note that this result gives the probability that any single $^{226}_{88}Ra$ nucleus will decay in 1 s.

(b) If a sample contains 3.0×10^{16} such nuclei at $t = 0$, determine its activity at this time.

Solution   (b) We can calculate the activity of the sample at $t = 0$ using $R_0 = \lambda N_0$, where R_0 is the decay rate at $t = 0$ and N_0 is the number of radioactive nuclei present at $t = 0$. Since $N_0 = 3.0 \times 10^{16}$, we have

$$R_0 = \lambda N_0 = (1.4 \times 10^{-11} \text{ s}^{-1})(3.0 \times 10^{16})$$
$$= 4.2 \times 10^5 \text{ decays/s}$$

Since $1 \text{ Ci} = 3.7 \times 10^{10}$ decays/s, the activity, or decay rate, at $t = 0$ is

$$R_0 = 11.1 \text{ } \mu\text{Ci}$$

(c) What is the decay rate after the sample is 2.0×10^3 yr old?

Solution   (c) We can use Equation 13.10 and the fact that $2.0 \times 10^3 \text{ yr} = (2.0 \times 10^3 \text{ yr})(3.15 \times 10^7 \text{ s/yr}) = 6.3 \times 10^{10}$ s:

$$R = R_0 e^{-\lambda t}$$
$$= (4.2 \times 10^5 \text{ decays/s}) e^{-(1.4 \times 10^{-11} \text{ s}^{-1})(6.3 \times 10^{10} \text{ s})}$$
$$= 1.7 \times 10^5 \text{ decays/s}$$

EXAMPLE 13.6   The Activity of Carbon

A radioactive sample contains 3.50 μg of pure $^{11}_6C$, which has a half-life of 20.4 min. (a) Determine the number of nuclei in the sample at $t = 0$.

Solution   The atomic mass of $^{11}_6C$ is approximately 11.0, and so 11.0 g contains Avogadro's number (6.02×10^{23}) of nuclei. Therefore 3.50 μg contains N nuclei, where

$$\frac{N}{6.02 \times 10^{23} \text{ nuclei/mol}} = \frac{3.50 \times 10^{-6} \text{ g}}{11.0 \text{ g/mol}}$$

$$N = 1.92 \times 10^{17} \text{ nuclei}$$

(b) What is the activity of the sample initially and after 8.00 h?

Solution   Since $T_{1/2} = 20.4 \text{ min} = 1224 \text{ s}$, the decay constant is

$$\lambda = \frac{0.693}{T_{1/2}} = \frac{0.693}{1224 \text{ s}} = 5.66 \times 10^{-4} \text{ s}^{-1}$$

Therefore, the initial activity or decay rate of the sample is

$$R_0 = \lambda N_0 = (5.66 \times 10^{-4} \text{ s}^{-1})(1.92 \times 10^{17})$$
$$= 1.08 \times 10^{14} \text{ decays/s}$$

We can use Equation 13.10 to find the activity at any time t. For $t = 8.00 \text{ h} = 2.88 \times 10^4$ s, we see that $\lambda t = 16.3$, and so

$$R = R_0 e^{-\lambda t} = (1.09 \times 10^{14} \text{ decays/s}) e^{-16.3}$$
$$= 8.96 \times 10^6 \text{ decays/s}$$

Table 13.3 lists values for activity versus the time in hours for this situation.

Table 13.3   Activity Versus Time for the Sample Described in Example 13.6

t (h)	R (decays/s)
0	1.08×10^{14}
1	1.41×10^{13}
2	1.84×10^{12}
3	2.39×10^{11}
4	3.12×10^{10}
5	4.06×10^9
6	5.28×10^8
7	6.88×10^7
8	8.96×10^6

Exercise 1   Calculate in Example 13.6 the number of radioactive nuclei that remain after 8.00 h.

Answer 1.58×10^{10} nuclei

EXAMPLE 13.7   A Radioactive Isotope of Iodine

A sample of the isotope ^{131}I, which has a half-life of 8.04 days, has an activity of 5 mCi at the time of shipment. Upon receipt of the ^{131}I in a medical laboratory, its activity is 4.2 mCi. How much time has elapsed between the two measurements?

Solution   We can make use of Equation 13.10 in the form

$$\frac{R}{R_0} = e^{-\lambda t}$$

Taking the natural logarithm of each side, we get

$$\ln\left(\frac{R}{R_0}\right) = -\lambda t$$

$$(1) \qquad t = -\frac{1}{\lambda} \ln\left(\frac{R}{R_0}\right)$$

To find λ, we use Equation 13.11:

$$(2) \qquad \lambda = \frac{0.693}{T_{1/2}} = \frac{0.693}{8.04 \text{ days}}$$

Substituting (2) into (1) gives

$$t = -\left(\frac{8.04 \text{ days}}{0.693}\right) \ln\left(\frac{4.2 \text{ mCi}}{5.0 \text{ mCi}}\right) = 2.02 \text{ days}$$

13.5   DECAY PROCESSES

As we stated in the preceding section, a radioactive nucleus spontaneously decays by means of one of three processes: alpha decay, beta decay, or gamma decay. Let us discuss these processes in more detail.

Alpha Decay

If a nucleus emits an α particle (^4_2He), it loses two protons and two neutrons. Therefore, the atomic number Z decreases by 2, the mass number A decreases by 4, and the neutron number decreases by 2. The decay can be written as

Alpha decay

$$^A_Z X \longrightarrow {}^{A-4}_{Z-2} Y + {}^4_2 He \tag{13.12}$$

where X is called the **parent nucleus** and Y the **daughter nucleus.** For example, ^{238}U and ^{226}Ra are both alpha emitters and decay according to the schemes

$$^{238}_{92} U \longrightarrow {}^{234}_{90} Th + {}^4_2 He \tag{13.13}$$

$$^{226}_{88} Ra \longrightarrow {}^{222}_{86} Rn + {}^4_2 He \tag{13.14}$$

The half-life for the ^{238}U decay is 4.47×10^9 years, and that for ^{226}Ra decay is 1.60×10^3 years. In both cases, note that the mass number of the daughter nucleus is 4 less than that of the parent nucleus. Likewise, the atomic number is reduced by 2. The differences are accounted for in the emitted α particle (the ^4He nucleus). Observe that alpha decay processes release energy because the decay products, especially the alpha particle, are more tightly bound than the parent nucleus (see Fig. 13.10.)

Figure 13.16 depicts the spontaneous decay of ^{226}Ra. As a general rule, (1) the sum of the mass numbers A must be the same on both sides of the equation, and (2) the net charge must be the same on both sides of the equation. In addition, the total relativistic energy and momentum must be conserved. If we call M_X the atomic mass of the parent, M_Y the mass of the daughter, and M_α the mass of the alpha particle, we can define the **disintegration energy** Q:

Disintegration energy Q

$$Q = (M_X - M_Y - M_\alpha)c^2 \qquad (13.15)$$

Note that atomic mass rather than nuclear mass can be used here, because the electronic masses cancel in an evaluation of the mass differences. Q is in joules when the masses are in kilograms, and c is the usual 3.00×10^8 m/s. However, when masses are expressed in the more convenient unit u, the value of Q can be calculated in MeV with the expression

$$Q = (M_X - M_Y - M_\alpha) \times 931.494 \text{ MeV/u} \qquad (13.16)$$

The disintegration energy Q appears in the form of kinetic energy of the daughter nucleus and the α particle. The quantity given by Equation 13.16 is sometimes referred to as the Q value of the nuclear reaction. In the case of the ^{226}Ra decay described in Figure 13.16, if the parent nucleus is at rest when it decays, the residual kinetic energy of the products is 4.87 MeV. Most of the kinetic energy is associated with the alpha particle because this particle is much less massive than the recoiling daughter nucleus, ^{222}Rn. That is, because momentum must be conserved, the lighter α particle recoils with a much higher speed than the daughter nucleus. Generally, light particles carry off most of the energy in nuclear decays.

Figure 13.16 Alpha decay of radium. The radium nucleus is initially at rest. After the decay, the radon nucleus has kinetic energy K_{Rn} and momentum $\mathbf{p}_{Rn}$, and the alpha particle has kinetic energy K_α and momentum $\mathbf{p}_\alpha$.

It is fairly easy to calculate the fraction of the disintegration energy carried off by the α particle by applying conservation of energy and momentum:

$$Q = K_Y + K_\alpha \qquad (13.17)$$

$$p_Y = p_\alpha \qquad (13.18)$$

where the subscript Y stands for the daughter nucleus. Since the total kinetic energy released in alpha decay (several MeV) is small compared to the rest energies of the α particle (3726 MeV) and the daughter nucleus (206.793 BeV for ^{222}Rn), we can use the classical expressions for momentum and kinetic energy in Equations 13.17 and 13.18 to show that

$$K_\alpha = \frac{M_Y}{M_Y + M_\alpha} Q \qquad (13.19)$$

where M_Y and M_α are the atomic masses of the daughter nucleus and the α particle (see Problem 43 at the end of the chapter).

Interestingly, if one assumed that ^{238}U (or other alpha emitters) decayed by emitting a proton or neutron, the mass of the decay products would exceed that of the parent nucleus, corresponding to negative Q values. These negative Q values indicate that such decays do not occur spontaneously.

EXAMPLE 13.8   The Energy Liberated When Radium Decays

The ^{226}Ra nucleus undergoes alpha decay according to Equation 13.12. Calculate the Q value for this process. Take the atomic masses to be 226.025 406 u for ^{226}Ra, 222.017 574 u for ^{222}Rn, and 4.002 603 u for $^{4}_{2}$He, as found in Appendix B.

Solution   Using Equation 13.16, we see that

$$Q = (M_X - M_Y - M_\alpha) \times 931.494 \frac{\text{MeV}}{\text{u}}$$

$$= (226.025\,406\text{ u} - 222.017\,574\text{ u}$$

$$- 4.002\,603\text{ u}) \times 931.494 \frac{\text{MeV}}{\text{u}}$$

$$= (0.005\,229\text{ u}) \times \left(931.494 \frac{\text{MeV}}{\text{u}}\right) = 4.87\text{ MeV}$$

It is left to Problem 43 to show that the kinetic energy of the α particle is about 4.8 MeV, whereas the recoiling daughter nucleus carries off only about 0.1 MeV of kinetic energy.

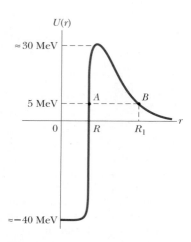

Figure 13.17   Potential energy versus separation for the alpha particle–nucleus system. Classically, the energy of the alpha particle is not great enough to overcome the barrier, and so the particle should not be able to escape the nucleus.

We now turn to the mechanism of α decay. Imagine that an α particle is somehow formed within the nucleus. Figure 13.17 is a plot of potential energy versus distance r from the nucleus for a typical alpha particle–nucleus system, where R is the range of the nuclear force. The curve represents the combined effects of (1) the Coulomb repulsive energy, which gives the positive peak for $r > R$, and (2) the nuclear attractive force, which causes the curve to be negative for $r < R$. As we saw in Example 13.8, a typical disintegration energy is about 5 MeV, which is the approximate kinetic energy of the α particle, represented by the lower dashed line in Figure 13.17. According to classical physics, the α particle is trapped in the potential well. How, then, does it ever escape from the nucleus?

The answer to this question was first provided by George Gamow in 1928 and, independently, by R. W. Gurney and E. U. Condon in 1929, using quantum mechanics. Briefly, the view of quantum mechanics is that there is always some probability that the particle can penetrate (tunnel through) the barrier (Section 7.2). Recall that the probability of locating the particle depends on its wavefunction ψ and that the probability of tunneling is measured by $|\psi|^2$. Figure 13.18 is a sketch of the wavefunction for a particle of energy E meeting a square barrier of finite height, a shape that approximates the nuclear barrier. Note that the wavefunction exists both inside and outside the barrier. Although the amplitude of the wavefunction is greatly reduced on the far side of the barrier, its finite value in this region indicates a small but finite probability that the particle can penetrate the barrier. As the energy E of the particle

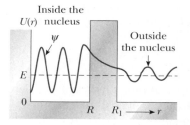

Figure 13.18   The nuclear potential energy is modeled as a square barrier. The energy of the alpha particle is E, which is less than the height of the barrier. According to quantum mechanics, the alpha particle has some chance of tunneling through the barrier, as indicated by the finite size of the wavefunction for $r > R_1$.

increases, its probability of escaping also increases. Furthermore, the probability increases as the width of the barrier decreases.

EXAMPLE 13.9 Probability for Alpha Decay

Apply the tunneling methods of Chapter 7 to compute the probability of escape from a $^{226}_{88}$Ra nucleus of an α particle with disintegration energy 5 MeV.

Solution The escape probability is none other than the transmission coefficient $T(E)$ for the Coulomb barrier shown in Figure 13.17. Equation 7.10 of Chapter 7 gives $T(E)$ approximately as

$$T(E) \approx \exp^{-(2/\hbar)\sqrt{2m}\int\sqrt{U(r)-E}\,dr}$$

The integral is taken over the classically forbidden region where $E < U$. For alpha decay, this region is bounded below by the nuclear radius R and above by $R_1 = 2Zke^2/E$ [from $E = U(R_1) = 2Zke^2/R_1$] (see Fig. 13.17). In this expression, Z is the atomic number of the *daughter* nucleus.

The tunneling integral to be evaluated is

$$\int \sqrt{U(r) - E}\,dr = \sqrt{E}\int_R^{R_1}\sqrt{\frac{R_1}{r} - 1}\,dr$$

$$= R_1\sqrt{E}\int_{R/R_1}^{1}\sqrt{\frac{1}{z} - 1}\,dz$$

where $z = r/R_1$. An exact value for this integral can be had with some effort, but a useful approximation is found readily by noting that R/R_1 is a small number ($R \sim 10$ fm; $R_1 \sim 50$ fm for $E \sim 5$ MeV). Thus, as a first estimate we set the lower limit to zero and change variables with $z = \cos^2\theta$, obtaining

$$\int_0^1 \sqrt{\frac{1}{z} - 1}\,dz = 2\int_0^{\pi/2} \sin^2\theta\,d\theta$$

$$= \int_0^{\pi/2} [1 - \cos 2\theta]\,d\theta = \frac{\pi}{2}$$

To improve upon this, we break the original integral into two and approximate the second, using $1/z \gg 1$ for z small, to get

$$\int_0^1 \sqrt{\frac{1}{z} - 1}\,dz - \int_0^{R/R_1}\sqrt{\frac{1}{z} - 1}\,dz = \frac{\pi}{2}$$

$$-\int_0^{R/R_1}\frac{dz}{\sqrt{z}} \approx \frac{\pi}{2} - 2\sqrt{\frac{R}{R_1}}$$

Combining this result with $R_1 = 2Zke^2/E$ gives, for the decay probability,

$$T(E) = \exp\{-4\pi Z\sqrt{(E_0/E)} + 8\sqrt{Z(R/r_0)}\}$$

The parameter $r_0 = \hbar^2/M_\alpha ke^2$ is a kind of Bohr radius for the alpha particle, with the value 7.25 fm, and E_0 is an energy unit analogous to the Rydberg in atomic physics:

$$E_0 = \frac{ke^2}{2r_0} = \frac{14.40 \text{ eV}\cdot\text{Å}}{(2)(7.25 \times 10^{-5}\text{ Å})} = 0.0993 \text{ MeV}$$

For the alpha decay of radium, the daughter nucleus is radon with atomic number $Z = 86$ and mass number $A = 222$. Equation 13.1 predicts the radius R of the radon nucleus to be

$$R = (1.2 \times 10^{-5}\text{ fm})(222)^{1/3} = 7.27 \text{ fm}$$

Then the decay probability for alpha disintegration at $E = 5$ MeV is

$$\exp\{-4\pi(86)\sqrt{(0.0993/5)} + 8\sqrt{86(7.27/7.25)}\}$$
$$= \exp\{-78.008\} = 1.32 \times 10^{-34}$$

This probability is quite small, but the actual number of disintegrations per second is much larger because of the many collisions the alpha particle makes with the nuclear barrier. This collision frequency f is the reciprocal of the transit time for the alpha particle crossing the nucleus; that is, $f = v/2R$, where v is the speed of the alpha particle inside the nucleus. Here, as in most cases, f is about 10^{21} collisions per second (see Problem 16 in Chapter 7), leading to a predicted decay rate $\approx 10^{-13}$ disintegrations per second, in reasonable agreement with the observed value of $\lambda = 1.4 \times 10^{-11}$ s^{-1}.

Beta Decay

When a radioactive nucleus undergoes beta decay, the daughter nucleus has the same number of nucleons as the parent nucleus, but the atomic number is changed by 1:

$$^A_Z X \longrightarrow {}^A_{Z+1}Y + e^-$$

(13.20) **Beta decay**

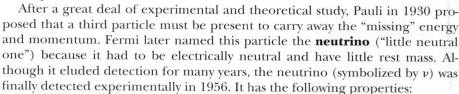

$$^A_Z X \longrightarrow\ _{Z-1}^{\ \ A}Y + e^+ \qquad\qquad (13.21)$$

Again, note that the nucleon number and total charge are both conserved in these decays. However, as we shall see later, these processes are not described completely by such expressions. We shall explain this shortly.

Two typical beta decay processes are

$$^{14}_{6}C \longrightarrow\ ^{14}_{7}N + e^-$$

$$^{12}_{7}N \longrightarrow\ ^{12}_{6}C + e^+$$

Notice that in beta decay a neutron changes into a proton (or vice versa). It is also important to point out that the electron or positron in these decays is not present beforehand in the nucleus but is created at the moment of decay out of the rest energy of the decaying nucleus.

Now consider the energy of the system before and after the decay. As with alpha decay, we assume energy is conserved and that the heavy recoiling daughter nucleus carries off *negligible* kinetic energy. (See Problem 43.) Experimentally, it is found that beta particles from a single type of nucleus are emitted with a continuous range of kinetic energies up to some maximum value, $K_{\max}$ (Fig. 13.19). The kinetic energy of the system after the decay is equal to the decrease in mass–energy of the system—that is, the Q value. However, because all parent nuclei of a given type have the same initial mass, *the Q value must be the same for each decay.* In view of this and the fact that the daughter nucleus carries off very little kinetic energy, why do the emitted beta particles have different kinetic energies? The law of conservation of energy seems to be violated! Further analysis shows that the decay processes given by Equations 13.20 and 13.21 also violate the principles of conservation of angular momentum (spin) and linear momentum!

After a great deal of experimental and theoretical study, Pauli in 1930 proposed that a third particle must be present to carry away the "missing" energy and momentum. Fermi later named this particle the **neutrino** ("little neutral one") because it had to be electrically neutral and have little rest mass. Although it eluded detection for many years, the neutrino (symbolized by ν) was finally detected experimentally in 1956. It has the following properties:

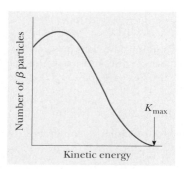

Figure 13.19 A typical beta decay curve. The maximum kinetic energy observed for the beta particles corresponds to the Q value for the reaction.

- Zero electric charge.
- A rest mass that is much smaller than that of the electron. Recent experiments show that the mass of the neutrino is not 0 but is less than $2.8\ \text{eV}/c^2$.
- A spin of $\frac{1}{2}$, which satisfies the law of conservation of angular momentum when applied to beta decay.
- Very weak interaction with matter, which makes it very difficult to detect.

Properties of the neutrino

For the general form of the beta decays considered earlier, we can now write

$$^A_Z X \longrightarrow\ _{Z+1}^{\ \ A}Y + e^- + \bar{\nu} \qquad\qquad (13.22)$$

$$^A_Z X \longrightarrow\ _{Z-1}^{\ \ A}Y + e^+ + \nu \qquad\qquad (13.23)$$

where the symbol $\bar{\nu}$ represents the **antineutrino,** the antiparticle to the neutrino. (We discuss antiparticles further in Chapter 15.) As in the case of alpha decay, the decays just listed are analyzed through conservation of energy and momentum, but we must use relativistic expressions because the kinetic energies of the electron and neutrino are not small compared to their rest energies.

Enrico Fermi, an Italian-American physicist, received his doctorate from the University of Pisa in 1922, then did post-doctoral work in Germany under Max Born. He returned to Italy in 1924, and in 1926 became a professor of physics at the University of Rome. He received the Nobel Prize for Physics in 1938 for his work with the production of transuranic radioactive elements (those more massive than uranium) by neutron bombardment.

Fermi first became interested in physics at the age of 14, after reading an old physics book in Latin. (He was an excellent scholar and could recite Dante's *Divine Comedy* and much of Aristotle's writings from memory.) His great ability to solve problems in theoretical physics and his skill for simplifying very complex situations made him somewhat of an oracle. Fermi was also a gifted experimentalist and teacher. During one of his early lecture trips to the United States, a car that he

B I O G R A P H Y
———
ENRICO FERMI
(1901–1954)

had purchased became disabled, and he pulled into a nearby gas station. After Fermi repaired the car with ease, the station owner offered him a job on the spot.

Fermi and his family emigrated to the United States, and he became a naturalized citizen in 1944. He taught first at Columbia University,

then at the University of Chicago. As part of the Manhattan Project during World War II, Fermi was commissioned to design and build a structure, called an atomic pile, in which a self-sustained chain reaction might occur. The structure, built in a squash court under the stadium of the University of Chicago, contained uranium in combination with graphite blocks to slow the neutrons to thermal speeds. Cadmium rods inserted in the pile were used to absorb neutrons and control the reaction rate. History was made at 3:45 P.M. on December 2, 1942, as the cadmium rods were slowly withdrawn and a self-sustained chain reaction was observed. Fermi's earth-shaking achievement—the world's first nuclear reactor—marked the beginning of the atomic age.

Fermi died of cancer in 1954 at the age of 53. One year later, the 100th element was discovered and named *fermium* in his honor.

(Fermi National Accelerator Laboratory)

A process that competes with e^+ decay is called **electron capture.** This occurs when a parent nucleus captures one of its own orbital atomic electrons and emits a neutrino. The final product after decay is a nucleus whose charge is $Z - 1$:

$$_Z^A\text{X} + e^- \longrightarrow _{Z-1}^A\text{X} + \nu \qquad (13.24)$$

Electron capture

In most cases it is an inner K-shell electron that is captured, and this is referred to as **K capture.** One example of this process is the capture of an electron by $_4^7\text{Be}$ to become $_3^7\text{Li}$:

$$_4^7\text{Be} + e^- \longrightarrow _3^7\text{Li} + \nu$$

Finally, it is instructive to mention the Q values for beta-decay processes. The Q values for e^- decay and electron capture are given by $Q = (M_X - M_Y)c^2$, while the Q values for e^+ decay are given by $Q = (M_X - M_Y - 2m_e)c^2$ where M_X and M_Y are the masses of neutral atoms. These relationships are useful for determining whether or not possible beta-decay processes are energetically allowed.

Carbon Dating

The beta decay of ^{14}C is commonly used to date organic samples. Cosmic rays in the upper atmosphere cause nuclear reactions that create ^{14}C. In fact, the ratio of ^{14}C to ^{12}C in the carbon dioxide molecules of our atmosphere has a constant value of approximately 1.3×10^{-12}. All living organisms have this same ratio of ^{14}C to ^{12}C, because they continuously exchange carbon dioxide with their surroundings. When an organism dies, however, it no longer absorbs ^{14}C from the atmosphere, and so the $^{14}C/^{12}C$ ratio decreases as a result of the beta decay of ^{14}C, which has a half-life of 5730 yr. It is therefore possible to measure the age of a material by measuring its activity per unit mass caused by the decay of ^{14}C. Using this technique, scientists have been able to identify samples of wood, charcoal, bone, and shell as having lived from 1000 to 25,000 yr ago. This knowledge has helped us reconstruct the history of living organisms—including humans—during that time span.

A particularly interesting example is the dating of the Dead Sea Scrolls, a group of manuscripts discovered by a shepherd in 1947. Translation showed them to be religious documents, including most of the books of the Old Testament. Because of their historical and religious significance, scholars wanted to know their age. Carbon dating performed on the material in which the scrolls were wrapped established their age at approximately 1950 yr.

EXAMPLE 13.10   The Age of Ice Man

In 1991 a German tourist discovered the well-preserved remains of a human, later dubbed "Ice Man," trapped in a glacier in the Italian Alps. Radioactive dating of a sample of Ice Man established an age of 5300 yr. Why did scientists date the sample with the isotope ^{14}C rather than ^{11}C, a beta emitter with a half-life of 20.4 min?

Reasoning Carbon-14 has a long half-life, 5730 yr, so the fraction of ^{14}C nuclei remaining after one half-life is high enough to enable accurate measurements of changes in the sample's activity. The ^{11}C isotope, which has a very short half-life, is not useful because its activity decreases to a vanishingly small value over the age of the sample, making it impossible to detect.

If a sample to be dated is not very old—say, about 50 yr—then the scientist should select the isotope of some other element whose half-life is comparable to the age of the sample. For example, if the sample contained hydrogen, one could measure the activity of ^{3}H (tritium), a beta emitter with a half-life of 12.3 yr. As a general rule, the expected age of the sample should be great enough to allow measurement of a change in activity but not so great that the activity is undetectable.

EXAMPLE 13.11   Radioactive Dating

An archaeologist finds a 25.0-g piece of charcoal in the ruins of an ancient city. The sample shows a ^{14}C activity of 250 decays/min. How long has the tree from which this charcoal came been dead?

Solution First calculate the decay constant for ^{14}C, which has a half-life of 5730 yr.

$$\lambda = \frac{0.693}{T_{1/2}} = \frac{0.693}{(5730 \text{ yr})(3.16 \times 10^7 \text{ s/yr})}$$
$$= 3.83 \times 10^{-12} \text{ s}^{-1}$$

The number of ^{14}C nuclei can be calculated in two steps. (1) The number of ^{12}C nuclei in 25.0 g of carbon is

$$N(^{12}C) = \frac{6.02 \times 10^{23} \text{ nuclei/mol}}{12.0 \text{ g/mol}} (25.0 \text{ g})$$
$$= 1.26 \times 10^{24} \text{ nuclei}$$

Knowing that the ratio of ^{14}C to ^{12}C in the live sample was 1.3×10^{-12}, we see that the number of ^{14}C nuclei in 25.0 g *before* decay is

$$N_0(^{14}C) = (1.3 \times 10^{-12})(1.26 \times 10^{24})$$
$$= 1.6 \times 10^{12} \text{ nuclei}$$

Hence the initial activity of the sample is

$$R_0 = N_0\lambda = (1.6 \times 10^{12} \text{ nuclei})(3.83 \times 10^{-12} \text{ s}^{-1})$$
$$= 6.13 \text{ decays/s} = 370 \text{ decays/min}$$

(2) We can now calculate the age of the charcoal, using Equation 13.10, which relates the activity R at any time t to the initial activity R_0:

$$R = R_0 e^{-\lambda t} \quad \text{or} \quad e^{-\lambda t} = \frac{R}{R_0}$$

Because it is given that $R = 250$ decays/min and because we found that $R_0 = 370$ decays/min, we can calculate t by taking the natural logarithm of both sides of the last equation:

$$-\lambda t = \ln\left(\frac{R}{R_0}\right) = \ln\left(\frac{250}{370}\right) = -0.39$$

$$t = \frac{0.39}{\lambda} = \frac{0.39}{3.84 \times 10^{-12}\,\text{s}^{-1}}$$

$$= 1.0 \times 10^{11}\,\text{s} = 3.2 \times 10^3\,\text{yr}$$

Gamma Decay

Very often, a nucleus that undergoes radioactive decay is left in an excited energy state. The nucleus can then undergo a second decay to a lower energy state, perhaps to the ground state, by emitting a high-energy photon:

$$^A_Z X^* \longrightarrow {}^A_Z X + \gamma \qquad (13.25)$$

Gamma decay

where X* indicates a nucleus in an excited state. The typical half-life of an excited nuclear state is 10^{-10} s. Photons emitted in such a deexcitation process are called gamma rays. Such photons have very high energy (in the range of 1 MeV to 1 GeV) relative to the energy of visible light (about 1 eV). Recall that the energy of photons emitted (or absorbed) by an atom equals the difference in energy between the two electronic states involved in the transition. Similarly, a gamma-ray photon has an energy hf that equals the energy difference ΔE between two nuclear energy levels. When a nucleus decays by emitting a gamma ray, the nucleus doesn't change its atomic mass A or atomic number Z.

A nucleus may reach an excited state as the result of a violent collision with another particle. It is also very common for a nucleus to be in an excited state after undergoing an alpha or beta decay. The following sequence of events represents a typical situation in which gamma decay occurs:

$$^{12}_5 B \longrightarrow {}^{12}_6 C^* + e^- + \bar{\nu} \qquad (13.26)$$

$$^{12}_6 C^* \longrightarrow {}^{12}_6 C + \gamma \qquad (13.27)$$

Figure 13.20 shows the decay scheme for ^{12}B, which undergoes beta decay with a half-life of 20.4 ms to either of two levels of ^{12}C. It can either (1) decay

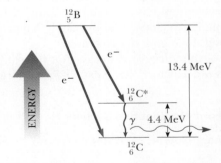

Figure 13.20 An energy-level diagram showing the initial nuclear state of a ^{12}B nucleus and two possible lower-energy states of the ^{12}C nucleus. The beta decay of the ^{12}B nucleus can result in either of two situations: The ^{12}C nucleus is in the ground state or in the excited state, in which case the nucleus is denoted as ^{12}C*. In the latter case, the beta decay to ^{12}C* is followed by a gamma decay to ^{12}C as the excited nucleus makes a transition to the ground state.

Table 13.4   Decay Processes

Alpha decay	${}_Z^A X \longrightarrow {}_{Z-2}^{A-4} X + {}_2^4 He$
Beta decay (e^-)	${}_Z^A X \longrightarrow {}_{Z+1}^A X + e^- + \bar{\nu}$
Beta decay (e^+)	${}_Z^A X \longrightarrow {}_{Z-1}^A X + e^+ + \nu$
Electron capture	${}_Z^A X + e^- \longrightarrow {}_{Z-1}^A X + \nu$
Gamma decay	${}_Z^A X^* \longrightarrow {}_Z^A X + \gamma$

directly to the ground state of ^{12}C by emitting a 13.4-MeV electron or (2) undergo e^- decay to an excited state of ^{12}C*, followed by gamma decay to the ground state. The latter process results in the emission of a 9.0-MeV electron and a 4.4-MeV photon. Table 13.4 summarizes the pathways by which a radioactive nucleus can undergo decay.

13.6   NATURAL RADIOACTIVITY

Four Radioactive Series

Radioactive nuclei are generally classified into two groups: (1) unstable nuclei found in nature, which give rise to what is called **natural radioactivity,** and (2) nuclei produced in the laboratory through nuclear reactions, which exhibit **artificial radioactivity.** There are three series of naturally occurring radioactive nuclei (Table 13.5). Each series starts with a specific long-lived radioactive isotope whose half-life exceeds that of any of its descendants. The three natural series begin with the isotopes ^{238}U **(Uranium Series),** ^{235}U **(Actinium Series),** and ^{232}Th **(Thorium Series),** and the corresponding stable end products are three isotopes of lead: ^{206}Pb, ^{207}Pb, and ^{208}Pb. The fourth series in Table 13.5 is an artificial radioactive series called the **Neptunium Series** because Neptunium is the longest-lived member of the series other than its stable end product, ^{209}Bi. The element ^{237}Np is a transuranic element (one having an atomic number greater than that of uranium) not found in nature. This element has a half-life of "only" 2.14×10^6 years, much less than the age of the Earth; consequently, any ^{237}Np present when the Earth was created would long since have decayed away.

Figure 13.21 shows the successive decays for the ^{232}Th series. Note that ^{232}Th first undergoes α decay to ^{228}Ra. Next, ^{228}Ra undergoes two successive β

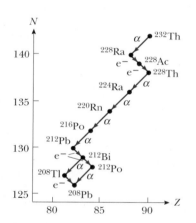

Figure 13.21   Successive decays for the ^{232}Th series.

Table 13.5   The Four Radioactive Series

Series	Starting Isotope	Half-Life (years)	Stable End Product
Uranium	${}_{92}^{238}$U	4.47×10^9	${}_{82}^{206}$Pb
Actinium	${}_{92}^{235}$U	7.04×10^8	${}_{82}^{207}$Pb
Thorium	${}_{90}^{232}$Th	1.41×10^{10}	${}_{82}^{208}$Pb
Neptunium	${}_{93}^{237}$Np[a]	2.14×10^6	${}_{83}^{209}$Bi

[a]This is the longest-lived member of the series. The starting isotope is actually plutonium, ${}_{94}^{241}$Pu.

decays to ^{228}Th. The series continues and finally branches when it reaches ^{212}Bi. At this point, there are two decay possibilities. The end of the decay series is the stable isotope ^{208}Pb. Each step in the decay sequence shown in Figure 13.21 may be characterized by a decrease of either 4 (for α decays) or 0 (for β or γ decays) in the mass number A, together with appropriate changes in Z. The two uranium series are somewhat more complex than the ^{232}Th series. Also, there are several naturally occurring radioactive isotopes, such as ^{14}C and ^{40}K, that are not part of the aforementioned decay series.

Radioactive series in nature constantly replenish our environment with radioactive elements that would otherwise have disappeared long ago. For example, because the Earth is approximately 5×10^9 years old, the supply of ^{226}Ra (whose half-life is only 1600 years) would have been depleted by radioactive decay long ago if it were not for the decay series that starts with ^{238}U.

Determining the Age of the Earth

The casual statement that the Earth is about 5 billion years old deserves more explanation not only because it has been a historically contentious subject but also because the science is interesting. Various methods, ranging from geological estimates based on weathering rates to estimates based on cooling rates of an initially hot Earth, have been used. However, Rutherford's method of using the half-life and measured amounts of decay products in rocks (he had already used it by 1904!) has proven the most accurate. Let's follow this interesting story.

The first major scientific theory of the origin of the Solar System was that of Pierre Simon de Laplace (1749–1827, French physicist) and William Herschel (1738–1822, German–British astronomer). They suggested that the solar system was formed from a hot, slowly rotating spherical cloud of gas, which cooled and contracted, flattening into more rapidly rotating rings to conserve angular momentum, with planets eventually condensing out of the rings. The Sun was formed from the hot leftover central part of the gas cloud. Hermann von Helmholtz (1821–1894, German physiologist and physicist), the great champion of the law of energy conservation, realized that there was no need to start with a rather artificial hot gas cloud. He modified the Laplace–Herschel theory by assuming that the primordial gas cloud was cold and that gravitational contraction produced heating in agreement with energy conservation. An enormous contribution to the quantitative study of the cooling of the Earth was made by Jean Baptiste Joseph Fourier (1768–1830, French mathematician) with his solution of the heat conduction equation in terms of sums of trigonometric functions (Fourier series). Taking into account varying seasonal inflow of solar energy as well as heat flowing out of the Earth's core, and using measurements of the temperature at different depths in mines and the thermal conductivity of rock as input parameters, Fourier concluded that the idea of a cooling Earth with an intially hot interior did not contradict known observations.

Geologists of the period generally agreed with these theories of an Earth solidifying from a molten state, the interior gradually cooling in an average way. Nineteenth-century geological estimates of the Earth's age based on continuous weathering processes, such as erosion, predicted ages of hundreds of millions of years. This hypothesis was readily adopted by Charles Darwin (1809–1882, English naturalist) who felt that the natural selection processes of his theory of evolution required long periods on the order of hundreds of millions of years.

Kelvin Weighs In.   Hardly any scientific story involves more disciplines and more famous scientific names than this one. Just sticking to the bare bones of the plot we have to mention William Thompson, Lord Kelvin (1824–1907, British physicist), one of the most important physicists of the 19th century. In the 1860s Kelvin refined Fourier's cooling model and estimated the Earth's age to be about 60 million years, starting from a uniform temperature of about 10,000°F, the highest melting point of rock. He also estimated that the Sun was about this age, or somewhat less, based on a model of solar energy production powered by gravitational contraction.

Kelvin's calculations started a serious disagreement between physicists who believed in the 60-million-year figure, and geologists, who felt the Earth was 10 times older. Even antievolutionists sided with physicists, whom they saw as providing clear scientific evidence that the Earth had not existed long enough for natural selection to occur. Neither side foresaw that the discovery of radioactivity would supply both the explanation of the failure of cooling models and an exquisitely accurate method of dating the Earth itself.

The End of The Age-of-the-Earth-Problem.   The discovery of radium and the measurement of its abundance and the energy it gives off allowed the Curies to explain Kelvin's erroneous age for the Earth. When the amount of thermal energy given off in one day by the radium in the Earth's crust is calculated, it turns out to be comparable to the energy lost per day by the Earth's core by conduction through the crust and eventual radiation into space. This means the Earth is warmer than it would be if energy were not added by radioactive decay. This led Kelvin to underestimate the actual cooling time when he did not incorporate the energy released by radioactive decay into the Earth's energy budget.

As mentioned previously, shortly after the prodigious discovery of radioactive series and the transmutation of series members into different chemical elements by Rutherford and Frederick Soddy (1877–1956, English physical chemist), Rutherford realized that radioactive techniques could be used to date the Earth's oldest rocks. The earliest methods involved measuring the amount of helium (from alpha decay) trapped in rocks, but these methods only provided a lower limit on the Earth's age because some helium had presumably leaked out of the rocks. In the 1940s, partly as a result of war work on the atomic bomb, very sensitive techniques were developed by Clair Cameron Patterson (1922–1995, American geochemist) and others of measuring tiny lead concentrations in samples. Since the three natural radioactive series all produce stable isotopes of lead, measurement of the amount of lead in very old rocks should indicate the age of the Earth. However, a major difficulty is that one needs to know the amount of primordial lead, that is, lead not produced by radioactive decay, present at the time of the formation of the Earth and Solar System. Because of their process of formation, iron meteorites containing almost no uranium are believed to contain true primordial levels of lead. Patterson used these and the oldest Earth rocks to arrive at the age of the Earth. In 1953 he found the still accepted value of (4.55 ± 0.07) billion years . A modest yet energetic scientist, Patterson is quoted as saying about his momentous discovery,

True scientific discovery renders the brain incapable at such moments of shouting vigorously to the world "Look at what I've done! Now I will reap the benefits of recognition and wealth." Instead such discovery instinctively forces the brain to thunder "*We* did it" in a voice no one else can hear, within its sacred, but lonely, chapel of scientific thought.

. . . "*We*" refers to what Patterson calls "the generations-old community of scientific minds." . . . To him it must have been an exercise in improving the state of the "community of scientific minds." His attitude recalls the remark of Newton: "If I have seen farther than others, it is because I have stood on the shoulders of giants." (From *Biographical Memoirs* by George R. Tilton)

Clair Cameron Patterson (1922–1995), American geochemist (*Courtesy of California Institute of Technology*)

SUMMARY

A nuclear species can be represented by $_Z^A X$, where A is the **mass number,** the total number of nucleons, and Z is the **atomic number,** the total number of protons. The total number of neutrons in a nucleus is the **neutron number,** N, where $A = N + Z$. Elements with the same Z but different A and N values are called **isotopes.**

Assuming that a nucleus is spherical, its radius is

$$r = r_0 A^{1/3} \tag{13.1}$$

where $r_0 = 1.2$ fm (1 fm $= 10^{-15}$ m).

Nuclei are stable because of the **nuclear force** between nucleons. This short-range force dominates the Coulomb repulsive force at distances of less than about 2 fm and is nearly independent of charge. Light nuclei are most stable when the number of protons equals the number of neutrons. Heavy nuclei are most stable when the number of neutrons exceeds the number of protons. In addition, the most stable nuclei have Z and N values that are both even. Nuclei with unusually high stability have Z or N values of 2, 8, 20, 28, 50, 82, and 126, called **magic numbers.**

Nuclei have a total angular momentum of magnitude $\sqrt{I(I+1)}\,\hbar$, where I is called the **nuclear-spin quantum number.** The magnetic moment of a nucleus is measured in terms of the **nuclear magneton** μ_n, where

$$\mu_n \equiv \frac{e\hbar}{2m_p} = 5.05 \times 10^{-27}\,\text{J/T} \tag{13.3}$$

When a nuclear moment is placed in an external magnetic field, it precesses about the field with a frequency that is proportional to the field strength.

The difference between the combined mass of the separate nucleons and that of the compound nucleus containing those nucleons, when multiplied by c^2, gives the **binding energy** E_b of the nucleus. We can calculate the binding energy of any nucleus of mass M_A with Z protons and N neutrons from

$$E_b(\text{MeV}) = [ZM(\text{H}) + Nm_n - M_A] \times 931.494\,\frac{\text{MeV}}{\text{u}} \tag{13.4}$$

where all masses are atomic masses and m_n is the mass of the neutron.

The **liquid-drop model** of nuclear structure treats the nucleons as molecules in a drop of liquid. The three main contributions influencing binding energy are the volume effect, the surface effect, and Coulomb repulsion. Summing such contributions results in the **semiempirical binding energy formula.**

The **independent-particle model** or **shell model** of nuclear structure assumes that each nucleon moves in a well-defined quantized orbit within the nucleus. The stability of certain nuclei—particularly those with magic numbers—can be explained with this model. Nuclei can move in well-defined orbits inside the densely packed nucleus because two nucleons will collide only if the energy of each state after the collision corresponds to one of the unoccupied nuclear states.

The **collective model** of the nucleus combines some features of the liquid-drop model and some features of the independent-particle model. It has been very successful in describing a variety of nuclear phenomena.

A radioactive substance decays by **alpha decay, beta decay,** or **gamma decay.** An alpha particle is a ^{4}He nucleus; a beta particle is either an electron (e^-) or a positron (e^+); a gamma particle is a high-energy photon.

If a radioactive material contains N_0 radioactive nuclei at $t = 0$, the number of nuclei N, remaining after a time t has elapsed is

$$N = N_0 e^{-\lambda t} \tag{13.9}$$

where λ is the **decay constant,** the probability per unit time that a nucleus will decay. The **decay rate,** or **activity,** of a radioactive substance is

$$R = \left| \frac{dN}{dt} \right| = R_0 e^{-\lambda t} \tag{13.10}$$

where $R_0 = N_0 \lambda$ is the activity, or number of decays per unit time at $t = 0$. The **half-life** $T_{1/2}$ is defined as the time required for half of a given number of radioactive nuclei to decay, where

$$T_{1/2} = \frac{0.693}{\lambda} \tag{13.11}$$

Alpha decay can occur because, according to quantum mechanics, alpha particles can tunnel through the coulomb barriers of nuclei. A nucleus undergoing beta decay emits either an electron (e^-) and an antineutrino ($\overline{\nu}$) or a positron (e^+) and a neutrino (ν). In electron capture, the nucleus of an atom absorbs one of its own orbital electrons and emits a neutrino. In gamma decay, a nucleus in an excited state decays to a lower energy state and emits a gamma ray.

QUESTIONS

1. Why are heavy nuclei unstable?
2. A proton precesses with a frequency ω_p in the presence of a magnetic field. If the intensity of the magnetic field is doubled, what happens to the precessional frequency?
3. Explain why nuclei that are well off the line of stability in Figure 13.4 tend to be unstable.
4. Consider two heavy nuclei, X and Y, with similar mass numbers. If X has the higher binding energy, which nucleus tends to be more unstable?
5. Discuss the differences between the liquid-drop model and the independent-particle model of the nucleus.
6. How many values of I_z are possible for $I = \frac{5}{2}$? for $I = 3$?
7. In nuclear magnetic resonance, how does increasing the dc magnetic field change the frequency of the ac field that excites a particular transition?
8. Would the liquid-drop model or the independent-particle model be more appropriate for predicting the behavior of a nucleus in a fission reaction? Which would be more successful at predicting the magnetic moment of a given nucleus? Which could better explain the gamma-ray spectrum of an excited nucleus?
9. If a nucleus has a half-life of one year, does that mean it will be completely decayed after two years? Explain.
10. What fraction of a radioactive sample has decayed after two half-lives have elapsed?

11. Two samples of the same radioactive nuclide are prepared, each having the same size. Sample A has twice the initial activity of sample B. How does the half-life of A compare with the half-life of B? After each sample has passed through five half-lives, what is the ratio of their activities?

12. Explain why the half-lives of radioactive nuclei are essentially independent of temperature.

13. The radioactive nucleus $^{226}_{88}$Ra has a half-life of approximately 1.6×10^3 yr. Given that the Solar System is about 5 billion years old, why do we still find this nucleus in nature?

14. A free neutron undergoes beta decay with a half-life of about 10 min. Can a free proton undergo a similar decay?

15. Explain how you can carbon-date the age of a sample.

16. What is the difference between a neutrino and a photon?

17. Use Equation 13.24 to explain why the neutrino must have a spin of $\frac{1}{2}$.

18. A nucleus such as ^{226}Ra that is initially at rest undergoes alpha decay. Which has more kinetic energy after the decay, the alpha particle or the daughter nucleus?

19. Can a nucleus emit alpha particles with different energies? Explain.

20. Explain why many heavy nuclei undergo alpha decay but do not spontaneously emit neutrons or protons.

21. If an alpha particle and an electron have the same kinetic energy, which undergoes greater deflection when passed through a magnetic field?

22. If photographic film is kept in a wooden box, alpha particles from a radioactive source outside the box cannot expose the film, but beta particles can. Explain.

23. Pick any beta-decay process, and show that the neutrino must have zero charge.

24. Suppose it could be shown that the intensity of cosmic rays at the Earth's surface was much greater 10,000 years ago. How would this difference affect what we accept as valid carbon-dated ages of ancient samples of once-living matter?

25. Why is carbon dating unable to provide accurate estimates of very old material?

26. Element X has several isotopes. What do these isotopes have in common? How do they differ?

27. Explain the main differences among alpha, beta, and gamma rays.

28. How many protons are in the nucleus $^{222}_{88}$Rn? How many neutrons? How many orbiting electrons are in the neutral atom?

PROBLEMS

Table 13.6 will be useful for many of these problems. A more complete list of atomic masses appears in Appendix B.

Table 13.6   Some Atomic Masses

Element	Atomic Mass (u)	Element	Atomic Mass (u)
$^{4}_{2}$He	4.002 603	$^{27}_{13}$Al	26.981 539
$^{7}_{3}$Li	7.016 003	$^{30}_{15}$P	29.978 310
$^{9}_{4}$Be	9.012 182	$^{40}_{20}$Ca	39.962 591
$^{10}_{5}$B	10.012 937	$^{42}_{20}$Ca	41.958 63
$^{12}_{6}$C	12.000 000	$^{43}_{20}$Ca	42.958 766
$^{13}_{6}$C	13.003 355	$^{56}_{26}$Fe	55.934 939
$^{14}_{7}$N	14.003 074	$^{64}_{30}$Zn	63.929 145
$^{15}_{7}$N	15.000 109	$^{63}_{29}$Cu	63.929 599
$^{15}_{8}$O	15.003 065	$^{93}_{41}$Nb	92.906 377
$^{17}_{8}$O	16.999 131	$^{197}_{79}$Au	196.966 543
$^{18}_{8}$O	17.999 160	$^{202}_{80}$Hg	201.970 617
$^{18}_{9}$F	18.000 937	$^{216}_{84}$Po	216.001 888
$^{20}_{10}$Ne	19.992 436	$^{220}_{86}$Rn	220.011 368
$^{23}_{11}$Na	22.989 768	$^{234}_{90}$Th	234.043 593
$^{23}_{12}$Mg	22.994 124	$^{238}_{92}$U	238.050 785

13.1   Some Properties of Nuclei

1. Find the radii of (a) a nucleus of $^{4}_{2}$H and (b) a nucleus of $^{238}_{92}$U. (c) What is the ratio of these radii?

2. The compressed core of a star formed in the wake of a supernova explosion consists of only neutrons and is called a *neutron star*. Calculate the mass of 10 cm^3 of a neutron star.

3. Consider the hydrogen atom to be a sphere of radius equal to the Bohr radius, a_0, given by Equation 3.29 in Chapter 3, and calculate the approximate value of the ratio of the nuclear mass density to the atomic mass density.

4. The Larmor precessional frequency is

$$f = \frac{\Delta E}{h} = \frac{2\mu B}{h}$$

Calculate the radio-wave frequency at which resonance absorption occurs for (a) free neutrons in a magnetic field of 1 T, (b) free protons in a magnetic field of 1 T, and (c) free protons in the Earth's magnetic field at a location where the magnetic field strength is 50 μT.

5. (a) Use energy methods to calculate the distance of closest approach for a head-on collision between an alpha particle with an initial energy of 0.5 MeV and a gold nucleus (^{197}Au) at rest. (Assume that the gold nu-

cleus remains at rest during the collision.) (b) What minimum initial speed must the alpha particle have in order to approach to a distance of 300 fm?

6. A neutron star is a nucleus composed entirely of neutrons and has a radius of about 10 km. Such a star forms when a larger star cools down and collapses under the influence of its own gravitational field. The compression continues until the protons and electrons merge to form neutrons. (a) Assuming that Equation 13.1 is valid, estimate the mass number A and mass M of a neutron star of radius $R = 10$ km. (b) Find the acceleration due to gravity at the surface of such a star. (c) Find the rotational kinetic energy of the star if it rotates about its axis 30 times/s. (Assume that the star is a sphere of uniform density.)

7. Consider a hydrogen atom with the electron in the $2p$ state. The magnetic field at the nucleus produced by the orbiting electron has a value of 12.5 T. The proton can have its magnetic moment aligned in either of two directions perpendicular to the plane of the electron's orbit. Because of the interaction of the proton's magnetic moment with the electron's magnetic field, there will be a difference in energy between the states with the two different orientations of the proton's magnetic moment. Find that energy difference in electron volts.

8. Using a reasonable scale, sketch an energy-level diagram for (a) a proton and (b) a deuteron, both in a magnetic field B. (c) What are the absolute values of the changes in energy that accompany the possible transitions between the levels shown in your diagrams?

9. Copper, as it occurs naturally, consists of two stable isotopes, ^{63}Cu and ^{65}Cu. What is the relative abundance of the two forms? In your calculations, take the atomic weight of copper to be 63.55 and take the masses of the two isotopes to be 62.95 u and 64.95 u.

10. (a) Find the radius of the $^{12}_{6}$C nucleus. (b) Find the force of repulsion between a proton at the surface of a $^{12}_{6}$C nucleus and the remaining five protons. (c) How much work (in MeV) must be done to overcome this electrostatic repulsion and put the last proton into the nucleus? (d) Repeat (a), (b), and (c) for $^{238}_{92}$U.

13.2   Binding Energy and Nuclear Forces

11. In Example 13.3, the binding energy of the deuteron was calculated to be 2.224 MeV. This corresponds to a value of 1.112 MeV/nucleon. What is the binding energy per nucleon for the heaviest isotope of hydrogen, ^{3}H (called *tritium*)?

12. Using the atomic mass of $^{56}_{26}$Fe given in Table 13.6, find its binding energy. Then compute the binding energy per nucleon and compare your result with Figure 13.10.

13. The $^{139}_{57}$La isotope of lanthanum is stable. A radioactive isobar (see Problem 14) of this lanthanum isotope, $^{139}_{59}$Pr, appears below the line of stable nuclei in Figure 13.4 and decays by e^{+} emission. Another radioactive isobar of ^{139}La, $^{139}_{55}$Cs, decays by e^{-} emission and appears above the line of stable nuclei in Figure 13.4. (a) Which of these three isobars has the highest neutron-to-proton ratio? (b) Which has the greatest binding energy per nucleon? (c) Which of the two radioactive nuclei (^{139}Pr or ^{139}Cs) do you expect to be heavier?

14. Two nuclei with the same mass number are known as *isobars*. If the two nuclei also have interchanged atomic and neutron numbers, such as $^{23}_{11}$Na and $^{23}_{12}$Mg, the nuclei are called *mirror isobars*. Binding-energy measurements on these nuclei can be used to obtain evidence of the charge independence of nuclear forces (that is, proton–proton, proton–neutron, and neutron–neutron forces are approximately equal). (a) Calculate the difference in binding energy for the two mirror nuclei $^{15}_{8}$O and $^{15}_{7}$N. (b) Calculate the difference in binding energy per nucleon for the mirror isobars $^{23}_{11}$Na and $^{23}_{12}$Mg. How do you account for the difference?

15. Calculate the binding energy per nucleon for the nuclei (a) $^{20}_{10}$Ne, (b) $^{40}_{20}$Ca, (c) $^{93}_{41}$Nb, and (d) $^{197}_{79}$Au.

16. Calculate the minimum energy required to remove a neutron from the $^{43}_{20}$Ca nucleus. (*Hint:* Use a formula analogous to Eq. 13.4.)

17. Using the graph in Figure 13.10, estimate how much energy is released when a nucleus of mass number 200 is split into two nuclei, each of mass number 100.

13.3   Nuclear Models

18. (a) In the liquid-drop model of nuclear structure, why does the surface-effect term $-C_2 A^{2/3}$ have a minus sign? (b) The binding energy of the nucleus increases as the volume-to-surface ratio increases. Calculate this ratio for both spherical and cubical shapes, and explain which is more plausible for nuclei.

19. Treat the nucleus as a sphere of uniform volume charge density ρ. (a) Derive an expression for the total energy required to assemble a sphere of charge corresponding to a nucleus of atomic number Z and radius R. (b) Using the result of (a), find an expression for the electrostatic potential energy U in terms of the mass number A for the case when $N = Z$ (the nucleus is on the line of stability). (c) Evaluate the result of (b) for the nucleus $^{30}_{15}$P.

20. (a) Use Equation 13.7 and the given values for the constants C_1, C_2, C_3, and C_4 to calculate the binding energy per nucleon for the isobars $^{64}_{29}$Cu and $^{64}_{30}$Zn. (b) Compare these values to the binding energy per nucleon calculated with Equation 13.4.

13.4 Radioactivity

21. Measurements on a radioactive sample show that its activity decreases by a factor of 5 during a 2-h interval. (a) Determine the decay constant of the radioactive nucleus. (b) Calculate the value of the half-life for this isotope.

22. The half-life of ^{131}I is 8.04 days. (a) Calculate the decay constant for this isotope. (b) Find the number of ^{131}I nuclei necessary to produce a sample with an activity of 0.5 μCi.

23. A freshly prepared sample of a certain radioactive isotope has an activity of 10 mCi. After an elapsed time of 4 h, its activity is 8 mCi. (a) Find the decay constant and half-life of the isotope. (b) How many atoms of the isotope were contained in the freshly prepared sample? (c) What is the sample's activity 30 h after it is prepared?

24. Tritium has a half-life of 12.33 yr. What percentage of the ^{3}H nuclei in a tritium sample will decay during a period of 5 years?

25. A sample of radioactive material is said to be *carrier-free* when no stable isotopes of the radioactive element are present. Calculate the mass of strontium in a carrier-free 5-mCi sample of ^{90}Sr whose half-life is 28.8 yr.

26. How many radioactive atoms are present in a sample that has an activity of 0.2 μCi and a half-life of 8.1 days?

27. A laboratory stock solution is prepared with an initial activity due to ^{24}Na of 2.5 mCi/mL, and 10 mL of the stock solution is diluted (at $t_0 = 0$) to a working solution with a total volume of 250 mL. After 48 h, a 5-mL sample of the working solution is monitored with a counter. What is the measured activity? (*Note:* 1 mL = 1 milliliter, and the half-life of ^{24}Na is 15.0 h.)

28. Start with Equation 13.10, and find the following useful forms for the decay constant and half-life:

$$\lambda = \frac{1}{t}\ln\left(\frac{R_0}{R}\right) \qquad T_{1/2} = \frac{(\ln 2)\,t}{\ln(R_0/R)}$$

29. The radioactive isotope ^{198}Au has a half-life of 64.8 h. A sample containing this isotope has an initial activity of 40 μCi. Calculate the number of nuclei that will decay in the time interval from $t_1 = 10$ h to $t_2 = 12$ h.

30. During the manufacture of a steel engine component, radioactive iron (^{59}Fe) is included in the total mass of 0.2 kg. The component is placed in a test engine when the activity due to this isotope is 20 μCi. After a 1000-h test period, oil is removed from the engine and found to contain enough ^{59}Fe to produce 800 disintegrations/min per liter of oil. The total volume of oil in the engine is 6.5 L. Calculate the total mass worn from the engine component per hour of operation. (The half-life for ^{59}Fe is 45.1 days.)

31. The activity of a sample of radioactive material was measured over 12 h, and the following *net* count rates were obtained at the times indicated:

Time (h)	Counting Rate (counts/min)
1	3100
2	2450
4	1480
6	910
8	545
10	330
12	200

(a) Plot the activity curve on semilog paper. (b) Determine the disintegration constant and the half-life of the radioactive nuclei in the sample. (c) What counting rate would you expect for the sample at $t = 0$? (d) Assuming the efficiency of the counting instrument to be 10%, calculate the number of radioactive atoms in the sample at $t = 0$.

32. A by-product of some fission reactors is the isotope $^{239}_{94}$Pu, which is an alpha emitter with a half-life of 24,000 years:

$$^{239}_{94}\text{Pu} \longrightarrow {}^{235}_{92}\text{U} + \alpha$$

Consider a sample of 1 kg of pure $^{239}_{94}$Pu at $t = 0$. Calculate (a) the number of $^{239}_{94}$Pu nuclei present at $t = 0$, (b) the initial activity in the sample, and (c) the time required for the activity to decrease to 1 decay/s.

33. A piece of charcoal has a mass of 25 g and is known to be about 25,000 years old. (a) Determine the number of decays per minute expected from this sample. (b) If the radioactive background in the counter without a sample is 20 counts/min and we assume 100% efficiency in counting, explain why 25,000 years is close to the limit of dating with this technique.

34. A fission reactor accident evaporates 5×10^6 Ci of ^{90}Sr($T_{1/2} = 27.7$ yr) into the air. The ^{90}Sr falls out over an area of 10^4 km^2. How long will it take the activity of the ^{90}Sr to reach the agriculturally "safe" level of 2 μCi/m^2?

35. What specific activity (see Problem 36), in disintegrations/min · g, would be expected for carbon samples from 2000-year-old bones? (Note that the ratio of ^{14}C to ^{12}C in living organisms is 1.3×10^{-12} and the half-life of ^{14}C is 5730 yr.)

36. In addition to the radioactive nuclei included in the natural decay series, there are several other radioactive nuclei that occur naturally. One is ^{147}Sm, which is 15% naturally abundant and has a half-life of approximately 1.3×10^{10} years. Calculate the number of decays per second per gram (due to this isotope) in a sample of natural samarium. The atomic weight of samarium is 150.4. (Activity per unit mass is called *specific activity*.)

37. A radioactive nucleus with decay constant λ decays to a stable daughter nucleus. (a) Show that the number of

daughter nuclei, N_2, increases with time according to the expression

$$N_2 = N_{01}(1 - e^{-\lambda t})$$

where N_{01} is the initial number of parent nuclei. (b) Starting with 10^6 parent nuclei at $t = 0$, with a half-life of 10 h, plot the number of parent nuclei and the number of daughter nuclei as functions of time over the interval 0 to 30 h.

38. Many radioisotopes have important industrial, medical, and research applications. One of these is ^{60}Co, which has a half-life of 5.2 yr and decays by the emission of a beta particle (energy 0.31 MeV) and two gamma photons (energies 1.17 MeV and 1.33 MeV). A scientist wishes to prepare a ^{60}Co sealed source that will have an activity of at least 10 Ci after 30 months of use. (a) What minimum initial mass of ^{60}Co is required? (b) At what rate will the source emit energy after 30 months?

39. The concept of radioactive half-life was described in Section 13.4, and Equation 13.11 gives the relationship between $T_{1/2}$ and λ. Another parameter that is often useful in the description of radioactive processes is the *mean life*, τ. Although the half-life of a radioactive isotope is accurately known, it is not possible to predict the time when any individual atom will decay. The mean life is a measure of the average length of existence of all the atoms in a particular sample. Show that $\tau = 1/\lambda$. (*Hint:* Remember that τ is essentially an average value, and use the fact that the number of atoms that decay between t and $t + dt$ is equal to dN. Furthermore, note that these dN atoms have a finite time of existence, t.)

40. Potassium as it occurs in nature includes a radioactive isotope ^{40}K, which has a half-life of 1.27×10^9 yr and a relative abundance of 0.0012%. These nuclei decay by two different pathways—89% by e^- emission and 11% by e^+ emission. Calculate the total activity in Bq associated with 1 kg of KCl *due to e^- emission*.

13.5 Decay Processes

41. Find the energy released in the alpha decay of $^{238}_{92}$U:

$$^{238}_{92}\text{U} \longrightarrow \, ^{234}_{90}\text{Th} + \, ^4_2\text{He}$$

You will find the mass values in Table 13.6 useful.

42. When, after a reaction or disturbance of any kind, a nucleus is left in an excited state, it can return to its normal (ground) state by emission of a gamma-ray photon (or several photons). Equation 13.25 describes this process. The emitting nucleus must recoil in order to conserve both energy and momentum. (a) Show that the recoil energy of the nucleus is

$$E_r = \frac{(\Delta E)^2}{2Mc^2}$$

where ΔE is the difference in energy between the excited and ground states of a nucleus of mass M. (b) Calculate the recoil energy of the ^{57}Fe nucleus when it decays by gamma emission from the 14.4-keV excited state. For this calculation, take the mass to be 57 u. (*Hint:* When writing the equation for conservation of energy, use $(Mv)^2/2M$ for the kinetic energy of the recoiling nucleus. Also, assume that $hf \ll Mc^2$.)

43. Equation 13.12 represents the decay of an unstable nucleus at rest by alpha emission. The disintegration energy Q given by Equation 13.15 must be shared by the alpha particle and the daughter nucleus in order to conserve both energy and momentum in the decay process. (a) Show that for nonrelativistic particles, Q and K_α, the kinetic energy of the α particle, are related by the expression

$$Q = K_\alpha \left(1 + \frac{M_\alpha}{M_Y}\right)$$

where M_Y is the mass of the daughter nucleus. (b) Use the result of (a) to find the energy of the α particle emitted in the decay of ^{226}Ra. (c) What is the kinetic energy of the daughter? (d) Apply the preceding expression for Q to the beta decay $^{210}_{83}\text{Bi} \longrightarrow \, ^{210}_{84}\text{Po} + e^-$ to verify the claim that the daughter nucleus carries off a negligible amount of kinetic energy in beta decay. Why is your answer an approximation?

44. Find the kinetic energy of an alpha particle emitted during the alpha decay of $^{220}_{86}$Rn. Assume that the daughter nucleus, $^{216}_{84}$Po, has zero recoil velocity.

45. Determine which of the following suggested decays can occur spontaneously:

(a) $^{40}_{20}\text{Ca} \longrightarrow e^+ + \, ^{40}_{19}\text{K} + \nu$

(b) $^{98}_{44}\text{Ru} \longrightarrow \, ^4_2\text{He} + \, ^{94}_{42}\text{Mo}$

(c) $^{144}_{60}\text{Nd} \longrightarrow \, ^4_2\text{He} + \, ^{140}_{58}\text{Ce}$

46. (a) Why is the following inverse β decay forbidden for a free proton?

$$\text{p} \longrightarrow \text{n} + e^+ + \nu$$

(b) Why is the same reaction possible if the proton is bound in a nucleus? For example, the following reaction occurs:

$$^{13}_7\text{N} \longrightarrow \, ^{13}_6\text{C} + e^+ + \nu$$

(c) How much energy is released in the reaction given in (b)? [Take the masses to be $m(e^+) = 0.000549$ u, $M(^{13}\text{C}) = 13.003355$ u, and $M(^{13}\text{N}) = 13.005739$ u, and see Problem 49.]

47. Use the Heisenberg uncertainty principle to make a reasonable argument against the hypothesis that free electrons can be present in a nucleus. Use relativistic expressions for the momentum and energy, and include appropriate assumptions and approximations.

48. The simplest example of beta decay is the beta decay of the free neutron. Free neutrons have a half-life of 10.2 min and decay according to the reaction

$$n \longrightarrow p + e^- + \bar{\nu}$$

   (a) Find the maximum kinetic energy the electron may have when a neutron decays at rest. (*Hint:* In this case, the energy of the antineutrino is approximately zero, the momentum of the proton equals the momentum of the electron, and the kinetic energy of the proton is negligible compared to that of the electron.) (b) Find the proton's kinetic energy in this case. (c) Find the maximum kinetic energy and momentum that the antineutrino may have and the proton's kinetic energy in this case. (*Hint:* Assume that the proton's kinetic energy is negligible compared to that of the $\bar{\nu}$.)

49. *Q values for β decay.* One must be careful in calculating Q values for beta decay. Atomic masses cannot always be used without correction, because electron masses do not always cancel as in alpha decay. Show that the correct expressions for beta decay are

   electron emission $Q = [M({}^A_Z X) - M(_{Z+1}^A Y)]c^2$

   positron emission $Q = [M({}^A_Z X) - M(_{Z-1}^A Y) - 2m_e]c^2$

   electron capture $Q = [M({}^A_Z X) - M(_{Z-1}^A Y)]c^2$

where $M({}^A_Z X)$ and $M(_{Z\pm 1}^A Y)$ are the atomic masses of the parent and daughter, respectively, and m_e is the mass of the electron.

13.6 Natural Radioactivity

50. Starting with ${}^{235}_{92}U$, the sequence of decays shown in Figure P13.50 is observed, ending with the stable isotope ${}^{207}_{82}Pb$. Enter the correct isotope symbol in each open square.

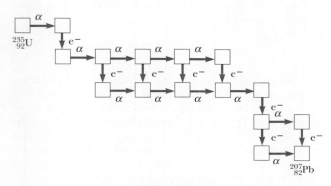

Figure P13.50

ADDITIONAL PROBLEMS

51. A ${}^3H$ nucleus beta-decays into ${}^3He$ by creating an electron and an antineutrino according to the reaction

$$ {}^3_1H \longrightarrow {}^3_2He + e^- + \bar{\nu} $$

   Use Appendix B to determine the total energy released in this reaction.

52. The nucleus ${}^{15}_8O$ decays by electron capture. Write (a) the basic nuclear process and (b) the decay process. (c) Determine the energy of the neutrino. Disregard the daughter's recoil.

53. In a piece of rock from the Moon, the ${}^{87}Rb$ content is assessed to be 1.82×10^{10} atoms per gram of material and the ${}^{87}Sr$ content is found to be 1.07×10^9 atoms per gram. (a) Determine the age of the rock. (b) Could the material in the rock actually be much older? What assumption is implicit in the use of the radioactive dating method? (The relevant decay is ${}^{87}Rb \rightarrow {}^{87}Sr + e^-$. The half-life of the decay is 4.8×10^{10} yr.)

54. (a) Can ${}^{57}Co$ decay by e^+ emission? Explain. (b) Can ${}^{14}C$ decay by e^- emission? Explain. (c) If either answer is yes, what is the range of kinetic energies available for the beta particle?

55. When a material of interest is irradiated by neutrons, radioactive atoms are produced continually and some decay according to their given half-lives. (a) If radioactive atoms are produced at a constant rate R and their decay is governed by the conventional radioactive decay law, show that the number of radioactive atoms accumulated after an irradiation time t is

$$ N = \frac{R}{\lambda} (1 - e^{-\lambda t}) $$

   (b) What is the maximum number of radioactive atoms that can be produced?

56. The ground state of ${}^{92}_{43}Tc$ (molar mass 92.9102) decays by electron capture and e^+ emission to energy levels of the daughter (molar mass in ground state is 92.9068) at 2.44 MeV, 2.03 MeV, 1.48 MeV, and 1.35 MeV. (a) For which of these levels are electron capture and e^+ decay allowed? (b) Identify the daughter and sketch the decay scheme, assuming that all excited states deexcite by direct gamma decay to the ground state.

57. In an experiment on the transport of nutrients in the root structure of a plant, two radioactive nuclides X and Y are used. Initially, 2.50 times more nuclei of type X are present than of type Y. Just three days later there are 4.20 times more nuclei of type X than of type Y. Isotope Y has a half-life of 1.60 d. What is the half-life of isotope X?

58. (a) The daughter nucleus formed in radioactive decay is often radioactive. Let N_{10} represent the number of parent nuclei at time $t = 0$, $N_1(t)$ the number of parent nuclei at time t, and λ_1 the decay constant of the parent. Suppose the number of daughter nuclei at time $t = 0$ is zero, let $N_2(t)$ be the number of daughter nuclei at time t, and let λ_2 be the decay constant of the daughter. Show that $N_2(t)$ satisfies the differential equation

$$\frac{dN_2}{dt} = \lambda_1 N_1 - \lambda_2 N_2$$

(b) Verify by substitution that this differential equation has the solution

$$N_2(t) = \frac{N_{10}\lambda_1}{\lambda_1 - \lambda_2} \left(e^{-\lambda_2 t} - e^{-\lambda_1 t}\right)$$

This equation is the law of successive radioactive decays. (c) ^{218}Po decays into ^{214}Pb with a half-life of 3.10 min, and ^{214}Pb decays into ^{214}Bi with a half-life of 26.8 min. On the same axes, plot graphs of $N_1(t)$ for ^{218}Po and $N_2(t)$ for ^{214}Pb. Let $N_{10} = 1000$ nuclei, and choose values of t from 0 to 36 min in 2-min intervals. The curve for ^{214}Pb at first rises to a maximum and then starts to decay. At what instant t_m is the number of ^{214}Pb nuclei a maximum? (d) By applying the condition for a maximum $\dfrac{dN_2}{dt} = 0$, derive a symbolic formula for t_m in terms of λ_1 and λ_2. Does the value obtained in (c) agree with this formula?

59. A certain African artifact is found to have a carbon-14 activity of (0.12 ± 0.01) Bq per gram of carbon. Assume the uncertainty is negligible in the half-life of ^{14}C (5730 yr) and in the activity of atmospheric carbon (0.25 Bq per gram). The age of the object lies within what range?

60. As part of his discovery of the neutron in 1932, James Chadwick determined the mass of the newly identified particle by firing a beam of fast neutrons, all having the same speed, at two different targets and measuring the maximum recoil speeds of the target nuclei. The maximum speeds arise when an elastic head-on collision occurs between a neutron and a stationary target nucleus. (a) Represent the masses and final speeds of the two target nuclei as m_1, v_1, m_2, and v_2 and assume Newtonian mechanics applies. Show that the neutron mass can be calculated from the equation

$$m_n = \frac{m_1 v_1 - m_2 v_2}{v_2 - v_1}$$

(b) Chadwick directed a beam of neutrons (produced from a nuclear reaction) on paraffin, which contains hydrogen. The maximum speed of the protons ejected was found to be 3.3×10^7 m/s. Since the velocity of the neutrons could not be determined directly, a second experiment was performed using neutrons from the same source and nitrogen nuclei as the target. The maximum recoil speed of the nitrogen nuclei was found to be 4.7×10^6 m/s. The masses of a proton and a nitrogen nucleus were taken as 1 u and 14 u, respectively. What was Chadwick's value for the neutron mass?

61. A rock sample contains traces of ^{238}U, ^{235}U, ^{232}Th, ^{208}Pb, ^{207}Pb, and ^{206}Pb. Careful analysis shows that the ratio of the amount of ^{238}U to ^{206}Pb is 1.164. (a) Assume that the rock originally contained no lead and determine the age of the rock. (b) What should be the ratios of ^{235}U to ^{207}Pb and of ^{232}Th to ^{208}Pb so that they would yield the same age for the rock? Ignore the minute amounts of the intermediate decay products in the decay chains. Note that this form of multiple dating gives reliable geological dates.

14

Nuclear Physics Applications

Chapter Outline

This chapter is concerned with nuclear reactions in which particles and nuclei collide and change into other nuclei and particles. We also consider the two means by which energy can be derived from nuclear reactions: *fission*, in which a large nucleus splits, or fissions, into two smaller nuclei, and *fusion*, in which two small nuclei fuse to form a larger one. In either case, a release of energy occurs that can then be used either destructively through bombs or constructively through production of electric power. Finally, we examine the interaction of radiation with matter and several devices for detecting radiation.

14.1   NUCLEAR REACTIONS

It is possible to change the structure of nuclei by bombarding them with energetic particles. Such collisions that change the identities of the target nuclei are called **nuclear reactions.** Rutherford was the first to observe them in 1919, using naturally occurring radioactive sources for the bombarding particles. Since then, thousands of nuclear reactions have been observed, especially after the development of charged-particle accelerators in the 1930s. With today's technology in particle accelerators, it is possible to achieve particle energies of 1000 GeV = 1 TeV and higher. These high-energy particles are used to create new particles whose properties are helping to solve the mystery of the nucleus.

Consider a reaction in which a target nucleus X is bombarded by a particle a, resulting in a nucleus Y and a particle b:

Nuclear reaction

$$a + X \longrightarrow Y + b \tag{14.1}$$

Sometimes this reaction is written in the more compact form

$$X(a, b)Y$$

As an example, consider the reaction $^7\text{Li}(p, \alpha)\,^4\text{He}$, or

$$^{1}_{1}\text{H} + \,^{7}_{3}\text{Li} \longrightarrow \,^{4}_{2}\text{He} + \,^{4}_{2}\text{He}$$

Cockroft and Walton were first to observe this reaction in 1932, using protons accelerated to 600 keV in an accelerator they had designed and built. A nuclear reaction such as this, and in fact any reaction, can occur only if it satisfies certain *conservation laws.* The conservation laws for nuclear reactions are

- *Conservation of mass number, A.* The total number of nucleons must be the same after the reaction as before. For the reaction under discussion, $A_{\text{before}} = 1 + 7 = A_{\text{after}} = 4 + 4$.
- *Conservation of charge, q.* Here the charged nuclear particles are protons, and $q_{\text{before}} = 1 + 3 = q_{\text{after}} = 2 + 2$.
- *Conservation of energy, linear momentum, and angular momentum.* These quantities are conserved because a nuclear reaction involves only internal forces between a target nucleus and a bombarding nucleus, and there are no external forces to upset these conservation principles.

Let us apply the conservation of energy to a reaction of the form of Equation 14.1 to compute the total kinetic energy released (or absorbed) in the reaction, which is called the **reaction energy,** Q. Assume that the target nucleus X is originally at rest, the bombarding particle a has kinetic energy K_a, and the reaction products b and Y have kinetic energies K_b and K_Y. Conserving energy,

$$M_X c^2 + K_a + M_a c^2 = M_Y c^2 + K_Y + M_b c^2 + K_b$$

As the total kinetic energy released in the reaction, Q, is equal to the difference between the kinetic energy of the final particles and that of the initial particle, we find

Q of a reaction

$$Q = (K_Y + K_b) - K_a = (M_X + M_a - M_Y - M_b)c^2 \tag{14.2}$$

A reaction for which Q is positive converts nuclear mass to kinetic energy of the products Y and b and is called an **exothermic reaction.** A reaction for which Q is negative requires some minimum input kinetic energy from the bombarding particle in order to occur. Such a reaction is called **endothermic.**

For an endothermic reaction to proceed, the incident particle must have a minimum kinetic energy called the **threshold energy,** K_{th}. Since K_{th} must not only supply $|Q|$, the excess mass−energy of the products, but also supply some kinetic energy to the products to conserve momentum, K_{th} is greater than $|Q|$.

For low-energy reactions, where the kinetic energies of all the interacting particles are small compared to their rest energies, we can apply the nonrela-

Table 14.1   *Q* **Values for Nuclear Reactions Involving Light Nuclei**

Reaction[a]	Measured *Q* Value (MeV)
$^2\text{H}(\text{n}, \gamma)^3\text{H}$	6.257 ± 0.004
$^2\text{H}(\text{d}, \text{p})^3\text{H}$	4.032 ± 0.004
$^6\text{Li}(\text{p}, \alpha)^3\text{H}$	4.016 ± 0.005
$^6\text{Li}(\text{d}, \text{p})^7\text{Li}$	5.020 ± 0.006
$^7\text{Li}(\text{p}, \text{n})^7\text{Be}$	-1.645 ± 0.001
$^7\text{Li}(\text{p}, \alpha)^4\text{He}$	17.337 ± 0.007
$^9\text{Be}(\text{n}, \gamma)^{10}\text{Be}$	6.810 ± 0.006
$^9\text{Be}(\gamma, \text{n})^8\text{Be}$	-1.666 ± 0.002
$^9\text{Be}(\text{d}, \text{p})^{10}\text{Be}$	4.585 ± 0.005
$^9\text{Be}(\text{p}, \alpha)^6\text{Li}$	2.132 ± 0.006
$^{10}\text{B}(\text{n}, \alpha)^7\text{Li}$	2.793 ± 0.003
$^{10}\text{B}(\text{p}, \alpha)^7\text{Be}$	1.148 ± 0.003
$^{12}\text{C}(\text{n}, \gamma)^{13}\text{C}$	4.948 ± 0.004
$^{13}\text{C}(\text{p}, \text{n})^{13}\text{N}$	-3.003 ± 0.002
$^{14}\text{N}(\text{n}, \text{p})^{14}\text{C}$	0.627 ± 0.001
$^{14}\text{N}(\text{n}, \gamma)^{15}\text{N}$	10.833 ± 0.007
$^{18}\text{O}(\text{p}, \text{n})^{18}\text{F}$	-2.453 ± 0.002
$^{19}\text{F}(\text{p}, \alpha)^{16}\text{O}$	-8.124 ± 0.007

[a]The symbols n, p, d, α, and γ denote the neutron, proton, deuteron, alpha particle, and photon, respectively. From C. W. Li, W. Whaling, W. A. Fowler, and C. C. Lauritsen, *Phy. Rev.*, 83:512, 1951.

tivistic expressions $K = \frac{1}{2}mv^2$ and $p = mv$ to find the threshold energy. It is left as a problem (Problem 9) to show that when momentum and energy are conserved in a low-energy negative Q reaction, the threshold energy is

$$K_{\text{th}} = -Q\left(1 + \frac{M_a}{M_X}\right) \qquad (14.3)$$

Finally, we have included a selected list of measured Q values for reactions involving light nuclei in Table 14.1.

EXAMPLE 14.1

(a) Calculate the Q value for the reaction observed by Cockcroft and Walton.

Solution   (a) The reaction is

$$^1_1\text{H} + {}^7_3\text{Li} \longrightarrow {}^4_2\text{He} + {}^4_2\text{He}$$

Using $Q = (M_{\text{Li}} + M_{\text{H}} - 2M_{\text{He}})c^2$ and   substituting atomic mass values from Appendix B, we find

$Q = (7.016\ 003\ \text{u} + 1.007\ 825\ \text{u}$

$\quad - 2(4.002\ 603\ \text{u}))(931.50\ \text{MeV/u}) = 17.3\ \text{MeV}$

(b) Find the kinetic energy of the products if 600-keV protons are incident.

Solution   (b) Since $Q = K_{\text{products}} - K_{\text{incident particle}}$,

$K_{\text{products}} = Q + K_{\text{incident particle}} = 17.3\ \text{MeV} + 0.6\ \text{MeV}$

$\qquad\qquad = 17.9\ \text{MeV}$

This means that the two alpha particles share 17.9 MeV of kinetic energy.

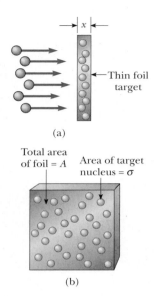

Figure 14.1 (a) A beam of particles incident on a thin foil target of thickness x. The view shown is of an edge of the target. (b) A front view of the target, where the circles represent the target nuclei.

14.2 REACTION CROSS SECTION

Since we deal in this chapter with the interactions between nuclei and matter, it is useful to introduce a quantity called the **cross section, which is a measure of the probability that a particular nuclear reaction will occur.**

When a beam of particles is incident on a target in the form of a thin foil, not every particle interacts with a target nucleus. The probability that an interaction will occur depends on the ratio of the "effective" area of the target nucleus to the area of the foil. The situation is analogous to throwing darts at a large wall upon which many inflated balloons are hanging. If the darts are thrown at random in the direction of the wall and the balloons are spread out so as not to touch each other, there is some chance that you will hit a balloon on any given throw. Furthermore, if you throw darts at a rate R_0, the rate R at which balloons burst will be less than R_0. In fact, the probability of hitting a balloon will equal R/R_0. The ratio R/R_0 will depend on the number of balloons N on the wall, the area σ of each balloon, and the area A of the wall. Since the total cross-sectional area of the balloons is $N\sigma$, the probability R/R_0 equals the ratio $N\sigma/A$.

With this analogy, we can now understand the concept of cross section as it pertains to nuclear events. Suppose a beam of particles is incident upon a thin-foil target, as in Figure 14.1a. Each target nucleus X has an effective area σ called the **cross section.** You can think of σ as an *effective* area of the nucleus at right angles to the direction of motion of the bombarding particles, as in Figure 14.1b, but note that the reaction cross reaction σ can be greater than, equal to, or less than the actual geometrical cross section of the target nucleus. It is assumed that the reaction X(a, b)Y will occur only if the incident particle strikes the area σ. Therefore, the probability that a collision will occur is proportional to σ. That is, the probability increases as σ increases. In the general case, the size of σ for a specific reaction may also depend on the energy of the incident particle.

Let us consider the concept of cross section in more detail. In what follows, we take the foil thickness to be x and its area to be A. Furthermore, we use the following notation.

R_0 = Rate at which incident particles strike the foil (particles/s)

R = Rate at which reaction events occur (reactions/s)

n = Number of target nuclei per unit volume (particles/m^3)

Since the total number of target nuclei in the foil is nAx, the total area exposed to the incident beam must be σnAx. It is assumed that the foil is sufficiently thin that nuclei are not "hidden" behind others. The ratio of the rate of interactions to the rate of incident particles, R/R_0, must equal the ratio of the area σnAx to the total area A of the foil, in analogy with the dart–balloon collision events. That is,

Reaction rate is proportional to cross section and target density

$$\frac{R}{R_0} = \frac{\sigma nAx}{A} = \sigma nx \qquad (14.4)$$

This result shows that the probability that a nuclear reaction will occur is proportional to the cross section σ, the density of target nuclei n, and the thickness of the target, x. A value for σ for a specific reaction can therefore be obtained by measuring R, R_0, n, and x and using Equation 14.4.

We can use the same reasoning to arrive at an expression for the number of particles that penetrate a foil without undergoing reaction. Suppose that N_0

particles are incident on a foil of thickness dx and dN is the number of particles that interact with the target nuclei (Fig. 14.2). The ratio of the number of interacting particles to the number of incident particles, dN/N, equals the ratio of the total target cross section, $nA\sigma\,dx$, to the total foil area A. That is,

$$-\frac{dN}{N} = \frac{nA\sigma\,dx}{A} = n\sigma\,dx$$

where the minus sign indicates that particles are being removed from the beam. Integrating this expression and taking $N = N_0$ at $x = 0$,

$$\int_{N_0}^{N} \frac{dN}{N} = -n\sigma \int_{0}^{x} dx$$

$$\ln\left(\frac{N}{N_0}\right) = -n\sigma x$$

$$N = N_0 e^{-n\sigma x} \qquad (14.5)$$

That is, if N_0 is the number of incident particles, the number that emerge from the slab, N, decreases exponentially with target thickness.

Nuclear cross sections, which have dimensions of area, are typically of the order of the square of the nuclear radius, which is about 10^{-28} m^2. For this reason, it is common to use the unit of 10^{-28} m^2 for measuring nuclear cross sections. This small unit is known, strangely enough, as the **barn** (b)[1] and is defined as

$$1 \text{ barn} = 10^{-28} \text{ m}^2 \qquad (14.6)$$

In reality, the concept of cross section in nuclear and atomic physics has little to do with the actual geometric area of target nuclei. The model we have used is simply a convenient one for describing the probability of occurrence of any nuclear reaction. In fact, cross sections vary with both the specific reaction considered and with the incident particle's kinetic energy over much more than several times the target nucleus's geometrical area. For example, the cross section for an antineutrino to interact with a proton via the nuclear weak interaction is only about 10^{-19} b in the reaction

$$\bar{\nu} + \text{p} \longrightarrow \text{e}^+ + \text{n} \qquad (\sigma = 10^{-19} \text{ b})$$

The cross sections for inelastic scattering of neutrons from iodine and xenon via the nuclear strong interaction, however, are about 4 barns in the following reactions:

$$\text{n} + {}^{127}\text{I} \longrightarrow {}^{127}\text{I}^* + \text{n} \qquad (\sigma = 4 \text{ b})$$

$$\text{n} + {}^{129}\text{Xe} \longrightarrow {}^{129}\text{Xe}^* + \text{n} \qquad (\sigma = 4 \text{ b})$$

(An inelastic scattering reaction is one in which the incident particle loses energy to the target nucleus, emerging from the reaction with less kinetic energy but leaving the target in an excited state, here denoted by the asterisk.)

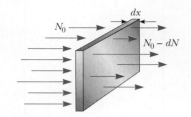

Figure 14.2 If N_0 is the number of particles incident on a target of thickness dx in some time interval, the number emerging from the target is $N_0 - dN$.

Number of particles transmitted through a target of thickness x

The barn

---

[1]"Barn" was introduced by the American physicists M. G. Holloway and C. P. Baker in 1942 in a humorous twist. It served the purpose of a code word in concealing war work on reaction probabilities and was appropriate because a cross section of 10^{-28} m^2 really is "as big as the broad side of a barn" for nuclear processes.

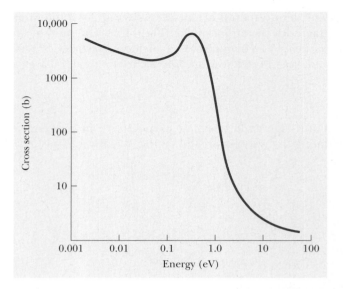

Figure 14.3   The neutron-capture cross section for cadmium. (*Reprinted with permission from Goldsmith, et al.*, Rev. Mod. Phys., *19:261, 1947. Copyright © 1947 by the American Physical Society.*)

Finally, an example of a nuclear reaction with a large scattering cross section (10^4 b), and with a strongly energy-dependent cross section as well, is the neutron capture reaction

$$\text{n} + {}^{113}\text{Cd} \longrightarrow {}^{114}\text{Cd} + \gamma \qquad (\sigma = 10^4 \text{ b})$$

In this reaction, which has a high probability only for low-energy neutrons, the cadmium target incorporates an extra neutron into its nucleus, is raised to an excited state, and emits its excess energy by gamma decay. Figure 14.3 shows the strong dependence of cross section on incident neutron energy in this case.

14.3   INTERACTIONS INVOLVING NEUTRONS

In order to understand the process of nuclear fission and the physics of the nuclear reactor, we must first understand the manner in which neutrons interact with nuclei. As mentioned earlier, because of their charge neutrality, neutrons are not subject to Coulomb forces. Since neutrons interact very weakly with electrons, matter appears fairly transparent to neutrons. In general, one finds that typical cross sections for neutron-induced reactions increase as the neutron energy decreases. Free neutrons undergo beta decay with a mean lifetime of about 10 min. On the other hand, neutrons in matter are usually absorbed by nuclei before they decay.

When a fast neutron (one with an energy greater than about 1 MeV) moves through matter, it undergoes scattering events with nuclei. In each such event, the neutron gives up some of its kinetic energy to a nucleus. The neutron continues to undergo collisions until its energy is of the order of the thermal energy $k_B T$, where k_B is Boltzmann's constant and T is the absolute temperature. A neutron with this energy is called a **thermal neutron.** At this low energy, there is a high probability that the neutron will be captured by a nucleus, as can be seen for the case of Cd, shown in Figure 14.3. This **neutron-capture** process can be written

$$\,_0^1 n + \,_Z^A X \longrightarrow \,_Z^{A+1} X + \gamma \qquad (14.7)$$

Although we do not indicate it here, the nucleus X is in an excited state X* for a very short time before it undergoes gamma decay. Also, the product nucleus $_Z^A X$ is usually radioactive and may decay by α or β emission.

The neutron-capture cross section associated with the preceding process depends on the nature of the target nucleus and the energy of the incident neutron. For some materials and for fast neutrons, the cross section is so small that elastic collisions dominate. Materials for which this occurs are called **moderators,** since they slow down (or moderate) the originally energetic neutrons very effectively. Boron, graphite, and water are a few examples of moderator materials.

During an elastic collision between two particles, the maximum kinetic energy is transferred from one particle to the other when they have the same mass. Consequently, a neutron loses all of its kinetic energy when it collides head-on with a proton, in analogy with the collision between a moving billiard ball and a stationary one. If the collision is oblique, the neutron loses only part of its kinetic energy. For this reason, materials which are abundant in hydrogen, such as paraffin and water, are good moderators for neutrons.

At some point, many of the neutrons in the moderator become thermal neutrons, which are neutrons in thermal equilibrium with the moderator material. Their average kinetic energy at room temperature is

$$K_{av} = \frac{3}{2} k_B T \approx 0.04 \text{ eV}$$

which corresponds to a neutron root-mean-square speed of about 2800 m/s. Thermal neutrons have a distribution of velocities just as the molecules in a container of gas do. A high-energy neutron, whose energy is several MeV, will thermalize (that is, reach K_{av}) in less than 1 ms when incident on a moderator such as graphite (carbon) or water.

EXAMPLE 14.2   Neutron Capture by Aluminum

An aluminum foil of thickness 0.30 mm is bombarded by energetic neutrons. The aluminum nuclei undergo neutron capture according to the process $^{27}\text{Al}(n, \gamma)$ ^{28}Al, with a measured capture cross section of 2.0×10^{-3} b $= 2.0 \times 10^{-31}$ m^2. Assuming the flux of incident neutrons to be 5.0×10^{12} neutrons/cm$^2 \cdot$s, calculate the number of neutrons captured per second by 1.0 cm^2 of the foil.

Solution   To solve this problem, we must first evaluate the density of nuclei n (which equals the density of atoms). Since the density of aluminum is 2.7 g/cm^3 and $A = 27$ for aluminum, we have

$$n = \frac{6.02 \times 10^{23} \text{ nuclei/mol}}{27 \text{ g/mol}} \times 2.7 \frac{\text{g}}{\text{cm}^3}$$

$$= 6.02 \times 10^{22} \text{ nuclei/cm}^3 = 6.02 \times 10^{28} \text{ nuclei/m}^3$$

Substituting this value into Equation 14.4, together with the given data for x and σ, gives

$$\frac{R}{R_0} = \sigma n x$$

$$= (2.0 \times 10^{-31} \text{ m}^2)(6.02 \times 10^{28} \text{ nuclei/m}^3)$$

$$(0.30 \times 10^{-3} \text{ m}) = 3.6 \times 10^{-6}$$

Since the rate of incident particles per unit area is $R_0 = 5.0 \times 10^{12}$ neutrons/cm$^2 \cdot$s,

$$R = (5.0 \times 10^{12} \text{ neutrons/cm}^2 \cdot \text{s})(3.6 \times 10^{-6})$$

$$= 1.8 \times 10^7 \text{ neutrons/cm}^2 \cdot \text{s}$$

Therefore, the number of neutrons captured by 1.0 cm^2/s is only 1.8×10^7 neutrons, whereas the incident number is 5.0×10^{12} neutrons. That is, only about 4 neutrons out of 1 million are captured!

14.4 NUCLEAR FISSION

Nuclear fission occurs when a very heavy nucleus such as ^{235}U splits, or fissions, into two particles of comparable mass. In such a reaction, **the total mass of the product particles is less than the original mass.** Fission is initiated by the capture of a thermal neutron by a heavy nucleus and involves the release of about 200 MeV per fission. This energy release occurs because the smaller fission-product nuclei are more tightly bound by about 1 MeV per nucleon than the original heavy nucleus.

The process of nuclear fission was first observed in 1938 by Otto Hahn (1879–1968, German chemist), Lise Meitner (1878–1968, Austrian physicist), and Fritz Strassmann (1902–1980, German chemist), following some basic studies by Fermi concerning the interaction of thermal neutrons with uranium. After bombarding uranium ($Z = 92$) with neutrons, Hahn and Strassmann performed a chemical analysis and discovered among the products two medium-mass elements, barium and lanthanum. Shortly thereafter, Lise Meitner and her nephew Otto Frisch (1904–1979, German-British physicist) explained what had happened and coined the term *fission*. The uranium nucleus could split into nearly equal fragments after absorbing a neutron. Measurements showed that about 200 MeV of energy was released in each fission event.

The fission of ^{235}U by thermal neutrons can be represented by

Fission of ^{235}U
$$\,^{1}_{0}n + \,^{235}_{92}U \longrightarrow \,^{236}_{92}U^* \longrightarrow X + Y + \text{neutrons} \qquad (14.8)$$

where ^{236}U* is an intermediate excited state that lasts for only about 10^{-12} s before splitting into X and Y. The resulting nuclei X and Y are called **fission fragments.** There are many combinations of X and Y that satisfy the requirements of conservation of mass–energy, charge, and nucleon number. Figure 14.4 shows the actual mass distribution of fragments in the fission of ^{235}U. The

(a) Lise Meitner

(b) Otto Frisch

While on vacation in the Alps, Meitner and her nephew realized that the large nucleus of uranium had split into smaller nuclei with the release of enormous amounts of energy. (*Photo by Emilio Segré, courtesy of AIP Emilio Segré Visual Archives; AIP Emilio Segré Visual Archives, Segré Collection.*)

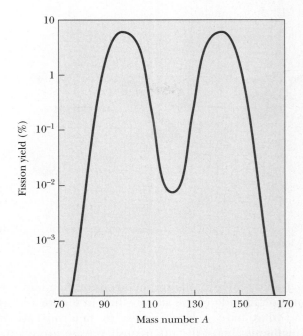

Figure 14.4   The distribution of fission products versus mass number for the fission of ^{235}U bombarded with slow neutrons. Note that the ordinate has a logarithmic scale.

process results in the production of several neutrons, typically two or three. On the average, about 2.5 neutrons are released per event. A typical reaction of this type is

$$\,_{0}^{1}\text{n} + \,_{92}^{235}\text{U} \longrightarrow \,_{56}^{141}\text{Ba} + \,_{36}^{92}\text{Kr} + 3\,_{0}^{1}\text{n} \qquad (14.9)$$

Of the 200 MeV or so released in this reaction, most goes into the kinetic energy of the heavy fragments barium and krypton.

The breakup of the uranium nucleus can be compared to what happens to a drop of water when excess energy is added to it. All the atoms in the drop have energy, but not enough to break up the drop. However, if enough energy is added to set the drop into vibration, it undergoes elongation and compression until the amplitude of vibration becomes large enough to cause the drop to break. In the uranium nucleus, a similar process occurs. Figure 14.5 shows various stages of the nucleus as the result of neutron capture. The sequence of events for ^{235}U can be described as follows:

1. The ^{235}U nucleus captures a thermal (slow-moving) neutron.
2. This capture results in the formation of ^{236}U*, and the excess energy of this nucleus causes it to undergo violent oscillations.
3. The ^{236}U* nucleus becomes highly distorted, and the force of repulsion between protons in the two halves of the dumbbell shape tends to increase the distortion.
4. The nucleus splits into two fragments, emitting several neutrons in the process.

As can be seen in Figure 14.4, the most probable fission events correspond to fission fragments with mass numbers $A \approx 140$ and $A \approx 95$. These

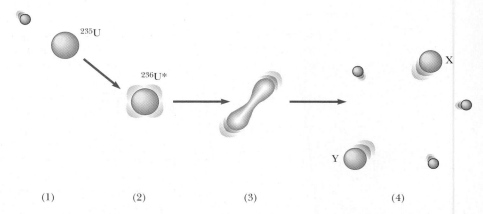

Figure 14.5 The stages in a nuclear fission event as described by the liquid-drop model of the nucleus.

fragments, which share the protons and neutrons of the mother nucleus, both fall on the neutron-rich side of the stability line in Figure 13.4 (Chapter 13). Since fragments that have a large excess of neutrons are unstable, the neutron-rich fragments almost instantaneously release two or three neutrons. The remaining fragments are still rich in neutrons and proceed to decay to more stable nuclei through a succession of beta decays. In the process of such decays, gamma rays are also emitted by nuclei in excited states.

Let us estimate the disintegration energy Q released in a typical fission process. From Figure 13.10 we see that the binding energy per nucleon for heavy nuclei ($A \approx 240$) is about 7.6 MeV, whereas in the intermediate mass range, the binding energy per nucleon is about 8.5 MeV. Taking the mass number of the mother nucleus to be $A = 240$, we see that the energy released per nucleon is estimated to be

$$Q = (240 \text{ nucleons})\left(8.5 \frac{\text{MeV}}{\text{nucleon}} - 7.6 \frac{\text{MeV}}{\text{nucleon}}\right) = 200 \text{ MeV}$$

About 85% of this energy appears in the form of kinetic energy in the heavy fragments. This energy is very large compared to the energy released in chemical processes. For example, the energy released in the combustion of one molecule of octane used in gasoline engines is about one-millionth the energy released in a single uranium fission event!

EXAMPLE 14.3   The Fission of Uranium

In addition to the barium–lanthanum reaction observed by Meitner and Frisch and the barium–krypton reaction of Equation 14.9, two other ways in which ^{235}U can fission when bombarded with a neutron are (1) by forming ^{140}Xe and ^{94}Sr and (2) by forming ^{132}Sn and ^{101}Mo. In each case, neutrons are also released. Find the number of neutrons released in each event.

Solution   By balancing mass numbers and atomic numbers, we find that these reactions can be written

$$^{1}_{0}n + {}^{235}_{92}U \longrightarrow {}^{140}_{54}Xe + {}^{94}_{38}Sr + 2({}^{1}_{0}n)$$
$$^{1}_{0}n + {}^{235}_{92}U \longrightarrow {}^{132}_{50}Sn + {}^{101}_{42}Mo + 3({}^{1}_{0}n)$$

Thus two neutrons are released in the first reaction and three in the second.

EXAMPLE 14.4   The Energy Released in the Fission of ^{235}U

Calculate the total energy released if 1.00 kg of ^{235}U undergoes fission, taking the disintegration energy per event to be $Q = 208$ MeV (a more accurate value than the estimate given before).

Solution   We need to know the number of nuclei in 1 kg of uranium. Since $A = 235$, the number of nuclei is

$$N = \frac{6.02 \times 10^{23} \text{ nuclei/mol}}{235 \text{ g/mol}} (1.00 \times 10^3 \text{ g})$$

$$= 2.56 \times 10^{24} \text{ nuclei}$$

Hence the disintegration energy is

$$E = NQ = (2.56 \times 10^{24} \text{ nuclei}) \left(208 \, \frac{\text{MeV}}{\text{nucleus}} \right)$$

$$= 5.32 \times 10^{26} \text{ MeV}$$

Since 1 MeV is equivalent to 4.14×10^{-20} kWh, we find that $E = 2.37 \times 10^7$ kWh. If this amount of energy were released suddenly, it would be equivalent to detonating 20,000 tons of TNT!

EXAMPLE 14.5   A Rough Mechanism for the Fission Process

Consider the fission reaction $^{235}_{92}\text{U} + \text{n} \rightarrow {}^{141}_{56}\text{Ba} + {}^{92}_{36}\text{Kr} + 2^1_0\text{n}$. Show that the model of the fission process in which an excited $^{235}_{92}$U nucleus elongates enough to overcome attractive nuclear forces and separates into two charged fragments can be used to estimate the energy released in this fission process. In this model, we assume

that the incident neutron provides a few MeV of excitation energy to separate the Ba and Kr nuclei within the uranium nucleus so that the two fragments ${}^{141}_{56}$Ba and ${}^{92}_{36}$Kr are driven apart by Coulomb repulsion. We also neglect the small amount of kinetic energy (several MeV) carried off by the neutrons produced in the reaction.

Solution   First we calculate the separation, r, of the Ba and Kr nuclei at which the nuclear force between them falls to zero. This is $r \cong r_{\text{Ba}} + r_{\text{Kr}}$, where r_{Ba} and r_{Kr} are the nuclear radii of Ba and Kr given by Equation 13.1. Thus

$$r_{\text{Ba}} = (1.2 \times 10^{-15} \text{ m})(141)^{1/3} = 6.2 \times 10^{-15} \text{ m}$$

$$r_{\text{Kr}} = (1.2 \times 10^{-15} \text{ m})(92)^{1/3} = 5.4 \times 10^{-15} \text{ m}$$

$$r \cong 12 \times 10^{-15} \text{ m}$$

Next calculate the Coulomb potential energy for two charges of $Z_1 = 56$ and $Z_2 = 36$ separated by a distance of 12 fm. The potential energy of the two nuclei on the brink of separation is

$$U = \frac{k(Z_1 e)(Z_2 e)}{r} = \frac{(1.440 \text{ eV} \cdot \text{nm})(56)(36)}{12 \times 10^{-15} \text{ m}}$$

$$= 240 \text{ MeV}$$

As the two fragments separate, this potential energy is converted to an amount of kinetic energy consistent with the total measured energy release of about 200 MeV. This shows that the simple fission mechanism suggested here is plausible.

14.5   NUCLEAR REACTORS

In the last section we saw that when ^{235}U fissions, an average of 2.5 neutrons are emitted per event. These neutrons can in turn trigger other nuclei to fission, with the possibility of a chain reaction (Fig. 14.6). Calculations show that if the chain reaction is not controlled (that is, if it does not proceed slowly), it can result in a violent explosion, with the release of an enormous amount of energy. For example, if the energy in 1 kg of ^{235}U were released, it would be equivalent to detonating about 20,000 tons of TNT! This, of course, is the principle behind the first nuclear bomb, an uncontrolled fission reaction.

A nuclear reactor is a system designed to maintain what is called a **self-sustained chain reaction.** This important process was first achieved in 1942 by Fermi at the University of Chicago, with natural uranium as the fuel (Fig. 14.7). Most reactors in operation today also use uranium as fuel. Natural uranium contains only about 0.7% of the ^{235}U isotope, with the remaining 99.3% being ^{238}U. This fact is important for the operation of a reactor, because ^{238}U almost never fissions. Instead, it tends to absorb neutrons,

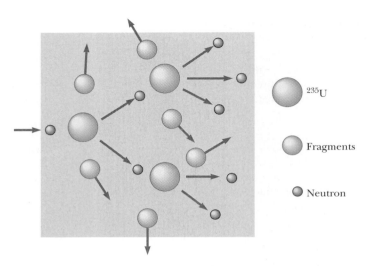

Figure 14.6 A nuclear chain reaction initiated by the capture of a neutron. (Many pairs of different isotopes are produced.)

producing neptunium and plutonium. For this reason, reactor fuels must be artificially enriched to contain at least a few percent ^{235}U.

Chain reaction   To achieve a self-sustained chain reaction, on average one of the neutrons emitted in ^{235}U fission must be captured by another ^{235}U nucleus and cause it to undergo fission. A useful parameter for describing the level of reactor

Figure 14.7 A sketch of the world's first reactor, which was composed of layers of graphite interspersed with uranium. (Because of wartime secrecy, there are no photographs of the completed reactor.) The first self-sustained chain reaction was achieved on December 2, 1942. Word of the success was telephoned immediately to Washington, D.C., in the form of this coded message: "The Italian navigator has landed in the New World and found the natives very friendly." The historic event took place in an improvised laboratory in the racquet court under the west stands of the University of Chicago's Stagg Field, and the Italian navigator guiding the work was Fermi.

operation is the **reproduction constant** K, defined as the average number of neutrons from each fission event that actually cause another fission event. As we have seen, K can have a maximum value of 2.5 in the fission of uranium. In practice, however, K is less than this because of several factors, to be discussed.

A self-sustained chain reaction is achieved when $K = 1$. Under this condition the reactor is said to be **critical.** When $K < 1$, the reactor is subcritical and the reaction dies out. When K is substantially greater than unity, the reactor is said to be supercritical and a runaway reaction occurs. In a nuclear reactor run by a utility company to furnish power, it is necessary to maintain a value of K slightly greater than unity.

Figure 14.8 shows the basic ingredients of a nuclear reactor core. The fuel elements consist of enriched uranium. The function of the remaining parts of the reactor and some aspects of its design will now be described.

Neutron Leakage

In any reactor, a fraction of the neutrons produced in fission leak out of the core before inducing other fission events. If the fraction leaking out is too great, the reactor will not operate. The percentage lost is large if the reactor is very small because leakage is a function of the ratio of surface area to volume. Therefore, a critical feature of the design of a reactor is to choose the correct surface-area-to-volume ratio so that a sustained reaction can be achieved.

Regulating Neutron Energies

Recall that the neutrons released in fission events are very energetic, having kinetic energies of about 2 MeV. It is necessary to slow these neutrons to thermal energies to allow them to be captured and produce fission of other ^{235}U nuclei, because the probability of neutron-induced fission increases with decreasing energy, as shown in Figure 14.9. The energetic neutrons are slowed down by a moderator substance surrounding the fuel, as shown in Figure 14.8.

Neutron Capture

In the process of being slowed down, neutrons may be captured by nuclei that do not fission. The most common event of this type is neutron capture by ^{238}U, which constitutes over 90% of the uranium in the fuel elements. The probability of neutron capture by ^{238}U is very high when the neutrons have high kinetic energies and very low when they have low kinetic energies. Thus the slowing down of the neutrons by the moderator serves the secondary purpose of making them available for reaction with ^{235}U and decreasing their chances of being captured by ^{238}U.

Control of Power Level

It is possible for a reactor to reach the critical stage ($K = 1$) after all the neutron losses just described are minimized. However, some method of control is needed to maintain a K value near unity. If K were to rise above this value, the heat produced in the runaway reaction would melt the reactor. To control the power level, control rods are inserted into the reactor core (see Fig. 14.8). These rods are made of materials such as cadmium that absorb neutrons very

Reproduction constant

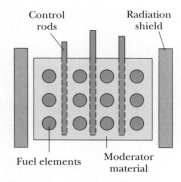

Figure 14.8 A cross section of a reactor core showing the control rods, fuel elements, and moderating material surrounded by a radiation shield.

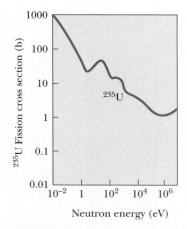

Figure 14.9 The cross section for neutron-induced fission of ^{235}U. The average cross section for room-temperature neutrons is about 500 b.

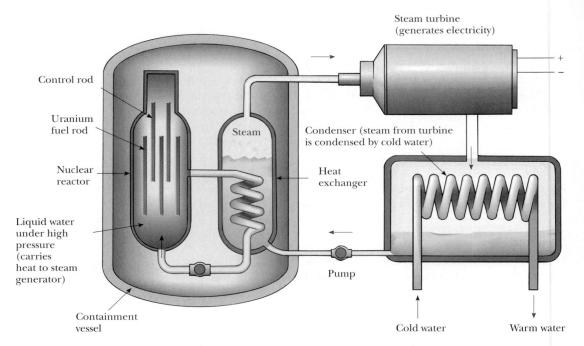

Figure 14.10   Main components of a pressurized-water reactor.

efficiently (see Fig. 14.3). By adjusting the number and position of these control rods in the reactor core, the K value can be varied and any power level within the design range of the reactor can be achieved.

Although there are several types of reactor that convert the kinetic energy of fission fragments to electrical energy, the most common type in use in the United States is the pressurized-water reactor (Fig. 14.10). Its main parts are common to all reactor designs. Fission events in the reactor core supply heat to the water contained in the primary (closed) loop, which is maintained at high pressure to keep it from boiling. This water also serves as the moderator. The hot water is pumped through a heat exchanger, and the heat is transferred to the water contained in the secondary loop. The hot water in the secondary loop is converted to steam, which drives a turbine-generator system to create electric power. Note that the water in the secondary loop is isolated from the water in the primary loop to prevent contamination of the secondary water and steam by radioactive nuclei from the reactor core.

Safety and Waste Disposal

The 1979 near-disaster at a nuclear power plant on Three Mile Island in Pennsylvania and the 1986 accident at the Chernobyl reactor in the Ukraine rightfully focused attention on reactor safety. The Three Mile Island accident was the result of inadequate control-room instrumentation and poor emergency-response training. There were no injuries or detectable health impacts from the event, even though more than one-third of the fuel melted. This unfortunately was not the case at Chernobyl, where the activity of the materials released immediately after the accident totaled approximately 1.2×10^{19} Bq and resulted in the evacuation of 135,000 people. Thirty individuals died during the accident or shortly thereafter and data from the Ukraine Radiological Institute suggest that more than 2,500 deaths could be attributed to the Chernobyl accident. In the period 1986–1997 there was a tenfold increase in the number of children contracting thyroid cancer from the ingestion of radioactive iodine in milk from cows that ate contaminated grass. One conclusion of an international conference studying the Ukraine accident was that the main causes of the Chernobyl accident were the coincidence of severe deficiencies in the reactor design and a violation of safety procedures. Most of these deficiencies have been addressed at plants of similar design in Russia and neighboring countries of the former Soviet Union.

Commercial reactors achieve safety through careful design, rigid operating protocol, and thorough emergency-response training of operators. It is only when these variables are compromised that reactors pose a danger. Radiation exposure and the potential health risks associated with such exposure are controlled by three layers of containment. The fuel and radioactive fission products are contained inside the reactor vessel. Should this vessel rupture, the reactor building acts as a second containment structure to prevent radioactive material from contaminating the environment. Finally, the reactor facilities must be in a remote location to protect the general public from exposure should radiation escape the reactor building.

A continuing concern about nuclear fission reactors is the safe disposal of radioactive material when the reactor core is replaced. Even when the uranium and plutonium are separated out and recycled, the remaining waste material contains long-lived, highly radioactive isotopes that must be stored over long time intervals in such a way that there is no chance of environmental contamination. At present, sealing radioactive wastes in waterproof containers and burying them in deep salt mines seems to be the most promising solution.

Transport of reactor fuel and reactor wastes poses additional safety risks. Accidents during transport of nuclear fuel could expose the public to harmful levels of radiation. To minimize these dangers, the Department of Energy requires stringent crash tests of all containers used to transport nuclear materials. Container manufacturers must demonstrate that their containers will not rupture even in high-speed collisions.

Despite these risks, there are advantages to the use of nuclear power to be weighed against the risks. For example, nuclear power plants do not produce air pollution and greenhouse gases as do fossil fuel plants, and the supply of uranium on the Earth is predicted to last longer than the supply of fossil fuels. For each source of energy, whether nuclear, hydroelectric, fossil fuel, wind, or solar, the risks must be weighed against the benefits and the regional availability of the energy source. Thus, thoughtful use of a variety of energy sources *and* increased emphasis on energy conservation methods appear to be logical components of a sensible energy policy.

14.6 NUCLEAR FUSION

In Chapter 13 we found that the binding energy for light nuclei (those having mass numbers less than 20) is much smaller than the binding energy for heavier nuclei. This suggests a process that is the reverse of fission, called **nuclear fusion.** Fusion occurs when two light nuclei combine to form a heavier nucleus. Because the mass of the final nucleus is less than the combined rest masses of the original nuclei, a loss of mass occurs, accompanied by a release of energy. The following are examples of such energy-liberating fusion reactions occuring in the Sun:

$$^1_1\text{H} + \,^1_1\text{H} \longrightarrow \,^2_1\text{H} + \,^0_1\text{e}^+ + \nu$$
$$^1_1\text{H} + \,^2_1\text{H} \longrightarrow \,^3_2\text{He} + \gamma \tag{14.10}$$

This second reaction is followed by one of the following reactions:

$$^1_1\text{H} + \,^3_2\text{H} \longrightarrow \,^4_2\text{He} + \,^0_1\text{e}^+ + \nu$$
$$^3_2\text{He} + \,^3_2\text{He} \longrightarrow \,^4_2\text{He} + \,^1_1\text{H} + \,^1_1\text{H}$$

These are the basic reactions in what is called the **proton–proton cycle,** believed to be one of the basic cycles by which energy is generated in the Sun and other stars that have an abundance of hydrogen. Most of the energy production takes place in the Sun's interior, where the temperature is approximately 1.5×10^7 K. As we will see later, such high temperatures are required to drive these reactions that they are called **thermonuclear fusion reactions.** The hydrogen (fusion) bomb, which was first exploded in 1952, is an example of an uncontrolled thermonuclear fusion reaction. All of the reactions in the proton–proton cycle are exothermic—that is, they involve a release of energy. An overall view of the proton–proton cycle is that four protons combine to form an alpha particle and two positrons, with the release of 25 MeV of energy.

Thermonuclear reactions

Fusion Reactions

The enormous amount of energy released in fusion reactions suggests the possibility of harnessing this energy for useful purposes here on Earth. A great deal of effort is currently directed toward developing a sustained and controllable thermonuclear reactor—a fusion power reactor. Controlled fusion is often called the ultimate energy source because of the availability of its fuel source: water. For example, if deuterium were used as the fuel, 0.12 g of it could be extracted from 1 gal of water at a cost of about 4 cents. Such rates would make the fuel costs of even an inefficient reactor almost insignificant. An additional advantage of fusion reactors is that comparatively few radioactive by-products are formed. For the proton–proton cycle described earlier in this section, the end product of the fusion of hydrogen nuclei is safe, nonradioactive helium. Unfortunately, a thermonuclear reactor that can deliver a net power output spread out over a reasonable time interval is not yet a reality, even though research has been in progress since the 1950s. Many difficulties must be resolved before a successful device is constructed.

We have seen that the Sun's energy is based, in part, upon a set of reactions in which hydrogen is converted to helium. Unfortunately, the proton–proton interaction is not suitable for use in a fusion reactor, because this reaction requires very high pressures and densities. The process works in the Sun only because of the extremely high density of protons in the Sun's interior.

The fusion reactions that appear most promising for a terrestrial fusion power reactor involve deuterium (^2_1H) and tritium (^3_1H):

$$^2_1\text{H} + \,^2_1\text{H} \longrightarrow \,^3_2\text{He} + \,^1_0\text{n} \qquad Q = 3.27 \text{ MeV}$$
$$^2_1\text{H} + \,^2_1\text{H} \longrightarrow \,^3_1\text{H} + \,^1_1\text{H} \qquad Q = 4.03 \text{ MeV} \tag{14.11}$$
$$^2_1\text{H} + \,^3_1\text{H} \longrightarrow \,^4_2\text{He} + \,^1_0\text{n} \qquad Q = 17.59 \text{ MeV}$$

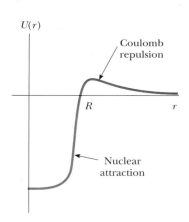

$U(r)$

Coulomb repulsion

R r

Nuclear attraction

Figure 14.11 Potential energy as a function of separation between two deuterons. The Coulomb repulsive force is dominant at long range, whereas the nuclear attractive force is dominant at short range, where R is of the order of several fermi.

As noted earlier, deuterium is available in almost unlimited quantities from our lakes and oceans and is very inexpensive to extract. Tritium, however, is radioactive ($T_{1/2} = 12.3$ yr) and undergoes beta decay to ^{3}He. As a result, tritium does not occur naturally to any great extent and must be artificially produced.

One of the major problems in obtaining energy from any fusion reaction is the fact that the Coulomb repulsion force between two charged nuclei must be overcome before they can fuse. The potential energy as a function of particle separation for two deuterons (each with charge $+e$) is shown in Figure 14.11. The potential energy is positive in the region $r > R$, where the Coulomb repulsive force dominates, and negative in the region $r < R$, where the strong nuclear force dominates. The fundamental problem, then, is to give the two nuclei enough kinetic energy to overcome this repulsive potential barrier. This can be accomplished by heating the fuel to extremely high temperatures (about 10^8 K, far greater than the interior temperature of the Sun). At these high temperatures, the atoms are ionized and the system consists of a collection of electrons and nuclei, commonly referred to as a **plasma.**

High temperatures are required to overcome the large Coulomb barrier

---

EXAMPLE 14.6   The Fusion of Two Deuterons

The separation between two deuterons must be about 1.0×10^{-14} m for the attractive nuclear force to overcome the repulsive Coulomb force. (a) Calculate the height of the potential barrier due to the repulsive force.

Solution   The potential energy associated with two charges separated by a distance r is

$$U = k \frac{q_1 q_2}{r}$$

where k is the Coulomb constant. For the case of two deuterons, $q_1 = q_2 = +e$, so

$$U = k \frac{e^2}{r} = \left(8.99 \times 10^9 \frac{\text{N} \cdot \text{m}^2}{\text{C}^2}\right) \frac{(1.60 \times 10^{-19}\ \text{C})^2}{1.0 \times 10^{-14}\ \text{m}}$$

$$= 2.3 \times 10^{-14}\ \text{J} = 0.14\ \text{MeV}$$

(b) Estimate the effective temperature required for a deuteron to overcome the potential barrier, assuming an energy of $\frac{3}{2} k_B T$ per deuteron (where k_B is Boltzmann's constant).

Solution   Since the total Coulomb energy of the pair of deuterons is 0.14 MeV, the Coulomb energy per deuteron is 0.07 MeV = 1.1×10^{-14} J. Setting this equal to the average thermal energy per deuteron gives

$$\tfrac{3}{2} k_B T = 1.1 \times 10^{-14}\ \text{J}$$

where k_B is equal to 1.38×10^{-23} J/K. Solving for T gives

$$T = \frac{2 \times (1.1 \times 10^{-14}\ \text{J})}{3 \times (1.38 \times 10^{-23}\ \text{J/K})} = 5.3 \times 10^8\ \text{K}$$

Example 14.6 suggests that deuterons must be heated to about 5×10^8 K to achieve fusion. This estimate of the required temperature is too high, however, because the particles in the plasma have a Maxwellian speed distribution, and therefore some fusion reactions are caused by particles in the high-energy "tail" of this distribution. Furthermore, even the particles without enough energy to overcome the barrier have some probability of tunneling through the barrier. When these effects are taken into account, a temperature of "only" 4×10^8 K appears adequate to fuse two deuterons.

The temperature at which the power generation rate exceeds the loss rate (due to mechanisms such as radiation losses) is called the **critical ignition temperature.** This temperature for the deuterium–deuterium (D–D) reaction is 4×10^8 K. According to $E \cong k_B T$, this temperature is equivalent to approximately 35 keV. It turns out that the critical ignition temperature for the deuterium–tritium (D–T) reaction is about 4.5×10^7 K, or only 4 keV.

Critical ignition temperature

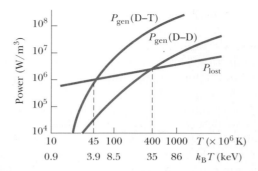

Figure 14.12 Power generated (or lost) versus temperature for the deuterium–deuterium and deuterium–tritium fusion reactions. When the generation rate P_{gen} exceeds the loss rate P_{lost}, ignition takes place.

Figure 14.12 is a plot of the power generated by fusion, P_{gen}, versus temperature for the two reactions. The straight line represents the power lost, via the radiation mechanism known as **bremsstrahlung,** versus temperature. This is the principal mechanism of energy loss, in which radiation (primarily x-ray) is emitted as the result of electron–ion collisions within the plasma.[2] The intersections of the P_{lost} line with the P_{gen} curves give the critical ignition temperatures.

Confinement time

In addition to the high temperature requirements, there are two other critical parameters that determine whether or not a thermonuclear reactor will be successful: the **ion density,** n, and **confinement time,** τ. **The confinement time is the period for which the interacting ions are maintained at a temperature equal to or greater than the ignition temperature.** The British physicist J. D. Lawson has shown that the ion density and confinement time must both be large enough to ensure that more fusion energy is released than is required to heat the plasma. In particular, **Lawson's criterion** states that a net energy output is possible under the following conditions:

Lawson's criterion

$$n\tau \geq 10^{14} \text{ s/cm}^3 \qquad \text{(D–T)}$$
$$n\tau \geq 10^{16} \text{ s/cm}^3 \qquad \text{(D–D)}$$

(14.12)

Figure 14.13 is a graph of $n\tau$ versus the so-called **kinetic temperature** $k_B T$ for the D–T and D–D reactions.

Lawson arrived at his criterion by comparing the energy required to heat the plasma with the energy generated by the fusion process. The energy E_h required to heat the plasma is proportional to the ion density n; that is, $E_h = D_1 n$. The energy generated by the fusion process, E_{gen}, is proportional to $n^2\tau$, or $E_{gen} = D_2 n^2 \tau$. This can be understood by realizing that the fusion energy released is proportional to both the rate at which interacting ions collide, n^2, and the confinement time, τ. Net energy is produced when the energy generated by fusion, E_{gen}, exceeds E_h. When the constants D_1 and D_2 are

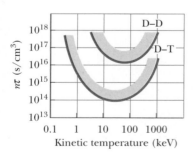

Figure 14.13 The Lawson number $n\tau$ at which net energy output is possible versus temperature for the D–T and D–D fusion reactions. The regions above the curves represent favorable conditions for fusion.

---

[2]Cyclotron radiation is another loss mechanism; it is especially important in the case of the D–D reaction.

calculated for different reactions, the condition that $E_{gen} \geq E_h$ leads to Lawson's criterion.[3]

In summary, the three basic requirements of a successful thermonuclear power reactor are

- The plasma temperature must be very high—about 4.5×10^7 K for the D–T reaction and about 4×10^8 K for the D–D reaction.
- The ion density must be high. A high density of interacting nuclei is necessary to increase the collision rate between particles.
- The confinement time of the plasma must be long. To meet Lawson's criterion, the product $n\tau$ must be large. For a given value of n, the probability of fusion between two particles increases as τ increases.

Requirements for a fusion power reactor

Current efforts are aimed at meeting Lawson's criterion at temperatures exceeding the critical ignition temperature. Although the minimum plasma densities have been achieved, the problem of confinement time is more difficult. How can a plasma be confined at 10^8 K for 1 s? Two basic techniques for confining plasmas are under investigation: magnetic field confinement and inertial confinement.

Magnetic Field Confinement

Many fusion-related plasma experiments use **magnetic field confinement** to contain the charged plasma. Figure 14.14a shows a device called a **tokamak,** first developed in Russia. A combination of two magnetic fields is used to confine and stabilize the plasma: (1) a strong toroidal field, produced by the current in the windings, and (2) a weaker "poloidal" field, produced by the toroidal current. The toroidal current heats the plasma in addition to confining it. The resultant helical field lines spiral around the plasma and keep it from touching the walls of the vacuum chamber. If the plasma comes into contact with the walls, its temperature is reduced and heavy impurities sputtered from the walls "poison" it and lead to large power losses. One of the major breakthroughs in the 1980s was in the area of auxiliary heating to reach ignition temperatures. Experiments have shown that injecting a beam of energetic neutral particles into the plasma is a very efficient method of heating the plasma to ignition temperatures (5 to 10 keV). Radio-frequency heating will probably be needed for reactor-size plasmas. Figure 14.14b shows a cutaway view of the Princeton Tokamak Fusion Test Reactor. When it was in operation, the Tokamak Fusion Test Reactor (TFTR) reported central ion temperatures of 510 million degrees Celsius, more than 30 times hotter than the center of the Sun. The $n\tau$ values in the TFTR for the D–T reaction were well above 10^{13} s/cm^3 and close to the value required by Lawson's criterion. By the late 1990s, tokamaks in England and Japan were reporting reaction rates of 10^{18} D–T fusions per second and $n\tau$ values of 5×10^{13} s/cm^3 at temperatures of 30 keV. Direct measurements showed that the output energy slightly exceeded the input energy to the plasma for brief periods.

---

[3]Note that Lawson's criterion neglects the energy needed to set up the strong magnetic field that is used to confine the hot plasma. This energy is expected to be about 20 times greater than the energy required to heat the plasma. Consequently, it is necessary to have a magnetic energy recovery system or to make use of superconducting magnets.

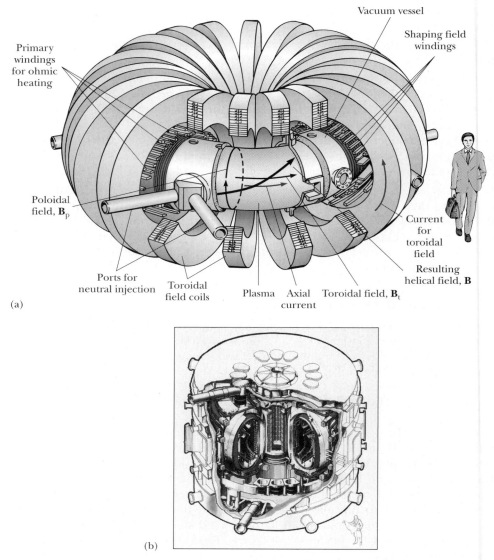

Primary
windings
for ohmic
heating

Vacuum vessel

Shaping field
windings

Poloidal
field, $\mathbf{B}_p$

Current
for
toroidal
field

Resulting
helical field, $\mathbf{B}$

Ports for
neutral injection

Toroidal
field coils

Plasma

Axial
current

Toroidal field, $\mathbf{B}_t$

(a)

(b)

Figure 14.14   (a) A schematic diagram of a tokamak used in magnetic confinement. The total magnetic field $\mathbf{B}$ is the superposition of the toroidal field $\mathbf{B}_t$ and the poloidal field $\mathbf{B}_p$. The plasma is trapped within the spiraling field lines as shown. (*Adapted from McGraw-Hill Encyclopedia of Science and Technology, New York, McGraw-Hill Book Co., 1987.*) (b) A cutaway view of the TFTR. (*Courtesy of Princeton Plasma Physics Laboratory*)

One of the new generation of fusion experiments is the National Spherical Torus Experiment (NSTX), shown in Figure 14.15. Rather than the donut-shaped plasma of a tokamak, the NSTX produces a spherical plasma that has a hole through its center. The major advantage of the spherical configuration is its ability to confine the plasma at a higher pressure in a given magnetic field. This approach could lead to development of smaller, more economical fusion reactors.

Figure 14.15   The National Spherical Torus Experiment (NSTX) that began operation in March 1999. (*Courtesy of Princeton Plasma Physics Laboratory, Princeton University*)

An international collaborative effort involving Canada, Europe, Japan, and Russia is currently under way to build a fusion reactor called ITER (International Thermonuclear Experimental Reactor). China and the United States began to participate in program activities in early 2003. This facility will address the remaining technological and scientific issues concerning the feasibility of fusion power. The design is completed, and site and construction negotiations are under way. If the planned device works as expected, the Lawson number for ITER will be about six times greater than the current record holder, the JT-60U tokamak in Japan. ITER will produce 1.5 GW of power, and the energy content of the alpha particles inside the reactor will be so intense that they will sustain the fusion reaction, allowing the auxiliary energy sources to be turned off once the reaction is initiated.

Inertial Confinement

The second technique for confining a plasma, called **inertial confinement,** makes use of a D–T target that has a very high particle density of 5×10^{25} particles/cm^3, or a mass density of about 200 g/cm^3. In this scheme, the confinement time is very short (typically 10^{-11} to 10^{-9} s), and so, because of their own inertia, the particles do not have a chance to move appreciably from their initial positions. Thus Lawson's criterion can be satisfied by combining a high particle density with a short confinement time.

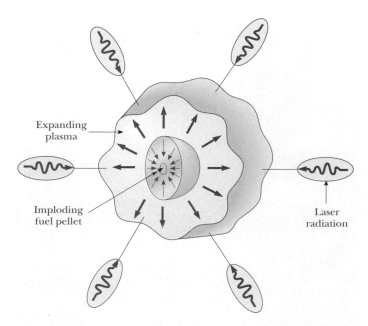

Figure 14.16 In the inertial confinement scheme, a D–T fuel pellet fuses when struck by several high-intensity laser beams simultaneously and symmetrically.

Laser fusion is the most common form of inertial confinement. A small D–T pellet about 1 mm in diameter is struck simultaneously by several focused, symmetrically incident, high-intensity laser beams, resulting in a large pulse of input energy that causes the surface of the fuel pellet to evaporate (Fig. 14.16). The escaping particles produce a reaction force on the core of the pellet, resulting in a strong, inwardly moving, compressive shock wave. This shock wave increases the pressure and density of the core and produces a corresponding increase in temperature. When the temperature of the core reaches ignition temperature, fusion reactions cause the pellet to explode. The process can be viewed as a miniature hydrogen bomb explosion.

Two of the leading laser fusion laboratories in the United States are the Omega facility at the University of Rochester and the Nova facility at Lawrence Livermore National Laboratory in California. Both facilities use neodymium glass lasers. The Omega facility focuses 60 laser beams on a target chamber about 3 m in diameter. Figure 14.17a shows the target at Omega, and Figure 14.17b shows the tiny, spherical D–T pellets used. Nova, operating at higher input power levels than Omega, is capable of injecting a power of 200 kJ into a 0.5-mm D–T pellet in 1 ns. With these high input powers, Nova has achieved $n\tau \approx 5 \times 10^{14}$ s/cm^3 and ion temperatures of 5.0 keV, values close to those required for D–T ignition.

Fusion Reactor Design

In the D–T fusion reaction

$$^2_1\text{H} + {}^3_1\text{H} \longrightarrow {}^4_2\text{He} + {}^1_0\text{n} \qquad Q = 17.6 \text{ MeV}$$

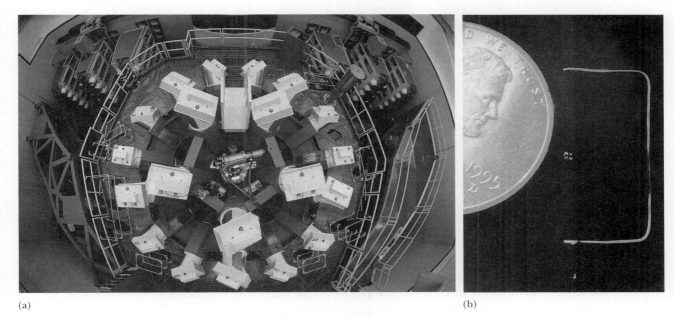

(a)   (b)

Figure 14.17 (a) A view of the Omega target bay from high above the target chamber shows critical components associated with the uv transport system. (b) An example of a plastic microballoon used as an inertial confinement fusion target. The microballoon has a diameter of 900 μm and is filled with an equimolar mix of deuterium and tritium. (*This work was supported by the U.S. Department of Energy (DOE), the University of Rochester, and the New York State Energy Research and Development Authority. The support of DOE does not constitute an endorsement by DOE of the views expressed in this text. Staff Photographer: Eugene Kowaluk*)

the alpha particle carries 20% of the energy and the neutron carries 80%, or about 14 MeV. While the charged alphas are primarily absorbed in the plasma and produce desired additional plasma heating, the 14-MeV neutrons pass through the plasma and must be absorbed in a surrounding blanket material to extract their large kinetic energy and generate electric power. A frequently proposed scheme is to use molten lithium metal as the neutron-absorbing material and to circulate the lithium in a closed heat-exchanging loop to produce steam and drive turbines, as in a conventional power plant. Figure 14.18 is a diagram of such a fusion reactor system. It is estimated that a blanket of lithium about 1 m thick would capture nearly 100% of the neutrons from the fusion of a small D–T pellet, not only absorbing the neutron kinetic energy but limiting the dangerous neutron flux to nearby workers.

The capture of neutrons by lithium is described by the reaction

$$_0^1 n + {}_3^6 Li \longrightarrow {}_1^3 H + {}_2^4 He$$

where the kinetic energies of the charged tritium and alpha particle products are converted to heat in the lithium. An extra advantage of using lithium as the energy transfer medium is the production of tritium, $_1^3 H$, which may be separated from the lithium and returned as fuel to the reactor. The process is indicated in the generic fusion reactor shown in Figure 14.18.

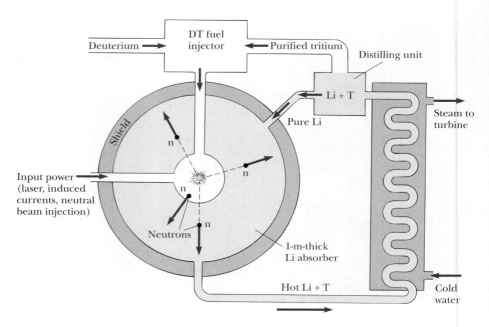

Figure 14.18   A generic fusion reactor.

Advantages and Problems of Fusion

If fusion power can be harnessed, it will offer several advantages over fission-generated power: (1) the low cost and abundance of the fuel (deuterium), (2) the impossibility of runaway accidents, and (3) a lesser radiation hazard than with fission. Some of the anticipated problem areas include (1) the as-yet-unestablished feasibility of fusion reactors, (2) the very high proposed plant costs, (3) the scarcity of lithium, (4) the limited supply of helium needed to cool the superconducting magnets used to produce strong confining fields (this problem may be significantly reduced by the development of high-temperature superconductors), (5) structural damage and induced radioactivity caused by neutron bombardment, and (6) the anticipated high degree of thermal pollution. If these basic problems and the engineering design factors can be resolved, nuclear fusion may become a feasible source of energy by the middle of the 21st century.

14.7   INTERACTION OF PARTICLES WITH MATTER

In this section we consider processes involving the interaction between energetic particles and matter. We shall deal mainly with the interaction of charged particles and photons with matter, since these are of primary importance in such factors as shielding characteristics, the design of particle detectors, and biological effects.

Heavy Charged Particles

A heavy charged particle, such as an alpha particle or a proton, moving through a solid, liquid, or gas travels a well-defined distance before coming to rest. This **Range of a particle**   distance is called the **range** of the particle. As the particle passes through the medium, it loses energy, primarily through the excitation and ionization of atoms in the medium. Some energy is also lost through elastic collisions with nuclei. The highly energetic particle loses its energy in many small increments, and to a good approximation one can treat the problem as a continuous loss of energy. At the end of its range, the particle is left with just the average thermal energy of its surroundings. The range depends on the charge, mass, and energy of the particle, the density of the medium through which it travels, and the ionization potential and atomic number of the atoms in the medium.

The ranges of α particles and protons in air as a function of their energy are plotted in Figure 14.19. Note that for a given energy, the proton has a range about 10 times that of the α particle. This is because the proton has less charge and so does not interact with the medium as strongly as the alpha particle. Furthermore, since the α particle is more massive, it travels at a lower average speed for a given energy. As a result, the slower-moving α particle loses its energy more readily, since it has more time to interact with atoms in the medium.

Figure 14.20 shows the rate of energy loss per unit length (energy loss rate), $-dE/dx$, versus the kinetic energy of the charged particle. This **energy loss rate** (also called the *stopping power*) is approximately proportional to the kinetic energy at low energies ($v \ll c$) and reaches a maximum at some point. At very high energies (as $v \rightarrow c$), the energy loss per unit length is approximately energy independent. At very low energies, the decrease in energy loss is a result of the fact that the particle is moving too slowly to produce ionization and loses its energy mainly by elastic collisions.

The energy loss rate is approximately proportional to the density of the medium through which the charged particles travel. This is explained by recognizing that the primary energy loss mechanism (ionization) involves the excitation of electrons in the medium, and the density of electrons increases with increasing density of the medium. For example, the range in aluminum of protons having an energy of 1 to 10 MeV is about 1/1600 of their range in air.

Fission fragments with energies of about 80 MeV have a range of only about 0.02 m in air, which is about 1000 times smaller than the range of protons with the same energy. This difference is due to the greater charge of the fission fragments and to the fact that the energy loss rate is proportional to the square of the charge.

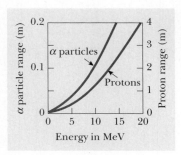

Figure 14.19   The range of alpha particles and protons in air under standard conditions. (*From E. Segrè, ed.*, Experimental Nuclear Physics, *Vol. 1, New York, Wiley, 1953*)

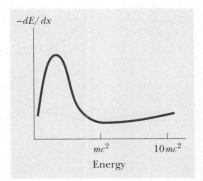

Figure 14.20   The energy loss $-dE/dx$ versus the energy of a charged particle of mass m moving through a medium.

Electrons

When an electron of energy less than about 1 MeV passes through matter, it loses its energy by the same processes that occur for heavy charged particles. However, the range of the electron is not as well defined. Since an electron is much smaller than a proton, there are large statistical variations in the electron's path length as a result of a phenomenon called **straggling.** It takes only a few large-deflection collisions to stop an electron in matter. Furthermore, electrons scatter much more readily than heavy charged particles with the same energy. As we mentioned earlier, a decelerated charged particle emits electromagnetic radiation called *bremsstrahlung*. This energy loss process is more important for electrons than for heavy charged particles, since low-mass electrons undergo greater accelerations when passing through matter. For example, the energy loss rate due to bremsstrahlung for 10-MeV electrons passing through lead is about equal to the loss rate due to ionization.

Photons

Since photons are uncharged, they are not as effective as charged particles in producing ionization or excitation in matter. Nevertheless, photons can be removed from a beam by either scattering or absorption in the medium. Figure 14.21 illustrates the three important processes that contribute to the total absorption of gamma rays in lead:

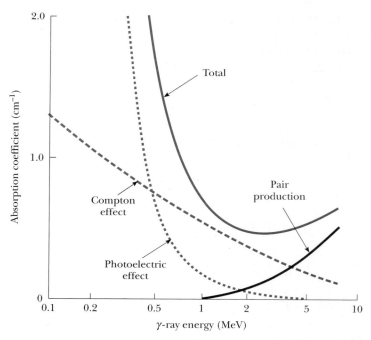

Figure 14.21 The absorption of gamma rays in lead. Shown are the three important absorption processes, which add up to the total absorption constant μ, measured in cm^{-1}. (*From W. Heitler,* The Quantum Theory of Radiation, *3rd ed., Oxford, Clarendon Press, 1954*)

- At low photon energies (less than about 0.5 MeV), the predominant process that removes photons from a beam is the photoelectric effect, in which a photon is absorbed and transfers all of its energy to an electron.
- At intermediate photon energies, the predominant process is Compton scattering, in which the photon transfers some of its energy to an electron. Both the photoelectric effect and Compton scattering were discussed in Chapter 3.
- At high energies, the predominant process is **pair production,** in which an electron–positron pair is created as a photon passes near a nucleus in the medium. Since the rest energy of an electron–positron pair is $2m_ec^2 = 1.02$ MeV (twice the rest energy of an electron), the gamma-ray photon must have at least this much energy to produce a pair. Pair production will be discussed further in Chapter 15.

If a beam of photons is incident on a medium, its intensity decreases exponentially with increasing depth of penetration into the medium. This reduction in intensity is referred to as **attenuation** of the beam. As in the case of neutrons (Eq. 14.5), the intensity in a medium varies according to the relation

$$I(x) = I_0 e^{-\mu x} \tag{14.13}$$

Absorption of photons in a medium

where I_0 is the incident photon intensity (measured in photons/m$^2 \cdot$s), x is the distance the beam travels in the medium, $I(x)$ is the intensity of the beam after it travels a distance x, and μ is a parameter called the **linear absorption coefficient** of the medium. This coefficient depends on the energy of the photon as well as on the properties of the medium. The variation of μ with energy is shown in Figure 14.21 for gamma rays in lead and in Tables 14.2 and 14.3 for x-rays and gamma rays in various media.

Linear absorption coefficient

Table 14.2   Linear Absorption Coefficients for X-Rays in Various Media

	Linear Absorption Coefficient (cm^{-1})				
λ (pm)	Air	Water	Aluminum	Copper	Lead
10		0.16	0.43	3.2	43
20		0.18	0.76	13	55
30		0.29	1.3	38	158
40		0.44	3.0	87	350
50	8.6×10^{-4}	0.66	5.4	170	610
60	1.3×10^{-3}	1.0	9.2	286	1000
70	1.95×10^{-3}	1.5	14	430	1600
80	2.73×10^{-3}	2.1	20	625	
90	3.64×10^{-3}	2.8	30	875	
100	4.94×10^{-3}	3.8	41	1200	
150	1.56×10^{-2}	12	124		
200	3.64×10^{-2}	28	275		
250	6.63×10^{-2}	51	524		

Table 14.3   Linear Absorption Coefficients for Gamma Rays in Various Media

Photon Energy (MeV)	Linear Absorption Coefficient (cm^{-1})[a]			
	Water	Aluminum	Iron	Lead
0.1	0.167	0.432	2.69	59.8
0.15	0.149	0.359	1.43	20.8
0.2	0.139	0.324	1.08	10.1
0.3	0.118	0.278	0.833	3.79
0.4	0.106	0.249	0.722	2.35
0.5	0.0967	0.227	0.651	1.64
0.6	0.0894	0.210	0.598	1.29
0.8	0.0786	0.184	0.525	0.946
1.0	0.0706	0.166	0.468	0.772
1.5	0.0576	0.135	0.380	0.581
2.0	0.0493	0.166	0.332	0.510
3.0	0.0396	0.0953	0.282	0.463
4.0	0.0339	0.0837	0.259	0.470
5.0	0.0302	0.0767	0.247	0.486
6.0	0.0277	0.0718	0.240	0.514
8.0	0.0242	0.0656	0.234	0.532
10.0	0.0221	0.0626	0.236	0.568

[a]Calculated using data in I. Kaplan, *Nuclear Physics*, Reading, MA., Addison-Wesley, 1962, Table 15.6.

EXAMPLE 14.7   Half-Value Thickness

The half-value thickness of an absorber is defined as the thickness that will reduce the intensity of a beam of particles by a factor of 2. Calculate the half-value thickness for lead, assuming an x-ray beam of wavelength 20 pm (1 pm = 10^{-12} m = 10^{-3} nm).

Solution   The intensity varies with distance traveled in the medium according to Equation 14.13:

$$I(x) = I_0 e^{-\mu x}$$

In this case, we are looking for a value of x such that $I(x) = I_0/2$. That is, we require that

$$\frac{I_0}{2} = I_0 e^{-\mu x}$$

or

$$e^{-\mu x} = \tfrac{1}{2}$$

Taking the natural logarithm of both sides of this equation gives

$$\mu x = \ln 2$$

or

$$x = \frac{\ln 2}{\mu}$$

Since $\mu = 55$ cm^{-1} for x rays in lead at a wavelength of 20 pm (Table 14.2),

$$x = \frac{\ln 2}{55 \text{ cm}^{-1}} = \frac{0.693}{55 \text{ cm}^{-1}}$$

$$= 1.26 \times 10^{-2} \text{ cm} = 0.126 \text{ mm}$$

Hence we conclude that lead is a very good absorber for x-rays.

14.8   RADIATION DAMAGE IN MATTER

Radiation passing through matter can cause severe damage. The degree and type of damage depend upon several factors, including the type and energy of the radiation and the properties of the matter. For example, the metals used in nuclear reactor structures can be severely weakened by high fluxes of energetic neutrons, because these high fluxes often lead to metal fatigue. The damage in such situations is in the form of atomic displacements, which often result in major alterations in the material properties. Materials can also be damaged by ionizing radiations, such as gamma rays and x-rays. For example, defects called color centers can be produced in alkali halide crystals by irradiating the crystals with x-rays. One extensively studied color center has been identified as an electron trapped in a Cl^- ion vacancy.

Radiation damage in biological organisms is due primarily to ionization effects in cells. The normal operation of a cell may be disrupted when highly reactive ions or radicals form as the result of ionizing radiation. For example, hydrogen and hydroxyl radicals produced from water molecules can induce chemical reactions that may break bonds in proteins and other vital molecules. Furthermore, the ionizing radiation may directly affect vital molecules by removing electrons from their structures. Large doses of radiation are especially dangerous, because damage to a great number of molecules in a cell may kill the cell. Although the death of a single cell is usually not a problem, the death of many cells may irreversibly damage the organism. Cells that divide rapidly, such as those of the digestive tract, reproductive organs, and hair follicles, are especially susceptible. Also, cells that do survive the radiation may become defective. These defective cells can produce more defective cells when they divide, leading to cancer.

In biological systems, it is common to separate radiation damage into two categories: somatic and genetic. **Somatic damage** is the damage associated with any body cell except the reproductive cells. At high dose rates, such damage can lead to cancer or seriously alter the characteristics of specific organisms. **Genetic damage** affects only reproductive cells. Damage to the genes in reproductive cells can lead to defective offspring. Clearly, we must be concerned about the effects of diagnostic treatments such as x rays and other forms of radiation exposure.

Several units are used to quantify the amount, or dose, of any radiation that interacts with a substance.

The **roentgen** (R) is that amount of ionizing radiation that produces an electric charge of 3.33×10^{-10} C in 1 cm^3 of air under standard conditions.	**The roentgen**

Equivalently, the roentgen is that amount of radiation that deposits an energy of 8.76×10^{-3} J into 1 kg of air. For most applications, the roentgen has been replaced by the rad (which is an acronym for radiation absorbed dose):

One **rad** is that amount of radiation that increases the energy of 1 kg of absorbing material by 1×10^{-2} J.	**The rad**

Although the rad is a perfectly good physical unit, it is not the best unit for measuring the degree of biological damage produced by radiation, because damage depends not only on the dose but on the type of the radiation. For example, a given dose of alpha particles causes about 10 times more biological damage than an equal dose of x rays. The **RBE** (relative biological

Table 14.4   RBE[a] for Several Types of Radiation

Radiation	RBE Factor
X-rays and gamma rays	1.0
Beta particles	1.0–1.7
Alpha particles	10–20
Slow neutrons	4–5
Fast neutrons and protons	10
Heavy ions	20

[a]RBE = relative biological effectiveness.

Table 14.5   Units for Radiation Dosage

Quantity	SI Unit	Symbol	Relation to Other SI Units	Older Unit	Conversion
Absorbed dose	gray	Gy	= 1 J/kg	rad	1 Gy = 100 rad
Dose equivalent	sievert	Sv	= 1 J/kg	rem	1 Sv = 100 rem

effectiveness) factor for a given type of radiation is the **number of rads of x radiation or gamma radiation that produces the same biological damage as 1 rad of the radiation being used.** Table 14.4 gives the RBE factors for several types of radiation. The values are only approximate, because they vary with particle energy and with the form of the damage. The RBE factor should be considered only a first-approximation guide to the actual effects of radiation.

Finally, the **rem** (radiation equivalent in man) is the product of the dose in rad and the RBE factor:

$$\text{Dose in rem} \equiv \text{dose in rad} \times \text{RBE} \qquad (14.14)$$

According to this definition, 1 rem of any type of radiation produces the same amount of biological damage as 1 rem of any other type. From Table 14.4 we see that a dose of 1 rad of fast neutrons represents an effective dose of 10 rem. On the other hand, 1 rad of gamma radiation is equivalent to a dose of 1 rem.

Low-level radiation from a natural source, such as cosmic rays or radioactive rocks and soil, delivers to each of us a dose of about 0.13 rem/yr and is called **background radiation.** It is important to note that background radiation varies with geography. The upper limit of radiation dose (apart from background radiation) recommended by the U.S. government is about 0.5 rem/yr. Many occupations involve much higher radiation exposures, and so an upper limit of 5 rem/yr has been set for combined whole-body exposure. Higher upper limits are permissible for certain parts of the body, such as the hands and forearms. A dose of 400 to 500 rem results in a mortality rate of about 50% (which means that half the people exposed to this radiation level die). The most dangerous forms of exposure are ingestion and inhalation of radioactive isotopes, especially those elements the body retains and concentrates, such as ^{90}Sr. In some cases, a dose of 1000 rem can result from ingesting only 1 mCi of radioactive material.

This discussion has focused on measurements of radiation dosage in units such as rads and rems because these units are still widely used. These units, however, have been formally replaced with new SI units. The rad has been replaced with the **gray** (Gy), equal to 100 rad. The rem has been replaced with the **sievert** (Sv), equal to 100 rem. Table 14.5 summarizes the older and the current SI units of radiation dosage.

14.9   RADIATION DETECTORS

Various devices have been developed for detecting radiation. They are used for a variety of purposes, including medical diagnoses, radioactive dating measurements, measurement of background radiation, and measurement of the mass, energy, and momentum of particles created in high-energy nuclear reactions.

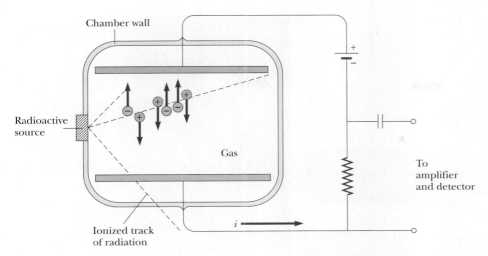

Figure 14.22   A schematic of an ion chamber. A charged particle stopping in the chamber creates electron–ion pairs, which are collected by the plates. This sets up a current *i* in the external circuit that is proportional to the particle's kinetic energy.

An **ion chamber** is a general class of detector that makes use of the electron–ion pairs generated by the passage of radiation through a gas to produce an electrical signal. It consists basically of two charged plates maintained at different potentials by a voltage supply (Fig. 14.22). The plates attract electrons or ions, depending on plate polarity, and cause a current pulse *i* that is proportional to the number of electron–ion pairs produced and to the particle energy if the particle comes to rest in the chamber. When an ion chamber is used in this way, both to detect the presence of an energetic charged particle and to measure its energy, it is called a **proportional counter.**

The **Geiger counter** (Fig. 14.23) is perhaps the most common form of ion chamber used to detect radiation. It can be considered the prototype of all counters that use the ionization of a medium as the basic detection process. It consists of a cylindrical metal tube filled with gas at low pressure and a long wire along the axis of the tube. The wire is maintained at a high positive potential (about 10^3 V) with respect to the tube. When a high-energy particle or photon enters the tube through a thin window at one end, some of the atoms of the gas are ionized. The electrons removed from these atoms are attracted toward the positive wire, and in the process they ionize other atoms in their path. This results in an avalanche of electrons that produces a current pulse at the output of the tube. After the pulse is amplified, it can either be used to trigger an electronic counter or be delivered to a loudspeaker that clicks each time a particle is detected. While a Geiger counter easily detects the presence of an energetic charged particle, the energy lost by the particle in the counter is *not* proportional to the current pulse produced in the avalanche process. Thus, although it is rugged, simple, and portable, a Geiger counter cannot be used to measure the energy of a particle.

A **semiconductor diode detector** is essentially a reverse-bias *p-n* junction. Recall from Chapter 12 that a *p-n* junction diode passes current readily when forward-biased and prohibits a current under reverse-bias conditions. As an

Ion chamber

Geiger counter

Semiconductor diode detector

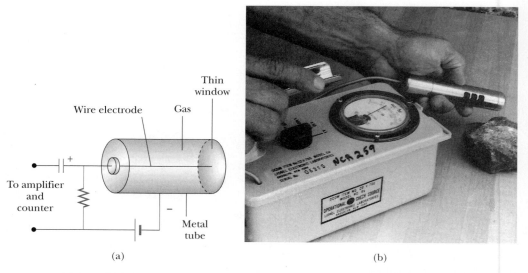

Figure 14.23   (a) A diagram of a Geiger counter. The voltage between the central wire and the metal tube is usually about 1000 V. (b) The use of a Geiger counter to measure the activity in a radioactive mineral. (*Henry Leap and Jim Lehman*)

energetic particle passes through the junction, electrons are excited into the conduction band and holes form in the valence band. The internal electric field sweeps the electrons toward the positive (*p*) side of the junction and the holes toward the negative (*n*) side. This movement of electrons and holes creates a pulse of current that is measured with an electronic counter. In a typical device, the duration of the pulse is about 10^{-8} s.

Scintillation counter

A **scintillation counter** (Fig. 14.24) usually uses a solid or liquid material whose atoms are easily excited by radiation. These excited atoms emit visible light when they return to their ground state. Common materials used as scintillators are transparent crystals of sodium iodide and certain plastics. If such a material is attached to one end of a device called a **photomultiplier** (PM) tube, the photons emitted by the scintillator can be converted to an electric signal. The PM tube consists of numerous electrodes, called dynodes, whose potentials increase in succession along the length of the tube, as shown in Figure 14.24. The top of the tube contains a photocathode, which emits electrons by the photoelectric effect. As one of these emitted electrons strikes the first dynode, the electron has sufficient kinetic energy to eject several other electrons. When these electrons are accelerated to the second dynode, many more electrons are ejected and a multiplication process occurs. The end result is 1 million or more electrons striking the last dynode. Hence one particle striking the scintillator produces a sizable electric pulse at the output of the PM tube, and in turn this pulse is sent to an electronic counter. Both the scintillator and diode detector are much more sensitive than a Geiger counter, mainly because of the higher density of the detecting medium. Both can also be used to measure particle energy if the particle stops in the detector.

Track detectors

Track detectors are devices that can be used to view the tracks of charged particles directly. High-energy particles produced in particle accelerators may have energies ranging from 10^9 to 10^{12} eV. Thus they cannot be stopped

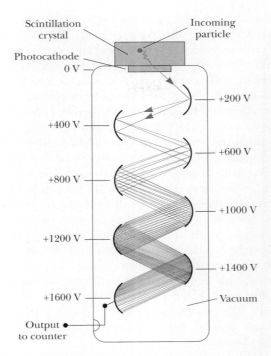

Figure 14.24   A diagram of a scintillation counter connected to a photomultiplier tube.

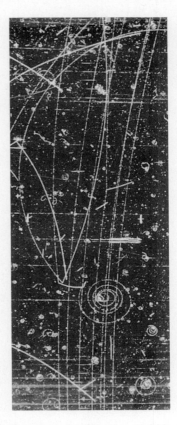

Figure   14.25 A   bubble-chamber photograph showing tracks of subatomic particles. (*Courtesy of Lawrence Radiation Laboratory, University of California, Berkeley*)

Photographic emulsion

Cloud chamber

Bubble chamber

Spark chamber

entirely, and their energy cannot be measured in the small detectors already mentioned. Instead, the energy and momentum of these energetic particles are found from the curvature of the path of the particle in a known magnetic field, hence the necessity of track detectors.

A **photographic emulsion** is the simplest track detector. A charged particle ionizes the atoms in an emulsion layer. The path of the particle corresponds to a family of points at which chemical changes have occurred in the emulsion. When the emulsion is developed, the particle's track becomes visible.

A **cloud chamber** contains a vapor that has been supercooled to just below its usual condensation point. An energetic particle passing through the chamber ionizes the vapor along its path. These ions serve as centers for condensation of the supercooled vapor. The track can be seen with the naked eye and can be photographed. A magnetic field can be applied to determine the charges of the particles as well as their momentum and energy.

A device called a **bubble chamber,** invented in 1952 by Donald A. Glaser, makes use of a liquid (usually liquid hydrogen) maintained near its boiling point. Ions produced by incoming charged particles leave bubble tracks, which can be photographed (Fig. 14.25). Because the density of the detecting medium of a bubble chamber is much higher than the density of the gas in a cloud chamber, the bubble chamber has a much higher sensitivity.

A **spark chamber** is a counting device that consists of an array of conducting parallel plates and is capable of recording a three-dimensional track record. Even-numbered plates are grounded, and odd-numbered plates are maintained at a high potential (about 10 kV). The spaces between the plates contain a noble gas at atmospheric pressure. When a charged particle passes through the chamber, ionization occurs in the gas, resulting in a large surge of current and a visible series of sparks along the particle path. These paths may be photographed or electronically detected and sent to a computer for path reconstruction and the determination of particle mass, momentum, and energy.

Wire chamber

Wire chambers or **drift chambers** are similar to spark chambers in their ability to record 3-D tracks and provide fast electronic readout to a computer for track reconstruction and display. A wire chamber consists of closely spaced parallel wires that collect the electrons created by a passing ionizing particle. A second grid, with its wires perpendicular to the first, allows the xy position of the particle in the plane of the two sets of wires to be determined. Finally, several such xy locating grids arranged parallel to each other along the z-axis can be used to determine the particle's trajectory in three dimensions.

Neutron detectors

Neutron detectors are a bit more difficult to construct than charged-particle detectors, because neutrons do not interact electrically with atoms in their passage through matter. Fast neutrons, however, can be detected by filling an ion chamber with hydrogen gas and detecting the ionization produced by high-speed recoiling protons produced in neutron–proton collisions. Slow neutrons with energies less than 1 MeV do not transfer sufficient energy to protons to be detected in this way but can be detected with an ion chamber filled with BF_3 gas. In this case, the boron nuclei disintegrate in a slow neutron-capture process, emitting highly ionizing alphas that are easily detected in the ion chamber.

14.10 USES OF RADIATION

Nuclear physics applications are extremely widespread in manufacturing, medicine, and biology. Even a brief discussion of all the possibilities would fill an entire book, and to keep such a book up-to-date would require frequent revisions. In this section we present a few of these applications and the underlying theories supporting them.

Tracing

Radioactive tracers are used to track chemicals participating in various reactions. One of the most valuable uses of radioactive tracers is in medicine. For example, iodine, a nutrient needed by the human body, is obtained largely through the intake of iodized salt and seafood. To evaluate the performance of the thyroid, the patient drinks a very small amount of radioactive sodium iodide containing ^{131}I, an artificially produced isotope of iodine (the natural, nonradioactive isotope is ^{127}I). The amount of iodine in the thyroid gland is determined as a function of time by measuring the radiation intensity at the neck area. How much or how little ^{131}I that is still in the thyroid is a measure of how well that gland is functioning.

A second medical application is indicated in Figure 14.26. A solution containing radioactive sodium is injected into a vein in the leg, and the time at which the radioisotope arrives at another part of the body is detected with a radiation counter. The elapsed time is a good indication of the presence or absence of constrictions in the circulatory system.

Tracers are also useful in agricultural research. Suppose the best method of fertilizing a plant is to be determined. A certain element in a fertilizer, such as

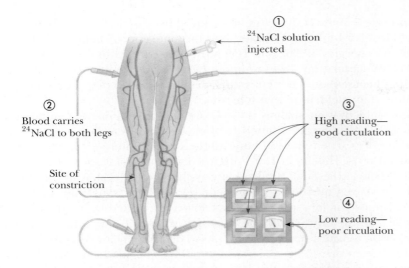

① ^{24}NaCl solution
injected

② Blood carries
^{24}NaCl to both legs

③ High reading—
good circulation

Site of
constriction

④ Low reading—
poor circulation

Figure 14.26   A tracer technique for determining the condition of the human circulatory system.

nitrogen, can be *tagged* (identified) with one of its radioactive isotopes. The fertilizer is then sprayed on one group of plants, sprinkled on the ground for a second group, and raked into the soil for a third. A Geiger counter is then used to track the nitrogen through the three groups.

Tracing techniques are as wide ranging as human ingenuity can devise. Present applications range from checking how teeth absorb fluoride to monitoring how cleansers contaminate food-processing equipment to studying deterioration inside an automobile engine. In the latter case, a radioactive material is used in the manufacture of the piston rings, and the oil is checked for radioactivity to determine the amount of wear on the rings.

Neutron Activation Analysis

For centuries, a standard method of identifying the elements in a sample of material has been chemical analysis, which involves determining how the material reacts with various chemicals. A second method is spectral analysis, which uses the fact that, when excited, each element emits its own characteristic set of electromagnetic wavelengths. These methods are now supplemented by a third technique, **neutron activation analysis.** Both chemical and spectral methods have the disadvantage that a fairly large sample of the material must be destroyed for the analysis. In addition, extremely small quantities of an element may go undetected by either method. Neutron activation analysis has an advantage over the other two methods in both respects.

When a material is irradiated with neutrons, nuclei in the material absorb the neutrons and are changed to different isotopes, most of which are radioactive. For example, ^{65}Cu absorbs a neutron to become ^{66}Cu, which undergoes beta decay:

$$\frac{1}{0}n + \frac{65}{29}Cu \longrightarrow \frac{66}{29}Cu \longrightarrow \frac{66}{30}Zn + e^- + \bar{\nu}$$

The presence of the copper can be deduced because it is known that ^{66}Cu has a half-life of 5.1 min and decays with the emission of beta particles having maximum energies of 2.63 and 1.59 MeV. Also emitted in the decay of ^{66}Cu is a 1.04-MeV gamma ray. By examining the radiation emitted by a substance after it has been exposed to neutron irradiation, one can detect extremely small amounts of an element in that substance.

Neutron activation analysis is used routinely in a number of industries, for example, in commercial aviation for the checking of airline luggage for hidden explosives. The following nonroutine use is of interest. Napoleon died on the island of St. Helena in 1821, supposedly of natural causes. Over the years, suspicion has existed that his death was not all that natural. After his death, his head was shaved and locks of his hair were sold as souvenirs. In 1961, the amount of arsenic in a sample of this hair was measured by neutron activation analysis, and an unusually large quantity of arsenic was found. (Activation analysis is so sensitive that very small pieces of a single hair could be analyzed.) Results showed that the arsenic was fed to him irregularly. In fact, the arsenic concentration patterns in the hair corresponded to the fluctuations in the severity of Napoleon's illness as determined from historical records.

Art historians use neutron activation analysis to detect forgeries. The pigments used in paints have changed throughout history, and old and new pigments react differently to neutron activation. The method can even reveal hidden works of art behind existing paintings because an older, hidden layer of paint reacts differently than the surface layer to neutron activation.

Radiation Therapy

Radiation causes the most damage to rapidly dividing cells. Therefore, it is useful in cancer treatment because tumor cells divide extremely rapidly.

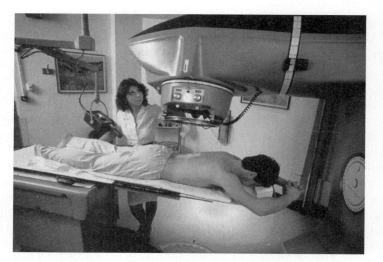

Figure 14.27   This large machine is being set to deliver a dose of radiation from ^{60}Co in an effort to destroy a cancerous tumor. Cancer cells are especially susceptible to this type of therapy because they tend to divide more often than cells of healthy tissue nearby. (*Martin Dohrn/Science Photo Library/Photo Researchers, Inc.*)

Several mechanisms can be used to deliver radiation to a tumor. In some cases, a narrow beam of x-rays or radiation from a source such as ^{60}Co is used, as shown in Figure 14.27. In other situations, thin radioactive needles called *seeds* are implanted in the cancerous tissue.

Food Preservation

Radiation is finding increasing use as a means of preserving food because exposure to high levels of radiation can destroy or incapacitate bacteria and mold spores. Techniques include exposing foods to gamma rays, high-energy electron beams, or x-rays. Food preserved this way can be placed in a sealed container (to keep out new spoiling agents) and stored for long periods of time. There is no evidence of adverse effect on the taste or nutritional value of food from irradiation. The safety of irradiated foods has been endorsed by the World Health Organization (WHO), the Centers for Disease Control and Prevention (CDC), the U.S. Department of Agriculture (USDA), and the Food and Drug Administration (FDA).

SUMMARY

A **nuclear reaction** can occur when a target nucleus X at rest is bombarded by a particle a, resulting in a nucleus Y and a particle b:

$$a + X \longrightarrow Y + b \qquad \text{or} \qquad X(a, b)Y \qquad (14.1)$$

The energy released in such a reaction, called the **reaction energy,** Q, is

$$Q = (M_X + M_a - M_Y - M_b)c^2 \qquad (14.2)$$

If N_0 beam particles are incident on a target of thickness x, the number N of beam particles that emerge from the target is

$$N = N_0 e^{-n\sigma x} \qquad (14.5)$$

where n is the density of nuclei in the target and σ is the **cross section.** Nuclear cross sections have dimensions of area and are usually measured in **barns** (b), where $1\ \text{b} = 10^{-28}\ \text{m}^2$.

The probability that neutrons will be captured as they move through matter generally increases with decreasing neutron energy. A **thermal neutron** is a slow-moving neutron that has a high probability of being captured by a nucleus according to the following **neutron-capture process:**

$$^{1}_{0}n + ^{A}_{Z}X \longrightarrow ^{A+1}_{Z}X + \gamma \qquad (14.7)$$

Energetic neutrons are slowed down readily in materials called **moderators.** These materials have small cross sections for neutron capture, and so neutrons lose their energy mainly through elastic collisions.

Nuclear fission occurs when a very heavy nucleus, such as ^{235}U, splits into two main fragments. Thermal neutrons can create fission in ^{235}U:

$$^{1}_{0}n + ^{235}_{92}U \longrightarrow ^{236}_{92}U^* \longrightarrow X + Y + \text{neutrons} \qquad (14.8)$$

where X and Y are the fission fragments and ^{236}U* is a nucleus in an excited state. On the average, 2.5 neutrons are released per fission event. The fragments and neutrons have a great deal of kinetic energy after the fission event.

The fragments then undergo a series of beta and gamma decays to various stable isotopes. The energy released per fission event is about 200 MeV.

The **reproduction constant** K is the average number of neutrons released from each fission event that cause another event. In a power reactor, it is necessary to maintain K slightly greater than 1. The value of K is affected by such factors as reactor geometry, the mean neutron energy, and the probability of neutron capture. Proper design of the reactor geometry is necessary to minimize neutron leakage from the reactor core. Neutron energies are regulated with a moderator material to slow down energetic neutrons and therefore increase the probability of neutron capture by other ^{235}U nuclei. The power level of the reactor is adjusted with control rods made of a material that absorbs neutrons very efficiently. The value of K can be adjusted by inserting the rods into the reactor core at varied depths.

Nuclear fusion is a process in which two light nuclei fuse to form a heavier nucleus. A great deal of energy is released in such a process. The major obstacle to obtaining useful energy from fusion is the large Coulomb repulsive force between the charged nuclei at close separations. Sufficient energy must be supplied to the particles to overcome this Coulomb barrier. The temperature required to produce fusion is of the order of 10^8 K, and at this temperature all matter occurs as plasma.

In a fusion reactor, the plasma temperature must reach at least the **critical ignition temperature,** the temperature at which the power generated by the fusion reactions exceeds the power lost in the system. The most promising fusion reaction is the D–T reaction, which has a critical ignition temperature of approximately 4.5×10^7 K. Two important parameters involved in fusion reactor design are **ion density** n and **confinement time** τ. The confinement time is the time period for which the interacting particles must be maintained at a temperature equal to or greater than the critical ignition temperature. **Lawson's criterion** states that for the D–T reaction, $n\tau \geq 10^{14}$ s/cm^3.

When energetic particles interact with a medium, they lose their energy by several processes. Heavy particles, such as alpha particles, lose most of their energy by excitation and ionization of atoms in the medium. The particles have a finite range in the medium, which depends on the energy, mass, and charge of the particle as well as the properties of the medium. Energetic electrons moving through a medium also lose their energy by excitation and ionization. However, they scatter more readily than heavy particles and therefore undergo fewer collisions before coming to rest.

Photons, or gamma rays, which have no charge, are not as effective as charged particles in producing ionization or excitation. However, they can be absorbed in a medium by several processes: (1) the photoelectric effect, predominant at low photon energies; (2) Compton scattering, predominant at intermediate photon energies; and (3) pair production, predominant at high photon energies.

If a beam of photons with intensity I_0 is incident on a medium, the intensity of the beam decreases exponentially with increasing depth of penetration into the medium, according to the relation

$$I(x) = I_0 e^{-\mu x} \tag{14.13}$$

where x is the distance traveled into the medium, $I(x)$ is the intensity of the beam after it travels a distance x in the medium, and μ is the **linear absorption coefficient** of the medium.

Devices used to detect radiation are the Geiger counter, the semiconductor diode detector, the scintillation counter, the photographic emulsion, the bubble chamber, the spark chamber, and the wire or drift chamber. Both the spark chamber and the wire chamber can easily make three-dimensional track measurements of charged particles, and both are interfaced to computers for rapid data collection and trajectory reconstruction.

SUGGESTIONS FOR FURTHER READING

1. A popular history and overview of fusion are given in R. Herman, *Fusion: The Search for Endless Energy*, Cambridge, England, Cambridge University Press, 1990.
2. The history of the German and U.S. A-bomb projects (including several interesting photographs) is given in several articles in *Physics Today*, 48:8, August 1995, Part 1.
3. The ITER is discussed in R. W. Conn, et al., "The International Thermonuclear Experimental Reactor," *Sci. Amer.*, 226:103, April 1992.
4. For an extensive bibliography and selected reprints on medical applications of nuclear physics, see *Medical Physics*, edited by Russel K. Hobbie, College Park, Md., American Association of Physics Teachers, 1986.

QUESTIONS

1. Explain the functions of a moderator and a control rod in a fission reactor.
2. Why is water a better shield against neutrons than lead or steel?
3. Discuss the advantages and disadvantages of fission reactors from the standpoints of safety, pollution, and resources. Make a comparison with power generated by the burning of fossil fuels.
4. Why would a fusion reactor produce less radioactive waste than a fission reactor?
5. Lawson's criterion states that the product of ion density and confinement time must exceed a certain number before a break-even fusion reaction can occur. Why should these two parameters determine the outcome?
6. Why is the temperature required for the D–T fusion less than that for the D–D fusion? Estimate the relative importance of Coulomb repulsion and nuclear attraction in each case.
7. What factors make a fusion reaction difficult to achieve?
8. Discuss the similarities and differences between fusion and fission.

9. Discuss the advantages and disadvantages of fusion power from the standpoints of safety, pollution, and resources.
10. Discuss three major problems associated with the development of a controlled-fusion reactor.
11. Describe two techniques that are being pursued in an effort to obtain power from nuclear fusion.
12. If two radioactive samples have the same activity measured in curies, will they necessarily create the same damage to a medium? Explain.
13. One method of treating cancer of the thyroid is to insert a small radioactive source directly into the tumor. The radiation emitted by the source can destroy cancerous cells. Very often, the radioactive isotope $^{131}_{53}$I is injected into the bloodstream in this treatment. Why do you suppose iodine is used?
14. Why should a radiologist be extremely cautious about x-ray doses when treating pregnant women?
15. The design of a PM tube might suggest that any number of dynodes may be used to amplify a weak signal. What factors do you suppose would limit the amplification in this device?

PROBLEMS

14.1 Nuclear Reactions

1. (a) Calculate the Q value corresponding to the reaction ^{18}O(p, n)^{18}F, and compare your result with the measured value in Table 14.1. (b) Calculate the threshold energy of the incident proton.
2. Supernova explosions are incredibly powerful nuclear reactions that tear apart the cores of relatively massive stars (greater than four times the Sun's mass). These blasts are produced by carbon fusion, which requires a temperature of about 6×10^8 K to overcome the strong Coulomb repulsion between carbon nuclei. (a) Estimate the repulsive energy barrier to fusion, using the required ignition temperature for carbon fusion. (In other words, what is the kinetic energy for a

carbon nucleus at 6×10^8 K?) (b) Calculate the energy (in MeV) released in each of these "carbon-burning" reactions:

$$^{12}C + {}^{12}C \longrightarrow {}^{20}Ne + {}^{4}He$$

$$^{12}C + {}^{12}C \longrightarrow {}^{24}Mg + \gamma$$

(c) Calculate the energy (in kWh) given off when 2 kg of carbon completely fuses according to the first reaction.

3. The following reaction, first observed in 1930, led to the discovery of the neutron by Chadwick:

$$^{9}_{4}Be(\alpha, n)^{12}_{6}C$$

Calculate the Q value of this reaction.

4. The following is the first known reaction (achieved in 1934) in which the product nucleus is radioactive:

$$^{27}_{13}Al(\alpha, n)^{30}_{15}P$$

Calculate the Q value of this reaction.

5. There are a few nuclear reactions in which the emerging particle and the product nucleus are identical. One example of this is the reaction

$$^{7}_{3}Li(p, \alpha)^{4}_{2}He$$

Calculate the Q value of this reaction.

6. Show that the following inverse reactions have the same *absolute value* of Q:

$$^{10}_{5}B(\alpha, p)^{13}_{6}C \quad \text{and} \quad ^{13}_{6}C(p, \alpha)^{10}_{5}B$$

7. (a) The first nuclear transmutation was achieved in 1919 by Rutherford, who bombarded nitrogen atoms with alpha particles emitted by the isotope ^{214}Bi. The reaction is

$$^{4}_{2}He + {}^{14}_{7}N \longrightarrow {}^{17}_{8}O + {}^{1}_{1}H$$

What is the Q value of the reaction? What is the threshold energy? (b) Cockroft and Walton performed the first nuclear reaction utilizing particle accelerators. In that case, accelerated protons were used to bombard lithium nuclei, producing the reaction

$$^{1}_{1}H + {}^{7}_{3}Li \longrightarrow {}^{4}_{2}He + {}^{4}_{2}He$$

Since the masses of the particles involved in the reaction were well known, these results were used to obtain an early proof of the Einstein mass–energy relation. Calculate the Q value of the reaction.

8. (a) One method of producing neutrons for experimental use is based on the bombardment of light nuclei by alpha particles. In one arrangement, alpha particles emitted by plutonium are incident on beryllium nuclei, and this results in the production of neutrons:

$$^{4}_{2}He + {}^{9}_{4}Be \longrightarrow {}^{12}_{6}C + {}^{1}_{0}n$$

What is the Q value for this reaction? (b) Neutrons are also often produced by small-particle accelerators. In

one design, deuterons (^2H) that have been accelerated in a Van de Graaff generator are used to bombard other deuterium nuclei, resulting in the reaction

$$^{2}_{1}H + {}^{2}_{1}H \longrightarrow {}^{3}_{2}He + {}^{1}_{0}n$$

Is this reaction exothermic or endothermic? Calculate its Q value.

9. When the nuclear reaction represented by Equation 14.2 is endothermic, the disintegration energy Q is negative. In order for the reaction to proceed, the incoming particle must have a minimum kinetic energy, called the *threshold energy*, K_{th}. (a) Show that for nonrelativistic particles

$$K_{th} = -Q\left(1 + \frac{M_a}{M_X}\right)$$

by using the fact that $K_{th} = -Q$ in the CM frame and by transforming this result for the CM frame back to the laboratory frame. Note that in the CM frame, a and X have equal and opposite momenta, $p = M_a v = M_X V$. In the lab frame, a has momentum $p_{lab} = M_a(v + V)$ and X is at rest. (b) Calculate the threshold energy of the incident alpha particle in the reaction

$$^{4}_{2}He + {}^{14}_{7}N \longrightarrow {}^{17}_{8}O + {}^{1}_{1}H$$

14.2   Reaction Cross Section

10. Consider a slab consisting of two layers of material with thicknesses x_1 and x_2 and target densities n_1 and n_2. If N_0 is the number of particles incident on the first layer of the slab in some time interval, determine the number N that emerge from the second layer in that interval. Assume that the cross section σ is the same for each material. What would you guess is the relationship for three or more layers?

11. The density of the liquid hydrogen target in a bubble chamber is 70 kg/m^3. If 20% of a beam of slow neutrons incident on the bubble chamber has reacted with the hydrogen by the time the beam has traveled 2 m through the hydrogen, what is the cross section, in barns, for the reaction of these slow neutrons with hydrogen atoms?

12. The density of lead is 11.35 g/cm^3, and its atomic weight is 207.2. Assume that 1.000 cm of lead reduces a beam of 1-MeV gamma rays to 28.65% of its initial intensity. (a) How much lead is required to reduce the beam to 10^{-4} of its initial intensity? (b) What is the effective cross section of a lead atom for a 1-MeV photon?

13. Neutrons are captured by a cadmium foil. Using the data in Figure 14.3, find (a) the ratio of 10-eV neutrons captured to 1-eV neutrons captured, (b) the ratio of 1-eV neutrons captured to 0.1-eV neutrons captured, and (c) the ratio of 0.1-eV neutrons captured to 0.01-eV neutrons captured. (d) In what range of energies can cadmium be used as an energy selector?

14. The atomic weight of cadmium is 112.41, and its density is 8.65 g/cm^3. Using Figure 14.3, estimate the attenuation distance of a thermal neutron beam in cadmium. (The attenuation distance is the distance traveled after which the intensity of the beam is reduced to $1/e$ of its initial value, where e is the base of the natural logarithms.)

15. A beam of 100-MeV protons from a Van de Graaff generator is incident on a gold foil 5.1×10^{-5} m thick. The beam current is 0.1 μA, and the beam has a cross-sectional area of 1 cm^2. If the scattering cross section is 500 b, calculate (a) the ratio N/N_0, (b) the number per second that pass through the foil, and (c) the number lost from the beam each second by scattering. (Gold has an atomic weight of 197 and a density of 19.3 g/cm^3.)

16. A tiny sphere is made of a material that absorbs all the photons incident on it. Many such spheres are embedded randomly in a transparent medium. (a) If the radius of one of these spheres is b, what is its cross section σ in terms of b for the absorption of photons? (b) If the radius of each sphere is $b = 2 \times 10^{-3}$ m, what is the cross section? (c) If 3×10^4 of these spheres are uniformly embedded in a cylinder of a transparent medium of height 2 m and cross-sectional area 0.5 m^2, and a light of beam intensity 0.75 W/m^2 is incident normally on one end of the cylinder, what is the intensity of the beam of light that emerges from the other end?

14.3   Interactions Involving Neutrons

17. The density of cadmium is 8.65 g/cm^3, and its cross section for thermal neutron absorption is 2450 b. Find the ratio of neutron absorption to neutron decay in cadmium for thermal neutrons at a temperature of 27°C. (*Hint:* The number of neutrons absorbed per second is

$$\frac{dN}{dt} = \frac{dN}{dx}\frac{dx}{dt} = \frac{dN}{dx}v$$

where v is the neutron speed and dN/dx is the number of neutrons lost per meter of travel. The half-life of the neutron is 636 s.)

18. In natural silver, the abundance of ^{107}Ag is 51.35%, and that of ^{109}Ag is 48.65%. The neutron-absorption cross section of ^{107}Ag is 31 b, and that of ^{109}Ag is 87 b. The activation products ^{108}Ag and ^{110}Ag formed by neutron absorption decay by beta emission with half-lives of 144 s and 24.5 s, respectively. A silver sample is removed from a fission reactor, and after some delay it is found that the ratio of ^{108}Ag to ^{110}Ag is 20:1. How long was the delay? (*Hint:* Assume that the sample has been in the reactor for a time sufficiently long that the decay rate equals the production rate for both ^{108}Ag and ^{110}Ag.)

19. A 1-MeV neutron is emitted in a fission reactor. If it loses half of its kinetic energy in each collision with a moderator atom, how many collisions must it undergo in order to achieve thermal energy (0.039 eV)?

20. A particle cannot generally be localized to distances much smaller than its de Broglie wavelength. This means that a slow neutron appears to be larger to a target particle than does a fast neutron, in the sense that the slow neutron will probably be found over a large volume of space. For a thermal neutron at room temperature (300 K), find (a) the linear momentum and (b) the de Broglie wavelength. Compare this effective neutron size with both nuclear and atomic dimensions.

14.4   Nuclear Fission

21. Find the energy released in the fission reaction

$$_{0}^{1}n + _{92}^{235}U \longrightarrow _{56}^{141}Ba + _{36}^{92}Kr + 3(_{0}^{1}n)$$

The required masses are

$$M(_{0}^{1}n) = 1.008665 \text{ u}$$
$$M(_{92}^{235}U) = 235.043915 \text{ u}$$
$$M(_{56}^{141}Ba) = 140.9139 \text{ u}$$
$$M(_{36}^{92}Kr) = 91.8973 \text{ u}$$

14.5   Nuclear Reactors

22. (a) How many grams of ^{235}U must undergo fission to operate a 1000-MW power plant for one day? (b) If the density of ^{235}U is 18.7 g/cm^3, how large a sphere of ^{235}U could you make from this much uranium?

23. In order to minimize neutron leakage from a reactor, the surface-area-to-volume ratio should be a minimum for a given shape. For a given volume V, calculate this ratio for (a) a sphere, (b) a cube, and (c) a rectangular parallelepiped of dimensions $a \times a \times 2a$. (d) Which of these shapes would have the minimum leakage? (e) Which would have the maximum leakage?

24. It has been estimated that there are 10^9 tons of natural uranium (mainly ^{238}U) at concentrations exceeding 100 parts per million, of which 0.7% is ^{235}U. If all the world's power needs (7×10^{12} J/s) were to be supplied by ^{235}U fission, how long would this supply last? (This estimate of uranium supply was taken from K. S. Deffeyes and I. D. MacGregor, *Sci. Amer.*, January 1980, p. 66.)

25. An electrical power plant operates on the basis of thermal energy generated in a pressurized-water reactor. The electrical power output of the plant is 1 GW, and its efficiency is 30%. (a) Find the total power generated by the reactor. (b) How much power is discharged to the environment as waste heat? (c) Calculate the rate of fission events in the reactor core. (d) Calculate the mass of ^{235}U used up in one year. (e) Using the results from (a), determine the rate at which fuel is converted to energy (in kg/s) in the reactor core, and compare your answer with the result from (d).

26. (a) Estimate the volume of space required to store the radioactive wastes that would be produced in one year if all the annual U.S. electricity production (which is about 2.2×10^{12} kWh/yr) came from uranium enriched to 3% ^{235}U. (Assume that the conversion efficiency is 30% and that the waste is in the form of a liquid with a density of 1 g/cm^3.) (b) If the waste could be formed into a cube, what would be the length of the cube's sides?

14.6   Nuclear Fusion

27. Consider the deuterium–tritium fusion reaction with the tritium nucleus at rest:

$$^2_1\text{H} + {}^3_1\text{H} \longrightarrow {}^4_2\text{He} + {}^1_0\text{n}$$

(a) From Equation 13.1, estimate the required distance of closest approach. (b) What is the Coulomb potential energy (in electron volts) at this distance? (c) If the deuteron has just enough energy to reach the distance of closest approach, what is the final velocity of the combined deuterium and tritium nuclei in terms of the initial deuteron velocity, v_0? (d) Use energy methods to estimate the minimum initial deuteron energy required to achieve fusion. (e) Why does the fusion reaction occur at much lower deuteron energies than that calculated in (d)?

28. The half-life of tritium is 12 yr. If the TFTR fusion reactor contains 50 m^3 of tritium at a density equal to 1.5×10^{14} particles/cm^3, how many curies of tritium are in the plasma? Compare this with a fission reactor supply of 4×10^{10} Ci.

29. (a) Calculate the amounts of energy carried off by the $^4_2$He nucleus and the neutron in the D–T fusion reaction. (Assume that the momentum is initially zero.) (b) Does this explain why only 20% of the energy produced can be used for critical ignition?

30. The Sun radiates energy at the rate of 4×10^{23} kW. Assuming that the reaction

$$4(^1_1\text{H}) \longrightarrow {}^4_2\text{He} + 2e^+ + 2\nu + \gamma$$

accounts for all the energy released, calculate (a) the number of protons fused per second and (b) the mass transformed into energy per second.

31. Suppose the target in a laser fusion reactor is a sphere of solid hydrogen with a diameter of 10^{-4} m and a density of 0.2 g/cm^3. Also assume that half of the nuclei are ^{2}H and half are ^{3}H. (a) If 1% of a 200-kJ laser pulse goes into heating this sphere, what temperature will the sphere reach? (b) If all of the hydrogen "burns" according to the D–T reaction, how many joules of energy will be released?

32. In a tokamak fusion reactor, suppose a 500-eV deuteron moves at an angle of 30° to the toroidal magnetic field. Assume that $B_t = 1$ T and $B_p = 0$ (see Fig. 14.14). (a) Calculate the components of velocity parallel and perpendicular to $\mathbf{B}_t$. (b) What is the radius of the spiral motion for the deuteron? (c) How far does

the deuteron travel *along* the magnetic field before it completes one revolution *around* the magnetic field?

33. The carbon cycle, first proposed by Bethe in 1939, is another cycle by which energy is released in stars and hydrogen is converted to helium. The carbon cycle requires higher temperatures than the proton–proton cycle. The series of reactions is

$$^{12}\text{C} + {}^1\text{H} \longrightarrow {}^{13}\text{N} + \gamma$$
$$^{13}\text{N} \longrightarrow {}^{13}\text{C} + e^+ + \nu$$
$$^{13}\text{C} + {}^1\text{H} \longrightarrow {}^{14}\text{N} + \gamma$$
$$^{14}\text{N} + {}^1\text{H} \longrightarrow {}^{15}\text{O} + \gamma$$
$$^{15}\text{O} \longrightarrow {}^{15}\text{N} + e^+ + \nu$$
$$^{15}\text{N} + {}^1\text{H} \longrightarrow {}^{12}\text{C} + {}^4\text{He}$$

(a) If the proton–proton cycle requires a temperature of 1.5×10^7 K, estimate the temperature required for the first step in the carbon cycle. (b) Calculate the Q value for each step in the carbon cycle and the overall energy released. (c) Do you think the energy carried off by the neutrinos is deposited in the star? Explain.

34. (a) Calculate the energy (in kWh) released if 1 kg of ^{239}Pu undergoes complete fission and the energy released per fission event is 200 MeV. (b) Calculate the energy (in MeV) released in the D–T fusion:

$$^2_1\text{H} + {}^3_1\text{H} \longrightarrow {}^4_2\text{He} + {}^1_0\text{n}$$

(c) Calculate the energy (in kWh) released if 1 kg of deuterium undergoes fusion. (d) Calculate the energy (in kWh) released by the combustion of 1 kg of coal if each $\text{C} + \text{O}_2 \rightarrow \text{CO}_2$ reaction yields 4.2 eV. (e) List the advantages and disadvantages of each of these methods of energy generation.

35. The ^{6}Li isotope is only about 7.5% naturally abundant. The remaining 92.5% of lithium is ^{7}Li. It is estimated that about 2×10^{13} g of lithium is available. If ^{6}Li is used as a tritium source in fusion reactors, with an energy release of 22 MeV per ^{6}Li nucleus, estimate the total energy available in D–T fusion. How does this number compare with the world's fossil fuel supply, which is estimated to be about 2.5×10^{23} J?

36. Find the Q value for each of the reactions in the proton–proton cycle (Eq. 14.10), and show that the overall Q value for the cycle is 25.7 MeV.

37. (a) Estimate the net power output of a fusion reactor that burns ten 50:50 3-mg D–T pellets every second. Assume that 30% of the fuel ignites and that a 5×10^{14} W laser pulse lasting 10 ns is needed to initiate burning. (b) What is the equivalent in liters of oil for a day's operation, if 2 L of oil gives off 100 MJ when burned?

38. In a laser fusion reaction, a pellet containing a 50:50 mixture of D and T experiences a density increase of 1000 times when struck by a laser pulse. (a) Find the particle density in the compressed state if the normal

density is 0.20 g/cm^3. (b) How long must the D and T nuclei be confined at this density if the reaction is to achieve a break-even condition?

39. Two nuclei with atomic numbers Z_1 and Z_2 approach each other with a total energy E. (a) If the minimum distance of approach for fusion to occur is $r = 10^{-14}$ m, find E in terms of Z_1 and Z_2. (b) Calculate the minimum energy for fusion for the D–D and D–T reactions (the first and third reactions in Eq. 14.11).

40. To understand why containment of a plasma is necessary, consider the rate at which a plasma would be lost if it were not contained. (a) Estimate the rms speed of deuterons in a plasma at 10^8 K. (b) Estimate the time interval for which such a plasma would remain in a 10-cm cube if no steps were taken to contain it.

41. Of all the hydrogen nuclei in the ocean, 0.0156% are deuterium. The oceans have a volume of 317 million cubic miles. (a) If all the deuterium in the oceans were fused to ^4_2He, how many joules of energy would be released? (b) Current world energy consumption is about 7×10^{12} W. If consumption were 100 times greater, how many years would the energy calculated in (a) last?

42. It has been pointed out that fusion reactors are safe from explosion because there is never enough energy in the plasma to do much damage. (a) Using a particle density of 10^{15} cm^{-3} and a kinetic temperature of 10 keV, calculate the amount of energy stored in the plasma of the TFTR reactor. (b) How many kilograms of water could be boiled by this much energy? (The plasma volume of the TFTR reactor is about 50 m^3.)

43. In order to confine a stable plasma, the magnetic energy density in the magnetic field must exceed the pressure $2nkT$ of the plasma by a factor of at least 10. In the following, assume a confinement time $\tau = 1$ s. (a) Using Lawson's criterion, determine the required ion density. (b) From the ignition temperature criterion for the D–T reaction, determine the required plasma pressure. (c) Determine the magnitude of the magnetic field that is required to contain the plasma.

14.7 Interaction of Particles with Matter

44. The densities and atomic weights of the materials in Table 14.3 are as follows:

Substance	Density (g/cm^3)	Atomic Weight
H$_2$O	1	18
Al	2.7	27
Fe	7.8	55.8
Pb	11.35	207.2

Compute the number of electrons per cubic centimeter for each material, and plot the gamma-ray linear absorption coefficient versus electron density. Draw three graphs corresponding to gamma-ray energies of 0.1, 1.0, and 10 MeV. What do you conclude from your graphs?

45. Three equally thick layers of aluminum, copper, and lead are used to reduce the intensity of an x-ray beam to one-third of its original intensity. The wavelength of the beam is 50 pm. (a) Find the thickness of a layer of material. (b) By what fraction would the initial beam intensity be reduced by the lead alone?

46. X-rays of wavelength 25 pm and gamma rays of energy 0.1 MeV have approximately the same absorption coefficient in lead. How do their energies compare?

47. What is the half-value thickness (Example 14.7) of water to x-rays of wavelength 20 pm? Since the human body is more than 90% water, what does your answer indicate about the use of x-rays as a diagnostic technique?

48. In a large-scale nuclear attack, typical radiation intensity from radioactive fallout might be 2000 rad in most places. In the following calculations, assume that one-third of the radiation is 10-MeV gamma radiation and that the linear absorption coefficient is the same for aluminum and concrete. (a) What thickness (in meters) of concrete would be needed to reduce the radiation intensity to 1 rad? (b) If a particular shelter were located at a "hot spot" receiving 100,000 rad, what thickness of concrete would be needed to reduce the radiation intensity to 1 rad?

14.8 Radiation Damage in Matter

49. Assume that an x-ray technician takes an average of eight x-rays per day and receives a dose of 5 rem/year as a result. (a) Estimate the dose in rem per x-ray taken. (b) How does this result compare with low-level background radiation?

50. In terms of biological damage, how many rad of heavy ions are equivalent to 100 rad of x-rays?

51. Two workers using an industrial x-ray machine accidentally insert their hands in the x-ray beam for the same length of time. The first worker inserts one hand in the beam, and the second worker inserts both hands. Which worker receives the larger dose in rad?

52. Calculate the radiation dose, in rad, supplied to 1 kg of water such that the energy deposited equals (a) the rest energy of the water and (b) its thermal energy. (Assume that each molecule has a thermal energy of $k_B T$.)

53. A person whose mass is 75 kg is exposed to a dose of 25 rad. How many joules of energy are deposited in the person's body?

14.9 Radiation Detectors

54. In a Geiger tube, the voltage between the electrodes is typically 1 kV and the current pulse generated by the detection of a β particle fully charges a 5-pF capacitor. (a) What is the energy amplification of this device for a 0.5-MeV beta ray? (b) How many electrons are avalanched by the initial electron?

55. In a PM tube, assume that there are seven dynodes with potentials of 100 V, 200 V, 300 V, . . . , 700 V. The average energy required to free an electron from the

dynode surface is 10 eV. For each incident electron, how many electrons are freed (a) at the first dynode and (b) at the last dynode?

ADDITIONAL PROBLEMS

56. A fission reaction that has been considered as a source of energy is the absorption of a proton by a boron-11 nucleus to produce three alpha particles,

$$^1_1\text{H} + ^{11}_5\text{B} \rightarrow 3\,^4_2\text{He}$$

This is an attractive possibility because large amounts of boron are present in the Earth's crust. A disadvantage is that the protons and boron nuclei must have large kinetic energies in order for the reaction to take place. This is in contrast to the initiation of uranium fission by slow neutrons. Chemical explosives might provide the high kinetic energies required. (a) What energy is released in each fission event here? (b) What is the reason the reactant particles must have high kinetic energies?

57. Consider a nucleus at rest, which then spontaneously splits into two fragments, of masses m_1 and m_2. Show that the fraction of the total kinetic energy that is carried by fragment m_1 is

$$\frac{K_1}{K_{\text{tot}}} = \frac{m_2}{m_1 + m_2}$$

and the fraction carried by m_2 is

$$\frac{K_2}{K_{\text{tot}}} = \frac{m_1}{m_1 + m_2}$$

assuming relativistic corrections can be ignored. (*Note:* If the parent nucleus was moving before the decay, then the fission products still divide the kinetic energy as shown, if all velocities are measured in the center-of-mass frame of reference, in which the total momentum of the system is zero.)

58. A stationary $^{236}_{92}\text{U}$ nucleus fissions spontaneously into two primary fragments, $^{87}_{35}\text{Br}$ and $^{149}_{57}\text{La}$. (a) Calculate the disintegration energy. The required atomic masses are 86.920 710 u for $^{87}_{35}\text{Br}$, 148.934 370 u for $^{149}_{57}\text{La}$, and 236.045 562 u for $^{236}_{92}\text{U}$. (b) How is the disintegration energy split between the two primary fragments? You may use the result of Problem 57. (c) Calculate the initial speed of each fragment.

59. The first nuclear bomb was a fissioning mass of plutonium-239, exploded before dawn on July 16, 1945, at Alamogordo, New Mexico. Enrico Fermi was 14 km away, lying on the ground facing away from the bomb. After the whole sky had flashed with unbelievable brightness, Fermi stood up and began dropping bits of paper to the ground. They first fell at his feet in the calm and silent air. As the shock wave passed, about 40 s after the explosion, the paper then in flight jumped about 5 cm away from ground zero. (a) Assume that the shock wave

in air propagated equally in all directions without absorption. Find the change in volume of a hemisphere of radius 14 km as it expands by 5 cm. (b) Find the work $P\Delta V$ done by the air in this sphere on the next layer of air farther from the center. (c) Assume the shock wave carried on the order of one-tenth of the energy of the explosion. Make an order-of-magnitude estimate of the bomb yield. (d) One ton of exploding trinitrotoluene (TNT) releases an energy of 4.2 GJ. What was the order of magnitude of the energy of the first nuclear bomb in equivalent tons of TNT? Fermi's immediate knowledge of the bomb yield agreed with that determined days later by analysis of elaborate measurements.

60. At time $t = 0$ a sample of uranium is exposed to a neutron source that causes N_0 nuclei to undergo fission. The sample is in a supercritical state, with a reproduction constant $K > 1$. A chain reaction occurs that produces fission throughout the mass of uranium. The chain reaction can be thought of as a succession of *generations*. The N_0 fissions produced initially are the zeroth generation of fissions. From this generation, $N_0 K$ neutrons go off to produce fission of new uranium nuclei. The $N_0 K$ fissions that occur subsequently are the first generation of fissions, and from this generation, $N_0 K^2$ neutrons go in search of uranium nuclei in which to cause fission. The subsequent $N_0 K^2$ fissions are the second generation of fissions. This process can continue until all the uranium nuclei have fissioned. (a) Show that the cumulative total of fissions, N, that have occurred up to and including the nth generation after the zeroth generation is given by

$$N = N_0 \left(\frac{K^{n+1} - 1}{K - 1} \right)$$

(b) Consider a hypothetical uranium bomb made from 5.50 kg of isotopically pure ^{235}U. The chain reaction has a reproduction constant of 1.10 and starts with a zeroth generation of 1.00×10^{20} fissions. The average time between one fission generation and the next is 10.0 ns. How long after the zeroth generation does it take the uranium in this bomb to fission completely? (c) Assume that the bulk modulus of uranium is 150 GPa. Find the speed of sound in uranium. You may ignore the density difference between ^{235}U and natural uranium. (d) Find the time interval required for a compressional wave to cross the radius of a 5.50-kg sphere of uranium. This time indicates how quickly the motion of explosion begins. (e) Fission must occur in a time interval that is short compared to that in part (d), for otherwise most of the uranium will disperse in small chunks without having fissioned. Can the bomb considered in part (b) release the explosive energy of all of its uranium? If so, how much energy does it release, in equivalent tons of TNT? Assume that 1 ton of TNT releases 4.20 GJ and that each uranium fission releases 200 MeV of energy.

15

Elementary Particles

Chapter Outline

The word *atom* is from the Greek *atomos*, which means "indivisible." At one time, atoms were thought to be the indivisible constituents of matter; that is, they were regarded as elementary particles. From as far back as the ancient Greek philosopher Democritus to the relatively recent works of John Dalton and Dmitri Mendeleev, the idea that everything consists of elementary atoms has been quite successful in explaining many properties of matter. When discoveries in the early part of the 20th century revealed that the atom is composed of other constituents, another simplification occurred with Bohr's atomic model and the invention of quantum mechanics. The variety of physical and chemical properties of approximately 100 elements has been explained in terms of rules governing just three constituents: electrons, protons, and neutrons. With the exception of the free neutron, these particles are very stable.

Beginning in about 1945, many new particles were discovered in experiments involving high-energy collisions between known particles. Such particles are highly unstable and have very short half-lives, ranging between 10^{-6} and 10^{-23} s. So far, more than 400 of these unstable, temporary particles have been catalogued.

Since the early 1960s, many powerful particle accelerators have been constructed throughout the world, making it possible to observe collisions of energetic particles under controlled laboratory conditions so as to reveal the subatomic world in finer detail. Until the 1960s, physicists were puzzled by the large number and variety of subatomic particles being discovered. They wondered whether the particles were like the varied animals in a zoo, with no systematic relationship connecting them, or whether a pattern was emerging that would provide a better understanding of the elaborate structure of the subnuclear world. Since about 1970, physicists have tremendously advanced our knowledge of the structure of matter by recognizing that all particles except electrons, photons, and a few others are made of smaller particles called **quarks.** Thus protons and neutrons, for example, are not truly elementary but are systems of tightly bound quarks. The quark model has reduced the array of particles to a manageable number and has successfully predicted new quark combinations that have subsequently been observed in many experiments. A kind of wave mechanics for quarks **(quantum chromodynamics)** has also been developed. This theory, although mathematically difficult, has deepened our understanding of elementary particles and has helped to tame the particle "zoo."

This chapter examines the properties and classifications of the known elementary particles, the interactions that govern their behavior, and the methods of producing elementary particles and measuring their properties. We also discuss the current theory of elementary particles, the standard model, in which elementary particles are divided into two catagories: particles of spin $\frac{1}{2}$—quarks and leptons—and force-carrying, or "field" particles with integral spin like the photon and gluon.

15.1   THE FUNDAMENTAL FORCES IN NATURE

To understand the properties of elementary particles, we must be able to describe the forces between them. Particles in nature are subject to four fundamental forces; in order of decreasing strength, these are the strong force, the electromagnetic force, the weak force, and the gravitational force.

The **strong force** is responsible for binding quarks tightly together to form protons, neutrons, and other heavy particles. It is extremely short-range and is negligible for separations greater than approximately 10^{-15} m. The nuclear force that binds neutrons and protons in nuclei is currently believed to be a residual effect of the more basic strong force between quarks, much as the molecular force binding electrically neutral atoms together in molecules is a residual electrical interaction.

The **electromagnetic force,** which binds electrons and protons within atoms and molecules to form ordinary matter, is approximately two orders of magnitude weaker than the strong force. It is a long-range force that decreases in strength as the inverse square of the separation between interacting particles.

The **weak force** is a short-range force that accounts for the beta decay of nuclei and the decay of heavier quarks and leptons. Its strength is only about 10^{-6} times that of the strong force. (As we shall discuss later, scientists now believe that the weak and electromagnetic forces are two manifestations of a unified force called the *electroweak force*.)

Finally, the **gravitational force** is a long-range force that has a strength only about 10^{-43} times that of the strong force. Although this familiar interaction holds the planets, stars, and galaxies together, its effect on elementary particles is negligible.

Classically, the entity that is responsible for transmitting a force from one particle to another is the field. For example, a positive electric charge produces an electric field in space, which in turn exerts an attractive force on a nearby negative charge. Furthermore, the field can carry energy and momentum from one particle to the other. According to quantum field theories, the energy and momentum of all fields are quantized, and the quantum that carries a "chunk" of momentum and energy from one type of particle to another is called a **field particle.** In particle physics, interactions between particles are described in terms of the exchange of field particles, or quanta, *which are all bosons.* In the case of the familiar electromagnetic interaction, for instance, the field particles are photons. In the language of modern physics, one can say that the electromagnetic force is *mediated*, or carried, by photons and that photons are the quanta of the electromagnetic field. Likewise, the strong force is mediated by field particles called *gluons*, the weak force is mediated by particles called the $W^{\pm}$ and Z^0 *bosons*, and the gravitational force is carried by quanta of the gravitational field called *gravitons*. These interactions, the particles they act on, their ranges, their relative strengths, and the corresponding field particles, are summarized in Table 15.1. Note that the fourth column of Table 15.1 presents another way of classifying interactions—by means of the observed lifetime of a decaying particle. In fact, with only a few exceptions, the lifetimes of decaying particles are excellent indicators of what interaction has caused the decay, with shorter lifetimes being associated with stronger forces. As shown in Table 15.1, particles decaying via the strong force are usually the shortest-lived; next come

Table 15.1   Particle Interactions

Interaction (Force)	Particles Acted on by Force	Relative Strength[a]	Typical Lifetimes for Decays via a Given Interaction	Range of Force	Force-Carrying Particle Exchanged
Strong	Quarks, hadrons	1	$\leq 10^{-20}$ s	Short (≈ 1 fm)	Gluon
Electromagnetic	Charged particles	$\approx 10^{-2}$	$\approx 10^{-16}$ s	Long (∞)	Photon
Weak	Quarks, leptons	$\approx 10^{-6}$	$\geq 10^{-10}$ s	Short ($\approx 10^{-3}$ fm)	$W^{\pm}$, Z^0 bosons
Gravitational	All particles	$\approx 10^{-43}$	?	Long (∞)	Graviton[b]

[a]For two u quarks at 3×10^{-17} m.

[b]Not experimentally detected.

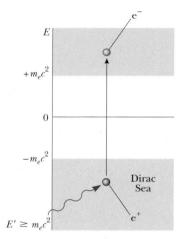

Figure 15.1 Dirac's model for the existence of antielectrons (positrons). The states lower in energy than $-m_ec^2$ are filled with electrons (the Dirac sea). One of these electrons can make a transition out of its state only if it is provided with energy equal to or larger than $2m_ec^2$. This leaves a vacancy in the Dirac sea, which can behave as a particle identical to the electron except for its positive charge.

those decaying via electromagnetic forces; and finally, particles decaying by the weak interaction have the longest observed lifetimes.

15.2 POSITRONS AND OTHER ANTIPARTICLES

In the 1920s the English theoretical physicist Paul Adrien Maurice Dirac (1902–1984) developed a version of quantum mechanics that incorporated special relativity. Dirac's theory automatically explained the origin of the electron's spin and its magnetic moment. However, it also presented a major difficulty. Dirac's relativistic wave equation required solutions corresponding to both positive and negative energies for free electrons.[1] We can easily see this from the expression for the total relativistic energy of an electron $E = \pm\sqrt{p^2c^2 + m_e^2c^4}$, which has both positive and negative roots. But if negative energy states existed, one would expect an electron in a state of positive energy to make a rapid transition to one of these lower energy states, emitting a photon in the process. Eventually all electrons in the universe would end up locked in negative energy states, contradicting the common observation of electrons with positive total energies. Dirac avoided this difficulty by postulating that all negative energy states are filled. Electrons that occupy the negative energy states are called collectively the "Dirac sea." Electrons in the Dirac sea are not directly observable, because the Pauli exclusion principle does not allow them to react to external forces—there are no available states to which an electron can make a transition in response to an external force. Therefore, an electron in such a state acts as an isolated system, unless enough energy ($\geq 2m_ec^2$) is supplied to excite the electron to a positive energy state. Such an excitation causes one of the negative energy states to become vacant, as in Figure 15.1, leaving a hole in the sea of filled states. *The hole can react to external forces and is observable.* The hole reacts in a way similar to that of the electron, except that it has a positive charge—it is the **antiparticle** to the electron. The electron's antiparticle, the **positron,** has a rest energy of 0.511 MeV and a positive charge of $+1.60 \times 10^{-19}$ C.

Carl Anderson (1905–1991) observed the positron experimentally in 1932, and in 1936 he was awarded a Nobel prize for his achievement. Anderson discovered the positron while examining tracks created in a cloud chamber by electron-like particles of positive charge. (These early experiments used cosmic rays—mostly energetic protons passing through interstellar space—to initiate high-energy reactions on the order of several GeV.) To discriminate between positive and negative charges, Anderson placed the cloud chamber in a magnetic field, causing moving charges to follow curved paths. He noted that some of the electron-like tracks deflected in a direction corresponding to a positively charged particle.

Since Anderson's discovery, positrons have been observed in a number of experiments. A common source of positrons is **pair production.** In this process, a gamma-ray photon with sufficiently high energy interacts with a nucleus, and an electron–positron pair is created from the photon. (The presence of the nucleus allows the principle of conservation of momentum to be satisfied.) Because the total rest energy of the electron–positron pair is

Pair production

[1]P. A. M. Dirac, *The Principles of Quantum Mechanics*, 3rd ed., New York, Oxford University Press, 1947, Chapter 11.

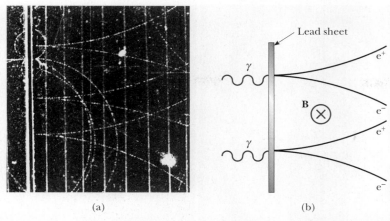

Figure 15.2   (a) Bubble-chamber tracks of electron–positron pairs produced by 300-MeV gamma rays striking a lead sheet. (*Courtesy Lawrence Berkeley Laboratory, University of California*) (b) A sketch of the pertinent pair-production events. Note that the positrons deflect upward while the electrons deflect downward in an applied magnetic field directed into the page.

$2m_e c^2 = 1.02$ MeV, the photon must have at least this much energy to create an electron–positron pair. Therefore, the electromagnetic energy of a photon is converted mainly to rest energy of the electron and positron in accordance with Einstein's relationship $E = mc^2$. If the gamma-ray photon has energy in excess of the rest energy of the electron–positron pair, the excess appears as kinetic energy of the two particles. Figure 15.2 shows tracks of electron–positron pairs created by 300-MeV gamma rays striking a lead sheet.

The reverse process can also occur. Under the proper conditions, an electron and a positron can annihilate each other to produce two gamma-ray photons with a combined energy of at least 1.02 MeV:

$$e^- + e^+ \longrightarrow 2\gamma$$

Electron–positron annihilation

A practical application of electron–positron annihilation occurs in the medical diagnostic technique called **positron emission tomography (PET).** The patient is injected with a glucose solution containing a radioactive substance that decays by positron emission, and the material is carried by the blood throughout the body. A positron emitted during a decay event in one of the radioactive nuclei in the glucose solution annihilates with an electron in the surrounding tissue, resulting in two gamma-ray photons emitted in opposite directions. A gamma detector surrounding the patient pinpoints the source of the photons and, with the assistance of a computer, displays an image of the sites at which the glucose accumulates. (Glucose is metabolized rapidly in cancerous tumors and accumulates at those sites, providing a strong signal for a PET detector system.) The images from a PET scan can indicate a wide variety of disorders in the brain, including Alzheimer's disease (Fig. 15.3). In addition, because glucose metabolizes more rapidly in active areas of the brain, a PET scan can indicate which areas of the brain are involved in activities in which a patient is engaging at the time of the scan, such as language use, music, and vision.

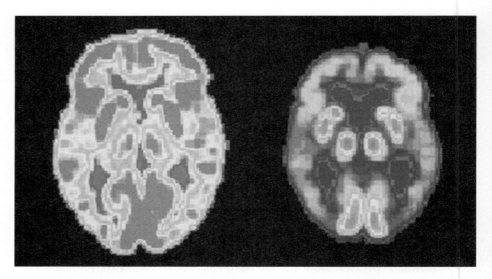

Figure 15.3   PET scans of the brain of a healthy older person *(left)* and that of a patient suffering from Alzheimer's disease *(right)*. (*National Institute of Health*)

Prior to 1955, on the basis of the Dirac theory, it was expected that every particle had a corresponding antiparticle, but antiparticles such as the antiproton and antineutron had not been detected experimentally. Since the relativistic Dirac theory had had some failures (it predicted the wrong-size magnetic moment for the proton) as well as many successes, it was important to determine whether the antiproton really existed. In 1955 a team led by Emilio Segrè (1905–1989, Italian-American physicist) and Owen Chamberlain (b. 1920, American physicist) used the Bevatron particle accelerator at the University of California, Berkeley, to produce both antiprotons and antineutrons. They thus established with certainty the existence of antiparticles. For this work Segrè and Chamberlain received the Nobel prize in 1959. **It is now accepted that every particle has a corresponding antiparticle of equal mass and spin and of equal and opposite charge, magnetic moment, and strangeness.** (The property of strangeness is explained in Section 15.6.) The only exceptions to these rules for particles and antiparticles are the neutral photon, pion, and eta, each of which is its own antiparticle.

EXAMPLE 15.1   Pair Production

When an electron and a positron meet at low speeds in free space, why are *two* 0.511-MeV gamma rays produced rather than *one* gamma ray with an energy of 1.022 MeV?

Reasoning   Gamma rays are photons, and photons carry momentum. If only one photon were produced, momentum would not be conserved, because the total momentum of the electron–positron system is approximately 0, whereas a single photon of energy 1.022 MeV would have a very large momentum. On the other hand, the two gamma-ray photons that are produced travel off in opposite directions, so their total momentum is 0.

15.3   MESONS AND THE BEGINNING OF PARTICLE PHYSICS

In the mid-1930s, physicists had a fairly simple view of the structure of matter. The building blocks were the proton, the electron, and the neutron. Three other particles were known or had been postulated at the time: the photon, the neutrino, and the positron. These six particles were considered the fundamental constituents of matter. With this marvelously simple picture of the world, however, no one could provide an answer to an important question: Since the many protons close together in a nucleus should strongly repel each other because of their like charges, what is the nature of the force that holds the nucleus together? Scientists recognized that this mysterious force must be much stronger than anything encountered in nature up to that time.

In 1935 the Japanese physicist Hideki Yukawa (1907–1981) proposed the first theory to successfully explain the nature of the nuclear force, an effort that later earned him the Nobel prize. To understand Yukawa's theory, it is useful to recall that in the modern view of electromagnetic interactions, *charged particles interact by exchanging photons.* Yukawa used this idea to explain the nuclear force by proposing a new particle whose exchange between nucleons in the nucleus produces the nuclear force. Furthermore, he established that the range of the force is inversely proportional to the mass of this particle and predicted that the mass would be about 200 times the mass of the electron. Since the new particle would have a mass between that of the electron and that of the proton, it was called a **meson** (from the Greek *meso*, "middle").

In an effort to substantiate Yukawa's predictions, physicists began an experimental search for the meson by studying cosmic rays entering the Earth's atmosphere. In 1937 Carl Anderson and his collaborators discovered a particle of mass $106 \text{ MeV}/c^2$, about 207 times the mass of the electron. However, subsequent experiments showed that the particle interacted very weakly with matter and hence could not be the carrier of the strong force. The puzzling situation inspired several theoreticians to propose that there were two mesons with slightly different masses. This idea was confirmed by the discovery in 1947 of the pi (π) meson, or simply **pion,** by Cecil Frank Powell (1903–1969) and Giuseppe P. S. Occhialini (b. 1907). The particle discovered by Anderson in 1937, the one thought to be a meson, is not really a meson. Instead, it takes part in the weak and electromagnetic interactions only, and it is now called the **muon.**

The pion, Yukawa's carrier of the nuclear force, comes in three varieties, corresponding to three charge states: π^+, π^-, and π^0. The π^+ and π^- particles have masses of $139.6 \text{ MeV}/c^2$, and the π^0 has a mass of $135.0 \text{ MeV}/c^2$. Both pions and muons are unstable particles. For example, the π^-, which has a mean lifetime of 2.6×10^{-8} s, first decays to a muon and a muon antineutrino. The muon, which has a mean lifetime of $2.2 \ \mu$s, then decays to an electron, a neutrino, and an electron antineutrino:

$$\pi^- \longrightarrow \mu^- + \overline{\nu}_\mu$$

$$\mu^- \longrightarrow e^- + \nu_\mu + \overline{\nu}_e \qquad (15.1)$$

The interaction between two particles can be represented in a useful diagram called a **Feynman diagram,** developed by the American physicist Richard P. Feynman (1918–1988). Figure 15.4 is such a diagram for the

Hideki Yukawa (1907–1981), a Japanese physicist, was awarded the Nobel prize in 1949 for predicting the existence of mesons. This photograph of Yukawa at work was taken in 1950 in his office at Columbia University. (© *Bettmann/CORBIS*)

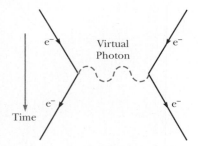

Figure 15.4 Feynman diagram showing how a photon carries the electromagnetic force between two interacting electrons. The blue arrow shows the direction of increasing time.

electromagnetic interaction between two electrons. A Feynman diagram is a qualitative graph of time on the vertical axis versus space on the horizontal axis. It is qualitative in the sense that actual values of time and space are not important, but the overall appearance of the graph conveniently serves to categorize different interaction processes. In this simple case, a photon is the field particle that mediates the electromagnetic force between the electrons. The photon transfers energy and momentum from one electron to the other in this interaction. The photon is called a **virtual photon** because it is emitted and reabsorbed without being detected. Virtual photons can violate the law of conservation of energy by ΔE for a very short time, Δt, provided that $\Delta E \Delta t \approx \hbar/2$ (from the minimum form of the uncertainty principle). Note that such quantum violations of energy conservation take place only in the short term and that system energy is conserved in the long run when the photon is reabsorbed by the other electron.

Now consider the pion exchange between a proton and a neutron that transmits the nuclear force according to Yukawa (Fig. 15.5). We can reason that the energy ΔE needed to create a pion of mass m_π is given by Einstein's equation $\Delta E = m_\pi c^2$. Again, the very existence of the pion would violate conservation of energy if it lasted for a time greater than $\Delta t \approx \hbar/2 \, \Delta E$, where ΔE is the energy of the pion and Δt is the time it takes the pion to travel from one nucleon to the other. Therefore,

$$\Delta t \approx \frac{\hbar}{2 \, \Delta E} = \frac{\hbar}{2 m_\pi c^2} \tag{15.2}$$

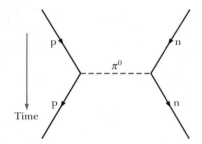

Figure 15.5 Feynman diagram representing a proton interacting with a neutron via the strong force. In this case, the pion mediates the strong force. The blue arrow shows the direction of increasing time.

Because the pion cannot travel faster than the speed of light, the maximum distance d it can travel in a time Δt is $c\Delta t$. Using Equation 15.2 and $d = c\Delta t$, we find this maximum distance to be

$$d \approx \frac{\hbar}{2 m_\pi c} \tag{15.3}$$

We know that the range of the nuclear force is approximately 1.0×10^{-15} m. Using this value for d in Equation 15.3, we calculate the rest energy of the pion to be

$$m_\pi c^2 \approx \frac{\hbar c}{2d} = \frac{197 \text{ MeV} \cdot \text{fm}}{2(1 \times 10^{-15} \text{ m})}$$

$$\approx 100 \text{ MeV}$$

This corresponds to a mass of $100 \text{ MeV}/c^2$ (approximately 250 times the mass of the electron), a value in reasonable agreement with the observed pion mass.

The concept we have just described is quite important. In effect, it says that a proton can change into a proton plus a pion as long as it returns to its original state in a very short time. Physicists often say that a nucleon undergoes "fluctuations" as it emits and absorbs pions. As we have seen, these fluctuations are a consequence of a combination of quantum mechanics (through the uncertainty principle) and special relativity (through Einstein's energy–mass relationship, $E = mc^2$). Also, as seen in Chapter 13, these virtual fluctuations can become real particles in collision processes if the incident particle can furnish the "missing" energy and momentum.

Richard Phillips Feynman was a brilliant theoretical physicist who shared the 1965 Nobel Prize in Physics with Julian S. Schwinger and Shinichiro Tomonaga for their fundamental work in the principles of quantum electrodynamics. Feynman's many important contributions to physics include the invention of simple diagrams to represent particle interactions graphically, the theory of the weak interaction of subatomic particles, a reformulation of quantum mechanics, the theory of superfluid helium, and his contribution to physics education through the magnificent three-volume text *The Feynman Lectures on Physics.*

Feynman did his undergraduate work at MIT and received his Ph.D. in 1942 from Princeton University, where he studied under John Archibald Wheeler. During World War II he worked on the Manhattan Project at Princeton and at Los Alamos, New Mexico. He then joined the faculty at Cornell University in 1945 and in 1950 was appointed professor of physics at California Institute of Technology, where he remained for the rest of his career.

It is well known that Feynman had a passion for finding new and better ways to formulate each problem or, as he would say, "turning it around." In the early part of his career, he was fascinated with electrodynamics and developed an intuitive view of quantum electrodynamics. Convinced that the electron could not interact with its own field, he said, "That was the beginning, and the idea seemed so obvious to me that I fell deeply in love with it. . . ." Often called the outstanding intuitionist of our age, he said in his Nobel acceptance speech, "Often, even in a physicist's sense, I did not have a demonstration of how to

BIOGRAPHY

RICHARD P. FEYNMAN
(1918–1988)

get all of these rules and equations, from conventional electrodynamics. . . . I never really sat down, like Euclid did for the geometers of Greece, and made sure that you could get it all from a single set of axioms."

In 1986 Feynman was a member of the presidential commission to investigate the explosion of the space shuttle *Challenger.* In this capacity, he performed a simple experiment for the commission members that showed that one of the shuttle's O-ring seals was the likely cause of the disaster. After placing a seal in a pitcher of ice water and squeezing it with a clamp, he demonstrated that the seal failed to spring back into shape once the clamp was removed.[1]

Feynman worked in physics with a style commensurate with his personality—that is, with energy, vitality, and humor. The following quotations from some of his colleagues hint at the great impact he made on the scientific community.[2]

Murray Gell-Mann: "A brilliant, vital, and amusing neighbor, Feynman was a stimulating (if sometimes exasperating) partner in discussions of profound issues—we would exchange ideas and silly jokes

in between bouts of mathematical calculation—we struck sparks off each other, and it was exhilarating."

David Pines: "Reading Feynman is a joy and a delight, for in his papers, as in his talks, Feynman communicated very directly, as though the reader were watching him derive the results at the blackboard."

David L. Goodstein: "He loved puzzles and games. In fact, he saw all the world as a sort of game, whose progress of 'behavior' follows certain rules, some known, some unknown. . . . Find places or circumstances where the rules don't work, and invent new rules that do."

Valentine L. Telegdi: "Feynman was not a theorist's theorist, but a physicist's physicist and a teacher's teacher."

Laurie M. Brown, one of his graduate students at Cornell, noted that Feynman, a playful showman, was "undervalued at first because of his rough manners [but] in the end triumphs through native cleverness, psychological insight, common sense and the famous Feynman humor. . . . Whatever else Dick Feynman may have joked about, his love for physics approached reverence."

[1] Feynman's own account of this inquiry can be found in *Physics Today,* 4:26, February 1988.

[2] For more on Feynman's life and contributions, see the articles in a special memorial issue of *Physics Today,* 42, February 1989. For a personal account of Feynman, see his popular autobiographical books, *Surely You're Joking Mr. Feynman,* New York, Bantam Books, 1985, and *What Do You Care What Other People Think,* New York, W. W. Norton & Co., 1987.

(© Shelly Grazin/CORBIS)

This section has dealt with early ideas about particles that carry the nuclear force, namely the pions, and the mediators of the electromagnetic force, photons. **Current ideas indicate that the nuclear force is more accurately described as an average or residual effect of the strong color force between quarks,** as will be explained in Section 15.10. The graviton, which is the mediator of the gravitational force, has yet to be observed. The $W^{\pm}$ and Z^0 particles that mediate the weak force were discovered in 1983 by the Italian physicist Carlo Rubbia (b. 1934) and his associates, using a proton–antiproton collider. Rubbia and Simon van der Meer, both at CERN near Geneva, Switzerland, shared the 1984 Nobel prize for the discovery of the $W^{\pm}$ and Z^0 particles and the development of the proton–antiproton collider. In this accelerator, protons and antiprotons that have a momentum of 270 GeV/c undergo head-on collisions with each other. In some of the collisions $W^{\pm}$ and Z^0 particles are produced, which in turn are identified by their decay products.

15.4 CLASSIFICATION OF PARTICLES

All particles other than field particles can be classified into two broad categories, hadrons and leptons, according to their interactions.

Hadrons

Particles that interact through the strong force are called **hadrons.** The two classes of hadrons, **mesons** and **baryons,** are distinguished by their masses and spins.

Mesons all have spin 0 or 1, with masses between that of the electron and that of the proton. All mesons are known to decay finally into electrons, positrons, neutrinos, and photons. The pion is the lightest of known mesons; it has a mass of approximately 140 MeV/c^2 and a spin of 0. Another is the K meson, with a mass of approximately 500 MeV/c^2 and a spin of 0.

Baryons, the second class of hadrons, have masses equal to or greater than the proton mass (*baryon* means "heavy" in Greek), and their spins are always odd half-integer values ($\frac{1}{2}$, $\frac{3}{2}$, $\frac{5}{2}$, etc.). Protons and neutrons are baryons, as are many other particles. With the exception of the proton, all baryons decay in such a way that the end products include a proton. For example, the baryon called the Ξ^- hyperon first decays to the Λ^0 baryon and a π^- in about 10^{-10} s. The Λ^0 then decays to a proton and a π^- in approximately 3×10^{-10} s.

It is important to note that hadrons are composite particles, not point particles, and have a measurable size of about 1 fm (10^{-15} m). Hadrons are composed of more elemental units called *quarks*, which are believed to be truly structureless point particles. Mesons consist of two quarks and baryons of three. For now, however, we defer discussion of the ultimate constituents of hadrons to Section 15.9 and continue with our empirical classification of particles. Table 15.2 lists important properties of the leptons and some hadrons. The symbols B, L_e, L_μ, L_τ, and S stand for baryon, electron, muon, and tau numbers and strangeness, respectively, and are explained in Sections 15.5 and 15.6.

Table 15.2   Some Particles and Their Properties

Category	Particle Name	Symbol	Anti-particle	Mass (MeV/c^2)	B	L_e	L_μ	L_τ	S	Lifetime (s)	Principal Decay Modes[a]
Leptons	Electron	e^-	e^+	0.511	0	+1	0	0	0	Stable	
	Electron-neutrino	ν_e	$\overline{\nu}_e$	$< 2.8 \times 10^{-6}$	0	+1	0	0	0	Stable	
	Muon	μ^-	μ^+	105.7	0	0	+1	0	0	2.19×10^{-6}	$e^- \overline{\nu}_e \nu_\mu$
	Muon-neutrino	ν_μ	$\overline{\nu}_\mu$	$< 3.5 \times 10^{-6}$	0	0	+1	0	0	Stable	
	Tau	τ^-	τ^+	1784	0	0	0	+1	0	3.3×10^{-13}	$\mu^- \overline{\nu}_\mu \nu_\tau$, $e^- \overline{\nu}_e \nu_\tau$
	Tau-neutrino	ν_τ	$\overline{\nu}_\tau$	$< 8.4 \times 10^{-6}$	0	0	0	+1	0	Stable	
Hadrons **Mesons**	Pion	π^+	π^-	139.6	0	0	0	0	0	2.60×10^{-8}	$\mu^+ \nu_\mu$
		π^0	Self	135.0	0	0	0	0	0	0.83×10^{-16}	2γ
	Kaon	K^+	K^-	493.7	0	0	0	0	+1	1.24×10^{-8}	$\mu^+ \nu_\mu$ $\pi^+ \pi^0$
		K_S^0	$\overline{K}_S^0$	497.7	0	0	0	0	+1	0.89×10^{-10}	$\pi^+ \pi^-$ $2\pi^0$
		K_L^0	$\overline{K}_L^0$	497.7	0	0	0	0	+1	5.2×10^{-8}	$\pi^\pm e^\mp \overline{\nu}_e$ $3\pi^0$ $\pi^\pm \mu^\mp \overline{\nu}_\mu$
	Eta	η	Self	548.8	0	0	0	0	0	$< 10^{-18}$	$2\gamma, 3\pi^0$
		η'	Self	958	0	0	0	0	0	2.2×10^{-21}	$\eta \pi^+ \pi^-$
Baryons	Proton	p	$\overline{p}$	938.3	+1	0	0	0	0	Stable	
	Neutron	n	$\overline{n}$	939.6	+1	0	0	0	0	624	$pe^- \overline{\nu}_e$
	Lambda	Λ^0	$\overline{\Lambda}^0$	1115.6	+1	0	0	0	-1	2.6×10^{-10}	$p\pi^-, n\pi^0$
	Sigma	Σ^+	$\overline{\Sigma}^-$	1189.4	+1	0	0	0	-1	0.80×10^{-10}	$p\pi^0, n\pi^+$
		Σ^0	$\overline{\Sigma}^0$	1192.5	+1	0	0	0	-1	6×10^{-20}	$\Lambda^0 \gamma$
		Σ^-	$\overline{\Sigma}^+$	1197.3	+1	0	0	0	-1	1.5×10^{-10}	$n\pi^-$
	Delta	Δ^{++}	$\overline{\Delta}$	1230	+1	0	0	0	0	6×10^{-24}	$p\pi^+$
		Δ^+	$\overline{\Delta}^-$	1231	+1	0	0	0	0	6×10^{-24}	$p\pi^0, n\pi^+$
		Δ^0	$\overline{\Delta}^0$	1232	+1	0	0	0	0	6×10^{-24}	$n\pi^0, p\pi^-$
		Δ^-	$\overline{\Delta}^+$	1234	+1	0	0	0	0	6×10^{-24}	$n\pi^-$
	Xi	Ξ^0	$\overline{\Xi}^0$	1315	+1	0	0	0	-2	2.9×10^{-10}	$\Lambda^0 \pi^0$
		Ξ^-	Ξ^+	1321	+1	0	0	0	-2	1.64×10^{-10}	$\Lambda^0 \pi^-$
	Omega	Ω^-	Ω^+	1672	+1	0	0	0	-3	0.82×10^{-10}	$\Xi^- \pi^0, \Xi^0 \pi^-$, $\Lambda^0 K^-$

[a]Notations in this column such as $p\pi^-$, $n\pi^0$ indicate two possible decay modes. In this case, the two possible decays are $\Lambda^0 \rightarrow p + \pi^-$ and $\Lambda^0 \rightarrow n + \pi^0$.

Leptons

Leptons (from the Greek *leptos*, meaning "small" or "light") are a group of particles that participate in the electromagnetic and weak interactions. All leptons have spins of $\frac{1}{2}$. Unlike hadrons, which have size and structure, **leptons appear to be truly elementary point-like particles with no structure.** Also unlike hadrons, the number of known leptons is small. Currently, scientists believe there are only six leptons: the electron, the muon, the tau, and—associated with each of these particles—three different neutrinos, the electron neutrino (ν_e), the muon neutrino (ν_μ), and the tau neutrino (ν_τ). We now classify the six known leptons into three pairs called families:

$$\begin{pmatrix} e^- \\ \nu_e \end{pmatrix} \quad \begin{pmatrix} \mu^- \\ \nu_\mu \end{pmatrix} \quad \begin{pmatrix} \tau^- \\ \nu_\tau \end{pmatrix}$$

Table 15.2 shows that the tau lepton is actually quite massive and has a mass about twice that of the proton. Also note in the table that each lepton has an antiparticle. It is quite unexpected and interesting that neutrinos and antineutrinos each have a distinct helicity, or relation between linear momentum and spin directions. The spin of a neutrino is opposite to its direction of travel, and the spin of an antineutrino is parallel to its direction of travel. Current evidence suggests that neutrinos have a small mass of several eV/c^2 as shown in Table 15.2. Problem 23 investigates an interesting time-of-flight method for setting an upper limit on neutrino mass. As we shall see, a firm knowledge of the neutrino's mass has great significance in physical models of energy production in stars and in grand unified theories of elementary particles.

The Solar Neutrino Mystery and Neutrino Oscillations

Conclusive measurements of the ability of neutrinos to change from one type or "flavor" to another, along with indirect measurements of neutrino masses have recently been made at the Sudbury Neutrino Observatory (SNO) in Canada and the Super Kamiokanda detector in Japan. These findings solve a puzzle over 40 years old and provide new confidence that physicists really do understand how energy is produced in the Sun's core.

The warming sunlight we receive on Earth should be accompanied by billions of neutrinos per square centimeter per second. In particular, electron neutrinos are produced in the Sun's fusion engine by typical reactions like

$$p + p \longrightarrow {}^2_1H + e^+ + \nu_e + 0.42 \text{ MeV}$$

and the boron decay

$$^8_5B \longrightarrow {}^8_4Be + e^+ + \nu_e + 14.6 \text{ MeV}$$

Measurements of the ν_e flux dating back to the 1960s have, however, been consistently mysterious because *only about one-third of the expected flux has been observed*. (The expected flux of solar electron neutrinos was calculated using the trusted standard solar model of the Sun's properties and energy production.) One proposed explanation of the mystery is that some ν_e's have changed into ν_μ's or ν_τ's during the journey from the Sun to the Earth, and these ν_μ's or ν_τ's would not show up in detectors designed to spot ν_e's.

The basic idea is that the change from ν_e to, say, ν_μ would not be permanent but would just be part of an ongoing oscillation between ν_e and ν_μ. The best part is that the frequency of oscillation would depend on the masses. Thus measurement of **neutrino oscillation** frequencies could indirectly determine neutrino masses, which are small and extremely

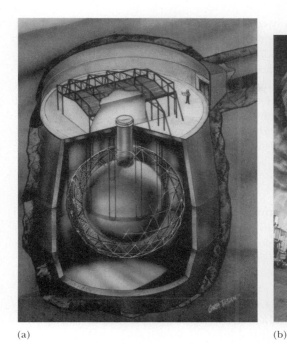

(a)   (b)

(a) Artist's concept of the underground SNO detector showing the 12 m diameter spherical acrylic vessel containing 1000 tons of D_2O. (b) A photo of the detector under construction. Photomultipliers mounted on the frame shown detect Cerenkov light flashes produced by neutrino reactions in the D_2O. (*Photos courtesy of Sudbury Neutrino Observatory*)

hard to measure directly. We can think of each flavor of neutrino — ν_e, ν_μ, ν_τ — as a linear combination of three other neutrinos — ν_1, ν_2, ν_3 — each having a different mass m_1, m_2, and m_3 and different energy for a given momentum. Since the time dependence of each state ν_1, ν_2, ν_3 is oscillatory and of the form $\exp(-iEt/\hbar)$, with time, the states ν_1, ν_2, ν_3 move out-of-phase with each other and interfere, producing an oscillation between flavors. Problem 15.22 explores how to calculate the probabilty that a ν_μ oscillates to a ν_τ.

The latest findings from SNO scientists, who used a new detection technique equally sensitive to electron, muon, and tau neutrinos, conclusively show that the missing electron neutrinos are accounted for by oscillation of ν_e's into ν_μ's and ν_τ's. Further, plans are under way at SNO to study how neutrino oscillation depends on the passage of neutrinos through matter. This can be done by observing day–night differences in neutrino flux from the Sun, because at night neutrinos must penetrate the whole Earth to reach the detector.

15.5   CONSERVATION LAWS

Conservation laws are important to an understanding of why certain decays and reactions occur and others do not. In general, the laws of conservation of energy, linear momentum, angular momentum, and electric charge provide us with a set of rules that all processes must absolutely follow. For example, conservation of electric charge requires that the total charge before a reaction equals the total charge after the reaction.

Certain new conservation laws are important in the study of elementary particles. Two of these laws, concerning baryon number and lepton number, are described in this section, and others will be discussed later in this chapter. Although the two described here have no theoretical foundation, they are supported by abundant empirical evidence and serve to indicate probable reactions, but do not absolutely hold 100% of the time.

Baryon Number

Conservation of baryon number

To apply conservation of baryon number, we assign a baryon number $B = +1$ for all baryons, $B = -1$ for all antibaryons, and $B = 0$ for all other particles. The **law of conservation of baryon number** states that *whenever a nuclear reaction or decay occurs, the sum of the baryon numbers before the process must equal the sum of the baryon numbers after the process.* An equivalent statement is that the net number of baryons remains constant in any process.

If baryon number is absolutely conserved, the proton must be absolutely stable. If it were not for the law of conservation of baryon number, the proton could decay to a positron and a neutral pion. However, such a decay has never been observed. At present, we can say only that the proton has a half-life of at least 10^{32} years (the estimated age of the Universe is only about 10^{10} years). In one recent version of a grand unified theory, physicists predicted that the proton is unstable. According to this theory, baryon number is not absolutely or perfectly conserved.

EXAMPLE 15.2   Checking Baryon Numbers

Determine whether or not each of the following reactions can occur on the basis of the law of conservation of baryon number.

$$(1) \qquad p + n \longrightarrow p + p + n + \bar{p}$$

$$(2) \qquad p + n \longrightarrow p + p + \bar{p}$$

Solution For reaction 1, recall that $B = +1$ for baryons and $B = -1$ for antibaryons. Hence the left side of reac-

tion 1 gives a total baryon number of $1 + 1 = 2$. The right side of reaction 1 gives a total baryon number of $1 + 1 + 1 + (-1) = 2$. Thus the reaction can occur, provided the incoming proton has sufficient energy.

The left side of reaction 2 gives a total baryon number of $1 + 1 = 2$. However, the right side gives $1 + 1 + (-1) = 1$. Because the baryon number is not conserved, the reaction cannot occur or at most has a small probability of occurrence.

Lepton Number

Conservation of lepton number

From observations of commonly occurring decays of the muon and tau we arrive at three conservation laws involving lepton numbers, one for each variety of lepton. The **law of conservation of electron-lepton number (lepton flavor conservation)** states that *the sum of the electron-lepton numbers before a reaction or decay must equal the sum of the electron-lepton numbers after the reaction or decay.*

The electron and the electron neutrino are assigned a positive lepton number, $L_e = +1$; the antileptons e^+ and $\bar{\nu}_e$ are assigned a negative lepton number, $L_e = -1$; all others (the muon and tau families) have $L_e = 0$. For example, consider the decay of the neutron

$$n \longrightarrow p + e^- + \overline{\nu}_e$$

Before the decay the electron-lepton number is $L_e = 0$; after the decay it is $0 + 1 + (-1) = 0$. Thus the electron-lepton number is conserved, although not perfectly in view of the proven neutrino oscillation previously mentioned. It is important to recognize that the baryon number must also be conserved. We can easily check this by noting that before the decay $B = +1$, and after the decay $B = +1 + 0 + 0 = +1$.

Similarly, when a decay involves muons, the muon-lepton number, L_μ, is conserved. The μ^- and the ν_μ are assigned positive numbers, $L_\mu = +1$; the antimuons μ^+ and $\overline{\nu}_\mu$ are assigned negative numbers, $L_\mu = -1$ and all others have $L_\mu = 0$. Finally, the tau-lepton number, L_τ, is conserved, and similar assignments can be made for the tau lepton and its neutrino. In all cases it is important to keep in mind that lepton flavor conservation is not absolute, and that lepton flavor-violating reactions have a small probability of occurrence.

EXAMPLE 15.3   Checking Lepton Numbers

Determine which of the following decay schemes can occur on the basis of conservation of electron-lepton number:

$$(1) \qquad \mu^- \longrightarrow e^- + \overline{\nu}_e + \nu_\mu$$

$$(2) \qquad \pi^+ \longrightarrow \pi^+ + \nu_\mu + \nu_e$$

Solution   Because decay 1 involves both a muon and an electron, L_μ and L_e must both be conserved. Before the

decay, $L_\mu = +1$ and $L_e = 0$. After the decay, $L_\mu = 0 + 0 + 1 = +1$, and $L_e = +1 - 1 + 0 = 0$. Thus both numbers are conserved, and on this basis the decay mode is possible.

Before decay 2 occurs, $L_\mu = 0$ and $L_e = 0$. After the decay, $L_\mu = -1 + 1 + 0 = 0$, but $L_e = +1$. Thus the decay is not possible, because the electron-lepton number is not conserved.

Exercise 1   Determine whether the decay $\mu^- \to e^- + \overline{\nu}_e$ can occur.

Answer   No. The muon-lepton number is $+1$ before the decay and 0 after the decay.

15.6   STRANGE PARTICLES AND STRANGENESS

Many particles discovered in the 1950s were produced by the nuclear interaction of pions with protons and neutrons in the atmosphere. Three of these particles—namely, the kaon (K), lambda (Λ), and sigma (Σ) particles—exhibited unusual properties in production and decay and hence were called *strange particles.*

One unusual property of these particles is their production in pairs. For example, when a pion collides with a proton, two neutral strange particles are produced with high probability (Fig. 15.6):

$$\pi^- + p \longrightarrow K^0 + \Lambda^0$$

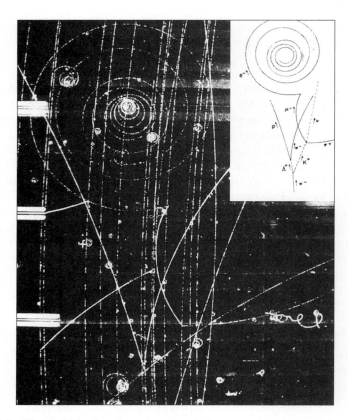

Figure 15.6 This bubble-chamber photograph shows many events, and the inset is a drawing of identified tracks. The strange particles Λ^0 and K^0 form at the bottom as the π^- interacts with a proton according to $\pi^- + p^+ \rightarrow \Lambda^0 + K^0$. (Note that the neutral particles leave no tracks, as indicated by the dashed lines.) The Λ^0 and K^0 then decay according to $\Lambda^0 \rightarrow \pi^- + p^+$ and $K^0 \rightarrow \pi^+ + \mu^- + \bar{\nu}_\mu$. (*Courtesy Lawrence Berkeley Laboratory, University of California, Photographic Services*)

On the other hand, the reaction $\pi^- + p \longrightarrow K^0 + n$ never occurred, even though no known conservation laws would have been violated and the energy of the pion was sufficient to initiate the reaction.

The second peculiar feature of strange particles is that although they are produced by the strong interaction at a high rate, they do not decay into particles that interact via the strong force at a very high rate. Instead, they decay very slowly, which is characteristic of the weak interaction, as shown in Table 15.1. Their half-lives are in the range 10^{-10} s to 10^{-8} s; most other particles that interact via the strong force have lifetimes on the order of 10^{-20} s and less.

To codify these unusual properties of strange particles, a law called conservation of strangeness was introduced with a new quantum number S, called **strangeness.** The strangeness numbers for some particles are given in Table 15.2. The production of strange particles in pairs is explained by assigning $S = +1$ to one of the particles and $S = -1$ to the other. All nonstrange particles are assigned strangeness $S = 0$. The **law of conservation of strange-**

Conservation of strangeness number

ness states that *whenever a nuclear reaction or decay occurs, the sum of the strangeness numbers before the process must equal the sum of the strangeness numbers after the process.*

The slow decay of strange particles can be explained by assuming that the strong and electromagnetic interactions obey the law of conservation of strangeness but the weak interaction does not. Because the decay reaction involves the loss of one strange particle, it violates strangeness conservation and hence proceeds slowly via the weak interaction.

EXAMPLE 15.4   Is Strangeness Conserved?

(a) Determine whether the following reaction occurs on the basis of conservation of strangeness.

$$\pi^0 + n \longrightarrow K^+ + \Sigma^-$$

Solution   The initial state has strangeness $S = 0 + 0 = 0$. Because the strangeness of the K^+ is $S = +1$ and the strangeness of the Σ^- is $S = -1$, the strangeness of the final state is $+1 - 1 = 0$. Thus strangeness is conserved and the reaction is allowed.

(b) Show that the following reaction does not conserve strangeness:

$$\pi^- + p \longrightarrow \pi^- + \Sigma^+$$

Solution   The initial state has strangeness $S = 0 + 0 = 0$, and the final state has strangeness $S = 0 + (-1) = -1$. Thus strangeness is not conserved.

Exercise 2   Show that the reaction $p + \pi^- \longrightarrow K^0 + \Lambda^0$ obeys the law of conservation of strangeness.

15.7   HOW ARE ELEMENTARY PARTICLES PRODUCED AND PARTICLE PROPERTIES MEASURED?

Examination of the bewildering array of entries in Table 15.2 leaves one yearning for firm ground. It is natural to wonder about a particle that exists for 10^{-20} s and has a mass of 1192.5 MeV/c^2. How is it possible to detect a particle that exists for only 10^{-20} s? Furthermore, how can the mass be measured? If a standard attribute of a particle is some type of permanence or stability, in what sense is a fleeting entity that exists for 10^{-20} s a particle? In this section we attempt to answer such questions and explain how elementary particles are produced and how their properties are measured.

Elementary particles, most of which are unstable and occur rarely naturally, are created abundantly in man-made collisions of high-energy particles with a suitable target. Since very high energy beams of incident particles are desirable, stable charged particles such as electrons or protons generally make up the incident beam, because it takes considerable time to accelerate particles to high energies with electromagnetic fields. Similarly, targets must be simple and stable, and the simplest target, hydrogen, serves nicely as both a target and a detector. In a liquid hydrogen bubble chamber, which is basically a large container filled with hydrogen near its boiling point, a charged particle traversing the chamber ionizes the atoms along its path, and the ionization causes a visible track of tiny bubbles. The liquid hydrogen also serves as an efficient source of target protons, with a proton density sufficient to ensure many incident particle–target collisions within a reasonable time.

Figure 15.6 shows a typical event in which a bubble chamber has served as both target and detector. In this figure, many parallel tracks of negative pions are visible entering the photograph from the bottom. One of the pions has hit a stationary proton in the hydrogen and produced two strange particles, the Λ^0 and K^0, according to the reaction

$$\pi^- + p \longrightarrow \Lambda^0 + K^0$$

Neither neutral strange particle leaves a track, but their subsequent decays into charged particles can be seen clearly, as indicated in Figure 15.6. A magnetic field directed into the plane of the photograph causes the track of each charged particle to curve, and from the measured curvature the particle's charge and linear momentum can be determined. If the mass and momentum of the incident particle are known, we can then usually calculate the product particle mass, kinetic energy, and speed from conservation of momentum and energy (see Section 1.10 and Example 1.18). Finally, combining a product particle's speed with a measurable decay track length, we can calculate the product particle's lifetime. Figure 15.6 shows that sometimes one can use this lifetime technique even for a neutral particle, which leaves no track. As long as the particle speed and the start and finish of the missing track are known, one can infer the missing track length and find the lifetime of the neutral particle.

Resonance Particles

With clever experimental technique and much effort, decay track lengths as short as 1 micron (10^{-6} m) can be measured. This means that lifetimes as short as 10^{-16} s can be measured with this technique in the case of high-energy parti-

A colliding beam detector at CERN. (*Philippe Plailly/Eurelios/Science Photo Library/Photo Researchers, Inc.*)

"Particles, particles, particles."

(© *Sydney Harris*)

cles traveling at about the speed of light. We arrive at this result by assuming that a decaying particle travels 1 micron in the laboratory at a speed of $0.99c$, yielding a laboratory lifetime of $\tau_{\text{lab}} = 10^{-6}$ m$/0.99c \approx 0.33 \times 10^{-14}$ s. Now, relativity helps us. Since the proper lifetime, as measured in the decaying particle's rest frame, is shorter than τ_{lab} by a factor of $\sqrt{1 - v^2/c^2}$, we can actually measure lifetimes of duration:

$$\tau_{\text{proper}} = \tau_{\text{lab}} \sqrt{1 - \frac{v^2}{c^2}} = (0.33 \times 10^{-14}\text{ s}) \sqrt{1 - \frac{(0.99c)^2}{c^2}} = 4 \times 10^{-16}\text{ s}$$

Unfortunately, even with Einstein's help, we are several orders of magnitude away from minimum hadron lifetimes of 10^{-23} s with the best efforts of the track-length method. How then, can we detect the presence of particles that exist for as short a time as 10^{-23} s? As we shall see shortly, the masses, lifetimes, and very existence of these very short-lived particles, known as **resonance particles,** can be inferred from peaks (resonances) in the cross section versus energy plots describing their decay products.

Let's consider this in more detail by looking at the case of the resonance particle called the delta plus (Δ^+), which has a mass of 1231 MeV$/c^2$ and a lifetime of about 6×10^{-24} s. The Δ^+ is produced in the reaction

$$\text{e}^- + \text{p} \longrightarrow \text{e}^- + \Delta^+ \tag{15.4}$$

which is followed in 6×10^{-24} s by the decay

$$\Delta^+ \longrightarrow \pi^+ + \text{n} \tag{15.5}$$

Because the Δ^+ lifetime is so short, it leaves no measurable track, and it might seem impossible to distinguish the reactions given in Equations 15.4 and 15.5 from the net direct reaction in which no Δ^+ is produced:

$$e^- + p \longrightarrow e^- + \pi^+ + n \qquad (15.6)$$

In fact, we can tell whether a Δ^+ was formed by measuring the momentum and energy of the suspected decay products (pion and neutron) and using the conservation of momentum and energy to see whether these values combine to give a Δ^+ mass of 1231 MeV/c^2.

To understand this in detail, consider the decay of the Δ^+ shown in Figure 15.7. The energy and momentum of the Δ^+ must satisfy the equation

$$E_\Delta^2 = (p_\Delta c)^2 + (m_\Delta c^2)^2$$

or

$$m_\Delta c^2 = \sqrt{E_\Delta^2 - (p_\Delta c)^2} \qquad (15.7)$$

where m_Δ is the Δ^+ mass. Although we cannot directly measure E_Δ and $\mathbf{p}_\Delta$, since the delta particle leaves no track, we can measure the energies and momenta of the outgoing particles, E_π, $\mathbf{p}_\pi$, E_n, and $\mathbf{p}_n$. Using conservation of momentum and energy, we can then find an expression for $m_\Delta c^2$ in terms of these measured quantities. Thus we have $E_\Delta = E_\pi + E_n$ and $\mathbf{p}_\Delta = \mathbf{p}_\pi + \mathbf{p}_n$, and substituting into Equation 15.7,

$$m_\Delta c^2 = \sqrt{(E_\pi + E_n)^2 - (\mathbf{p}_\pi + \mathbf{p}_n)^2 c^2} \qquad (15.8)$$

Equation 15.8 holds for all events in which a Δ^+ particle actually formed and decayed. That is, for many different measured values of E_π, $\mathbf{p}_\pi$, E_n, and $\mathbf{p}_n$ corresponding to many repeated experiments, we will always find the same value of the quantity $m_\Delta c^2 = 1231$ MeV within experimental uncertainty *if the decay of a Δ^+ is involved*. On the other hand, if no Δ^+ is involved and the direct reaction $e^- + p \rightarrow e^- + \pi^+ + n$ occurs, $\sqrt{(E_\pi + E_n)^2 - (\mathbf{p}_\pi + \mathbf{p}_n)^2 c^2}$ will not equal 1231 MeV but will sweep over a broad range of values, some larger and some smaller than $m_\Delta c^2$, as the experiment is repeated. The typical method for showing the existence of a resonance particle involves calculating the quantity $Z = \sqrt{(E_\pi + E_n)^2 - (\mathbf{p}_\pi + \mathbf{p}_n)^2 c^2}$ for a large number of events in which a π^+

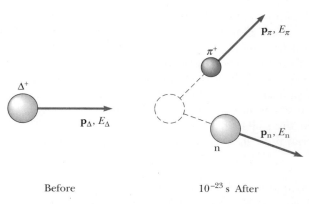

Before 10^{-23} s After

Figure 15.7   The decay of a Δ^+ particle into a positive pion and a neutron.

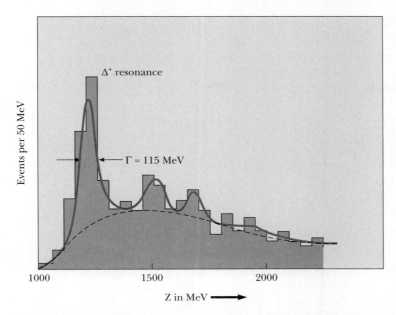

Figure 15.8   Experimental evidence for the existence of the Δ^+ particle. The sharp peak near 1230 MeV was produced by the events in which a Δ^+ formed and promptly decayed into a π^+ and neutron.

and a neutron are produced and then drawing a histogram of those events. By plotting the number of events with a given value of Z against this value of Z, we obtain a slowly varying curve with sharp peaks superimposed, the peaks showing the existence of resonance particles. Figure 15.8 shows such an experimental histogram for the Δ^+ particle. The broad background (dashed curve) is produced by direct events in which no Δ^+ was created; the sharp peak near $Z = 1230$ MeV, containing many events, was produced by all the events in which a Δ^+ formed and decayed into a pion and neutron with just the right energies and momenta to make up a delta particle. Peaks corresponding to two other resonance particles of larger mass can also be seen in Figure 15.8.

Histograms like Figure 15.8 can tell us not only the mass of a short-lived particle but also the lifetime of the particle from the full width at half maximum of the resonance peak, denoted by Γ in Figure 15.8. Because Γ is twice the uncertainty in energy of the delta, ($\Gamma = 2\,\Delta E$), we can use the energy–time uncertainty relation, $\Delta E\,\Delta t \approx \hbar/2$, to infer the lifetime of the delta, Δt:

$$\Delta t \approx \frac{\hbar}{2\,\Delta E} = \frac{\hbar}{\Gamma} \tag{15.9}$$

The measured width of $\Gamma = 115$ MeV leads to a value for the lifetime of the delta particle:

$$\Delta t \approx \frac{\hbar}{\Gamma} = \frac{6.6 \times 10^{-16}\ \text{eV·s}}{115 \times 10^6\ \text{eV}} = 5.7 \times 10^{-24}\ \text{s}$$

In this incredibly short lifetime, a delta particle moving at the highest possible speed of c travels only 10^{-15} m, or about one nuclear diameter.

Energy Considerations in Particle Production

It is ironic that the highly sophisticated branch of physics known as elementary particle physics relies for its very existence on the most brutish of experimental methods: smashing an incident particle moving at relativistic speed into a stationary target particle and observing what pieces come flying out! Yet this process is at the heart of experimental particle physics, and in this section we shall determine the threshold energy that is required for the production of new particles in a collision. Since the energy needed to manufacture new particles comes from the kinetic energy of the incident particle, and incident particle energies are quite large (the Fermi National Laboratory produces 1000-GeV protons), we must use relativistic equations in the calculation of these threshold energies.

Consider the specific particle production process

$$m_1 + m_2 \longrightarrow m_3 + m_4 + m_5 \tag{15.10}$$

Here m_1 is the mass of the incident particle, m_2 is the mass of the target particle that is at rest in the laboratory, and m_3, m_4, and m_5 are product-particle masses. Figure 15.9a shows such a particle reaction in the laboratory reference frame. The energies shown in Figure 15.9—E_1, E_2, and so on—are total energies (kinetic energy + rest energy) and the momenta are labeled

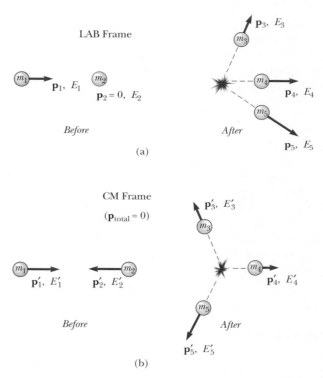

Figure 15.9   (a) The reaction $m_1 + m_2 \rightarrow m_3 + m_4 + m_5$ viewed in the laboratory frame, where m_2 is initially at rest and the energies are total relativistic energies (kinetic plus rest energy). (b) The same reaction viewed from the center-of-mass (CM) frame, in which the total momentum is always zero.

$\mathbf{p}_1$, $\mathbf{p}_2$, and so on. At first glance it might seem that the total initial energy of m_1 and m_2, $E_1 + E_2$, could be converted completely to rest energy of particles 3, 4, and 5. However, as we have already seen in the calculation of nuclear reaction thresholds in Chapter 14, some of the initial energy must go into energy of motion of the product particles in order to conserve momentum in the collision. This means that not all of the initial energy can go into creating new masses. The issue is to find out how much of the initial energy can go toward creating masses m_3, m_4, and m_5. That is, we wish to calculate the minimum or **threshold kinetic energy K_{th}** that m_1 must have in order to create particles with masses of m_3, m_4, and m_5 and also conserve momentum.

An important clue to finding K_{th} comes from understanding that if we could somehow arrange for a zero initial momentum, all of the initial energy could be converted into new particles. What we can do is solve for K_{th} in the frame in which the initial momentum is zero—the center-of-mass frame— and then transform that result back into the frame in which the experiment is actually carried out, the laboratory frame. Here's how it goes.

Figure 15.9b shows the same reaction as 15.9a but viewed from the center-of-mass (CM) frame, in which, by definition, the total momentum is always zero. Thus, in Figure 15.9b, m_1 and m_2 have equal and opposite momenta, and the vector sum of the momenta of m_3, m_4, and m_5 is zero. Actually, Figure 15.9b shows the case of an incident particle with more than threshold energy. A moment's reflection reveals that when m_1 has the minimum, or threshold, energy in the CM frame, all of the initial energy, $E'_1 + E'_2$, should be converted to the masses of particles 3, 4, and 5. This occurs when the product particles are created at rest in the CM frame, which is now possible since the total momentum is zero in the CM frame. Figure 15.10 shows the reaction $m_1 + m_2 \rightarrow m_3 + m_4 + m_5$ in the CM frame when m_1 has the threshold kinetic energy and m_3, m_4, and m_5 are created at rest.

To calculate a numerical expression for K_{th}, we make use of the invariant quantity $E^2 - p^2 c^2$, introduced in Chapter 2. Recall that $E^2 - p^2 c^2$ is called an invariant because it has the same numerical value for a system of particles in *any inertial frame* (see Problem 16 in Chapter 2). Applying the invariance of $E^2 - p^2 c^2$ to the CM and laboratory frames, we have, for the period before the collision,

$$E_{\text{CM}}^2 - p_{\text{CM}}^2 c^2 = E_{\text{lab}}^2 - p_{\text{lab}}^2 c^2 \tag{15.11}$$

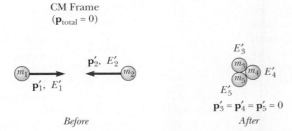

Figure 15.10   The reaction $m_1 + m_2 \rightarrow m_3 + m_4 + m_5$ in the CM frame when m_1 has the threshold kinetic energy required to produce m_3, m_4, and m_5. For the threshold condition, m_3, m_4, and m_5 are created at rest.

Here, E_{CM} and E_{lab} stand for the sum of the energies of particles m_1 and m_2 in the CM and laboratory frames, respectively. Likewise, p_{CM} and p_{lab} stand for the sum of the momenta of particles m_1 and m_2 in the CM and laboratory frames. Because $p_{CM} = 0$, $E_{lab} = E_1 + m_2 c^2$, and $p_{lab} = p_1$, Equation 15.11 becomes

$$E_{CM\,before}^2 = (E_1 + m_2 c^2)^2 - p_1^2 c^2 \qquad (15.12)$$

We can eliminate p_1 from Equation 15.12 by using $E_1^2 = p_1^2 c^2 + m_1^2 c^4$ to obtain

$$E_{CM\,before}^2 = 2E_1 m_2 c^2 + (m_2^2 + m_1^2) c^4 \qquad (15.13)$$

Observe that Equation 15.13 is in a useful form, since we can readily solve for E_1 or K_{th} ($E_1 = K_{th} + m_1 c^2$) in terms of all the masses if we can find an expression for $E_{CM\,before}$ in terms of the product masses m_3, m_4, and m_5. We can find such an expression by conserving relativistic energy in the CM frame:

$$E_{CM\,before} = E_{CM\,after} = E_3' + E_4' + E_5' = (m_3 + m_4 + m_5) c^2 \qquad (15.14)$$

Substituting this expression into Equation 15.13 and using $E_1 = K_{th} + m_1 c^2$, after a bit of algebra we obtain our final result:

$$K_{th} = \frac{(m_3 + m_4 + m_5)^2 c^2 - (m_1 + m_2)^2 c^2}{2 m_2} \qquad (15.15)$$

Equation 15.15 gives the threshold kinetic energy of an incident particle m_1 colliding with a stationary target m_2 required to produce three product particles of mass m_3, m_4, and m_5. For more than three product particles, inspection of our derivation shows that Equation 15.15 can be generalized as

$$K_{th} = \frac{(m_3 + m_4 + m_5 + m_6 + \cdots)^2 c^2 - (m_1 + m_2)^2 c^2}{2 m_2} \qquad (15.16)$$

EXAMPLE 15.5   How to Make a Virtual Particle Real

Consider the scattering of two protons. Assume that the protons interact by exchanging virtual field particles or field quanta. Thus, if the protons are attracted gravitationally, they exchange a graviton; if they are repelled electrically, they exchange a photon. In the present case we wish to consider the nuclear force between two protons that we assume is carried by the field particle called the pi meson or pion. When an incoming proton scatters from a stationary proton at low energy, a virtual pion with a mass of about 140 MeV/c^2 can blink into existence, transport energy and momentum from one proton to the other, and then blink out of existence in a time so short that violation of energy conservation is not observable. However, if an incident proton interacts with a stationary target at high enough energy, the incident proton may supply enough energy to make a virtual pion real according to the reaction

$$p + p \longrightarrow p + p + \pi^0 \qquad (15.17)$$

where π^0 represents a neutral pion. In this example, we wish to find the threshold energy for π^0 production ac-

cording to the reaction given by Equation 15.17. The neutral pion has a rest energy of 135 MeV.

Solution   Use Equation 15.15 with $m_1 = m_2 = m_3 = m_4 = m_p = 938.3$ MeV/c^2 and $m_5 = m_\pi = 135$ MeV/c^2. Equation 15.15 becomes

$$K_{th} = \frac{(2m_p + m_\pi)^2 c^2 - (2m_p)^2 c^2}{2m_p}$$

$$= \frac{(4m_p m_\pi + m_\pi^2) c^2}{2m_p} = 2m_\pi c^2 + \frac{m_\pi^2 c^2}{2m_p}$$

$$= 2(135\ \text{Mev}/c^2) c^2 + \frac{(135\ \text{MeV}/c^2)^2 c^2}{2(938.3\ \text{MeV}/c^2)}$$

$$= 280\ \text{MeV}$$

Thus, if stationary protons are bombarded with protons of at least 280 MeV kinetic energy, particle accelerators can produce neutral pion beams that can be used for additional pion studies.

Exercise 3   Since equally strong nuclear forces exist between proton and proton, neutron and neutron, and proton and neutron, we might expect the pion to come in three different charge states, $+e$, $-e$, and 0, and this is indeed the case. Three different pions exist—π^+, π^-, and π^0—with masses $m_{\pi-} = 139.6$ MeV/c^2, $m_{\pi+} = 139.6$ MeV/c^2, and $m_{\pi0} = 135$ MeV/c^2. Two reactions involving the production of the π^+ and π^- are

$$p + p \longrightarrow p + n + \pi^+ \tag{15.18}$$

$$p + p \longrightarrow p + p + \pi^+ + \pi^- \tag{15.19}$$

Calculate the threshold energy for these reactions.

Answers $K_{th} = 292$ MeV for Equation 15.18; $K_{th} = 600$ MeV for Equation 15.19.

15.8   THE EIGHTFOLD WAY

One of the tools scientists use is the detection of patterns in data, patterns that contribute to our understanding of nature. One of the best examples of the use of this tool is the development of the periodic table, which provides a fundamental understanding of the chemical behavior of the elements. The periodic table explains how more than 100 elements can be formed from three particles—the electron, the proton, and the neutron. The table of nuclides contains hundreds of nuclides, but all can be built from protons and neutrons.

The number of observed particles and resonances observed by particle physicists is also in the hundreds. Is it possible that a small number of entities exist from which all of these can be built? Taking a hint from the success of the periodic table and the table of nuclides, let us explore the historical search for patterns among the particles.

Many classification schemes have been proposed for grouping particles into families. Consider, for instance, the baryons listed in Table 15.2 that have spins of $\frac{1}{2}$: p, n, Λ^0, Σ^+, Σ^0, Σ^-, Ξ^0, and Ξ^-. If we plot strangeness versus charge for

The American physicists Richard Feynman *(left)* and Murray Gell-Mann *(right)* were awarded the Nobel prize in 1965 and 1969, respectively, for their theoretical studies dealing with subatomic particles. *(Courtesy of Michael R. Dressler)*

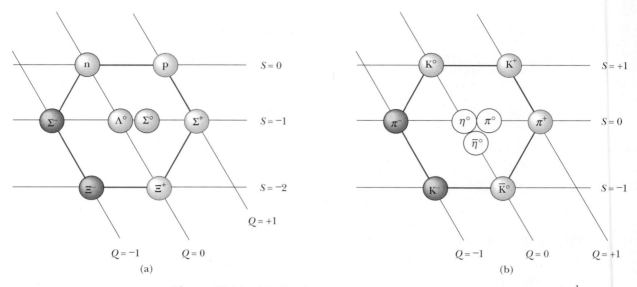

Figure 15.11   (a) The hexagonal eightfold-way pattern for the eight spin-$\frac{1}{2}$ baryons. This strangeness-versus-charge plot uses a sloping axis for the charge number Q but a horizontal axis for the strangeness (S) values. (b) The eightfold-way pattern for the nine spin-0 mesons.

these baryons using a sloping coordinate system, as in Figure 15.11a, we observe a fascinating pattern: Six of the baryons form a hexagon, and the remaining two are at the hexagon's center.

As a second example, consider the following nine spin-zero mesons listed in Table 15.2: π^+, π^0, π^-, K^+, K^0, K^-, η, η', and the antiparticle $\overline{K}^0$. Figure 15.11b is a plot of strangeness versus charge for this family. Again, a hexagonal pattern emerges. In this case, each particle on the perimeter of the hexagon lies opposite its antiparticle, and the remaining three (which form their own antiparticles) are at the center of the hexagon. These and related symmetric patterns were developed independently in 1961 by Murray Gell-Mann and Yuval Ne'eman (b. 1925). Gell-Mann called the patterns the **eightfold way,** after the eightfold path to nirvana in Buddhism.

Groups of baryons and mesons can be displayed in many other symmetric patterns within the framework of the eightfold way. For example, the family of spin-$\frac{3}{2}$ baryons known in 1961 contained nine particles arranged in a pattern like that of the pins in a bowling alley, as in Figure 15.12. (The particles Σ^{*+}, Σ^{*0}, Σ^{*-}, Ξ^{*0}, and Ξ^{*-} are excited states of the particles Σ^+, Σ^0, Σ^-, Ξ^0, and Ξ^-. In these higher-energy states, the spins of the three quarks making up the particle are aligned so that the total spin of the particle is $\frac{3}{2}$. When this pattern was proposed, an empty spot occurred in it (at the bottom position), corresponding to a particle that had never been observed. Gell-Mann predicted that the missing particle, which he called the omega minus (Ω^-), should have spin $\frac{3}{2}$, charge -1, strangeness -3, and rest energy of approximately 1680 MeV. Shortly thereafter, in 1964, scientists at the Brookhaven National Laboratory found the missing particle through careful analyses of bubble-chamber photographs (Fig. 15.13) and confirmed all its predicted properties.

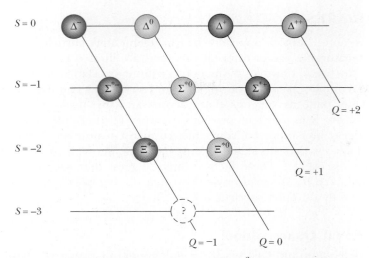

Figure 15.12   The pattern for the higher-mass, spin-$\frac{3}{2}$ baryons known at the time the pattern was proposed. The three Σ^* and two Ξ^* particles are excited states of the corresponding spin-$\frac{1}{2}$ particles in Figure 15.11. These excited states have higher mass and spin $\frac{3}{2}$. The absence of a particle in the bottom position was evidence of a new particle yet to be discovered, the Ω^-.

The prediction of the missing particle in the eightfold way has much in common with the prediction of missing elements in the periodic table. Whenever a vacancy occurs in an organized pattern of information, experimentalists have a guide for their investigations.

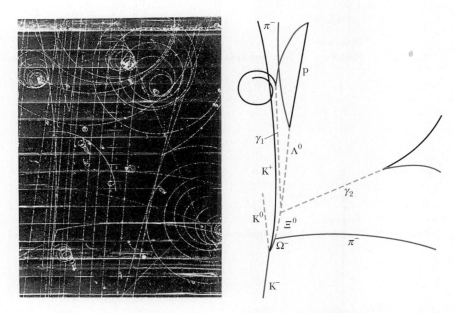

Figure 15.13   Discovery of the Ω^- particle. The photograph on the left shows the original bubble-chamber tracks. The drawing on the right isolates the tracks of the important events. The K^- particle at the bottom collides with a proton to produce the first detected Ω^- particle plus a K^0 and a K^+. (*Courtesy of Brookhaven National Laboratory*)

15.9   QUARKS

As we have noted, leptons appear to be truly elementary particles because they have no measurable size or internal structure, are limited in number, and do not seem to break down into smaller constituents. Hadrons, on the other hand, are complex particles having size and structure. The existence of the eightfold-way patterns suggests that baryons and mesons—in other words, hadrons—have a more elemental substructure. Furthermore, we know that hadrons decay into other hadrons and are many in number. Table 15.2 lists only the hadrons that are stable against hadronic decay; hundreds of others have been discovered. These facts strongly suggest that hadrons cannot be truly elementary. In this section we show that the complexity of hadrons can be explained by a simpler substructure.

The Original Quark Model

In 1963 Gell-Mann and George Zweig independently proposed that hadrons have a more elemental substructure. According to their model, all hadrons are composite systems of two or three fundamental constituents called **quarks** (pronounced to rhyme with *forks*). (Zweig called them "aces." Gell-Mann borrowed the word *quark* from the passage "Three quarks for Muster Mark" in James Joyce's *Finnegan's Wake*.) In the original quark model, there were three types of quarks designated by the symbols u, d, and s. These were given the arbitrary names **up, down,** and *sideways* or, now more commonly, **strange.**

Table 15.3   Properties of Quarks and Antiquarks

				Quarks				
Name	**Symbol**	**Spin**	**Charge**	**Baryon Number**	**Strangeness**	**Charm**	**Bottomness**	**Topness**
Up	u	$\frac{1}{2}$	$+\frac{2}{3}e$	$\frac{1}{3}$	0	0	0	0
Down	d	$\frac{1}{2}$	$-\frac{1}{3}e$	$\frac{1}{3}$	0	0	0	0
Strange	s	$\frac{1}{2}$	$-\frac{1}{3}e$	$\frac{1}{3}$	-1	0	0	0
Charmed	c	$\frac{1}{2}$	$+\frac{2}{3}e$	$\frac{1}{3}$	0	$+1$	0	0
Bottom	b	$\frac{1}{2}$	$-\frac{1}{3}e$	$\frac{1}{3}$	0	0	$+1$	0
Top	t	$\frac{1}{2}$	$+\frac{2}{3}e$	$\frac{1}{3}$	0	0	0	$+1$
				Antiquarks				
Name	**Symbol**	**Spin**	**Charge**	**Baryon Number**	**Strangeness**	**Charm**	**Bottomness**	**Topness**
Anti-up	$\bar{u}$	$\frac{1}{2}$	$-\frac{2}{3}e$	$-\frac{1}{3}$	0	0	0	0
Anti-down	$\bar{d}$	$\frac{1}{2}$	$+\frac{1}{3}e$	$-\frac{1}{3}$	0	0	0	0
Anti-strange	$\bar{s}$	$\frac{1}{2}$	$+\frac{1}{3}e$	$-\frac{1}{3}$	$+1$	0	0	0
Anti-charmed	$\bar{c}$	$\frac{1}{2}$	$-\frac{2}{3}e$	$-\frac{1}{3}$	0	-1	0	0
Anti-bottom	$\bar{b}$	$\frac{1}{2}$	$+\frac{1}{3}e$	$-\frac{1}{3}$	0	0	-1	0
Anti-top	$\bar{t}$	$\frac{1}{2}$	$-\frac{2}{3}e$	$-\frac{1}{3}$	0	0	0	-1

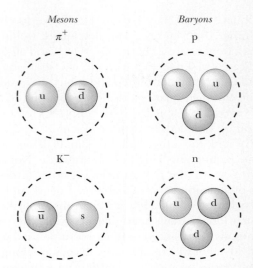

Figure 15.14 Quark compositions of two mesons and two baryons. Note that each meson on the left contains two quarks, while each baryon on the right contains three quarks.

Table 15.4   Quark Compositions of Several Hadrons

Particle	Quark Composition
Mesons	
π^+	$u\bar{d}$
π^-	$\bar{u}d$
K^+	$u\bar{s}$
K^-	$\bar{u}s$
K^0	$d\bar{s}$
Baryons	
p	uud
n	udd
Λ^0	uds
Σ^+	uus
Σ^0	uds
Σ^-	dds
Ξ^0	uss
Ξ^-	dss
Ω^-	sss

An unusual property of quarks is that they had to be assigned fractional electronic charges. The u, d, and s quarks have charges of $+2e/3$, $-e/3$, and $-e/3$, respectively, where e is the charge of a positron. Each quark has a baryon number of $\frac{1}{3}$ and a spin of $\frac{1}{2}$. The u and d quarks have strangeness 0, and the s quark has strangeness -1. Associated with each quark is an antiquark of opposite charge, baryon number, and strangeness. Table 15.3 gives other properties of quarks and antiquarks.

The compositions of all hadrons that were known when Gell-Mann and Zweig presented their models could be completely specified by three simple rules:

- A meson consists of one quark and one antiquark, which gives it a baryon number of 0, as required. Mesons are thus quark–antiquark combinations (quarkonium) bound together by a swarm of gluons, which are the field particles that transmit the strong force between quarks.
- A baryon consists of three quarks and is a sort of quark "molecule" held together by gluons.
- An antibaryon consists of three antiquarks.

Table 15.4 lists the quark compositions of several mesons and baryons. Note that just two of the quarks, u and d, are contained in all hadrons encountered in ordinary matter (protons and neutrons). The third quark, s, is needed only to construct strange particles with a strangeness number of either $+1$ or -1. Figure 15.14 is a pictorial representation of the quark compositions of several particles.

Charm and Other Developments

Although the original quark model was highly successful in classifying particles into families, some discrepancies occurred between predictions of the model and certain experimental decay rates. Consequently, several physicists proposed a fourth quark in 1967. The fourth quark, denoted by c, was given a property

charm. A *charmed* quark would have charge $+2e/3$, but its charm would distinguish it from the other three quarks. The new quark would have a charm of $C = +1$, its antiquark would have a charm of $C = -1$, and all other quarks would have $C = 0$, as indicated in Table 15.3. Charm, like strangeness, would be conserved in strong and electromagnetic interactions but not in weak interactions.

Evidence that the charmed quark exists began to accumulate in 1974 when a new heavy particle called the J/Ψ particle (or simply Ψ) was discovered independently by two groups, one led by Burton Richter at the Stanford Linear Accelerator (SLAC) and the other led by Samuel Ting at the Brookhaven National Laboratory. Richter and Ting were awarded the Nobel prize in 1976 for this work. Although massive, the J/Ψ particle was not a three-quark baryon but had the properties of a combination of a charmed quark and its antiquark ($c\bar{c}$). It was much more massive (~3100 MeV/c^2) than the other known mesons, and its lifetime was much longer than those that decay via the strong force. Soon, related charmed mesons were discovered that corresponded to such quark combinations as $\bar{c}d$ and $c\bar{d}$, all of which have large masses and long lifetimes. The existence of these new mesons provided firm evidence for the fourth quark flavor.

In 1975 researchers at Stanford University reported strong evidence for the tau (τ) lepton, with a mass of 1784 MeV/c^2. This was the fifth type of lepton to be discovered, which led physicists to propose that two new quarks, **top** (t) and **bottom** (b), might exist. (Some physicists prefer the designators *truth* and *beauty*.) To distinguish these quarks from the old ones, quantum numbers called **topness** and **bottomness** (with allowed values $+1$, 0, -1) were assigned to all quarks and antiquarks (see Table 15.3). In 1977 researchers at the Fermi National Laboratory, under the direction of Leon Lederman, reported the discovery of a very massive new meson, Y, whose composition is considered to be $b\bar{b}$. In March 1995 researchers at the Fermi National Laboratory announced the discovery of the top quark (supposedly the last of the quarks), with a mass of 173 GeV/c^2. The researchers identified $t\bar{t}$ pairs from their decay into W bosons and bottom quarks and the subsequent decay of those particles into leptons, neutrinos, and hadrons. The original $t\bar{t}$ pairs were produced by the collision of 0.9 TeV protons with 0.9 TeV antiprotons. An interesting essay by Melissa Franklin and David Kestenbaum at the end of this chapter explains the operation of the accelerator and the particle detectors used to find the "top."

You are probably wondering whether such discoveries will ever end. How many "building blocks" of matter really exist? At the present, physicists believe that the fundamental particles in nature are 12 fermions, including 6 quarks and 6 leptons (together with their antiparticles) and the field particles listed in Table 15.1. Table 15.5 lists properties of these 12 fermions as well as properties of the field particles excepting the graviton.

Despite extensive experimental efforts, no isolated quark has ever been observed. Physicists now believe that quarks are permanently confined inside hadrons because of an exceptionally strong force that prevents them from escaping. This force, called the strong or "color" force, *increases* with separation distance, eventually approaching a constant value—implying a potential energy that grows linearly with quark separation and the requirement of infinite energy to produce two truly free quarks

Table 15.5   Properties of the Fundamental Point Particles

Particle		Approximate Rest Mass in Terms of Proton Mass[a]	Interactions Experienced[b]	Interaction Mediated	Electric Charge in Units of $|e|$	Color Charge
Leptons	e^-, μ, τ	$\frac{1}{1836}, \frac{1}{9}, 1.9$	EM, W	None	-1	No
	ν_e, ν_μ, ν_τ	$<3 \times 10^{-9}, <4 \times 10^{-9},$ $<9 \times 10^{-9}$	W	None	0	No
Quarks	u, c, t	$\frac{1}{2.8}, 1.6, 185$	EM, W, S	None	$+\frac{2}{3}$	Yes
	d, s, b	$\frac{1}{2.8}, \frac{1}{1.7}, 5.3$	EM, W, S	None	$-\frac{1}{3}$	Yes

All are fermions with spin $\frac{1}{2}$.

Field Particles

Photon	γ	0	EM	EM	0	No
Intermediate bosons	$W^\pm$	86	W, EM	W	± 1	No
	Z^0	97	W	W	0	No
Gluons	g	0	S	S	0	Yes

All are bosons with spin 1.

[a]For quarks these are inferred values because free quarks have never been observed.

[b]EM = electromagnetic, W = weak, S = strong

15.10   COLORED QUARKS, OR QUANTUM CHROMODYNAMICS

Shortly after the concept of quarks was proposed, scientists recognized that certain particles had quark compositions that violated the Pauli exclusion principle. Because quarks are fermions with spins of $\frac{1}{2}$, they are expected to follow the exclusion principle. One example of a particle that violates the exclusion principle is the Ω^- (s s s) baryon, which contains three s quarks with parallel spins, giving it a total spin of $\frac{3}{2}$. Other examples of baryons that have identical quarks with parallel spins are the Δ^{++} (u u u) and the Δ^- (d d d). To resolve this problem, Moo-Young Han and Yoichiro Nambu suggested in 1965 that quarks possess a new property, called **color** or **color charge.** This property is similar in many respects to electric charge except that it occurs in three varieties, called red, green, and blue. (The antiquarks have the colors antired, antigreen, and antiblue.) To satisfy the exclusion principle, all three quarks in a baryon must have *different* colors. Just as a combination of actual colors of light can produce the neutral color white, a combination of three quarks with different colors is said to be white, or colorless. A meson consists of a quark of one color and an antiquark of the corresponding anticolor. The result is that both baryons and mesons are always colorless (or white).

Although the concept of color in the quark model was originally conceived to satisfy the exclusion principle, it also provided an improved theory for explaining certain experimental results. For example, the modified theory correctly predicts the lifetime of the π^0 meson. The general theory of how quarks interact with each other is called **quantum chromodynamics** (QCD), to parallel quantum

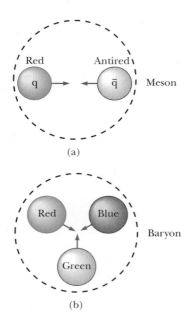

Figure 15.15 (a) A red quark is attracted to an antired quark. This forms a meson whose quark structure is (q$\overline{q}$). (b) Three differently colored quarks attract each other to form a baryon.

electrodynamics, the theory of interaction between electric charges. In QCD the quark is said to carry a **color charge,** in analogy with electric charge, and the color charge is responsible for the strong force between quarks. The strong force between quarks is often called the **color force.** The color force between quarks is analogous to the electric force between charges; like colors repel and opposite colors attract. Therefore, two red quarks repel each other, but a red quark is attracted to an antired quark. The attraction between quarks of opposite color to form a meson (q$\overline{q}$) is indicated in Figure 15.15a. Differently colored quarks also attract each other, but with less intensity than the oppositely colored quark and antiquark. For example, the red, blue, and green quarks in a cluster all attract each other to form baryons, as indicated in Figure 15.15b. Thus every baryon contains three quarks of three different colors.

As stated earlier, the strong force between quarks is carried by massless particles that travel at the speed of light called **gluons** (g). According to QCD, there are eight gluons, *all carrying two color charges, a color and an anticolor,* such as a "blue-antired" gluon. When a quark emits or absorbs a gluon, its color changes. For example, a blue quark that emits a blue-antired gluon becomes a red quark, and a red quark that absorbs this gluon becomes a blue quark. Because gluons carry color charge, they should clump together to form entities called **glue-balls,** but no glue-ball particles have yet been experimentally detected.

Experimental Evidence for Quarks

As already mentioned, an isolated quark has never been observed and probably never will be, according to QCD. This leaves us with the question of how quarks can be detected and what experimental evidence we have for the existence of quarks.

Experiments at the Stanford Linear Accelerator (SLAC) in the late 1960s established two important results concerning quarks. (1) The recoils of high-energy electrons scattered from protons could be completely modeled by assuming *three* point-like quarks within the proton. (2) Although gluons were not directly detected, it was found that the incident electrons (which carry no color charge) interacted with only about half the mass making up the proton. This is exactly what is expected if approximately half the material in a proton consists of gluons that are electrically neutral and do not interact electromagnetically with the electron (see Table 15.5).

It is interesting to consider what happens to a quark–antiquark pair (q$\overline{q}$) formed in a high-energy collision as the pair attempts to separate. Because the color force is so strong, all quarks must form colorless baryons (three quarks) or mesons (two quarks). The process goes like this. As two quarks separate, the increase in energy of the color field between them is so large that additional quark–antiquark pairs are created in streams (or jets) following the original (q$\overline{q}$) pair. When all the available energy is used up, the quarks cluster into color-neutral or colorless combinations, which continue to separate indefinitely as two jets of hadrons, as in Figure 15.16a. Figure 15.16b shows actual experimental data corresponding to three jets of hadrons emerging from a collision in which a quark and antiquark were produced and a gluon was radiated. Because gluons carry color charge, they, like quarks,

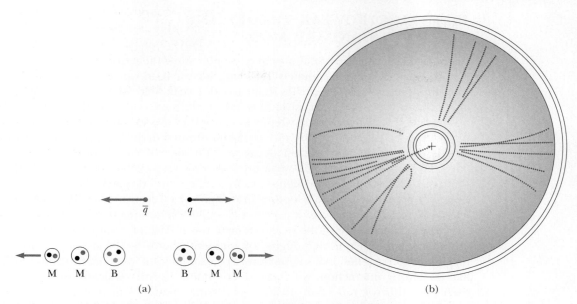

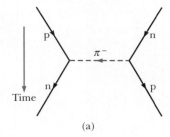

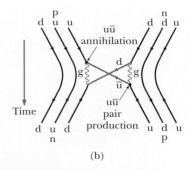

Figure 15.16   (a) As the quark–antiquark pair separates, the potential energy in the color field is transformed into additional q$\overline{q}$ pairs, which quickly condense into color-less mesons (M) and baryons (B). These separate as jets of hadrons. (b) Three jets of hadrons produced by a quark, an antiquark, and a gluon. This figure is from the JADE detector at the German laboratory DESY in Hamburg. (*Adapted from Gordon Kane,* The Particle Garden, *Figure 6.2, p. 100*)

must also appear as a jet of hadrons. Thus *experimental signatures of quarks and gluons are narrow "jets" of hadrons.*

Explanation of Nuclear Force in Terms of Quarks

Although the color force between two color-neutral hadrons is negligible at large separations, the strong color force between their constituent quarks does not exactly cancel at small separations. This residual strong force is, in fact, the short-range nuclear force that binds protons and neutrons to form nuclei. In Section 15.3 we explained the nuclear interaction of a proton with a neutron using Yukawa's early theory of pion exchange. According to QCD, a more basic explanation of nuclear force can be given in terms of quarks and gluons, as shown by contrasting Feynman diagrams of the same process in Figure 15.17. Each quark within the neutron and proton is continually emitting and absorbing virtual gluons and creating and annihilating virtual (q$\overline{q}$) pairs. When the neutron and proton approach within 1 fm of each other, these virtual gluons and quarks can be exchanged between the two nucleons, and such exchanges produce the nuclear force. Figure 15.17b depicts one likely possibility or contribution to the general process shown in Figure 15.17a. A down quark emits a virtual gluon (represented by a wavy line, g), which creates, a little later, a u$\overline{u}$ pair. Both the recoiling d quark and the $\overline{u}$ are transmitted to the proton, where the $\overline{u}$ annihilates a proton u quark (with the creation of a gluon) and the d is captured.

Figure 15.17   (a) A nuclear interaction between a proton and a neutron explained in terms of Yukawa's pion-exchange model. (b) The same interaction as in (a), explained in terms of quarks and gluons. Note that the exchanged $\overline{u}$d quark pair makes up a π^- meson.

15.11 ELECTROWEAK THEORY AND THE STANDARD MODEL

Recall that the weak interaction is an extremely short-range force with an interaction distance of approximately 10^{-18} m. Such a short-range interaction implies that the quantized particles that carry the weak field (the spin-1 W^+, W^-, and Z^0 bosons) are quite massive, as is indeed the case (see Table 15.5). These bosons are especially amazing when we realize they are structureless, point-like particles as massive as krypton atoms! As mentioned earlier, the weak interaction is responsible for neutron decay and the beta decay of other, heavier baryons. More important, the weak interaction is responsible for the decay of the c, s, b, and t quarks into lighter, more stable u and d quarks as well as the decay of the massive μ and τ leptons into (lighter) electrons. Thus, *the weak interaction is very important because it governs the stability of the basic matter particles.*

A mysterious feature of the weak interaction is its lack of symmetry, especially in comparison with the high degree of symmetry shown by the strong, electromagnetic, and gravitational interactions. For example, the weak interaction, unlike the strong interaction, is not symmetric under mirror reflection or charge exchange. (Mirror reflection means that all the quantities in a given particle reaction are exchanged as in a mirror reflection—left for right, an inward motion toward the mirror for an outward motion. Charge exchange means that all the electric charges in a particle reaction are converted to their opposites—all positives to negatives and vice versa.) When we say that the weak interaction is not symmetric, we mean that the reaction with all quantities changed occurs less frequently than the direct reaction. For example, the decay of the K^0, which is governed by the weak interaction, is not symmetric under charge exchange, since

$$K^0 \longrightarrow \pi^- + e^+ + \nu_e$$

occurs much more frequently than

$$K^0 \longrightarrow \pi^+ + e^- + \bar{\nu}_e$$

In 1979 Sheldon Glashow, Abdus Salam, and Steven Weinberg won a Nobel prize for developing a theory that unifies the electromagnetic and weak interactions. This **electroweak theory** postulates that *the weak and electromagnetic interactions have the same strength at very high particle energies.* Thus the two interactions are viewed as two different manifestations of a single, unifying electroweak interaction. The photon and the three massive bosons ($W^\pm$ and Z^0) play key roles in the electroweak theory. Perhaps the most spectacular of the theory's many concrete predictions was the prediction of the masses of the W and Z particles at about 80 GeV/c^2 and 91 GeV/c^2, respectively. The 1984 Nobel prize was awarded to Carlo Rubbia and Simon van der Meer for their work leading to the discovery of these particles with just those masses at CERN.

The combination of the electroweak theory and QCD for the strong interaction is called **the Standard Model** in high-energy physics. It includes almost all the constituents of matter—six leptons, six quarks, and three forces and their field particles—but not the gravitational force at this time. Physicists, in continuing pursuit of unification, hope that string theory, mentioned in the next section, will provide the unification of gravity with the Standard Model.

The Standard Model does not answer all questions. It requires as input parameters over a dozen measured numbers such as the lepton and quark masses. Further, it does not explain the mysterious ratios of these masses, why there are only six types of leptons, or why there are six types of *both* leptons and quarks. A specific question regarding field particles is why the photon has no mass while the W and Z bosons do have mass. Because of this difference, the electromagnetic and weak forces are quite distinct at low energies but become similar, or symmetric, at very high energies, where the rest energies of the W and Z bosons are insignificant fractions of their total energies. This behavior in the transition from high to low energies is called **symmetry breaking.** In 1964 Peter Higgs (b. 1929, Scottish physicist) introduced a mechanism for electroweak symmetry breaking by proposing a new field, called the **Higgs field,** which permeates all of space and gives particles their mass. Roughly, the Higgs field may be viewed as causing a kind of drag force on particles as they interact with it, giving particles their characteristic inertia. As with all other classical fields, quantization of the Higgs field results in a force-carrying particle called the **Higgs boson.** The Standard Model, including the Higgs mechanism, provides a logically consistent explanation of the massive nature of the W and Z bosons. Unfortunately, the Higgs boson has not yet been found, but physicists think its mass should be less than 1 TeV (10^{12} eV).

To determine whether the Higgs boson exists, two quarks of at least 1 TeV of energy must collide, but calculations show that this requires injecting 40 TeV of energy within the volume of a proton. Scientists are convinced that because of the limited energy available in conventional accelerators using fixed targets, it is necessary to build colliding-beam accelerators called **colliders.** The concept of colliders is straightforward. Particles with equal masses and kinetic energies, traveling in opposite directions in an accelerator ring, collide head-on to produce the required reaction and form new particles. Because the total momentum of the interacting particles is zero, all of their kinetic energy is available for the reaction. The Large Electron–Positron Collider (LEP) at CERN and the Stanford Linear Collider in California collide electrons and positrons. The Super Proton Synchrotron at CERN accelerates protons and antiprotons to

Peter W. Higgs (b. 1930). (*Courtesy of Peter Tuffy/Edinburgh University*)

energies of 270 GeV, while the world's highest-energy proton accelerator, the Tevatron, at the Fermi National Laboratory in Illinois produces protons at almost 1000 GeV (1 TeV). CERN expects a 2007 completion of the Large Hadron Collider (LHC), a proton–proton collider that will provide a center-of-mass energy of 14 TeV and allow an exploration of Higgs boson physics. The accelerator is being constructed in the same 27-km-circumference tunnel as CERN's LEP Collider, and many countries are participating in the project.

15.12   BEYOND THE STANDARD MODEL

Grand Unification Theory and Supersymmetry

Following the success of the electroweak theory, scientists attempted to combine it with QCD in a **grand unification theory,** or **GUT.** In this model, the next step is taken of merging the electroweak force with the strong color force to form a grand unified force. GUT considers leptons and quarks to be specific states of a single particle called a *leptoquark* and it is this identity that leads to the same number of flavors for leptons and quarks. Also, because leptons and quarks are states of the same particle, GUT predicts that quarks and leptons should be able to change into each other given sufficient time. Thus GUT predicts that quark-filled protons are unstable and will decay with a lifetime of about 10^{32} years to a positron, which is a lepton, and other nonbaryons. Attempts to detect such proton decays have so far been unsuccessful.

The search for unification has also led to another beautiful symmetry principle, **Supersymmetry (SUSY).** According to this principle the fundamental equations of nature are unchanged by the exchange of a fermion for a boson in these equations. SUSY suggests that every elementary particle has a **superpartner,** called a **sparticle,** although no sparticle has yet been observed. It is believed that supersymmetry is a broken symmetry (like the broken electroweak symmetry at low energies) and that the masses of the superpartners are too large to be produced in current accelerators. Continuing with the fun and whimsey in naming particles and their properties, superpartners are given the names *squarks* and *sleptons* (the boson superpartners of quarks and leptons) and *photinos, winos,* and *gluinos* (the fermion superpartners of the field bosons—photons, $W^{\pm}$ s, and gluons).

String Theory—A New Perspective

String theory is an effort to *unify* the four fundamental forces by modeling all particles as various vibrational modes of a single entity—an incredibly small string. The typical length of such a string is on the order of 10^{-35} m, called the **Planck length.** In string theory each quantized mode of vibration of the string corresponds to a different elementary particle in the Standard Model.

One of the complicating factors in string theory is that it requires spacetime to have 10 dimensions. Despite the theoretical and conceptual difficulties in dealing with 10 dimensions, string theory holds promise *in incorporating gravity with the other forces.* Four of the 10 dimensions are visible to us—3 space dimensions and 1 time dimension—and the other 6 are said to be *compactified.* That is, the 6 dimensions are curled up so tightly that they are not visible in the macroscopic world.

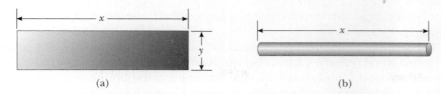

(a)   (b)

Figure 15.18   (a) A piece of paper is cut into a rectangular shape. As a rectangle, the shape has two dimensions. (b) The paper is rolled up into a soda straw. From far away, it appears to be one-dimensional. The curled-up second dimension is not visible when viewed from a distance large compared to the diameter of the straw.

As an analogy, consider a soda straw. We can build a soda straw by cutting a rectangular piece of paper (Fig. 15.18a), which clearly has two dimensions, and rolling it up into a small tube (Fig. 15.18b). From far away, the soda straw looks like a one-dimensional straight line. The second dimension has been curled up and is not visible. String theory claims that 6 space-time dimensions are curled up in an analogous way, with the curling being on the size of the Planck length—impossible to see from our viewpoint.

Another complicating factor with string theory is that it is difficult for string theorists to guide experimentalists as to what to look for in an experiment. The Planck length is so small that direct experimentation on strings is impossible. Until the theory has been further developed, string theorists are restricted to applying the theory to known results and testing for consistency.

Other theorists are working on **M-theory,** which is an 11-dimensional theory based on membranes rather than strings. In a way reminiscent of the correspondence principle, M-theory is claimed to reduce to string theory if one compactifies from 11 dimensions to 10 dimensions.

SUMMARY

The strong, electromagnetic, weak, and gravitational forces are the four fundamental forces in nature. All the interactions in which these forces take part are mediated by **field particles.** The electromagnetic interaction is mediated by the photon; the weak interaction is mediated by the $W^{\pm}$ and Z^0 bosons; the gravitational interaction is mediated by gravitons; the strong interaction is mediated by gluons.

A charged particle and its **antiparticle** have the same mass but opposite charge, and other properties may have opposite values, such as lepton number and baryon number. It is possible to produce particle–antiparticle pairs in nuclear reactions if the available energy is greater than $2mc^2$, where m is the mass of the particle (or antiparticle).

Particles other than field particles are classified as hadrons or leptons. **Hadrons** interact via all four fundamental forces. They have size and structure and are not elementary particles. There are two types—**baryons** and **mesons.** Baryons, which generally are the most massive particles, have nonzero **baryon number** and a spin of $\frac{1}{2}$ or $\frac{3}{2}$. Mesons have baryon number zero and either zero or integral spin.

Leptons have no structure or size and are considered truly elementary. They interact only via the weak, gravitational, and electromagnetic forces. Six types of leptons exist: the electron e^-, the muon μ^-, the tau τ^-; and their neutrinos ν_e, ν_μ, and ν_τ.

In reactions and decays, quantities such as energy, linear momentum, angular momentum, electric charge, baryon number, and lepton number are generally conserved. Conservation of relativistic energy and momentum yields the following expression for the threshold energy for particle production:

$$K_{\text{th}} = \frac{(m_3 + m_4 + m_5 + \cdots)^2 c^2 - (m_1 + m_2)^2 c^2}{2m_2}$$

Here, K_{th} is the kinetic energy of the incident particle of mass m_1, m_2 is the mass of the stationary target particle, and the remaining m's are product-particle masses.

Certain particles have properties called **strangeness** and **charm.** These unusual properties are conserved only in the decays and nuclear reactions that occur via the strong force.

Theorists in elementary particle physics have postulated that all hadrons are composed of smaller units known as **quarks,** and experimental evidence agrees with this model. Quarks have fractional electric charge and come in six **flavors:** up (u), down (d), strange (s), charmed (c), top (t), and bottom (b). Each baryon contains three quarks, and each meson contains one quark and one antiquark.

According to the theory of **quantum chromodynamics,** quarks have a property called **color charge,** and the force between quarks is referred to as the **strong force** or the **color force.** The strong force is now considered to be a fundamental force. The nuclear force, which was originally considered to be fundamental, is now understood to be a secondary effect of the strong force, due to gluon exchanges between hadrons.

The electromagnetic and weak forces are now considered to be manifestations of a single force called the **electroweak force.** The combination of quantum chromodynamics and the electroweak theory is called the **Standard Model.**

SUGGESTIONS FOR FURTHER READING

Particle Physics

1. J. Bahcall, "The Solar Neutrino Problem," *Sci. Amer.*, May 1990.
2. H. Breuker et al., "Tracking and Imaging Elementary Particles," *Sci. Amer.*, August 1991.
3. F. Close, *The Cosmic Onion: Quarks and the Nature of the Universe*, New York, The American Institute of Physics, 1986. A timely monograph on particle physics, including lively discussions of the Big Bang theory.
4. H. Fritzsch, *Quarks, The Stuff of Matter*, London, Allen and Lane, 1983. An excellent introductory overview of elementary particle physics.
5. S. Glashow, *Interactions: A Journey Through the Mind of a Particle Physicist*, New York, Warner Books, 1988.
6. H. Harari, "The Structure of Quarks and Leptons," *Sci. Amer.*, April 1983.

7. L. M. Lederman, "The Value of Fundamental Science," *Sci. Amer.*, November 1984.
8. N. B. Mistry, R. A. Poling, and E. H. Thorndike, "Particles with Naked Beauty," *Sci. Amer.*, July 1983.
9. C. Quigg, "Elementary Particles and Forces," *Sci. Amer.*, April 1985.
10. M. Riordan, "The Discovery of Quarks," *Science* 29, May, 1992.
11. M. J. G. Veltman, "The Higgs Boson," *Sci. Amer.*, November 1986.
12. E. Witten, "Reflections on the Fate of Spacetime," *Physics Today*, April 1996. This article discusses the string theory of elementary particles.
13. M. Perl, "The Leptons After 100 Years," *Physics Today*, October 1997. An excellent overview of leptons by Nobel prize winner and discoverer of the tau lepton, Martin L. Perl.

14. G. Kane, *Supersymmetry*, Cambridge, MA, Helix Books, 2000. A clear and nonmathematical description of supersymmetry.

15. B. Greene, *The Elegant Universe*, New York, Vintage Books, 2000. A clear and nonmathematical description of string theory.

QUESTIONS

1. Name the four fundamental interactions and the field particle that mediates each.
2. Describe the quark model of hadrons, including the properties of quarks.
3. What are the differences between hadrons and leptons?
4. Describe the properties of baryons and mesons and the important differences between them.
5. Particles known as resonances have very short lifetimes, of the order of 10^{-23} s. From this information, would you guess they are hadrons or leptons? Explain.
6. Kaons all decay into final states that contain no protons or neutrons. What is the baryon number of kaons?
7. The Ξ^0 particle decays by the weak interaction according to the decay mode $\Xi^0 \rightarrow \Lambda^0 + \pi^0$. Would you expect this decay to be fast or slow? Explain.
8. Identify the particle decays listed in Table 15.2 that occur by the weak interaction. Justify your answers.
9. Identify the particle decays listed in Table 15.2 that occur by the electromagnetic interaction. Justify your answers.
10. Two protons in a nucleus interact via the strong interaction. Are they also subject to the weak interaction?
11. Discuss the following conservation laws: energy, linear momentum, angular momentum, electric charge, baryon number, lepton number, and strangeness. Are all of these laws based on fundamental properties of nature? Explain.
12. An antibaryon interacts with a meson. Can a baryon be produced in such an interaction? Explain.
13. Describe the essential features of the Standard Model of particle physics.
14. How many quarks are in (a) a baryon? (b) an antibaryon? (c) a meson? (d) an antimeson? How do you account for the fact that baryons have half-integral spins while mesons have spins of 0 or 1? (*Hint:* Quarks have spin $\frac{1}{2}$.)
15. In the theory of quantum chromodynamics, quarks come in three colors. How would you justify the statement that "all baryons and mesons are colorless"?
16. Which baryon did Murray Gell-Mann predict in 1961? What is the quark composition of this particle?
17. What is the quark composition of the Ξ^- particle? (See Table 15.4.)
18. The $W^\pm$ and Z bosons were first produced at CERN in 1983 (by the collision of a beam of protons and a beam of antiprotons at high energy). Why was this an important discovery?

PROBLEMS

15.1 The Fundamental Forces in Nature

1. When a high-energy proton or pion traveling near the speed of light collides with a nucleus, it travels an average distance of 3×10^{-15} m before interacting. From this information, estimate the time required for the strong interaction to occur.
2. The neutral ρ^0 meson decays by the strong interaction into two pions according to $\rho^0 \rightarrow \pi^+ + \pi^-$, with a half-life of about 10^{-23} s. The neutral K^0 meson also decays into two pions according to $K^0 \rightarrow \pi^+ + \pi^-$ but with a much longer half-life, about 10^{-10} s. How do you explain these observations?

15.2 Positrons and Other Antiparticles

3. Two photons are produced when a proton and an antiproton annihilate each other. What are the minimum frequency and the corresponding wavelength of each photon?
4. A photon produces a proton–antiproton pair at rest according to the reaction $\gamma \rightarrow p + \bar{p}$. If a nearby nucleus of mass 100 u (initially at rest) carries off the photon's momentum, what is the frequency of the photon? What is its wavelength?

15.3 Mesons and the Beginning of Particle Physics

5. One of the mediators of the weak interaction is the Z^0 boson, which has a mass of 91 GeV/c^2. Use this information to find an approximate value for the range of the weak interaction.

15.5 Conservation Laws

6. High-energy muons occasionally collide with electrons and produce two neutrinos according to the reaction $\mu^+ + e^- \rightarrow 2\nu$. What kind of neutrinos are these?
7. (a) Show that baryon number and charge are conserved in the following reactions of a pion with a proton.

$$\pi^- + p \longrightarrow K^- + \Sigma^+ \qquad (1)$$
$$\pi^- + p \longrightarrow \pi^- + \Sigma^+ \qquad (2)$$

(b) The first reaction is observed, but the second never occurs. Explain these observations.

8. The following reactions or decays involve one or more neutrinos. Supply the missing neutrinos (ν_e, ν_μ, or ν_τ).
   (a) $\pi^- \to \mu^- + ?$
   (b) $K^+ \to \mu^+ + ?$
   (c) $? + p \to n + e^+$
   (d) $? + n \to p + e^-$
   (e) $? + n \to p + \mu^-$
   (f) $\mu^- \to e^- + ? + ?$

9. Determine which of the following reactions can occur. For those that cannot occur, determine the conservation law (or laws) that is violated.
   (a) $p \to \pi^+ + \pi^0$
   (b) $p + p \to p + p + \pi^0$
   (c) $p + p \to p + \pi^+$
   (d) $\pi^+ \to \mu^+ + \nu_\mu$
   (e) $n \to p + e^- + \overline{\nu}_e$
   (f) $\pi^+ \to \mu^+ + n$

15.6   Strange Particles and Strangeness

10. Determine whether strangeness is conserved in each of the following decays and reactions:
    (a) $\Lambda^0 \to p + \pi^-$
    (b) $\pi^- + p \to \Lambda^0 + K^0$
    (c) $\overline{p} + p \to \overline{\Lambda}^0 + \Lambda^0$
    (d) $\pi^- + p \to \pi^- + \Sigma^+$
    (e) $\Xi^- \to \Lambda^0 + \pi^-$
    (f) $\Xi^0 \to p + \pi^-$

11. The following decays are forbidden. Determine a conservation law that each violates.
    (a) $\mu^- \to e^- + \gamma$
    (b) $n \to p + e^- + \nu_e$
    (c) $\Lambda^0 \to p + \pi^0$
    (d) $p \to e^+ + \pi^0$
    (e) $\Xi^0 \to n + \pi^0$

12. The following reactions are forbidden. Determine a conservation law that each violates.
    (a) $p + \overline{p} \to \mu^+ + e^-$
    (b) $\pi^- + p \to p + \pi^+$
    (c) $p + p \to p + \pi^+$
    (d) $p + p \to p + p + n$
    (e) $\gamma + p \to n + \pi^0$

15.7   How Are Elementary Particles Produced and Particle Properties Measured?

13. In 1959 Emilio Segrè and Owen Chamberlain were awarded the Nobel prize for demonstrating the existence of the antiproton. In a series of experiments started in 1955, using the Bevatron accelerator at Berkeley, they produced both antiprotons and antineutrons in the reactions

$$p + p \longrightarrow p + p + p + \overline{p}$$

$$p + p \longrightarrow p + p + n + \overline{n}$$

(a) Calculate the threshold kinetic energy of the incident proton (fixed-target proton) required for the production of an antiproton, $\overline{p}$. (b) For the same initial conditions, calculate the threshold kinetic energy needed for the production of an antineutron, $\overline{n}$.

14. Calculate the threshold for production of strange particles in the following reactions. Assume that the first particle is moving and the second is at rest.

(a) $p + p \longrightarrow n + \Sigma^+ + K^0 + \pi^+$

(b) $\pi^- + p \longrightarrow \Sigma^0 + K^0$

15. *Efficiency of fixed-target accelerators vs. colliding-beam accelerators.* The efficiency is the percentage of the initial kinetic energy that is converted to the mass of new product particles in a given reaction. Calculate the efficiencies of the reactions in Example 15.5, Exercise 3, and Problem 13. If one could arrange these experiments so that the two colliding protons approached each other with equal speed at threshold, the total momentum would be zero and none of the initial kinetic energy would go into kinetic energy of the products. All of the initial kinetic energy would then go into creating new mass, and the efficiency would be 100%. That is why most current experiments involving the production of heavy particles are carried out with colliding-beam accelerators. Although there are formidable experimental difficulties in storing, focusing, and causing two oppositely circulating low-density beams of particles to collide, the great gain in efficiency makes colliders worthwhile.

16. Consider the reaction $p + p \to p + p + X$. (a) For a fixed-target accelerator capable of producing incident protons with a kinetic energy of 1000 GeV, find the heaviest particle X that can be produced. (b) If colliding protons, each with a kinetic energy of 500 GeV, are available in a collider, what is the mass of the heaviest particle X that can be produced in the same reaction?

15.9   Quarks

17. A Σ^0 particle traveling through matter strikes a proton, and a Σ^+ and a gamma ray, as well as a third particle, emerge. Use the quark models of the Σ^+ and the gamma ray to determine the identity of the third particle.

18. The quark compositions of the K^0 and Λ^0 particles are $d\overline{s}$ and uds, respectively. Show that the charge, baryon number, and strangeness of these particles equal the sums of these numbers for the quark constituents.

19. Neglecting binding energies, estimate the masses of the u and d quarks from the masses of the proton and neutron.

20. Figure P15.20 shows a neutron–proton interaction in which a virtual π^+ is exchanged. Draw a Feynman diagram showing this interaction in terms of quarks and gluons.

21. A neutron undergoing beta decay is believed to emit a virtual W^- that decays into an e^- and an antielectron neutrino, $\bar{\nu}_e$. Draw a Feynman diagram showing this decay in terms of quarks, the W^-, the e^-, and the $\bar{\nu}_e$.

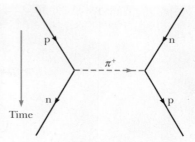

Figure P15.20

ADDITIONAL PROBLEMS

22. *Neutrino flavor oscillation* Neutrino oscillation is based on the idea that the ν_e, ν_μ, and ν_τ neutrinos are mixtures of three other neutrinos called ν_1, ν_2, and ν_3. For example,

$$\nu_\tau = U_{\tau 1}\nu_1 + U_{\tau 2}\nu_2 + U_{\tau 3}\nu_3$$

where each of the neutrinos ν_1, ν_2, and ν_3 is represented as a stationary state solution to the Schrodinger equation and each has a different mass m_1, m_2, and m_3. For algebraic simplicity consider only oscillations of the ν_μ, ν_τ system. We limit the nonzero U coefficients to those connecting ν_2 and ν_3 and write the Us in terms of a single variable θ, called the mixing angle:

$$\nu_\mu = \cos\theta\nu_2 - \sin\theta\nu_3$$

$$\nu_\tau = \sin\theta\nu_2 + \cos\theta\nu_3$$

The mixing angle controls the amount of ν_2 and ν_3 present in ν_μ and ν_τ. Note that there is no mixing for $\theta = 0$. In a typical experiment a ν_μ (beam) with a given momentum, p, is created at $t = 0$. The time evolution of the ν_μ as it moves downstream is given by

$$\nu_\mu(t) = \cos\theta\nu_2 e^{-iE_2 t/\hbar} - \sin\theta\nu_3 e^{-iE_3 t/\hbar}$$

where ν_2 and ν_3 are stationary states with energies E_2 and E_3 respectively.
(a) Write a similar expression for $\nu_\tau(t)$.
(b) Express, mathematically, the initial condition that the neutrino created at $t = 0$ is entirely a μ neutrino. (*Hint:* one condition is $|\nu_\tau(0)|^2 = 0$)
(c) Show that the probability of finding a τ neutrino at time t (or that the μ neutrino has oscillated into a τ neutrino) is

$$P(\nu_\mu \longrightarrow \nu_\tau) = \frac{\sin^2 2\theta}{2}\left(1 - \cos\frac{E_2 - E_3}{\hbar}t\right)$$

(d) Show that for a fixed momentum p and small masses m_2 and m_3 ($m_2 c$, $m_3 c \ll p$)

$$E_2 - E_3 \cong \frac{(m_2^2 - m_3^2)c^3}{2p}$$

The results of (c) and (d) show that observation of neutrino oscillation fixes the difference of the squared masses and determines the mixing angle.

23. The most recent naked-eye supernova was Supernova Shelton 1987A (Fig. P15.23). It was 170,000 ly away in the next galaxy to ours, the Large Magellanic Cloud. About 3 h before its optical brightening was noticed, two continuously running neutrino detection experiments simultaneously registered the first neutrinos from an identified source other than the Sun. The Irvine-Michigan-Brookhaven experiment in a salt mine in Ohio registered eight neutrinos over a 6-second period, and the Kamiokande II experiment in a zinc mine in Japan counted 11 neutrinos in 13 s. (Because the supernova is far south in the sky, these neutrinos entered the detectors from below. They passed through the Earth before they were by chance absorbed by nuclei in the detectors.) The neutrino energies were between about 8 MeV and 40 MeV. If neutrinos have no mass, then neutrinos of all energies should travel together at the speed of light—the data are consistent with this possibility. The arrival times could show scatter simply because neutrinos were created at different moments as the core of the star collapsed into a neutron star. If neutrinos have nonzero mass, then lower-energy neutrinos should move comparatively slowly. The data are consistent with a 10-MeV neutrino requiring at most about 10 s more than a photon would require to travel from the supernova to us. Find the upper limit that this observation sets on the mass of a neutrino. (Other evidence sets an even tighter limit.)

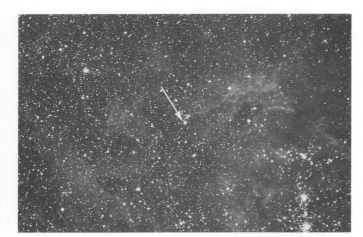

Figure P15.23   The giant star Sanduleak −69° 202 in the "before" picture became Supernova Shelton 1987A in the "after picture." (*Anglo-Australian Telescope Board*)

24. An unstable particle, initially at rest, decays into a proton (rest energy 938.3 MeV) and a negative pion (rest energy 139.5 MeV). A uniform magnetic field of 0.250 T exists perpendicular to the velocities of the created particles. The radius of curvature of each track is found to be 1.33 m. What is the rest mass of the original unstable particle?

25. Calculate the kinetic energies of the proton and pion resulting from the decay of a Λ^0 at rest:

$$\Lambda^0 \longrightarrow p^+ + \pi^-$$

26. A Σ^0 particle at rest decays according to

$$\Sigma^0 \longrightarrow \Lambda^0 + \gamma$$

Find the gamma-ray energy.

27. If a K^0 meson at rest decays in 0.90×10^{-10} s, how far will a K^0 meson travel if it is moving at $0.96c$ through a bubble chamber?

28. A pi meson at rest decays according to $\pi^- \rightarrow \mu^- + \bar{\nu}_\mu$. What is the energy carried off by the neutrino? (Assume that the neutrino moves off with the speed of light.) $m_\pi c^2 = 139.5$ MeV, $m_\mu c^2 = 105.7$ MeV, $m_\nu = 0$.

29. Two protons approach each other with 70.4 MeV of kinetic energy and engage in a reaction in which a proton and positive pion emerge at rest. What third particle, obviously uncharged and therefore difficult to detect, must have been created?

30. What processes are described by the Feynman diagrams in Figure P15.30? What is the exchanged particle in each process?

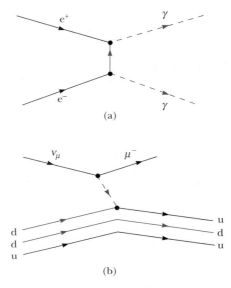

Figure P15.30

31. Identify the mediators for the two interactions described in the Feynman diagrams in Figure P15.31.

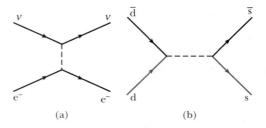

Figure P15.31

32. The Λ^0 is an unstable particle that decays into a proton and a negatively charged pion. Determine the kinetic energies of the proton and pion if the Λ^0 is at rest when it decays. The rest mass of the Λ^0 is

1115.7 MeV/c^2, the mass of the π^- is 139.5 MeV/c^2, and the mass of the proton is 938.3 MeV/c^2.

33. A free neutron beta-decays by creating a proton, an electron, and an antineutrino according to the reaction n $\rightarrow$ p + e$^-$ + $\bar{\nu}$. For simplicity assume that a free neutron beta-decays by creating a proton and electron according to the reaction

$$n \longrightarrow p + e^-$$

and assume that the neutron is initially at rest in the laboratory. (a) Determine the energy released in this reaction. (b) Determine the speeds of the proton and electron after the reaction. (Energy and momentum are conserved in the reaction.) (c) Are any of these particles moving at relativistic speeds? Explain.

34. The quark composition of the proton is uud, while that of the neutron is udd. Show that the charge, baryon number, and strangeness of these particles equal the sums of those same numbers for their quark constituents.

35. The particle decay $\Sigma^+ \rightarrow \pi^+ + n$ is observed in a bubble chamber. Figure P15.35 represents the curved tracks of the particles Σ^+ and π^+ and the invisible track of the neutron, in the presence of a uniform magnetic field of 1.15 T directed out of the page. The measured radii of curvature are 1.99 m for the Σ^+ particle and 0.580 m for the π^+ particle. (a) Find the momenta of the Σ^+ and the π^+ particles, in units of MeV/c. (b) The angle between the momenta of the Σ^+ and the π^+ particles at the moment of decay is 64.5°. Find the momentum of the neutron. (c) Calculate the total energy of the π^+ particle, and of the neutron, from their known masses ($m_\pi = 139.6$ MeV/c^2, $m_n = 939.6$ MeV/c^2) and the relativistic energy–momentum relation. What is the total energy of the Σ^+ particle? (d) Calculate the mass and speed of the Σ^+ particle.

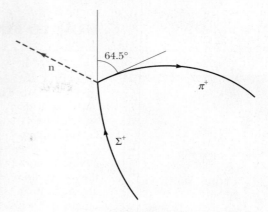

Figure P15.35

36. Supernova 1987A, located about 170,000 ly from the Earth, is estimated to have emitted a burst of neutrinos carrying energy ~10^{46} J (Fig. P15.23). Suppose the average neutrino energy was 6 MeV and your body presented cross-sectional area 5000 cm^2. To an order of magnitude, how many of these neutrinos passed through you?

37. A rocket engine for space travel using photon drive and matter–antimatter annihilation has been suggested. Suppose the fuel for a short-duration burn consists of N protons and N antiprotons, each with mass m. (a) Assume all of the fuel is annihilated to produce photons. When the photons are ejected from the rocket, what momentum can be imparted to it? (b) If half of the protons and antiprotons annihilate each other and the energy released is used to eject the remaining particles, what momentum could be given to the rocket? Which scheme results in the greatest change in speed for the rocket?

HOW TO FIND A TOP QUARK

Melissa Franklin

Harvard University

David Kestenbaum

Harvard University

In 1932 Carl Andersen observed a new particle in his cosmic ray experiment at Pike's Peak. This was the muon, the first in a second generation of leptons. "Who ordered that?," exclaimed the Nobel prize winner I. I. Rabi. Nature had spoken, and the theory had to listen. The Standard Model was expanded to include a second lepton generation.

In 1977 when the discovery of the bottom quark ushered in the addition of a third family of quarks, the Standard Model demanded a "top" quark to balance it. This time the new addition had been "ordered," but would nature deliver? In 1994, when the top quark was finally sighted, it marked not only a great triumph for the theory, but also a remarkable achievement for modern detectors and accelerators. This is the story of the accelerator and detector at Fermi National Accelerator Laboratory ("Fermilab") that brought you the top quark.

THE ACCELERATOR

In quantum mechanics a particle is described as a wave with wavelength inversely proportional to the particle momentum. Using waves with smaller and smaller wavelengths enables scientists to probe smaller and smaller structures. Therefore, being able to accelerate particles up to large momenta enables us to probe small structures such as the proton. In order to create new high-mass particles in the laboratory, we need high-energy particle beams, since the available energy to create new states is approximately $\sqrt{2mE}$ in a fixed target experiment, or $2E$ for a collider experiment, where E is the energy of the particle beam and m is the mass of the target particle. For these two reasons it is necessary to understand techniques used to accelerate charged particles.

A particle accelerator has two main elements, the beam and the accelerating structure. All acceleration techniques use only electromagnetic fields. Particles are usually accelerated in evacuated pipes to reduce scattering. The accelerating fields must be carefully controlled, and methods for aligning the beam are necessary. Finally, since the magnets get very hot, they need to be cooled.

There are different types of accelerating techniques. One uses a DC electric field, and another uses either standing or traveling radio frequency (RF) waves to accelerate particles. The DC technique is limited by arcing. It is used for initial acceleration of particles, followed by the RF wave technique, which leads to bunched beams.

The source of charged particles for the beam is either a hot metal filament for electrons or a bottle of hydrogen for protons. The particles must be guided down the beam pipe. This involves steering or bending the beam, focusing it, and breaking it up to form bunches of particles rather than a continuous beam.

The proton accelerator at Fermilab consists of many parts (Fig. 1). First, hydrogen is ionized to form H^- ions which are then introduced into an accelerating structure ("PreAcc" in Fig. 1) which produces a DC electric field of 750 kV. After the ions are accelerated in this field, they enter a linear accelerator (LINAC) which uses 201 MHz–805 MHz RF waves supplied to accelerating cavities. The RF signal produces an alternating electric field in a cavity pointed first along the beam and then opposite to it. The ions are accelerated in the forward direction by shielding them from the RF signal in drift tubes while the **E** field is of the wrong sign. The ions are accelerated down the 145-m long LINAC to an energy of 400 MeV. The particles introduced into the first accelerating cavity do not all arrive simultaneously and have a spread of energies. A particle arriving with exactly the right energy and at the right time will be accelerated to the desired energy. Because of the alternating nature of the electric field, particles arriving slightly after

An aerial view of the Fermi National Accelerator Laboratory, Batavia, Illinois. The largest circle is the main accelerator. Three experimental lines extend at a tangent from the accelerator. The 16-story twin-towered Wilson Hall is seen at the base of the experimental lines. (*Fermi National Accelerator Laboratory*)

or before the "ideal" particle will be accelerated in such a way that they tend toward the ideal particle trajectory. This feature is called *phase stability* and is a necessary feature for our accelerators. Both electrons in the H$^-$ ion are then stripped off by a foil, and the protons are introduced into a circular accelerator called the Synchrotron Booster.

The Booster's structure also uses RF electromagnetic fields to accelerate the protons by 550 keV per turn. The protons, kept inside the ring by magnetic fields, are now trav-

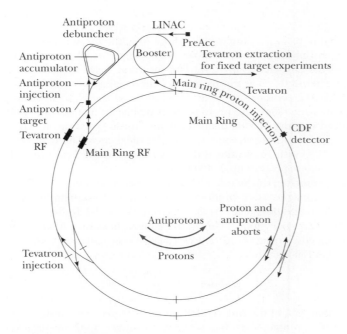

Figure 1   Fermilab accelerator complex.

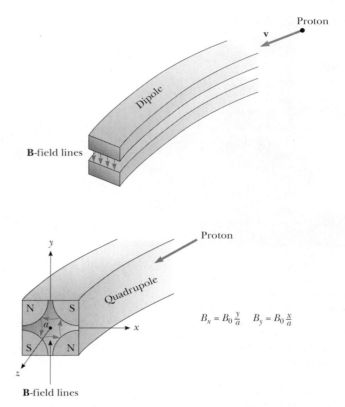

Figure 2   A dipole magnet in the synchrotron exerts a force on the proton toward the center of the ring. A quadrupole magnet exerts a force on protons traveling off axis.

eling at a speed of $0.37c$. Each time around the ring, the protons pass through accelerating cavities and pick up energy. Dipole magnets (shown schematically in Fig. 2) are used to bend the particles in a circle. The magnetic fields in these magnets are increased from 0.05 to 0.7 T as the particles change energy from 400 MeV to 8 GeV in such a way that the protons stay in the same orbit in the circular beam pipe. When their energy reaches 8 GeV, the protons are extracted using fast-rising magnetic fields and introduced into the 2000-m diameter main proton synchrotron. The 6000-m tunnel houses two accelerating structures, one atop the other. The first, called the Main Ring, built in 1969 with conventional dipole magnets whose field ranges from 0.04 to 0.67 T, can accelerate protons up to 400 GeV.

The superconducting magnet accelerator, the Tevatron, built in 1979, makes use of superconducting magnets that produce strong, uniform magnetic fields. The field in the superconducting dipole magnets ranges from 0.66 to 4.4 T, thus enabling the Tevatron to accelerate particles to 900 GeV in the same radius as the Main Ring. In general, we can describe the relationship between accelerator radius, energy achieved, and magnetic field as follows:

$$\text{Radius} = 44 \ \frac{E \ (\text{GeV})}{B \ (\text{kG})} \ m$$

This assumes that 75% of the ring is filled with dipole magnets. In the case of the Tevatron, there are 774 dipoles, each 6 m long, comprising 74% of the Tevatron circumference. This shows that the limitation for accelerators is either building magnets with a

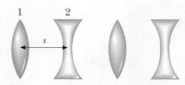

Figure 3 Strong focusing using quadrupole magnets separated by a distance s and rotated $90°$ in series. Quadrupole magnets in series with equal and opposite focal lengths f have a total focal length given by $1/f_{12} = s/f^2$.

high uniform magnetic field or having enough money to make an accelerator with a huge radius.

The synchrotron would be unable to keep the protons inside the beampipe during acceleration without a method of focusing the beam. The concept of Strong Focusing developed in 1952 is crucial. Quadrupole magnets shown in Figure 2 have zero field on axis, but have a field gradient off axis. As a particle wanders away from the axis in the x direction, it experiences a force that pushes it back toward the axis. A particle that wanders off in the y direction experiences a force away from the axis. Strong Focusing relies on a series of focusing and defocusing magnets that have an overall effect of focusing the beam, provided the magnets are spaced sufficiently closely together, as shown in Figure 3. The field gradients in the quadrupoles in the Tevatron reach as high as 67 T/m. The synchrotron consists of strings of magnets, something like F-B-B-B-B-D-B-B-B-B-F etc., where F stands for focusing, B stands for bending, and D stands for defocusing magnets.

Particles in the Tevatron are accelerated by RF cavities driven at 53 MHz. The protons are accelerated through a beampipe maintained at a vacuum of 10^{-9} torr to a final energy of 900 ± 0.72 GeV.

The top quark was discovered at a proton–antiproton collider. The antiprotons are produced by extracting the proton beam at 120 GeV and shining it on a copper target. Approximately 2.5×10^{12} protons are extracted from the Main Ring every 2.4 s, and 3×10^7 antiprotons are produced with a distribution of energy peaked at 8.9 GeV. A smaller 250-m radius ring first debunches the antiprotons into a continuous beam and then "cools" the antiprotons until they form an intense, monoenergetic and focused beam. The method used, called stochastic cooling, was invented by Simon Van der Meer in 1972. The antiprotons are cooled and stored in the accumulator, a ring concentric with the debuncher held at a vacuum of 2×10^{-10} torr for a few hours until enough antiprotons are accumulated for reentry into the Main Ring.

The antiprotons are introduced into the Main Ring in bunches approximately 50 cm long and are accelerated to 900 GeV in the same accelerating structure used for accelerating the protons. Bunches of roughly 3×10^{11} protons and 1×10^{11} antiprotons circulating in opposite directions in the Tevatron collide at two interaction regions containing the particle detectors every 3.5 μs.

THE DETECTOR

Because the mass of the top quark is almost 200 times the mass of a proton, it is very rarely produced, even in collisions where the center of mass energy is 1.8 TeV. Only a few top quark pairs are produced from a typical 50 billion collisions per day. The Standard Model predicts that the top quark (with a lifetime of 10^{-24} s) should decay in a manner similar to the familiar β-decay which converts a neutron to a proton by emitting a W boson which then decays to an electron and neutrino. Each top quark decays to a W and a b-quark. We focus on the case where one W decays either to a

The tunnel of the main accelerator at Fermilab. The upper ring of magnets is the 400-GeV accelerator. The lower ring is superconducting magnets for the Tevatron. (*Fermi National Accelerator Laboratory*)

$\mu \nu_\mu$ or $e \nu_e$ and the other W decays to a quark and an antiquark, for example, an up quark and an anti-down quark (Fig. 4). Quarks are confined to hadronic bound states; hence the outgoing quarks in this interaction manifest themselves as collimated "jets" of hadrons.

To find the top quark pairs produced, the detector must then be able to identify electrons, muons, neutrinos, light quark jets, and b-quark jets and to measure their momentum and energy.

All detectors function by registering the effects of energy lost by particles as they pass through matter. Fortunately electrons, hadrons, jets, neutrinos, and muons all behave quite differently, which allows us to identify each type.

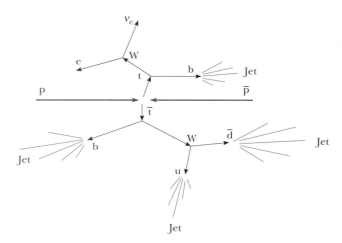

Figure 4 Schematic of top quark pair production and subsequent top quark decays.

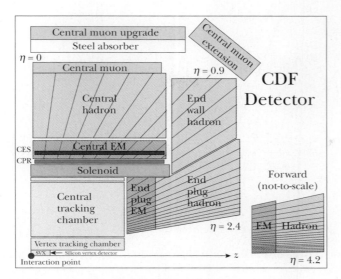

Figure 5   Schematic of a quadrant of the CDF detector.

The experiment that first observed the production of top quarks, called the Collider Detector at Fermilab (CDF), was designed for a variety of physics measurements. The detector surrounds the point where the protons and antiprotons collide (called the interaction point), and measures the momentum, energy, and identity of the particles produced in the collision. Figure 5 is a schematic view of one quadrant of the CDF detector. To measure the momentum of high-energy particles, we require a large region with a **B** field to measure their curvature. To measure their energy, we use a calorimeter that contains enough material to bring the particles to rest. Hence, detectors must be large and heavy; CDF, which weighs over 6000 tons, is 12 m high and 32 m long. From each collision, it records over 50 000 pieces of information. CDF, like most collider detectors, surrounds the collision point with tracking chambers, followed by calorimeters and muon chambers.

Charged Particle Tracking   The momentum of charged particles emerging from a collision can be measured in a drift chamber. In its simplest form, a drift chamber is a gas-filled rectangular tube with a high-voltage wire running along its axis, similar to a Geiger counter. The wire (called a "sense" wire) is held at a high potential relative to the tube to create an electric field. By adding wires of different voltages, the field can be made roughly constant over the width of the tube. When a charged particle passes through the tube, it ionizes some of the atoms in the gas. The electrons from ionization drift in the field toward the sense wire (Fig. 6). With a suitably chosen electric field strength and gas, the drift velocity can be made roughly constant. Near the wire, the electric field strength increases proportional to $1/r$. In this region the electrons gain enough energy to ionize more atoms in the gas. This process continues, producing an "avalanche" of electrons that appears as a small electronic pulse.

By measuring the drift time, we can determine the position of the original particle. Drift velocities are typically 5 cm/μs, and although the drift time can be measured to less than 1 ns, the position resolution (typically 250 μm) is dominated by variations in the arrival time of the avalanche as a result of collisions with the gas molecules. A single planar drift chamber measures position in one dimension, but if a drift chamber with two planes of wires is used (one, for example, with its sense wire along the x direction, the other along the y direction) the particle's position can be reconstructed in two di-

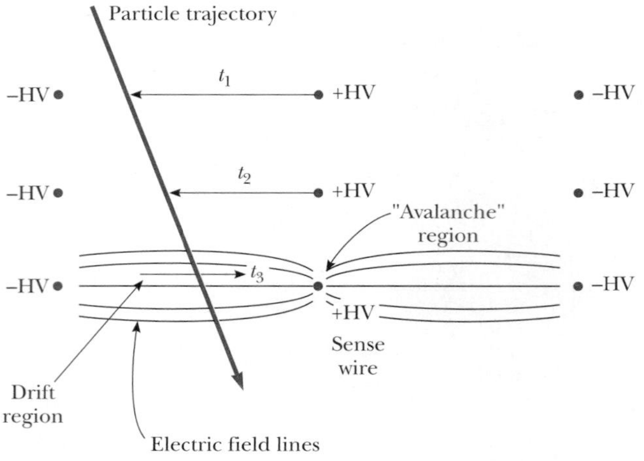

Figure 6   Schematic of a planar drift chamber showing **E** field lines, avalanche region, electron drift region, sense wires (+HV), incident particle trajectory, and the times measured between traversal of the particle in each cell and avalanche at each sense wire.

mensions. If a series of drift chambers are placed in a magnetic field, the particle's momentum can be determined by measuring the curvature of its track.

CDF uses a 1-m-radius, 3-m-long cylindrical drift chamber surrounding the collision point, containing over 4000 sense wires. The wires are grouped into nine layers. Five layers run parallel to the beam line and alternate with the other four, which are tilted at ± 3 degrees to the beam line to provide 3D position information. Another 30 000 wires along the length of the cylinder create drift fields (of about 1 kV/cm) around each sense wire. The entire chamber is immersed in a 1.4-T magnetic field, which is directed parallel to the beam line and is generated by a superconducting solenoid. Momentum is measured more accurately for low-momentum tracks with large curvatures than for high-momentum tracks which bend only a few degrees in the field. A 20-GeV/c particle traveling in the radial direction can be measured to about 4%, but a 1-GeV/c particle's momentum can be determined to 0.2%.

For extremely short-lived particles, we need a detector with higher resolution than the drift chamber can provide. Although b-hadrons have a lifetime of only 1.5 ps, they can often be identified by the short distance they travel from the collision before decaying. B-hadrons from top decays that have an average momentum of 40 GeV/c transverse to the beam and travel an average of 3.4 mm before decaying. B-hadrons can be identified by tracing back the tracks from the decay to a vertex which is displaced from the collision point (Fig. 7). We use a high-precision silicon microstrip detector (an 8-cm-radius, 60-cm-long cylindrical device) which is placed inside the drift chamber and immediately surrounds the beampipe.

Strips of silicon separated by about 20 μm provide the necessary resolution for finding displaced vertices. The semiconducting strips are reverse-biased with voltages applied to the n and p sides. This produces an electric field in the depletion region, while preventing much current flow. A charged particle passing through the detector can create electron–hole pairs that move in the electric field and appear as small pulses on the n and p electrodes. The CDF microstrip detector uses four layers of silicon strips placed at small radii inside the main drift chamber. The combination of the two tracking devices provides enough resolution to pick out tracks and identify displaced vertices from b-hadron decays. For a b-hadron traveling in the radial direction, the position of the displaced vertex can be resolved to about 130 μm.

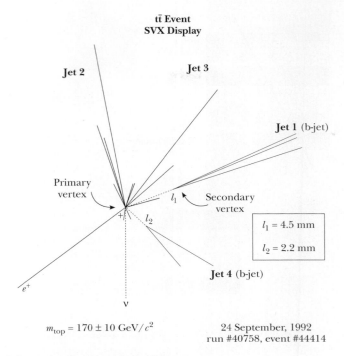

$t\bar{t}$ **Event**
SVX Display

Jet 2

Jet 3

Jet 1 (b-jet)

Primary
vertex

Secondary
vertex

l_1

l_2

$l_1 = 4.5$ mm

$l_2 = 2.2$ mm

Jet 4 (b-jet)

e^+

ν

$m_{\text{top}} = 170 \pm 10 \text{ GeV}/c^2$

24 September, 1992
run #40758, event #44414

Figure 7   Schematic of top quark candidate event.

Calorimetry: Measuring the Energy of Particles   We now have a snapshot of the event which tells us the direction and momentum of many of the particles. Outside the tracking chambers and the solenoid coil are calorimeters designed to measure particle energies. The energy of the hadron jets is crucial to the determination of the top quark mass. Most particles lose energy in the calorimeters by producing a shower of lower-energy particles. Figure 8 illustrates the various effects of these particles in the detector. Calorimeters contain alternate layers of an "absorber" like lead or iron, with "active" layers of detectors for sampling the particles in the shower as it develops. Since we would like to contain the entire shower, high-energy particles require deep calorimeters, and since we want to know roughly where the energy was deposited, the calorimeters are segmented into hundreds of read-out towers. Unlike tracking chambers which are sensitive only to charged particles, calorimeters can also provide information about neutral particles like photons and neutrons. As we will see, they also aid in distinguishing electrons from hadrons.

Electrons, because of their low mass, are accelerated by the protons in the lead nuclei, and lose much of their energy through *bremsstrahlung*, or "braking radiation." Because energy loss due to bremsstrahlung radiation for a particle falls as M^{-2}, this process is neglible for the heavy particles. Hadrons, instead, lose most of their energy through inelastic collisions with the lead nuclei, producing secondary hadrons.

In both these cases, the loss of energy results in the production of other, lower-energy particles. Photons from bremsstrahlung will subsequently convert to electron–positron pairs which continue in the same manner. Hadrons also shower, producing lower-energy hadrons. Eventually, all the particles in the "shower" will come to rest. By measuring the energy deposited in the active layers of the calorimeter, a good measurement of the energy of the particle is made. The more times we sample the shower the better our estimate of the energy of the incident particle.

In practice, although we use the same method to measure both electron and hadron energies, we often use two separate devices, an "electromagnetic" calorimeter

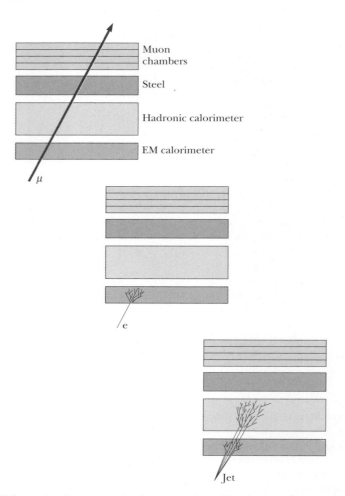

Figure 8   Schematic of muons (μ), electrons (e) and jets interacting in a detector.

(in our case containing about 7.5 cm of lead) followed by a "hadronic" calorimeter with much more material (about 80 cm of steel). Each uses about 30 active layers to sample the number of particles in the shower. The sampling layers are slabs of plastic scintillator. The scintillation light they produce is converted to electronic pulses by photomultiplier tubes. By summing these pulses, we obtain a measure of the total energy deposited in the region.

In the CDF, electrons traverse 18 radiation lengths of lead in the electromagnetic calorimeter. A pion traverses about 5 nuclear absorption lengths of iron in the hadronic calorimeter. Electron and hadron showers are distinguishable by the fact that electrons shower early. In lead a radiation length is 0.56 cm while an absorption length is about 10 cm. As a result, an electron will lose most of its energy in a very short amount of material compared to a pion or a proton.

Identifying Muons and Neutrinos   We have discussed some of the ways in which jets, b-hadrons and electrons can be identified and are left now with the problem of finding muons and neutrinos. A muon is just a heavy version of the electron. Recall that high-energy electrons lose energy mostly through bremsstrahlung, a process which is negligible for the heavier muons. As a result, muons lose energy almost entirely through the relatively slow process of ionization. A 20-GeV muon from a top quark decay will travel through the drift chamber, leaving a track, but then pass through the calorimeters depositing only a few GeV of energy. To identify muons, we place small drift chambers outside the calorimeters and look for the track of a particle. Because most other particles stop in the calorimeter, a track in the muon drift chambers establishes the particle as a muon candidate. We already know the muon's momentum from the track it left in the main drift chamber.

Neutrinos are like muons without charge and little or no mass. Because they have no charge, they cannot lose energy through ionization. Neutrinos interact only through the weak interaction. Only one out of a billion neutrinos from a top quark decay will interact in the detector. Neutrinos can be detected only by measuring the missing momentum in the collision.

Top quark pairs are produced predominantly by quark–antiquark interactions. Although proton–antiproton collisions take place at the CDF in their center-of-mass frame, the quarks inside them carry only a fraction of their momentum. As a result, the top quark system is typically boosted along the beamline. Fortunately, the quarks and antiquarks collide with very little momentum transverse to the beam direction. Conservation of momentum requires that the final transverse momentum should then also be near zero. If a large imbalance is detected, we can infer the presence of a neutrino and estimate its momentum transverse to the beam. Since undetected particles also result in such an imbalance, this technique requires a "hermetic" detector with few cracks, intercepting most of the particles from the collision.

RECONSTRUCTING THE TOP

For over two years, this process continued, beginning with a bottle of hydrogen, and ending with a few thousand magnetic tapes full of data. After a long period of analyz-

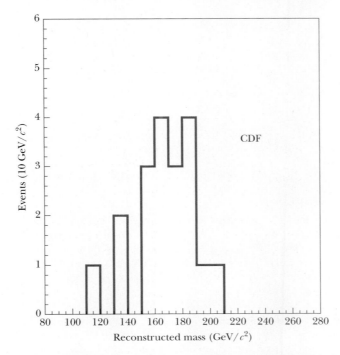

Figure 9 Top quark candidates' invariant mass distribution. (*Phys. Rev. Lett.* 74:2626, 1995)

ing the data, selecting electrons, muons, and b-hadrons, a small pool of 50 events were identified containing about 28 top quark pairs. Using the measured momentum and energies, the mass of the top was reconstructed to be 176 ± 13 GeV/c^2 (Fig. 9). The top quark is the heaviest known fundamental particle; why it is so heavy is one of the fundamental questions remaining to be answered.

Suggestions for Further Reading

D. A. Edwards and M. J. Syphers, *An Introduction to the Physics of High Energy Accelerators*, New York, John Wiley & Sons, 1993.

T. Ferbel, *Experimental Techniques in High Energy Physics*, Frontiers in Physics Lecture Note Series, Reading, Mass., Addison-Wesley, 1987.

M. Reiser, *Theory and Design of Charged Particle Beams*, New York, John Wiley & Sons, 1994.

Best Known Values for Physical Constants

There is an uncertainty in the last digit, typically of 1 to 7 parts.

speed of light, c	299 792 458 m/s (exact)
gravitational constant, G	$6.672\ 6 \times 10^{-11}$ N·m²·kg⁻²
Planck's constant, h	$6.626\ 068\ 76 \times 10^{-34}$ J·s
Boltzmann constant, k_B	$1.380\ 650\ 3 \times 10^{-23}$ J·K⁻¹
Stefan-Boltzmann constant, σ	$5.670\ 400 \times 10^{-8}$ W·m⁻²·K⁻⁴
Rydberg constant, R	$10\ 973\ 731.568\ 549$ m⁻¹
fine-structure constant, α	$1/137.035\ 989$
Bohr radius, a_0	$5.291\ 772\ 083 \times 10^{-11}$ m
Avogadro's number, N_A	$6.022\ 141\ 99 \times 10^{23}$ mol⁻¹
mass of neutron, m_n	$1.674\ 927\ 16 \times 10^{-27}$ kg $939.565\ 330$ MeV/c^2
mass of proton, m_p	$1.672\ 621\ 58 \times 10^{-27}$ kg $938.271\ 998$ MeV/c^2
mass of electron, m_e	$9.109\ 381\ 88 \times 10^{-31}$ kg $0.510\ 998\ 902$ MeV/c^2
elementary charge, e	$1.602\ 176\ 462 \times 10^{-19}$ C
permeability of vacuum, μ_0	$4\pi \times 10^{-7}$ N/A² (exact) $12.566\ 370\ 6 \times 10^{-7}$ N/A²
permittivity of vacuum, ϵ_0	$1/\mu_0 c^2$ C²/N·m² (exact) $8.854\ 187\ 817 \times 10^{-12}$ C²/N·m²
Coulomb constant, k	$1/(4\pi\epsilon_0)$ (exact) $8.987\ 551\ 78 \times 10^{9}$ N·m²/C²
Bohr magneton, μ_B	$9.274\ 008\ 99 \times 10^{-24}$ J/T $5.788\ 381\ 749 \times 10^{-5}$ eV/T
Nuclear magneton, μ_n	$5.050\ 783\ 17 \times 10^{-27}$ J/T $3.152\ 451\ 238 \times 10^{-8}$ eV/T
Atomic mass unit, u	$931.494\ 013$ MeV/c^2 $1.660\ 538\ 73 \times 10^{-27}$ kg

Values based on 1998 CODATA recommended values.

Table of Selected Atomic Masses*

$1 \text{ u} = 931.494\ 013 \text{ MeV}/c^2 = 1.660\ 538\ 73 \times 10^{-27} \text{ kg}$

Atomic Number Z	Element	Symbol	Mass Number A	Atomic Mass** (u)	Percent Abundance, or Decay Mode (if radioactive)†	Half-Life (if radioactive)
0	(Neutron)	n	1	1.008665	β^-	10.6 min
1	Hydrogen	H	1	1.007825	99.985	
	Deuterium	D	2	2.014102	0.015	
	Tritium	T	3	3.016049	β^-	12.33 yr
2	Helium	He	3	3.016029	0.00014	
			4	4.002603	≈ 100	
3	Lithium	Li	6	6.015122	7.5	
			7	7.016004	92.5	
4	Beryllium	Be	7	7.016929	EC, γ	53.3 days
			8	8.005305	2α	6.7×10^{-17} s
			9	9.012182	100	
5	Boron	B	10	10.012937	19.8	
			11	11.009306	80.2	
6	Carbon	C	11	11.011433	β^+, EC	20.4 min
			12	12.000000	98.89	
			13	13.003355	1.11	
			14	14.003242	β^-	5730 yr
7	Nitrogen	N	13	13.005739	β^+	9.96 min
			14	14.003074	99.63	
			15	15.000109	0.37	
8	Oxygen	O	15	15.003066	β^+, EC	122 s
			16	15.994915	99.76	
			18	17.999160	0.204	
9	Fluorine	F	19	18.998403	100	
10	Neon	Ne	20	19.992440	90.51	
			22	21.991386	9.22	
11	Sodium	Na	22	21.994437	β^+, EC, γ	2.602 yr
			23	22.989770	100	
			24	23.990964	β^-, γ	15.0 h
12	Magnesium	Mg	24	23.985042	78.99	
13	Aluminum	Al	27	26.981538	100	
14	Silicon	Si	28	27.976926	92.23	2.62 h
			31	30.975363	β^-, γ	
15	Phosphorus	P	31	30.973761	100	
			32	31.973907	β^-	14.28 days

Atomic Number Z	Element	Symbol	Mass Number A	Atomic Mass** (u)	Percent Abundance, or Decay Mode (if radioactive)†	Half-Life (if radioactive)
16	Sulfur	S	32	31.972071	95.0	
			35	34.969032	β^-	87.4 days
17	Chlorine	Cl	35	34.968853	75.77	
			37	36.965903		
18	Argon	Ar	40	39.962383	99.60	
19	Potassium	K	39	38.963707	93.26	
			40	39.963999	β^-, EC, γ, β^+	1.28×10^9 yr
20	Calcium	Ca	40	39.962591	96.94	
21	Scandium	Sc	45	44.955910	100	
22	Titanium	Ti	48	47.947947	73.7	
23	Vanadium	V	51	50.943964	99.75	
24	Chromium	Cr	52	51.940512	83.79	
25	Manganese	Mn	55	54.938049	100	
26	Iron	Fe	56	55.934942	91.8	
27	Cobalt	Co	59	58.933200	100	
			60	59.933822	β^-, γ	5.271 yr
28	Nickel	Ni	58	57.935348	68.3	
			60	59.930790	26.1	
			64	63.927969	0.91	
29	Copper	Cu	63	62.929601	69.2	
			64	63.929766	β^-, β^+	12.7 h
			65	64.927794	30.8	
30	Zinc	Zn	64	63.929146	48.6	
			66	65.926036	27.9	
31	Gallium	Ga	69	68.925581	60.1	
32	Germanium	Ge	72	71.922076	27.4	
			74	73.921178	36.5	
33	Arsenic	As	75	74.921597	100	
34	Selenium	Se	80	79.916522	49.8	
35	Bromine	Br	79	78.918338	50.69	
36	Krypton	Kr	84	83.911508	57.0	
			89	88.917563	β^-	3.2 min
37	Rubidium	Rb	85	84.911792	72.17	
38	Strontium	Sr	86	85.909265	9.8	
			88	87.905617	82.6	28.8 yr
			90	89.907738	β^-	
39	Yttrium	Y	89	88.905849	100	
40	Zirconium	Zr	90	89.904702	51.5	
41	Niobium	Nb	93	92.906376	100	
42	Molybdenum	Mo	98	97.905407	24.1	
43	Technetium	Te	97	96.906364		2.6×10^6 yr
			98	97.907215	β^-, γ	4.2×10^6 yr
44	Ruthenium	Ru	102	101.904349	31.6	
45	Rhodium	Rh	103	102.905504	100	
46	Palladium	Pd	106	105.903484	27.3	
47	Silver	Ag	107	106.905093	51.83	
			109	108.904756	48.17	
48	Cadmium	Cd	114	113.903359	28.7	
49	Indium	In	115	114.903879	95.7; β^-	5.1×10^{14} yr

Atomic Number Z	Element	Symbol	Mass Number A	Atomic Mass** (u)	Percent Abundance, or Decay Mode (if radioactive)[†]	Half-Life (if radioactive)
50	Tin	Sn	120	119.902199	32.4	
51	Antimony	Sb	121	120.903822	57.3	
52	Tellurium	Te	130	129.906223	34.5; β^-	2×10^{21} yr
53	Iodine	I	127	126.904468	100	
			131	130.906118	β^-, γ	8.04 days
54	Xenon	Xe	132	131.904155	26.9	
			136	135.907220	8.9	
55	Cesium	Cs	133	132.905447	100	
56	Barium	Ba	137	136.905822	11.2	
			138	137.905242	71.7	
			144	143.922845	β^-	11.9 s
57	Lanthanum	La	139	138.906349	99.911	
58	Cerium	Ce	140	139.905435	88.5	
59	Praseodymium	Pr	141	140.907648	100	
60	Neodymium	Nd	142	141.907719	27.2	
61	Promethium	Pm	145	144.912743	EC, α, β^-	17.7 yr
			146	145.914708		5.53 yr
62	Samarium	Sm	152	151.919729	26.6	
63	Europium	Eu	153	152.921227	52.1	
64	Gadolinium	Gd	158	157.924101	24.8	
65	Terbium	Tb	159	158.925343	100	
66	Dysprosium	Dy	164	163.929171	28.1	
67	Holmium	Ho	165	164.930319	100	
68	Erbium	Er	166	165.930290	33.4	
69	Thulium	Tm	169	168.934211	100	
70	Ytterbium	Yb	174	173.938858	31.6	
71	Lutecium	Lu	175	174.940768	97.39	
72	Hafnium	Hf	180	179.946549	35.2	
73	Tantalum	Ta	181	180.947996	99.988	
74	Tungsten (wolfram)	W	184	183.950932	30.7	
75	Rhenium	Re	187	186.955750	62.60, β^-	4×10^{10} yr
76	Osmium	Os	191	190.960928	β^-, γ	15.4 days
			192	191.961479	41.0	
77	Iridium	Ir	191	190.960591	37.3	
			193	192.962923	62.7	
78	Platinum	Pt	195	194.964774	33.8	
79	Gold	Au	197	196.966551	100	
80	Mercury	Hg	202	201.970625	29.8	
81	Thallium	Tl	205	204.974412	70.5	
82	Lead	Pb	203	202.973375	β^-	51.9 h
			204	203.973028	β^-, 1.48	1.4×10^{17} yr
			206	205.974449	24.1	
			207	206.975880	22.1	
			208	207.976636	52.3	22.3 yr
			210	209.984163	α, β^-, γ	36.1 min
			211	210.988735	β^-, γ	
			212	211.991871	β^-, γ	10.64 h
			214	213.999798	β^-, γ	26.8 min

Atomic Number Z	Element	Symbol	Mass Number A	Atomic Mass** (u)	Percent Abundance, or Decay Mode (if radioactive)[†]	Half-Life (if radioactive)
83	Bismuth	Bi	209	208.980384	100	
			211	210.987258	α, β^-, γ	2.15 min
84	Polonium	Po	207	206.981570	β^+	5.8 h
			208	207.981222	α	2.9 yr
			209	208.982404	α, γ	102 yr
			210	209.982848	α, γ	138.38 days
			214	213.995176	α	164 μs
			218	218.008966		3.11 min
85	Astatine	At	210	209.987131	β^+	8.1 h
			211	210.987470	EC, α	7.21 h
86	Radon	Rn	211	210.990575	α, β^+	14.6 h
			220	220.011368	α, γ	55.6 s
			222	222.017570	α, γ	3.8235 days
87	Francium	Fr	223	223.019731	α, β^-, γ	22.0 min
88	Radium	Ra	223	223.018501	α	11.43 d
			224	224.020186	α	3.66 d
			225	225.023604	β^-	14.8 days
			226	226.025402	α, γ	1.599×10^3 yr
			228	228.031064	β^-	5.75 yr
89	Actinium	Ac	227	227.027747	α, β^-, γ	21.773 yr
90	Thorium	Th	228	228.028715	α, γ	1.91 yr
			230	230.033128	α, γ	7.54×10^4 yr
			232	232.038051	100, α, γ	1.40×10^{10} yr
91	Protactinium	Pa	231	231.035880	α, γ	3.25×10^4 yr
			233	233.040242	β^-	27.0 days
92	Uranium	U	231	231.036264	EC	4.2 days
			232	232.037129	α, γ	68.9 yr
			233	233.039628	α, γ	1.592×10^5 yr
			234	234.040947	0.006, α, γ	2.45×10^5 yr
			235	235.043924	0.72; α, γ	7.038×10^8 yr
			236	236.045563	α, γ	2.342×10^7 yr
			238	238.050785	99.275; α, γ	4.468×10^9 yr
			239	239.054289	β^-	23.5 min
93	Neptunium	Np	237	237.048168	α	2.14×10^6 yr
			239	239.052933	β^-, γ	2.36 days
94	Plutonium	Pu	239	239.052158	α, γ	2.41×10^4 yr
			242	242.058737	α	3.75×10^5 yr
			244	244.064198	α	8.00×10^7 yr
95	Americium	Am	241	241.056824	α	432.7 yr
			243	243.061372	α, γ	7.37×10^3 yr
96	Curium	Cm	243	243.0614		29.1 yr
			244	244.0627		18.1 yr
			245	245.0655		8.48×10^3 yr
			246	246.0672		4.76×10^3 yr
			247	247.070346	α	15.6×10^6 yr
			248	248.072343	α	3.48×10^5 yr
97	Berkelium	Bk	247	247.070298	α, γ	1.4×10^3 yr
			248	248.073107	α	9 yr
			249	249.0750		3.26×10^2 yr

Atomic Number Z	Element	Symbol	Mass Number A	Atomic Mass** (u)	Percent Abundance, or Decay Mode (if radioactive)†	Half-Life (if radioactive)
98	Californium	Cf	249	249.074845	α	351 yr
			251	251.079579	α	9.0×10^2 yr
99	Einsteinium	Es	252	252.082945	α, β^+	472 days
			254	254.088019	α, β^-, γ	276 days
100	Fermium	Fm	253	253.085173	EC, α, γ	3.0 days
			257	257.095099	α	100 days
101	Mendelevium	Md	258	258.098572	α	51.5 days
			260	260.1037		27.8 days
102	Nobelium	No	255	255.093258	α, β^+	3.1 min
			259	259.100932	EC, α	58 min
103	Lawrencium	Lr	260	260.105314	EC, α	180 s
104	Rutherfordium	Rf	261	261.108685	EC, α	65 s
105	Dubnium	Db	262	262.113763	EC, α	34 s
106	Seaborgium	Sg	266	266.121955	α	~21 s
107	Bohrium	Bh	262	262.123028	α	102 ms
			264	264.124746	α	0.44 sec
108	Hassium	Hs	269	269.134086	α	9 s
			277			16.5 min
109	Meitnerium	Mt	266	266.137950	α	0.8 ms
			268	268.138809	α	70 ms
110	Darmstadtium	Ds	271	271.146081	α	1.1 ms
111	Unununium‡	Uuu	272	272.1535		1.5×10^{-3} s
112	Ununbium‡	Uub	285			15.4 min
114	Ununquadium‡	Uuq	289			30.4 s
116	Ununhexium‡	Uuh	289			0.60×10^{-3} s

*Data are taken from *Chart of the Nuclides,* 12th ed., General Electric, 1977, from C. M. Lederer and V. S. Shirley, eds., *Table of Isotopes,* 7th ed., John Wiley & Sons, Inc., New York, 1979, and from G. Audi and H. Wapstra, "The 1993 Atomic Mass Evaluation," *Nuclear Physics* A565, 1, 1993, the International Union of Pure and Applied Chemistry (IUPAC), and *Nuclear Wallet Cards* on the Web at **http://www.nndc.bnl.gov/wallet.**

**The masses given are those for the neutral atom, including the Z electrons.

†The abbreviation EC stands for "electron capture."

‡Elements 111, 112, 114, and 116 have not yet been named. The IUPAC provisional names are shown.

Nobel Prizes

All Nobel Prizes in physics are listed (and marked with a P), as well as relevant Nobel Prizes in Chemistry (C). The key dates for some of the scientific work are supplied; they often antedate the prize considerably.

1901   (P) *Wilhelm Roentgen* for discovering x-rays (1895).

1902   (P) *Hendrik A. Lorentz* for predicting the Zeeman effect and *Pieter Zeeman* for discovering the Zeeman effect, the splitting of spectral lines in magnetic fields.

1903   (P) *Antoine-Henri Becquerel* for discovering radioactivity (1896) and *Pierre* and *Marie Curie* for studying radioactivity.

1904   (P) *Lord Rayleigh* for studying the density of gases and discovering argon.
(C) *William Ramsay* for discovering the inert gas elements helium, neon, xenon, and krypton, and placing them in the periodic table.

1905   (P) *Philipp Lenard* for studying cathode rays, electrons (1898–1899).

1906   (P) *J. J. Thomson* for studying electrical discharge through gases and discovering the electron (1897).

1907   (P) *Albert A. Michelson* for inventing optical instruments and measuring the speed of light (1880s).

1908   (P) *Gabriel Lippmann* for making the first color photographic plate, using interference methods (1891).
(C) *Ernest Rutherford* for discovering that atoms can be broken apart by alpha rays and for studying radioactivity.

1909   (P) *Guglielmo Marconi* and *Carl Ferdinand Braun* for developing wireless telegraphy.

1910   (P) *Johannes D. van der Waals* for studying the equation of state for gases and liquids (1881).

1911   (P) *Wilhelm Wien* for discovering Wien's law giving the peak of a blackbody spectrum (1893).
(C) *Marie Curie* for discovering radium and polonium (1898) and isolating radium.

1912   (P) *Nils Dalén* for inventing automatic gas regulators for lighthouses.

1913   (P) *Heike Kamerlingh Onnes* for the discovery of superconductivity and liquefying helium (1908).

1914   (P) *Max T. F. von Laue* for studying x-rays from their diffraction by crystals, showing that x-rays are electromagnetic waves (1912).
(C) *Theodore W. Richards* for determining the atomic weights of sixty elements, indicating the existence of isotopes.

1915   (P) *William Henry Bragg* and *William Lawrence Bragg*, his son, for studying the diffraction of x-rays in crystals.

1917   (P) *Charles Barkla* for studying atoms by x-ray scattering (1906).

1918   (P) *Max Planck* for discovering energy quanta (1900).

1919   (P) *Johannes Stark,* for discovering the Stark effect, the splitting of spectral lines in electric fields (1913).

1920 (P) *Charles-Édouard Guillaume* for discovering invar, a nickel-steel alloy with low coefficient of expansion.

(C) *Walther Nernst* for studying heat changes in chemical reactions and formulating the third law of thermodynamics (1918).

1921 (P) *Albert Einstein* for explaining the photoelectric effect and for his services to theoretical physics (1905).

(C) *Frederick Soddy* for studying the chemistry of radioactive substances and discovering isotopes (1912).

1922 (P) *Niels Bohr* for his model of the atom and its radiation (1913).

(C) *Francis W. Aston* for using the mass spectrograph to study atomic weights, thus discovering 212 of the 287 naturally occurring isotopes.

1923 (P) *Robert A. Millikan* for measuring the charge on an electron (1911) and for studying the photoelectric effect experimentally (1914).

1924 (P) *Karl M. G. Siegbahn* for his work in x-ray spectroscopy.

1925 (P) *James Franck* and *Gustav Hertz* for discovering the Franck-Hertz effect in electron-atom collisions.

1926 (P) *Jean-Baptiste Perrin* for studying Brownian motion to validate the discontinuous structure of matter and measure the size of atoms.

1927 (P) *Arthur Holly Compton* for discovering the Compton effect on x-rays, their change in wavelength when they collide with matter (1922), and *Charles T. R. Wilson* for inventing the cloud chamber, used to study charged particles (1906).

1928 (P) *Owen W. Richardson* for studying the thermionic effect and electrons emitted by hot metals (1911).

1929 (P) *Louis Victor de Broglie* for discovering the wave nature of electrons (1923).

1930 (P) *Chandrasekhara Venkata Raman* for studying Raman scattering, the scattering of light by atoms and molecules with a change in wavelength (1928).

1932 (P) *Werner Heisenberg* for creating quantum mechanics (1925).

1933 (P) *Erwin Schrödinger* and *Paul A. M. Dirac* for developing wave mechanics (1925) and relativistic quantum mechanics (1927).

(C) *Harold Urey* for discovering heavy hydrogen, deuterium (1931).

1935 (P) *James Chadwick* for discovering the neutron (1932).

(C) *Irène* and *Frédéric Joliot-Curie* for synthesizing new radioactive elements.

1936 (P) *Carl D. Anderson* for discovering the positron in particular and antimatter in general (1932) and *Victor F. Hess* for discovering cosmic rays.

(C) *Peter J. W. Debye* for studying dipole moments and diffraction of x-rays and electrons in gases.

1937 (P) *Clinton Davisson* and *George Thomson* for discovering the diffraction of electrons by crystals, confirming de Broglie's hypothesis (1927).

1938 (P) *Enrico Fermi* for producing the transuranic radioactive elements by neutron irradiation (1934–1937).

1939 (P) *Ernest O. Lawrence* for inventing the cyclotron.

1943 (P) *Otto Stern* for developing molecular-beam studies (1923), and using them to discover the magnetic moment of the proton (1933).

1944 (P) *Isidor I. Rabi* for discovering nuclear magnetic resonance in atomic and molecular beams.

(C) *Otto Hahn* for discovering nuclear fission (1938).

1945 (P) *Wolfgang Pauli* for discovering the exclusion principle (1924).

1946 (P) *Percy W. Bridgman* for studying physics at high pressures.

1947 (P) *Edward V. Appleton* for studying the ionosphere.

1948   (P) *Patrick M. S. Blackett* for studying nuclear physics with cloud-chamber photographs of cosmic-ray interactions.

1949   (P) *Hideki Yukawa* for predicting the existence of mesons (1935).

1950   (P) *Cecil F. Powell* for developing the method of studying cosmic rays with photographic emulsions and discovering new mesons.

1951   (P) *John D. Cockcroft* and *Ernest T. S. Walton* for transmuting nuclei in an accelerator (1932).
        (C) *Edwin M. McMillan* for producing neptunium (1940) and *Glenn T. Seaborg* for producing plutonium (1941) and further transuranic elements.

1952   (P) *Felix Bloch* and *Edward Mills Purcell* for discovering nuclear magnetic resonance in liquids and gases (1946).

1953   (P) *Frits Zernike* for inventing the phase-contrast microscope, which uses interference to provide high contrast.

1954   (P) *Max Born* for interpreting the wave function as a probability (1926) and other quantum-mechanical discoveries and *Walther Bothe* for developing the coincidence method to study subatomic particles (1930–1931), producing, in particular, the particle interpreted by Chadwick as the neutron.

1955   (P) *Willis E. Lamb, Jr.* for discovering the Lamb shift in the hydrogen spectrum (1947) and *Polykarp Kusch* for determining the magnetic moment of the electron (1947).

1956   (P) *John Bardeen, Walter H. Brattain,* and *William Shockley* for inventing the transistor (1956).

1957   (P) *T.-D. Lee* and *C.-N. Yang* for predicting that parity is not conserved in beta decay (1956).

1958   (P) *Pavel A. Čerenkov* for discovering Čerenkov radiation (1935) and *Ilya M. Frank* and *Igor Tamm* for interpreting it (1937).

1959   (P) *Emilio G. Segrè* and *Owen Chamberlain* for discovering the antiproton (1955).

1960   (P) *Donald A. Glaser* for inventing the bubble chamber to study elementary particles (1952).
        (C) *Willard Libby* for developing radiocarbon dating (1947).

1961   (P) *Robert Hofstadter* for discovering internal structure in protons and neutrons and *Rudolf L. Mössbauer* for discovering the Mössbauer effect of recoilless gamma-ray emission (1957).

1962   (P) *Lev Davidovich Landau* for studying liquid helium and other condensed matter theoretically.

1963   (P) *Eugene P. Wigner* for applying symmetry principles to elementary-particle theory and *Maria Goeppert Mayer* and *J. Hans D. Jensen* for studying the shell model of nuclei (1947).

1964   (P) *Charles H. Townes, Nikolai G. Basov,* and *Alexandr M. Prokhorov* for developing masers (1951–1952) and lasers.

1965   (P) *Sin-itiro Tomonaga, Julian S. Schwinger,* and *Richard P. Feynman* for developing quantum electrodynamics (1948).

1966   (P) *Alfred Kastler* for his optical methods of studying atomic energy levels.

1967   (P) *Hans Albrecht Bethe* for discovering the routes of energy production in stars (1939).

1968   (P) *Luis W. Alvarez* for discovering resonance states of elementary particles.

1969   (P) *Murray Gell-Mann* for classifying elementary particles (1963).

1970   (P) *Hannes Alfvén* for developing magnetohydrodynamic theory and *Louis Eugène Félis Néel* for discovering antiferromagnetism and ferrimagnetism (1930s).

1971   (P) *Dennis Gabor* for developing holography (1947).

(C) *Gerhard Herzberg* for studying the structure of molecules spectroscopically.

1972   (P) *John Bardeen, Leon N. Cooper,* and *John Robert Schrieffer* for explaining super-conductivity (1957).

1973   (P) *Leo Esaki* for discovering tunneling in semiconductors, *Ivar Giaever* for discovering tunneling in superconductors, and *Brian D. Josephson* for predicting the Josephson effect, which involves tunneling of paired electrons (1958–1962).

1974   (P) *Anthony Hewish* for discovering pulsars and *Martin Ryle* for developing radio interferometry.

1975   (P) *Aage N. Bohr, Ben R. Mottelson,* and *James Rainwater* for discovering why some nuclei take asymmetric shapes.

1976   (P) *Burton Richter* and *Samuel C. C. Ting* for discovering the J/psi particle, the first charmed particle (1974).

1977   (P) *John H. Van Vleck, Nevill F. Mott,* and *Philip W. Anderson* for studying solids quantum-mechanically.

(C) *Ilya Prigogine* for extending thermodynamics to show how life could arise in the face of the second law.

1978   (P) *Arno A. Penzias* and *Robert W. Wilson* for discovering the cosmic background radiation (1965) and *Pyotr Kapitsa* for his studies of liquid helium.

1979   (P) *Sheldon L. Glashow, Abdus Salam,* and *Steven Weinberg* for developing the theory that unified the weak and electromagnetic forces (1958–1971).

1980   (P) *Val Fitch* and *James W. Cronin* for discovering CP *(charge-parity) violation* (1964), which possibly explains the cosmological dominance of matter over antimatter.

1981   (P) *Nicolaas Bloembergen* and *Arthur L. Schawlow* for developing laser spectroscopy and *Kai M. Siegbahn* for developing high-resolution electron spectroscopy (1958).

1982   (P) *Kenneth G. Wilson* for developing a method of constructing theories of phase transitions to analyze critical phenomena.

1983   (P) *William A. Fowler* for theoretical studies of astrophysical nucleosynthesis and *Subramanyan Chandrasekhar* for studying physical processes of importance to stellar structure and evolution, including the prediction of white dwarf stars (1930).

1984   (P) *Carlo Rubbia* for discovering the W and Z particles, verifying the electroweak unification, and *Simon van der Meer,* for developing the method of stochastic cooling of the CERN beam that allowed the discovery (1982–1983).

1985   (P) *Klaus von Klitzing* for the quantized Hall effect, relating to conductivity in the presence of a magnetic field (1980).

1986   (P) *Ernst Ruska* for inventing the electron microscope (1931), and *Gerd Binnig* and *Heinrich Rohrer* for inventing the scanning-tunneling electron microscope (1981).

1987   (P) *J. Georg Bednorz* and *Karl Alex Müller* for the discovery of high temperature superconductivity (1986).

1988   (P) *Leon M. Lederman, Melvin Schwartz,* and *Jack Steinberger* for a collaborative experiment that led to the development of a new tool for studying the weak nuclear force, which affects the radioactive decay of atoms.

1989   (P) *Norman Ramsay* for various techniques in atomic physics; and *Hans Dehmelt* and *Wolfgang Paul* for the development of techniques for trapping single charge particles.

1990 (P) *Jerome Friedman, Henry Kendall,* and *Richard Taylor* for experiments important to the development of the quark model.

1991 (P) *Pierre-Gilles de Gennes* for discovering that methods developed for studying order phenomena in simple systems can be generalized to more complex forms of matter, in particular to liquid crystals and polymers.

1992 (P) *George Charpak* for developing detectors that trace the paths of evanescent subatomic particles produced in particle accelerators.

1993 (P) *Russell Hulse* and *Joseph Taylor* for discovering evidence of gravitational waves.

1994 (P) *Bertram N. Brockhouse* and *Clifford G. Schull* for pioneering work in neutron scattering.

1995 (P) *Martin L. Perl* for discovery of the tau particle, and *Frederick Reines* for first detection of a neutrino.

1996 (P) *David M. Lee, Douglas C. Osheroff,* and *Robert C. Richardson* for developing a superfluid using helium-3.

1997 (P) *Steven Chu, Claude Cohen-Tannoudji,* and *William D. Phillips* for developing methods to cool and trap atoms with laser light.

1998 (P) *Robert B. Laughlin, Horst L. Störmer,* and *Daniel C. Tsui* for discovering a new form of quantum fluid with fractionally charged excitations.

1999 (P) *Gerardus 't Hooft* and *Martinus J. G. Veltman* for studies in the quantum structure of electroweak interactions in physics.

2000 (P) *Zhores I. Alferov* and *Herbert Kroemer* for developing semiconductor heterostructures used in high-speed electronics and optoelectronics and *Jack St. Clair Kilby* for participating in the invention of the integrated circuit.

2001 (P) *Eric A. Cornell, Wolfgang Ketterle,* and *Carl E. Wieman* for the achievement of Bose–Einstein condensation in dilute gases of alkali atoms.

2002 (P) *Raymond Davis Jr.* and *Masatoshi Koshiba* for the detection of cosmic neutrinos and *Riccardo Giaconni* for contributions to astrophysics that led to the discovery of cosmic x-ray sources.

2003 (P) *Alexei Abrikosov, Anthony Leggett,* and *Vitaly Ginzburg* for their contributions to the study of superconductivity and superfluidity, which shed light on the outlandish properties of matter at extremely low temperatures.

Answers to Odd Problems

CHAPTER 1

3. In the rest frame, $p_i = p_f = 0.9$ kg·m/s. In the moving frame, $p_i' = p_f' = 1.9$ kg·m/s.

5. $0.866c$

7. $0.048c = 1.44 \times 10^6$ m/s

9. $0.436L$

11. Moving clock runs slower by $\cong 3.2$ ns.

13. (a) $= 7.1\ \mu s$ (b) $\cong 1.1 \times 10^4$ muons

15. (c) 1.5×10^7 m/s

17. (a) Galaxy A is approaching at $v = 0.198c$.
(b) Galaxy B is receding at $v = 0.237c$.

19. $+0.41c$ and $-0.41c$

21. $-0.50c$

23. (a) $-0.33\ \mu s$ (b) 140 m   (c) No. Event 2 occurs $0.33\ \mu s$ earlier than event 1.

25. (a) $39.2\ \mu s$ (b) Accurate to one digit. More precisely, he aged $1.78\ \mu s$ less on each orbit.

27. $0.789c$

29. (a) $\dfrac{2v}{1 + v^2/c^2}$ (b) $\dfrac{3v + v^3/c^2}{1 + 3v^2/c^2}$

33. (a) Yes   (b) Yes

35. 5.45 yr; Goslo

37. B occurred 4.44×10^{-7} s before A.

39. (a) 26.6 Mm   (b) 3.87 km/s   (c) -8.34×10^{-11}
(d) 5.29×10^{-10} (e) $+4.46 \times 10^{-10}$

CHAPTER 2

1. (a) 5.01×10^{-21} kg m/s   (b) 2.89×10^{-19} kg m/s
(c) 1.03×10^{-18} kg m/s   (d) for a, 9.38 MeV/c; for b, 540 MeV/c; for c, 1930 MeV/c

9. (a) 3.07 MeV   (b) $0.986c$

11. (a) 2.71×10^{-17} kg m/s   (b) 2.9995×10^8 m/s

13. (a) $0.999997c$ (b) 3.744×10^5 MeV

15. (a) $\cong 0.412c$ (b) $0.422c$

17. 4.9 MeV

19. 7.42 MeV

21. 1.011 MeV, 1.422 MeV/c, $\theta = 45.3°$

23. 2.51×10^{-28} kg and 8.84×10^{-28} kg

27. 1.47 km

29. (a) 3.65 MeV/c^2 (b) $0.589c$

31. $M = \dfrac{2m}{3} \sqrt{\dfrac{4 - (u^2/c^2)}{1 - (u^2/c^2)}}$

33. 6.28×10^7 kg

CHAPTER 3

1. (c) $\Delta\omega = 8.8 \times 10^{10}$ rad/s; $\Delta\omega/\omega = 2.3 \times 10^{-5}$
(d) If the electron's plane of rotation is parallel to **B**, the magnetic flux ϕ_B is always zero so that **E** and F are zero and there is no Δv for the electron.

3. (a) $E_{total} = 2.0$ J; $f = 0.56$ Hz   (b) 5.4×10^{33}
(c) 3.7×10^{-34} J

5. (a) $x = 5(1 - e^{-x})$, where $x = hc/\lambda_{max}k_B T$
(b) Solution to (a) is $x = 4.965$ or $\lambda_{max}T = 2.90 \times 10^{-3}$ mK

7. (a) 10 modes   (b) 0.5 cm^{-2} (c) Yes   (d) For short wavelengths, n is almost a continuous function of λ.

9. (a) $\lambda = 600$ nm   (b) $\lambda = 0.03$ m   (c) $\lambda = 10$ m

11. 9.45×10^{44} photons/s

13. 2.04 eV

15. (a) 1.0×10^{15} Hz   (b) 2.0 V

17. (a) lithium and beryllium   (b) lithium $= 1.83$ eV, beryllium $= 0.23$ eV

19. 1.55 V

21. (a) 1.6 eV   (b) 4.0×10^{-15} V·s   (c) 775 nm
(d) 3%

23. $E = 2.48$ eV, $p = 1.32 \times 10^{-27}$ kg m/s

25. (a) 3.25×10^{-4} nm   (b) 2.78×10^5 eV   (c) $\cong 22$ keV

29. (a) $\theta = 41.5°$ (b) 0.679 MeV

31. (a) 9.11×10^4 eV   (b) 8.90 keV   (c) 55.4°

35. (a) 17.4 keV   (b) 0.0760 nm   (c) 16.3 keV
(d) 1.1 keV

37. 4.49°, 9.00°, 13.6°

39. (a) $d = 2.80 \times 10^{-8}$ cm   (b) 6.13×10^{23} formula units/mole

41. (a) 1500 m   (b) 1.1×10^{17}

43. (a) 3.77×10^{-5} eV   (b) 3.10 eV   (c) No, because the maximum energy transferred ($\theta = 180°$) is insufficient.

45. $h = 6.5 \times 10^{-34}$ J·S

CHAPTER 4

1. 1.60×10^{-19} C

3. (a) 9.58×10^7 C/kg   (b) proton
(c) 2.19×10^6 m/s $\sim 0.01c$
(d) For a speed of $0.01c$ there is no need to use relativity.

5. (b) speed $= 2.39 \times 10^8$ m/s $= 0.795c$. Using $p = mv$ gives
$r = 0.0769$ m and $y = 0.00408$ m. Using $p = \gamma mv$ gives
$r = 0.1267$ m and $y = 0.00243$ m.

7. mass $= 8.418 \times 10^{-11}$ g, $mg/E = 25.96 \times 10^{-19}$ C,
$\overline{q} = 1.661 \times 10^{-19}$ C

9. 6.00×10^{-15} m

11. 656.112 nm, 486.009 nm, 433.937 nm

13. (a) 3   (b) no

15. (a) $E_n = (-13.6 \text{ eV}) Z^2/n^2 = (-54.4 \text{ eV})/n^2$
(b) 54.4 eV

17. $r_{\text{He}^+} = 0.0265$ nm, $r_{\text{Li}^{2+}} = 0.0177$ nm,
$r_{\text{Be}^{3+}} = 0.0132$ nm

19. (a) 1.89 eV   (b) 658 nm   (c) 4.56×10^{14} Hz

21. (a) $\lambda_{\max} = 1874.606$ nm, $\lambda_{\min} = 820.140$ nm
(b) $E_{\min} = 0.6627$ eV, $E_{\max} = 1.515$ eV

23. (a) 0.0529 nm   (b) 1.99×10^{-24} kg·m/s
(c) 1.05×10^{-34} kg·m^2/s $= \hbar$ (d) 13.6 eV
(e) -27.2 eV   (f) -13.6 eV

25. (a) 1.60×10^{14} Hz   (b) For $n = 3$,
$f = 2.44 \times 10^{14}$ Hz. For $n = 4$, $f = 1.03 \times 10^{14}$ Hz.

29. (b) 3.23×10^{-8} %

33. (a) 3.1×10^{-15} m   (b) -18.9 MeV

35. $r_{\text{positronium}} = 2r_{\text{hydrogen}}$, $E_{\text{positronium}} = \frac{1}{2} E_{\text{hydrogen}}$

39. (a) 4.9000134 eV   (b) 1.31×10^6 m/s $= 4.38 \times 10^{-3}c$
(c) 3.60 m/s   (d) 3.68×10^{-11} eV

41. $\lambda = 634$ nm, red

43. (a) $\sim -10^6$ m/s^2 (b) ~ 1 m

CHAPTER 5

1. 3.97×10^{-13} m

3. 1.79×10^{-36} m

5. For electrons: (a) 0.0150 eV   (b) 150 eV
(c) 1240 MeV. For alphas: (a) 2.06×10^{-6} eV
(b) 0.0206 eV   (c) 201 MeV.

7. 9.05 fm $= 9.1 \times 10^{-15}$ m

9. 1.1×10^{-34} m

11. (a) ~ 100 MeV   (b) The kinetic energy is too large to expect that the electron could be confined to a region the size
of the nucleus.

13. 50th plane

17. $v_p = c(1 + m_e/\hbar^2 h^2)^{1/2}$
$v_g = c(1 + m_e/\hbar^2 h^2)^{-1/2}$;
$v_p v_g = c^2$. If $v_p > c$, $v_g < c$.

19. 4.6×10^{-14} m

21. $\Delta x = \sqrt{\hbar/2m} \cdot \sqrt[4]{H/2g} = 1.8 \times 10^{-16}$ m.

23. (a) $\Delta v \geq 0.5$ m/s   (b) 3.5 m

25. The intrinsic energy width, ΔE, is given by
$\Delta E \geq 3.29 \times 10^{-6}$ eV. It can't be measured with a
gamma detector that has a resolution of ± 5 eV.

27. 550 eV

29. 2.25 to 1

31. 9.5×10^{27} m

33. $\Delta m/m = 5.6 \times 10^{-8}$

35. (a) $f(x) = \dfrac{A}{\alpha\sqrt{2}} e^{-x^2/4\alpha^2} e^{ik_0 x}$

(b) Re $f(x) = \dfrac{A}{\alpha\sqrt{2}} e^{-x^2/4\alpha^2} \cos k_0 x$ (c) $\Delta k = \dfrac{1}{2\alpha}$

37. 2.27 pA

CHAPTER 6

1. (a) Not acceptable—diverges as $x \to \infty$.
(b) Acceptable.   (c) Acceptable.   (d) Not acceptable—not single-valued.   (e) Not acceptable—wavefunction is discontinuous (as is its slope).

3. (a) 0.126 nm   (b) 5.26×10^{-24} kg·m/s   (c) 95 eV

5. $U(x) = \dfrac{\hbar^2}{2mL^2}\left\{\dfrac{4x^2}{L^2} - 6\right\}$ is a parabola opening upward

with its vertex at $\left(0, -\dfrac{3\hbar^2}{mL^2}\right)$.

9. $\lambda = 2.02 \times 10^{-4}$ nm, $E = 6.14$ MeV. This is the gamma-ray
region of the electromagnetic spectrum.

11. $\psi_1(x) = \sqrt{2/L}\cos(\pi x/L)$, $P_1(x) = (2/L)\cos^2(\pi x/L)$;
$\psi_2(x) = \sqrt{2/L}\sin(2\pi x/L)$, $P_2(x) = (2/L)\sin^2(2\pi x/L)$;
$\psi_3(x) = \sqrt{2/L}\cos(3\pi x/L)$, $P_3(x) = (2/L)\cos^2(3\pi x/L)$

13. (a) 5.13×10^{-3} eV   (b) 9.40 eV
(c) Much smaller electron mass

15. (a) $U = -\dfrac{7}{3}\dfrac{ke^2}{d}$ (b) $\dfrac{\hbar^2}{36 md^2}$

(c) $d = \dfrac{\hbar^2}{42 mke^2} = 0.050$ nm

(d) Lithium spacing $= a = 0.28$ nm

17. (b) $P_1 = 0.200$ (c) $P_2 = 0.351$
(d) $E_1 = 0.377$ eV; $E_2 = 1.51$ eV

19. $n = 4.27 \times 10^{28}$; excitation energy $= 4.69 \times 10^{-32}$ J

21.

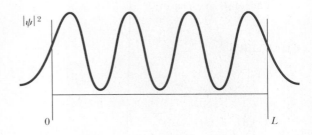

23. $\psi(x) = A \sin kx$ for $0 \le x \le L$ and $\psi(x) = Ce^{-\alpha}x$ for

$x > L$, where $k^2 = \dfrac{2mE}{\hbar^2}$ and $\alpha^2 = \dfrac{2m(U-E)}{\hbar^2}$. Allowed

energies satisfy $\dfrac{kL}{\sin(kL)} = \left[\dfrac{2mUL^2}{\hbar^2}\right]^{1/2}$, which has

solutions only if $\dfrac{2mUL^2}{\hbar^2} > 1$.

29. (a) $12^{1/2}$ nm$^{-1/2}$ (b) 0.693 nm

(c) $\langle x \rangle = 13/12$ nm, $\langle x \rangle$ is somewhat greater than the most probable position.

31. $\langle x \rangle = 0$, $\langle x^2 \rangle = x_0^2/2$, $\Delta x = x_0/\sqrt{2}$, $P = 1 - e^{-\sqrt{2}} = 0.757$

33. (a) $\langle p \rangle = 0$ (b) $\langle p^2 \rangle = \dfrac{m\hbar\omega}{2}$

(c) $\Delta p = \left(\dfrac{m\hbar\omega}{2}\right)^{1/2}$

35. c and d, with eigenvalue $\hbar k$

37. $\Psi(x, t) = \dfrac{1}{\sqrt{2}}[\psi_1(x)e^{-iE_1 t/\hbar} + \psi_2(x)e^{-iE_2 t/\hbar}]$, where

E_1, E_2 are the ground- and first excited-state energies, respectively.

CHAPTER 7

1. (b) $k^2 = 2mE/\hbar^2$ to the left and $k^2 = 2m(U-E)/\hbar^2$ to the right of the step. $E/U = \frac{1}{2}$.   (c) 1.44 fm

3. $R = 0.146$, $T = 0.854$ for both, provided their energies are the same.

5. (b) (1) 0.90   (2) 0.36   (3) 0.41   (4) ≈ 0

9. $T = 0.33552$, $R = 0.66448$, $T(E) = 0.333596$ (exact value), Thickness $= 0.02334$ Å for protons.

11. (a) The matter wave reflected from the front of the well $(x = 0)$ suffers a 180° change in phase; that reflected from the rear is not phase shifted.

(b) $\Psi(x_1, t) = \psi(x) e^{-i\omega t}$, where

$\psi(x) = Ae^{ik'x} + Be^{-ik'x}$ (region 1)
$= Ce^{-ikx} + De^{ikx}$ (region 2)
$= Fe^{ik'x} + Ge^{-ik'x}$ (region 3)

with $k = \sqrt{2mE/\hbar^2}$
$k' = \sqrt{2m(E-U)/\hbar^2}$
and $G = 0$ for particle incident from the left.

13. $R = -\dfrac{E_0}{E}\left(1 - \dfrac{E_0}{E}\right)^{-1}$; $T = \left(1 - \dfrac{E_0}{E}\right)^{-1}$; with $E_0 = -\dfrac{mS^2}{2\hbar^2}$

17. 9.35×10^{20} Hz

19. The first transmission resonance occurs at about $E = 0.08293$ eV, where $T(E)$ is nearly unity. Transmission drops to about 50% at $E = 0.08254$ eV and again at $E = 0.08332$ eV, for a width of about 0.0008 eV.

CHAPTER 8

1. $E/E_0 = 6, 9, 9, 12, 14, 14$, where $E_0 = \dfrac{\pi^2\hbar^2}{8mL^2}$

3. (a) $E = \dfrac{11\pi^2\hbar^2}{2mL^2}$ (b) n_1, n_2, n_3 can have values (1, 1, 3); (1, 3, 1); (3, 1, 1).

(c) $\psi_{113} = A \sin(\pi x/L)\sin(\pi y/L)\sin(3\pi z/L)$
$\psi_{131} = A \sin(\pi x/L)\sin(3\pi y/L)\sin(\pi z/L)$
$\psi_{311} = A \sin(3\pi x/L)\sin(\pi y/L)\sin(\pi z/L)$

5. (a) $E_{111} \approx 1.54$ MeV   (b) The states ψ_{211}, ψ_{121}, ψ_{112} have the same energy $= 2E_{111} \approx 3.08$ MeV. The states ψ_{221}, ψ_{122}, and ψ_{212} have the same energy $E = 3E_{111} \approx 4.63$ MeV.

(c) Both states are threefold degenerate.

7. $A = \sqrt{8/V}$, where $V = L_1 L_2 L_3$ is the volume of the box.

9. $\ell = 4$

11. (a) $\ell \approx 4.58 \times 10^{65}$ (b) $\Delta L/L \approx 1/\ell = 2.18 \times 10^{-66}$

13. (a) For $n = 3$, ℓ can have the values of 0, 1, or 2. For $\ell = 0$, $m_\ell = 0$; for $\ell = 1$, $m_\ell = 1, 0, -1$; for $\ell = 2$, $m_\ell = 2, 1, 0, -1, -2$.   (b) $E = -6.04$ eV

15. (b) helium $= 164.1$ nm (ultraviolet), positronium $= 1312.6$ nm (infrared)

17. (a) 2.58×10^{-34} J·s   (b) 3.65×10^{-34} J·s

21. (a) 9.88×10^{14} m$^{-3/2}$ (b) 9.75×10^{29} m^{-3}

(c) 3.43×10^{10} m^{-1}

23. (a) 137.036   (b) $2\pi \times 137$ (c) $137/2\pi$

(d) $4\pi(137)$

25. $(3 + \sqrt{5})a_0 \approx 5.236a_0$ for 2s state and $4a_0$ for 2p state.

27. $13.1a_0$ for 3s; $12.0a_0$ for 3p; $9.00a_0$ for 3d. Bohr theory predicts $9a_0$.

29. $\Delta r = 0.866a_0$, more than half the average itself.

31. $\psi(0) = 0$; $E_n = -\dfrac{mA^2}{2\hbar^2}\left[\dfrac{1}{n^2}\right]$ $(n = 1, 2, \ldots)$

33. $E = 4.3123$ MeV, $\ell = 1$ for 1st excited state (degeneracy 3), $E = 7.0797$ MeV, $\ell = 2$ for 2nd excited state (degeneracy 5), $E = 8.3882$ MeV, $\ell = 0$ for 3rd excited state (non-degenerate).

CHAPTER 9

1. $f = 9.79 \times 10^9$ Hz

3. (a) 2   (b) 8   (c) 18   (d) 32   (e) 50

5. $\dfrac{dB_z}{dz} = 0.387$ T/m

7. $v/c = 278.6$

9. $|\mathbf{s}| = \dfrac{\sqrt{15}}{2}\hbar$, $\theta = 140.8°, 105.0°, 75.0°, 39.2°$. Ω^- does obey the Pauli exclusion principle.

11. $j = \frac{5}{2}, \frac{3}{2}$, and $m_j = \pm\frac{5}{2}, \pm\frac{3}{2}, \pm\frac{1}{2}$ $(j = \frac{5}{2})$, $m_j = \pm\frac{3}{2}, \pm\frac{1}{2}$ $(j = \frac{3}{2})$

13. (a) $n = 4$, $\ell = 3$, $j = \frac{5}{2}$ (b) $|\mathbf{J}| = \dfrac{\sqrt{35}}{2}\hbar$

(c) $J_z = m_j\hbar$, where $m_j = \pm\frac{5}{2}, \pm\frac{3}{2}, \pm\frac{1}{2}$

15. $\Delta E_{calc} = 3.18 \times 10^{-5}$ eV; $\Delta E_{meas} = 5 \times 10^{-5}$ eV

17. (a) 394.8 eV, $(n_1, n_2, n_3) = (1, 1, 1), (1, 1, 2)$, $(1, 2, 1)$, and $(2, 1, 1)$

(b) All eight particles go into the $(n_1, n_2, n_3) = (1, 1, 1)$ state, and so the energy is $8 \times E_{111} = 225.6$ eV.

19. 6s wavefunction has five nodes and $E = -8.64$ eV. Ionization energy is 8.64 eV vs. 9.22 eV from Table 9.2. Most probable distance is $1.37a_0$ compared with $1.00a_0$ for the (outermost) electron in hydrogen, suggesting that the gold atom is not much larger even though it has 79 electrons.

21. (a) $1s^2 2s^2 2p^4$ (b) $(1, 0, 0, \pm\frac{1}{2})$, $(2, 0, 0, \pm\frac{1}{2})$, $(2, 1, 0, \pm\frac{1}{2})$, $(2, 1, \pm1, \pm\frac{1}{2})$

23. [Kr] $4d^{10}$ has the lesser energy, but [Kr] $4d^9 5s^1$ has two more unpaired spins. Thus, Hund's rule is violated in this case (since filled subshells are especially stable). The element is Pd.

25. Between 10^4 K and 10^5 K

27. (a) $\sqrt{f} = \sqrt{\frac{5}{36}\left(\frac{13.6\ eV}{h}\right)}(Z-7)$

(b) Theoretical: slope $= 0.214 \times 10^8\ Hz^{1/2}$, intercept $= 7$
Experimental: slope $= 0.21 \times 10^8\ Hz^{1/2}$, intercept $= 6.3$
(c) $Z - I = Z - 6.3$.

CHAPTER 10

3. $s_{\bar{v}} = 1.58$ cm, $s_{\text{rms}} = 1.45$ cm, $s_{\text{mp}} = 1.78$ cm
5. $P(E) = 0.385\ e^{-(0.408\,E/E1)}$
7. (b) $2e^{-2p\mathcal{E}/k_B T}$ (c) $T = 2.83$ K

(d) $\bar{E} = \frac{2p\mathcal{E}}{1 + \frac{1}{2}e^{2p\mathcal{E}/k_B T}}$ (e) $E_{\text{TOTAL}} = N\bar{E}$

(f) C is a maximum at $2p\mathcal{E}/k_B T = 2.65$ or $T = 0.055$ K
9. $\bar{v} = 1.51 \times 10^3$ m/s, so the classical Doppler shift formula can be used:

$$\frac{\Delta f}{f_0} = 1.01 \times 10^{-5}$$

11. $\bar{n}(0E)$ through $\bar{n}(5E) = 2$, $\bar{n}(6E) = 1.89$, $\bar{n}(7E) = 1.78$, $\bar{n}(8E) = 1.55$, $\bar{n}(9E) = 1.22$, $\bar{n}(10E) = 0.777$, $\bar{n}(11E) = 0.444$, $\bar{n}(12E) = 0.222$, $\bar{n}(13E) = 0.111$, and $\bar{n}(14E) = 0$, $E_F(0) = 9E$
13. (b) T_E(lead) ≈ 100 K, T_E(aluminum) ≈ 300 K, and T_E(silicon) ≈ 500 K
15. (a) 1.80×10^{29} free electrons/m^3 (b) 3
17. $E_F = 33.4$ MeV, $\bar{E} = 20$ MeV
19. 93.8%
21. (a) 2.54×10^{28} electrons/m^3 (b) 3.15 eV
(c) 1.05×10^6 m/s
23. 3.40×10^{17} electrons

CHAPTER 11

1. (a) 1.28 eV   (b) $\sigma = 0.272$ nm, $\varepsilon = 4.65$ eV
(c) 6.55 nN
3. $R_0 = (I/\mu)^{1/2} = 1.13$ Å, same as Example 11.1
5. (a) 8.10 cm, 3.70 GHz,   (b) 4.53×10^{-45} kg·m^2
7. 5.69×10^{12} rad/s
$\ell = 0$ $E_{\text{rot}} = 0$
$\ell = 1$ $E_{\text{rot}} = 2.62 \times 10^{-3}$ eV
9. (a) $\ell = 2$ $E_{\text{rot}} = 7.86 \times 10^{-3}$ eV
$\ell = 3$ $E_{\text{rot}} = 1.57 \times 10^{-3}$ eV
(b) $K = 480$ N/m, $f = 8.66 \times 10^{13}$ Hz
(c) $E_0 = 0.179$ eV, $A_0 = 0.0109$ nm, $E_1 = 0.538$ eV, $A_1 = 0.0189$ nm   (d) rotational: 473 μm; vibrational 3.46 μm
11. $E_{\text{rot}} = \hbar^2/mR_0^2\{\ell(\ell+1)\}$, $\ell = 0, 1, 2, \ldots$
13. (a) $R_\ell \approx R_0 + [\ell(\ell+1)\hbar^2/(\mu\omega_0)^2]/R_0^3$
(b) $U_0 \approx \ell(\ell+1)\hbar^2/2\mu R_0^2$,
$\omega_\ell^2 \approx \omega_0^2 + 3[\ell(\ell+1)\hbar^2]/\mu^2 R_0^4$

15. $\Delta E = \hbar\omega[1 - (v+1)\hbar\omega/2U_0]$, $v_{\text{MAX}} = \frac{2U_0}{\hbar\omega} - 1$,
$E_{\text{MAX}} = U_0 - \frac{(\hbar\omega)^2}{16U_0}$

17. When the energy of the rotation-vibration ground state $(v = 0, 1 = 0)$ is taken into account ($E = 0.268$ eV from Problem 16), the numerical results for $1 = 1, 2$ agree with the predictions of Equation 11.10 with $\frac{h^2}{I_{\text{CM}}}$ $\frac{h^2}{I_{\text{CM}}}$ eV. Discrepancies do arise for larger values of 1. $E = 0.283$ eV $(v = 0, \ell = 1)$ and $E = 0.313$ eV $(v = 0, \ell = 2)$
19. Ground state: $E = (2x^2\hbar^2/mL^2)$, where x is the smallest root of the equation $\tan x = -(2\hbar^2/mSL)x$. First excited state: $E = (2\pi^2\hbar^2/mL^2)$. As $S \to \infty$, $x \to \pi$ and the two energies coincide; as $S \to 0$, $x \to \pi/2$, and the energies reduce to the ground and first excited states of an infinite well with no barrier.
21. (a) $R_0 = 1.44$ bohrs
(b) $K = 1.03$ Ry/bohr^2 = 801 N/m
23. The crossing takes about 6.1 fs (= 6.1×10^{-15} s) for a crossing frequency of 1.64×10^{14} Hz. The bonding and antibonding states have energies $E = 6.379$ eV and $E = 6.718$ eV, respectively, for a splitting $\Delta E = 0.339$ eV and a characteristic frequency $\Delta E/h = 8.20 \times 10^{13}$ Hz, or about half the crossing frequency.

CHAPTER 12

3. 7.84 eV/ion pair
7. (a) 7.12 eV/ion pair   (b) 6.39 eV/atom pair
11. (c) 3.9×10^5 cm/s   (d) 0.36 Ωm
13. (a) 3.80×10^{-14} s   (b) 52.7 nm   (c) $L/d \approx 200$
15. (a) 2.75×10^{14} Hz   (b) 1090 nm (IR)
19. (b) 0.36 nm   (c) 6.7×10^{-4} nm. The controlling factor is cavity length.
21. (b) 10.7 kA
25. (a) 2.98 mA   (b) 67.1 Ω (c) 8.39 Ω

CHAPTER 13

1. (a) 1.9 fm   (b) 7.44 fm   (c) 3.92
3. 8.57×10^{13}
5. (a) 4.55×10^{-13} m   (b) 6.03×10^6 m/s
7. 2.2×10^{-6} eV
9. 30% ^{63}Cu
11. 2.657 MeV/nucleon
13. (a) ^{139}Cs with $N/Z = 1.53$
(b) ^{139}La, with 8.353 MeV/nucleon
(c) ^{139}Cs, with a mass of 138.913 u
15. (a) 8.03 MeV/nucleon   (b) 8.55 MeV/nucleon
(c) 8.66 MeV/nucleon   (d) 7.92 MeV/nucleon
17. 160 MeV
19. (a) $U = 3k\frac{(Ze)^2}{5R}$ (b) $2.88 \times 10^{-14}\ (A^{5/3})$ J
(c) 52.1 MeV
21. (a) 0.805 h^{-1} (b) 0.861 h
23. (a) $\lambda = 1.55 \times 10^{-5}$ s^{-1}; $T_{1/2} = 12.4$ h
(b) 2.39×10^{13} atoms   (c) 1.87 mCi

25. 36.3×10^{-6} g

27. 0.055 mCi

29. 9.46×10^9

31. (b) $\lambda = 0.25$ h^{-1}; $T_{1/2} = 2.77$ h   (c) 4×10^3 counts/min
(d) 9.59×10^6

33. (a) 18.3 counts/min

35. 11.8 decays/min · g

41. 2.26 MeV

43. (b) 4.79 MeV   (c) 0.08 MeV   (d) The Po daughter carries off about 3×10^{-6} of the kinetic energy in the beta decay.

45. (a) $Q = -1.82$ MeV, so reaction *cannot* occur.
(b) $Q = -1.68$ MeV, so reaction *cannot* occur.
(c) $Q = 1.86$ MeV, so reaction *can* occur.

47. Using the uncertainty principle, electrons in a nucleus are found to have an energy of about 100 MeV. Since the most energetic electrons emitted in beta decay have energies less than 10 MeV, electrons are not present in the nucleus.

51. 18.6 keV

53. (a) 3.96×10^9 yr   (b) It could be no older. The rock could be younger if some Sr were initially present.

55. (b) R/λ

57. 2.66 d

59. 5400 yr to 6800 yr

61. (a) 4.00 Gyr   (b) 0.0199 and 4.60

CHAPTER 14

1. (a) -2.4386 MeV   (b) 2.5751 MeV

3. 5.70 MeV

5. 17.35 MeV

7. (a) $Q = -1.19$ MeV, $K_{th} = 1.53$ MeV   (b) 17.35 MeV

9. (b) 1.53 MeV

11. 0.0266 b

13. (a) 0.0373   (b) 0.0663   (c) ≈ 1 (d) 0.1 to 10 eV

15. (a) 0.86   (b) 6.1×10^{11} protons/s
(c) 8.7×10^{10} protons/s

17. 2.25×10^{12}

19. 25

21. 200.6 MeV

23. (a) 4.84 $V^{-1/3}$ (b) 6 $V^{-1/3}$ (c) 6.30 $V^{-1/3}$
(d) Sphere   (e) Parallelepiped

25. (a) 3333 MW   (b) 2333 MW   (c) 1.04×10^{20} events/s
(d) 1.34×10^3 kg   (e) 3.7×10^{-8} kg/s

27. (a) 2.70×10^{-15} m   (b) 720 keV
(c) $v_F = v_0 m_D/(m_D + m_T)$ (d) 1.2 MeV
(e) Possibly by tunneling

29. (a) $K_n = 14.1$ MeV, $K_\alpha = 3.45$ MeV   (b) Yes; since the neutron is uncharged, it is not confined by the **B** field, and only K_α contributes directly to achieving critical ignition.

31. (a) 1.9×10^9 K   (b) 355 kJ

33. (a) 52×10^6 K   (b) $1.943 + 1.709 + 7.551 + 7.297 + 2.242 + 4.966 = 25.75$ MeV
(c) Most energy is lost, since neutrinos have such low cross sections for interaction with matter.

35. 5.3×10^{23} J

37. (a) 3000 MW   (b) 5.2×10^6 L

39. (a) $2.3 \times 10^{-19} Z_1 Z_2$ J   (b) 0.14 MeV

41. (a) 2.63×10^{33} J   (b) 119 billion years

43. (a) 10^{14}/cm^3 (b) 2.2×10^5 J/m^3 (c) 2.35 T

45. (a) 1.4×10^{-3} cm   (b) $I/I_0 = 0.426$

47. 3.85 cm

49. (a) 0.0025 rem   (b) 38 times background of 0.13 rem/yr

51. The two workers receive the same dose in rad.

53. 18.8 J

55. (a) 10   (b) 10^8

59. (a) $\sim 1 \times 10^8$ m^3 (b) $\sim 1 \times 10^{13}$ J   (c) $\sim 10^{14}$ J
(d) ~ 10 kilotons

CHAPTER 15

1. 10^{-23} s

3. $f_{min} = 2.26 \times 10^{23}$ Hz, $\lambda_{max} = 1.32 \times 10^{-15}$ m

5. 2.2×10^{-18} m

9. (a) Does not occur (violates baryon number)
(b) Occurs
(c) Does not occur (violates baryon number)
(d) Occurs
(e) Occurs
(f) Does not occur (violates baryon number and muon–lepton number)

11. (a) Electron and muon–lepton number
(b) Electron–lepton number
(c) Strangeness and charge
(d) Baryon number
(e) Strangeness

13. (a) 5.63 GeV   (b) 5.64 GeV

15. For Example 15.5, $E = 48\%$; for Exercise 3, $E = 48\%$, $E = 46\%$; for Problem 13, $E = 33\%$.

17. A neutron, udd

19. $m_u = 312.3$ MeV/c^2, $m_d = 313.6$ MeV/c^2

21.

23. 19 eV/c^2

25. $K_p = 5.4$ MeV, $K_\pi = 32.3$ MeV

27. 9.3 cm

29. A neutron

31. (a) A Z^0 boson   (b) A gluon

33. (a) 0.782 MeV   (b) Proton speed $= 0.001266c = 380$ km/s, electron speed $= 0.9185c = 2.76 \times 10^8$ m/s (c) The electron is relativistic, the proton is not.

35. (a) $p_\Sigma = 686$ MeV/c, $p_\pi = 200$ MeV/c (b) 627 MeV/c
(c) $E_\pi = 244$ MeV, $E_n = 1\,130$ MeV, $E_\Sigma = 1\,370$ MeV
(d) 1 190 MeV/c^2, $0.500c$

37. (a) $2Nmc$ (b) $3^{1/2}Nmc$ (c) Method (a)

Index

Page numbers followed by "f" indicate figures; page numbers followed by "n" indicate footnotes; page numbers followed by "t" indicate tables.